Annual Reports in
MEDICINAL CHEMISTRY

VOLUME **45**

Annual Reports in

MEDICINAL CHEMISTRY

VOLUME **45**

Sponsored by the Division of Medicinal Chemistry of the American Chemical Society

Edited-in-Chief

JOHN E. MACOR
Neuroscience Discovery Chemistry
Bristol-Myers Squibb R&D
Wallingford, CT, United States

Section Editors

ROBICHAUD • STAMFORD • BARRISH/WEINSTEIN
PRIMEAU • LOWE • DESAI • MCALPINE

ELSEVIER

Amsterdam • Boston • Heidelberg • London • New York • Oxford
Paris • San Diego • San Francisco • Singapore • Sydney • Tokyo
Academic Press is an imprint of Elsevier

ACADEMIC
PRESS

Academic Press is an imprint of Elsevier
Linacre House, Jordan Hill, Oxford OX2 8DP, UK
84 Theobald's Road, London WC1X 8RR, UK
Radarweg 29, PO Box 211, 1000 AE Amsterdam, The Netherlands
30 Corporate Drive, Suite 400, Burlington, MA 01803, USA
525 B Street, Suite 1900, San Diego, CA 92101-4495, USA

First edition 2010

Notice
No responsibility is assumed by the publisher for any injury and/or damage to persons
or property as a matter of products liability, negligence or otherwise, or from any use
or operation of any methods, products, instructions or ideas contained in the material
herein. Because of rapid advances in the medical sciences, in particular, independent
verification of diagnoses and drug dosages should be made

ISBN: 978-0-12-380902-5
ISSN: 0065-7743

For information on all Academic Press publications
visit our website at www.elsevierdirect.com

Printed and bound in USA

10 11 12 13 14   10 9 8 7 6 5 4 3 2 1

Working together to grow
libraries in developing countries
www.elsevier.com | www.bookaid.org | www.sabre.org

ELSEVIER   BOOK AID
              International   Sabre Foundation

CONTENTS

PART II: Cardiovascular and Metabolic Diseases

Section Editor: Andy Stamford, Merck Research Laboratories, Kenilworth, New Jersey

8. **Inhibitors of HIF Prolyl Hydroxylases**   **123**

  Michael H. Rabinowitz, Terrance D. Barrett, Mark D. Rosen and
  Hariharan Venkatesan

PART III: Inflammation/Pulmonary/Gastrointestinal Diseases

Section Editors: Joel C. Barrish and David Weinstein, Bristol-Myers Squibb R&D, Princeton,
                   New Jersey

9. **Opioid Receptor Antagonists for Gastrointestinal Dysfunction**   **143**

  R. William Hipkin and Roland E. Dolle

10. **CFTR Modulators for the Treatment of Cystic Fibrosis**   **157**

  Sabine Hadida, Fredrick Van Goor and Peter D. J. Grootenhuis

11. **Targeting B-cells in Inflammatory Disease**   **175**

  Kevin S. Currie

PART IV: Oncology

Section Editor: Shelli R. McAlpine, Department of Chemistry and Biochemistry, San Diego State University, San Diego, California

PART V: Infectious Diseases

Section Editor: John Primeau, Astra Zeneca, Waltham, Massachusetts

PART VI: Topics in Biology

Section Editor: John Lowe, JL3Pharma LLC, Stonington, Connecticut

PART VII: Topics in Drug Design and Discovery

Section Editor: Manoj C. Desai, Medicinal Chemistry Gilead Sciences, Inc.,
                   Foster City, California

23. Role of Physicochemical Properties and Ligand Lipophilicity Efficiency in Addressing Drug Safety Risks   381
Martin P. Edwards and David A. Price

24. Reducing the Risk of Drug Attrition Associated with Physicochemical Properties   393
Paul D. Leeson and James R. Empfield

25. Compound Collection Enhancement and Paradigms for High-Throughput Screening—an Update   409
Stevan W. Djuric, Irini Akritopoulou-Zanze, Philip B. Cox and
Scott Galasinski

PART VIII: Trends and Perspectives

Color Plate Section at the end of this book

CONTRIBUTORS

Annual Reports in Medicinal Chemistry has reached Volume 45. Its longevity is a testament to the vitality of the field of medicinal chemistry, and I hope that *Annual Reports in Medicinal Chemistry* continues to be the review resource for medicinal chemists. Volume 45 upholds the traditions of *Annual Reports in Medicinal Chemistry* with 28 chapters covering the themes of Central Nervous System Disease, Cardiovascular and Metabolic Diseases, Inflammation/Pulmonary/GI, Oncology, Infectious Disease, Topics in Biology, Topics in Drug Design and Discovery and finally our review of new drugs introduced in 2009 in the To-Market-to-Market section.

Putting Volume 45 of *Annual Reports in Medicinal Chemistry* together was more difficult this year than in years past as a result of the continued consolidation of pharmaceutical companies, driven by larger companies and their business needs. While these business needs *may* be fulfilled by these acquisitions, such events have continued to result in a significant reduction in the depth of the talent pool of medicinal chemists. These reductions in employed medicinal chemists and overall drug discovery research capacity will undoubtedly have long-term effects on the pharmaceutical industry's ability to find new innovative medicines, and in the end, the patient will bear the burden of these business decisions. For the book this year, a significant number of Chapter Authors and Section Editors have been directly affected by this business environment, and I want to applaud them for their dedication to *Annual Reports in Medicinal Chemistry* in completing their tasks despite the significant distractions.

Small molecule medicinal chemistry continues to be challenged to prove its integral value to the discovery of new medicines. Numerous new research modalities have been introduced in the last 25 or so years, which have claimed to limit the need for medicinal chemistry and chemists, only for researchers to find that these "revolutionary" approaches to the discovery of drugs are really just new useful tools to be added to the arsenal of drug discoverers. Computational structure-based drug design, high-throughput screening, and combinatorial chemistry are the three examples of modalities which were touted to radically change the way drugs were discovered (accompanied by the need for fewer medicinal chemists), only for the industry to eventually recognize those techniques as part of a maturing drug discovery process. Medicinal chemistry and medicinal chemists still discover the molecules that eventually become the innovative drugs of the future. Most recently, biologics have made their appearance on the scene as a revolutionary type of new drug. Without a doubt, they are an important part of the present and future of the pharmaceutical industry's goal of helping patients. However, biologics are not without their limitations, and there are still many, many diseases that will be best treated with small molecules. The need for medicinal chemists today should not be seen as diminished, especially since the regulatory environment requires so much more of a molecule today than in years past. In fact, I am confident that the medicinal chemistry community will continue to evolve and provide solutions to problems in drug discovery that have not even yet been fully defined, be those issues in small or large molecules. However, given that the medicinal chemistry community and drug discovery in general has been negatively impacted by mergers and business decisions, I worry going forward whether the industry can maintain a critical mass to attack the diseases we have yet to conquer.

Creating Volume 45 of *Annual Reports in Medicinal Chemistry* has been another stellar team effort comprised entirely of volunteers. I would first like to thank the chapter authors in Volume 45 for their dedication and talent. They enthusiastically provided a consistent quality to the book. Helping to bring all of these together were the Section Editors: Joel C. Barrish and David Weinstein, Manoj Desai, John Lowe, Shelli McAlpine, John Primeau, Albert Robichaud, and Andrew Stamford. I would like to thank them for their time and constant dedication. I would particularly like to welcome Professor McAlpine to the Editorial Staff. Professor McAlpine who assumed the responsibility of the Oncology Section of the book is the only non-industrial Section Editor on *Annual Reports in Medicinal Chemistry* in quite some time, and should bring a different perspective to the book. I'd also like to specifically thank Joel C. Barrish for his efforts in compiling and editing the Immunology/Pulmonary/GI section of the book for the last few years. Joel will be leaving his role in *Annual Reports in Medicinal Chemistry* to assume key leadership roles in the Medicinal Chemistry Division of the American Chemical Society (the sponsor of this book). David Weinstein from Bristol-Myers Squibb has helped Joel this year and will assume the role as Section Editor or the Immunology/Pulmonary/GI section of the book starting with Volume 46.

Helping the Section Editors and me were a team of reviewers/proof readers that have done another spectacular job behind the scenes as well. I would like to acknowledge these reviewers/proof readers by listing their names below as a demonstration of our appreciation for their time and effort.

AstraZeneca—Greg Bisacchi

Bristol-Myers Squibb—Kenneth Boy, Joanne Bronson, Robert Cherney, Andrew Degnan, Gene Dubowchik, Murali Dhar, Carolyn Dzierba, W. Richard Ewing, Samuel Gerritz, Matthew Hill, John Kadow, George Karageorge, Michael Lawrence, Ivar McDonald, Harold Mastalerz, Nicholas Meanwell, Natesan Murugesan, Richard Olson, Brad Pearce, William Pitts, Kenneth Santone, Lawrence Snyder, John Starrett, Drew Thompson, Joseph Tino, Dolatrai Vyas, Michael Walker, and Christopher Zusi

Gilead Sciences—Randall Britton Corkey, Randall Halcomb, Richard Mackman, Paul Roethle

The "To-Market-to-Market" review is always one of my favorite chapters in the book as it details all of the new drugs introduced in the previous year (in this case 2009). Once again, Michelle Schmidt and Shridar Hedge have done an outstanding job compiling and writing this chapter. I cannot thank them enough. Finally, I would like to thank Ms. Catherine Hathaway, my Administrative Assistant, who helps in many different ways.

In summary, I hope that Volume 45 of *Annual Reports in Medicinal Chemistry* continues to be a key reference for your medicinal chemistry pursuits. As Editor-in-Chief, I continue to look for ways to optimize, improve, and evolve the series. Please do not hesitate to contact me with suggestions for improving the series [john.macor@bms.com]. Thank you.

John E. Macor
Bristol-Myers Squibb, R&D,
Wallingford, CT, USA

PART I:
Central Nervous System Diseases

Editor: Albert J. Robichaud
Lundbeck Research USA, Inc.
Paramus
New Jersey

CHAPTER 1

Voltage-Gated Calcium Channel Antagonists for the Central Nervous System

James C. Barrow and **Joseph L. Duffy**

1. INTRODUCTION

Voltage-gated calcium channels are Ca^{2+}-selective pores in the plasma membrane of conductive cells [1]. Under physiological conditions, the individual channel subtypes are activated in response to specific

Merck Research Laboratories, Rahway, NJ 07065, USA

Annual Reports in Medicinal Chemistry, Volume 45
ISSN 0065-7743, DOI 10.1016/S0065-7743(10)45001-0

membrane voltage depolarization from resting potentials and permit the rapid entry of extracellular calcium ions into the cell [2]. The resulting increase in intracellular Ca^{2+} concentration can alter neuron excitability and firing mode, or trigger intracellular biochemical processes including membrane fusion and exocytosis of synaptic vesicles containing neurotransmitters. In this way, voltage-gated calcium channels represent a critical step in nerve impulse initiation and transmission and provide a rich potential opportunity for pharmacological intervention.

In this chapter, we provide a brief overview of voltage-gated calcium channel subtypes and function, with an emphasis on neuronal transmission. The structure and kinetics of calcium channels is an extensive field of study, and outstanding reviews have recently appeared [3,4]. This chapter is focused on the potential opportunities for small-molecule intervention in neuronal signaling via antagonism of voltage-gated calcium channel subtypes, with a brief review of the chemotypes that have been recently reported to modify channel function.

1.1 Nomenclature and tissue distribution

Voltage-gated calcium channels may be categorized into two groups based on the membrane potential required for activation [2]. The low-voltage-activated channels include those that are activated at potentials from about –40 to –50 mV. The high-voltage-activated channels are activated at more positive potentials of –35 to +15 mV. The nomenclature assigned to these channels has evolved, and the three primary nomenclature systems listed in Table 1 may still be found in the current literature. Calcium channels are most commonly grouped in families by their kinetic and pharmacologic current characteristics [5,6]. The low-voltage-activated family is responsible for the transient current (T-type). The high-voltage-activated channel families include those responsible for the long-lasting current (L-type), the current found in Purkinje cells (P/Q-type), the neuronal current (N-type), and the residual current (R-type).

Ten specific mammalian calcium channel subtypes have been identified based on the sequence of their membrane pore-forming α_1 subunit. As illustrated in Table 1, calcium channels may be individually identified by their α_1 subunit, or more commonly using a systematic nomenclature based on a gene subfamily (Ca_v1–Ca_v3) and the order of discovery of the α_1 subunit (e.g., $Ca_v1.1$–$Ca_v 1.4$) [7].

The pore-forming α_1 subunit alone is sufficient to form a functional calcium channel in the cell membrane [8]. However, high-voltage-activated calcium channels are hetero-oligomeric structures that may contain as many as three additional smaller subunits, labeled $\alpha_{2\delta}$ (itself a heterodimer), β, and γ. These three subunits associate in an auxiliary

Table 1 Nomenclature of voltage-gated calcium channels

Family	Channel	α_1 Subunit	Distribution
L-type	$Ca_v1.1$	α_{1S}	Sketetal muscle
	$Ca_v1.2$	α_{1C}	Cardiac and smooth muscle, endocrine cells, CNS, neuron cell bodies
	$Ca_v1.3$	α_{1D}	Cochlea, retina, neuroendocrine cells in the pancreas
	$Ca_v1.4$	α_{1F}	Retina, lymphoid tissue, spinal chord
P/Q-type	$Ca_v2.1$	α_{1A}	Neuron presynaptic terminals, heart, brain
N-type	$Ca_v2.2$	α_{1B}	Neuron presynaptic terminals
R-type	$Ca_v2.3$	α_{1E}	Neuron cell bodies and presynaptic terminals, heart, pituitary
T-type	$Ca_v3.1$	α_{1G}	Brain, heart
	$Ca_v3.2$	α_{1H}	Kidney, liver, brain, heart
	$Ca_v3.3$	α_{1I}	Brain

fashion and are believed to modify both the voltage-gating properties of the channel and the trafficking of the channel to the cell surface. Low-voltage-activated (T-type) calcium channels have not been demonstrated to associate with auxiliary subunits in native tissues [9].

1.2 State-dependent calcium channel antagonism

Voltage-gated calcium channels fluctuate between three primary functional states, which provide the opportunity for selective pharmacological intervention [10]. Initially, when the membrane is hyperpolarized, the channel is in the resting (closed) state. In response to the appropriate depolarized membrane potential, the channel may undergo a conformational change to the activated state (open), permitting Ca^{2+} ions to enter the cell through the channel. Finally, the channel may enter an inactivated conformational state from either the activated or the resting state. In this inactivated state, the channel remains unresponsive to a depolarizing potential until it transitions to the closed state (recovers). Increased neuronal firing rates, such as those in chronic pain syndromes or epileptic episodes, are believed to drive a greater proportion of calcium channels into the inactivated state [11]. These distinct conformational states afford the possibility of discovering state-dependent calcium channel antagonists that block neuronal transmission only under conditions of hyperexcitability. By contrast, state-independent antagonists block calcium channels under most conditions of electrical excitability, and this latter mechanism forms the basis of a variety of naturally occurring peptidyl neurotoxins [12]. State-dependent blockers may be identified using labor-intensive patch-clamp electrophysiology techniques or by using more recently developed high-throughput Fluorometric Imaging Plate Reader (FLIPR) assays [13–15].

1.3 Potential central nervous system indications for calcium channel antagonists

The most often studied central nervous system (CNS) indication for voltage-gated calcium channel antagonists is for the treatment of pain [16]. Both N-type and T-type channels have been implicated in pain processing [17]. Indeed, ziconotide (synthetic ω-conotoxin, Prialt®) is a 25-amino acid peptide N-type calcium channel blocker that is used clinically to treat intractable pain via intrathecal injection. Ziconotide is a state-independent N-type calcium channel blocker. This pharmacology may underlie the narrow therapeutic window found with this treatment for analgesia over ancillary CNS effects [10,12], as well as the hemodynamic effects resulting from systemic administration of the peptide [18]. Genetic evidence in mice also supports the role of N- and T-type calcium channels in nociception. N-type calcium channel knockout mice display reduced response to noxious mechanical, thermal, chemical, and inflammatory visceral stimuli [19]. Recent knockout and genetic studies in rodents have similarly implicated T-type calcium channels in pain pathways [20,21], but reports of efficacy of T-type selective small-molecule inhibitors in preclinical pain models are limited [22].

Calcium channels have been shown to play a role in epilepsy as well [23]. Currently used antiepileptic drugs exhibit a wide spectrum of activity, including modulation of voltage-gated sodium and calcium channels. T-type calcium channels have been demonstrated to play an important role in absence epilepsy, a specific form of epilepsy characterized by brief lapses in consciousness correlated with spike-and-wave discharges in the electroencephalogram [14,24–28]. Ethosuximide 1 has been shown to block T-type calcium channels and is used clinically to treat absence epilepsy [25]. Several selective small-molecule T-type calcium channel antagonists have demonstrated efficacy in rodent epilepsy models (*vide infra*).

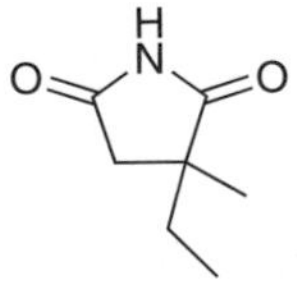

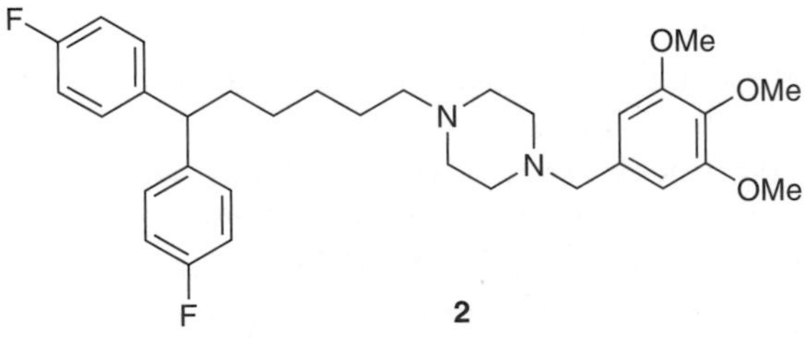

Recent studies highlight the potential for calcium channel antagonists in the treatment of addiction. In rodent behavioral models, a mixed N- and T-type antagonist NP078585 (**2**) reduced ethanol intoxication and attenuated the reinforcing and rewarding properties of ethanol [29]. These effects were absent in $Ca_v2.2$ knockout mice, suggesting the N-type calcium channel as the primary driver of the observed effects. Other studies implicate a role for T-type calcium channel in addiction. Urbano and coworkers studied T-type calcium channel expression in mice after a cocaine "binge" and showed increases in T-type currents from thalamic relay ventrobasal neurons [30].

The block of calcium channels has been investigated as a treatment for the prevention of neuronal damage during ischemic stroke [31]. These studies are based on the finding that cerebral ischemia results in unregulated increases in intracellular Ca^{2+} concentration, which leads to uncontrolled release of excitatory neurotransmitters from presynaptic neurons and a cascade of intracellular events leading to cell death. Preclinical results with N-type peptidyl antagonists and small-molecule blockers have been encouraging [32–34]. However, the translation of these findings to clinical results utilizing L-type antagonists has been disappointing. The poor outcomes may be due to inadequate preclinical models for ischemic stroke or the lack of understanding of an optimal clinical trial design for this type of therapy [31].

The $Ca_v3.1$ and $Ca_v3.3$ subtypes of the T-type calcium channel are highly expressed in the thalamus and cortex and are important for the regulation of thalamocortical signaling which is an important component of sleep/wake regulation [35,36]. Selective T-type calcium channel antagonists have recently been shown to modify sleep architecture in rodents, resulting in altered feeding patterns and resistance to high-fat diet-induced weight gain [37]. In addition to their role in normal sleep rhythms, abnormalities in these signaling pathways can lead to a variety of conditions and have been called "thalamocortical dysrhythmias" [38]. Depending on the particular thalamocortical circuit involved, these abnormalities can lead to neuropathic pain, epilepsy, schizophrenia, tremor, or tinnitus [39,40]. To date, all these studies have been in preclinical species, and clinical validation has not yet been reported.

The exact roles of P/Q-type channels ($Ca_v2.1$) and R-type channels ($Ca_v2.3$) in the CNS are less well understood. Genetic evidence indicates that $Ca_v2.1$ channels play a central role in motor function and sensory processing [41]. Mice deficient in the α_{1A} subunit of the $Ca_v2.1$ channel display ataxia, dystonia, absence seizures, cerebellar degeneration, and altered nociception, and die within 3–4 weeks following birth [42]. Human $Ca_v2.1$ mutations have been associated with absence epilepsy, episodic ataxia, and familial hemiplegic migraine [43,44]. Mice deficient in the α_{1E} subunit of the $Ca_v2.3$ channel exhibit a mild phenotype, with alterations in inflammatory pain response and learning [45]. As a result, no medicinal chemistry programs directed toward selective P/Q- and R-type calcium channels have been reported.

2. T-TYPE CALCIUM CHANNEL BLOCKERS

Several diverse structural classes of T-type calcium channel antagonists have been reported over the last decade [46,47]. This section focuses on recent literature reports of CNS T-type antagonists since 2008.

2.1 Historical compounds

The two most frequently studied compounds with T-type calcium channel antagonist properties are ethosuximide 1 and mibefradil 3. However, the modest potency of ethosuximide (~200 μM) [48] and the poor selectivity of mibefradil [49] make these compounds suboptimal tools for the investigation of these channels. Guided by a pharmacophore model [50], several analogs of 3 were prepared. Compound 4 represents the most potent compound identified (IC$_{50}$ 8 nM, patch-clamp assay) with good selectivity over the L-type calcium channel [51]. Compound 4 showed a modest brain-to-plasma ratio (0.25) after oral dosing to rats at 50 mg/kg. However, no *in vivo* efficacy assay results have been reported with this compound.

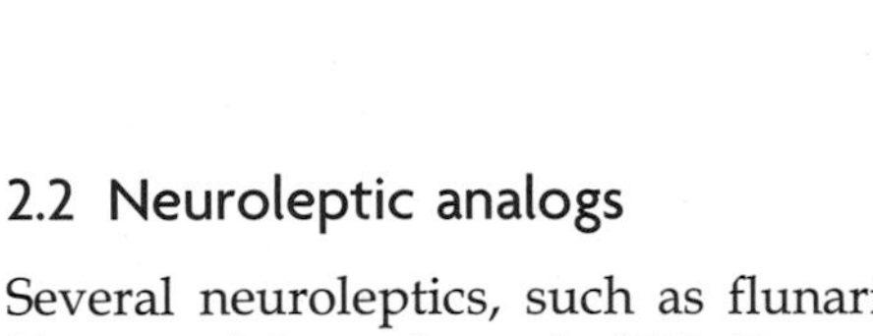

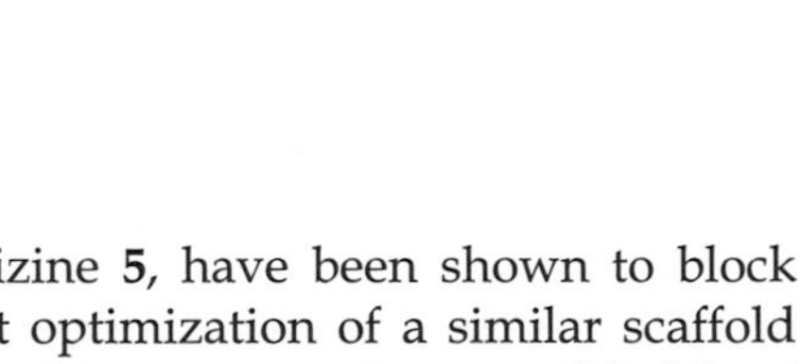

2.2 Neuroleptic analogs

Several neuroleptics, such as flunarizine 5, have been shown to block T-type calcium channels [52]. Recent optimization of a similar scaffold using a pharmacophore model resulted in compounds exemplified by 6 (IC$_{50}$ 280 nM, patch-clamp assay), which were shown to block the Ca$_v$3.1 subtype of the T-type calcium channel [53]. No selectivity information was reported for 6. Antagonists based on scaffolds identified using similar modeling and screening approaches have been reported, such as 7 (IC$_{50}$ 0.93 μM) and 8 (IC$_{50}$ 6.9 μM), although potency was somewhat reduced with these templates [54].

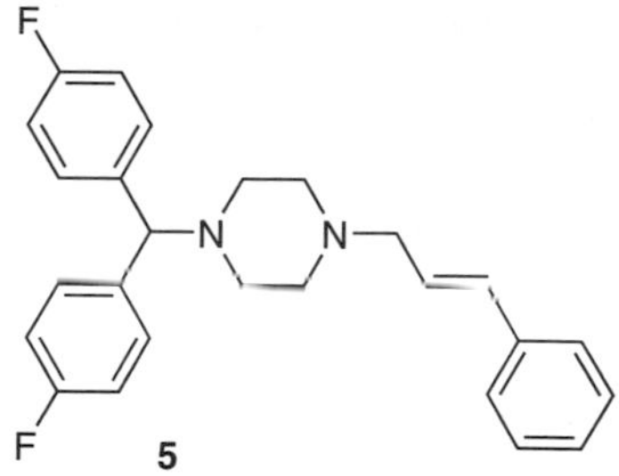

2.3 Quinazolines and quinazolinones

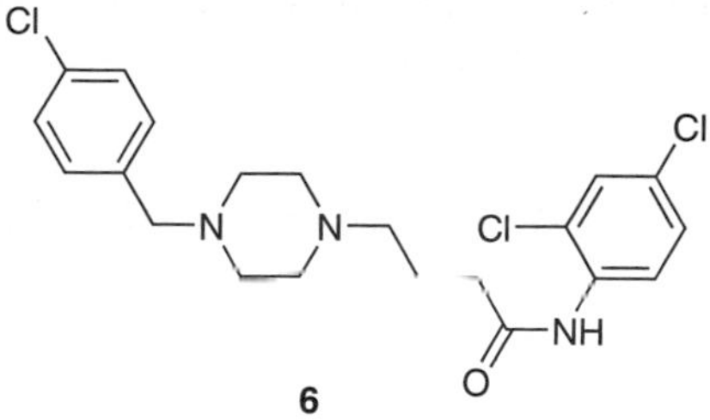

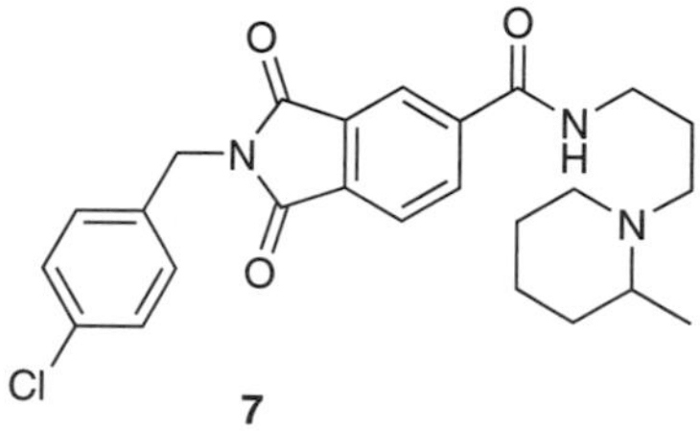

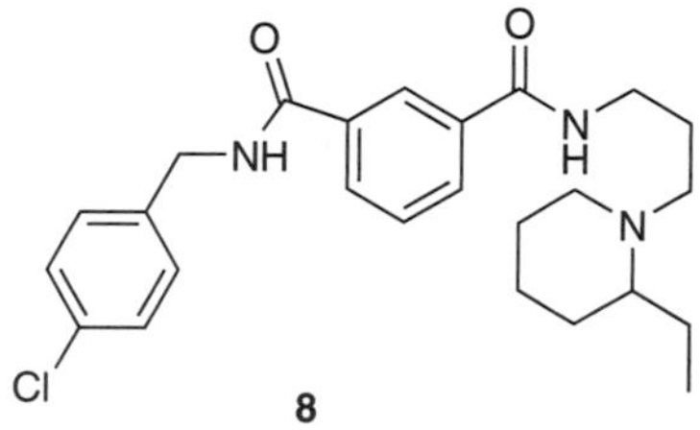

A series of 2-amino-3,4-dihydro quinazolines have been extensively explored as selective T-type calcium channel antagonists. A recent disclosure included KYS05090 (**9**) with an IC_{50} of 41 nM on the $Ca_v3.1$ subtype of the T-type channel and 120-fold selectivity versus the N-type calcium channel $Ca_v2.2$ [55]. A pharmacophore model was recently published based on this and related structures [56], but no other selectivity or *in vivo* activity have been disclosed since the original report.

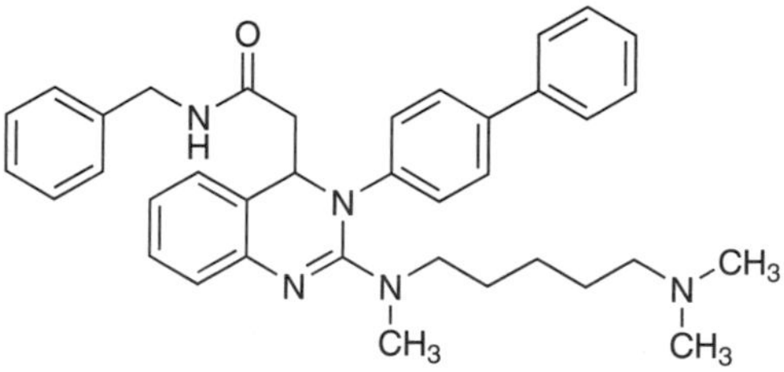

9

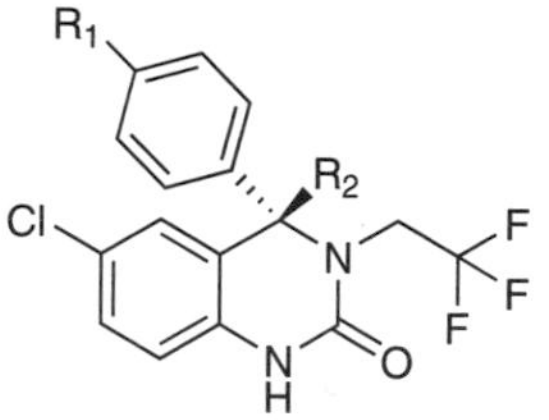

10 $R^1 = F$, $R^2 = Et$
11 $R^1 = CN$, $R^2 = cyclopropyl$

A series of 4,4-disubstituted quinazolin-2-ones derived from HIV nonnucleoside reverse transcriptase inhibitor leads have shown good *in vitro* potency and *in vivo* efficacy [28]. Using FLIPR assays on cell lines with different resting membrane potentials, TTA-Q3 (**10**) and TTA-Q6 (**11**) were shown to be state-dependent blockers of the $Ca_v3.3$ channel. In the depolarized state, compound **10** blocked calcium influx (IC_{50} 61 nM), and compound **11** was more potent (IC_{50} 14 nM). However, in the hyperpolarized state assay, the potency of **10** was more than 20-fold weaker (IC_{50} 1200 nM) than for **11** (IC_{50} 590 nM). Both compounds have good selectivity versus L-type and N-type calcium channels as measured by FLIPR assays, with good bioavailability across three species (rat, dog, rhesus macaque). *In vivo* efficacy was demonstrated in the WAG/Rij rat epilepsy model as well as the rat harmaline model of tremor. Compound **11** showed significant decreases in active wake activity after oral dosing to rats, despite being dosed immediately prior to the rat's inactive period. This finding highlights the potential for T-type calcium channel antagonists to affect arousal.

2.4 Piperidines

Compounds **12** and **13**, potent T-type calcium channel antagonists, were identified by modification of a nonselective piperidine lead, principally by employing piperidine 3- and 4-fluoro substituents to reduce basicity [26,27]. These changes resulted in improved selectivity versus other ion channels. The resulting ligands display little state dependence as determined by a FLIPR functional assay on the $Ca_v3.3$ channel, and this finding was later confirmed by electrophysiology (Table 2). The potency and state independence were similar when the $Ca_v3.1$ and $Ca_v3.2$ subtypes were examined, demonstrating the difficulty of finding subtype selective compounds within a family of calcium channels. Both **12** and **13** have good *in vivo* pharmacokinetic parameters and were efficacious in epilepsy and tremor models. These compounds were also extensively examined for cardiovascular effects, with no significant changes in blood pressure or ECG intervals (PR, QRS, and QT) noted in dogs. A recent report confirmed the potency and selectivity of **12** in thalamic slice preparations [57].

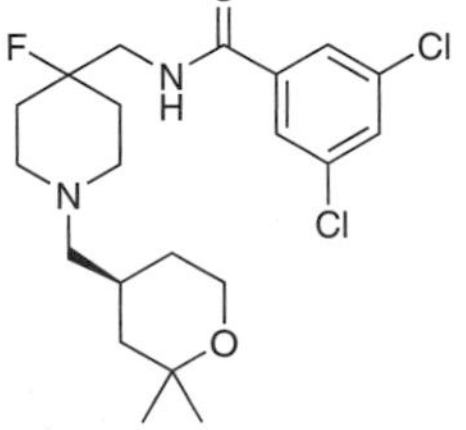

12 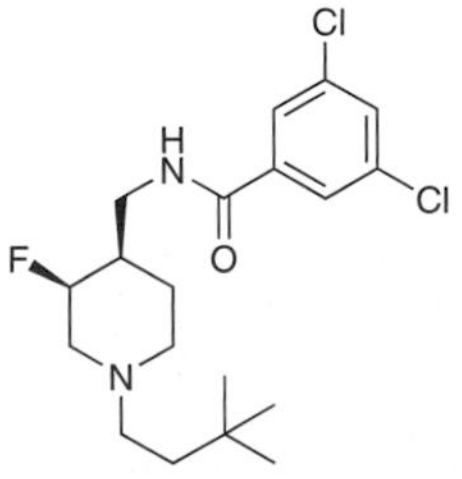   **13**

Table 2 Potency of 12 and 13 versus T-type calcium channel subtypes

	12	13
Cav3.3 FLIPR (−60 mV)	297 nM	79 nM
Cav3.3 FLIPR (−20 mV)	94 nM	32 nM
Cav3.3 EP (−80 mV)	115 nM	25 nM
Cav3.3 EP (−100 mV)	84 nM	43 nM
Cav3.2 EP (−100 mV)	196 nM	NT
Cav3.1 EP (−100 mV)	93 nM	62 nM

2.5 Phenylacetamides

A phenylacetamide series of state-dependent T-type calcium channel antagonists are exemplified by TTA-A1 (14), which has a $Ca_v3.3$ FLIPR IC_{50} of 22 nM in a depolarized state assay and an IC_{50} of 312 nM in a hyperpolarized cell line [14]. The potency and physical properties of this compound made it a useful radioligand in the binding assay. Thus, distinct binding sites for 14 compared to quinazolinones 10 and 11 were identified using this ligand. An optimal compound from this structural series, TTA-A2 (15), showed an IC_{50} of 9 nM in a depolarized cell line as well as good selectivity and pharmacokinetics. Compound 15 was dosed orally to mice at 10 mg/kg daily for 14 weeks, showing significant effects on sleep/wake activity patterns. Interestingly, significant weight loss was observed with mice on a high-fat, high-carbohydrate diet, but not those fed normal chow. This unexpected pharmacological outcome was hypothesized to be the result of better circadian timing of sleep and wake cycles with feeding patterns [37].

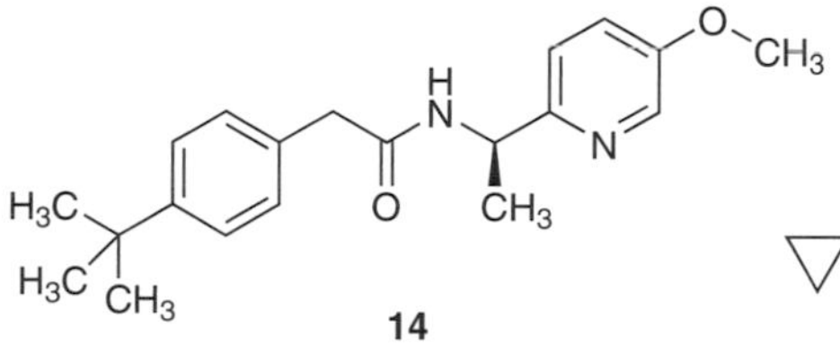

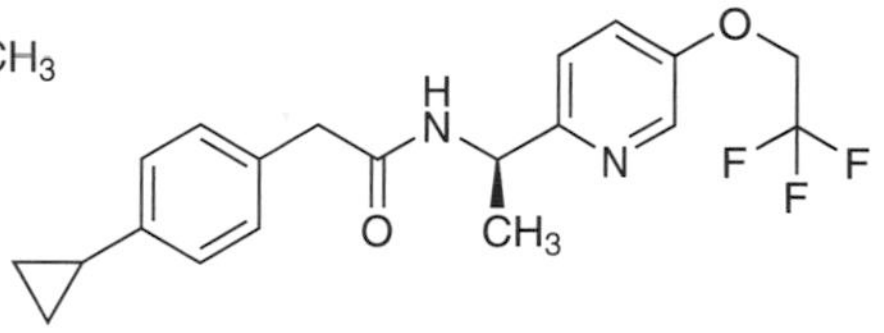

2.6 Imidazopyridazines

Imidazopyridazines DM1 (16) and DM2 (17), originally prepared as cyclooxygenase inhibitors, have demonstrated T-type antagonist properties in voltage clamp studies using the Cav3.1 subtype, as well as seizure suppression in WAG/Rij rats at 1, 3, and 10 mg/kg. The potency of these compounds toward the block of other ion channels was not reported [58].

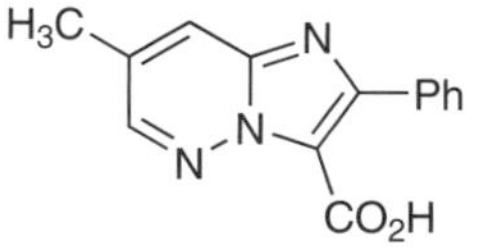

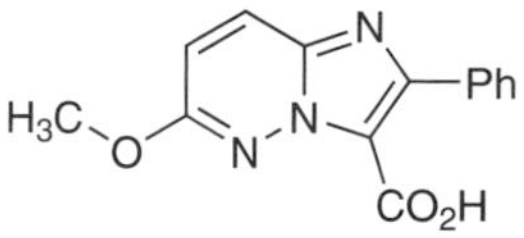

3. N-TYPE CALCIUM CHANNEL BLOCKERS

A structurally diverse array of N-type calcium channel antagonists has been reported, and two recent extensive reviews have summarized the structures and associated data through 2008 [59,60]. Among these

reported structures, there are pharmacophores that have been dispropor-
tionally represented, exemplified by the most recent reports included
below.

3.1 Peptidomimetic analogs

Extensive structural studies and structure–activity relationship (SAR)
investigations have been performed on the selective peptidyl N-type
calcium channel blocker ω-conotoxins GVIA and MVIIA in an attempt
to identify a minimal pharmacophore required for potency [61,62]. Compound **18** has been reported as a small-molecule mimetic of GVIA, pre-
senting the critical tripeptide pharmacophore of Lys2, Tyr13, and Arg17,
and was found to displace ^{125}I-labelled ω-conotoxin GVIA from rat brain
membrane (IC$_{50}$ 1.8 μM). Recently, the dipeptidyl mimetic **19** was
reported (IC$_{50}$ 5.8 μM), and this truncated structure resulted in only a
moderate loss of binding potency [63]. However, the peptide displacement binding assay used to examine these compounds does not afford
the same insight into the state dependence of block of the channel as does
the calcium functional assay or whole-cell patch-clamp electrophysiolo-
gical methods.

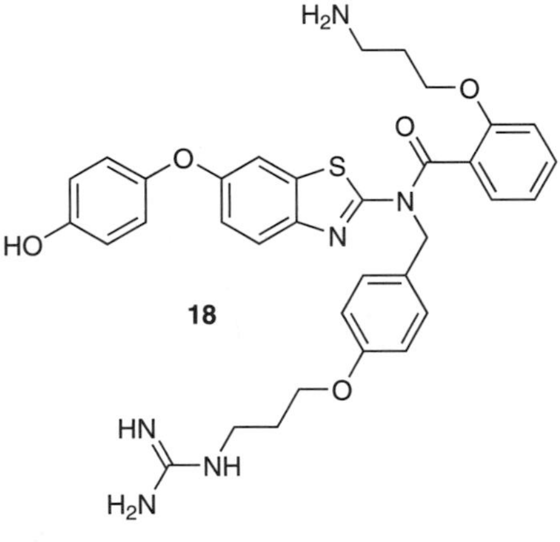

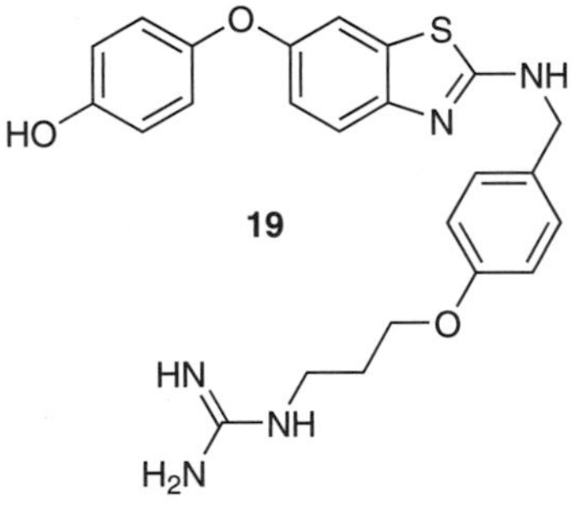

3.2 Aliphatic benzhydryl derivatives

The benzhydryl substituent is common to many reported N-type calcium
channel blockers. A recently reported series derived from the neuroepi-
leptic flunarizine (**5**), which exhibits N-type calcium channel potency in
a whole-cell patch-clamp assay (IC$_{50}$ 0.08 μM) as well as L-type potency

(IC$_{50}$ 0.31 µM) [64], in addition to the T-type activity noted above (IC$_{50}$ 0.53 µM vs. Ca$_v$3.1) [52]. A successful optimization effort based on **5** was recently reported to improve the subtype selectivity of this pharmaco-phore, resulting in NP118809 (**20**). This compound afforded similar potency against the N-type channel (IC$_{50}$ 0.11 µM) and was >100-fold selective over the L-type channel and the cardiac potassium channel human ether-a-go-go (hERG). The compound was found to have rat pharmacokinetic properties suitable for preclinical efficacy studies and afforded significant analgesic efficacy in the phase IIA portion of the rat formalin model of inflammatory pain (25 mg/kg i.p.). Substantial rever-sal of both mechanical allodynia and thermal hyperalgesia following oral dosing of 30 mg/kg of **20** in the rat spinal nerve ligation (SNL) model of neuropathic pain was recently reported [65]. Further optimization stu-dies on **20** gave the related benzhydryl N-type blockers **21** (IC$_{50}$ 0.05 µM) and **22** (IC$_{50}$ 0.15 µM), which also showed significant efficacy in the SNL model following oral administration at 30 mg/kg.

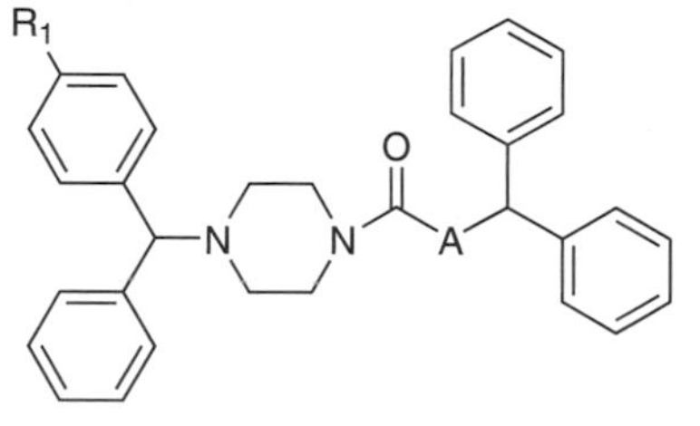

20: R^1 = H, A = CH$_2$
21: R^1 = Cl, A = CH$_2$ (*R* isomer)
22: R^1 = H, A = NH

3.3 Sulfonamide and sulfone derivatives

A variety of structurally related aminopiperidine sulfonamide deriva-tives have been independently reported by several groups as N-type calcium channel antagonists [59,66]. A common feature of many of these compounds is the trifluoromethylphenylsulfonyl substituent, as illustrated in **23–25**. The exocyclic sulfonamides are the most widely represented, exemplified by **23**, with reported highly selective potency for the N-type calcium channel (IC$_{50}$ 0.39 µM) over the L-type channel (IC$_{50}$ > 20 µM) in patch-clamp assays [67]. A modification of this phar-macophore has afforded a series of endocyclic piperidine and piperazine sulfonamides, illustrated by **24**, although no specific biological data was provided [68]. A further extension of this pharmacophore has recently appeared, wherein the sulfonamide group has been replaced by the gem-dimethyl sulfone moiety as in the piperidine **25**, which retains potency in a FLIPR assay (IC$_{50}$ 0.15 µM) against the N-type calcium channel [69].

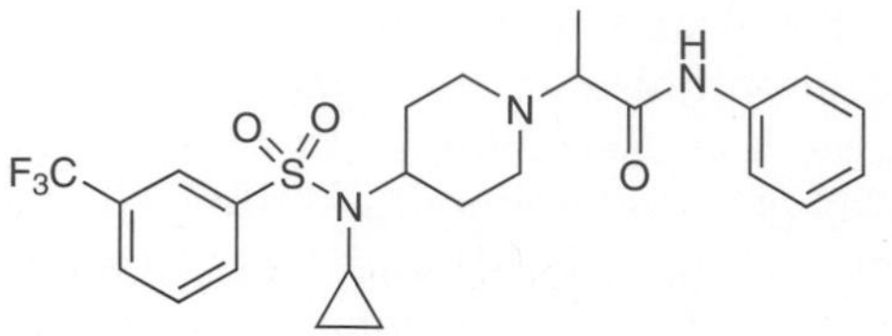

23

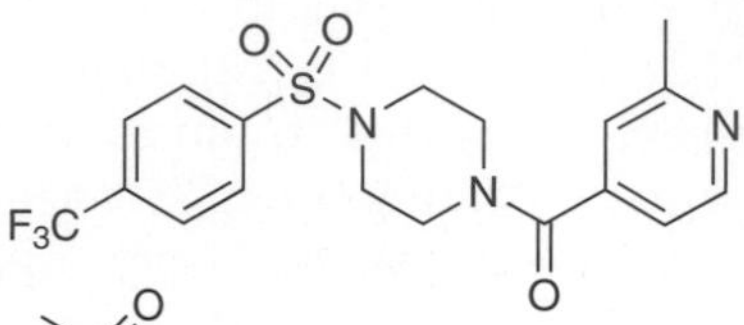

24

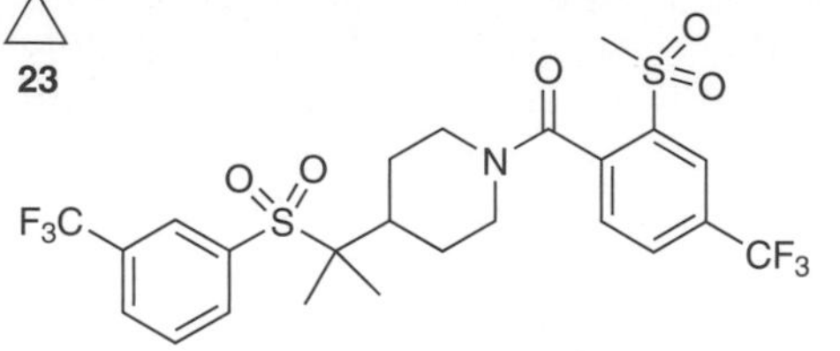

25

3.4 Oxindole derivatives

A novel oxindole pharmacophore was recently reported to be a Ca_v2 blocker [70]. The initial lead compound **26**, identified via high-throughput screening (HTS), was optimized to provide "TROX-1" (**27**). Compound **27** is an orally bioavailable, state-dependent antagonist of N-type ($Ca_v2.2$) channels (IC_{50} 0.27 μM, FLIPR assay), but was also found to be a state-dependent blocker of the related P/Q- and R-type channels. The analgesic efficacy of **27** is similar to nonsteroidal anti-inflammatory drugs (NSAIDs) in rat models of inflammatory pain and similar to pregabalin and duloxetine in a rat model of neuropathic pain. The efficacy was abrogated in mice deficient in $Ca_v2.2$, suggesting that the analgesic activity is derived primarily from blockade of $Ca_v2.2$ channels. Importantly, at efficacious doses, **27** produced no observable changes in motor coordination (rats), cardiovascular function (dogs), or hemodynamic parameters (dogs), and established a therapeutic exposure window of ~30 fold between significant preclinical analgesia and ancillary pharmacologic effects. These findings indicate that a state-dependent Ca_v2 blocker may afford a substantially greater therapeutic window for neurologic and hemodynamic effects than ziconotide [71].

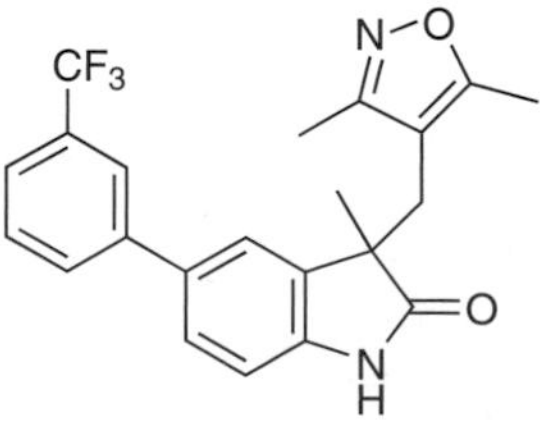

26

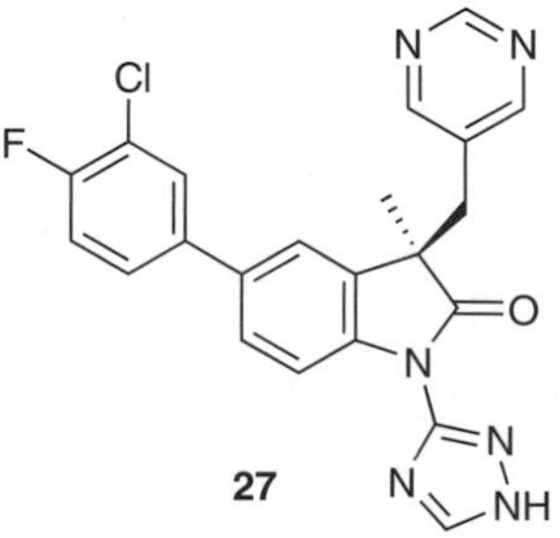

27

4. CONCLUSIONS

Recent work shows that antagonism of voltage-gated calcium channels offers a rich potential to alter neuronal transmission via small-molecule intervention. However, investigation of this target class is complicated by the differential tissue distribution of the channel subtypes, different activation potentials, and different activation states for each channel. The chemotypes reported in the recent literature illustrate the wide variety of pharmacophores that are available to further explore these targets. The variety of calcium channel subtypes, while complex as a target class, affords the possibility for selective intervention in neuronal signaling pathways. However, the development of calcium channel antagonists as safe and selective therapeutics will rely on the discovery of compounds that antagonize only the aberrant channel activity while sparing normal neuronal function. Recent advances in optical assays for ion channel targets, such as the FLIPR technology, provide a more rapid format to examine compounds for ion channel selectivity and state-dependent properties of channel blockade [72]. These assays will support HTS activities and SAR progression that would have been difficult or impossible with more laborious patch-clamp methods. A refined understanding of the pharmacology associated with this target class will result from the iterative cycle of compound discovery and biochemical investigation with the ultimate proof of concept left to await human clinical trial results.

REFERENCES

[1] T. P. Snutch in *Encyclopedia of Neuroscience*, (ed. R. S. Larry), Academic Press, Oxford, 2009, p. 427.
[2] W. A. Catterall, E. Perez-Reyes, T. P. Snutch, and J. Striessnig, *Pharmacol. Rev.*, 2005, **57**, 411.
[3] C. J. Doering and G. W. Zamponi, *Curr. Pharm. Des.*, 2005, **11**, 1887.
[4] W. A. Catterall in *Encyclopedia of Neuroscience*, (ed. R. S. Larry), Academic Press, Oxford, 2009, p. 543.
[5] A. A. Kochegarov, *Cell Calcium*, 2003, **33**, 145.
[6] M. Spedding and P. M. Vanhoutte, *J. Cardiovasc. Pharmacol.*, 1993, **22**, 906.
[7] E. A. Ertel, K. P. Campbell, M. M. Harpold, F. Hofmann, Y. Mori, E. Perez-Reyes, A. Schwartz, T. P. Snutch, T. Tanabe, L. Birnbaumer, R. W. Tsien, and W. A. Catterall, *Neuron*, 2000, **25**, 533.
[8] W. A. Catterall, *Annu. Rev. Cell Dev. Biol.*, 2000, **16**, 521.
[9] S. J. Dubel, C. Altier, S. Chaumont, P. Lory, E. Bourinet, and J. Nargeot, *J. Biol. Chem.*, 2004, **279**, 29263.
[10] R. J. Winquist, J. Q. Pan, and V. K. Gribkoff, *Biochem. Pharmacol.*, 2005, **70**, 489.
[11] P. G. Patil, D. L. Brody, and D. T. Yue, *Neuron*, 1998 **20**, 1027.
[12] J. P. Bingham, E. Mitsunaga, and Z. L. Bergeron, *Chem. Biol. Interact.*, 2010, **183**, 1.
[13] G. Dai, R. J. Haedo, V. A. Warren, K. S. Ratliff, R. M. Bugianesi, A. Rush, M. E. Williams, J. Herrington, M. M. Smith, O. B. McManus, and A. M. Swensen, *Assay Drug Dev. Technol.*, 2008, **6**, 195.

[14] V. N. Uebele, C. E. Nuss, S. V. Fox, S. L. Garson, R. Cristescu, S. M. Doran, R. L. Kraus, V. P. Santarelli, Y. Li, J. C. Barrow, Z. Q. Yang, K. A. Schlegel, K. E. Rittle, T. S. Reger, R. A. Bednar, W. Lemaire, F. A. Mullen, J. E. Ballard, C. Tang, G. Dai, O. B. McManus, K. S. Koblan, and J. J. Renger, *Cell Biochem. Biophys.*, 2009, **55**, 81.

[15] M. Xia, J. P. Imredy, K. S. Koblan, P. Bennett, and T. M. Connolly, *Anal. Biochem.*, 2004, **327**, 74.

[16] Y. Q. Cao, *Pain*, 2006, **126**, 5.

[17] S. M. Todorovic, and V. Jevtovic-Todorovic, *CNS Neurol. Disord. Drug Targets*, 2006, **5**, 639.

[18] G. P. Miljanich, *Curr. Med. Chem.*, 2004, **11**, 3029.

[19] C. Kim, K. Jun, T. Lee, S. S. Kim, M. W. McEnery, H. Chin, H. L. Kim, J. M. Park, D. K. Kim, S. J. Jung, J. Kim, and H. S. Shin, *Mol. Cell Neurosci.*, 2001, **18**, 235.

[20] E. Bourinet, A. Alloui, A. Monteil, C. Barrere, B. Couette, O. Poirot, A. Pages, J. McRory, T. P. Snutch, A. Eschalier, and J. Nargeot, *EMBO J.*, 2005, **24**, 315.

[21] S. Choi, H. S. Na, J. Kim, J. Lee, S. Lee, D. Kim, J. Park, C. C. Chen, K. P. Campbell, and H. S. Shin, *Genes Brain Behav.*, 2007, **6**, 425.

[22] M. E. Barton, E. L. Eberle, and H. E. Shannon, *Eur. J. Pharmacol.*, 2005, **521**, 79.

[23] A. C. Errington, T. Stohr, and G. Lees, *Curr. Top. Med. Chem.*, 2005, **5**, 15.

[24] D. Varela, H. Khosravani, S. E. Heron, C. Bladen, T. C. Williams, M. Newman, S. F. Berkovic, I. E. Scheffer, J. C. Mulley, and G. W. Zamponi, *Biophys. J.*, 2007, **92**, 601A.

[25] I. Bidaud, I. Vitko, S. Dubel, E. Bourinet, J. Nargeot, E. Perez-Reyes, and P. Lory, *Epilepsia*, 2005, **46**, 59.

[26] W. D. Shipe, J. C. Barrow, Z. Q. Yang, C. W. Lindsley, F. V. Yang, K. A. S. Schlegel, Y. Shu, K. E. Rittle, M. G. Bock, G. D. Hartman, C. Tang, J. E. Ballard, Y. Kuo, E. D. Adarayan, T. Prueksaritanont, M. M. Zrada, V. N. Uebele, C. E. Nuss, T. M. Connolly, S. M. Doran, S. V. Fox, R. L. Kraus, M. J. Marino, V. K. Gratifelds, H. M. Vargas, P. B. Bunting, M. Hasbun-Manning, R. M. Evans, K. S. Koblan, and J. J. Renger, *J. Med. Chem.*, 2008, **51**, 3692.

[27] Z. Q. Yang, J. C. Barrow, W. D. Shipe, K. A. S. Schlegel, Y. S. Shu, F. V. Yang, C. W. Lindsley, K. E. Rittle, M. G. Bock, G. D. Hartman, V. N. Uebele, C. E. Nuss, S. V. Fox, R. L. Kraus, S. M. Doran, T. M. Connolly, C. Y. Tang, J. E. Ballard, Y. S. Kuo, E. D. Adarayan, T. Prueksaritanont, M. M. Zrada, M. J. Marino, V. K. Graufelds, A. G. DiLella, I. J. Reynolds, H. M. Vargas, P. B. Bunting, R. F. Woltmann, M. M. Magee, K. S. Koblan, and J. J. Renger, *J. Med. Chem.*, 2008, **51**, 6471.

[28] J. C. Barrow, K. E. Rittle, T. S. Reger, Z.-Q. Yang, P. Bondiskey, G. B. McGaughey, M. G. Bock, G. D. Hartman, C. Tang, J. Ballard, Y. Kuo, T. Prueksaritanont, C. Nuss, S. Doran, S. V. Fox, S. L. Garson, R. L. Kraus, Y. Li, M. J. Marino, V. Kuzmick Graufelds, V. N. Uebele, and J. J. Renger, *ACS Med. Chem. Lett.*, 2010, **1**, 75.

[29] P. M. Newton, L. Zeng, V. Wang, J. Connolly, M. J. Wallace, C. Kim, H. S. Shin, F. Belardetti, T. P. Snutch, and R. O. Messing, *J. Neurosci.*, 2008, **28**, 11712.

[30] F. J. Urbano, V. Bisagno, S. I. Wikinski, O. D. Uchitel, and R. R. Llinas, *Biol. Psychiatry*, 2009, **66**, 769.

[31] V. K. Gribkoff, and R. J. Winquist, *Expert Opin. Invest. Drugs*, 2005, **14**, 579.

[32] K. Valentino, R. Newcomb, T. Gadbois, T. Singh, S. Bowersox, S. Bitner, A. Justice, D. Yamashiro, B. B. Hoffman, and R. Ciaranello, *Proc. Natl. Acad. Sci. USA*, 1993, **90**, 7894.

[33] K. Shahlaie, B. G. Lyeth, G. G. Gurkoff, J. P. Muizelaar, and R. F. Berman, *J. Neurotrauma*, 2010, **27**, 175.

[34] C. A. Hicks, M. A. Ward, and M. J. O'Neill, *Eur. J. Pharmacol.*, 2000, **408**, 241.

[35] L. Cueni, M. Canepari, and J. P. Adelman, A. Luthi, *Pfluegers Arch.*, 2009, **457**, 1161.

[36] D. Contreras, *CNS Neurol. Disord. Drug Targets*, 2006, **5**, 571.

[37] V. N. Uebele, A. L. Gotter, C. E. Nuss, R. L. Kraus, S. M. Doran, S. L. Garson, D. R. Reiss, Y. X. Li, J. C. Barrow, T. S. Reger, Z. Q. Yang, J. E. Ballard, C. Y. Tang, J. M. Metzger, S. P. Wang, K. S. Koblan, and J. J. Renger, *J. Clin. Invest.*, 2009, **119**, 1659.

[38] R. R. Llinas, U. Ribary, D. Jeanmonod, E. Kronberg, and P. P. Mitra, *Proc. Natl. Acad. Sci. USA*, 1999, **96**, 15222.

[39] R. R. Llinas and M. Steriade, *J. Neurophysiol.*, 2006, **95**, 3297.

[40] Y. Zhang, R. R. Llinas, and J. E. Lisman, *Front. Neural. Circuits*, 2009, **3**, 20.

[41] E. Takahashi, K. Niimi, and C. Itakura, *Behav. Brain Res.*, 2010, **207**, 273.

[42] K. Jun, E.S. Piedras-Renteria, S. M. Smith, D. B. Wheeler, S. B. Lee, T. G. Lee, H. Chin, M. E. Adams, R. H. Scheller, R. W. Tsien, and H. S. Shin, *Proc. Natl. Acad. Sci. USA*, 1999, **96**, 15245.

[43] R. A. Ophoff, G. M. Terwindt, M. N. Vergouwe, E. R. van Eijk, P. J. Oefner, S. M. Hoffman, J. E. Lamerdin, H. W. Mohrenweiser, D. E. Bulman, M. Ferrari, J. Haan, D. Lindhout, G. J. van Ommen, M. H. Hofker, M. D. Ferrari, and R. R. Frants, *Cell*, 1996, **87**, 543.

[44] P. Imbrici, S. L. Jaffe, L. H. Eunson, N. P. Davies, C. Herd, R. Robertson, D. M. Kullmann, and M. G. Hanna, *Brain*, 2004, **127**, 2682.

[45] M. Osanai, H. Saegusa, A. A. Kazuno, S. Nagayama, Q. Hu, S. Zong, T. Murakoshi, and T. Tanabe, *Biochem. Biophys. Res. Commun.*, 2006, **344**, 920.

[46] E. Takahashi and K. Niimi, *Recent Pat. CNS Drug Discov.*, 2009, **4**, 96.

[47] J. C. Barrow and T. M. Connolly in *Voltage-Gated Ion Channels as Drug Targets*, (ed. D. J. Triggle, M. Gopalakrishnan, D. Rampe, and W. Zheng), Wiley-VCH Verlag GmbH and Co. KgaA, Weinheim, 2006, p. 84.

[48] D. A. Coulter, J. R. Huguenard, and D. A. Prince, *Ann. Neurol.*, 1989, **25**, 582.

[49] P. R. Strege, C. E. Bernard, Y. J. Ou, S. J. Gibbons, and G. Farrugia, *Am. J. Phys. Gastroenterol. Liver Phys.*, 2005, **289**, G249.

[50] M. R. Doddareddy, H. K. Jung, J. Y. Lee, Y. S. Lee, Y. S. Cho, H. Y. Koh, and A. N. Pae, *Bioorg. Med. Chem.*, 2004, **12**, 1605.

[51] H. K. Lee, Y. S. Lee, E. J. Roh, H. Rhim, J. Y. Lee, and K. J. Shin, *Bioorg. Med. Chem. Lett.*, 2008, **18**, 4424.

[52] C. M. Santi, F. S. Cayabyab, K. G. Sutton, J. E. Mcrory, J. Mezeyova, K. S. Hamming, D. Parker, A. Stea, and T. P. Snutch, *J. Neurosci.*, 2002, **22**, 396.

[53] J. H. Park, J. K. Choi, E. Lee, J. K. Lee, H. Rhim, S. H. Seo, Y. Kim, M. R. Doddareddy, A. N. Pae, J. Y. Kang, E. J. Roh, *Bioorg. Med. Chem.*, 2007, **15**, 1409.

[54] Y. Oh, Y. Kim, S. H. Seo, J. K. Lee, H. Rhim, A. N. Pae, K. S. Jeong, H. Choo, and Y. S. Cho, *Bull. Korean Chem. Soc.*, 2008, **29**, 1881.

[55] H. N. Seo, J. Y. Choi, Y. J. Choe, Y. Kim, H. Rhim, S. H. Lee, J. Kim, D. J. Joo, and J. Y. Lee, *Bioorg. Med. Chem. Lett.*, 2007, **17**, 5740.

[56] J. A. Jeong, H. Cho, S. Y. Jung, H. B. Kang, J. Y. Park, J. Kim, D. J. Choo, and J. Y. Lee, *Bioorg. Med. Chem. Lett.*, 2010, **20**, 38.

[57] F. M. Dreyfus, A. Tscherter, A. C. Errington, J. J. Renger, H. S. Shin, V. N. Uebele, V. Crunelli, R. C. Lambert, and N. Leresche, *J. Neurosci.*, 2010, **30**, 99.

[58] M. G. Rimoli, E. Russo, M. Cataldi, R. Citraro, P. Ambrosino, D. Melisi, A. Curcio, S. De Lucia, P. Patrignani, G. De Sarro, and E. Abignente, *Neuropharmacology*, 2009, **56**, 637.

[59] T. Yamamoto and A. Takahara, *Curr. Top. Med. Chem.*, 2009, **9**, 377.

[60] B. Bear, J. Asgian, A. Termin, and N. Zimmermann, *Curr. Opin. Drug Discovery Dev.*, 2009, **12**, 543.

[61] J. B. Baell, S. A. Forsyth, R. W. Gable, R. S. Norton, and R. J. Mulder, *J. Comput. Aided Mol. Des.*, 2001, **15**, 1119.

[62] J. B. Baell, P. J. Duggan, and Y. P. Lok, *Aust. J. Chem.*, 2004, **57**, 179.

[63] P. J. Duggan, R. J. Lewis, Y. Phei Lok, N. G. Lumsden, K. L. Tuck, and A. Yang, *Bioorg. Med. Chem. Lett.*, 2009, **19**, 2763.

[64] G. W. Zamponi, Z. P. Feng, L. Zhang, H. Pajouhesh, Y. Ding, F. Belardetti, H. Pajouhesh, D. Dolphin, L. A. Mitscher, and T. P. Snutch, *Bioorg. Med. Chem. Lett.*, 2009, **19**, 6467.

[65] H. Pajouhesh, Z. P. Feng, Y. Ding, L. Zhang, H. Pajouhesh, J. L. Morrison, F. Belardetti, E. Tringham, E. Simonson, T. W. Vanderah, F. Porreca, G. W. Zamponi, L. A. Mitscher, and T. P. Snutch, *Bioorg. Med. Chem. Lett.*, 2010, **20**, 1378.

[66] P. K. Chakravarty, P. S. Shao, and F. Ye, *Patent Application WO 2007/075524*, 2007.

[67] B. Shao and J. Yao, *Patent Application WO 2009/040659*, 2009.

[68] P. J. Beswick, A. Campbell, A. P. Cridland, R. J. Gleave, J. P. Heer, N. H. Nicholson, L. W. Page, and S. Vile, *Patent Application WO 2010/007072*, 2010.

[69] P. K. Chakravarty, Y. Ding, J. L. Duffy, H. Pajouhesh, P. S. Shao, S. Tyagarajan, and F. Ye, *Patent Application WO 2009/045382*, 2009.

[70] S. B. Hoyt, MEDI-017, 238th ACS National Meeting, Washington DC, August, 2009.

[71] C. Abbadie, O. B. McManus, S.-Y. Sun, R. M. Bugianesi, G. Dai, R. J. Haedo, J. B. Herrington, G. J. Kaczorowski, M. M. Smith, A. M. Swensen, V. A. Warren, B. Williams, S. P. Arneric, C. Eduljee, T. P. Snutch, E. W. Tringham, N. Jochnowitz, A. D. Liang, E. MacIntyre, E. McGowan, S. Mistry, V. V. White, S. B. Hoyt, C. London, K. A. Lyons, P. B. Bunting, S. Volksdorf, and J. L. Duffy, *J. Pharmacol. Exp. Ther.*, 2010, **334**, 545.

[72] E. Molokanova and A. Savchenko, *Drug Discov. Today*, 2008, **13**, 14–22.

CHAPTER 2

Recent Developments in Glycine Transporter-1 Inhibitors

Christopher L. Cioffi, Shuang Liu and **Mark A. Wolf**

1. INTRODUCTION

Schizophrenia is a chronic and devastating mental illness that affects approximately 1% of the world's population [1]. Men and women are affected equally and the incident rate is nearly the same for all countries and ethnic groups throughout the world. The disease is ranked among the top ten causes of disability worldwide [1].

The symptoms of schizophrenia generally appear in late adolescence or early adulthood and can be grouped into three specific categories. These include positive symptoms (auditory and visual hallucinations, disorganized thoughts and speech, delusions and irrational fears), negative symptoms (social withdrawal, anhedonia, blunted affect, lack of energy, and catatonia), and cognitive dysfunction (diminished capacity for learning and memory, attention, vigilance, and social cognition). Negative symptoms and especially cognitive impairment are considered

AMRI, 30 Corporate Circle, Albany, NY 12203, USA

Annual Reports in Medicinal Chemistry, Volume 45
ISSN 0065-7743, DOI 10.1016/S0065-7743(10)45002-2

to be the primary contributors to poor social functioning, the inability to work, lack of independent living skills, and poor quality of life for nearly all patients [1,2].

2. CURRENT ANTIPSYCHOTICS

Current antipsychotics used to treat patients are divided into two classes: the first generation antipsychotics (FGA) or typicals (e.g., chlorpromazine, haloperidol, thioridazine, and loxapine) and the second generation antipsychotics (SGA) or atypicals (i.e., clozapine, olanzapine, quetiapine, risperidone, aripiprazole, ziprasidone, and asenapine).

Activity of the typical antipsychotics is widely attributed to dopamine (DA) D_2 receptor antagonist activity in the mesolimbic pathway. These agents are regarded as efficacious at ameliorating positive symptoms, but generally ineffective against negative symptoms and cognitive deficiencies. Many typicals display extrapyramidal side effects (EPS) and pose the risk of tardive dyskinesia and hyperprolactinemia upon chronic usage [3].

Atypical antipsychotics possess strong serotonin (5-HT_{2A}) receptor antagonism in addition to D_2 antagonism, which is believed to attenuate EPS. However, these agents are largely ineffective against negative symptoms and cognitive deficits, with the exception of clozapine in regards to negative symptoms. Many atypicals exhibit their own side effects that may include metabolic syndrome (weight gain and diabetes) and associated cardiovascular risks [4].

Safe and effective treatment for the entire spectrum of symptoms remains a critical unmet need within the patient population [5]. Alternative avenues of pharmacological intervention are being explored, including nondopaminergic central nervous system (CNS) targets that may play critical roles in disease amelioration.

3. NMDA RECEPTOR HYPOFUNCTION

The D_2 antagonist activity of current antipsychotics led to the "dopamine hypothesis," which states that the pathophysiology of schizophrenia is due to excessive dopaminergic neurotransmission and dysfunctional D_2 signaling [6]. This hypothesis has prevailed for nearly 60 years; however, it falls short as a complete explanation due to the deficiencies current antipsychotics exhibit against negative and cognitive symptoms.

A growing body of evidence is emerging that suggests glutamatergic neurotransmission plays a key role in the etiology of the disease.

The "glutamate hypothesis" originated from the discovery that the non-competitive N-methyl-D-aspartate (NMDA) glutamate receptor antagonists ketamine and phencyclidine (PCP) induce positive, negative, and cognitive symptoms in healthy individuals and exacerbate symptoms in stable schizophrenic patients [7]. These observations suggest that NMDA receptor hypofunction plays a critical role in schizophrenia [7,8]. Thus, agents that can enhance receptor function may effectively ameliorate the full symptomology of psychosis. Direct agonism of the glutamate receptor is not a viable approach as it leads to excitotoxicity [9]. Consequently, several strategies to indirectly potentiate the receptor and avoid toxic side effects are being investigated [9–12]. One approach involves elevating synaptic glycine levels via inhibition of glycine transporter-1 (GlyT-1).

4. GLYCINE TRANSPORTERS

The NMDA receptor is an ionotropic glutamate receptor involved in fast excitatory neurotransmission. It plays a key role in a variety of CNS functions, most notably long-term potentiation (LTP) and neuronal plasticity, and is regulated by several mechanisms. One such mechanism involves the amino acid glycine (**1**).

Glycine is an inhibitory neurotransmitter at strychnine-sensitive glycine receptors (GlyA site) and an excitatory neurotransmitter at the glycine modulatory site located on the NR1 subunit of the NMDA receptor (GlyB site). It is an obligatory coagonist that allows glutamate to bind to and stimulate the receptor [13]. Thus, a possible method to potentiate NMDA receptor function would be to increase synaptic glycine concentrations. Studies show that patients experience improvement of negative symptoms when administered glycine (0.8 g/kg/day) or D-serine (a GlyB site agonist, 0.03 g/kg/day) in conjunction with clozapine [14]. Not surprisingly, large doses of these amino acids were required due to poor pharmacokinetics (PK) and CNS penetration. However, these encouraging results provided an impetus to discover alternative approaches for increasing synaptic glycine levels.

Recently, two high-affinity transporters that regulate synaptic glycine concentrations were identified: GlyT-1 and GlyT-2 [13]. Both share a high level of homology across species and approximately 50% homology with each other. Glycine transporters belong to the Na^+/Cl^- solute carrier 6 (SLC6) family, which includes the DA, serotonin (5-HT), norepinephrine (NE), leucine, taurine, proline, and GABA transporters. GlyT-2 is expressed in the brainstem and spinal cord and is colocalized with strychnine-sensitive glycine receptors. GlyT-1 is primarily expressed in neuronal and glial cells of the forebrain and is largely colocalized with NMDA receptors [13]. Thus, inhibition of GlyT-1 provides an opportunity to elevate glycine levels within close proximity to the GlyB site.

Studies reveal that a homozygous GlyT-1 (−/−) knockout in mice is neonatally lethal. However heterozygous GlyT-1 (+/−) mice survive to adulthood and display enhanced NMDA receptor function in the hippocampus, better memory retention, and no disruption in sensory gating when dosed with amphetamine [15].

A small placebo-controlled study, whereby 20 stably treated schizophrenic patients were given sarcosine (N-methyl glycine, 2), a weak but selective GlyT-1 inhibitor, at a dose of 2 g/day, showed that patients exhibited improvement in negative and cognitive symptoms [16]. These data, in conjunction with the results from the glycine and D-serine trials, have led to significant activity to develop GlyT-1 inhibitors with better drug-like and CNS-penetrant properties.

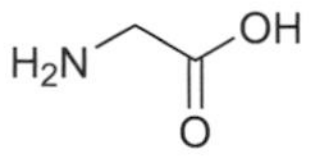

1 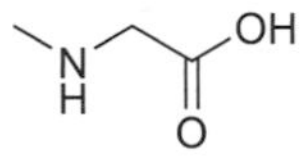   **2**

Several diverse, potent, and selective GlyT-1 inhibitors have appeared in the literature and many are reported to be efficacious in animal psychosis models. Several of these have advanced into Phase I and Phase II clinical studies. Recent Phase II results from a double-blind, 320-patient study with the investigational GlyT-1 inhibitor RG1678 (**33**) [17] demonstrated that the compound improved negative symptoms and social functioning of stable patients currently on atypical antipsychotic therapy and was well tolerated at all doses tested [18].

GlyT-1 inhibitors are categorized as either sarcosine-based or nonsarcosine-based inhibitors and several comprehensive reviews have appeared [13,19–23]. This chapter will highlight recent developments of some of these agents.

5. GLYCINE TRANSPORTER-1 INHIBITORS

5.1 Sarcosine-based GlyT-1 inhibitors

The design of many first-generation GlyT-1 inhibitors started with sarcosine and appended large N-linked hydrophobic moieties to it. In many cases, motifs were chosen to exploit the homology that GlyT-1 shares with other members of the SLC6 family. Thus, numerous early sarcosine-based inhibitors contain structural features shared with inhibitors of related transporters.

The first sarcosine-based inhibitor, NFPS ((R)-N-[3-(4'-fluorophenyl)-3-(4'-phenylphenoxy)propyl]sarcosine or ALX-5407, **3**), bears close resemblance to the selective serotonin reuptake inhibitor (SSRI) fluoxetine (**4**) [24]. The inhibitor exhibits good potency with regard to blockade of [³H] glycine uptake in QT6 cells stably expressing GlyT-1 (IC_{50} = 3 nM) and is also selective (GlyT-2, IC_{50} >75 µM). NFPS has been used extensively in the field as a pharmacological tool for gaining a better understanding of the role of GlyT-1 in the CNS and for providing a proof of mechanism in several animal models predictive of antipsychotic activity [24–36].

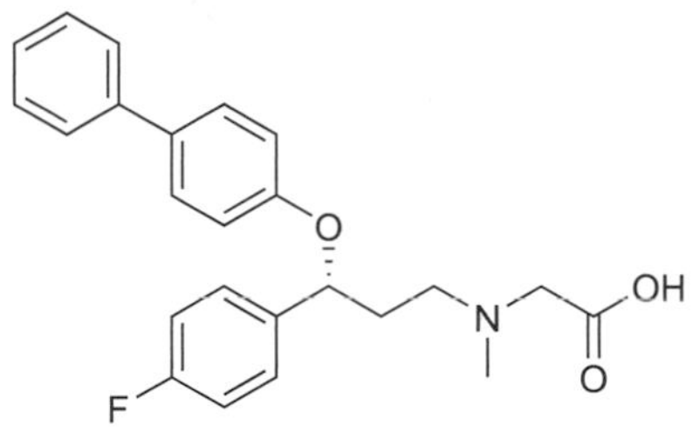

3 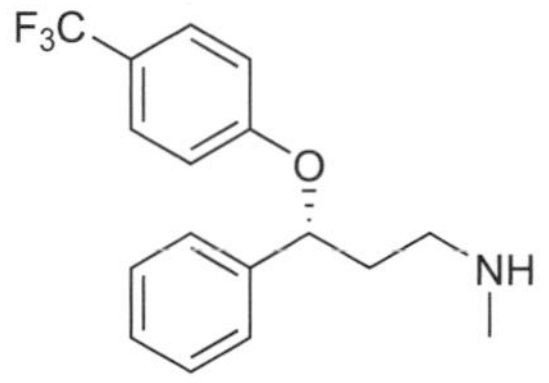   4

Since then, several potent and selective sarcosine-based inhibitors have been reported in the literature. Representative examples include **5** (LY2365109) [37,38], **6** ((R)-N[3-phenyl-3-(4'-(4-toluoyl)phenoxy)-propyl] sarcosine or (R)-NPTS) [39,40], **7** [41], **8** [42], **9** (JNJ-17305600) [43], and **10** [44]. Members of these series have demonstrated efficacy in several psychosis models [38,40] and JNJ-17305600 is reportedly in Phase I clinical trials for schizophrenia (data not available) [23].

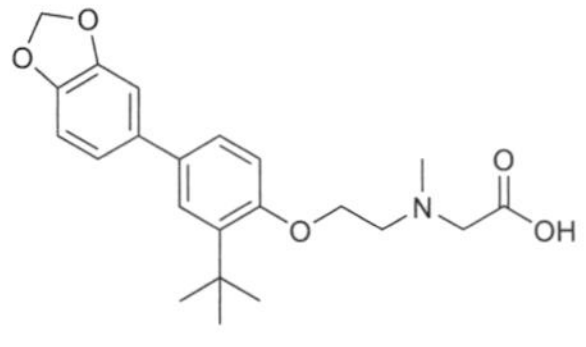

5 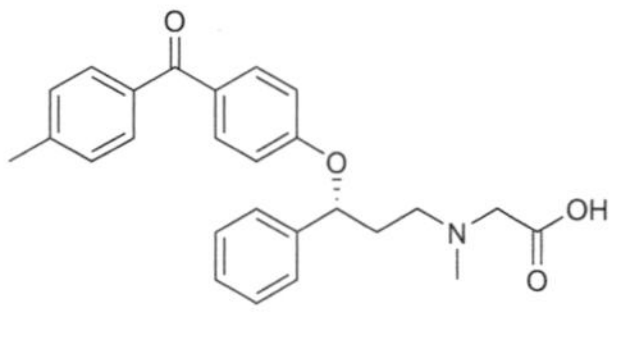   6 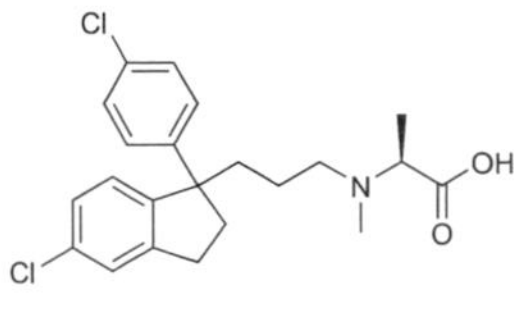   7

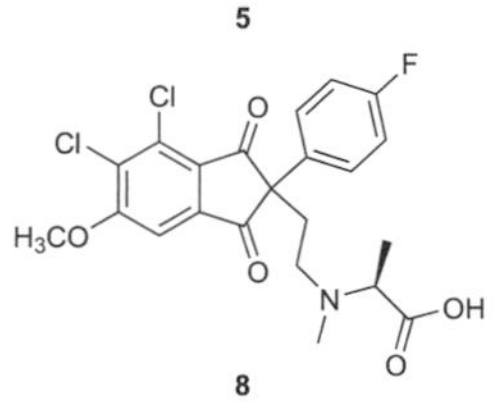

8 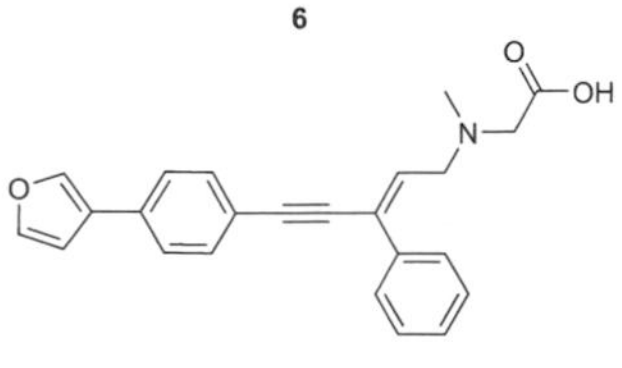   9 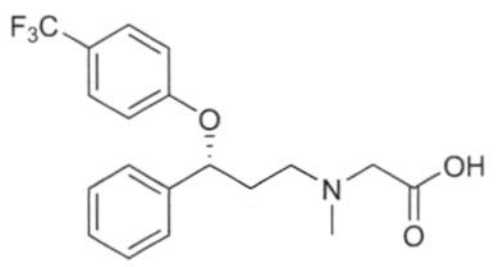   10

The most advanced sarcosine-based inhibitor reported to date is Org-25935 (**11**) [45]. This ligand increased striatal and cerebral spinal fluid (CSF) glycine levels (commonly used to measure *in vivo* efficacy) in both rats and cynomolgus monkeys in a dose- and time-dependent manner. Rats dosed with 6 mg/kg (i.p.) exhibited significantly increased CSF glycine levels relative to vehicle-treated rats and CSF drug levels reached 58 ± 5 ng/mL. Dose-dependent increases of CSF glycine levels in both male and female cynomolgus monkeys were observed during 3 weeks of daily oral administration of Org-25935 with basal levels returning 3 weeks after the final dose [45].

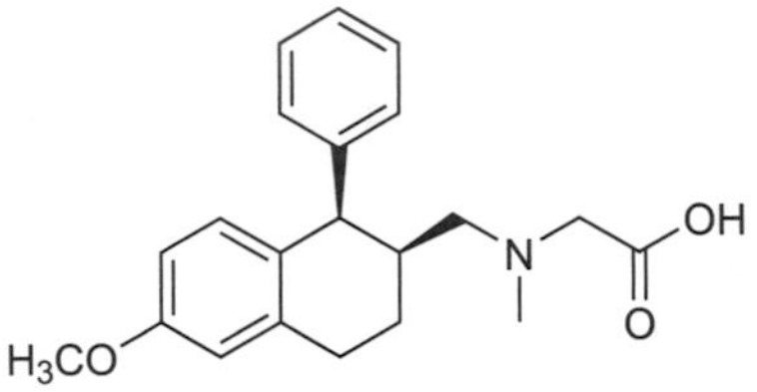

11

A Phase II study was recently completed whereby Org-25935 was compared against placebo for the ability to improve negative symptoms in 246 subjects maintained on a stable dose of an atypical antipsychotic (data not disclosed) [46]. A second Phase II study in progress (200 patients) is designed to assess the efficacy of Org-25935 as a stand-alone therapy versus placebo, using olanzapine as the active control [46]. Org-25935 is also being investigated in separate Phase II studies as a treatment for panic disorder and for recidivism in subjects with alcohol dependence [46].

Despite the progress reported for a few sarcosine-based inhibitors (e.g., **11**), the class has come under scrutiny due to reported toxicity associated with two ligands, NFPS and LY2365109. Both induce hypoactivity, impaired respiration, and ataxia in rodents [38], and NFPS exhibits slow dissociation kinetics, rendering binding to GlyT-1 essentially irreversible and leading to elevated glycine levels in the rat prefrontal cortex (PFC) for periods >24 h [24]. Furthermore, NFPS and (*R*)-NPTS are noncompetitive with glycine [47], and the binding profile of these inhibitors is thought to contribute to excessive and prolonged elevated glycine levels in rodents, which could lead to overstimulation of strychnine-sensitive glycine receptors [20,21,47]. However, it has not been determined whether the toxicity associated with NFPS and LY2365109 is mechanism based or compound specific.

Concurrent to these findings, the development of a second-generation of GlyT-1 inhibitors was pursued. These inhibitors differ from their predecessors in that they do not contain a sarcosine motif. The pharmaceutical industry has focused much of its effort on this emerging class, leading to several compounds currently being studied in clinical trials.

5.2 Nonsarcosine-based inhibitors

Efforts in the area of second-generation nonsarcosine GlyT-1 inhibitors enjoy a stunning breadth of structural diversity. The field encompasses a highly competitive landscape and many of the scaffolds have been disclosed only in patent literature.

Several patent applications claiming a class of inhibitors derived from the DA transporter (DAT) inhibitor methylphenidate (**12**) have been published [48–50]. A variety of structural modifications represented by analogs **13–15** have been claimed and pharmacological data has been reported for analog SSR-504734 (**13**) [51–53].

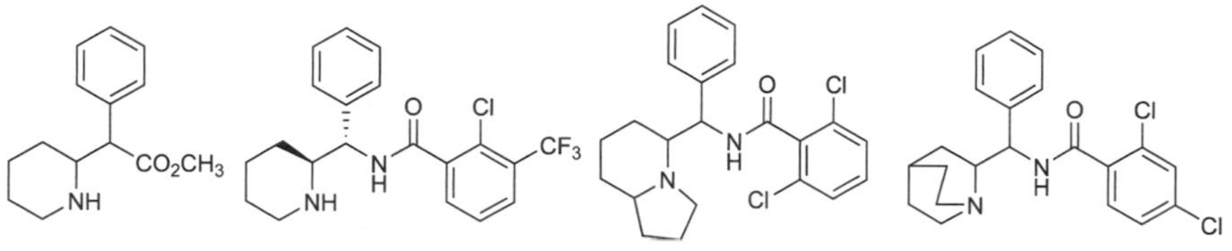

SSR-504734 is a potent, selective, and reversible inhibitor ($IC_{50} = 18$ nM) that is competitive with glycine [47,51]. The inhibitor rapidly and reversibly blocked the uptake of [^{14}C]glycine in mouse cortical homogenates, which was sustained for up to 7 h. Complete cessation of blockade and return to glycine basal levels occurred prior to 24 h, which is in stark contrast to NFPS (>24 h). SSR-504734 potentiated a nearly twofold increase of NMDA receptor-mediated excitatory postsynaptic currents (EPSCs) in rat hippocampal slices and produced an increase in contralateral rotations in mice when microinjected into the striatum. Microdialysis experiments indicated that the inhibitor induced a rapid and sustained increase in extracellular glycine levels in the PFC of freely moving rats [51]. The compound also demonstrated efficacy in a variety of psychosis models [51–53]. SSR-504734 was reportedly in clinical trials for schizophrenia but discontinued after Phase I (data not disclosed) [54].

A similar pharmacological profile was reported for the GlyT-1 inhibitor SSR-103800 (structure not disclosed) [55,56]. The compound is a potent ($IC_{50} = 1.9$ nM), selective, and reversible inhibitor that demonstrated efficacy in the aforementioned *in vitro* and *in vivo* studies. In addition, SSR-103800 did not produce catalepsy in rodents at doses up to 30 mg/kg, suggesting a reduced side-effect potential relative to typical antipsychotics [56]. The compound is reportedly in Phase I clinical trials for schizophrenia [54].

Other patent applications for scaffolds derived from methylphenidate have been published (**16–18**); however, no pharmacological data has been reported [57–59].

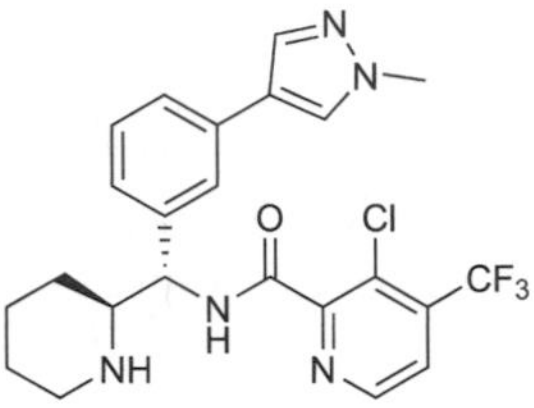

16 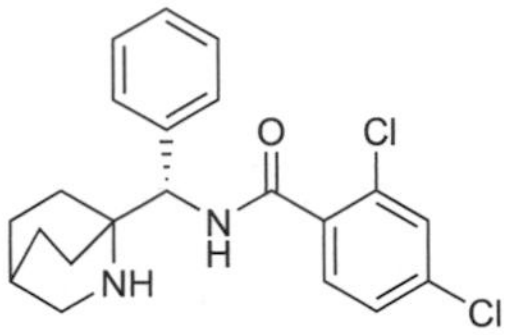   17 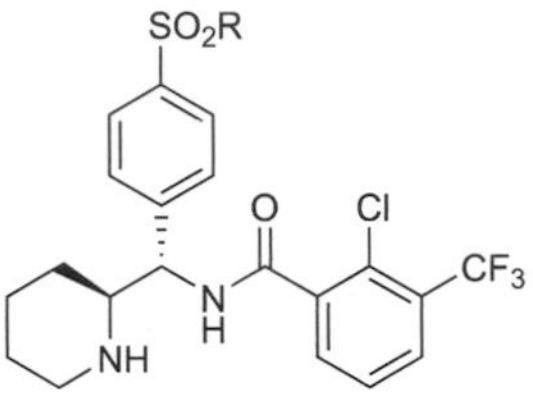   18

Sulfonamide **19**, a selective GlyT-1 inhibitor (IC$_{50}$ = 135 nM), was discovered via a high-throughput screening (HTS) effort [60]. Rapid structure–activity relationship (SAR) exploration led to **20** (IC$_{50}$ = 2.6 nM). There was a strong preference for the (S)-enantiomer and the α-methyl moiety eliminated rodent Pgp liabilities. Compound **20** increased PFC glycine concentrations and demonstrated significant enhancement of prepulse inhibition (PPI) in rodents. In addition, it was shown that [^{35}S]-radiolabeled **20** could be used as a radioligand for *in vivo* occupancy studies [61]. Drawbacks of this series included poor solubility and time-dependent cytochrome P450 (CYP) inhibition (TDI) [62]. Compound **21** exhibited good potency (IC$_{50}$ = 4.4 nM), improved solubility, and was devoid of TDI [63], although it suffered from poor PK and brain penetrance.

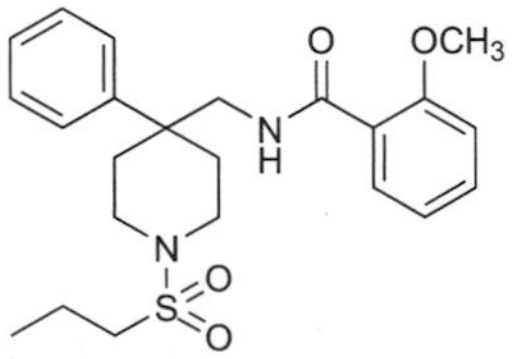

19 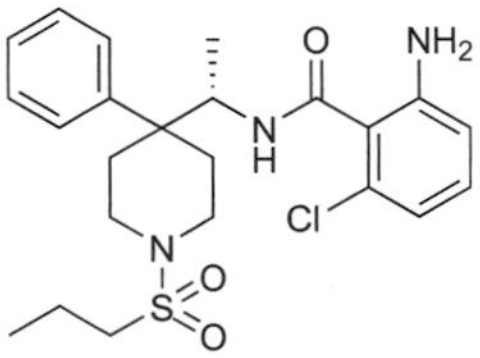   20 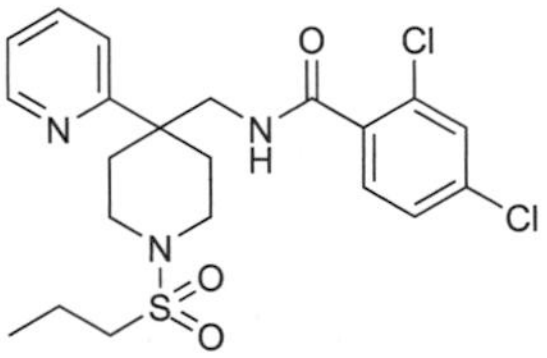   21

Metabolic oxidation of the sulfonamide propyl chain was found to be problematic. Truncation to the ethyl analog **22** resulted in good potency

(IC_{50} = 26 nM), selectivity, improved dog PK, and brain exposure in rodents [64]. Further optimization involved replacement of the alkyl chain with N-methyl imidazoles and triazoles; however, mitigation of sulfonamide side-chain metabolism resulted in oxidation of the piperidine ring. Consequently, cyclohexane sulfone analogs such as **23** (DCCCyB) were prepared [65]. In 2008, it was disclosed that **23** selectively bound to GlyT-1 receptors in rhesus monkeys, determined using a positron emission tomography (PET) study, and that **23** had successfully completed a Phase I clinical trial [66].

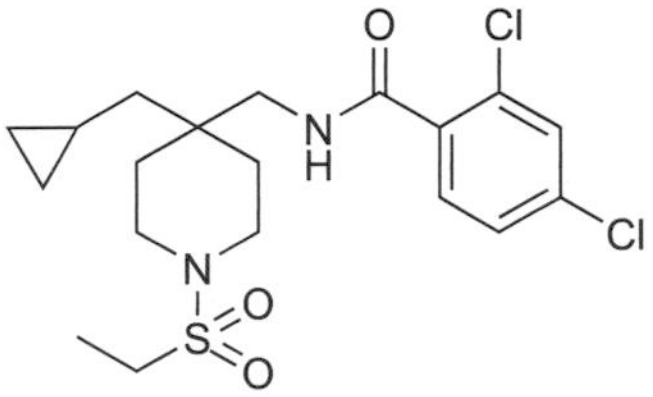

22 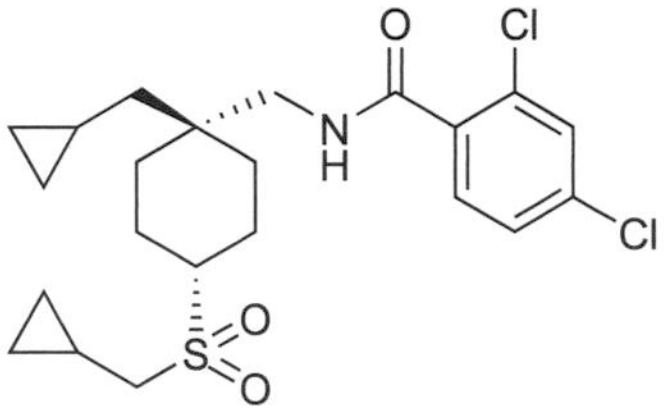　　　　　　　　　　**23**

Compound **24** represents a series of potent spiropiperidine GlyT-1 inhibitors that showed significant binding to the μ opioid and nociceptin/orphanin FQ peptide (NOP) receptors [67]. The hydroxy analog **25** demonstrated improved selectivity and N-propyl analog **26** exhibited increased microsomal stability but reduced GlyT-1 potency (EC_{50} = 450 nM) [68].

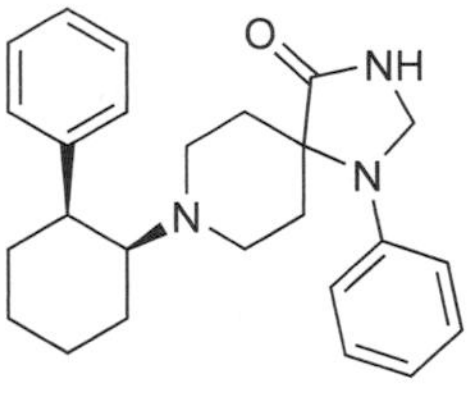

(+/–)　　　　　　　　(+/–)　　　　　　　　(+/–)

24　　　　　　　　**25**　　　　　　　　**26**

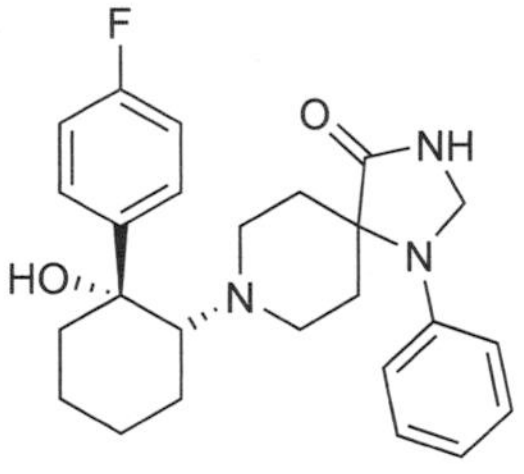

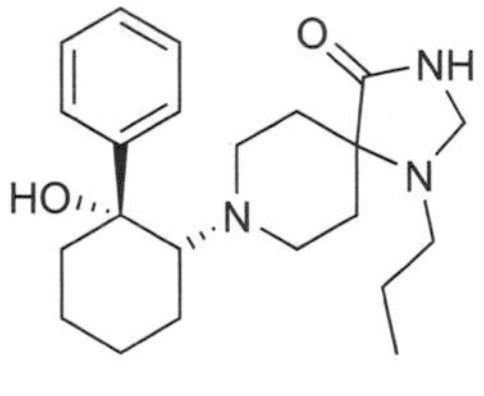

Removal of the anilino nitrogen of **24** was tolerated (**27**, EC_{50} = 61 nM) and selectivity remained against μ opioid and NOP receptors [69]. The 2-hydroxy analog **28** displayed good affinity for GlyT-1 (EC_{50} = 70 nM) and complete selectivity against GlyT-2, μ opioid, and NOP receptors but low micromolar affinity at the human ether-a-go-go related gene (hERG) channel (EC_{50} = 1.8 μM) [70]. A potent and selective pyran derivative **29** (EC_{50} = 91 nM) was reportedly devoid of hERG activity ($EC_{50} > 20$ μM) [71]. Further SAR exploration led to a 1-phenyl series exemplified by **30** ($EC_{50} = 97$ nM), with high oral bioavailability (60%) and CNS exposure in the mouse [71].

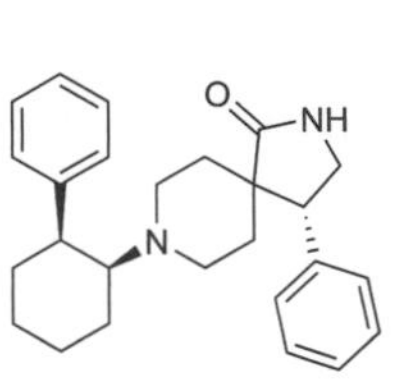

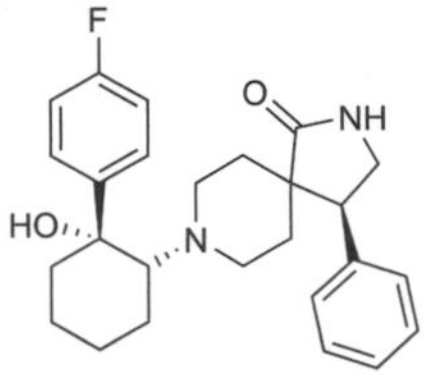

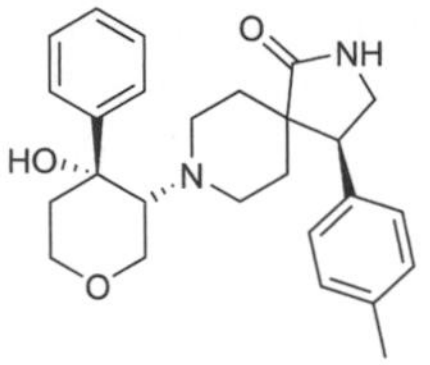

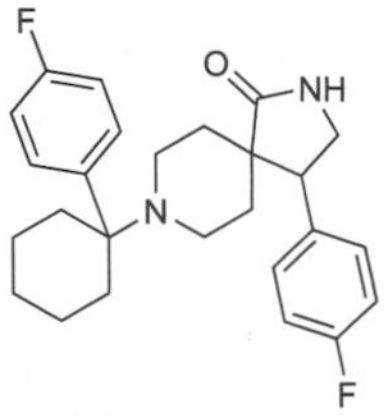

27   28   29   30

Benzoylpiperazine **31** is reported as a potent and selective inhibitor of GlyT-1 ($EC_{50} = 15$ nM) [72].

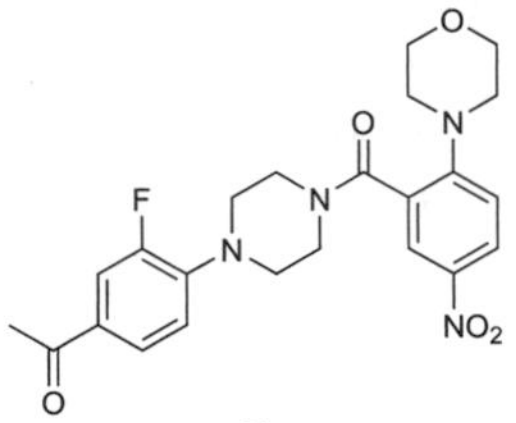

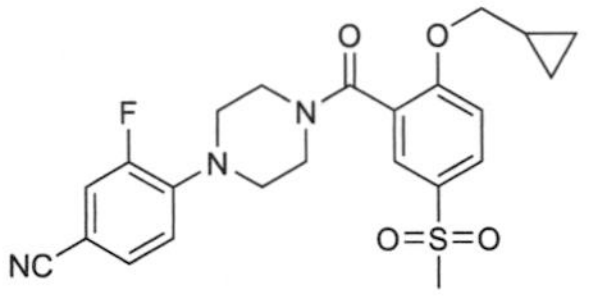

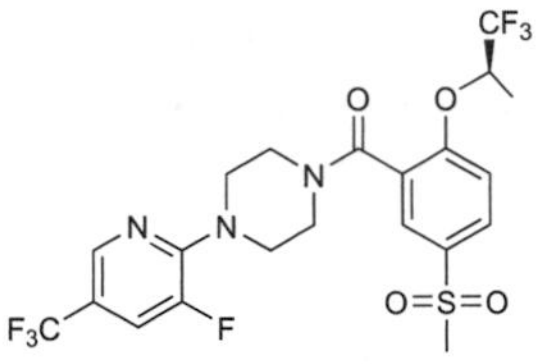

31   32   33

Further optimization of this scaffold led to compound **32**, which emerged as a prominent analog of the series, demonstrating good potency ($EC_{50} = 16$ nM) and PK in the mouse. Brain levels were above the inhibitory activity at GlyT-1 despite a low brain to plasma (b/p) ratio (0.1) [72]. Microdialysis experiments in the mouse striatum showed a 2.3-fold increase of glycine over basal levels. However, the inhibitor did display appreciable activity at the hERG channel ($IC_{50} = 0.6$ μM) and efforts to eliminate this issue are ongoing [72]. The structure of **33** was revealed to be RG1678 [17] and a patent application claiming specific crystal forms of **33** has been published [73].

Potent and selective bis-amide GlyT-1 inhibitors have also been reported [74]. Analog **34**, derived from a benzodiazepine HTS hit, demonstrated good potency ($EC_{50} = 4.7$ μM) and selectivity. Truncating the N-acyl linker led to the more potent analog **35** ($EC_{50} = 25$ nM), which exhibited high microsomal stability but poor aqueous solubility [74]. Patent applications for scaffolds represented by **36** [75] and **37** [76] have also been published; however, no pharmacological data has been reported.

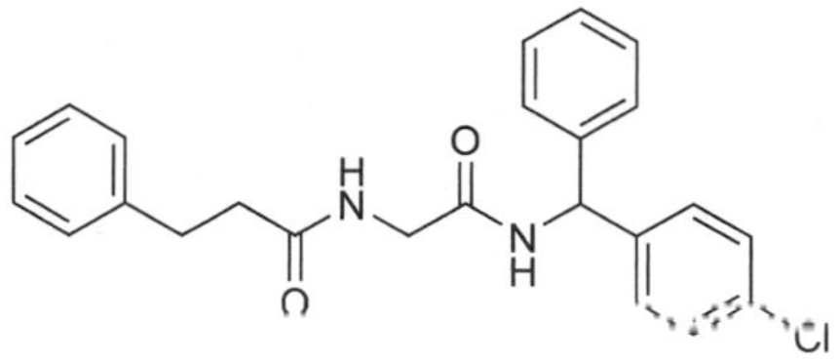

34

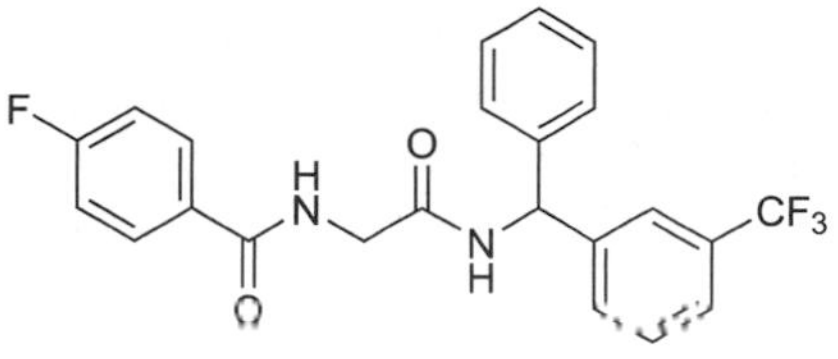

35

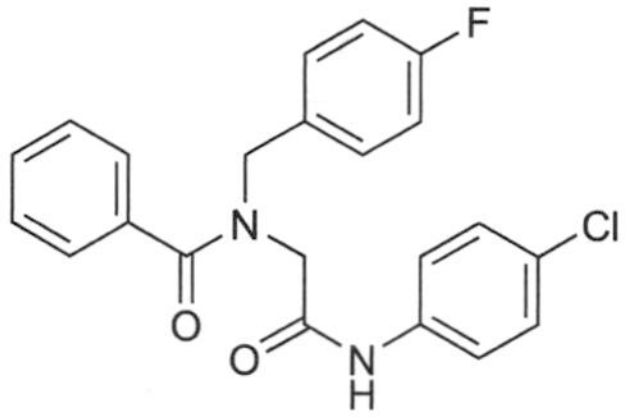

36

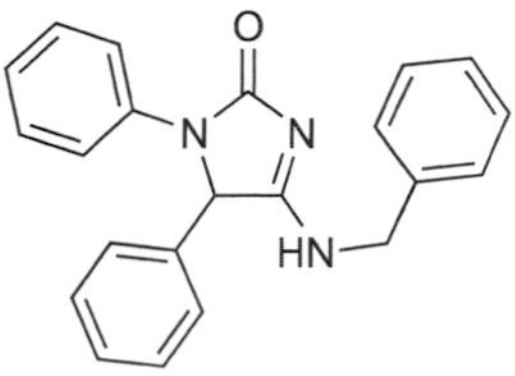

37

Naphthyl inhibitor **38** ($K_i = 500$ nM) was discovered via a displacement-based HTS assay utilizing radiolabeled (R)-NPTS [77]. However, compound **38** demonstrated high affinity for 5-HT$_{1B}$ ($K_i < 1$ nM), hERG ($K_i = 1.2$ μM), significant inhibition of CYP 2D6, and poor microsomal stability.

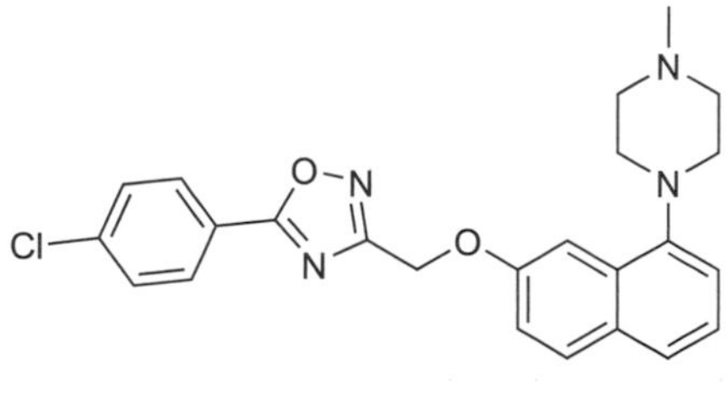

38

Modification of **38** improved the selectivity versus 5-HT$_{1B}$; however, hERG activity, CYP 2D6 inhibition, and poor microsomal stability remained. The series was discontinued due to a lack of progress with overcoming these challenges [77].

Amide **39** is a potent and selective inhibitor of GlyT-1 both *in vitro* ($K_i = 1.79$ nM) and *in vivo* (CSF-glycine ED$_{200}$ 3.9 mg/kg, rat, p.o.) but with limited permeability across Madin–Darby canine kidney (MDCK) cell membranes [78]. Optimization led to fused [3.1.0] and [3.3.0] azabicyclic analogs **40** (PF-3463275) [78] and **41** [79], respectively. Analogs of both systems demonstrated excellent potency ($K_i = 1.7$–95 nM), and improved permeability, PK, and *in vivo* efficacy. Spatial working memory studies in rodents and nonhuman primates found that PF-3463275 reversed ketamine-induced deficits [80]. Phase IB studies evaluating the safety and efficacy of **40** the structure for compound 40 should have a 4-fluoro-3-chloro substitution on the benzyl ring and not the 3,4-dichloro substitution shown in adjunctive treatment of cognitive deficits in schizophrenia were recently completed (data not disclosed) [46].

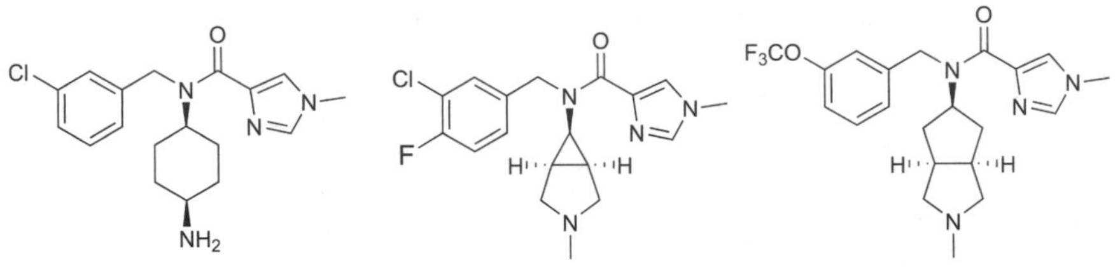

39   **40**   **41**

Several patent applications claiming variations represented by scaffold **42** have been published [81]. A developed PET ligand from this scaffold class, [^{11}C]-GSK931145 (**43**) [82], was used in a Phase I study designed to evaluate the relationship between plasma concentrations and brain occupancy of GSK1018921 in healthy individuals (structure and data not disclosed) [46].

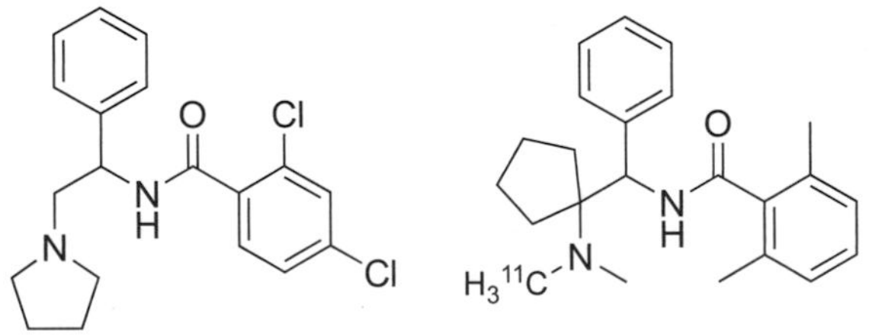

42   **43**

The 2, 4-dimethylpyrrolidine analog **44** (as a mixture of four diastereomers) is a selective inhibitor of GlyT-1 (IC$_{50}$ = 800 nM). Modification led to optimized analog **45**, which demonstrated good potency and selectivity but poor microsomal clearance and oral bioavailability [83].

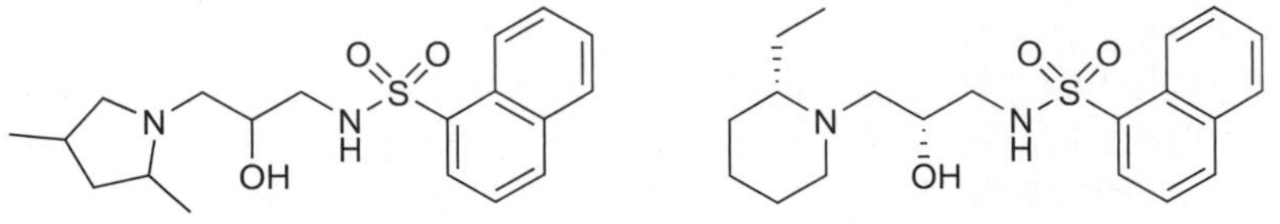

44   **45**

Several patent applications for compounds represented by spirocycle **46** [84–88] and urea **47** [89] have been published; however, no pharmacological data has been reported.

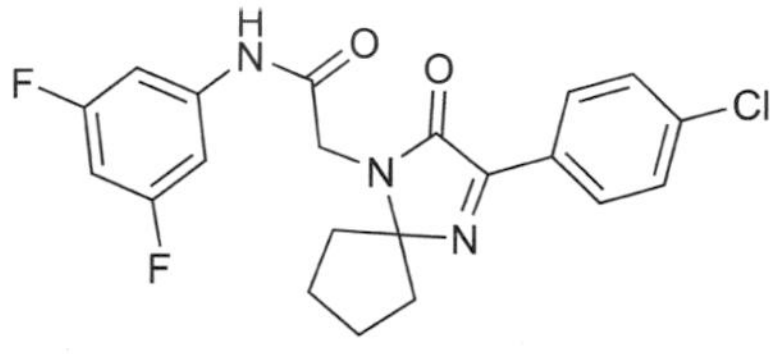

46 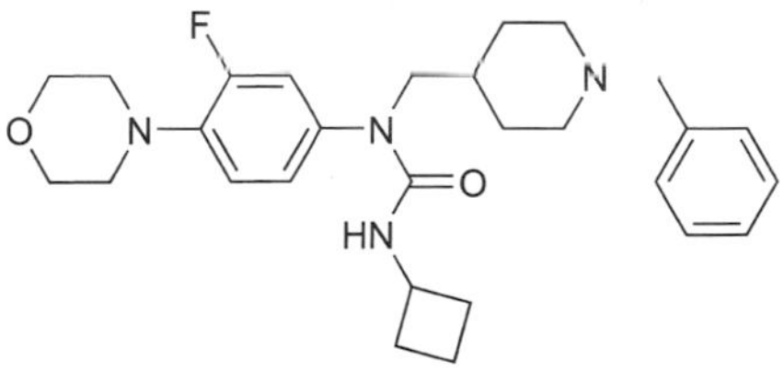   **47**

Quinoline-derived GlyT-1 inhibitors, represented by analog **48**, have also been claimed in published patent applications [90]. Functional IC_{50} data has been provided for several examples, with values ranging between 10 and 1000 nM.

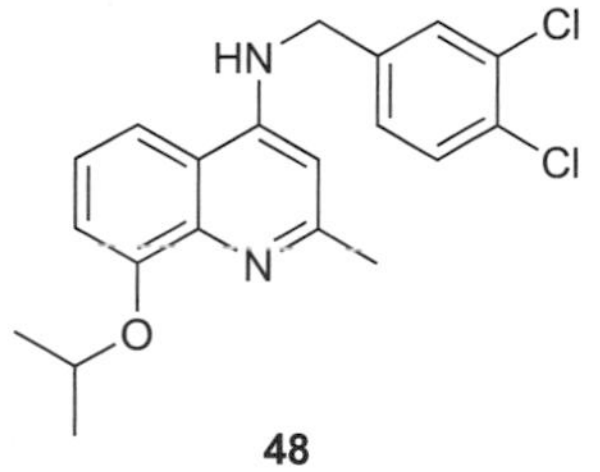

48

6. CONCLUSION

The growing body of evidence implicating NMDA glutamate receptor hypofunction as an underlying cause of schizophrenia has led to a paradigm shift in the search for safer and more efficacious antipsychotics. Inhibition of GlyT-1 provides an attractive means to potentiate NMDA receptor function because of its distribution within the forebrain and colocalization with the receptor. The approach has been validated in a variety of pre-clinical models predictive of antipsychotic activity and recent human clinical trials. Inhibition of GlyT-1 has garnered a great deal of interest by the pharmaceutical industry resulting in a competitive landscape with several structurally diverse sarcosine-based and nonsarcosine-based inhibitors reported. The field will undoubtedly gain further momentum with the recent proof-of-concept Phase II data provided by

Roche for their inhibitor RG1678. This trial provides compelling evidence that GlyT-1 inhibitors show promise as a novel class of antipsychotics that are effective at ameliorating the negative symptoms and cognitive dysfunction associated with schizophrenia.

REFERENCES

[1] N. C. Andreasen, *Brain Res. Rev.*, 2000, **31**, 106.
[2] M. Karayiorgou, *Clin. Neurosci. Res.*, 2001, **1**, 158.
[3] W. T. Carpenter, R. R. Conley, R. W. Buchanan, S. J. Enna, and J.T. Coyle(eds) in *Pharmacological Management of Neurological and Psychiatric Disorders*, McGraw-Hill, New York, NY, 1998, p. 27.
[4] S. Miyamoto, G. E. Duncan, C. E. Marx, and J. A. Liberman, *Mol. Psychiatry*, 2005, **10**, 79.
[5] C. A. Tamminga and J. M. Davis, *Schizophr. Bull.*, 2007, **33**, 937.
[6] A. Sawa and S. H. Snyder, *Science*, 2002, **296**, 692.
[7] J. T. Coyle, *Cell. Mol. Neurobiol.*, 2006, **26**, 365.
[8] M. J. Millan, *Psychopharmacology*, 2005, **179**, 30.
[9] M. J. Marino, L. J. S. Knutsen, and M. Williams, *J. Med. Chem.*, 2008, **51**, 1077.
[10] C. Hui, B. Wardwell, and G. E. Tsai, *Recent Pat. CNS Drug Discov.*, 2009, **4**, 220.
[11] J. M. Stone and L. S. Pilowsky, *CNS Neurol. Diord. Drug Targets*, 2007, **6**, 265.
[12] J. A. Gray and B. L. Roth, *Mol. Psychiatry*, 2007, **12**, 904.
[13] L. G. Harsing Jr., Z. Juranyi, I. Gacsalyi, P. Tapolcsanyi, A. Czompa, and P. Matyus, *Current Med. Chem.*, 2006, **13**, 1017.
[14] D. C. Javitt, *Biol. Psychiatry*, 2008, **63**, 6.
[15] J. Gomeza, W. Armsen, H. Betz, and V. Eulenburg, *Handb. Exp. Pharmacols.*, 2006, **175**, 457.
[16] H. Lane, Y. Liu, C. Huang, Y. Chang, C. Liau, C. Perng, and G. E. Tsai, *Biol. Psychiatry*, 2008, **63**, 9.
[17] E. Pinard, A. Alanine, D. Alberati, E. Borroni, H. Fischer, D. Hainzl, S. Jolidon, J-L. Moreau, R. Narquizian, M. Nettekoven, R. Norcross, H. Stalder, A. Thomas, and J. G. Wettstein, Abstract 273, 2010 Spring ACS National Meeting and Exposition, San Francisco, CA, USA, March 21—25, 2010, http://www.acsmedchem.org/mediabstracts2010.pdf
[18] http://www.roche-trials.com
[19] S. E. Wolkenberg and C. Sur, *Curr. Top. Med. Chem.*, 2010, **10**, 170.
[20] T. M. Bridges, R. Williams, and C. W. Lindsey, *Curr. Opin. Mol. Ther.*, 2008, **10**, 591.
[21] D. C. Javitt, *Curr. Opin. Drug Discov. Devel.*, 2009, **12**, 468.
[22] S. M. Lechner, *Curr. Opin. Pharmacol.*, 2006, **6**, 75.
[23] C. Thomsen, *Drug Discov. Today:Ther. Strateg.*, 2006, **3**, 539.
[24] B. N. Atkinson, S. C. Bell, M. De Vivo, L. R. Kowalski, S. M. Lechner, V. I. Ognyanov, C. S. Tham, C. Tsai, J. Jia, D. Ashton, and M. A. Klitenick, *Mol. Pharmacology*, 2001, **60**, 1414.
[25] K. R. Aubrey and R. J. Vandenberg, *Br. J. Pharmacol.*, 2001, **134**, 1429.
[26] A. J. Berger, S. Dieudonne, and P. Ascher,*J. Neurophysiol.*, 1998, **80**, 3336.
[27] R. Bergeron, T. M. Meyer, J. T. Coyle, and R. W. Greene, *Proc. Natl. Acad. Sci. USA*, 1998, **95**, 15730.
[28] L. Chen, M. Muhlhauser, and C. R. Yang, *J. Neurophysiol.*, 2003, **89**, 691.
[29] G. G. Kinney, C. Sur, M. Burno, P. J. Mallorga, J. B. Williams, D. J. Figueroa, M. Wittman, W. Lemaire, and P. J. Conn, *J. Neurosci.*, 2003, **23**, 7586.
[30] D. C. Javitt, H. Sershen, A. Hashim, and A. Lajtha, *Neuropsychopharmacology*, 1997, **17**, 202.
[31] E. Toth and A. Lajtha, *Neurochem. Res.*, 1986, **11**, 393.

[32] J. McCaughran, E. Mahjubi, E. Decena, and R. Hitzemann, *Psychopharmacology (Berl.)*, 1997, **134**, 131.

[33] G. Le Pen, J. Kew, D. Alberati, E. Borroni, M. P. Heitz, and J. L. Moreau, *Biol. Psychiatry*, 2003, **54**, 1162.

[34] B. K. Lipska and D. R. Weinberger, *Brain Res. Dev. Brain Res.*, 1993, **75**, 213.

[35] L. G. Harsing, I. Gacsalyi, G. Szabo, E. Schmidt, N. Sziray, C. Sebban, B. Tesolin-Decros, A. Egyed, M. Spedding, and G. Levay, *Pharmacol. Biochem. Behav.*, 2003, **74**, 811.

[36] D. Manahan-Vaughan, V. Wildforster, and C. Thomsen, *Eur. J. Neurosci.*, 2008, **28**, 1342.

[37] T. Man, G. Milot, W. J. Porter, J. K. Reel, H. C. E. Rudyk, M. J. Valli, M. W. Walter, *Patent Application WO 2005/100301*, 2005.

[38] K. W. Perry, J. F. Falcone, M. J. Fell, J. W. Ryder, H. Yu, P. L. Love, J. Katner, K. D. Gordon, M. R. Wade, T. Man, G. G. Nomikos, L. A. Phebus, A. J. Cauvin, K. W. Johnson, C. K. Jones, B. J. Hoffmann, G. E. Sandusky, M. W. Walter, W. J. Porter, L. Yang, K. M. Merchant, H. E. Shannon, and K. A. Svensson, *Neuropharmacology*, 2008, **55**, 743.

[39] J. A. Lowe III, S. E. Drozda, K. Fisher, C. Strick, L. Lebel, C. Schmidt, D. Hiller, and K. S. Zandi, *Bioorg. Med. Chem. Lett.*, 2003, **13**, 1291.

[40] M. Martina, Y. Gorfinkel, S. Halman, J. A. Lowe, P. Periyalwar, C. J. Schmidt, and R. Bergeron, *J. Physiol.*, 2004, **557**, 489.

[41] E. K. Moltzen, G. P. Smith, C. Krog-Jensen, and K. P. Bogeso, *Patent Application WO 2002/08216-A1*, 2002.

[42] C. G. Thomson, K. Duncan, S. R. Fletcher, I. T. Huscroft, G. Pillai, P. Raubo, A. J. Smith, and D. Stead, *Bioorg. Med. Chem. Lett.*, 2006, **16**, 1388.

[43] I. Egle J. Frey, and M. Isaac, *Patent Application WO 2001/32602-A1*, 2001.

[44] A. Brown, I. Carlyle, J. Clark, W. Hamilton, S. Gibson, G. McGarry, S. McEachen, D. Rae, S. Thorn, and G. Walker, *Bioorg. Med. Chem. Lett.*, 2001, **11**, 2007.

[45] N. Andrews, J. Ge, G. Walker, J. Schipper, and H. M. Marston, *Eur. Neuropsychopharmacology.*, 2007, **17**, S497.

[46] http://clinicaltrials.gov

[47] M. Mezler, W. Hornberger, R. Meuller, M. Schmidt, W. Amberg, W. Braje, M. Ochse, H. Schoemaker, and B. Behl, *Mol. Pharmacol.*, 2008, **74**, 1705.

[48] G. Dargazanli, G. Estenne-Bouhtou, F. Medaisko, and N. Rakotoarisoa, *Patent Application WO 2005/037785-A2*, 2005.

[49] G. Dargazanli and G. Estenne-Bouhtou, F. Medaisko, and M. C. Renones, *Patent Application WO 2008/037881-A2*, 2008.

[50] G. Dargazanli, G. Estenne-Bouhtou, P. Magat, B. Marabout, and P. Roger, *Patent Application WO 2005/037783-A2*, 2005.

[51] R. Depoortere, G. Dargazanli, G. Estenne-Bouhtou, A. Coste, C. Lanneau, C. Desvignes, M. Poncelet, M. Heaulme, V. Santucci, M. Decobert, A. Cudennec, C. Voltz, D. Boulay, J. P. Terranova, J. Stemmelin, P. Roger, B. Marabout, M. Sevrin, X. Vige, B. Biton, R. Steinberg, D. Francon, R. Alonso, P. Avenet, F. Oury-Donat, G. Perrault, G. Griebel, P. George, P. Soubrie, and B. Scatton, *Neuropsychopharmacology*, 2005, **30**, 1963.

[52] P. Singer, J. Feldon, and B. K. Yee, *Psychopharmacology (Berl.)*, 2009, **202**, 371.

[53] M. D. Black, G. B. Varty, M. Arad, S. Barak, A. De Levie, D. Boulay, P. Pichat, G. Griebel, and I. Weiner,*Psychopharmacology (Berl.)*, 2009, **202**, 385.

[54] http://adisinsight.com

[55] D. Boulay, P. Pichat, G. Dargazanli, G. Estenne-Bouhtou, J. P. Terranova, N. Rogacki, J. Stemmelin, A. Coste, C. Lanneau, C. Desvignes, C. Cohen, R. Alonso, X. Vige, B. Biton, R. Steinberg, M. Sevrin, F. Oury-Donat, P. George, O. Bergis, G. Griebel, P. Avenet, and B. Scatton, *Pharmacol. Biochem. Behav.*, 2008, **91**, 47.

[56] D. Boulay, O. Bergis, P. Avenet, and G. Griebel, *Neuropsychopharmacology*, 2010, **35**, 416.

[57] Y. Sekiguchi, T. Okubo, T. Shibata, K. Abe, S. Yamamoto, and S. Kashiwa, *Patent Application WO 2008/018639-A2*, 2008.

[58] J. S. Albert, C. Alhambra, T. A. Brugel, and J. G. Varnes, *Patent Application WO 2009/013535-A1*, 2009.

[59] W. P. Blackaby, I. T. Huscroft, L. E. Keown, R. T. Lewis, P. A. Raubo, L. J. Street, C. G. Thomson, and J. Thomson, *Patent Application WO 2006/134341-A1*, 2006.

[60] C. W. Lindsley, Z. Zhao, W. H. Leister, J. O'Brien, W. Lemaire, D. L. Williams Jr., T.-B. Chen, R. S. Chang, M. Burno, M. A. Jacobson, C. Sur, G. G. Kinney, D. J. Pettibone, P. R. Tiller, S. Smith, N. N. Tsou, M. E. Duggan, and P. J. Conn, *Chem. Med. Chem.*, 2006, **1**, 807.

[61] Z. Zeng, J. A. O'Brien, W. Lemaire, S. S. O'Malley, P. J. Miller, Z. Zhao, M. A. Wallace, C. Raab, C. W. Lindsley, C. Sur, and D. L. Williams Jr., *Nucl. Med. Biol.*, 2008, **35**, 315.

[62] Z. Zhao, J. A. O'Brien, W. Lemair, D. L. Williams Jr., M. A. Jacobson, C. Sur, D. J. Pettibone, P. R. Tiller, S. Smith, G. D. Hartman, S. E. Wolkenberg, and C. W. Lindsley, *Bioorg. Med. Chem. Lett.*, 2006, **16**, 5968.

[63] Z. Zhao, W. H. Leister, J. A. O'Brien, W. Lemaire, D. L. Williams Jr., M. A. Jacobson, C. Sur, G. G. Kinney, D. J. Pettibone, P. R. Tiller, S. Smith, G. D. Hartman, C. W. Lindsley, and S. E. Wolkenberg, *Bioorg. Med. Chem. Lett.*, 2009, **19**, 1488.

[64] S. E. Wolkenberg, Z. Zhao, D. D. Wisnoski, W. H. Leister, J. O'Brien, W. Lemaire, D. L. Williams Jr., M. A. Jacobson, C. Sur, G. G. Kinney, D. J. Pettibone, P. R. Tiller, S. Smith, C. Gibson, B. K. Ma, S. L. Polsky-Fisher, C. W. Lindsley, and G. D. Hartman, *Bioorg. Med. Chem. Lett.*, 2009, **19**, 1492.

[65] J. L. Thomson, W. P. Blackaby, A. S. R. Jennings, S. C. Goodacre, A. Pike, S. Thomas, T. A. Brown, A. Smith, G. Pillai, L. J. Street, and R. T. Lewis, *Bioorg. Med. Chem. Lett.*, 2009, **19**, 2235.

[66] C. Drahl, *Chem. Eng. News*, 2008, **86**, 38.

[67] E. Pinard, S. M. Ceccarelli, H. Stalder, and D. Alberati, *Bioorg. Med. Chem. Lett.*, 2006, **16**, 349.

[68] S. M. Ceccarelli, E. Pinard, H. Stalder, and D. Alberati, *Bioorg. Med. Chem. Lett.*, 2006, **16**, 354.

[69] D. Alberati, S. M. Ceccarelli, S. Jolidon, E. A. Krafft, A. Kurt, A. Maier, E. Pinard, H. Stalder, D. Studer, A. W. Thomas, and D. Zimmerli, *Bioorg. Med. Chem. Lett.*, 2006, **16**, 4305.

[70] D. Alberati, D. Hainzl, S. Jolidon, E. A. Krafft, A. Kurt, A. Maier, E. Pinard, A. W. Thomas, and D. Zimmerli, *Bioorg. Med. Chem. Lett.*, 2006, **16**, 4311.

[71] D. Alberati, D. Hainzl, S. Jolidon, A. Kurt, E. Pinard, A. W. Thomas, and D. Zimmerli, *Bioorg. Med. Chem. Lett.*, 2006, **16**, 4321.

[72] E. Pinard, D. Alberti, E. Borroni, H. Fischer, D. Hainzl, S. Jolidon, J.-L. Moreau, R. Narquizian, M. Nettekoven, R. D. Norcross, H. Stalder, and A. W. Thomas, *Bioorg. Med. Chem. Lett.*, 2008, **18**, 5134.

[73] A. Bubendorf, A. Deynet-Vucenovic, R. Diodone, O. Grassmann, K. Lindenstruth, E. Pinard, F. E. Roher, and U. Schwitter, *Patent Application WO 2008/080821-A1*, 2008.

[74] S. Jolidon, D. Alberati, A. Dowle, H. Fischer, D. Hainzl, R. Narquizian, R. Norcross, and E. Pinard, *Bioorg. Med. Chem. Lett.*, 2008, **18**, 5533.

[75] S. Jolidon, R. Narquizian, and E. Pinard, *Patent Application WO 2008/025694-A1*, 2008.

[76] S. Jolidon, R. Narquizian, R. Norcross, and E. Pinard, *Patent Application WO 2007/101802-A1*, 2007.

[77] J. Lowe, S. Drozda, W. Qian, M.-C. Peakman, J. Liu, J. Gibbs, J. Harms, C. Schmidt, K. Fisher, C. Strick, A. Schmidt, M. Vanase, and L. Lebel, *Bioorg. Med. Chem. Lett.*, 2007, **17**, 1675.

[78] J. A. Lowe III, X. Hou, C. Schmidt, F. D. Tingley III, S. McHardy, M. Kalman, S. DeNinno, M. Sanner, K. Ward, L. Lebel, D. Tunucci, and J. Valentine, *Bioorg. Med. Chem. Lett.*, 2009, **19**, 2974.

[79] J. A. Lowe III, S. L. DeNinno, S. E. Drozda, C. J. Schmidt, K. M. Ward, F. D. Tingley III, M. Sanner, D. Tunucci, and J. Valentine, *Bioorg. Med. Chem. Lett.*, 2010, **20**, 907.

[80] B. M. Roberts, C. L. Schaffer, P. A. Seymour, C. J. Schmidt, G. V. Williams, and S. A. Castner, *Neuroreport*, 2010, **21**, 390.

[81] C. L. Branch, H. Marshall, J. McCritchie, R. A. Porter, and S. Spada, *Patent Application WO 2007/080159-A2*, 2007.

[82] G. Gentile, H. J. Herdon, J. Passchier, and R. A. Porter, *Patent Application WO 2007/147838-A1*, 2007.

[83] S. S. Rahman, S. Coulton, H. J. Herdon, C. F. Joiner, J. Jin, and R. A. Porter, *Bioorg. Med. Chem. Lett.*, 2007, **17**, 1741.

[84] N. M. Ahmad, J. Y. Q. Lai, H. R. Marshall, D. J. Nash, and R. A. Porter, *Patent Application WO 2008/092877-A2*, 2008.

[85] N. M. Ahmad, J. Y. Q. Lai, D. Andreotti, H. R. Marshall, D. J. Nash, and R. A. Porter, *Patent Application WO 2008/092876-A1*, 2008.

[86] N. M. Ahmad, A. A. Jaxa-Chamiec, J. Y. Q. Lai, and R. A. Porter, *Patent Application WO 2008/092873-A1*, 2008.

[87] D. G. Cooper and R. A. Porter, *Patent Application WO 2009/034062-A1*, 2009.

[88] N. M. Ahmad, C. L. Branch, A. A. Jaxa-Chamiec, J. Y. Q. Lai, and D. J. Nash, *Patent Application WO 2008/092878-A1*, 2008.

[89] C. L. Branch, H. Marshall, D. J. Nash, and R.A.. Porter, *Patent Application WO 2005/058882-A1*, 2005.

[90] W. Amberg, M. Ochse, W. Braje, B. Behl, W. Hornberger, M. Mezler, and C. W. Hutchins, *Patent Application WO 2009/024611-A2*, 2009.

CHAPTER 3

Modulators of Transient Receptor Potential Ion Channels

Rajagopal Bakthavatchalam and **S. David Kimball**

1. INTRODUCTION

The transient receptor potential (TRP) superfamily is comprised of 28 mammalian ligand-, stimulus-, and second messenger-gated cation channels and is classified into six subfamilies: TRPA (Ankyrin), TRPV (Vanilloid), TRPM (Melastatin), TRPC (Canonical), TRPP (Polycystin), and TRPML (Mucolipin) [1]. These channels contain putative transmembrane-spanning domains (S1–S6) with the ion-conducting pore located between S5 and S6 [2]. The TRP ion channels conduct both monovalent and divalent cations, and display a wide diversity of activation and regulation by temperature, taste, mechanical input, and noxious chemical stimuli central to the signaling of sensory neurons. Additionally, there is

Hydra Biosciences, 790 Memorial Drive, Cambridge, MA 02139, USA

Annual Reports in Medicinal Chemistry, Volume 45
ISSN 0065-7743, DOI 10.1016/S0065-7743(10)45003-4

mounting evidence that the TRP ion channels play a vital role in the pathophysiology of many illnesses. This chapter will discuss TRP channel modulators that have been identified since this topic was previously reviewed [3–5].

2. NEW MODULATORS OF TRP CHANNELS

2.1 TRPA1 agonists

TRPA1, also known as ANKTM1, is expressed in the polymodal C- and A-δ fiber sensory neurons of the dorsal root ganglia (DRG) and trigeminal ganglia (TG). It is coexpressed with TRPV1 in a subset of TRPV1-containing neurons. Recombinant TRPA1 is activated by noxious cold (<17 °C) and a variety of natural compounds that are nocifensive or produce burning sensations [6,7]. Several exogenous electrophiles such as cinnamaldehyde (1, CA, $EC_{50} = 19\,\mu M$), oxindole (2, SC, $EC_{50} = 0.8\,\mu M$), oxindole alkyne (3, SCA, $EC_{50} = 0.1\,\mu M$), acrolein (4, $EC_{50} = 5\,\mu M$), allyl isothiocyanate (5, AITC, $EC_{50} = 33.5\,\mu M$), and others have been shown to activate TRPA1 *in vitro* through covalent binding to cysteine residues present in the cytoplasmic N-terminus of TRPA1. The endogenous ligand 4-hydroxynonenal (6, 4-HNE), which is produced under conditions of tissue damage or oxidative stress, can also activate TRPA1 ($EC_{50} = 13\,\mu M$) at physiologically relevant concentrations [8–10]. Other small-molecule electrophiles such as formaldehyde and iodoacetamide (7) are also known to activate TRPA1 [8,13].

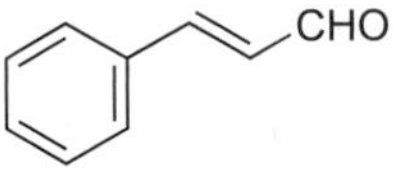

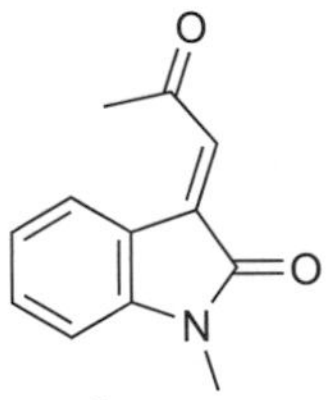

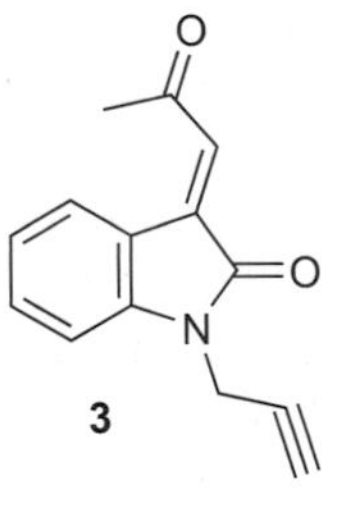

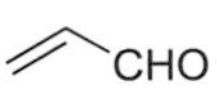

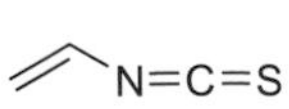

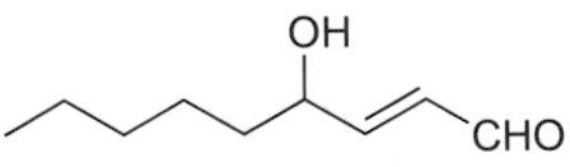

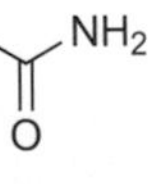

The recently disclosed terpenoids miogadial (8, MD), miogatrial (9, MT), and polygotrial (10, PG) have been shown to activate TRPA1 more potently than AITC (5) in heterogeneously expressed HEK293 or CHO cells [11].

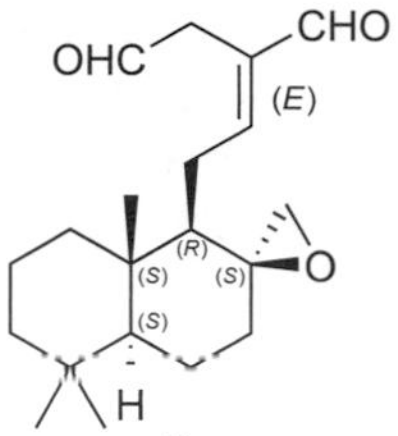

A recent patent application describes dibenz[*b,f*][1,4]oxazepine (**11**, military code CR), a powerful lachrymator used as a riot control agent, as a potent activator of human TRPA1 (EC$_{50}$ = 0.3 nM) in HEK cells [12]. The parent dibenz[*b,f*][1,4]oxazepine ring system has been shown to form a covalent adduct with benzylthiol, mimicking the putative alkylation of N-terminal cysteine residues involved in the activation of TRPA1. Related tricyclic analogs (**12–16**) are also thought to be activators of TRPA1.

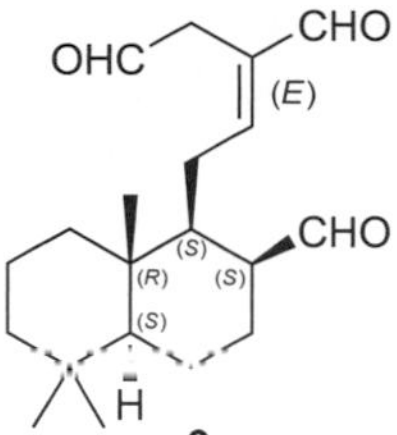

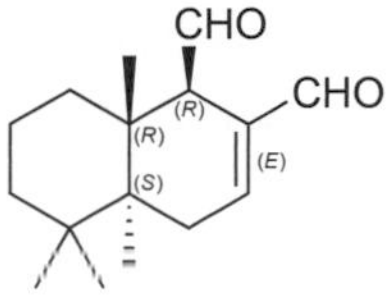

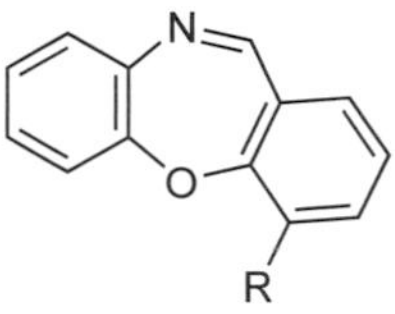

11 (R = H)
12 (R = CO$_2$Me)
13 (R = CONH$_2$)

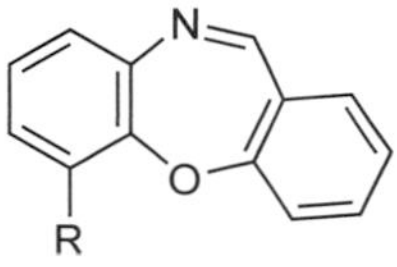

14 (R = CO$_2$Me)
15 (R = CONH$_2$)

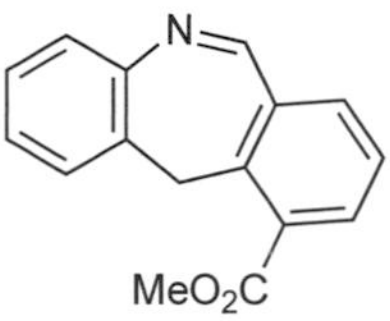

16

A number of compounds activate TRPA1 without any apparent ability to form covalent adducts, including nonelectrophilic fenamate nonsteroidal anti-inflammatory drugs (NSAIDs), such as flufenamic acid (**17**, FFA), niflumic acid (**18**, NFA), and mefenamic acid (**19**, MFA) [13]. Phenols such as thymol (**20**) and 2-*tert*-butyl-5-methylphenol (**21**) have been shown to activate human TRPA1 with micromolar EC$_{50}$ values in stably transfected HEK293 cells [14].

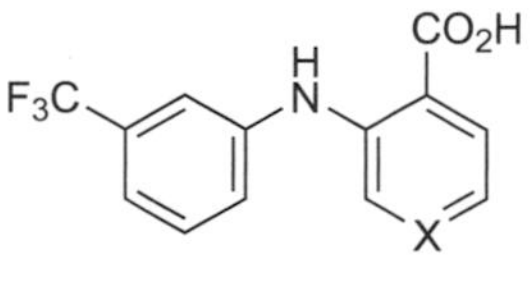

17 (X = CH)
18 (X = N)

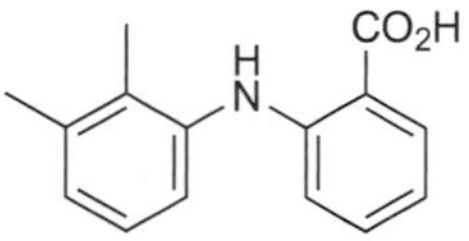

19

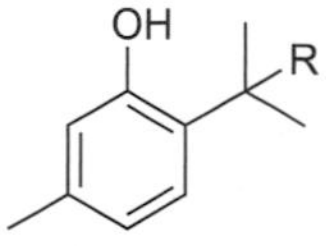

20 (R = H)
21 (R = Me)

While most TRPA1 agonists are primarily useful as research tools, some may also have potential therapeutic utility. For example, it has been reported that TRPA1 is highly expressed in enterochromaffin (EC) cells, and activation of these cells with TRPA1 agonists such as AITC (**5**), CA (**1**), and acrolein (**4**) causes 5-hydroxytryptamine (5-HT) secretion *in vitro* [15]. A recent patent application claims indolinone-derived TRPA1 agonist (**3**) as a useful agent to treat constipation-type irritable bowel syndrome (IBS) [16].

2.2 TRPA1 antagonists

In recent years there has been considerable effort directed at identifying potent selective antagonists for TRPA1 [17]. Studies of TRPA1 knockout (KO) mice have revealed the importance of TRPA1 in pain signaling to chemical and mechanical stimuli as well as in mediating inflammatory pain [8,17–19]. The xanthine HC-030031 (**22**) has been extensively profiled by several research groups [17,20–23]. HC-030031 (**22**) blocks AITC (**5**) and CA (**1**) induced Ca^{2+} increases in TRPA1-expressing cells with IC_{50} values of 6.2 and 4.9 μM, respectively [17,23]. HC-030031 is reported to show good selectivity for TRPV1, TRPV3, TRPV4, hERG, and NaV1.2 channels and has demonstrated efficacy in pain models such as the acute formalin model, Complete Freund's Adjuvant (CFA)-induced inflammatory pain, and the spinal nerve ligation (SNL) model of neuropathic pain [17,23]. HC-030031 (**22**) and related congeners (**23–25**) are claimed as TRPA1 inhibitors [24–26]. Related quinazolinediones (**26–28**) have also been reported to be potent TRPA1 antagonists [27].

AP18 (**29**), a structurally distinct low-molecular-weight oxime derivative, has been reported as a potent TRPA1 antagonist [28]. AP18 (**29**) blocks CA (**1**) activation of TRPA1 with IC_{50} values of 3.1 and 4.5 μM for human and mouse channels, respectively, and is highly selective for TRPA1 over TRPV1, TRPV2, TRPV3, TRPV4, and TRPM8. AP18 (**29**) significantly suppresses nociceptive behavior caused by injection of CA (**1**) into the mouse hind paw. The antagonist AP18 (**29**) reverses CFA-induced mechanical hyperalgesia in wild-type (WT) mice. However, TRPA1 KO mice appear to develop compensatory mechanical hyperalgesia to CFA treatment, which is not reversible by this TRPA1 antagonist. Taken together, these data suggest that the activity of AP18 (**29**) is on-target [29].

The compound designated A-967079 (**30**), a recently disclosed TRPA1 antagonist, blocks AITC-induced TRPA1 activation in human and rat cell lines with IC_{50} values of 67 and 289 nM, respectively [30]. Similarly, A-967079 inhibits AITC-induced TRPA1 current in human and rat cells

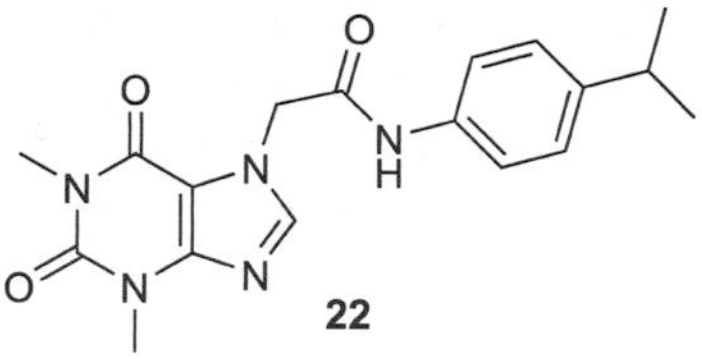

22

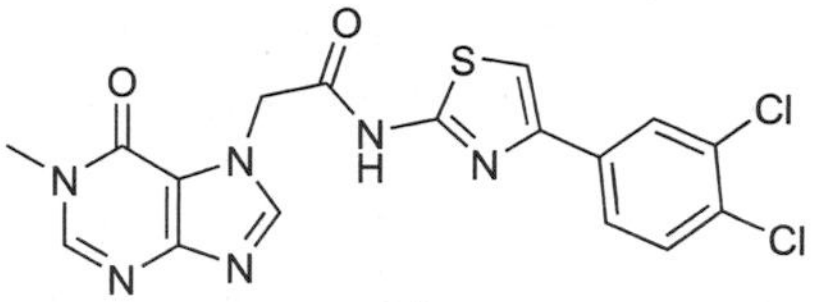

23

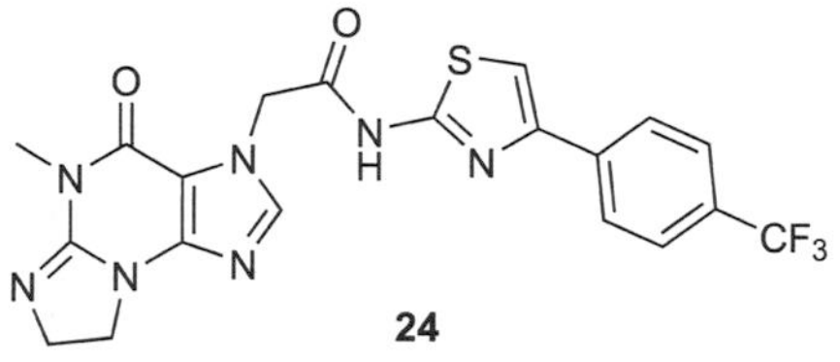

24

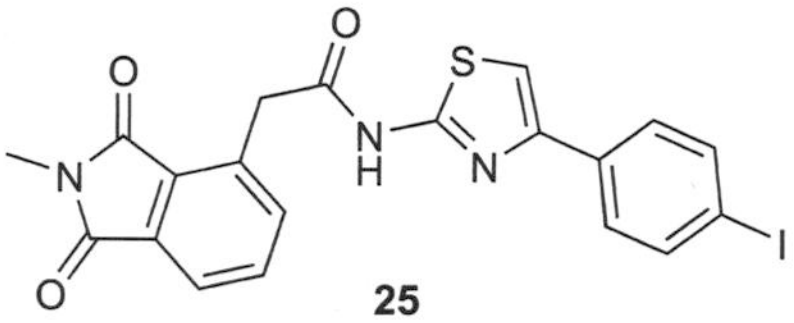

25

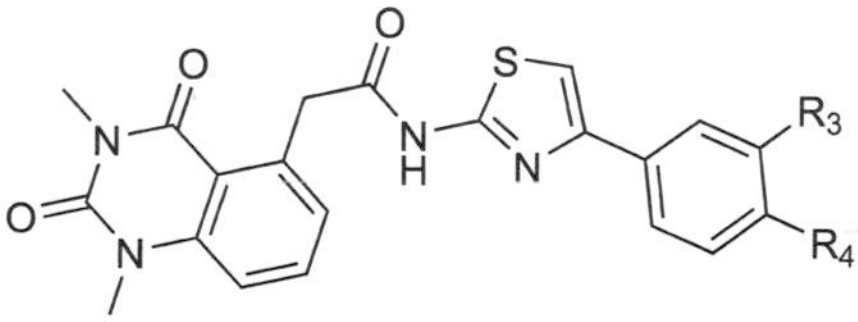

26 (R^3 = H, R^4 = Cl)
27 (R^3 = CF$_3$, R^4 = H)
28 (R^3 = F, R^4 = CF$_3$)

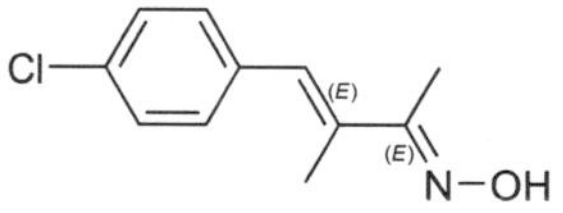

29

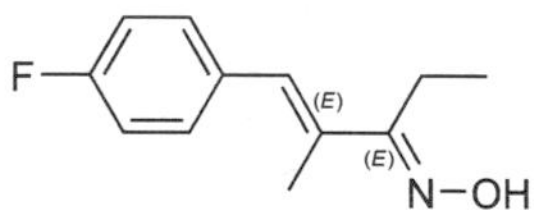

30

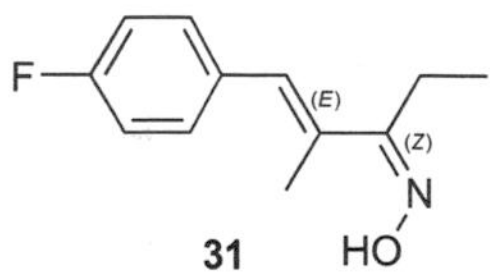

31

with IC$_{50}$ values of 51 and 101 nM, respectively. The stereochemistry of the oxime double bond (*E,E*-isomer) in A-967079 appears to be important for TRPA1 activity, as the *E,Z*-isomer (**31**) is ten-fold less potent (IC$_{50}$ = 701 nM) [31]. A-967079 is efficacious in reducing pain and the response of wide dynamic range (WDR) neurons in the monoiodoacetate (MIA) model of osteoarthritis as well as the CFA model of inflammatory pain [31].

Stereochemistry also plays an important role in the recently disclosed 3,4-dihydropyrimidinethione analogs [32]. Racemic dihydropyrimidine derivative (**32**) is reported to inhibit activation by the TRPA1 agonist 11*H*-dibenzo[*b,e*]azepine-10-carboxylic acid methyl ester (**16**) with an IC$_{50}$ value of 128 nM for human TRPA1 in HEK293 cells.

The (R)-isomer (33) is claimed to be active ($IC_{50} = 76\,nM$), whereas the (S)-isomer (34, $IC_{50} = 10\,\mu M$) is completely inactive in this experiment.

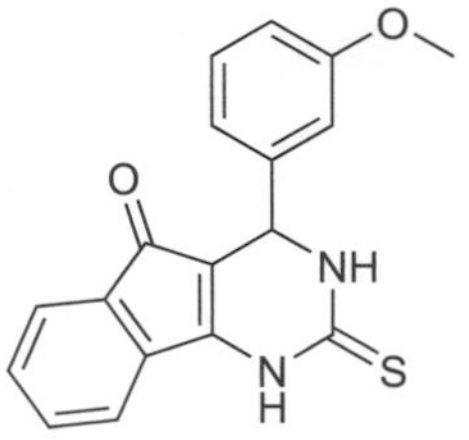

32 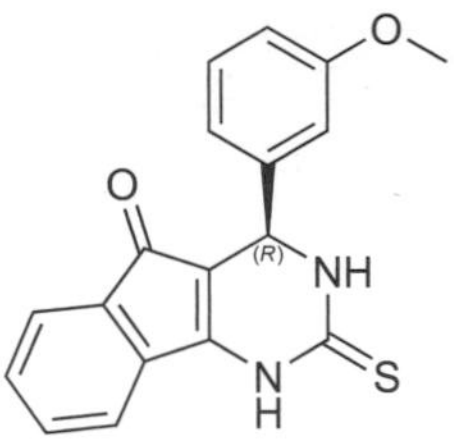   33 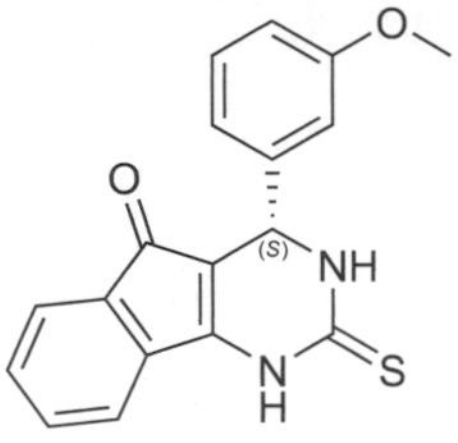   34

Trichloro(sulfanyl)ethyl benzamides, AMG9090 (35), AMG5545 (36), and AMG2504 (37) have been shown to potently block activation of human TRPA1 in CHO cells by AITC and noxious cold (4°C). However, these compounds (35–37) show species-specific pharmacology at the rat channel, as nitro-substituted compound 37 is a weak antagonist, while compounds 35 and 36 are partial agonists in CHO cells stably expressing rat TRPA1 [33].

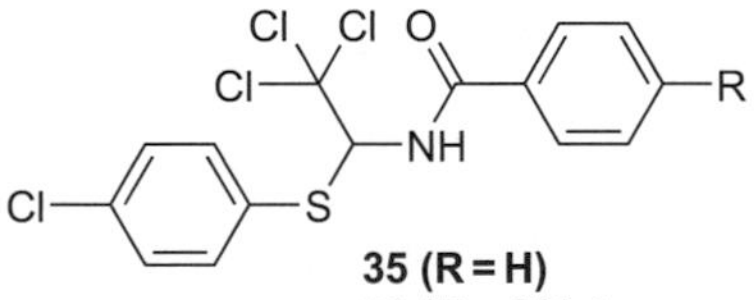

35 (R = H)
36 (R = OMe)
37 (R = NO$_2$)

2.3 TRPV1 antagonists

TRPV1 is a nonselective cation channel predominantly expressed in sensory neurons and activated by capsaicin, heat (>42°C), pH (<5.4), and noxious stimuli. Several compounds have advanced into clinical development and have been extensively covered in recent reviews [5,34–39]. Among these, MK-2295 (structure not disclosed), AMG 517 (38), SB-705498 (39), and GRC-6211 (structure not disclosed) have reportedly encountered safety issues [35]. In phase II clinical trials with MK-2295, patients experienced an increase in core body temperature

[36]. A similar hyperthermia was reported with AMG 517 (**38**), and phase IB clinical trials were discontinued [34]. The clinical trials of two urea analogs, SB-705498 (**39**) and ABT-102 (**40**), have been discontinued without any disclosed data [38,40].

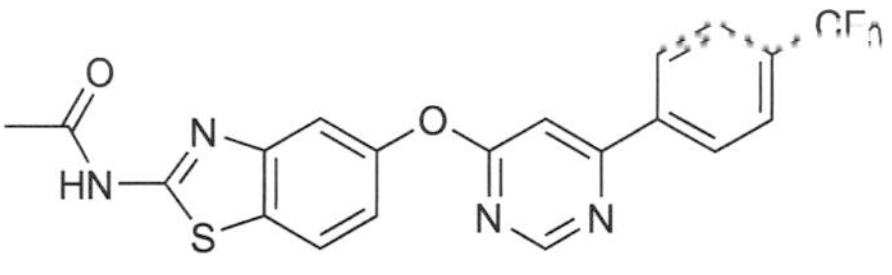

38 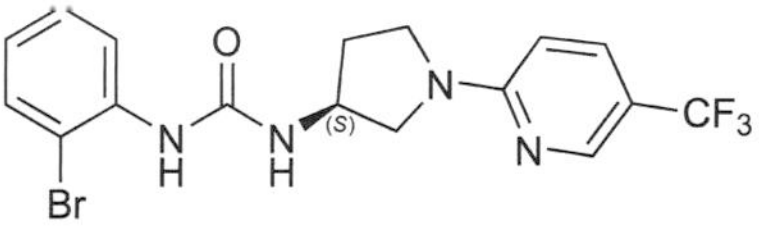   **39**

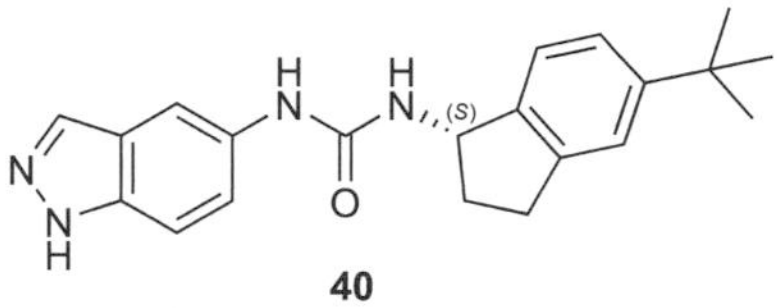

40

2.4 TRPV2 modulators

Modulation of the TRPV2 channel is species specific. TRPV2 is activated by heat (>52°C) and 2-aminoethoxydiphenyl borate (2-APB) (**41**) in both rat and mouse; however, human TRPV2 does not respond to either heat or 2-APB [41]. Recently, cannabinoids such as cannabindiol (**42**) and δ-9-tetrahydrocannabinol (**43**, THC) have been disclosed as agonists for rat TRPV2 with EC_{50} values of 3.7 and 15.5 μM, respectively [42,43]. Probenecid (**44**) is also an agonist of rat TRPV2 ($EC_{50} = 31.5$ μM) [44,45]. Ruthenium Red and SKF96365 (**45**) block 100 μM 2-APB (**41**) activation of mouse TRPV2 with IC_{50} values of 7 and 21 μM, respectively [46].

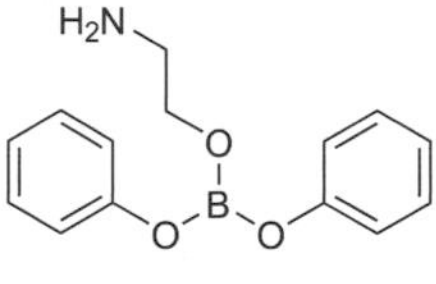

41 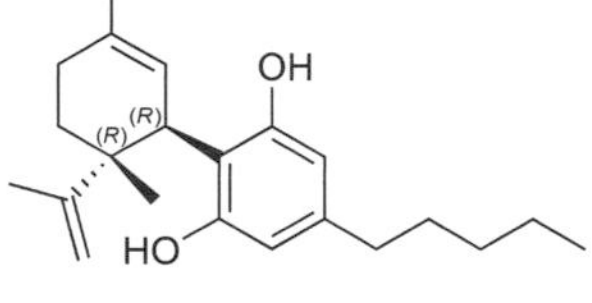   **42** 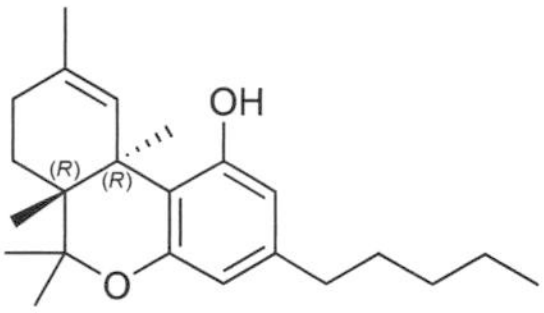   **43**

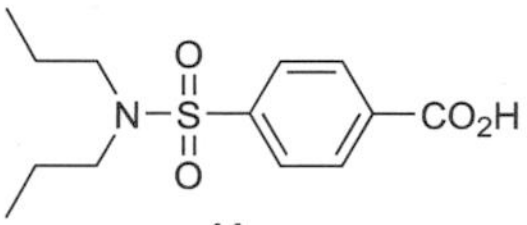

44 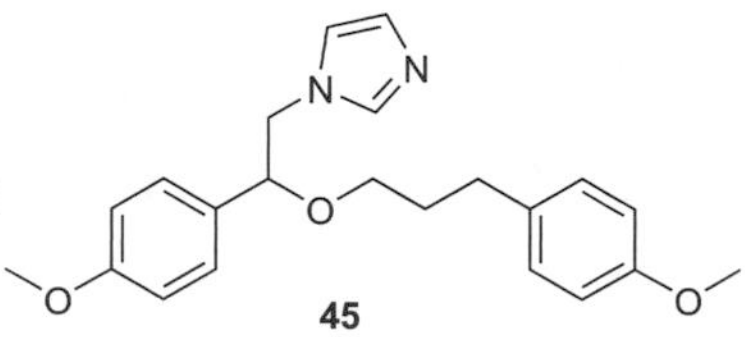   **45**

2.5 TRPV3 modulators

TRPV3 is found in pain-signaling neural pathways such as DRG, TG, and spinal cord, as well as in keratinocytes, brain, skin, and tongue [47]. TRPV3 is activated by warm temperatures (>32°C) and 2-APB (**41**) as well as by natural products such as camphor (**46**), thymol (**20**), carvacrol (**47**), eugenol (**48**), vanillin (**49**), menthol (**50**), and synthetic ethyl vanillin (**51**) [48–50]. Incensole acetate (**52**, IA), a plant-derived diterpenoid, is reported to increase Ca^{2+} influx in HEK293 cells stably expressing mouse TRPV3 with an EC_{50} value of 16 μM [51]. IA (**52**) shows anxiolytic (elevated plus maze test) and antidepressive (Porsolt forced swim test) effects in WT mice but not in TRPV3 KO mice.

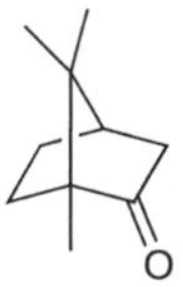

46 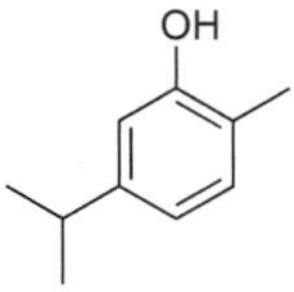   **47** 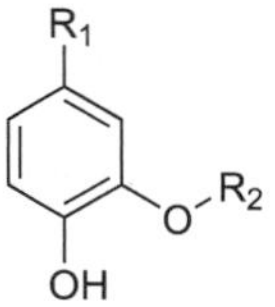   **48** (R = CH₂–CH = CH₂; R^2 = Me)   **50**   **52**
                       49 (R = CHO; R^2 = Me)
                       51 (R = CHO; R^2 = Et)

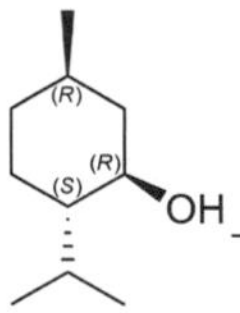

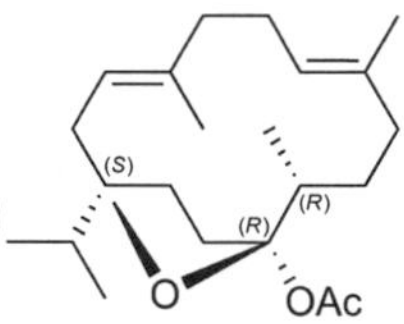

A number of TRPV3 antagonists have been described in the recent patent literature [52–57]. The benzthiazole (**53**) blocks stably transfected TRPV3 in a whole-cell patch clamp experiment in HEK293 cells following activation by 2-APB (**41**) [52]. Compound **53** showed greater than forty-fold selectivity against TRPV1, TRPV6, NaV1.2, and hERG. At 200 mg/kg i.p., compound **53** is efficacious in models of thermal hyperalgesia and formalin-induced pain [52].

Quinazolin-4-one analogs exemplified by compound **54** have also been disclosed as inhibitors of TRPV3 [53]. Compound **54** blocks both human and rat TRPV3-mediated current in whole-cell patch clamp and is selective relative to TRPA1, TRPV1, TRPM8, NaV1.2, and hERG. Compound **54** is poorly bioavailable in the rat ($F = 1\%$), yet shows efficacy in a carrageenan-induced model of acute inflammatory pain at 200 mg/kg i.p. Related analogs (**55** and **56**) have also been reported as TRPV3 blockers [54,55], as have chromane analogs (**57** and **58**) [56,57].

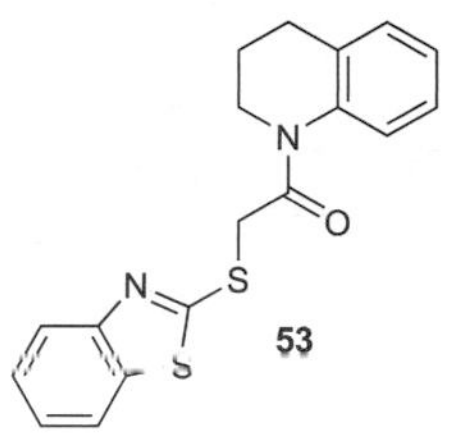

2.6 TRPV4 modulators

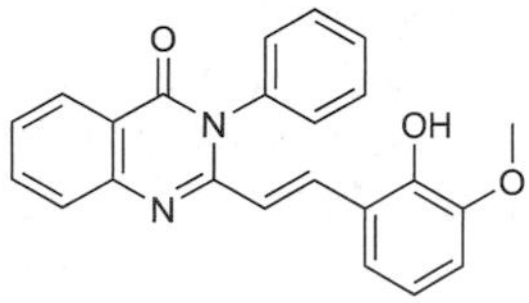

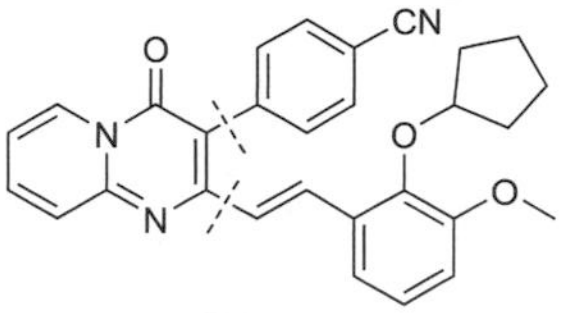

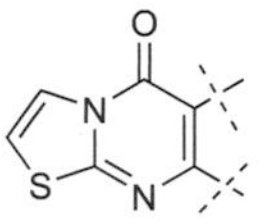

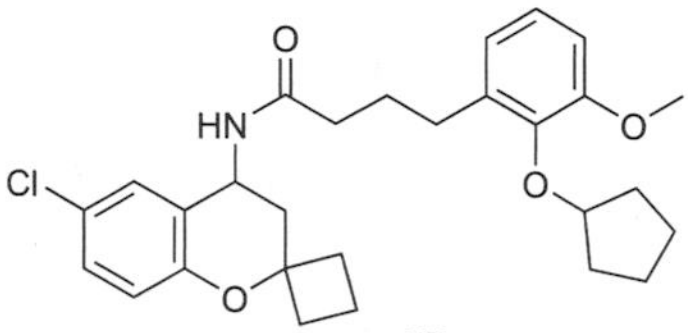

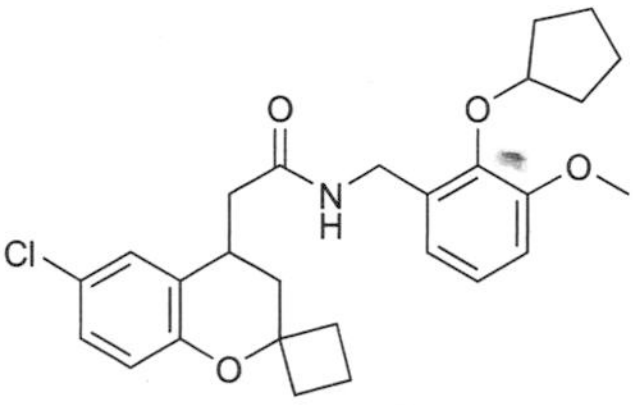

TRPV4 was identified a decade ago as an osmotransducer that is expressed in lung, heart, kidney, airway muscle cells, sensory neurons, brain, skin, gut, sympathetic nerves, inner ear, endothelium, and fat tissue [58–61]. TRPV4 is activated by heat (27–34°C), endogenous substances such as anandamide (**59**, AA) and the arachidonic acid metabolite 5,6-epoxyeicosatrienoic acid (**60**, 5,6-EET), a plant dimeric diterpenoid bisandrographolide A (**61**, BAA), and the semisynthetic phorbol ester 4α-phorbol-12,13-didecanoate (**62**, 4α-PDD) [62].

Activation of TRPV4 by phorbol derivative (**62**) results in an increase of intracellular Ca^{2+} concentration in mTRPV4 with an EC_{50} value of 370 nM. A related analog (**63**) was reported to be several-fold more potent than compound **62**. GSK1016790A (**64**), a highly potent TRPV4 agonist, has been found to activate Ca^{2+} influx in HEK cells expressing human and mouse TRPV4 cells with EC_{50} values of 2.1 and 18 nM, respectively [63]. Similarly, GSK1016790A showed dose-dependent activation of TRPV4 whole-cell currents at concentrations above 1 nM. GSK1016790A (**64**) is reported to display good selectivity for TRPV4 relative to human TRPM8 and TRPA1, but shows only ten-fold selectivity versus TRPV1. TRPV4 activation by GSK1016790A (**64**) causes contraction in TRPV4$^{+/+}$ mouse bladders *ex vivo*, but does not affect the bladder of TRPV4$^{-/-}$ mice [63], suggesting a role for TRPV4 agonists in the treatment of overactive bladder syndrome. Another sulfonamide, RN-1747 (**65**), is reported to be a TRPV4 agonist with selectivity versus TRPV1, TRPV3, and TRPM8 [64].

Few TRPV4 antagonists have been reported to date. RN-1734 (**66**) has been identified as a micromolar TRPV4 antagonist with good selectivity relative to TRPV1, TRPV3, and TRPM8 [64]. Diazabicyclo [2,2,1]hept-2-yl analogs (**67**, **68**) have been described as TRPV4 antagonists, but no biological data have been provided for these compounds [65,66].

59

60

61

62 (R = n-C_9H_{19})
63 (R = n-C_5H_{11})

64

65

66

67 (R = H)
68 (R = F)

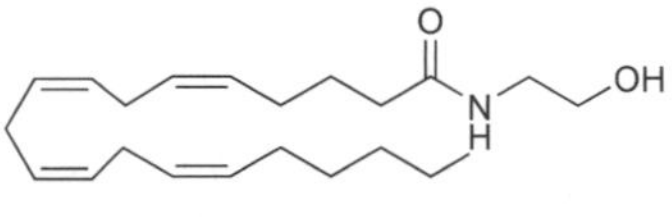

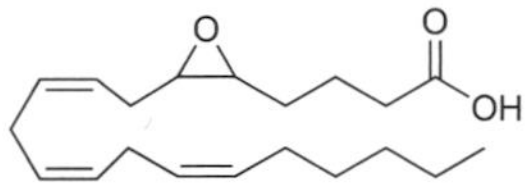

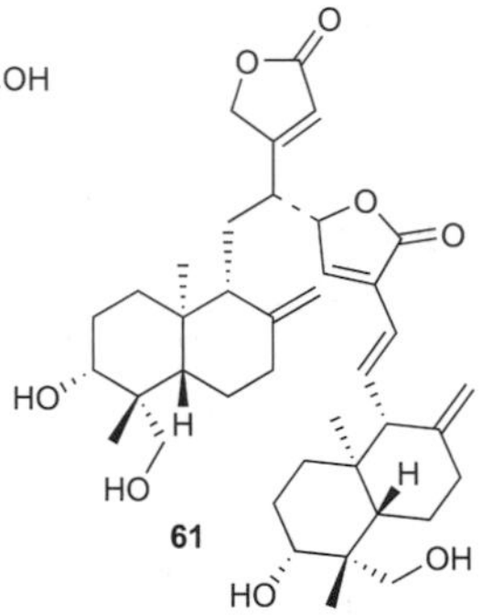

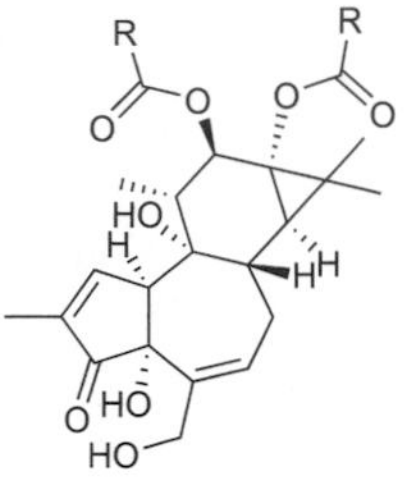

2.7 TRPV5 and TRPV6

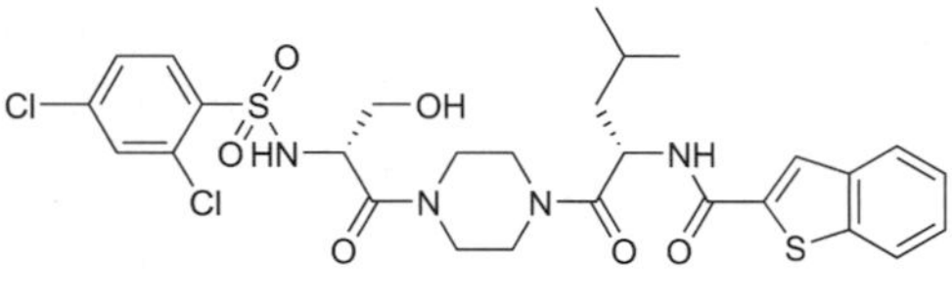

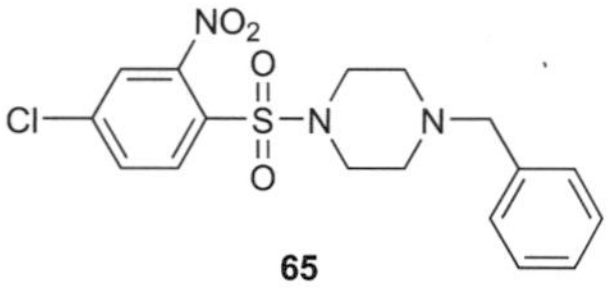

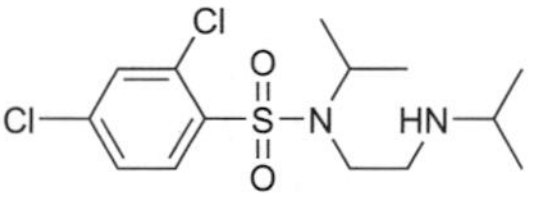

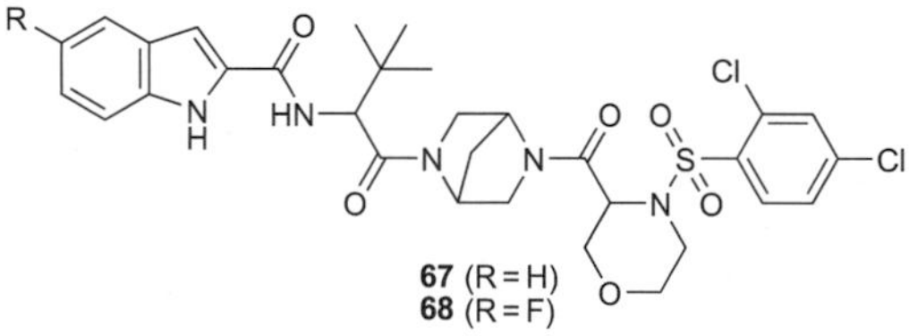

To date, no small-molecule modulators have been reported for either of these channels.

2.8 TRPM (1–8 subtypes)

TRP-melastatin (TRPM) channels are believed to be involved as signal integrators in a number of physiological and pathological conditions and are therefore compelling targets for drug discovery [67]. The TRPM subfamily consists of eight members (1–8), some of which contain an enzymatic domain in the C-terminus.

TRPM2 is a nonselective Ca^{2+}-permeable channel that contains an adenosine diphosphoribose (ADPR) pyrophosphatase domain and is activated by ADPR, cyclic ADP-ribose (cADPR), and H_2O_2; this activation is potentiated by heat (40°C). TRPM2 appears to be involved in oxidative stress-induced cell death and inflammation processes. In addition, TRPM2 activation has been implicated in the release of insulin in rat pancreatic islets at physiological body temperature [68]. Heat-induced insulin release was almost completely blocked by 2-APB (100 µM). In addition to 2-APB (**41**), FFA (**17**), antifungal agents miconazole (**69**) and clotrimazole (**70**), and the phospholipase A2 inhibitor N-(p-amylcinnamoyl)anthranilic acid (**71**, ACA) inhibit TRPM2 [69].

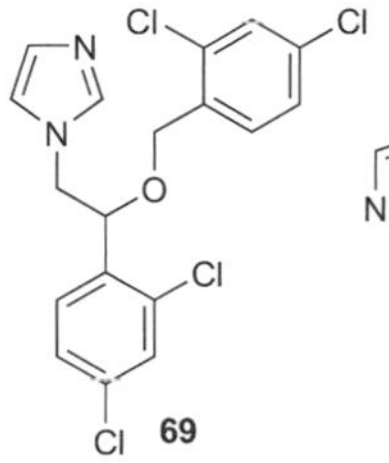

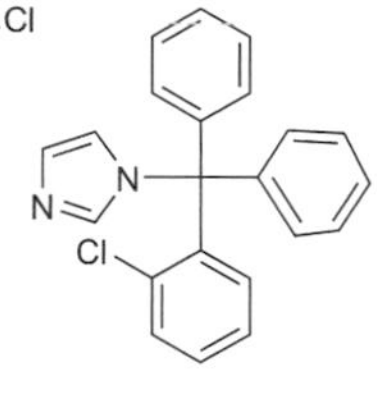

Human TRPM3 is a divalent permeable channel that is expressed predominantly in kidney but also in brain, testis, and spinal cord [70]. In TRPM3-expressing HEK293 cells the channel is rapidly and reversibly activated by the neuroactive steroid pregnenolone sulfate (**72**, PS) [71], which also activates endogenous TRPM3 channels in insulin-producing β cells. Administration of PS (**72**) leads to a rapid calcium influx and enhanced insulin secretion from pancreatic β cells and pancreatic islets. To date, no specific TRPM3 antagonist has been reported.

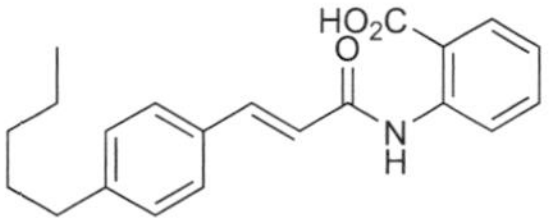

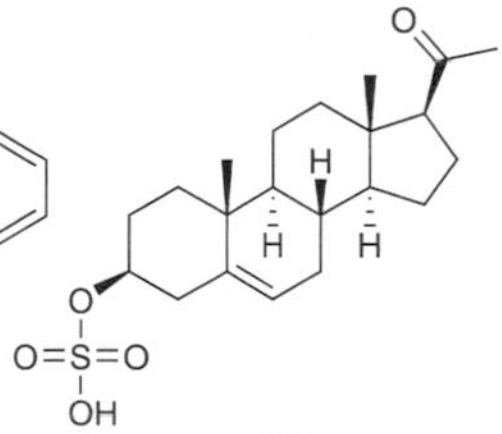

TRPM8 protein was originally identified in the prostate but later found to be present also in DRG, TG, lung, and bladder [72,73]. TRPM8 is activated by cold temperatures and compounds such as menthol (**50**, $EC_{50} = 192$ µM) and icilin (**73**, $EC_{50} = 7$ µM). Several menthol derivatives such as WS-5 (**74**, $EC_{50} = 26$ µM), WS-12 (**75**, EC_{50} 12 µM), CPS-125 (**76**, $EC_{50} = 32$ µM), CPS-368 (**77**, $EC_{50} = 104$ µM), and CPS-369 (**78**, $EC_{50} = 65$ µM) are also reported as agonists of TRPM8 [74]. TRPM8$^{-/-}$ mice are impaired in their ability to detect cold temperatures and unlike the WT animals, the number of neurons in these mice responding to cold (18°C) and menthol (100 µM) is greatly decreased. Studies with TRPM8$^{-/-}$ mice also revealed that TRPM8 may be involved in mediating both cold and mechanical allodynia in rodent models of neuropathic pain [75].

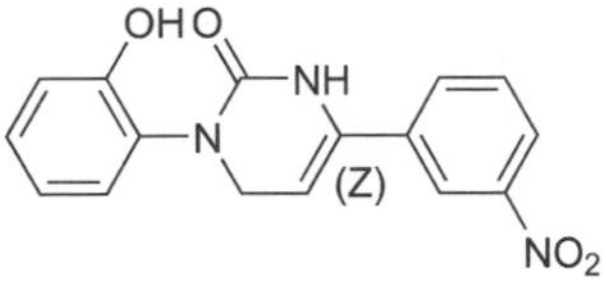

73 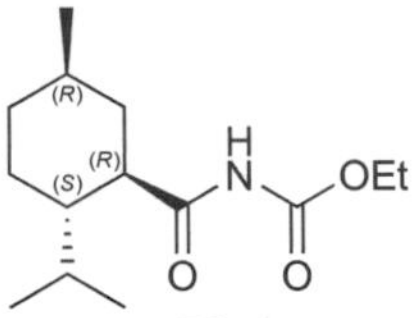   74 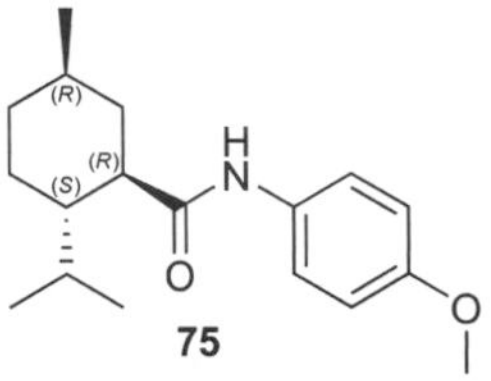   75

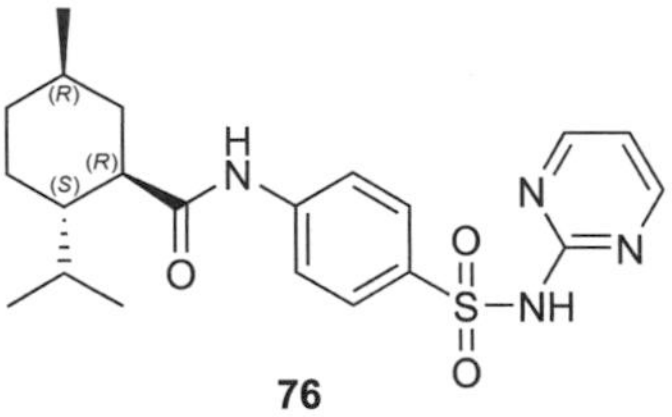

76 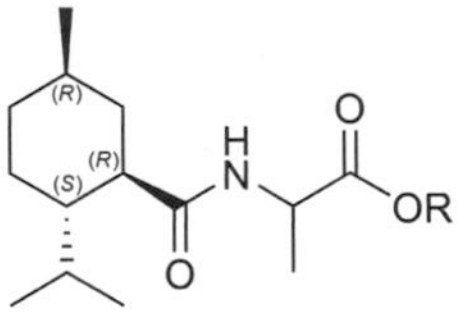   77 (R = Me)
78 (R = Et)

Several drug discovery efforts are currently directed at identifying selective inhibitors of TRPM8 as potential analgesic drugs for the treatment of inflammatory and neuropathic pain. The activation of TRPM8 has also been implicated in overactive bladder (OAB) and painful bladder syndromes (PBS). Benzamide (79, AMTB) has been reported recently as a TRPM8 antagonist that blocks icilin activation of human TRPM8 with modest potency (IC$_{50}$ = 0.6 µM) [76]. AMTB (79) is selective for TRPM8 versus TRPV1 and TRPV4; selectivity data for other ion channels have not been reported. Several chemical classes of TRPM8 antagonists have been reported recently [77,78]. The carboxylic acid derivative (80) is described as a potent TRPM8 antagonist in HEK293 cells expressing human (4 nM) and rat (5 nM) TRPM8 [77]. This compound blocks both menthol and cold-activated currents with IC$_{50}$ values of 1 and 2 nM, respectively. Efficacy *in vivo* has been reported for compound 80 in models of icilin-induced wet-dog shakes in rats, CFA-induced paw radiant heat hypersensitivity, and CCI-induced cold hypersensitivity. Benzthiophene derivatives (81 and 82) were also claimed as potent inhibitors of human TRPM8. It is interesting to note that the (R)-enantiomer (81) is approximately ten-fold less potent than the (S)-enantiomer (82) as an antagonist of TRPM8 in HEK 293 cells [78].

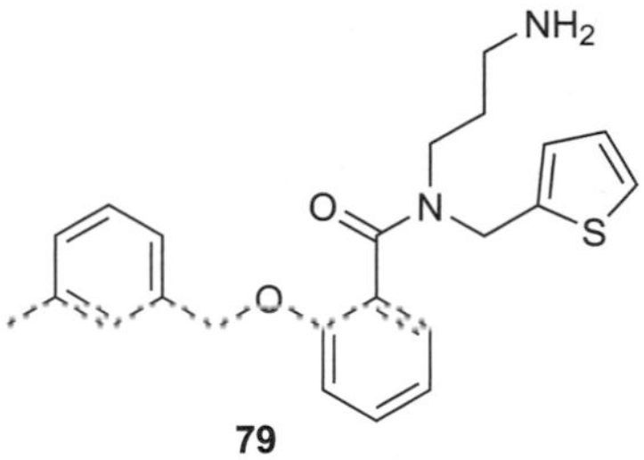

79

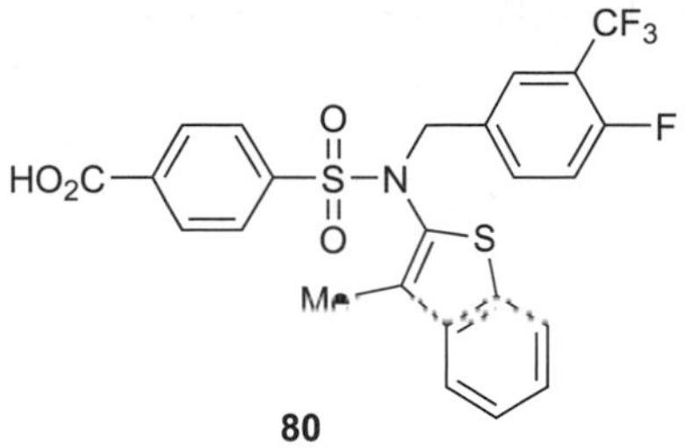

80

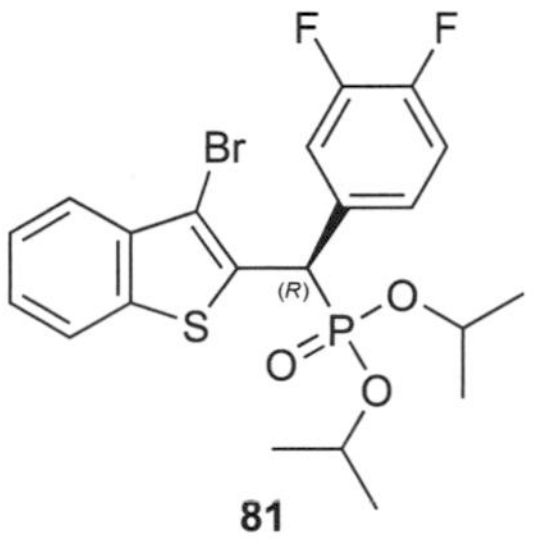

81

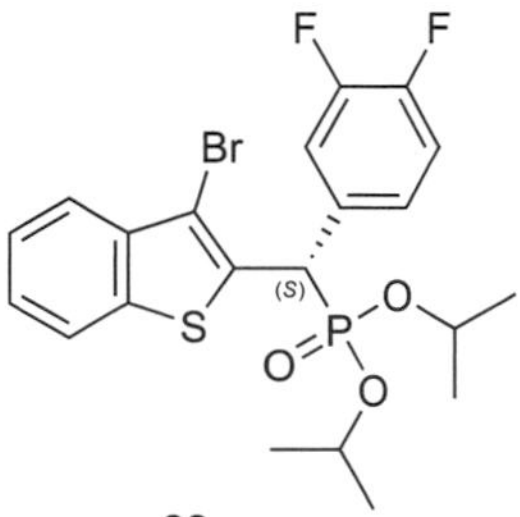

82

No small-molecule modulators of TRPM1, TRPM3, TRPM4, TRPM5, TRPM6, or TRPM7 have been described.

2.9 TRPC, TRPP, and TRPML families

TRPC5 is a Gq-linked cation channel that is highly expressed in the brain, particularly in the hippocampus and amygdala. Behavioral studies in WT and KO mice have linked TRPC5 to innate fear behavior [79,80]. A recent patent application discloses modulators of the TRPC5 channel, including the diaminotriazine (**83**).

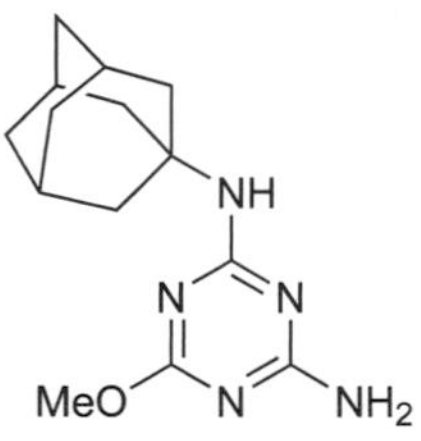

83

No small-molecule modulators of the remaining TRPC, TRPP, or TRPML ion channel families have yet been described.

3. CONCLUSIONS

The transient receptor potential ion channel superfamily comprises a significant fraction of known ion channels. The role of these channels both in normal physiology and in the pathophysiology of disease is being elucidated, and several important TRP drug targets have already been identified. TRPV1 is the first of the TRP ion channels to be systematically studied as a potential small-molecule drug target. While TRPV1 antagonists have shown efficacy in human trials, concerns of on-target effects of hyperthermia and an inability to detect noxious heat have hampered their clinical development. The apparent physiological role of TRPA1 as an irritant receptor positions it as a key player in the modulation of pain, inflammation, and pulmonary disease. TRPV3 and TRPM8 are likewise promising drug targets due to their pivotal roles in pain signaling. These targets are now being probed preclinically with newly discovered agonists and antagonists. Our evolving understanding of the physiological role of TRP ion channels and the biological impact of small-molecule modulators of these channels *in vivo* should enable the rapid discovery of important drugs for the treatment of human disease.

REFERENCES

[1] I. S. Ramsey, M. Delling, and D. E. Clapham, *Annu. Rev. Physiol.*, 2006, **68**, 619.

[2] B. Nilius, G. Owsianik, T. Voets, and J. A. Peters, *Physiol. Rev.*, 2007, **87**, 165.

[3] A. Vasudevan and M. E. Kort, *Ann. Rep. Med. Chem.*, 2008, **43**, 81.

[4] W. E. Childers, A. M. Gilbert, J. D. Kennedy, and G. T. Whitesides, *Expert Opin. Ther. Patents*, 2008, **18**, 1027.

[5] J. Vriens, G. Appendino, and B. Nilius, *Mol. Pharm.*, 2009, **75**, 1262.

[6] A. Hinman, H. Chaung, D. M. Bautista, and D. Julius, *Proc. Natl. Acad. Sci. USA*, 2006, **103**, 19564.

[7] L. J. Macpherson, A. E. Dubin, M. J. Evans, F. Marr, P. G. Schultz, B. F. Cravatt, and A. Patapoutian, *Nature*, 2007, **445**, 541.

[8] L. J. Macpherson, B. Xiao, K. Y. Kwan, M. J. Petrus, A. E. Dubin, S. Hwang, B. Cravatt, D. P. Corey, and A. Patapoutian, *J. Neurosci.*, 2007, **27**, 11412.

[9] M. Trevisani, J. Siemens, S. Materazzi, D. M. Bautista, R. Nassini, B. Campi, N. Imamachi, E. Andre, R. Patacchini, G. S. Cottrell, R. Gatti, A. I. Basbaum, N. W. Bunnett , D. Julius, and P. Geppetti, *Proc. Natl. Acad. Sci. USA*, 2007, **104**, 13519.

[10] C. Tai, S. Zhu, and N. Zhou, *J. Neurosci.*, 2008, **28**, 1019.

[11] Y. Iwasaki, M. Tanabe, Y. Kayama, M. Abe, M. Kashio, K. Koizumi, Y. Okumura, Y. Morimitsu, M. Tominaga, Y. Ozawa, and T. Watanabe, *Life Sci.*, 2009, **85**, 60.

[12] H. J. M. Gijsen and M. H. Mercken, *Patent Application WO 2009/071631-A*, 2009.

[13] H. Hu, J. Tian, Y. Zhu, C. Wang, R. Xiao, J. M. Herz, J. D. Wood, and M. X. Zhu, *Pflugers Arch-Eur. J. Physiol.*, 2010, **459**, 579 .

[14] P. S. Lee, Q. Yang, R. W. Bryant, and T. Buber, *Patent Application WO 2008/013861-A*, 2008.

[15] H. Doihara, K. Nozawa, E. Kawabata-Shoda, R. Kojima, T. Yokoyama, and H. Ito, *Mol. Cell. Biochem.*, 2009, **331**, 239.

[16] H. Kaku, T. Takuwa, M. Kageyama, K. Nozawa, and H. Doihara, *Patent Application WO 2009/123080-A*, 2009.

[17] C. R. McNamara, J. Mandel-Brehm, D. M. Bautista, J. Siemens, K. L. Deranian, M. Zhao, N. J. Hayward, J. A. Chong, D. Julius, M. M. Moran, and C. M. Fanger, *Proc. Natl. Acad. Sci. USA*, 2007, **104**, 13525.

[18] D. M. Bautista, S. E. Jordt, T. Nikai, P. R. Tsuruda, A. J. Read, J. Poblete, E. N. Yamoah, A. I. Basbaum, and D. Julius, *Cell*, 2006, **124**, 1269.

[19] K. Y. Kwan, A. J. Allchorne, M. A. Vollrath, A. P. Christensen, D. S. Zhang, C. J. Woolf, and D. P. Corey, *Neuron*, 2006, **50**, 277.

[20] X. Cai, *Expert Rev. Neurother.*, 2008, **8**, 1675.

[21] A. I. Caceres, M. Brackmann, M. D. Elia, B. F. Bessac, D. del Camino, M. D'Amours, J. S. Witek, C. M. Fanger, J. A. Chong, N. J. Hayward, R. J. Homer, L. Cohn, X. Huang, M. M. Moran, and S. E. Jordt, *Proc. Natl. Acad. Sci. USA*, 2009, **106**, 9099.

[22] P. C. Kerstein, D. del Camino, M. M. Moran, and C. L. Stucky, *Mol. Pain*, 2009, **5**, 19.

[23] S. R. Eid, E. D. Crown, E. L. Moore, H. A. Liang, K. C. Choong, S. Dima, D. A. Henze, S. A. Kane, and M. O. Urban, *Mol. Pain*, 2008, **4**, 48.

[24] H. Ng, M. Weigele, M. M. Moran, J. Chong, C. Fanger, G. R. Larsen, G. Del Camino, N. Hayward, S. P. Adams, and A. Ripka, *Patent Application WO 2009/002933-A*, 2009.

[25] M. Muthuppalniappan, S. Kumar, A. Thomas, N. Khairatkar-Joshi, and I. Mukhopadhyay, *Patent Application WO 2009/118596-A*, 2009.

[26] S. S. Chaudhari, A. Thomas, N. P. Patil, V. G. Deshmukh, N. Khairatkar-Joshi, and I. Mukhopadhyay, *Patent Application WO 2009/144548-A*, 2009.

[27] M. Muthuppalniappan, A. Thomas, S. Kumar, S. Margal, N. Khairatkar-Joshi I. Mukhopadhyay, and S. Gullapalli, *Patent Application US 2009/0325987-A*, 2009.

[28] A. Patapoutian and T. J. Jegla, *Patent Application WO 2007/098252-A*, 2007.

[29] M. Petrus, A. M. Peier, M. Bandell, S. W. Hwang, T. Huynh, N. Olney, T. Jegla, and A. Patapoutian, *Mol. Pain*, 2007, **3**, 40.

[30] R. J. Perner, M. E. Kort, S. Didomenico, J. Chen, and A. Vasudevan, *Patent Application WO 2009/089082-A*, 2009.

[31] S. McGaraughty, K. L. Chu, R. J. Perner, S. Didomenico, M. E. Kort, and P. R. Kym, *Mol. Pain*, 2010, **6**, 14.

[32] H. J. M. Gijsen, D. J. Berthelot, and M. A. J. Cleyn, *Patent Application WO 2009/147079-A*, 2009.

[33] L. Klionsky, R. Tamir, B. Gao, W. Wang, D. C. Immke, N. Nishimura, and N. R. Gavva, *Mol. Pain*, 2007, **3**, 39.

[34] F. Viana and A. Ferrer-Montiel, *Expert Opin. Ther. Patents*, 2009, **19**, 1787.

[35] N. R. Gavva, *Open Drug Discov. J.*, 2009, **1**, 1.

[36] M. Crutchlow, American Society for Clinical Pharmacology and Therapeutics Meeting, Maryland, USA, March 18–21, 2009.

[37] J. Lázár, L. Gharat, N. Khairathkar-Joshi, P. M. Blumberg, and A. Szallasi, *Expert Opin. Drug Discov.*, 2009, **4**, 159.

[38] M. J. Gunthrope and B. A. Chiz, *Drug Discov. Today*, 2009, **14**, 56.

[39] D. N. Cortright and A. Szallasi, *Curr. Pharm. Design*, 2009, **15**, 1736.

[40] A. Gomtsyan, E. K. Bayburt, R. G. Schmidt, C. S. Surowy, P. Honore, K. C. Marsh, S. M. Hannick, H. A. McDonald, J. M. Wetter, J. P. Sullivan, M. F. Jarvis, C. R. Faltynek, and C. H. Lee, *J. Med. Chem.*, 2008, **51**, 392.

[41] M. P. Neeper, Y. Liu, T. L. Hutchinson, Y. Wang, C. M. Flores, and N. Qin, *J. Biol. Chem.*, 2007, **282**, 15894.

[42] N. Qin, M. P. Neeper, Y. Liu, T. L. Hutchinson, M. L. Lubin, and C. M. Flores, *J. Neurosci.*, 2008, **28**, 6231.

[43] N. Qin, Y. Liu, C. M. Flores, T. Hutchinson, and M. P. Neeper, *US Patent 7,575,882-B2*, 2009.

[44] S. Bang, K. Y. Kim, S. Yoo, S. H. Lee, and S. W. Hwang, *Neurosci Lett.*, 2007, **425**, 120.

[45] S. W. Hwang, S. S. Bang, and S. H. Lee, *Patent Application WO 2009/0215107-A*, 2009.

[46] V. Juvin, A. Penna, J. Chemin, Y. L. Lin, and F. A. Rassendren, *Mol. Pharmacol.*, 2007, **72**, 1258.

[47] H. Xu, I. S. Ramsey, S. A. Kotecha, M. M. Moran, J. A. Chong, D. Lawson, P. Ge, J. Lilly, I. Silos-Santiago, Y. Xie, P. S. DiStefano, R. Curtis, and D. E. Clapham, *Nature*, 2002, **418**,181.

[48] A. Moqrich, S. W. Hwang, T. J. Earley, M. J. Petrus, A. N. Murray, K. S. R. Spencer, M. Andahazy, G. M. Story, and A. Patapoutian, *Science*, 2005, **307**, 1468.

[49] A. K. Vogt-Eisele, K. Weber, M. A. Sherkheli, G. Vielhaber, J. Panten, G. Gisselmann, and H. Hatt, *Br. J. Pharmacol.*, 2007, **151**, 530.

[50] H. Xu, M. Delling, J. C. Jun, and D. E. Clapham, *Nat. Neurosci.*, 2006, **9**, 628.

[51] A. Moussaieff, N. Rimmerman, T. Bregman, A. Straiker, C. C. Felder, S. Shoham, Y. Kashman, S. M. Huang, H. Lee, E. Shohami, K. Mackie, M. J. Caterina, J. M. Walker, E. Fride, and R. Mechoulam, *FASEB J.*, 2008, **22**, 3024.

[52] J. A. Chong, C. Fanger, M. M. Moran, D. J. Underwood, X. Zhen, A. Ripka, M. Weigele, W. C. Lumma, and G. R. Larsen, *Patent Application WO 2006/122156-A*, 2006.

[53] J. A. Chong, C. Fanger, G. R. Larsen, W. C. Lumma, M. M. Moran, A. Ripka, D. J. Underwood, M. Weigele, and X. Zhen, *Patent Application WO 2007/056124-A*, 2007.

[54] V. S. P. Lingam, A. Thomas, D. A. More, J. Y. Khatik, N. Khairatkar-Joshi, and V. G. Kattigei, *Patent Application WO 2009/109987-A*, 2009.

[55] V. S. P. Lingam, S. C. Sachin, A. Thomas, N. Khairatkar-Joshi, and V. G. Kattigei, *Patent Application WO 2009/130560-A*, 2009.

[56] V. S. P. Lingam, A. Thomas, A. L. Gharat, D. V. Ukirde, A. S. Mindhe, N. Khairatkar-Joshi, V. G. Kattigei, and S. K. Phatangare *Patent Application WO 20090/084034-A*, 2009.

[57] V. S. P. Lingam, A. Thomas, J. Y. Khatik, N. Khairatkar-Joshi, and V. G. Kattigei, *Patent Application WO 2010/0004379-A*, 2010.

[58] B. Nilius, J. Vriens, J. Prenen, G. Droogmans, and T. Voets, *Am. J. Physiol. Cell Physiol.*, 2004, **286**, C195.

[59] W. Liedtke, Y. Choe, M. A. Marti-Renom, M. A. Bell, C. S. Denis, A. Sali, A. J. Hudspeth, J. M. Friedman, and S. Heller, *Cell*, 2000, **103**, 525–535.

[60] R. Strotmann, C. Harteneck, K. Nunnenmacher, G. Schultz, and T. D. Plant, *Nat. Cell Biol.*, 2000, **2**, 695.

[61] Y. Jia, X. Wang, L. Varty, C. A. Rizzo, R. Yang, C. C. Correll, P. T. Phelps, R. W. Egan, and J. A. Hey, *Am. J. Physiol. Lung Cell Mol. Physiol.*, 2004, **287**, L272.

[62] T. K. Klausen, A. Pagani, A. Minassi, A. Ech-Chahad, J. Prenen, G. Owsianik, E. K. Hoffmann, S. F. Pedersen, G. Appendino, and B. Nilius, *J. Med. Chem.*, 2009, **52**, 2933.

[63] K. S. Thorneloe, A. C. Sulpizio, Z. Lin, D. J. Figueroa, A. K. Clouse, G. P. McCafferty, T. P. Chendrimada, E. S. Lashinger, E. Gordon, L. Evans, B. A. Misajet, D. J. Demarini, J. H. Nation, L. N. Casillas, R. W. Marquis, B. J. Votta, S. A. Sheardown, X. Xu, D. P. Brooks, N. J. Laping, and T. D. Westfall, *J. Pharmacol. Exp. Ther.*, 2008, **326**, 432.

[64] F. Vincent, A. Acevedo, M. T. Nguyen, M. Dourado, J. DeFalco, A. Gustafson, P. Spiro, D. E. Emerling, M. G. Kelly, and M. A. J. Duncton, *Biochem. Biophys. Res. Commun.*, 2009, **389**, 490.

[65] M. Cheung, H. S. Eidam, R. M. Fox, and E. S. Manas, *Patent Application WO 2009/146177-A*, 2009.

[66] M. Cheung, H. S. Eidam, R. M. Fox, and E. S. Manas, *Patent Application WO 2009/146182-A*, 2009.

[67] A. Fleig and R. Penner, *Trends Pharmacol. Sci.*, 2004, **25**, 633.

[68] K. Togashi, Y. Hara, T. Tominaga, T. Higashi, Y. Konishi, Y. Mori, and M. Tominaga, *EMBO J.*, 2006, **25**, 1804.

[69] K. Togashi, H. Inada, and M. Tominaga, *Br. J. Pharma.*, 2008, **153**, 1324.
[70] N. Lee, J. Chen, L. Sun, S. Wu, K. R. Gray, A. Rich, M. Huang, J. H. Lin, J. N. Feder, E. B. Janovitz, P. C. Levesque, and M. A. Blanar, *J. Biol. Chem.*, 2003, **278**, 20890.
[71] T. F. J. Wagner, S. Loch, S. Lambert, I. Straub, S. Mannebach, I. Mathar, M. Dufer, A. Lis , V. Flockerzi, S. E. Philipp, and J. Oberwinkler, *Nature Cell Biol.*, 2008, **10**, 1421.
[72] L. Tsavaler, M. H. Shapero, S. Morkowski, and R. Laus, *Cancer Res.*, 2001, **61**, 3760.
[73] D. D. McKemy, W. M. Neuhausser, and D. Julius, *Nature*, 2002, **416**, 52
[74] M. A. Sheikhleh, G. Gisselmann, A. K. Vogt-Eisele, J. F. Doerner, and H. Hatt, *Pak. J. Pharm. Sci.*, 2008, **21**, 370.
[75] R. W. Colburn, M. L. Lubin, D. J. Stone, Y. Wang, D. Lawrence, M. R. D'Andrea, M. R. Brandt, Y. Liu, C. M. Flores, and N. Qin, *Neuron*, 2007, **54**, 379.
[76] E. S. R. Lashinger, M. S. Steiginga, J. P. Hieble, L. A. Leon, S. D. Gardner, R. Nagilla, E. A. Davenport, B. E. Hoffman, N. J. Laping, and X. Su, *Am. J. Physiol. Renal Physiol.*, 2008, **295**, F803.
[77] S. T. Branum, R. W. Colburn, S. L. Dax, C. M. Flores, M. C. Jetter, Y. Liu, D. Ludovici, M. J. Macielag, J. M. Matthews, J. J. Mcnally, L. M. Reany, R. K. Russell, N. Qin, K. M. Wells, S. C. Youells, and M. A. Youngman, *Patent Application WO 2009/012430-A*, 2009.
[78] R. W. Colburn, S. L. Dax, C. M. Flores, J. Matthews, M. A. Youngman, C. Teleha, and L. M. Reany, *Patent Application WO 2007/134107-A*, 2007.
[79] A. Riccio, Y. Li, J. Moon, K. S. Kim, K. S. Smith, U. Rudolph, S. Gapon, G. L. Yao, E. Tsvetkov, S. J. Rodig, A. Van't Veer, E. G. Meloni, W. A. Carlezon Jr., V. Y. Bolshakov, and D. E. Clapham, *Cell*, 2009, **137**, 761.
[80] J. Chong, C. Fanger, M. M. Moran, E. Singer, T. Strassmaier, and H. Ng, *Patent Application WO 2010/017368-A2*, 2010.

CHAPTER **4**

Defining Neuropharmacokinetic Parameters in CNS Drug Discovery to Determine Cross-Species Pharmacologic Exposure–Response Relationships

Christopher L. Shaffer

Department of Pharmacokinetics, Pharmacodynamics and Metabolism, Neuroscience Research Unit, Pfizer
Global Research and Development, Groton Laboratories, Pfizer Inc., Groton, CT 06340, USA

Annual Reports in Medicinal Chemistry, Volume 45
ISSN 0065-7743, DOI 10.1016/S0065-7743(10)45004-6

1. INTRODUCTION

Drug discovery and development are scientifically complex and financially risky [1]. Estimates suggest that just 0.02% of molecules evaluated in discovery and 11% of compounds entering clinical development receive regulatory approval [2,3]. Although this encompasses all therapeutic areas, pharmaceuticals targeting central nervous system (CNS) indications have both significantly longer clinical development times and much lower probabilities of success [3,4]. Two proposed reasons for these lower clinical success rates are the challenges in penetrating the blood–brain barrier (BBB) and the dearth of clinically translatable biomarkers and animal models [5]; the latter points to 60% of CNS agents failing in Phase 2 [3]. While significant advances have been made in designing CNS-penetrant compounds [6–8], one key to lowering neurotherapeutic clinical attrition, particularly in Phase 2, other than purely developing more predictive disease models is understanding more precisely preclinically what central concentrations of neuroeffector agents elicit both desired and deleterious effects, and accurately forecasting these exposures in humans [9].

Accordingly, in early discovery stages of CNS pharmacologic agents, animal pharmacokinetic studies evaluate the brain penetration of lead molecules [10]. These *in vivo* studies, supplemented with physicochemical, *in vitro*, and *in silico* data, influence medicinal chemistry strategies to optimize cerebral access [11–20]. However, brain tissue partitioning alone is insufficient to understand fully the centrally mediated effects of these agents [21–23]. Thus, preclinical "neuropharmacokinetic" studies, which define temporal CNS intercompartmental concentration relationships via unbound brain, cerebrospinal fluid (CSF), and unbound plasma area under the compound concentration–time curve (AUC) ratios, are useful to relate a molecule's different realms of CNS penetration and unbound concentration(s) to a specific *in vivo* effect. These *in vivo* unbound brain compound concentrations may be used for important translational medicine purposes, such as determining *in vitro–in vivo* pharmacological correlations; projecting receptor occupancy (RO); and defining a target-based *in vivo* exposure–response relationship across multiple species models of mechanism, efficacy, and/or safety. This chapter will highlight the importance of both unbound brain compound concentrations and the unbound brain-to-unbound plasma compound concentration relationship, as defined by rat neuropharmacokinetic studies, to optimizing the cross-species (i.e., animals and humans) translational pharmacology of small molecules targeting transmembrane proteins within the CNS.

2. CONCEPTUAL OVERVIEW OF CNS PENETRATION

2.1 The blood—brain and blood—CSF barriers

The brain is highly protected anatomically to physical injury by the skull and to chemical insult by both the BBB and the blood–CSF barrier (BCSFB). The BBB, a tight-junctioned monolayer of cerebral capillary endothelial cells containing a plethora of transport proteins and metabolizing enzymes, physically separates the vasculature from brain tissue, itself comprised of interstitial (or extracellular) fluid (ISF) and parenchyma [21,24]. The BCSFB, distinct from the BBB and constituted primarily by the choroid plexus, separates blood from CSF [24,25]. The BCSFB possesses fenestrated endothelial cells adjacent to tight-junctioned epithelial cells and contains various transport proteins, some different from those in the BBB [24,25]. Due to the much greater cerebrovasculature surface area of the BBB versus that of the BCSFB, the BBB mediates the vast majority of the brain access of blood-borne molecules [21,24,25]. Central penetration at these barriers is mediated by metabolic, passive, and active components, with the latter comprised of uptake and efflux transporters [24]. Finally, ISF and CSF are separated by the highly passively permeable ependyma, a cellular monolayer lining brain ventricles believed to be devoid of transporter proteins [21,26]. This greatly simplified, three-compartment neuroanatomical understanding (Figure 1)

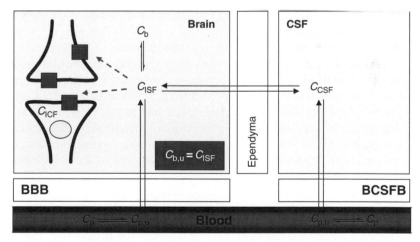

Figure 1 A schematic conceptualization of the three-compartment model of CNS penetration demonstrating the importance of intercompartmental unbound compound concentration relationships to target pharmacology interactions [21,22,25–28]. An exaggerated synapse is shown in the brain compartment to emphasize the locale of transmembrane proteins (squares) versus intracellular (oval) targets, and the matrix compound concentrations dictating their respective ligand–target interactions. (See Color Plate 4.1 in the Color Plate Section.)

forms the fundamental basis for assessing small-molecule CNS penetration and is an essential foundation for neuropharmacokinetic concepts.

2.2 Unbound compound concentrations and the free drug hypothesis

Based on the described neuroanatomy, the three key matrices for evaluating compound concentrations to determine the extent and/or rate of CNS penetration are blood, CSF, and brain tissue. Due to the nearly universal analysis of plasma to determine systemic concentrations of small molecules, total plasma compound concentration (C_p) will henceforth be substituted for total blood concentration, which is the product of C_p and compound blood-to-plasma ratio.

2.2.1 Intercompartmental compound concentration relationships

According to the free drug hypothesis [27], only unbound ligand is available for a pharmacologic interaction. For intratissue molecular targets, this theory is often accommodated by converting an *in vivo*-derived C_p to an unbound plasma compound concentration ($C_{p,u}$), using an experimentally determined unbound plasma fraction ($f_{u,p}$), and assuming equilibrium between $C_{p,u}$ and unbound compound concentration at its site of action (either extracellular or intracellular) [29]. For CNS targets, such a supposition is convenient but naive since it clearly discounts the uniqueness of the BBB and BCSFB.

Accepting that only an unbound neurotherapeutic at its pharmacologic target elicits its medicinal action(s), then the most appropriate CNS matrix for determining this effect-causing compound concentration is dependent on target cellular locale. Specifically, intracellular fluid (ICF) compound concentration (C_{ICF}) dictates pharmacological interaction with an intracellular target, whereas ISF compound concentration (C_{ISF}) is vital for a transmembrane protein with an extracellular ligand binding site (Figure 1); both ICF and ISF lack binding constituents so their compound concentrations are considered completely free [24,25]. However, for intramembrane targets, such as the inhibitory site of γ-secretase [30], the most appropriate neuromatrix compound concentration affecting their interaction is murkier. Determining C_{ICF} or intramembrane compound concentrations experimentally are currently impossible, although C_{ICF} may be projected [31]. However, measuring C_{ISF} is achievable via intracerebral microdialysis, a labor-intensive methodology with its own caveats [25,32]. Thus, experimentally derived C_{ISF} is highly desirable since it simultaneously provides CNS penetration insight and exposure–response relationships directly for transmembrane proteins and indirectly (presuming, precariously [21], C_{ISF} equilibrium with C_{ICF} and cellular constituents) for intracellular and intramembrane targets. However, due to the low-throughput nature of intracerebral microdialysis studies preclinically and its essential unavailability clinically, C_{ISF} is usually estimated by assuming its equilibrium with compound concentrations in CSF (C_{CSF}) [25,33]. Although CSF contains soluble proteins, this protein concentration is so minimal that C_{CSF}, like C_{ISF} and C_{ICF}, is considered an unbound compound concentration [25]. In a discovery setting, CSF collection, particularly in rodents, is more amenable to screening cycle times than intracerebral microdialysis, and in the clinic this is typically the only neuromatrix providing central compound concentrations. Lastly, the ratio between C_{CSF} and $C_{p,u}$ ($C_{CSF}:C_{p,u}$) is important if C_{CSF} is used as a C_{ISF} surrogate as this affords the ratio between C_{ISF} and $C_{p,u}$ ($C_{ISF}:C_{p,u}$), allowing the projection of C_{ISF} from experimentally determined $C_{p,u}$ from blood samples collected either at a single time point or serially in animals or humans [34].

2.2.2 The utility of unbound brain compound concentrations

Due to the difficulties in attaining C_{ISF} in animals undergoing pharmacology studies, a more facile method has been employed in which unbound brain compound concentration ($C_{b,u}$) is projected from a compound's experimentally determined total brain concentration (C_b) and unbound brain homogenate fraction ($f_{u,b}$) [15,33,35–40]. The theory underlying this technique is $C_{b,u}$ equals C_{ISF}, and hence the CNS intercompartmental relationships of a compound may be derived more efficiently experimentally between its $C_{p,u}$, C_{CSF}, and $C_{b,u}$. This methodology has been criticized [16,31] since homogenized brain tissue is not truly indicative of intact brain. However, it was recently reported [39] that for eight of nine compounds, with varying physicochemical properties and degrees of CNS penetration, evaluated at steady-state conditions in rats, $C_{b,u}$ values were within threefold of intracerebral microdialysis-determined C_{ISF}. Similar results in rats dosed acutely have been disclosed for an α_7 nicotinic acetylcholine agonist [41]. These datasets, combined with experimental practicality, support using $C_{b,u}$ as a high-confidence C_{ISF} surrogate generally in CNS drug discovery programs.

2.3 Defining brain penetration and its caveats

A molecule is commonly deemed "brain penetrant" if its brain-to-plasma concentration ratio ($C_b:C_p$) is >0.04 using nonperfused brain tissue, as cerebral blood volume approximates 4% of total brain volume [6]. Though this ratio determines technically if a compound enters brain tissue, it is treacherous in early discovery to use this single parameter to select one brain-penetrant compound over another as this value alone by

Table 1 Rat CNS intercompartmental parameters for compounds **1**, **2**, and **3**

$AUC_{0-\infty}$ Ratio[a]	Compound		
	1	**2**	**3**
$C_b:C_p$	6.4	99	45
$C_{CSF}:C_{p,u}$	6.2	1.3	0.6
$C_{b,u}:C_{CSF}$	0.4	1.2	1.7
$C_{b,u}:C_{p,u}$	2.2	1.6	1.0

[a]Section 3.1 details how these data are derived.
C_b, total brain value; C_p, total plasma value; C_{CSF}, cerebrospinal fluid value; $C_{p,u}$, unbound plasma value; $C_{b,u}$, unbound brain value.

no means determines if adequate brain penetration is achieved for ligand–target interaction. Instead, respective CNS intercompartmental relationships should be considered for each compound.

To make this point more explicit, it is worth evaluating these different AUC-derived intercompartmental relationships in rats of three structurally distinct compounds, none being multidrug resistance 1 P-glycoprotein (P-gp) substrates [19], but each targeting both the serotonin (5-HT) transporter (SerT) and the 5-HT$_{2A}$ receptor (Table 1) [37]. If compound progression was solely based on $C_b:C_p$, then **2** would be selected since it demonstrated the greatest brain tissue partitioning. However, a holistic view of the dataset, in which all realms (i.e., C_{CSF} directly and C_{ISF} indirectly via $C_{b,u}$) of central penetration are evaluated, concludes that all three compounds in terms of $C_{b,u}$, a surrogate for C_{ISF} which is the neuromatrix dictating compound interaction with the specifically cited synaptic transmembrane targets, and $C_{p,u}$ are effectively identical. This example clearly shows the limited understanding of intercompartmental relationships provided purely by $C_b:C_p$, and that increasing this ratio does not necessarily equate to a "better" CNS compound. In the majority of cases, designing to increase passive brain penetration (e.g., increasing calculated logarithm of the partition coefficient (clogP)) concomitantly decreases $f_{u,b}$, and together these parameters determine $C_{b,u}$. Clearly, the CNS parameter $C_{b,u}:C_{p,u}$ is the differentiator between compounds as $C_{b,u}$ is the key component driving pharmacologic interactions at the intended target.

3. PRECLINICAL NEUROPHARMACOKINETIC STUDIES

Building on prior topics and believing $C_{b,u}$ is a high-confidence C_{ISF} surrogate, efficient *in vivo* studies may be conducted in CNS drug discovery to evaluate the neuropharmacokinetics of ligands while helping to define the relationship between *in vivo* compound concentrations and mechanism-based effects. Thus, one study provides insights into both the CNS penetration properties of a compound and the targeted exposures for subsequent studies in small and large animals.

3.1 Establishing temporal CNS intercompartmental relationships in rats

Since rats are used ubiquitously in CNS research, rat neuropharmacokinetic studies performed in discovery provide temporal cerebral intercompartmental compound concentrations that help guide doses and evaluation times for rat pharmacology studies. Neuropharmacokinetic studies may employ any dosing route, but subcutaneous administration is often preferred since it bypasses first-pass metabolism, affording low variability pharmacokinetics data. These studies, like most pharmacology models, are typically conducted with a single dose, lending this methodology to caveats from a CNS penetration perspective as it may be argued that acute dosing does not account for steady-state considerations in determining true CNS intercompartmental relationships [12,21,39]. Although this is a fair criticism for compounds with intrinsically low rates and extents of CNS penetration, an acute dose assessment should be similar to a steady-state evaluation for highly passively permeable compounds. Nonetheless, such single-dose neuropharmacokinetic studies are critical for understanding exposure–response relationships in animal pharmacology models, themselves nearly always evaluated acutely in discovery settings. For high-interest compounds, subsequent steady-state pharmacology and CNS penetration studies should be performed to assess specific pharmacological phenomena as chronic clinical dosing regimens are typical.

From such rat neuropharmacokinetic studies, in which animals ($N = 2/$ time point, ≥ 2 doses) are euthanized at specific time points postdose for plasma, CSF, and brain collection for compound concentration analysis, composite neuromatrix-specific compound concentration–time curves are generated (Figure 2).

These data afford important pharmacokinetic parameters providing insight into both the rate and extent of compound CNS penetration and the rates of elimination from each neurocompartment. Knowing matrix-specific diffusional and elimination rates is important as compounds may have intercompartmental concentration-related delays and/or much longer half-lives in brain tissue versus CSF or plasma. Such phenomena are critical especially if hysteresis is observed in a C_p–response relationship [43]. Furthermore, a combination of AUC-derived values for each neuromatrix and compound-specific rat $f_{u,p}$ and $f_{u,b}$ values, determined

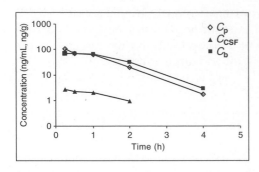

	Plasma	CSF	Brain
C_{max} [a]	105	2.7	70.3
T_{max} [b]	0.25	0.25	0.25
$AUC_{0-\infty}$ [c]	131	5.4	150
$t_{1/2}$ [b]	0.7	1.3	0.8

[a] ng/mL.
[b] h.
[c] ng•h/mL.

	$AUC_{0-\infty}$ Ratios
$C_b:C_p$	1.1
$C_{CSF}:C_{p,u}$	0.9
$C_{b,u}:C_{CSF}$	0.6
$C_{b,u}:C_{p,u}$	0.5

Figure 2 Mean plasma (C_p), CSF (C_{CSF}), and brain (C_b) compound concentration–time profiles (graph) and matrix-specific neuropharmacokinetic parameters (table) of a compound in rats following subcutaneous administration [42]. Abbreviations: C_{max}, maximal compound concentration; T_{max}, time of C_{max}; $t_{1/2}$, compound half-life.

by equilibrium dialysis using plasma or brain homogenate [23], respectively, allows derivation of the net intercompartmental concentration relationships between plasma (unbound and total), CSF, and brain (unbound and total) using the following equations:

$$AUC_{CSF}/(AUC_p \bullet f_{u,p}) = C_{CSF}:C_{p,u} \quad (1)$$

$$(AUC_b \bullet f_{u,b})/AUC_{CSF} = C_{b,u}:C_{CSF} \quad (2)$$

$$(AUC_{CSF}/AUC_{p,u}) \bullet (AUC_{b,u}/AUC_{CSF}) = C_{b,u}:C_{p,u} \quad (3)$$

Equation 1 defines the ratio of $C_{CSF}:C_{p,u}$, which provides an idea of partitioning at the BCSFB. Equation 2 calculates the $C_{b,u}:C_{CSF}$ ratio that, assuming $C_{b,u}$ equals C_{ISF}, defines compound concentrations between ISF and CSF. This ratio is very important because if C_{CSF} is determined in a higher species (principally humans), C_{ISF} may be estimated from it using this rodent-derived $C_{b,u}:C_{CSF}$ assuming an identical ratio across species. Lastly, Equation 3 projects the $C_{b,u}:C_{p,u}$ relationship, which, again presuming equivalent $C_{b,u}$ and C_{ISF}, provides the concentration relationship across the BBB, and, in the absence of measured C_{CSF}, allows directly calculating $C_{b,u}$ from $C_{p,u}$. For CNS compounds, careful characterization of these three ratios increases the chances of adequate compound delivery to the pharmacologic target and understanding cross-species exposure–response relationships.

Importantly, these ratios are *net* values comprised of metabolic, passive, and active components that dictate CNS intercompartmental relationships. As such, they simply compensate for all aspects of BBB and BCSFB penetration, either known (e.g., P-gp-mediated efflux [34]) or unknown (e.g., a yet-to-be-identified uptake transporter [44]). Ideally, a compound's $C_{b,u}{:}C_{p,u}$ would be unity ("equilibrium"), allowing measured C_p to provide $C_{b,u}$ without any conversion. However, this is rarely the case due to a combination of the following: experimental and interanimal variability; intrinsic compound permeability, nonspecific binding, metabolic lability, and active transporter susceptibility; and natural bulk flow gradients affecting CNS fluid dynamics [25]. Based on literature data, ratios within two- to threefold of each other are considered "equilibriums" [20,33,35,39], whereas those with greater variability are deemed "disequilibriums" [19,34]. This ratio helps project $C_{b,u}$ from C_p, measured at any time during a rat assay, via the equation:

$$C_p \bullet f_{u,p} \bullet (C_{b,u}{:}C_{p,u}) = C_{b,u} \qquad (4)$$

For rat pharmacology studies lacking pharmacokinetic samples, these neuropharmacokinetic studies estimate $C_{b,u}$ at a specific dose and time during pharmacodynamic assessment. This $C_{b,u}$ may be used for multiple purposes, primarily the comparison of target-mediated $C_{b,u}$-causing *in vivo* effects to *in vitro* parameters (e.g., K_i or EC_{50}) useful in driving structure–activity relationships (SARs) in synthetic design cycles.

3.2 Projecting large animal unbound brain compound concentrations

Based on rat neuropharmacokinetic concepts, it has been proposed [37,42,45] that $C_{b,u}$ may be projected directly from C_p for large animal species (i.e., dog, monkey, and human) in which C_b are rarely measured. For such species, in which serial blood sampling is common for determining plasma pharmacokinetics, compound $C_{b,u}$ may be extrapolated using its C_p, species-specific $f_{u,p}$, and rat-derived $C_{b,u}{:}C_{p,u}$ assuming a fixed ratio across species:

$$C_{p,species} \bullet f_{u,p,species} \bullet (C_{b,u}{:}C_{p,u})_{rat} = C_{b,u,species} \qquad (5)$$

A common cross-species $C_{b,u}{:}C_{p,u}$ is not universal as occasionally compounds are found to have different degrees of cross-species CNS penetration due to reasons ranging from species differences in P-gp transport to variable intrabrain distribution [46]. However, for compounds in which brain penetration is predominately mediated by passive diffusion, such species differences are more often the exception than the rule. Finally, this methodology implies that a compound's $f_{u,b}$ is identical

Table 2 Observed versus predicted $C_{b,u}$ in dogs for compounds **4** and **5**

Compound	Dog	Observed			Predicted
		C_p (ng/mL)	C_b (ng/g)	$C_{b,u}$ (nM)	$C_{b,u}$ (nM)
4	A	335	395	26	27
4	B	273	423	28	22
5	C	277	574	32	28
5	D	628	1031	58	62
5	E	294	542	30	29

in rats and large animals, which is supported by a recent dataset [14] and the conserved physiological constituents of mammalian brains.

One evaluation of this methodology for dogs is summarized in Table 2 for two structurally distinct compounds targeting the same glutamatergic receptor [45]. For each compound, individual dog C_p and C_b at their respective times of maximal C_p were determined experimentally, as well as rat $C_{b,u}{:}C_{p,u}$ and dog $f_{u,p}$ and $f_{u,b}$. The observed $C_{b,u}$ was compared to the $C_{b,u}$ predicted from C_p using Equation 5. For these two compounds, dog $C_{b,u}$ was projected accurately using this approach. Such an evaluation provides confidence in using this technique to predict primate $C_{b,u}$ to aid in higher order species translational pharmacology.

4. PRECLINICAL AND CLINICAL APPLICATIONS

This section contains examples applying the presented concepts in which measured C_b and $f_{u,b}$, satellite rat neuropharmacokinetic studies, and/or measured C_p and $f_{u,p}$ were used to determine compound-dependent $C_{b,u}$ required for a specific mechanism-mediated *in vivo* effect. The presented studies demonstrate the utility of $C_{b,u}$ in defining target pharmacology relationships for small molecules affecting transmembrane proteins.

4.1 *In vivo* minimally effective $C_{b,u}$ versus *in vitro* functional data

For an agonist (**6**) preferentially activating α_6/α_3-subtype-containing nicotinic acetylcholine receptors, neuronal transmembrane ion channels whose orthosteric binding sites are bathed by ISF, its $C_{b,u}$, for mice or rats undergoing pharmacodynamic evaluation, was determined either directly by C_b measurement at test conclusion and species-specific $f_{u,b}$ or indirectly by projection using rat neuropharmacokinetic data [47]. For six of the seven mechanistic and behavioral rodent assays evaluated, minimally effective $C_{b,u}$ of **6** were within twofold of its functional EC_{50} (Figure 3). For each

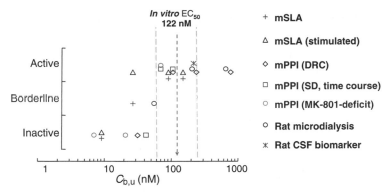

Figure 3   Rodent pharmacodynamic effects versus $C_{b,u}$ for **6**. Dashed lines represent a twofold separation from the *in vitro* functional assay EC_{50} (122 nM, dashed arrow). mSLA, mouse spontaneous locomotor activity; mPPI, mouse prepulse inhibition; DRC, dose–response curve; SD, single dose. (See Color Plate 4.3 in the Color Plate Section.)

program series, the effective $C_{b,u}$ for a sampling of first-generation compounds had similar correlations with their respective EC_{50} values, thus providing high confidence in the *in vitro* screening of subsequent molecules. The early establishment of such *in vitro–in vivo* pharmacological correlations allowed high-confidence *in vitro* pharmacology-based SAR for subsequent chemical matter.

4.2 *In vivo* $C_{b,u}$ versus *in vivo* cross-species receptor occupancy

For CNS transmembrane proteins whose binding sites are positioned extracellularly, C_{ISF} dictates ligand–receptor engagement (Figure 1). Accordingly, the extent of these *in vivo* interactions may be projected by RO, which estimates the percentage of total receptors occupied by a specific compound concentration at a certain time. For centrally acting compounds, such RO–concentration relationships may be determined more definitively using a receptor-specific radioligand [48] or more speculatively using a simple equation (Equation 6) [34,49] incorporating ligand concentration at the site (or surrogate of it, such as $C_{b,u}$) of compound–receptor interaction at a time point (C_t) and molecule binding affinity (K_i) for the specific protein:

$$RO = (C_t/(C_t + K_i)) \bullet 100 \qquad (6)$$

The validity of this concept, in which $C_{b,u}$ is used to project RO accurately across species, was evaluated for a dual-pharmacology antagonist (**1**) simultaneously targeting the synaptic transmembrane proteins SerT and 5-HT$_{2A}$ [37]. Briefly, SerT RO was determined in rats and humans by *ex vivo* binding and positron emission tomography (PET) studies,

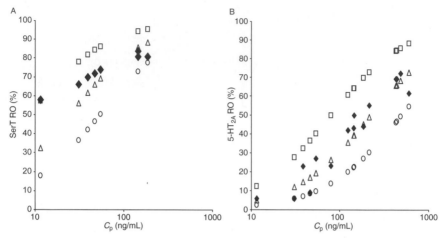

Figure 4 SerT (panel A) and 5-HT$_{2A}$ (panel B) receptor occupancy (RO) in rats versus total plasma exposure (C_p) for **1**. Plotted RO are observed (♦) or projected from $C_{p,u}$ (○), C_{CSF} (□), and $C_{b,u}$ (△), itself estimated from $C_{p,u}$ and rat-derived $C_{b,u}:C_{p,u}$.

respectively, while 5-HT$_{2A}$ RO was determined in rats by *ex vivo* binding, and in humans and nonhuman primates by PET. Rat neuropharmacokinetic-derived intercompartmental parameters of $C_{CSF}:C_{p,u}$ and $C_{b,u}:C_{p,u}$ for **1** (Table 1) allowed the projection of C_{CSF} and $C_{b,u}$ from C_p, measured in actual rats undergoing RO assessment; each neuromatrix value and respective rat K_i were substituted into Equation 6 to estimate rat RO. This analysis showed that $C_{p,u}$ underpredicted, C_{CSF} overpredicted, and $C_{b,u}$ most accurately predicted *ex vivo*-derived SerT and 5-HT$_{2A}$ RO in rats (Figure 4). This observation is consistent with reported rat [38,40] and mouse [36] single-dose RO studies for transmembrane CNS receptors.

This rationale was subsequently applied to human and nonhuman primate C_p, collected from subjects undergoing PET imaging, using Equations 5 and 6 to derive $C_{b,u}$ and RO, respectively; all RO calculations incorporated species-specific K_i values for each target. For these primates, projected $C_{b,u}$, using species-specific $f_{u,p}$ and rat-derived $C_{b,u}:C_{p,u}$, closely predicted PET-determined ROs (Table 3). Together, this cross-species dataset supports using rat-derived $C_{b,u}:C_{p,u}$ to forecast higher order species $C_{b,u}$ from species-specific C_p.

4.3 Defining cross-species exposure–response continuums

Accepting both $C_{b,u}$ as a valid C_{ISF} surrogate and the ability to project $C_{b,u}$ accurately in large animal species, $C_{b,u}$ values may be utilized to develop a mechanism-based cross-species exposure–response continuum for CNS

Table 3 Mean (±SD) observed versus predicted SerT and 5-HT$_{2A}$ receptor occupancy (RO) in humans and nonhuman primates for compound 1

Species	C_p (ng/mL)	$C_{b,u}$ (nM)	SerT RO (%)		5-HT$_{2A}$ RO (%)	
			RO_{pred}	RO_{obs}	RO_{pred}	RO_{obs}
Human[a]	61.2	9.6	—	—	31	30
	13.4	2.1	—	—	9	5
	51.4 ± 30.3	8.1 ± 4.8	64 ± 17	78 ± 15	—	—
	9.5 ± 3.7	1.5 ± 0.6	28 ± 9	47 ± 28	—	—
Nonhuman primate[b]	120 ± 75	24 ± 15	—	—	60 ± 12	47 ± 12

[a] N = 2 and 3 for 5-HT$_{2A}$ and SerT RO studies, respectively.
[b] N = 5.
—, not determined.

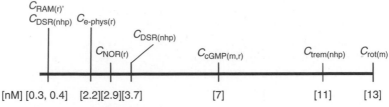

Figure 5 The $C_{b,u}$-normalized cross-species exposure–response continuum for **7** across multiple pharmacology models [42]. A listed value (C_X) is the $C_{b,u}$ (nM) affecting a response in assay X. RAM(r), rat radial arm maze; DSR(nhp), nonhuman primate delayed spatial response task; e-phys(r), rat electrophysiology model; NOR(r), rat novel object recognition; cGMP(m,r), mouse/rat cerebellar cGMP; trem(nhp), nonhuman primate tremor; rot(m), mouse rotarod.

compounds. As an example, this concept has been applied to an α-amino-3-hydroxy-5-methyl-4-isoxazolepropionic acid (AMPA) receptor potentiator (**7**) that underwent a series of preclinical *in vivo* assays evaluating AMPA-mediated pharmacology [42]. Supplementing pharmacological assessments of **7**, rat neuropharmacokinetic studies and C_b determinations from actual rodents undergoing specific pharmacodynamic evaluations were used to determine the $C_{b,u}$ required for a specific *in vivo* effect. These *in vivo*-derived, model-dependent $C_{b,u}$ values for **7** were used to define its exposure–response relationship across multiple species. Furthermore, rat-defined $C_{b,u}:C_{p,u}$ for **7** was used to project its $C_{b,u}$ in nonhuman primate cognition and safety assays to determine the cross-species translatability of such effects. Based on these datasets, a $C_{b,u}$-normalized cross-species exposure–response continuum for **7** across multiple models of cognition, mechanism and safety was ascertained to understand better its exposure separations between enhancing cognition and disrupting motor coordination (Figure 5). Such cross-species definitions preclinically, which normalize exposures in each species model to a $C_{b,u}$ value and allow cross-species translational pharmacology, provide insight into what responses are expected at what systemic exposures clinically, and attempt to position CNS compounds optimally for high-confidence clinical evaluation.

5. CONCLUSIONS

The current challenges in neuroscience drug discovery and development are largely attributed to overcoming the BBB for adequate compound exposure and the shortage of clinically translatable animal disease models. Thus, this chapter has attempted to highlight what may be done using laboratory animals to enhance the progression of small-molecule neurotherapeutics, most specifically those targeting transmembrane proteins, from the laboratory to patients. Of course, the concepts and methodologies explored herein are by no means foolproof. Instead, a greater appreciation of the reviewed literature and its uses, chiefly the application of rat-derived neuropharmacokinetic parameters across species and the importance of both $C_{h,u}$ and $C_{b,u}:C_{p,u}$, should provide an ever greater chance at delivering adequate concentrations of a centrally acting compound to its intended target to test assuredly a pharmacological mechanism in any species, particularly humans. Consequently, Phase 2 studies may be interpreted appropriately without the worry of erroneous target exposure.

REFERENCES

[1] J. A. DiMasi, R. W. Hansen, and H. G. Grabowski, *J. Health Econ.*, 2003, **22**, 151–185.

[2] J. Caldwell, *J. Pharmaceut. Sci.*, 1996, **2**, 117–119.

[3] I. Kola and J. Landis, *Nat. Rev. Drug Discov.*, 2004, **3**, 711–715.

[4] Tufts Center for the Study of Drug Development, *Tufts CSDD Impact Report*, 2005, 1–4.

[5] M. N. Pangalos, L. E. Schechter, and O. Hurko, *Nature Rev. Drug Discov.*, 2007, **6**, 521–532.

[6] S. A. Hitchcock and L. D. Pennington, *J. Med. Chem.*, 2006, **49**, 7559–7583.

[7] T. Wager, R. Chandrasekaran, X. Hou, M. Troutman, P. Verhoest, A. Villalobos, and Y. Will, *ACS Chem. Neurosci.*, 2010, **1**, 420–434.

[8] T. Wager, X. Hou, P. Verhoest, and A. Villalobos, *ACS Chem. Neurosci.*, 2010, **1**, 435–449.

[9] J. Gabrielsson and A. R. Green, *J. Pharmacol. Exp. Ther.*, 2009, **331**, 767–774.

[10] A. Reichel, *Curr. Drug Metabol.*, 2006, **7**, 183–203.

[11] U. Fagerholm, *Drug Discov. Today*, 2007, **12**, 1076–1082.

[12] S. G. Summerfield, A. J. Stevens, L. Cutler, M.d.C. Osuna, B. Hammond, S-P. Tang, A. Hersey, D. J. Spalding, and P. Jeffrey, *J. Pharmacol. Exp. Ther.*, 2006, **316**, 1282–1290.

[13] S. G. Summerfield, K. Read, D. J. Begley, T. Obradovic, I. J. Hidalgo, S. Coggon, A. V. Lewis, R. A. Porter, and P. Jeffrey, *J. Pharmacol. Exp. Ther.*, 2007, **322**, 205–213.

[14] S. G. Summerfield, A. J. Lucas, R. A. Porter, P. Jeffrey, R. N. Gunn, K. R. Read, A. J. Stevens, A. C. Metcalf, M. C. Osuna, P. J. Kilford, J. Passchier, and A. D. Ruffo, *Xenobiotica*, 2008, **38**, 1518–1535.

[15] J. C. Kalvass, T. S. Maurer, and G. M. Pollack, *Drug Metab. Dispos.*, 2007, **35**, 660–666.

[16] S. Becker and X. Liu, *Drug Metab. Dispos.*, 2006, **34**, 855–861.

[17] X. Liu, B. J. Smith, C. Chen, E. Callegari, S. L. Becker, X. Chen, J. Cianfrogna, A. C. Doran, S. D. Doran, J. P. Gibbs, N. Hosea, J. Liu, F. R. Nelson, M. A. Szewc, and J. Van Deusen, *J. Pharmacol. Exp. Ther.*, 2005, **313**, 1254–1262.

[18] H. Wan, M. Rehngren, F. Giordanetto, F. Bergstrom, and A. Tunek, *J. Med. Chem.*, 2007, **50**, 4606–4615.

[19] A. Doran, R. S. Obach, B. J. Smith, N. A. Hosea, S. Becker, E. Callegari, C. Chen, X. Chen, E. Choo, J. Cianfrogna, L. M. Cox, J. P. Gibbs, M. A. Gibbs, H. Hatch, C.E.C.A. Hop, I. N. Kasman, J. LaPerle, J. Liu, X. Liu, M. Logman, D. Maclin, F. M. Nedza, F. Nelson, E. Olson, S. Rahematpura, D. Raunig, S. Rogers, K. Schmidt, D. K. Spracklin, M. Szewc, M. Troutman, E. Tseng, M. Tu, J. W. Van Deusen, K. Venkatakrishnan, G. Walens, E. Q. Wang, D. Wong, A. S. Yasgar, and C. Zhang, *Drug Metab. Dispos.*, 2005, **33**, 165–174.

[20] M. Fridén, S. Winiwater, G. Jerndal, O. Bengtsson, H. Wan, U. Bredberg, M. Hammarlund-Udenaes, and M. Antonsson, *J. Med. Chem.*, 2009, **52**, 6233–6243.

[21] M. Hammarlund-Udenaes, M. Fridén, S. Syvänen, and A. Gupta, *Pharmaceut. Res.*, 2008, **25**, 1737–1750.

[22] X. Liu and C. Chen, *Curr. Opin. Drug Discov. Dev.*, 2005, **8**, 505–512.

[23] J. C. Kalvass and T. S. Maurer, *Biopharm. Drug Dispos.*, 2002, **23**, 327–338.

[24] G. Lee, S. Dallas, M. Hong, and R. Bendayan, *Pharmacol. Rev.*, 2001, **53**, 569–596.

[25] D. D. Shen, A. A. Artru, and K. K. Adkison, *Adv. Drug Deliv. Rev.*, 2004, **56**, 1825–1857.

[26] E. C. M. de Lange and M. Danhof, *Clin. Pharmacokinet.*, 2002, **41**, 691–703.

[27] J.-P. Tillement, S. Urien, P. Chaumet-Riffaud, P. Riant, F. Bree, D. Morin, E. Albengres, and J. Barre, *Fundam. Clin. Pharmacol.*, 1988, **2**, 223–238.

[28] X. Liu, C. Chen, and B. J. Smith, *J. Pharmacol. Exp. Ther.*, 2008, **325**, 349–356.

[29] G. R. Wilkinson, in *Goodman and Gilman's The Pharmacological Basis of Therapeutics*, 10th Edition, (ed. J. G. Hardman, L. E. Limbird, and A. G. Gilman), McGraw-Hill, New York, 2001, pp. 3–29.

[30] M. S. Wolfe, *Chem. Rev.*, 2009, **109**, 1599–1612.

[31] M. Fridén, A. Gupta, M. Antonsson, U. Bredberg, and M. Hammarlund-Udenaes, *Drug Metab. Dispos.*, 2007, **35**, 1711–1719.

[32] E. C. M. de Lange, M. Danhof, A. G. de Boer, and D. D. Breimer, *Brain Res.*, 1994, **666**, 1–8.

[33] X. Liu, B. J. Smith, C. Chen, E. Callegari, S. L. Becker, X. Chen, J. Cianfrogna, A. C. Doran, S. D. Doran, J. P. Gibbs, N. Hosea, J. Liu, F. R. Nelson, M. A. Szewc, and J. Van Deusen, *Drug Metab. Dispos.*, 2006, **34**, 1443–1447.

[34] K. Venkatakrishnan, E. Tseng, F. R. Nelson, H. Rollema, J. L. French, I. V. Kaplan, W. E. Horner, and M. A. Gibbs, *Drug Metab. Dispos.*, 2007, **35**, 1341–1349.

[35] T.S. Maurer, D. B. DeBartolo, D. A. Tess, and D. O. Scott, *Drug Metab. Dispos.*, 2005, **33**, 175–181.

[36] J. C. Kalvass, E. R. Olson, M. P. Cassidy, D. E. Selley, and G. M. Pollack, *J. Pharmacol. Exp. Ther.*, 2007, **323**, 346–355.

[37] C. L. Shaffer, *Drug Metabol. Rev.*, 2008, **40**, 14.

[38] J. Watson, S. Wright, A. Lucas, K. L. Clarke, J. Viggers, S. Cheetham, P. Jeffrey, R. Porter, and K. D. Read, *Drug Metab. Dispos.*, 2009, **37**, 753–760.

[39] X. Liu, K. Van Natta, H. Yeo, O. Vilenski, P. E. Weller, P. D. Worboys, and M. Monshouwer, *Drug Metab. Dispos.*, 2009, **37**, 787–793.

[40] X. Liu, O. Vilenski, J. Kwan, S. Apparsundaram, and R. Weikert, *Drug Metab. Dispos.*, 2009, **37**, 1548–1556.

[41] S. M. Osgood, S. L. Becker, R. Gorczyca, L. M. Buchholz, H. Rollema, and C. L. Shaffer, 146, 57th ASMS Conference on Mass Spectrometry and Allied Topics, Philadelphia, PA, 2009

[42] C. L. Shaffer, R. J. Scialis, S. Lotarski, D. K. Bryce, J. Liu, M. J. Majchrzak, C. Christoffersen, W. E. Hoffmann, B. Campbell, R. S. Hurst, S. McLean, A. H. Ganong, M. Hajos, P. A. Seymour, F. S. Menniti, and C. J. Schmidt, 883.17, Society for Neuroscience, Chicago, IL, 2009.

[43] N. H. G. Holford and L. B. Sheiner, *Clin. Pharmacokinet.*, 1981, **6**, 429–453.

[44] E. Boström, U. S. H. Simonsson, and M. Hammarlund-Udenaes, *Drug Metab. Dispos.*, 2006, **34**, 1624–1631.

[45] C. L. Shaffer, ACS ProSpectives Conference Series: Discovery and Selection of Successful Drug Candidates with special emphasis on Structure-based Drug Design, Cambridge, MA, 2009.

[46] S. Syvänen, Ö. Lindhe, M. Palner, B. R. Kornum, O. Rahman, B. Långström, G. M. Knudsen, and M. Hammarlund-Udenaes, *Drug Metab. Dispos.*, 2009, **37**, 635–643.

[47] R. S. Hurst, R. J. Mather, K. M. Ward, A. Shrikhande, D. E. Johnson, F. D. Tingley, T. Tritto, R. Kozak, B. Ellerbrock, S. M. Osgood, C. L. Shaffer, B. N. Rogers, G. Arora, J. W. Coe, S. L. Deninno, J. A. Lowe, T. Nguyen, I. Sakurada, and L. Zhang, 228.10, Society for Neuroscience, Chicago, IL, 2009.

[48] S. Grimwood and P. R. Hartig, *Pharmacol. Therapeut.*, 2009, **122**, 281–301.

[49] P. Taylor and P. A. Insel, in *Principles of Drug Action: The Basis of Pharmacology*, 3rd Edition, (ed. W. B. Pratt and P. Taylor), Churchill Livingstone, New York, 1990, p. 1–102.

PART II:
Cardiovascular and Metabolic Diseases

Editor: Andy Stamford
Merck Research Laboratories
Kenilworth
New Jersey

CHAPTER 5

High-Affinity Niacin Receptor GPR109A Agonists

Hong C. Shen and **Steven L. Colletti**

1. INTRODUCTION

Niacin (nicotinic acid, **1**) at high doses (>1 g/day) favorably modulates the human lipid profile by elevating high-density lipoprotein cholesterol (HDL-C) and decreasing low-density lipoprotein cholesterol (LDL-C), very low-density lipoprotein cholesterol (VLDL-C), triglyceride (TG),

Department of Medicinal Chemistry, Merck Research Laboratories, Merck & Co., Inc., Rahway, NJ 07065-0900, USA

Annual Reports in Medicinal Chemistry, Volume 45
ISSN 0065-7743, DOI 10.1016/S0065-7743(10)45005-8

and lipoprotein a (Lp(a)) [1,2]. Specifically, the ability of niacin to increase HDL-C (~20%) is greater than any other drug on the market. As such niacin is considered a broad-spectrum lipid-lowering drug. On the other hand, along with the structurally closely related nicotinamide, niacin as vitamin B_3 provides precursors to the coenzyme nicotinamide adenine dinucleotide, a vital electron receptor involved in the oxidation of fuel metabolites [3].

In a landmark study reported in 1955, niacin became the first drug to favorably affect plasma lipids [4]. Clinical data from the Coronary Drug Project (CDP) released in 1975 unequivocally demonstrated the effectiveness of niacin as the first drug to reduce atherosclerotic cardiovascular events [5–8]. Over a 6-year dosing period, the niacin treatment cohort, which was administered a 2-g daily dose, experienced 26% fewer nonfatal myocardial infarctions and 24% fewer cerebrovascular events compared to the placebo group. In a follow-up study conducted 9 years after the completion of this trial, total mortality was reduced by 11% in patients originally treated with niacin in comparison to placebo-treated patients [9]. In the subsequent HATS [10] and ARBITER2 trials [11], the combination of niacin and simvastatin slowed the progression of atherosclerosis, reduced the number of cardiac events, and established further advantages over statin treatment alone. In addition, the Stockholm Ischaemic Heart Disease Study also achieved mortality reduction with the combination therapy of niacin and clofibrate [12].

The major adverse effect of niacin treatment is intense cutaneous flushing (vasodilation), which manifests as an uncomfortable burning sensation and itchiness of the face and upper body, thereby limiting patient compliance to therapy [13]. Moreover, a short half-life, dyspepsia, hyperuricemia, and modest hyperglycemia were also reported [14–16].

To attenuate vasodilation induced by niacin, four strategies have emerged. The first involves a once-daily extended-release formulation of niacin which substantially ameliorates the magnitude of flushing [17]. Based on the discovery that niacin causes prostaglandin D_2 (PGD_2) release in skin [18], and that antagonism of the PGD_2 receptor DP1 suppresses niacin-induced vasodilation [19], the second strategy entails the combination of laropiprant (**2**), a DP1 receptor antagonist and an extended release form of niacin. This combination has been shown to provide lipid effects similar to those of niacin alone, but with significantly diminished flushing symptoms [20]. As such, this combination was approved as Tredaptive in Europe in April 2008. The third approach is to take advantage of niacin derivatives with potential for reduced flushing. For example, in an *in vitro* assay to assess flushing, compound **3** caused less PGD_2 release by THP-1 macrophages than did niacin (34% vs. 100% respectively). On the other hand, compound **3** was comparable to niacin for inhibiting HDL uptake by Hep-G2 cells (13% vs. 19%

respectively), suggesting that **3** and niacin may have similar HDL-raising activity [21]. In the same assays, compound **4** provided similar PGD_2 release (91%) with respect to niacin but more HDL uptake inhibition (30%). The fourth strategy involves development of agonists of the high-affinity niacin receptor and is the subject of this chapter.

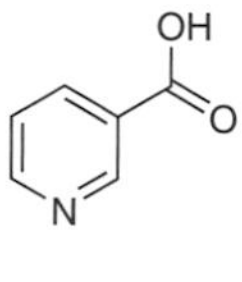

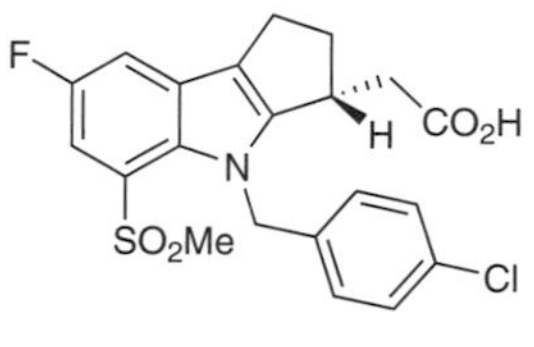

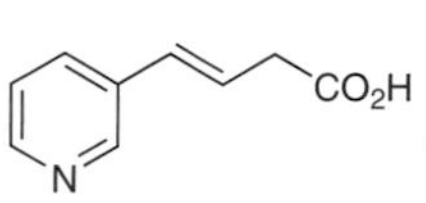

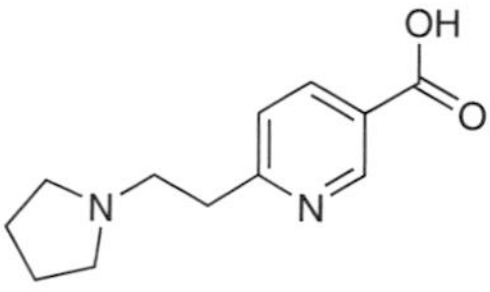

| 1 | 2 | 3 | 4 |

2. DISCOVERY OF GPR109A

The murine niacin receptor, protein upregulated in macrophages by interferon-γ (PUMA-G), was identified as a G-protein-coupled receptor (GPCR) expressed in adipocytes and spleen [22,23]. This finding suggested a role of this receptor in macrophage function [24]. In 2003, the high- and low-affinity niacin receptors, G-protein-coupled receptor 109A (GPR109A, also known as HM74A) and G-protein-coupled receptor 109B (GPR109B, also known as HM74) respectively, were reported as G_i-coupled orphan GPCRs [25–28]. Despite the high homology (95%) between these two receptors, niacin showed over 1000-fold higher binding affinity and functional activity against GPR109A over GPR109B. In addition, GPR109A is expressed primarily on adipocytes, spleen, and immune cells.

It was soon discovered that β-hydroxybutyrate, a ketone body-derived fatty acid, activates GPR109A and PUMA-G receptors in transfected Chinese hamster ovary (CHO) cells with half-maximal effect at 0.7–0.8 mmol/L [29]. Indeed, the physiological concentration of β-hydroxybutyrate approaches 0.2–0.4 mmol/L after an overnight fast, and 1–2 mmol/L after 2–3 days of fasting. Furthermore, infusion of β-hydroxybutyrate successfully suppressed lipolysis. These observations strongly support the hypothesis that β-hydroxybutyrate may serve as an endogenous ligand for GPR109A. On the other hand, niacin, albeit more potent with a half-maximal effect of 100 nmol/L, does not achieve endogenous levels sufficient to provide GPR109A-mediated pharmacological effects.

While the precise mechanism regarding the pharmacological action of niacin remains elusive, it is known that niacin binds to GPR109A on adipocytes and decreases the hydrolysis of adipocyte TG, thereby resulting in a transient reduction of plasma concentrations of free fatty acids (FFA). It has been hypothesized that the reduction of FFA decreases the availability of FFA in the liver, thereby reducing TG synthesis and their subsequent packaging into VLDL-C. The reduction of TG-rich VLDL-C may lead to decreased HDL-C metabolism via limiting the cholesterol ester transfer protein (CETP)-mediated exchange of cholesterol from HDL-C to VLDL-C and of TG from VLDL to HDL [30]. It is also postulated that niacin may inhibit the uptake and removal of Apo-AI-containing HDL-C in hepatocytes and augment reverse cholesterol transport [31,32]. This allows efflux of more cholesterol from the vascular wall. In addition, activation of peroxisome proliferator-activated receptor-γ (PPAR-γ) in a monocytoid cell line was reported as a result of treatment of 1 mmol/L of niacin, thereby increasing the efflux-related receptors including CD36 and ATP-binding cassette protein A1 [33]. Moreover, since GPR109A is also expressed on immune cells, it is possible that some of its antiatherosclerotic effects are derived from modulation of vascular inflammation.

With respect to vasodilation, niacin-elicited vasodilation requires the activation of GPR109A in skin Langerhans cells [34,35], which then triggers the release of arachidonic acid from membrane phospholipids and its subsequent metabolism to PGD_2. The production of PGD_2 then activates DP1 receptors in dermal blood vessels to cause vasodilation [36].

Thus GPR109A mediates the niacin-induced antilipolysis effect in adipocytes and vasodilation effect in skin. This discovery represents significant progress in understanding the mechanism of action for niacin, but also poses a challenge to develop a therapeutic agent to separate the lipid and vasodilation effects.

3. MODELING

A modeling study of the GPR109A niacin-binding site was first reported in 2005 [37]. A site-directed mutagenesis study coupled with generation of chimeric receptors comprising GPR109A and GPR109B led to the identification of putative niacin-binding pocket. This model was based on the bovine rhodopsin crystal structure template 1HZX. It was determined that Asn86/Trp91 [transmembrane helix (TMH) 2/extracellular loop (ECL) 1], Arg111 (TMH3), Ser178 (ECL2), Phe276 (TMH7), and Tyr284 (TMH7) were critical for binding of niacin. In this binding pocket, the carboxylate group of niacin forms a salt bridge with Arg111 at TMH3. In addition, Phe276/Tyr284 at TMH7 and Trp91 at the junction TMH2/ECL1 may contribute to the binding of niacin via π–π interactions. Lastly, the pyridine nitrogen atom appears to form a hydrogen bond with Ser187 at ECL2. Thus, the niacin-binding pocket in GPR109A is apparently distinct from most other rhodopsin family receptors.

A subsequent modeling study by a different group was based on the bovine rhodopsin homology model using the 1L9H construct [38]. However, the site-directed mutagenesis data and docking suggested that the niacin-binding pocket is different to that described in the aforementioned report, but similar to the ligand-binding pocket of most class A GPCRs. The presumed binding pocket consists of residues mainly from TMH3, TMH5, TMH6, and ECL2. Interestingly, Arg111 has also been recognized as a critical residue for binding by functioning as a basic anchor point for the acid component of ligands. An example of this model-directed structure–activity relationship (SAR) efforts is described in Section 5.3 *vide infra*.

4. *IN VITRO* ASSAYS AND *IN VIVO* ANIMAL MODELS

Five major *in vitro* assays have been employed to evaluate GPR109A agonists: a ^{3}H-niacin-binding displacement assay and a guanine nucleotide exchange functional assay on the GPR109A-expressing CHO cell membranes [25], a calcium mobilization assay [37], a forskolin-stimulated cyclic adenosine monophosphate (cAMP) accumulation assay in CHO-hGPR109A cells [27], and a lipolysis inhibition assay in human adipocytes [29].

To evaluate the *in vivo* therapeutic index (TI) of GPR109A agonists regarding FFA reduction and vasodilation, mouse [39], rat, and dog [40,41] models have been developed. GPR109A agonists caused dose- and time-dependent vasodilation in these models. The TI can be defined as the threshold drug level that induces flushing over the IC_{50} for FFA reduction. In mice and rats, FFA reduction and vasodilation were measured as a function of plasma drug levels in two separate experiments. Changes in ear perfusion were established using laser Doppler flowmetry to measure the magnitude of vasodilation and were used to estimate the threshold plasma level of drug required to trigger vasodilation. Compared to the rat model, the dog model allowed measurement of FFA suppression, vasodilation, and compound blood levels in the same conscious dogs after oral compound administration. Vasodilation was established by measuring changes in red color values in the ear using a spectrocolorimeter. Because FFA reduction and flushing were measured in the same animal, it was possible to establish TI on a dose basis as well. Thus the TI was defined as the threshold dose at which vasodilation occurred in a subset of the animals divided by the dose which gave maximal FFA suppression. Based on these criteria, the TI of niacin is 0.5–0.7 and the TI of acipimox is 1–1.4 in rats and dogs, which indicates that it is difficult to mechanistically separate FFA reduction and vasodilation effects for these two drugs. This is consistent with the evidence (Section 2, *vide supra*) that both FFA suppression and vasodilation are GPR109A-dependent events [34].

5. MEDICINAL CHEMISTRY

The discovery of the high-affinity niacin receptor, GPR109A, has spawned significant interest from academia as well as the pharmaceutical industry in searching for novel and potent agonists of this receptor [42–44]. Medicinal chemistry efforts toward development of GPR109A agonists have led to several classes of potent modulators including full orthosteric, partial orthosteric, and positive allosteric agonists.

5.1 Niacin analogs

Prior to the discovery of niacin receptors, medicinal chemistry efforts were mainly directed toward small heterocyclic carboxylic acids that are structurally similar to niacin. Systematic study of nitrogen-containing five- and six-membered heterocyclic carboxylic acids revealed that activity at GPR109A was significantly reduced for any of the variants of niacin shown in general structures (**A** and **B**) [45,46]. These heterocycles include pyrazole, isoxazole, thiazole, pyrazine, and pyrimidine.

Acipimox (**5**) was introduced in Europe to treat hyperlipidemia in 1985 [47,48]. Acipimox is a weak agonist of GPR109A with micromolar binding and functional activity. Like niacin, acipimox raises HDL-C and triggers vasodilation in humans. However, it remains unclear whether acipimox causes mild hyperglycemia as is observed with niacin [49,50].

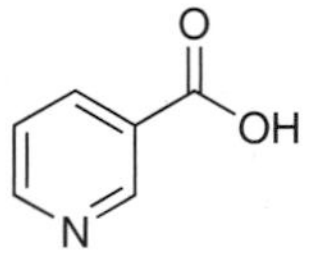

1	**A**	**B**	**5**
h-^{3}H-niacin binding			h-^{3}H-niacin binding
$IC_{50} = 0.081\ \mu M$			$IC_{50} = 5.1\ \mu M$
hGTPγS			h-cAMP (GPR109A)
$EC_{50} = 1.0\ \mu M$ (100%)			$EC_{50} = 2.0\ \mu M$

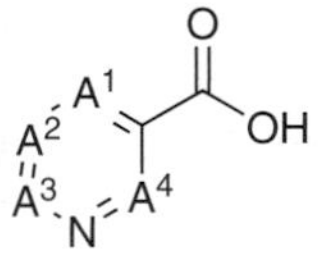

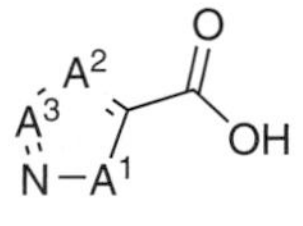

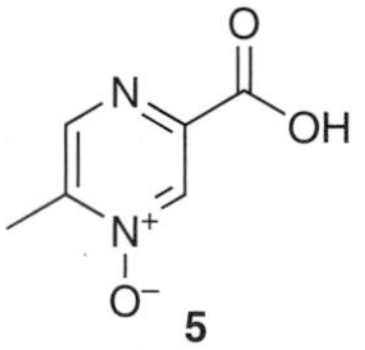

5.2 Acifran analogs

Acifran (**6**) was developed by Ayerst Laboratories in the early 1980s to lower lipids [51–53]. It was later found that acifran was a dual agonist of GPR109A and GPR109B. Analogs of acifran are typically also dual agonists of GPR109A and GPR109B. Several of these analogs were tested for their ability to decrease TG in a fructose fed rat model, yet none was more efficacious than acifran except analog **7** (racemic), which was equipotent [54].

Analog *ent-(+)*-**8** is 20-fold more potent than racemic acifran and is threefold more potent than niacin in the cAMP whole-cell assay [55]. The selectivity was also improved with respect to acifran with 10-fold selectivity for GPR109A over GPR109B. The importance of the absolute configuration of *ent-(+)*-**8**, which is tentatively assigned *S*, is demonstrated by the 70-fold better cAMP whole-cell activity of the eutomer compared to that of the distomer.

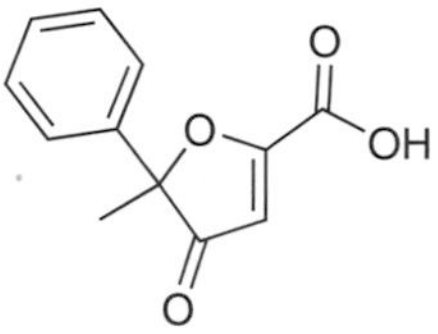

rac-6: acifran
h-³H-niacin binding
$IC_{50} = 1.1\,\mu M$
cAMP whole cell for GPR109A
$EC_{50} = 1.3\,\mu M$
cAMP whole cell for GPR109B
$EC_{50} = 4.2\,\mu M$

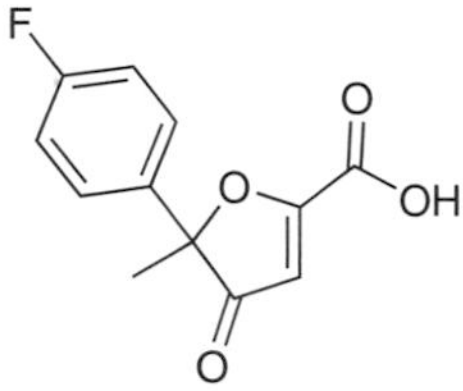

rac-7

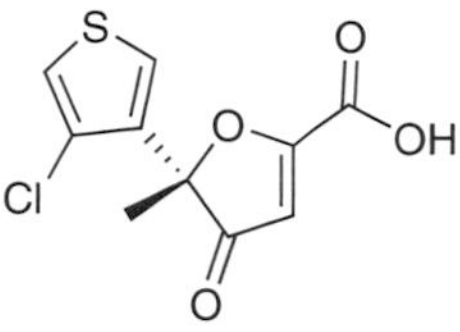

ent-(+)-8
cAMP whole cell for GPR109A
$EC_{50} = 0.066\,\mu M$
cAMP whole cell for GPR109B
$EC_{50} = 0.69\,\mu M$

5.3 Anthranilides

Initial efforts in the pharmaceutical industry to develop small-molecule agonists of GPR109A culminated in two major classes of GPR109A agonists: anthranilides and xanthines. The anthranilide class was described by general structure **C** [56–59]. In the corresponding patent applications [56], R^2 was defined as various 9- or 10-membered saturated, partially saturated, or unsaturated bicyclic ring systems, and Z was a 2- to 4-carbon and/or heteroatom linker. R^1 could be hydrogen, halogen, or C_{1-3} alkyl groups. In the second patent application [57], R^2 represents five- and six-membered aryl, heteroaryl, heterocyclic, or alicyclic rings with a 0- to 3-atom linker. This patent also covered analogs containing directly linked biaryls or diaryls with one atom between the two aryl groups. Representative compounds are naphthyl **9** and biaryl **10**.

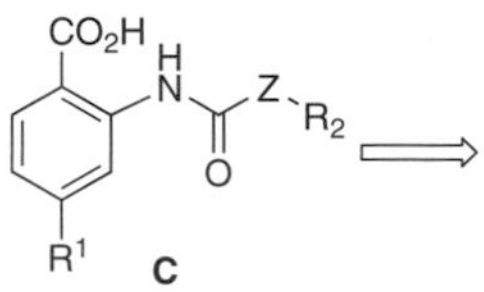

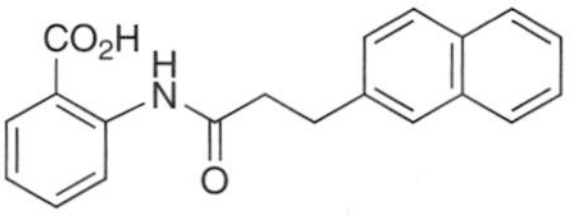

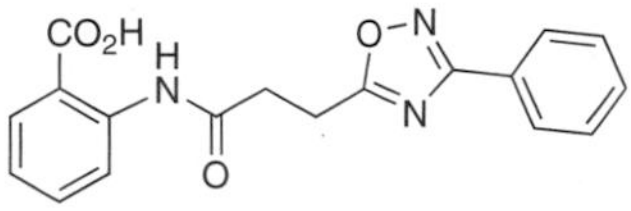

Subsequently, another group also explored the anthranilide series. Based on the homology of GPR109A with bovine rhodopsin [38], Arg111 (TMH3) and Arg251 (TMH6) form a salt bridge with the carboxylate of the high-throughput screening (HTS) hit **11**, and Ser178 (ECL2) may participate in a hydrogen bond with the ligand amide carbonyl group. Hydrophobic residues including Ile254 (TMH6), Phe255 (TMH6), and Phe276 (TMH6) were proposed to provide favorable interactions with the anisole moiety of **11**. The proximity of polar residues such as Asn171 (ECL2), Ser179 (ECL2), and His259 (ECL3) suggested that a small polar tail may improve the activity against GPR109A. Hence a large hydrophobic body (e.g., analog **12**) and a small polar tail incorporated in biaryl (**13**, **14**) [60,61] or fused bicyclic (**15**) [62] anthranilides bearing a terminal hydroxyl group were developed as potent and selective full agonists for GPR109A. The SAR study successfully reduced serum shift and cytochrome P450 2C8 (CYP2C8) and 2C9 inhibition liabilities, while improving pharmacokinetic (PK) properties and maintaining excellent receptor activity. Several highly potent biaryl anthranilide derivatives were discovered as agonists of GPR109A with 10- to 100-fold better activity than niacin *in vitro*.

Compound **13** had good antilipolytic efficacy in mice and did not induce vasodilation at the maximum feasible dose of 100 mg/kg (i.p.), corresponding to a plasma C_{max} of 490 μM and a TI of over 70. Based on the same criteria, the TI for niacin was ~1. Hence compound **13** demonstrated a significantly improved TI with respect to niacin in mice. This observation for the first time suggested that development of GPR109A agonists with good efficacy on the proximal biomarker, FFA, and an improved TI with respect to niacin-induced flushing was possible.

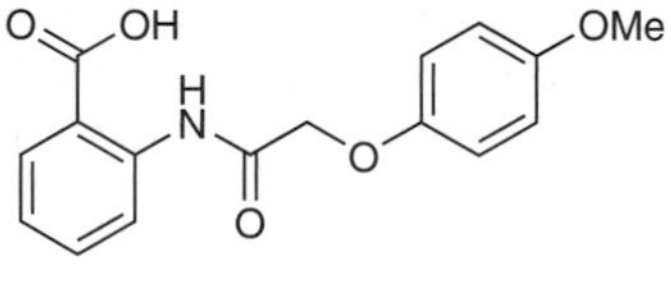

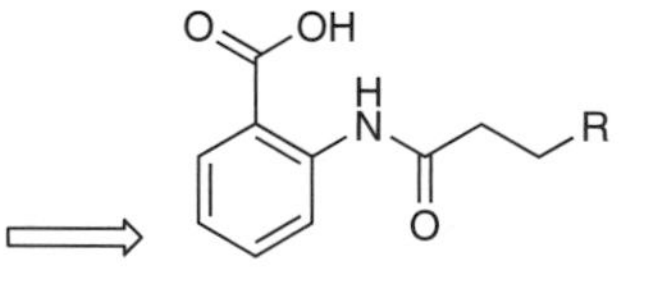

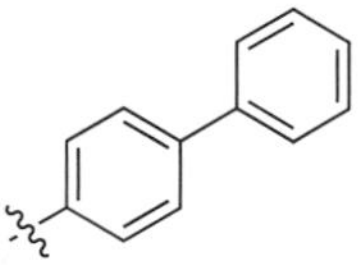

11
h-^{3}H-niacin binding
$IC_{50} > 2.5$ μM
hGTP$_\gamma$S
$EC_{50} = 6.0$ μM

12 R =
h-^{3}H-niacin binding
$IC_{50} = 0.094$ μM
hGTP$_\gamma$S
$EC_{50} = 0.59$ μM

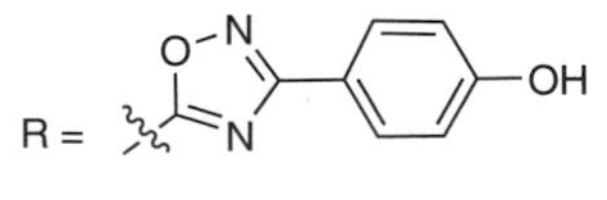

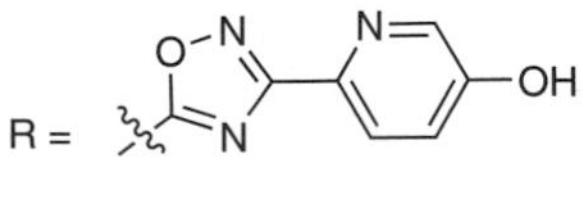

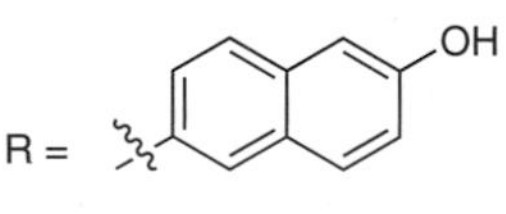

13
h-^{3}H-niacin binding
$IC_{50} = 0.010$ μM
(40-fold serum shift)
hGTP$_\gamma$S
$EC_{50} = 0.12$ μM

14
h-^{3}H-niacin binding
$IC_{50} = 0.004$ μM
(30-fold serum shift)
hGTP$_\gamma$S
$EC_{50} = 0.027$ μM

15

Anthranilide derivatives in which an aryl group was directly connected to the anthranilide unit were also reported [63]. Similarly, anthranilides connected to a five-membered heterocycle such as thiazole, thiophene, or pyrazole were also covered by the patent literature [64]. Representative examples of these two classes are analogs **16** and **17**.

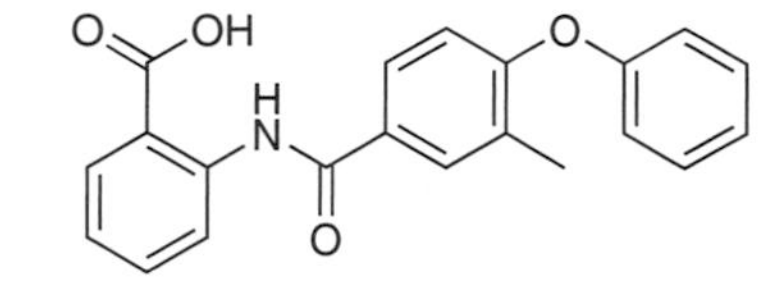

16 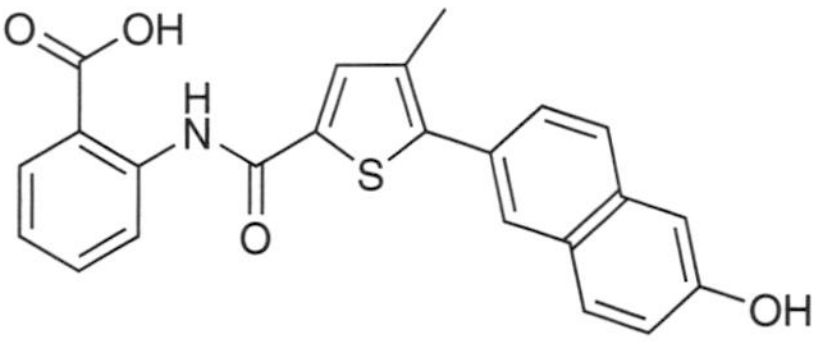   **17**

Anthranilic acid derivatives of genus **D** that possess carbamate and urea linkers (W contains an oxygen or nitrogen atom directly linked to the carbonyl carbon, e.g., compounds **18** and **19** respectively) were claimed as GPR109A agonists in a patent application [65].

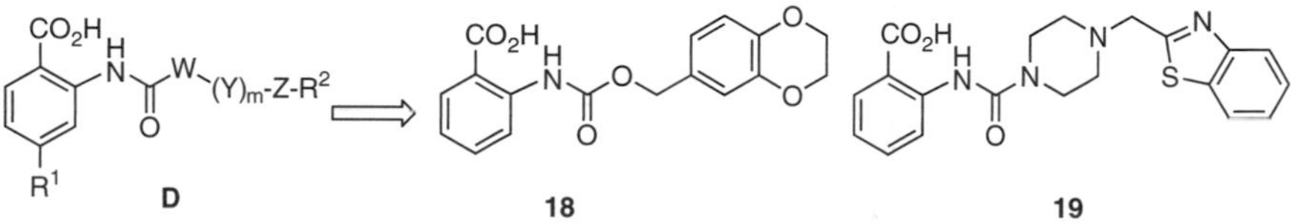

D   **18**   **19**

The SAR of ureas containing an anthranilide component (**E**) revealed the importance of the piperazine and quinoxaline moieties [66,67]. With modest *in vitro* activity and good PK properties, analog **20** achieved good efficacy in reducing FFA and was devoid of flushing at the dose of 100 mg/kg (i.p.) in the mouse model. Compound **21** bearing a terminal hydroxyl group matched the potency of niacin indicating the added favorable interaction of the hydroxyl group with the receptor.

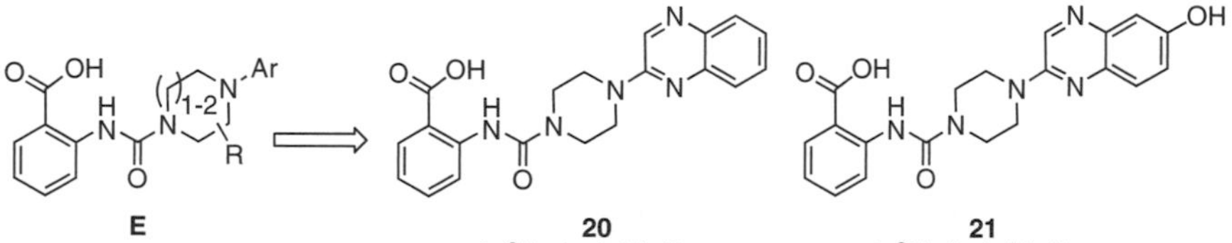

E

20
h-^{3}H-niacin binding
IC_{50}=0.55 μM
hGTPγS
EC_{50}=1.6 μM (94%)

21
h-^{3}H-niacin binding
IC_{50}=0.14 μM
hGTPγS
EC_{50}=0.47 μM (83%)

Hybridization of bicyclic and biaryl anthranilides led to the discovery of novel tricyclic compounds as potent GPR109A agonists [68]. One such compound, **22**, displayed good activity against GPR109A, a good mouse PK profile, a superior TI over niacin regarding FFA reduction and vaso-dilation effects in rats, and minimal CYP2C8 and 2C9 inhibition liability.

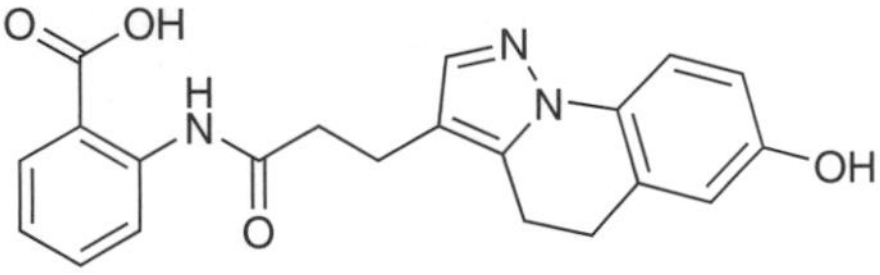

22

h-^{3}H-niacin binding

$IC_{50} = 0.013 \ \mu M$

hGTP$_\gamma$S

$EC_{50} = 0.064 \ \mu M$

5.4 Thiophenes/furans

Anthranilide bioisosteric replacement has also been pursued as a strategy for identifying novel GPR109A agonists. For example, five-membered heterocycles (**F**) in particular thiophene or furan regioisomers as bioisos-teres of benzene have been disclosed in patent applications by a number of groups (e.g., **23** and **24**) [69–72].

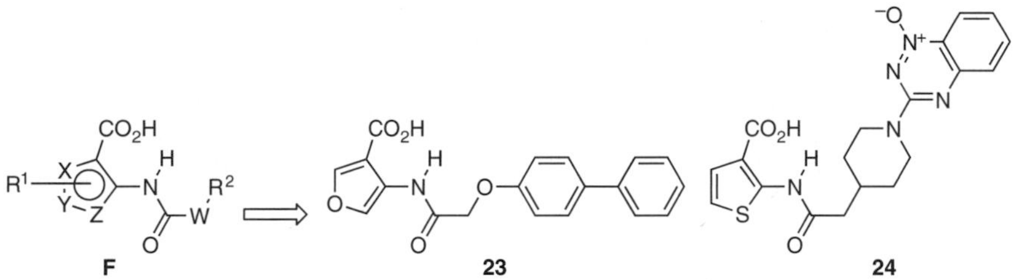

F   **23**   **24**

5.5 Cycloalkenes

Replacement of the phenyl group of anthranilic acid by a cycloalkene represents another strategy to obtain intellectual property. This endeavor resulted in three consecutive patent applications [73–75] and a publica-tion [76]. The advantages of tetrahydro anthranilic acid as a surrogate for anthranilic acid include reduced CYP2C8 and 2C9 inhibition and improved oral exposure in mice (analogs **25** vs. **12**). Ultimately, a pre-clinical candidate, MK-6892 (**26**), was selected from this series due to its excellent receptor activity, good PK properties across species, favorable off-target and ancillary pharmacological profiles, and a superior TI compared to niacin with respect to FFA reduction versus vasodilation in rats and dogs [77]. Furthermore, the cycloalkene ring is tolerant of various substitution patterns including aryl groups as shown by racemic analog 27 [75,78].

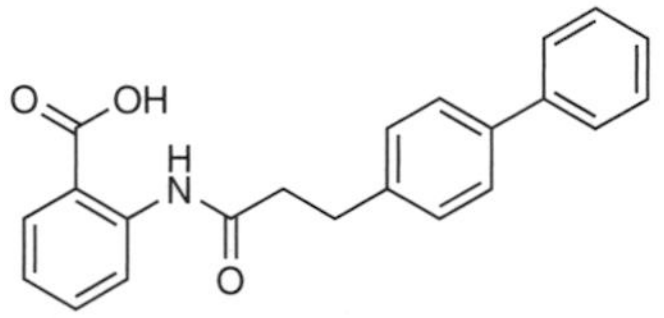

12
h-^{3}H-niacin binding
$IC_{50} = 0.094$ μM
hGTPγS
$EC_{50} = 0.59$ μM
Mouse PK: Cl = 4.8 mL/min/kg
AUCN = 1.4 μM.h.kg/mg
F% = 13%

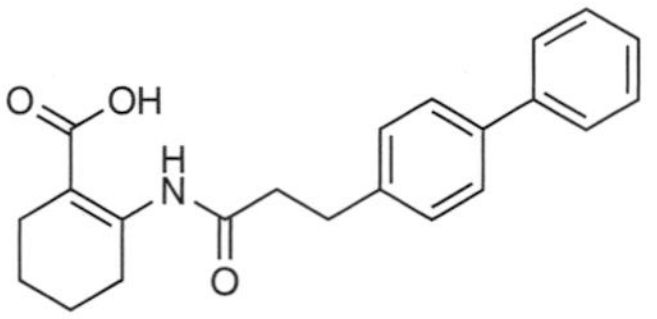

25
h-^{3}H-niacin binding
$IC_{50} = 0.021$ μM
hGTPγS
$EC_{50} = 0.38$ μM
Mouse PK: Cl = 1.2 mL/min/kg
AUCN = 28 μM.h.kg/mg
F% = 64%

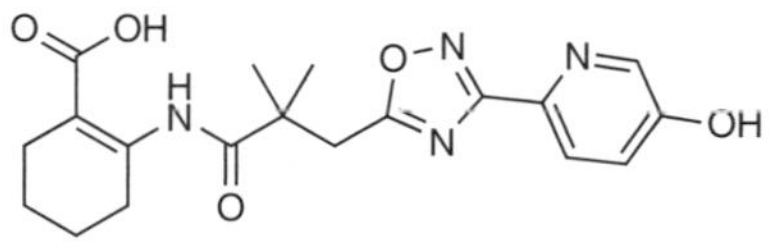

26
h-^{3}H-niacin binding
$IC_{50} = 0.004$ μM
hGTPγS
$EC_{50} = 0.016$ μM
Dog PK: Cl = 1.8 mL/min/kg
AUCN = 21 μM.h.kg/mg
F% = 90%

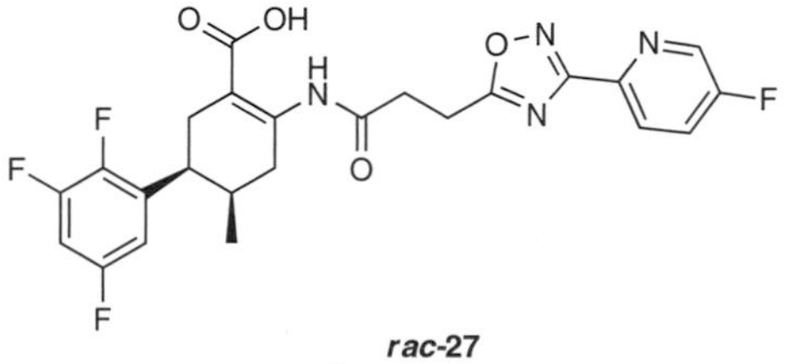

rac-27
h-^{3}H-niacin binding
$IC_{50} = 0.006$ μM
hGTPγS
$EC_{50} = 0.075$ μM
Rat PK: Cl = 2.3 mL/min/kg
AUC = 13.2 μM.h
F% = 44%

In addition, substitution of the central ethylene linker by an amino group was reported as exemplified by **28** and exemplified by **29** [79]. This may have beneficial effects such as reduced IC_{50} shift in the presence of serum and improved physical properties.

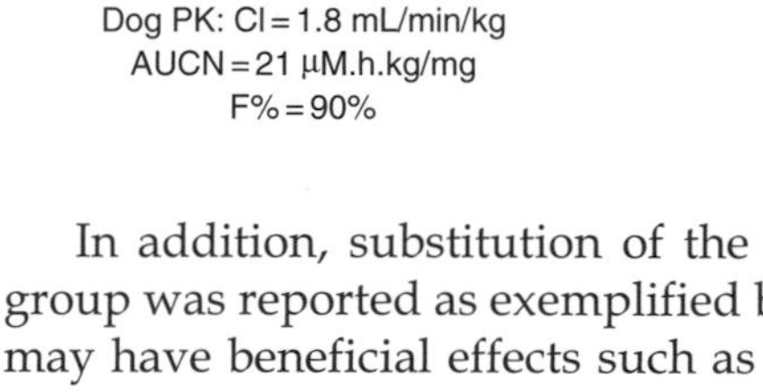

28

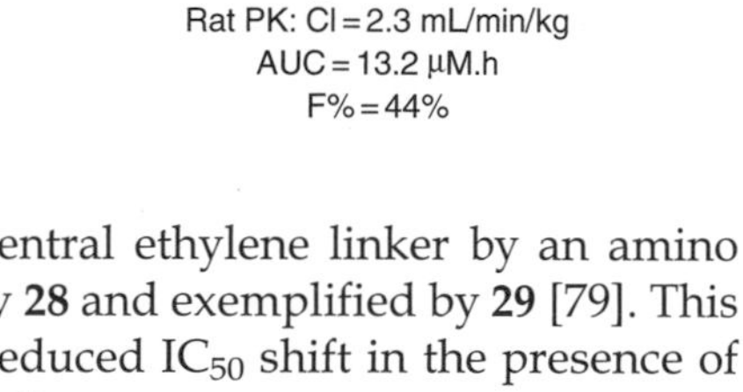

29

5.6 Hydroxypyrazoles

The hydroxypyrazole class GPR109A agonists embody another effective and novel approach to replacement of the anthranilide moiety [80,81]. In this case, a strategy of tethering the amino and carboxyl groups of anthranilides (G) has been applied to generate fused bicyclic systems (H) bearing a hydroxyl group to mimic the carboxylate hydroxyl group. In addition, a ring such as pyrazolo is utilized to simulate the intramolecular hydrogen bonding of the anthranilide group. In particular, cycloalkyl (ranging from cyclopentyl to cycloheptyl)-fused hydroxypyrazole and benzene-fused hydroxypyrazole analogs were claimed in this patent application. The calculated pK_a of the hydroxyl group (~4) matches that of a carboxylic acid [82]. Several representative examples selected from the patent are shown below– (30–32).

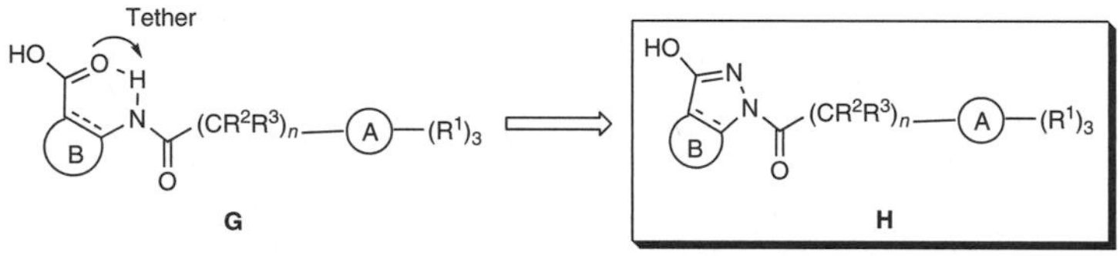

B = phenyl, pyridinyl, cyclohexene, or cyclopentene; A = biaryl, tricyclic, bicyclic.

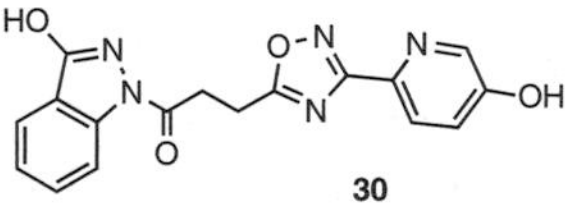

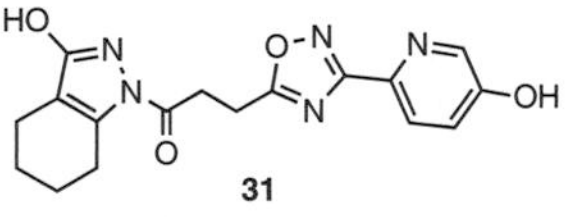

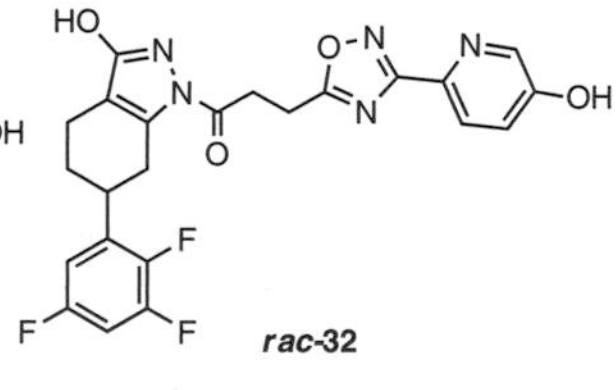

5.7 Xanthine derivatives

A xanthine series (I) exemplified by 33 and 34 is another important class of GPR109A agonists [83–88]. It is worth noting that members of this series lack the carboxylic acid to interact with Arg111 of GPR109A, unlike compounds in the aforementioned series. However, it is conceivable that the N–H present in the bicyclic core of this class of compounds is sufficiently acidic (pK_a~5.6) [82] to mimic the function of a carboxylic acid present in the anthranilide class.

Structurally related to the xanthine series, a triazole ring was employed to tether either one of the two carbonyl groups of the xanthine with the side chain, resulting in two patent applications [89,90]. This approach led to novel tricyclic analogs (e.g., **35–38**) containing a purinone motif (**J** and **K**). It should be noted that a clinical candidate, INCB19602, may fall into the scope of these patent applications [91].

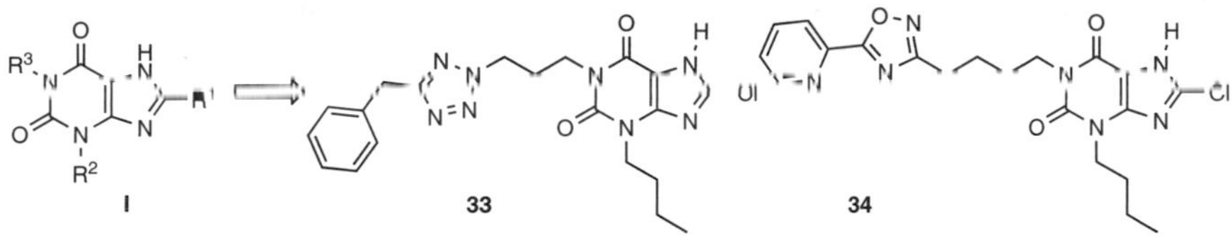

5.8 Barbituric acid derivatives

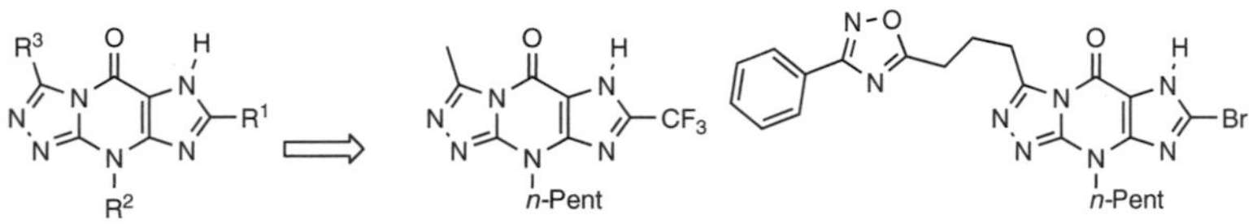

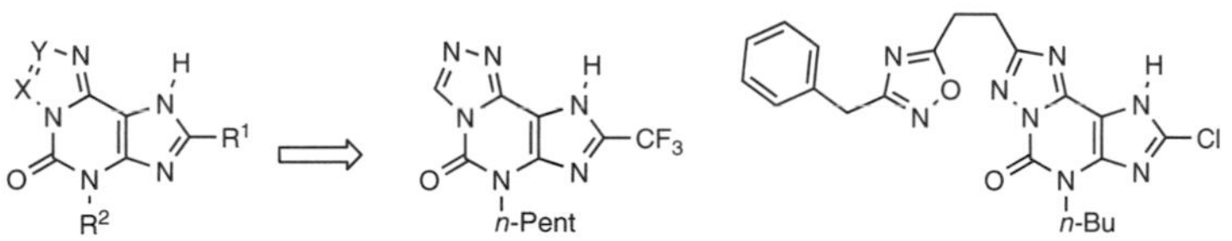

Fusion of a barbituric acid motif and a pyrone ring afforded compounds containing a novel pyranopyrimidine core (**L**) that were discovered as GPR109A agonists [92,93]. This core appears to be distinct from other fused bicyclic cores such as xanthine and anthranilide based on their poor overlap. Furthermore, several compounds, exemplified by **39** and **40**, provided remarkable potency in the cAMP assay. The critical acidic functional group is present as the N–H of barbituric acid motif which has a calculated pK_a of 8 [82].

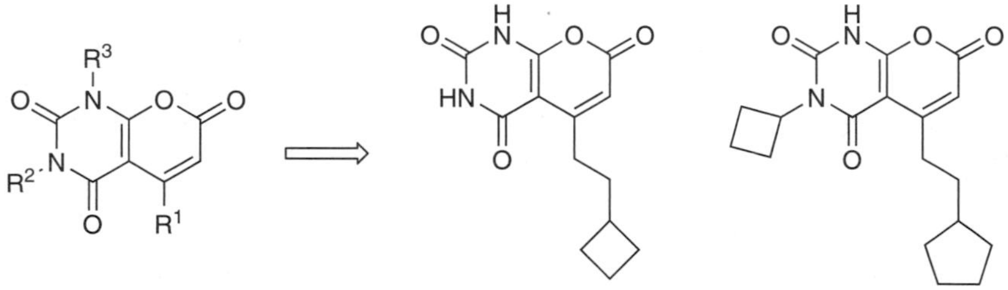

L
R^2 and/or $R^3 = H$

39
h-cAMP
$EC_{50} = 1.2$ nM

40
h-cAMP
$EC_{50} = 13$ nM

5.9 Pyridopyrimidinone derivatives

Apparently related to the aforementioned xanthine analogs, pyridopyrimi-dinone ($X^1=X^2=X^3$=CH or CR) and aza-pyridopyrimidinone (X^1 or X^3=N, X^2=CH or CR) derivatives (**M**), exemplified by compounds **41** and **42** respectively, were discovered as GPR109A agonists [94,95]. So far no biolo-gical data has been provided for this class of molecules. The calculated pK_a of the N–H group present in compounds **41** and **42** ranges from 8 to 9 [82].

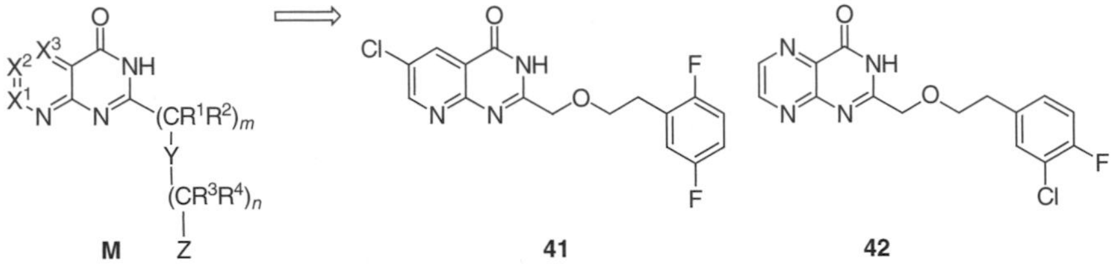

5.10 Pyrazole carboxylic acids

Before the discovery of niacin receptors, 5-alkyl-substituted pyrazole 3-carboxylic acids were known to result in hypolipidemia in rats [96]. Several pyrazole carboxylic acid derivatives (**N**) elicited full lipolysis inhibition *in vitro* and *in vivo*. Not only did these compounds fail to cause flushing, they also antagonized the ability of niacin to induce flushing. In contrast to flushing compounds, these nonflushing agonists did not cause internalization of the receptor in cells expressing GPR109A. In addition, they did not activate extracellular signal-regulated kinase 1 and 2 (ERK1/2) mitogen-activated protein kinase phosphorylation [97].

In the guanosine 5'-O-[γ-thio]triphosphate (GTPγS)-binding assay, a series of pyrazole carboxylic acids were claimed as partial agonists with respect to niacin [98]. It was then postulated that compared to full ago-nists, partial agonists may exhibit reduced adverse effects. Interestingly, compound **43**, a partial agonist in the GTPγS assay, appeared to be a full agonist in the cAMP whole-cell assay [99]. In addition, a series of 4-fluoro-5-alkyl pyrazole 3-carboxylic acids were also identified as selec-tive agonists for GPR109A. One such compound **44** at a dose of 10 mg/kg to rats elicited a reduction of FFA essentially equivalent to that of niacin. As an extension of the cycloalkyl-fused pyrazole carboxylic acids, aryl substituents on the cycloalkyl group, particularly the cyclohexyl group, were also examined [100]. Compound **45** showed no significant flushing when administered to rats at the dose of 30 mg/kg, corresponding to plasma levels of 135 μM. On the other hand, this compound dose-dependently decreased FFA at doses of 3–100 mg/kg.

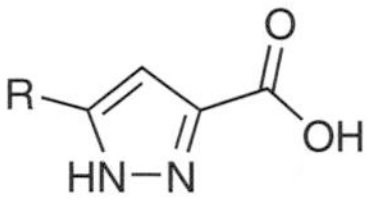

N

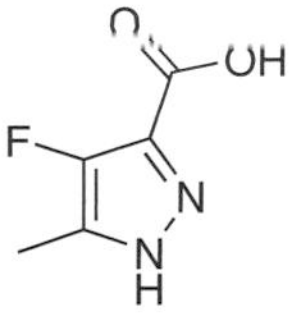

44
h-cAMP
$pEC_{50} = 7.4$ (95–100%)*

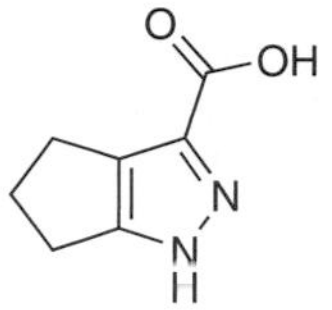

43
h-^{3}H-niacin binding
$IC_{50} = 0.16$ μM
r-GTPγS
$EC_{50} = 6.4$ μM (52%)*
h-cAMP
$EC_{50} = 0.86$ μM (106%)*

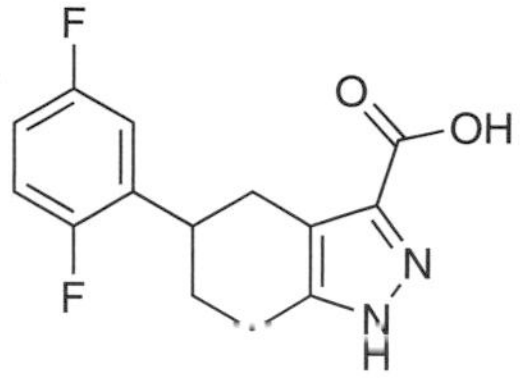

45
h-^{3}H-niacin binding
$IC_{50} = 0.09$ μM
r-GTPγS
$EC_{50} = 1.1$ μM (90%)*

*Relative intrinsic activity
with respect to niacin

5.11 Pyrazole tetrazoles

The replacement of the acid in the pyrazole analog **43** with a tetrazole bioisostere yielded analog **46**, a partial agonist in the cAMP assay [101]. Compound **46** effectively reduced plasma FFA in mice, but did not induce vasodilation at the maximum feasible dose (400 mg/kg). Unlike niacin, **46** did not activate ERK 1/2 mitogen-activated protein kinase (MAPK) pathways *in vitro*. Furthermore, preadministration of **46** in mice blocked niacin-induced, but not PGD$_2$-induced vasodilation. Taken together, these observations constitute the hallmark of a partial agonist. Therefore, the partial agonist strategy apparently achieved the separation of the antilipolysis and vasodilatory effects in mice.

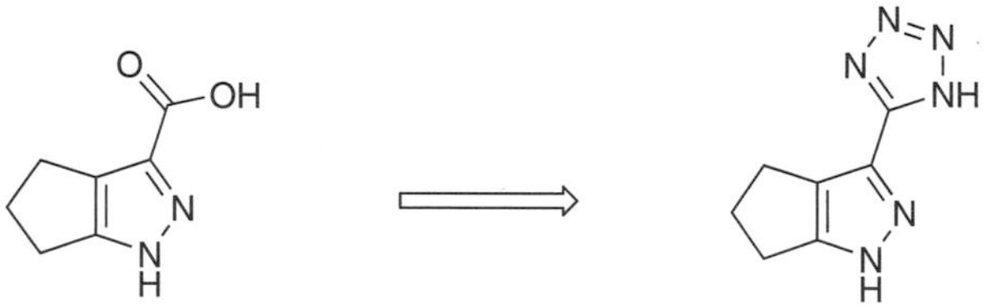

43
h-cAMP
$EC_{50} = 0.86$ μM (106%)*
*Relative intrinsic activity
with respect to niacin

46
h-cAMP
$EC_{50} = 1.65$ μM (59%)*
*Relative intrinsic activity
with respect to niacin

Compound **46** was further developed as a clinical candidate designated MK-0354 [102], which demonstrated robust dose-dependent reduction of plasma FFA in humans over 5 h after single doses up to 4000 mg and multiple doses up to 3600 mg. Single doses of 300 mg of MK-0354 and 1 g of an extended release form of niacin produced similar reduction of FFA. Importantly, no receptor tachyphylaxis was detected as suppression of FFA after seven daily doses was similar to that after a single dose. In a Phase II, multicenter, randomized, double-blind, placebo-controlled clinical study, 2.5 g of MK-0354 was dosed qd for 4 weeks and little flushing was observed at this dose. It was anticipated that the extent and duration of FFA reduction as a result of a 2.5 g dose of MK-0354 would exceed or at least match the effect of 2 g of an extended release form of niacin, the maximal clinical dose of the drug. However, there were minimal effects on lipids with 0.4% increase of HDL-C (95% confidence interval: –5.2 to 6.0), 9.8% reduction of LDL-C (95% confidence interval: –16.8 to –2.7), and 5.8% reduction of TG (95% confidence interval: –22.6 to 11.9). It was thus concluded that MK-0354 failed to elicit HDL-C elevation, and LDL-C and TG reduction. It is unclear why MK-0354 did not achieve lipid effects similar to those observed for niacin. It may imply that niacin-induced global lipid effects are based on different mechanisms, whereas MK-0354 is unable to affect these mechanistic pathways.

Analogs of MK-0354 have been covered in patent applications. For example, different aryl substitutions were introduced onto the core of MK-0354 represented by compound *ent*-**47**, and the ring fused to the pyrazole ring was also enlarged from a five- to a six-membered ring (**48**) [103]. Furthermore, different substitution on MK-0354 [104–106] and analogs containing a small fused ring such as fused cyclopropyl analog (+)-**49** were also explored [107]. A potent partial agonist of GPR109A, compound (+)-49 blocked niacin-induced vasodilation but did not interfere with the PGD$_2$-elicited normal flushing response in mice. This compound displayed excellent PK profiles in four different animal species and reduced FFA in human adipocytes and in mice [108].

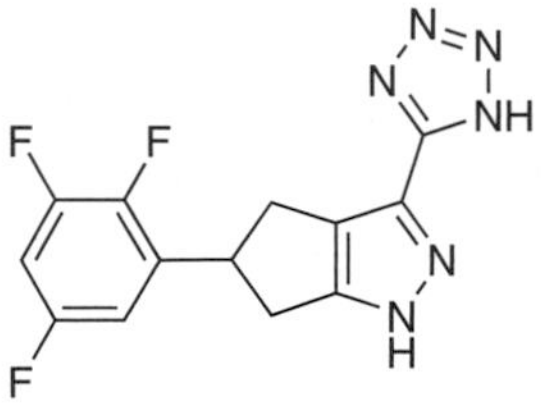

ent-47
h-^{3}H-niacin binding
IC$_{50}$ = 0.04 µM

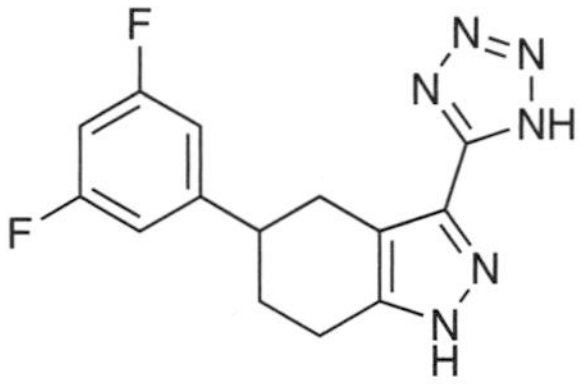

48

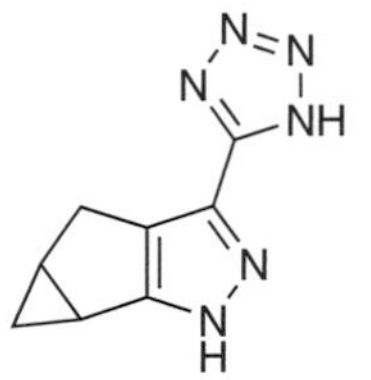

(+)-49
h-cAMP
EC$_{50}$ = 0.21 µM (87%)*
*Relative intrinsic activity
with respect to niacin

Monkey PK: Cl = 13.9 mL/min/kg
AUCN = 1.26 µM.h/mg/kg
F% > 100%

5.12 Pyrazolopyrimidines

GPCR ligands comprise orthosteric and allosteric modulators [109,110]. All the previously described GPR109A agonists are characterized as orthosteric agonists. The first reported allosteric agonists for GPR109A are pyrazolopyrimidines [111]. In addition to its intrinsic activity, a representative analog 50 significantly enhanced niacin binding to the receptor, thereby potentiating the niacin-induced cAMP reduction in a dose-dependent fashion. However, pharmacodynamic effects of an allosteric agonist alone or of the combination of an allosteric agonist and an orthosteric agonist remain to be seen. It should be noted that this class of allosteric agonists does not have an acidic proton as found in their orthosteric agonist counterparts.

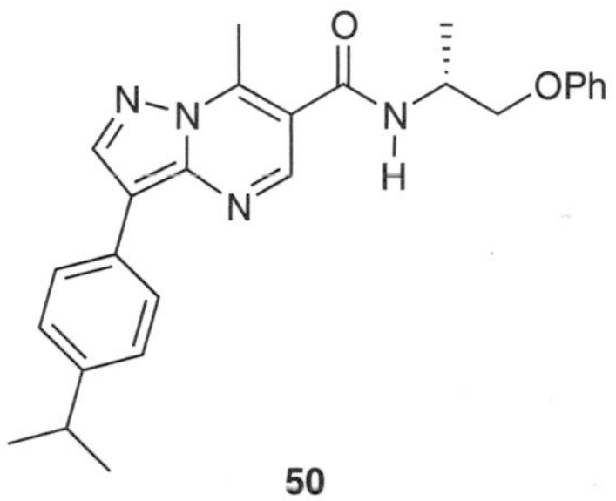

50
^{3}H-niacin binding
$IC_{50} = 0.17\ \mu M$
hGTPγS
$EC_{50} = 0.12\ \mu M\ (75\%)$

6. CONCLUSIONS

Since the discovery of GPR109A, the high-affinity niacin receptor, multiple classes of GPR109A agonists have been discovered and developed, of which many seem to possess excellent potency in FFA suppression and good pharmacological properties with regard to reduced vasodilation. This chapter highlighted eleven structural classes of orthosteric GPR109A agonists and one class of allosteric agonists. All series of orthosteric agonists invariably have an acidic proton which likely interacts with Arg111 at TMH3 through a salt bridge.

   With well-established animal models to evaluate vasodilation and FFA reduction, several compounds were profiled *in vivo* and they indeed displayed improved TIs relative to niacin. Furthermore, these models appeared to correlate to humans as two candidates including both

MK-0354 and INCB-19062 showed excellent FFA reduction accompanied by minimal flushing in humans.

At least four clinical candidates including GSK-256073 [112], MK-0354 [102], MK-1903 [113], and INCB-19062 [91], and one preclinical candidate MK-6892 [77] have been reported. Neither the structure of GSK-256073, nor the clinical data, has been reported. In Phase II clinical trials, neither GPR109A partial agonist MK-0354, nor the full agonist MK-1903 showed substantial lipoprotein effects, and both candidates were discontinued. INCB-19062 is targeted to a type II diabetes indication based upon the related role of FFA to insulin sensitization in type II diabetes, and the robust FFA lowering effect observed in a Phase I clinical trial devoid of FFA rebound.

Based on the reported clinical data to date, it appears that GPR109A may not play a significant role in the HDL-C elevation induced by niacin. As such, the pathway responsible for niacin to raise HDL-C remains elusive. In addition, it is unclear whether the beneficial effects of niacin documented from clinical outcomes are exclusively due to lipid alteration. It is conceivable that GPR109A is still involved in the vascular inflammation pathway related to the antiatherosclerotic effect of niacin. In order to further investigate the therapeutic potential of GPR109A agonists in cardiovascular diseases or metabolic disorders, it is necessary to further elucidate the mechanism of action of niacin and the pharmacology of GPR109A.

REFERENCES

[1] L. A. Carlson, *J. Intern. Med.*, 2005, **258**, 94.
[2] H. Vosper, *Br. J. Pharmacol.*, 2009, **158**, 429.
[3] C. A. Elvehjem, R. J. Madden, F. M. Strong, and D. W. Woolley, *J. Biol. Chem.*, 1938, **123**, 137.
[4] R. Altschul, A. Hoffer, and J. D. Stephen, *Arch. Biochem.*, 1955, **54**, 558.
[5] The Coronary Drug Research Group. *J. Am. Med. Assoc.*, 1975, **231**, 360.
[6] Coronary Drug Project. *Circulation*, 1973, **47**, 1.
[7] K. G. Berge and P. L. Canner, *Eur. J. Clin. Pharmacol.*, 1991, **40** (suppl 1), S49.
[8] S. Tavintharan and M. L. Kashyap, *Curr. Atheroscler. Rep.*, 2001, **3**, 74.
[9] P. L. Canner, K. G. Berge, N. K. Wenger, J. Stamler, L. Friedman, R. J. Prineas, and W. Friedewald, *J. Am. Coll. Cardiol.*, 1986, **8**, 1245.
[10] B. G. Brown, X. Q. Zhao, A. Chait, L. D. Fisher, M. C. Cheung, J. S. Morse, A. A. Dowdy, E. K. Marino, E. L. Bolson, P. Alaupovic, J. Frohlich, and J. J. Albers, *N. Engl. J. Med.*, 2001, **345**, 1583.
[11] A. J. Taylor, L. E. Sullenberger, H. J. Lee, J. K. Lee, and K. A. Grace, *Circulation*, 2004, **110**, 3512.
[12] L. A. Carlson and G. Rosenhamer, *Acta Med. Scand.*, 1988, **203**, 405.
[13] J. G. Hiatt, S. G. Shamsie, and G. Schectman, *Am. J. Manage. Care*, 1999, **5**, 437.
[14] J. M. McKenney, J. D. Proctor, S. Harris, and V. M. Chinchili, *J. Am. Med. Assoc.*, 1994, **271**, 672.
[15] A. Christensen, R. Achor, K. G. Berge, and H. Mason, *J. Am. Med. Assoc.*, 1961, **177**, 546.
[16] J. R. Guyton, *Curr. Opin. Lipidol.*, 2007, **18**, 415 and references therein.

[17] J. M. Morgan, D. M. Capuzzi, J. R. Guyton, R. M. Centor, R. Goldberg, D. C. Robbins, D. DiPette, S. Jenkins, and S. Marcovina, *J. Cardiovasc. Pharamacol. Ther.*, 1996, **1**, 195.

[18] J. D. Morrow, J. A. Awad, J. A. Oates, and L. J. Roberts II, *J. Invest. Dermatol.*, 1992, **98**, 812.

[19] K. Cheng, T. J. Wu, K. K. Wu, C. Sturino, K. Metters, K. Gottesdiener, S. D. Wright, Z. Wang, G. O'Neill, E. Lai, and M. G. Waters, *Proc. Natl. Acad. Sci. USA*, 2006, **103**, 6682.

[20] J. F. Paolini, Y. B. Mitchel, R. Reyes, U. Kher, E. Lai, D. J. Watson, J. M. Norquist, A. G. Meehan, H. E. Baye, M. Davidson, and C. M. Ballantyne, *Am. J. Cardiol.*, 2008, **101**, 625.

[21] Bachovchin, W. W. and Lai, H. *Patent Application WO 2008/016968*, 2008.

[22] A. Lorenzen, C. Stannek, H. Lang, V. Andrianov, I. Kalvinsh, and U. Schwabe, *Mol. Pharmacol.*, 2001, **59**, 349.

[23] A. Lorenzen, C. Stannek, A. Burmeister, I. Kalvinsh, and U. Schwabe, *Biochem. Pharmacol.*, 2002, **64**, 645.

[24] A. Schaub, A. Futterer, and K. Pfeffer, *Eur. J. Immunol.*, 2001, **31**, 3714.

[25] A. Wise, S. M. Foord, N. J. Fraser, A. A. Barnes, N. Elshourbagy, M. Eilert, D. M. Ignar, P. R. Murdock, K. Steplewski, A. Green, A. J. Brown, S. J. Dowell, P. G. Szekeres, D. G. Hassall, F. H. Marshall, S. Wilson, and N. B. Pike, *J. Biol. Chem.*, 2003, **278**, 9869.

[26] S. Tunaru, J. Kero, A. Schaub, C. Wufka, A. Blaukat, K. Pfeffer, and S. Offermanns, *Nat. Med.*, 2003, **9**, 352.

[27] T. Soga, M. Kamohara, J. Takasaki, S. I. Matsumoto, T. Saito, T. Ohishi, H. Hiyama, A. Matsuo, H. Matsushime, and K. Furuichi, *Biophys. Res. Commun.*, 2003, **303**, 364.

[28] F. Karpe and K. Frayn, *Lancet*, 2004, **363**, 1892.

[29] A. K. Taggart, J. Kero, X. Gan, T. Q. Cai, K. Cheng, M. Ippolito, N. Ren, R. Kaplan, K. Wu, T. J. Wu, L. Jin, C. Liaw, R. Chen, J. Richman, D. Connolly, S. Offermanns, S. D. Wright, and M. G. Waters, *J. Biol. Chem.*, 2005, **280**, 26649.

[30] S. Offermanns, *Trends Pharmacol. Sci.*, 2006, **27**, 384.

[31] T. Sakai, V. S. Kamanna, and M. L. Kashyap, *Arterioscler. Thromb. Vasc. Biol.*, 2001, **21**, 1783.

[32] F. Y. Jin, V. S. Kamanna, and M. L. Kashyap, *Arterioscler. Thromb. Vasc. Biol.*, 1997, **17**, 2020.

[33] T. Rubic, M. Trottmann, and R. L. Lorenz, *Biochem. Pharmacol.*, 2004, **67**, 411.

[34] Z. Benyo', A. Gille, C. L. Bennett, B. E. Clausen, and S. Offermanns, *Mol. Pharmacol.*, 2006, **70**, 1844.

[35] Z. Benyo', A. Gille, J. Kero, M. Csiky, M. C. Suchánoková, R. Nüsing, A. Moers, K. Pfeffer, and S. Offermanns, *J. Clin. Invest.*, 2005, **115**, 3634.

[36] D. Maciejewski-Lenoir, J. G. Richman, Y. Hakak, I. Gaidarov, D. P. Behan, and D. T. Connolly, *J. Invest. Dermatol.*, 2006, **126**, 2637.

[37] S. Tunaru, J. Lättig, J. Kero, G. Krause, and S. Offermanns, *Mol. Pharmacol.*, 2005, **68**, 1271.

[38] Q. Deng, J. L. Frie, D. M. Marley, R. T. Beresis, N. Ren, T. Q. Cai, A. K. P. Taggart, K. Cheng, E. Carballo-Jane, J. Wang, X. Tong, M. G. Waters, J. R. Tata, and S. L. Colletti, *Bioorg. Med. Chem. Lett.*, 2008, **18**, 4963.

[39] K. Cheng, T. J. Wu, K. K. Wu, C. Sturino, K. Metters, K. Gottesdiener, S. D. Wright, Z. Wang, G. O'Neil, E. Lai, and M. G. Waters, *Proc. Natl. Acd. Sci. USA*, 2006, **103**, 6682 and references therein.

[40] E. Carballo-Jane, L. S. Gerckens, S. Luell, A. S. Parlapiano, M. Wolff, S. L. Colletti, J. R. Tata, A. K. P. Taggart, M. G. Waters, J. R. Richman, M. E. McCann, and M. J. Forrest, *J. Pharmacol. Toxicol. Methods*, 2007, **56**, 308.

[41] E. Carballo-Jane, T. Ciecko, S. Luell, J. W. Woods, E. I. Zycband, and M. G. Waters, *Inflamm. Res.*, 2007, **56**, 254.

[42] G. Semple, P. D. Boatman, and J. G. Richman, *Curr. Opin. Drug Discov. Dev.*, 2007, **10**, 452.

[43] P. D. Boatman, J. G. Richman, and G. Semple, *J. Med. Chem.*, 2008, **51**, 7653.

[44] H. C. Shen and S. L. Colletti, *Exp. Opin. Ther. Patents*, 2009, **19**, 957.

[45] T. Gharbaoui, P. J. Skinner, Y. J. Shin, C. Averbuj, J. K. Jung, B. R. Johnson, T. Duong, M. Decaire, J. Uy, M. C. Cherrier, P. J. Webb, S. Y. Tamura, N. Zou, N. Rodriguez, P. D. Boatman, C. R. Sage, A. Lindstrom, J. Xu, T. O. Schrader, B. M. Smith, R. Chen, J. G. Richman, D. T. Connolly, S. L. Colletti, J. R. Tata, and G. Semple, *Bioorg. Med. Chem. Lett.*, 2007, **17**, 4914.

[46] L. A. Carlson, C. Hedbom, E. Helgstrand, B. Sjoberg, and N. E. Stjernstrom, *Acta. Pharm. Suec.*, 1972, **9**, 289.

[47] M. J. O'Kane, T. R. Trinick, M. B. Tynan, E. R. Trimble, and D. P. Nicholls, *Br. J. Clin. Pharmacol.*, 1992, **33**, 451.

[48] P. Tornvall and G. A. Walldius, *J. Intern. Med.*, 1991, **230**, 415.

[49] A. T. Santomauro, G. Boden, M. E. Silva, D. M. Rocha, R. F. Santos, M. J. Ursich, P. G. Strassmann, and B. L. Wajchenberg, *Diabetes*, 1999, **48**, 1836.

[50] K. Cusi, S. Kashyap, A. Gastaldelli, M. Bajaj, and E. Cersosimo, *Am. J. Physiol. Endocrinol. Metab.*, 2007, **292**, E1775.

[51] I. Jirkovsky and M. N. Cayen, *J. Med. Chem.*, 1982, **25**, 1154.

[52] J. C. LaRosa, V. T. Miller, K. D. Edwards, M. R. DeBovis, and D. B. Stoy, *Artery*, 1987, **14**, 338.

[53] D. B. Hunninghake, K. D. Edwards, G. S. Sopko, and R. L. Tosiello, *Clin. Pharmacol. Ther.*, 1985, **38**, 313.

[54] K. Mahboubi, T. Witman-Jones, J. E. Adamus, J. T. Letsinger, D. Whitehouse, A. R. Moorman, D. Sawicki, N. Bergenhem, and S. A. Ross, *Biochem. Biophys. Res. Commun.*, 2006, **340**, 482.

[55] J. K. Jung, B. R. Johnson, T. Duong, M. Decaire, J. Uy, T. Gharbaoui, P. D. Boatman, C. R. Sage, R. Chen, J. G. Richman, D. T. Connolly, and G. Semple, *J. Med. Chem.*, 2007, **50**, 1445.

[56] M. Campbell, R. J. Hatley, J. P. Heer, A. M. Mason, I. L. Pinto, S. S. Rahman, and I. E. D. Smith, *Patent Application WO 2005/016867*, 2005.

[57] A. M. Mason, I. L. Pinto, and S. S. Rahman, *Patent Application WO 2006/085111*, 2006.

[58] R. Hatley, A. M. Mason, I. L. Pinto, and I. E. D. Smith, *Patent Application WO 2006/085112*, 2006.

[59] M. Campbell, R. J. Hatley, J. P. Heer, A. M. Mason, N. H. Nicholson, I. L. Pinto, S. S. Rahman, and I. E. D. Smith, *Patent Application WO 2005/016870*, 2005.

[60] H. C. Shen, F. X. Ding, A. Taggart, K. Cheng, E. Carballo-Jane, N. Ren, Q. Chen, J. Wang, M. Wolff, G. Waters, M. Hammond, J. R. Tata, and S. L. Colletti, *J. Med. Chem.*, 2007, **50**, 6303.

[61] S. L. Colletti, J. R. Tata, H. C. Shen, F. X. Ding, J. L. Frie, J. E. Imbriglio, and W. Chen, *Patent Application WO 2006/057922*, 2006.

[62] S. L. Colletti, R. T. Beresis, W. Chen, J. R. Tata, H. C. Shen, D. M. Marley, Q. Deng, J. L. Frie, and F. X. Ding, *Patent Application WO 2006/052555*, 2006.

[63] H. Dehmlow, U. Grether, N. A. Kratochwil, R. Narquizian, C. Panousis, and J. U. Peters, *Patent Application WO 2006/0281810*, 2006.

[64] S. L. Colletti, J. E. Imbriglio, R. T. Beresis, and J. L. Frie, *Patent Application WO 2007/035478*, 2007.

[65] M. Campbell, A. M. Mason, I. L. Pinto, D. R. Pollard, and I. E. D. Smith, *Patent Application WO 2006/085108*, 2006.

[66] H. C. Shen, M. Szymonifka, Q. Deng, E. Carballo-Jane, K. Cheng, K. Wu, T. J. Wu, J. Wang, X. Tong, N. Ren, A. Taggart, T. Cai, G. Waters, M. Hammond, J. R. Tata, and S. L. Colletti, *Bioorg. Med. Chem. Lett.*, 2007, **17**, 6723.

[67] S. L. Colletti, H. Shen, J. R. Tata, and M. J. Szymonifka, *Patent Application WO 2007/027532*, 2007.

[68] H. C. Shen, F. X. Ding, Q. Deng, L. C. Wilsie, M. L. Krsmanovic, A. K. Taggart, E. Carballo-Jane, N. Ren, E. T. Cai, T. J. Wu, K. K. Wu, K. Cheng, Q. Chen, M. S.

Wolff, X. Tong, T. G. Holt, M. G. Waters, M. L. Hammond, J. R. Tata, and S. L. Colletti, *J. Med. Chem.*, 2009, **52**, 2587.

[69] I. L. Pinto and J. K. Simpson, *Patent Application WO 2006/085113*, 2006.

[70] S. L. Colletti, J. R. Tata, W. Chen, R. T. Beresis, F. X. Ding, D. R. Schmidt, H. Shen, and S. Raghavan, *Patent Application WO 2007/120575*, 2007.

[71] H. Dehmlow, U. Grether, N. A. Kratochwil, R. Narquizian, and C. Panousis, *Patent Application US 2007/0072873*, 2007.

[72] G. Cao, C. B. Xue, R. Anand, T. Huang, L. Kong, J. Glenn, and H. Feng, *Patent Application WO 2007/015744*, 2007.

[73] S. Raghavan, S. L. Colletti, F. X. Ding, H. Shen, J. R. Tata, A. R. Lins, A. L. Smenton, W. Chen, D. R. Schmidt, and G. S. Tria, *Patent Application WO 2007/002557*, 2007.

[74] S. Raghavan, S. L. Colletti, F. X. Ding, H. Shen, J. R. Tata, A. R. Lins, A. L. Smenton, W. Chen, D. R. Schmidt, and G. S. Tria, *Patent Application US 2006/0293364*, 2006.

[75] S. Raghavan, D. R. Schmidt, S. L. Colletti, and A. L. Smenton, *Patent Application WO 2007/092364*, 2007.

[76] S. Raghavan, G. S. Tria, H. C. Shen, F. X. Ding, A. K. P. Taggart, N. Ren, L. C. Wilsie, M. L. Krsmanovic, T. G. Holt, M. S. Wolff, M. G. Waters, M. L. Hammond, J. R. Tata, and S. L. Colletti, *Bioorg. Med. Chem. Lett.*, 2008, **18**, 3163.

[77] H. C. Shen, F. X. Ding, S. Raghavan, Q. Deng, S. Luell, M. J. Forrest, E. Carballo-Jane, L. C. Wilsie, M. L. Krsmanovic, A. K. Taggart, K. K. Wu, T. J. Wu, K. Cheng, N. Ren, T.-Q. Cai, Q. Chen, J. Wang, M. S. Wolff, X. Tong, T. G. Holt, M. G. Waters, M. L. Hammond, J. R. Tata, and S. L. Colletti, *J. Med. Chem.*, 2010, **53**, 2666.

[78] D. Schmidt, A. Smenton, S. Raghavan, H. Shen, F. X. Ding, E. Carballo-Jane, S. Luell, T. Ciecko, T. G. Holt, M. Wolff, A. Taggart, L. Wilsie, M. Krsmanovic, N. Ren, D. Bolm, K. Cheng, P. E. McCann, M. G. Waters, J. Tata, and S. Colletti, *Bioorg. Med. Chem. Lett.*, 2010, **20**, 3426.

[79] J. Imbriglio, S. L. Colletti, J. R. Tata, R. T. Beresis, D. Marley, S. Raghavan, D. R. Schmidt, A. R. Lins, A. L. Smenton, W. Chen, H. Shen, F. X. Ding, and R. Bodner, *Patent Application WO 2007/075749*, 2007.

[80] R. T. Beresis and S. L. Colletti, *Patent Application WO 2008/051403*, 2008.

[81] H. C. Shen, *Expert Opin. Ther. Patents*, 2009, **19**, 1149.

[82] pK$_a$ is calculated using ACD/Labs Extension for CS Chemdraw (version 11.0).

[83] I. L. Pinto, S. S. Rahman, and N. H. Nicholson, *Patent Application WO 2005/077950*, 2005.

[84] R. J. D. Hatley and I. L. Pinto, *Patent Application WO 2006/045564*, 2006.

[85] R. J. D. Hatley and I. L. Pinto, *Patent Application WO 2006/045565*, 2006.

[86] R. J. D. Hatley, A. M. Mason, and I. L. Pinto, *Patent Application WO 2007/017261*, 2007.

[87] R. J. D. Hatley, J. P. Heer, J. Liddle, A. M. Mason, I. L. Pinto, S. S. Rahman, and I. E. D. Smith, *Patent Application WO 2007/017262*, 2007.

[88] J. P. Heer and I. E. D. Smith, *Patent Application WO 2007/17265*, 2007.

[89] C. Zheng, C. B. Xue, G. Cao, M. Xia, A. Wang, H. F. Ye, and B. Metcalf, *Patent Application WO 2007/150025*, 2007.

[90] B. Metcalf, K. Zhang, C. Zheng, C. B. Xue, G. Cao, and A. Wang, *Patent Application WO 2007/150026*, 2007.

[91] Incyte corporation presentation at JP Morgan Healthcare Conference, January 8, 2008. http://findarticles.com/p/articles/mi_m0EIN/is_2008_jan_8/ai_n24224669 (accessed June 24, 2010).

[92] A. Palani, J. Su, D. Xiao, X. Huang, A. U. Rao, X. Chen, H. Tang, J. Qin, Y. R. Huang, R. G. Aslanian, and B. A. Mckittrick, *Patent Application WO 2006/124490*, 2006.

[93] A. Palani, J. Su, D. Xiao, X. Huang, A. U. Rao, X. Chen, H. Tang, J. Qin, Y. R. Huang, R. G. Aslanian, B. A. Mckittrick, and S. J. Degrado, *Patent Application US 2008/019978*, 2008.

[94] A. Conte, H. Dehmlow, U. Grether, N. A. Kratochwil, H. Kuehne, R. Narquizian, C. G. Panousis, J. U. Peters, and F. Ricklin, *Patent Application WO 2008/0234277*, 2008.

[95] A. Conte, H. Dehmlow, U. Grether, N. A. Kratochwil, H. Kuehne, R. Narquizian, C. G. Panousis, J. U. Peters, F. Ricklin, and S. Roever, *Patent Application US 2007/0275987*, 2007.

[96] K. Seki, J. Isegawa, M. Fukuda, and M. Ohki, *Chem. Pharm. Bull.*, 1984, **32**, 1568.

[97] J. G. Richman, M. Knemitsu-Parks, I. Gaidarov, J. S. Cameron, P. Griffin, H. Zheng, N. C. Guerra, L. Cham, D. Maciejewski-Lenoir, D. P. Behan, D. Boatman, R. Chen, P. Skinner, P. Ornelas, G. M. Waters, S. D. Wright, G. Semple, and D. T. Connolly, *J. Biol. Chem.*, 2007, **282**, 18028.

[98] T. van Herk, J. Brussee, A. M. van den Nieuwendijk, P. A. van der Klein, A. P. Ijzerman, C. Stannek, A. Burmeister, and A. Lorenzen, *J. Med. Chem.*, 2003, **46**, 3945.

[99] P. J. Skinner, M. C. Cherrier, P. J. Webb, Y. J. Shin, T. Gharbaoui, A. Lindstrom, V. Hong, S. Y. Tamura, H. T. Dang, C. C. Pride, R. Chen, J. G. Richman, D. T. Connolly, and G. Semple, *Bioorg. Med. Chem. Lett.*, 2007, **17**, 5620.

[100] D. Schmidt, A. Smenton, S. Raghavan, E. Carballo-Jane, S. Lubell, T. Ciecko, T. G. Holt, M. Wolff, A. Taggart, L. Wilsie, M. Kramanovic, N. Ren, D. Blom, K. Cheng, P. E. McCann, M. G. Waters, J. Tata, and S. Colletti, *Bioorg. Med. Chem. Lett.*, 2009, **19**, 4768.

[101] G. Semple, P. J. Skinner, T. Gharbaoui, Y. J. Shin, J. K. Jung, M. C. Cherrier, P. J. Webb, S. Y. Tamura, P. D. Boatman, C. R. Sage, T. O. Schrader, R. Chen, S. L. Colletti, J. R. Tata, M. G. Waters, K. Cheng, A. K. Taggart, T. Q. Cai, E. Carballo-Jane, D. P. Behan, D. T. Connolly, and J. G. Richman, *J. Med. Chem.*, 2008, **51**, 5101.

[102] E. Lai, M. G. Waters, J. R. Tata, W. Radziszewski, I. Perevozskaya, W. Zheng, L. Wenning, D. T. Connolly, G. Semple, A. O. Johnson-Levonas, J. A. Wagner, Y. Mithel, and F. Paolini, *J. Clin. Lipidol.*, 2008, **2**, 375.

[103] J. E. Imbriglio, S. L. Colletti, J. R. Tata, R. Liang, S. Raghavan, D. R. Schmidt, A. R. Smenton, and S. Y. Chan, *Patent Application WO 2006/113150*, 2006.

[104] G. Semple, T. Gharbaoui, Y. J. Shin, M. Decaire, C. Averbuj, and P. J. Skinner, *Patent Application WO 2005/011677*, 2005.

[105] G. Semple, T. Schrader, P. J. Skinner, S. L. Colletti, G. Tawfik, J. E. Imbriglio, J. K. Jung, R. Liang, S. Raghavan, D. Schmidt, and J. R. Tata, *Patent Application WO 2005/044816*, 2005.

[106] J. E. Imbriglio, S. Chang, R. Liang, S. Raghavan, D. Schmidt, A. Smenton, S. Tria, T. O. Schrader, J. K. Jung, C. Esser, A. K. P. Taggart, K. Cheng, E. Carballo-Jane, M. G. Waters, J. R. Tata, and S. L. Colletti, *Bioorg. Med. Chem. Lett.*, 2009, **19**, 2121.

[107] D. P. Boatman, T. O. Schrader, G. Semple, P. J. Skinner, and J. K. Jung, *Patent Application WO 2006/069242*, 2006.

[108] D. P. Boatman, T. O. Schrader, M. Kasem, B. R. Johnson, P. J. Skinner, J. K. Jung, J. Xu, M. C. Cherrier, P. J. Webb, G. Semple, C. R. Sage, J. Knudsen, R. Chen, A. K. Taggart, E. Carballo-Jane, and J. G. Richman, *Bioorg. Med. Chem. Lett.*, 2010, **20**, 2797.

[109] A. Christopoulos, *Nat. Rev. Drug Discov.*, 2002, **1**, 198.

[110] A. Christopoulos and T. Kenakin, *Pharmacol. Rev.*, 2002, **54**, 323.

[111] H. C. Shen, A. Taggart, L. Wilsie, G. Waters, M. Hammond, J. R. Tata, and S. L. Colletti, *Bioorg. Med. Chem. Lett.*, 2008, **18**, 4948.

[112] Pipeline summary of GSK from www.gsk.com.

[113] http://www.nasdaq.com/aspx/company-news-story.aspx?storyid=200912231651 dowjonesdjoinline000524&title=arena-Pharma-merck-to-discontinue-mk-1903-development (accessed June 24, 2010).

CHAPTER 6

Recent Advances in Acetyl-CoA Carboxylase Inhibitors

Matthew P. Bourbeau, John G. Allen and **Wei Gu**

1. INTRODUCTION

Recent studies suggest that approximately 22% of American adults exhibit symptoms of metabolic syndrome, constituting a significant public health crisis [1]. Metabolic syndrome is associated with an increased risk of type 2 diabetes (T2D) and cardiovascular disease (CVD), specifically insulin resistance, high blood pressure, central obesity, decreased high-density lipoprotein (HDL) cholesterol, and elevated triglycerides [2–4]. Although the pathogenesis of metabolic syndrome remains poorly understood, abnormal fatty acid metabolism is considered to be a key element in its development [5,6]. Studies have demonstrated positive associations between insulin resistance and intracellular accumulations of triglycerides and fatty acid metabolites in insulin-responsive tissues such as muscle and liver [7]. Fatty acid metabolites such as acyl-CoA, diacylglycerol (DAG), and ceramides have been shown to interfere with insulin

Amgen, Inc., 1 Amgen Center Dr., Thousand Oaks, CA 91320, USA

Annual Reports in Medicinal Chemistry, Volume 45
ISSN 0065-7743, DOI 10.1016/S0065-7743(10)45006-X

signaling [8]. Accumulation of intracellular fatty acid metabolites depends on fatty acid uptake and the rates of fatty acid synthesis and oxidation. Therefore, therapeutic approaches to block fatty acid uptake, inhibit fatty acid synthesis, or increase fatty acid oxidation may improve insulin sensitivity, reduce body weight, and decrease the risks of T2D and associated CVD. Although drugs such as nicotinic acid and fibric acids have shown efficacy in lowering the lipid risk factors of metabolic syndrome and reduction of cardiovascular events, their overall single agent efficacy is not robust [3]. Thus, there remains a significant need for the development of effective therapeutics which can modify multiple risk factors for the treatment of this metabolic disorder.

The acetyl-CoA carboxylases (ACCs) are a drug target class currently under investigation for the treatment of metabolic syndrome. The ACCs are enzymes that control the rate-limiting reactions for long-chain fatty acid synthesis and mitochondrial fatty acid oxidation. Inhibition of ACC activity may have the potential to significantly modify lipid metabolism and consequently improve insulin sensitivity and promote weight loss. The ACCs are large proteins (>250 kDa), consisting of three active domains: the biotin carboxylase (BC) domain, the biotin carboxyl carrier protein (BCCP) domain, and the carboxyl transferase (CT) domain. There are two known ACC isoforms, commonly referred to as ACC1 and ACC2 [9]. Both enzymes catalyze the conversion of acetyl-CoA to malonyl-CoA. ACC1 is expressed primarily in liver and fat and initiates the fatty acid synthesis pathway [10]. In contrast, ACC2 is primarily expressed in muscle where it is associated with the mitochondrial membrane. ACC2 serves as a negative regulator of fatty acid uptake by mitochondria via inhibition of carnityl palmitate transfer protein 1 (CPT1) by malonyl-CoA. CPT1 conjugates free fatty acids to carnitine for transport across the mitochondrial membrane, resulting in subsequent β-oxidation [11]. As such, the ACCs are fundamentally important for both the synthesis and the utilization of fatty acids. Inhibitors of ACC activity are expected to reduce fatty acid synthesis in liver and adipose tissues, as well as increase fatty acid oxidation in liver and muscle tissues. Thus, ACC inhibition may have the beneficial effects of increased energy expenditure, reduced body weight, reduced adiposity, and improved insulin sensitivity.

The groups of Wakil and Cooney independently reported generation and characterization of ACC2 knockout (KO) mice with some overlapping but largely distinct phenotypes [12–14]. The KO mice generated by the two groups differed in their genetic background and in the design of the targeting construct. In both cases, ACC2 KO mice were viable and had normal life spans relative to wild-type mice. Consistent with the known function of ACC2, the KO mice generated by both groups had significantly reduced levels of malonyl-CoA and increased fatty acid oxidation in heart and skeletal muscle [12,14]. When placed on a high-fat/high-calorie diet, the KO mice generated by Wakil's group showed increased insulin sensitivity relative to wild-type mice, as well as improved lipid profiles and reduced body weight and adiposity [13,15]. In contrast, the high-fat-fed ACC2 KO mice generated by Cooney's lab had no beneficial metabolic phenotype [14]. Moreover, utilization of glucose for lipid synthesis was increased in Cooney's ACC2 KO mice, suggesting that a compensatory alteration of fuel metabolism may have contributed to the differential phenotype [14]. ACC1 KO was reported to be lethal in mice during embryonic development [16]. As a result, the phenotype of ACC1/ACC2 dual genetic KO has not been characterized.

There is some evidence of a role for ACCs in both human diabetes and obesity. Obese and type 2 diabetic subjects have been shown to have increased ACC2 activity and reduced fatty acid oxidation in skeletal muscle. Interestingly, treatment with thiazolidinedione, an antidiabetic agent, resulted in a moderate reduction of ACC activity [17]. Human subjects placed on a 3-month exercise regimen showed decreased ACC2 expression and increased fatty acid oxidation [18]. The glucose-lowering therapeutic agent metformin may work in part through decreasing ACC2 activity via activation of adenosine monophosphate-activated protein kinase (AMPK). AMPK has been shown to inhibit ACC2 activity by phosphorylation [19]. Therefore, increased ACC2 activity is associated with obesity and insulin resistance in humans and the treatment of these conditions is associated with reduced ACC2 activity.

In response to this biological data, there has been an active interest in the medicinal chemistry community to develop ACC inhibitors for the treatment of metabolic disorders. This chapter begins by briefly describing notable efforts in the ACC literature prior to 2009. The bulk of the chapter focuses on publications and patents that have appeared in 2009–early 2010. Finally, available data showing the efficacy of ACC inhibition in long-term studies is discussed.

2. MEDICINAL CHEMISTRY EFFORTS REPORTED PRIOR TO 2009

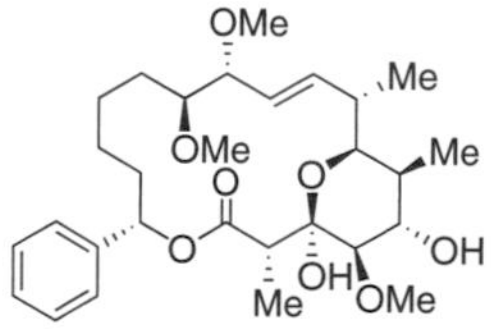

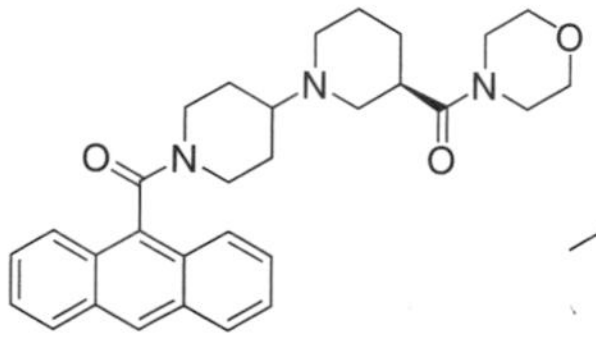

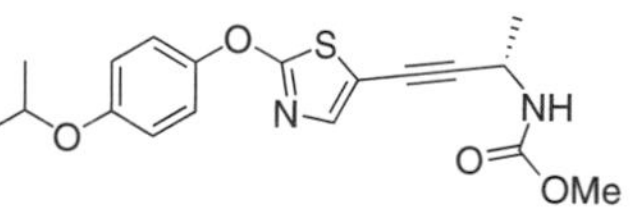

Several reports of ACC inhibitors have appeared over the last 10 years [20,21]. The natural product soraphen (1) has been shown to be an inhibitor of human ACC2 [22] and human ACC1 [23,24] with IC_{50}'s of 2

and 1–5 nM, respectively. Soraphen has also been shown to increase the oxidation rate of ^{14}C-labeled palmitate in Wistar rats and also to increase the percentage of lipid metabolism as measured by respiratory quotient (RQ) [22]. Soraphen has been cocrystallized with the BC domain of yeast ACC and is the only ACC inhibitor that has been published that binds in this domain [25]. CP-640186 (**2**) has been shown to be an inhibitor of rat liver ACC1 and rat skeletal muscle ACC2 with IC_{50}'s of 53 and 63 nM, respectively [26]. CP-640186 also inhibited fatty acid synthesis in HepG2 cells ($IC_{50} = 1.8$ μM). Oral dosing of CP-640186 to Sprague–Dawley rats lowered malonyl-CoA levels in liver (ED_{50} 55 mg/kg), heart (ED_{50} 8 mg/kg), soleus (ED_{50} 6 mg/kg), and quadriceps (ED_{50} 15 mg/kg). Rats dosed orally with 100 mg/kg of CP-640186 also showed an 89% reduction in RQ without increased consumption of O_2, indicating a shift in metabolic preference from carbohydrates to lipids. CP-640186 has been cocrystallized with the CT domain of yeast ACC [27].

More recently, a series of alkynyl thiazoles was reported, of which some analogs exemplified by **3** (ACC2 $IC_{50} = 38$ nM, ACC1 IC_{50} >30,000 nM) showed high selectivity for ACC2 versus ACC1 [28]. Sprague–Dawley rats dosed orally with **3** showed a dose-dependent reduction of malonyl-CoA in liver and muscle, indicating that inhibition of ACC2 alone may be sufficient to achieve a metabolic response. Extensive structure–activity relationship (SAR) studies around the phenyl and acetylene groups in this series have also been reported [29–31]. Additionally, gene expression analysis in animals treated with **3** suggested that the peroxisome proliferator-activated receptor α (PPARα) pathway was upregulated by ACC2 inhibition [32]. Although other compounds from the same group did not show PPARα activity, such an effect of **3** could potentially confound the interpretation of the pharmacology of its ACC inhibition.

3. CURRENT MEDICINAL CHEMISTRY EFFORTS

3.1 Spirochromanone-containing compounds

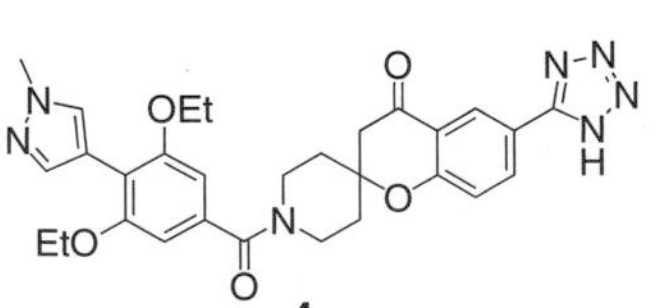

4

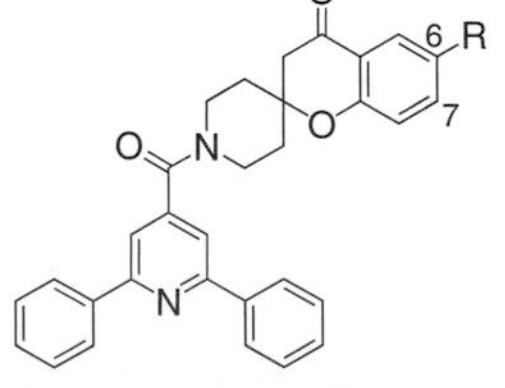

5: R = 4-morpholinyl
6: R = NHCO(CH₂)₂CH₃
7: R = 2-oxo-1-pyrrolidinyl

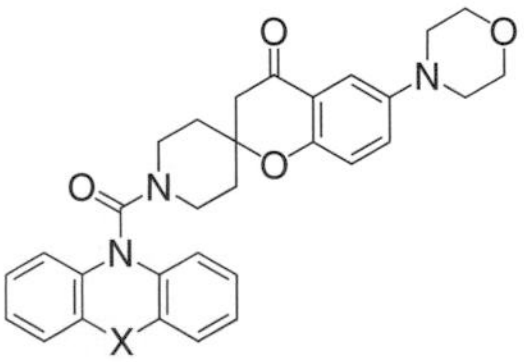

8: X = O
9: X = S

A number of recent publications have featured dual inhibitors of ACC1 and ACC2 with a conserved spirochromanone core. This spirochromanone core was first reported in two patent applications in 2007; however, only % inhibition of ACC1 and ACC2 was reported for a small number of examples [33,34]. A more recent patent application from these two groups expanded the scope of the structures exemplified (i.e., **4**, 100% inhibition of human ACC1 and ACC2 at 1 μM), but again only % inhibition data was included [35].

A more detailed report that published recently examined substitution effects at the 6 and 7 positions of the spirochromanone ring as well as changes to the amide region of the molecule [36]. 2,6-Diphenylpyridines **5–7** inhibited rat muscle ACC activity with IC_{50} values in the 6–16 nM range. Shifting the substitution from the 6 to the 7 position of the spirochromanone ring led to reduced ACC inhibition (4- to 10-fold). Replacement of the 2,6-diphenylpyridine amine with tricycles such as phenoxazine or phenothiazine resulted in compounds with significantly reduced ACC inhibitory activity (**8**, $IC_{50} = 99$ nM; **9**, 64% of control at 10 μM compound concentration). Compounds **5** and **6** were reported to have poor permeability, whereas compound **7** had moderate permeability (PAMPA permeability 26.76×10^{-6} cm/s at pH 6.8). Compound **7** administered intraperitoneally to C57BL/6J mice at doses of 30 and 60 mg/kg resulted in a dose-dependent decrease of RQ, indicative of increased fatty acid oxidation.

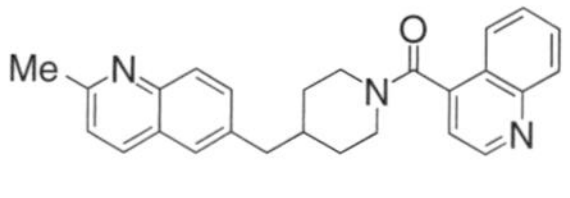

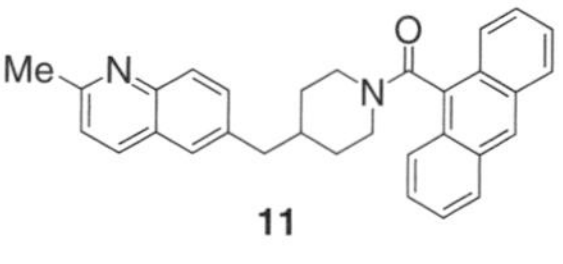

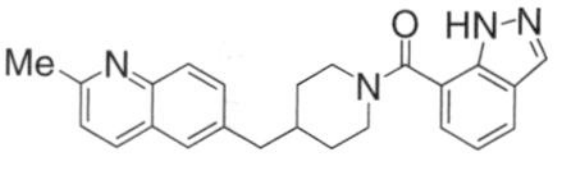

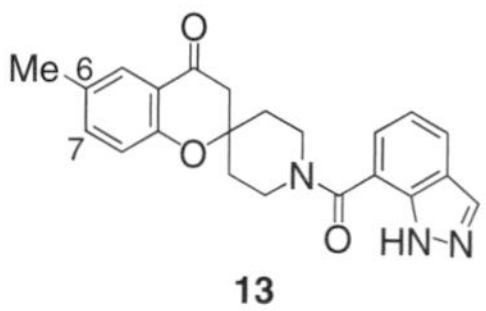

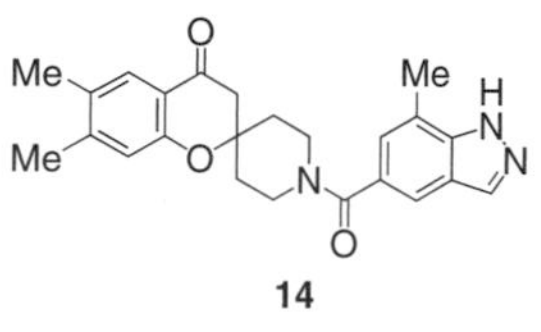

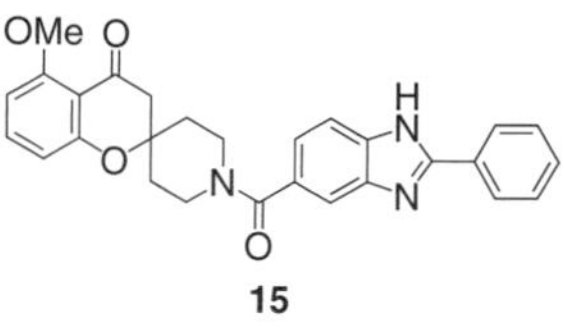

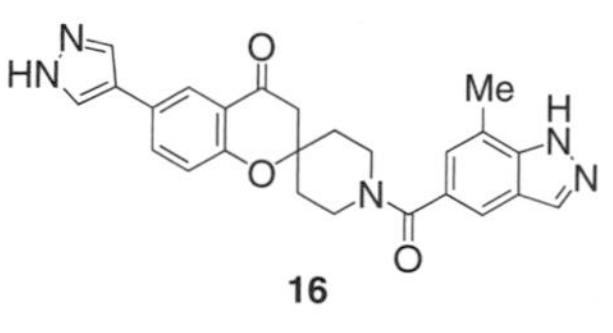

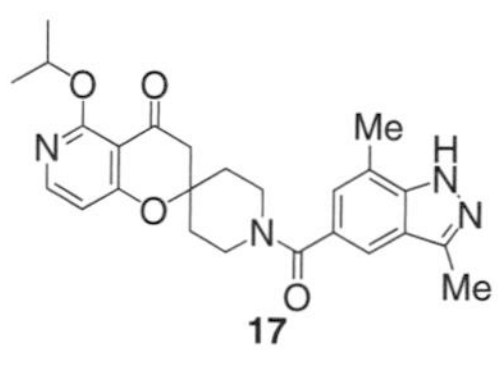

Another group has also reported studies of the spirochromanone scaffold in a series of patent applications and papers [37,38]. Work in this area was initiated by the discovery by high-throughput screening of **10**, a modestly potent inhibitor of rat ACC1 ($IC_{50} = 8910$ nM). Docking of **10** into the cocrystal structure of **2** with yeast ACC suggested that the interaction of the quinoline moiety with the protein was suboptimal [27]. Indeed, when the quinoline was replaced with an anthracene (**11**), the IC_{50} improved to 345 nM. Efforts were then undertaken to improve the physical properties of the series. An amide scan identified indazole **12**, which inhibited ¯rat ACC1 ($IC_{50} = 3380$ nM) with reasonable ligand efficiency (LE) of 0.26, a parameter that was monitored throughout the development of this series [39].

Compound **12** had reasonable rat pharmacokinetic (PK) properties (iv dose 1 mg/kg, $Cl = 44.3$ mL/min/kg, $Vdss = 2.4$ L/kg; po dose 5 mg/kg, $C_{max} = 617$ ng/mL, %$F = 45$, $t^{1/2} = 1.7$ h), but SAR around the methyl quinoline portion of the molecule was relatively flat. X-ray crystallographic analysis of **11** suggested that Gly B-1958 (numbering based on yeast ACC) was available to donate a hydrogen bond to the ligand, so efforts were made to incorporate an additional H-bond acceptor into the methyl quinoline moiety while introducing conformational constraints to eliminate unnecessary rotatable bonds. These investigations resulted in the discovery of spirochromanone **13**, which showed improved affinity and LE (rat ACC1 $IC_{50} = 633$ nM, LE = 0.30; human ACC2 $IC_{50} = 641$ nM). X-Ray crystallographic analysis revealed a H-bonding interaction of **13** with Gly B-1958 as predicted.

Further investigation of 6,5-fused heterocyclic amides determined that C5-linked indazoles such as **14** (rat ACC1 $IC_{50} = 22$ nM, LE = 0.34; human ACC2 $IC_{50} = 48$ nM) offered a significant improvement in activity relative to the original C7-linked indazoles. The indazole could be replaced with other heterocycles such as benzimidazoles, and additional aromatic substitution at the 2 position of the benzimidazole was also tolerated, as shown by benzimidazole **15** (rat ACC1 $IC_{50} = 16$ nM, LE = 0.30; human ACC2 $IC_{50} = 16$ nM). Additionally, it was found that a number of five-membered heterocyclic rings could be tolerated at the 6 position of the spirochromanone ring, exemplified by pyrazole **16** (rat ACC1 $IC_{50} = 7.4$ nM, LE = 0.34). Finally, 5 and 6 azaspirochromanones were prepared. The most potent of these was 6-azaspirochromanone **17**, which showed potent inhibition of both rat ACC1 and human ACC2 (rat ACC1 $IC_{50} = 12$ nM, LE = 0.33; human ACC2 $IC_{50} = 20$ nM). Compound **17** was further profiled in rat and dog PK studies. In rat, compound **17** showed moderate iv clearance, half-life, and volume of distribution (iv 1 mg/kg: $Cl = 31$ mL/min/kg, $t^{1/2} = 1.4$ h, $Vdss = 3.3$ L/kg). Oral bioavailability of the compound was highly dependent on the formulation method and was highest when a spray-dried dispersion was used (po 3 mg/kg: methylcellulose formulation, $\%F = 32$; lipid emulsion, $\%F = 8$; spray-dried dispersion, $\%F = 52$). Compound **17** also showed reasonable PK parameters in dogs (iv 1 mg/kg; Cl = 3.6 mL/min/kg, $t\frac{1}{2} = 6.1$ h, Vdss = 1.6 L/kg). No oral bioavailability data was reported for **17** in dogs.

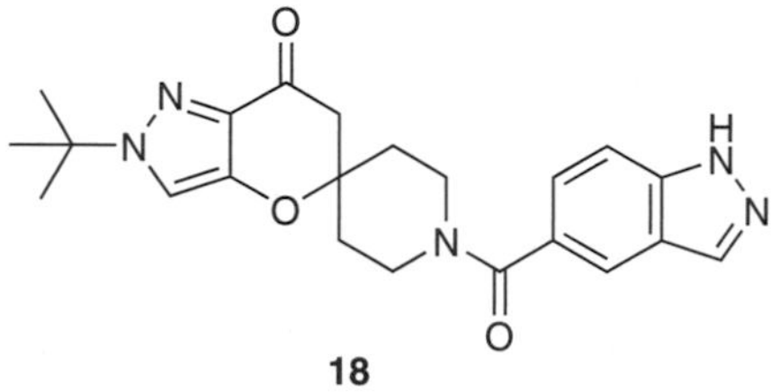

18

In two more recent patent applications, work in the spirochromanone series was extended to include spiropyranopyrazolones [40,41]. Of the compounds exemplified, most data was reported for compound **18** (rat ACC1 $IC_{50} = 17.2$ nM; human ACC2 $IC_{50} = 6.7$ nM). While no PK data was reported, **18** dosed to Sprague–Dawley rats on a normal diet caused a dose-dependent reduction of malonyl-CoA in liver and quadriceps, suggesting that it acts on both ACC1 and ACC2 *in vivo* (Table 1). A greater effect was seen in the liver than in the quadriceps at lower doses. A possible explanation for this observation is that the concentration of **18** in the liver may be higher than in the muscle tissue. Another explanation may be that there is a difference in the enzyme kinetics of ACC1 and ACC2.

Table 1 Percent decrease in malonyl-CoA relative to control animals after oral dosing of **18**

Dose (mg/kg)	Muscle malonyl-CoA^a (quadriceps)	Liver malonyl-CoA^a
1	2 (5)	35 (2.6)
3	24 (6.8)	54 (1.2)
10	49 (2.4)	71 (3.5)
30	57 (2.0)	64 (3.6)

^{a} ± standard error of the mean in parentheses.

3.2 Biphenyl and phenylpyridine cores

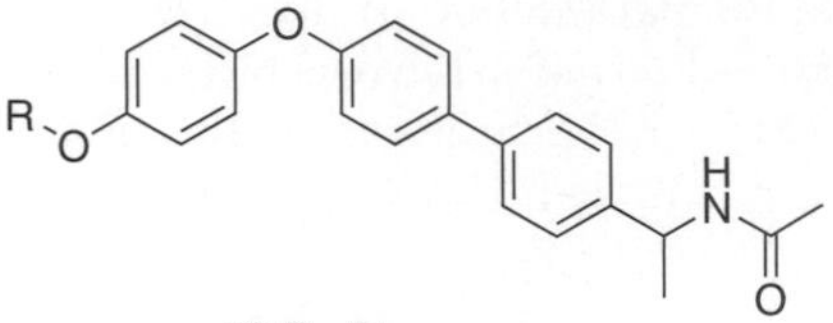

19 R = Ph
20 R = CH(CH₃)₂
21 R = (CH₂)₂OCH₃
22 R = OEt

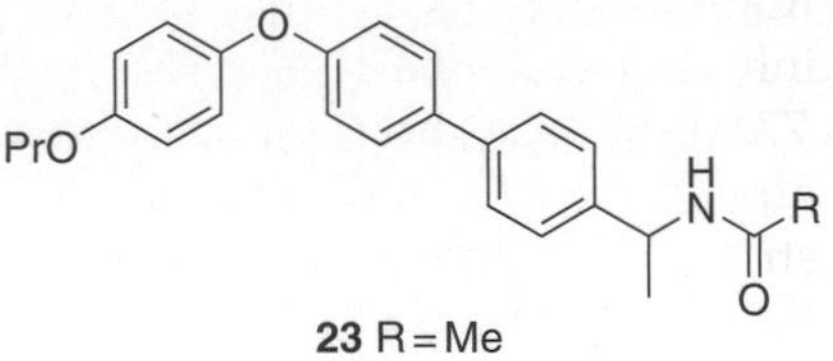

23 R = Me
24 R = Et
25 R = i-Pr

A series of biphenylmethyl acetamides and pyridylbenzyl amides has also been reported as ACC inhibitors [42]. Interestingly, relatively small changes to the structure of these inhibitors were shown to produce large changes in the ACC1/ACC2 selectivity ratio. For example, phenyl ether 19 and isopropyl ether 20 exhibit IC_{50} values in the low nanomolar range for both ACC1 and ACC2 (19; ACC1 $IC_{50} = 14$ nM, ACC2 $IC_{50} = 7$ nM: 20, ACC1 $IC_{50} = 17$ nM; ACC2 $IC_{50} = 33$ nM). In contrast, 2-methoxylethyl ether 21 showed potent inhibition of ACC1 ($IC_{50} = 89$ nM) with no measurable effect on ACC2. Ethyl ether 22, on the other hand, showed a modest preference for inhibition of ACC2 versus ACC1 (ACC2 $IC_{50} = 119$ nM, ACC1 $IC_{50} = 1630$ nM). While selective inhibitors of ACC2 versus ACC1 such as 3 had previously been reported [28], this is the first published example of compounds that selectively inhibit ACC1 versus ACC2.

Changes to the amide portion of this scaffold have also been reported. In general, substitution with groups larger than methyl led to compounds with decreased potency (23; ACC1 $IC_{50} = 17$ nM, ACC2 $IC_{50} = 14$ nM: 24; ACC1 $IC_{50} = 47$ nM, ACC2 $IC_{50} = 202$ nM: 25; ACC1 $IC_{50} = 188$ nM, ACC2 $IC_{50} = 214$ nM).

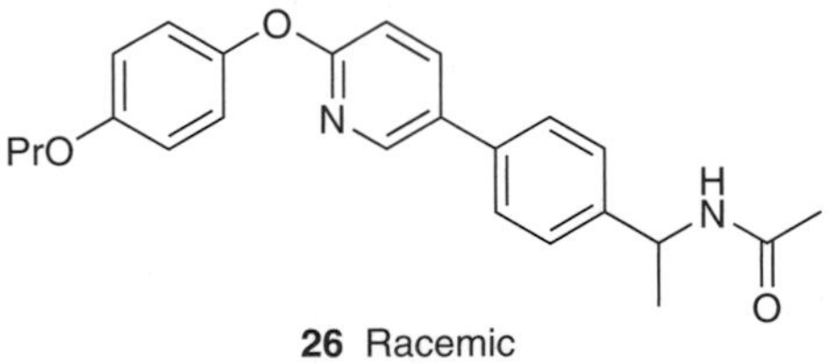

26 Racemic
27   S
28   R

Switching from the biphenyl core to a 3-phenylpyridine core resulted in compounds that were similar in potency to the biphenyl compounds, but with increased inhibition of ACC2 versus ACC1, as seen for **26** (ACC1 $IC_{50} = 110$ nM; ACC2 $IC_{50} = 18$ nM). Upon separation of the enantiomers of **26**, it was found that the S enantiomer **27** (ACC1 $IC_{50} = 35$ nM, ACC2 $IC_{50} = 8$ nM) was a significantly more potent ACC inhibitor than the R enantiomer **28** (ACC1 $IC_{50} > 30,000$ nM, ACC2 IC_{50} 777 nM). This increased relative potency of the S enantiomer was also observed in the related alkynylthiazole series (cf. **3**) [28, 31]. Indeed, the structural similarity between the two series raises the possibility that these two classes of inhibitors might occupy the same ACC-binding pocket. However, no ACC cocrystal structures for either class of inhibitors have been reported.

3.3 Piperazine core

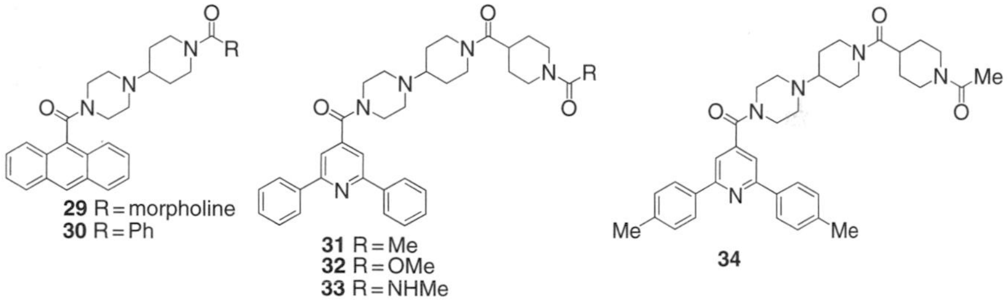

29 R = morpholine
30 R = Ph

31 R = Me
32 R = OMe
33 R = NHMe

34

A new series of ACC inhibitors containing a piperazine core was recently reported [43]. Using **2** as a design template, morpholine urea **29** was found to be a moderately potent inhibitor of ACC partially purified from human liver ($IC_{50} = 239$ nM). Compound **29** also showed poor microsomal stability. Various ureas and amides such as phenyl amide **30** ($IC_{50} = 66$ nM) were moderately potent ACC inhibitors; however, the effects of these structural changes on microsomal stability was not reported. Following literature precedent, replacement of the anthracene by a 2,6-diphenyl pyridine failed to provide analogs with improved potency [36]. However, replacement of the R-group of **30** by an acyl piperidine as in **31** ($IC_{50} = 126$ nM) allowed incorporation of the 2,6-diphenyl pyridine moiety and conferred potency similar to that of **29** and **30**. The acyl piperidine was tolerant of a variety of different polar functionalities, as exemplified by carbamate **32** ($IC_{50} = 102$ nM) and urea **33** ($IC_{50} = 95$ nM). Finally, it was found that substitution of the 2,6-diphenylpyridine could further improve potency, as exemplified by compound **34** ($IC_{50} = 76$ nM). Detailed profiling of **34** showed that it inhibited recombinant human ACC1 ($IC_{50} = 101$ nM) and recombinant human ACC2 ($IC_{50} = 23$ nM). Compound **34** also inhibited fatty acid synthesis in HepG2 cells ($IC_{50} = 340$ nM) and increased fatty acid oxidation in the same cell line ($ED_{50} = 580$ nM). Furthermore, **34** showed good stability in human and rat liver microsomes (87% and 96% remaining, respectively, after 15-min incubation with 1 mg protein/mL) and had reasonable PK properties when orally administered to rats at a dose of 10 mg/kg (C_{max} 107 ng/mL, $T_{max} = 4.0$ h, $AUC = 1000$ ng/mL/h).

3.4 Other series

Two recently published patent applications [44,45] extended previously described work by the same group [21]. Enzymatic data in the patent applications is reported in ranges (<1, 1–10, >10 µM, etc), making it difficult to thoroughly assess the compounds covered in the claims. Bissulfonamide **35** is a representative compound and is reported to inhibit both ACC1 and ACC2 with IC_{50}'s < 5 µM.

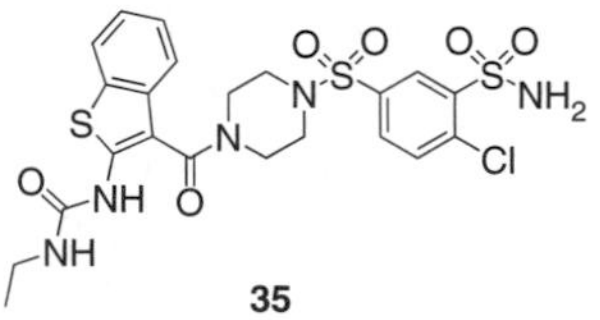

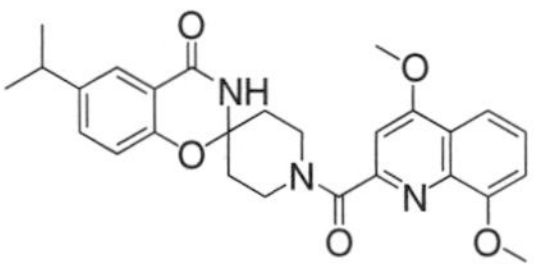

Another recent patent application reported a series of spiro 2,3-dihydro-4*H*-1,3-benzoxazin-4-one ACC1/2 dual inhibitors [46]. These compounds are structurally related to spirochromanones such as **4** and **17**, and the scope of substitutions examined in the series is similar to the previously reported studies [33,34]. Representative example **36** potently inhibited human ACC1 and ACC2 ($IC_{50} = 2$ nM for both enzymes). Compound **36** was shown to inhibit fatty acid synthesis *in vivo* (10 mg/kg, 55% inhibition; 30 mg/kg, 78% inhibition), as determined by inhibition of incorporation of radiolabeled acetyl-CoA into palmitate. Additionally, this patent application also included antiproliferation data for a number of compounds in four different cancer cell lines (A2780, HCT-116, MDA-MB-231, PC-3). At a concentration of 10 µM, compound **36** inhibited proliferation (<52% of control) in all four cell lines. To the best of our knowledge, this is the first antiproliferation data that has been reported for inhibitors of ACC.

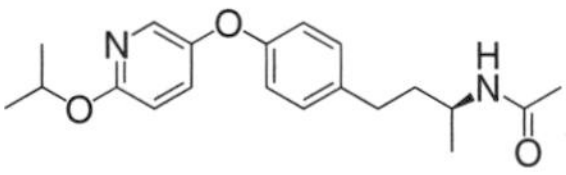

ACC inhibitors from another recently published patent application are exemplified by compound **37** [47] and resemble phenyl pyridine **26** [42] where one of the aromatic rings has been replaced by an aliphatic acetamide. Compound **37** showed some selectivity for human ACC2 versus ACC1 (ACC1 $IC_{50} = 190$ nM; ACC2 $IC_{50} = 30$ nM). Interestingly, compound **37** was also assayed against rat ACC1 and ACC2, and was a less potent inhibitor of rat ACC2 compared to human ACC2 (ACC1 $IC_{50} = 170\,nM$, ACC2 $IC_{50} = 400\,nM$). No rationale was presented to explain this species difference in ACC2 inhibitor potency, which has not been documented elsewhere in the ACC literature.

4. ACC INHIBITORS IN LONG-TERM EFFICACY STUDIES

Despite the relatively large number of publications in the area of ACC inhibition, there have been few disclosures reporting the effects of long-term administration of ACC inhibitors on metabolic syndrome-related endpoints. Recently, two disclosures reported the results of studies of the balanced ACC1/ACC2 inhibitor 2 in mouse models of metabolic syndrome [48,49]. In the first study, 2 was administered in chow to ob/ob mice at doses of 50 and 100 mg/kg for 8 weeks. Treatment with 2 resulted in increased glucose and triglyceride levels and decreased insulin levels in blood. The AUC and peak glucose response in an oral glucose tolerance test (OGTT) were also increased. After 6 weeks, the 100 mg/kg dose group was lowered to 20 mg/kg. This resulted in normalization of triglyceride and glucose levels, but insulin levels remained low. Body weight decreased in the 100 mg/kg dosing group, but food intake remained approximately the same for all the animals in the study. It was concluded that 2 worsened the diabetic phenotype in ob/ob mice, but the authors indicated that the origin of this effect was unclear [48].

The second study with 2 was performed in diet-induced obese (DIO) C57Bl6/J mice. Compared to control DIO mice, mice treated with 2 showed significantly less weight gain over the course of the 6-week study. However, mice treated with 2 had liver triglyceride and total cholesterol levels that were identical to untreated DIO mice. In a glucose infusion clamp study, the glucose infusion rate was slightly higher for the treated animals (~10%) than for the untreated DIO mice. The study concluded that treatment with 2 had a moderate effect on improving the diabetic phenotype of DIO mice [49].

In addition to these studies with 2, the effects of the dual ACC1/ACC2 inhibitor 1 in *in vivo* efficacy studies in high-fat-fed C57Bl6/J mice have been reported [50]. In mice fed a high-fat chow diet, 1 administered in-diet (50 and 100 mg/kg/day) dose-dependently reduced body weight gain relative to a high-fat chow-fed control group. Mice treated with 1 also had significantly higher levels of β-hydroxybutyrate than the high-fat-fed control group, indicating a shift in energy utilization to increased fatty acid metabolism in these mice. Additionally, fasting plasma insulin levels were lower in mice treated with 1 than in the high-fat control group, and in a glucose infusion rate study, mice dosed with 1 showed an improved response relative to the high-fat mice. Interestingly, the level of hepatic malonyl-CoA was found to be lower in the group dosed with 50 mg/kg of **1** than it was in the animals that were fed normal chow, while animals receiving 100 mg/kg of **1** had identical malonyl-CoA levels to the animals fed normal chow. The study concluded that treatment with **1** resulted in favorable effects on diabetic symptoms in DIO mice.

5. CONCLUSIONS

Significant progress has been made in the development of potent inhibitors of the two ACC isoforms. A number of different chemotypes have been designed, some of which possess selectivity for ACC1 or ACC2. Published mouse KO studies from two research groups, using different constructs and different backgrounds, have provided somewhat conflicting results [12–14]. However, limited *in vivo* efficacy studies suggest that a beneficial effect may be seen in DIO mouse models of diabetes when treated with an ACC1/2 dual inhibitor and thus ACC inhibition still remains a promising therapeutic approach. It remains to be seen whether ACC1 or ACC2 selective inhibitors will show beneficial *in vivo* effects and whether inhibition of ACC will be a useful treatment for metabolic disease in humans.

REFERENCES

[1] E. S. Ford, W. H. Giles, and W. H. Dietz, *J. Amer. Med. Assoc.*, 2002, **283**, 356.
[2] D. E. Moller, *Nature*, 2001, **414**, 821.
[3] S. M. Grundy, *Nat. Rev. Drug Discov.*, 2006, **5**, 295.
[4] D. E. Moller and K. D. Kaufman, *Annu. Rev. Med.*, 2005, **56**, 45.
[5] P. Bjorntorp, *Curr. Opin. Lipidol.*, 1994, **5**, 166.
[6] J. D. McGarry, *Science*, 1992, **258**, 766.
[7] D. B. Savage, K. F. Petersen, and G. I. Shulman, *Physiol. Rev.*, 2007, **87**, 507.
[8] G. I. Shulman, *J. Clin. Invest.*, 2000, **106**, 17.
[9] S. E. Ploakis, R. B. Guchhait, E. E. Zwergel, M. D. Lane, and T. G. Cooper, *J. Biol. Chem.*, 1974, **249**, 6657.
[10] J. D. McGarry and D. W. Foster, *Annu. Rev. Biochem.*, 1980, **49**, 395.
[11] B. B. Rasmussen, U. C. Holmback, E. Volpi, B. Morio-Liondore, D. Paddon-Jones, and R. R. Wolfe, *J. Clin. Invest.*, 2002, **110**, 1687.
[12] L. Abu-Elheiga, M. M. Matzuk, K. A. Abo-Hashema, and S. J. Wakil, *Science*, 2001, **291**, 2613.
[13] L. Abu-Elheiga, W. Oh, P. Kordari, and S. J. Wakil, *Proc. Natl. Acad. Sci. USA*, 2003, **100**, 10207.
[14] K. L. Hoehn, N. Turner, M. M. Swarbrick, D. Wilks, E. Preston, Y. Phua, H. Joshi, S. M. Furler, M. Larance, B. D. Hegarty, S. J. Leslie, R. Pickford, A. J. Hoy, E. W. Kraegen, D. E. James, and G. J. Cooney, *Cell Metab.*, 2010, **11**, 70.
[15] C. S. Choi, D. B. Savage, L. Abu-Elheiga, Z. X. Zhu, S. Kim, A. Kulkarni, A. Distefano, Y. J. Hwang, R. M. Reznick, R. Codella, D. Zhang, G. W. Cline, S. J. Wakil, and G. I Shulman, *Proc. Natl. Acad. Sci. USA*, 2007, **104**, 16480.

[16] L. Abu-Elheiga, M. M. Matzuk, P. Kordari, W. Oh, T. Shaikenov, Z. Gu, and S. J. Wakil, *Proc. Natl. Acad. Sci. USA*, 2005, **102**, 12011.

[17] G. K. Bandyopadhyay, J. G. Yu, J. Ofrecio, and J. M. Olefsky, *Diabetes*, 2006, **55**, 2277.

[18] P. Schrauwen, D. P. C. van Aggel-Leijssen, G. Hui, A. J. M. Wagenmakers, H. Vidal, W. H. M. Saris, and M. A. van Baak, *Diabetes*, 2002, **51**, 2220.

[19] N. Musi, M. F. Hirshman, J. Nygren, M. Svanfeldt, P. Bavenholm, O. Rooyackers, G. Zhou, J. M. Williamson, O. Ljunqvist, S. Efendic, D. E. Moller, A. Thorell, and L. J. Goodyear, *Diabetes*, 2002, **51**, 2074.

[20] J. W. Corbett, *Expert Opin. Ther. Pat.*, 2009, **19**, 943.

[21] J. W. Corbett and J. H. Harwood, Jr., *Recent Pat. Cardiol.*, 2007, **2**, 162.

[22] M. Gubler and J. Mizrahi, *Patent Application WO 2003/011867*, 2003.

[23] J. H. Harwood, Jr., *Expert Opin. Ther. Targets*, 2005, **9**, 267.

[24] L. Abu-Elheiga, D. B. Almarza-Ortega, A. Baldini, and S. J. Wakil, *J. Biol. Chem.*, 1996, **272**, 10669.

[25] Y. Shen, S. L. Volrath, S. C. Weatherly, T. D. Elich, and L. Tong, *Mol. Cell*, 2004, **16**, 881.

[26] H. J. Harwood, Jr., S. F. Petras, L. D. Shelly, L. M. Zaccaro, D. A. Perry, M. R. Makowski, D. M. Hargrove, K. A. Martin, W. R. Tracy, J. G. Chapman, W. P. Magee, D. K. Dalvie, V. F. Soliman, W. H. Martin, C. J. Mularski, and S. A. Eisenbeis, *J. Biol. Chem.*, 2003, **278**, 37099.

[27] H. Zhang, B. Tweel, J. Li, and L. Tong, *Structure*, 2004, **12**, 1683.

[28] Y. G. Gu, M. Weitzberg, R. F. Clark, X. Xu, Q. Li, T. Zhang, T. M. Hansen, G. Liu, Z. Xin, X. Wang, R. Wang, T. McNally, B. A. Zinker, E. U. Frevert, H. S. Camp, B. A. Beutel, and H. L. Sham, *J. Med. Chem.*, 2006, **49**, 3770.

[29] R. F. Clark, T. Zhang, Z. Xin, G. Liu, Y. Wang, T. M. Hansen, X. Wang, R. Wang, X. Zhang, E. U. Frevert, H. S. Camp, B. A. Beutel, H. L. Sham, and Y. G. Gu, *Bioorg. Med. Chem. Lett.*, 2006, **16**, 6078.

[30] R. F. Clark, T. Zhang, X. Wang, R. Wang, H. S. Camp, B. A. Beutel, H. L. Sham, and Y. G. Gu, *Bioorg. Med. Chem. Lett.*, 2007, **17**, 1961.

[31] Y. G. Gu, M. Weitzberg, R. F. Clark, X. Xu, Q. Li, N. L. Lubbers, Y. Yang, D. W. Beno, D. L. Widomski, T. Zhang, T. M. Hansen, R. F. Keyes, J. F. Waring, S. L. Carroll, X. Wang, R. Wang, C. H. Healan-Greenberg, E. A. Bloome, B. A. Beutel, H. L. Sham, and H. S. Camp, *J. Med. Chem.*, 2007, **50**, 1078.

[32] J. F. Waring, Y. Wang, C. H. Healan-Greenberg, A. L. Adler, R. Dickenson, T. McNally, X. Wang, M. Weitzberg, X. Xiangdong, A. Lisowski, S. E. Warder, Y. G. Gu, B. A. Zinker, E. A. Bloome, and H. S. Camp, *J. Pharmacol. Exp. Ther.*, 2007, **324**, 507.

[33] Y. J. Yamakawa, H. Jona, K. Niiyama, K. Yamada, T. Lino, M. Ohkubo, H. Imamura, J. Shibata, J. Kusunoki, and L. Yang, *Patent Application WO2007/011809*, 2007.

[34] Y. J. Yamakawa, H. Jona, K. Niiyama, K. Yamada, T. Lino, M. Ohkubo, H. Imamura, J. Kusunoki, and L. Yang, *Patent Application WO 2007/011811*, 2007.

[35] T. Lino, H. Jona, H. Kurihara, M. Nakamura, K. Niiyama, J. Shibata, T. Shimamura, H. Watanabe, T. Yamakawa, and L. Yang, *Patent Application WO 2008/088692*, 2008.

[36] P. Shinde, S. K. Srivastava, R. Odedara, D. Tuli, S. Munshi, J. Patel, S. P. Zambad, R. Sonawane, R. C. Gupta, V. Chauthaiwale, and C. Dutt, *Bioorg. Med. Chem. Lett.*, 2009, **19**, 949.

[37] J. W. Corbett, R. L Elliott, and A. S. Bell, *Patent Application WO 2008/065508*, 2008.

[38] J. W. Corbett, K. D. Freeman-Cook, R. Elliott, F. Vajdos, F. Rajammohan, D. Kohls, E. Marr, H. Zhang, L. Tong, M. Tu, S. Murdande, S. D. Doran, J. A. Houser, W. Song, C. J. Jones, S.B. Coffey, L. Buzon, M. L. Minich, J. K. Dirico, S. Tapley, R. K. McPherson, E. Sugarman, H. J. Harwood, and W. Esler, *Bioorg. Med. Chem. Lett.*, 2010, **20**, 2383.

[39] A. L. Hopkins, C. R. Groom, and A. Alex, *Drug Discov. Today*, 2004, **9**, 430.

[40] J. W. Corbett, R. L. Elliott, K. D. Freeman-Cook, D. A. Griffith, and D. P. Phillion, *Patent Application WO 2009/144554*, 2009.

[41] K. D. Freeman-Cook and B. M. Samas, *Patent Application WO 2009/144555*, 2009.

[42] T. S. Haque, N. Liang, R. Golla, R. Seethala, Z. Ma, W. R. Ewing, C. B. Cooper, M. A. Pelleymounter, M. A. Poss, and D. Cheng, *Bioorg. Med. Chem. Lett.*, 2009, **19**, 5872.

[43] T. Chonan, T. Oi, D. Yamamoto, M. Yashiro, D. Wakasugi, H. Tanaka, A. Ohoka-Sugita, F. Io, H. Koretsune, and A. Hiratate, *Bioorg. Med. Chem. Lett.*, 2009, **19**, 6645.

[44] E. Chang and M. McNeill, *Patent Application US 2009/0035375*, 2009.

[45] E. Chang, T. Duong, and A. Vassar, *Patent Application US 2009/0253725*, 2009.

[46] R. Anderson, S. Breazeale, T. Elich, and S.-F. Lee, *Patent Application US 2010/0009982*, 2010.

[47] G. Zoller, D. Schmoll, M. Muller, G. Haschke, and I. Focken, *Patent Application WO 2010/ 003624*, 2010.

[48] J. L. Treadway, R. K. McPherson, S. F. Petras, L. D. Shelly, K. S. Frederick, K. Sagawa, D. A. Perry, and H. J. Harwood, Presented at 64th Scientific Sessions Meeting, June 4–8, 2004, Orlando, FL, Abstract Number 679-P The abstract and the poster can be viewed at http://professional.diabetes.org.

[49] M. Schreurs, M. H. Oosterveer, T. H. van Dijk, R. Havinga, D.-J. Reijngoud, and F. Kuipers, Presented at 43rd EASD Annual Meeting, September 17–21, 2007. Amsterdam. The abstract can be viewed at http://www.easd.org.

[50] M. Schreurs, T. H. van Dijk, A. Gerding, R. Havinga, D.-J. Reijngoud, and F. Kuipers, *Diabetes Obes. Metab.*, 2009, **11**, 987.

CHAPTER 7

Enzymatic Targets in the Triglyceride Synthesis Pathway

Dmitry O. Koltun and **Jeff Zablocki**

1. INTRODUCTION

Obesity has reached epidemic proportions globally, with more than 1 billion adults overweight—at least 300 million of them clinically obese—and is a major contributor to the global burden of chronic disease and disability. Confined to older adults for most of the 20th century, obesity now affects children even before puberty. There is a strong correlation between type 2 diabetes and obesity: 90% of type 2 diabetics are either obese or overweight [1] . Obesity is commonly associated with elevated serum triglyceride (TG) levels. Increased TG levels, particularly when accompanied by low high-density lipoprotein (HDL) levels, have consistently been shown to be a surrogate marker of insulin resistance, a strong predisposing condition for type 2 diabetes [2].

Circulating TG levels represent a balance between TG synthesis and utilization. These are greatly affected by lifestyle factors (nutritional habits and exercise) and by insulin sensitivity. Circulating TGs are secreted from the liver in the form of very low-density lipoprotein (VLDL) or from the intestine in the form of chylomicrons. The underlying

Gilead Sciences, Inc., 333 Lakeside Dr., Foster City, CA 94404, USA

Annual Reports in Medicinal Chemistry, Volume 45
ISSN 0065-7743, DOI 10.1016/S0065-7743(10)45007-1

mechanism of TG synthesis involves maintaining the appropriate balance between the essential saturated and unsaturated fatty acid (FA) building blocks, and the coupling of three of these FA residues with one glycerol molecule. Finally, TGs must be transferred onto nascent VLDL or chylomicron particles. This chapter focuses on three enzymatic targets in the TG synthesis pathway: stearoyl-CoA desaturases (SCDs, also known as Δ9-desaturases) that catalyze the synthesis of monounsaturated FAs from saturated FAs, acylCoA:diacylglyceride O-acyltransferases (DGATs) that catalyze the last committed step in the synthesis of TGs, and microsomal triglyceride transfer protein (MTP) that is responsible for the process of loading of TGs onto nascent VLDL particles or chylomicrons.

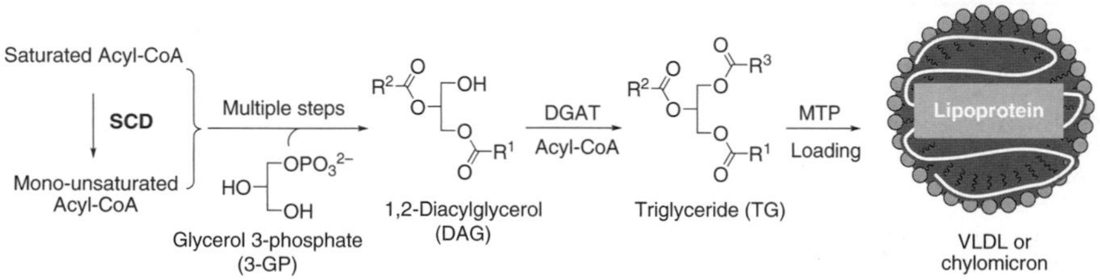

2. STEAROYL-CoA DESATURASES

SCDs are a family of microsomal Fe-based metalloenzymes. They act on long-chain saturated acyl CoAs and introduce a cis-double bond at the C-9 or C-10 position. For example, SCDs convert stearic acid into oleic acid, and palmitic acid into palmitoleic acid. Monounsaturated FAs constitute a major component of TGs, cholesteryl esters, and phospholipids. The reaction requires molecular O_2 and NADH and generates H_2O in the process [3,4].

Two isoforms (SCD1 and SCD5, also known as SCD2 in rodents) have been identified in humans. SCD1 is most highly expressed in liver, adipose tissue, and skeletal muscle [5] while SCD2 is found primarily in the brain [6]. For this reason, in papers that utilize liver microsomes or cells as sources of SCD for assays, the terms SCD1 and SCD are often used interchangeably.

Deletion, mutation, or inhibition of SCD1 in mice and rats results in decreased hepatic TGs [7–11], resistance to weight gain, and improvements in insulin sensitivity and glucose uptake. In one study, naturally occurring lean, hypermetabolic SCD-deficient asebia mice were crossed with obese leptin-deficient *ob/ob* mice and the resulting offspring were lean, hypermetabolic and had normal liver histology [8]. Thus, SCD inhibition may offer a novel approach to treating obesity, insulin resistance, and diabetes, as well as fatty liver diseases, such as steatosis and nonalcoholic steato-hepatatis (NASH) [12–16]. However, mechanism-based side effects have been reported and appear to be due to a deficit of unsaturated FAs in peripheral tissues like skin and pancreas [9,17]. Reported skin side effects included loss of hair (alopecia) and eye fissure [18]. Additionally, using antisense oligonucleotides to inhibit SCD in a mouse model of hyperlipidemia and atherosclerosis (LDLr$^{-/-}$ Apob$^{100/100}$), it has been demonstrated that SCD inhibition increased atherosclerosis independently of improvements in obesity and insulin resistance and argued against SCD1 inhibition as a safe therapeutic approach for treatment of the metabolic syndrome [19]. The development of atherosclerosis could not be prevented with dietary oleic acid. However, in a follow-up study, fish oils were able to fully prevent the development of atherosclerosis following SCD inhibition. It is proposed that fish oils antagonize the inflammatory effects of saturated FAs on the Toll4 receptor system [20].

The pursuit of SCD as a "druggable" target began in the early 2000s. In 2005, a series of patent applications emerged that are exemplified by compound 1. The common motif in the generic structures of these patent applications is a pair of linked nitrogen heterocycles, one aromatic and one saturated, for example, pyridazine and piperazine, flanked by amide-linked hydrophobic groups [21]. This family of patents inspired a number of second-generation designs, such as A-939572 (2), featuring a bioisosteric replacement of acylpiperazine with hydroxypiperidine. Compound 2 was found to be a potent SCD inhibitor (IC$_{50}$ 37 and <4 nM in human and mouse liver microsomal assays, respectively). Compound 2 also possessed very good mouse pharmacokinetic (PK) properties ($F = 92\%$, Cl$_p = 0.4$ L/h/kg) and demonstrated efficacy in lowering the ratio of unsaturated to saturated FAs (desaturation index) in liver and plasma when administered to ob/ob mice [22]. Unfortunately, in the same study, side effects consistent with asebia and the SCD knock-out mouse phenotype were observed that included alopecia and eye fissure [23].

Other designs structurally related to those reported in the 2005 series of patent applications included thiazole derivative 3 and MF-438 (4). In the case of 3, the six-membered heteroaromatic ring in 1 was contracted to a five-membered ring, and the adjacent amide was replaced by a 1,2,4-oxadiazole bioisostere [24]. In 4, a 1,3,4-thiadiazole was utilized [25]. Compound 3 showed an SCD IC$_{50}$ of 1 nM in both rat microsomal and human HEPG2 cell-based assays. The compound preferentially distributed into liver and adipose tissues compared to skin (3.7- and 4.9-fold respectively at 6 h postdose) and displayed efficacy for SCD inhibition $in~vivo$. C57BL6 mice fed a high-fat diet that were dosed with 3 (0.2 mg/kg, qd, 28 days) displayed significantly reduced body weight gain compared to a high-fat-fed control group and had body weight gain similar to that of regular chow-fed mice. However, despite apparent tissue selectivity, skin and eye side effects consistent with the SCD knock-out phenotype were also observed beginning at approximately day 7 of the study.

Another group of researchers built on SAR from previously reported compounds including 1–4 and generated a lead compound, pyridazine 5. Similarly to 2, the piperazine ring of 1 was transformed into piperidine and the keto moiety was retained. The amide moiety in the left-hand portion of the molecule was retained, but similar to 3 addition of a hydroxy group was found to be beneficial. Introduction of a 3-pyridyl group contributed to potency (SCD $IC_{50} = 37$ nM, mouse microsomal assay), oral bioavailability ($F = 68\%$, mouse), solubility ($>100\,\mu$M, pH 1.2), and oral efficacy of 5 for lowering the desaturase activity ($ID_{50} = 3$ mg/kg) [26]. The same group reported a second-generation design (6), featuring a spirocyclic ring system that retained an *ortho*-CF_3 substituent common to many SCD inhibitors. Compound 6 possessed superior potency for SCD inhibition *in vitro* ($IC_{50} = 60$ pM, mouse microsomal assay) and *in vivo* (ID_{50} 0.5 and 0.8 mg/kg, 2–3 and 6–7 h after administration respectively) [27].

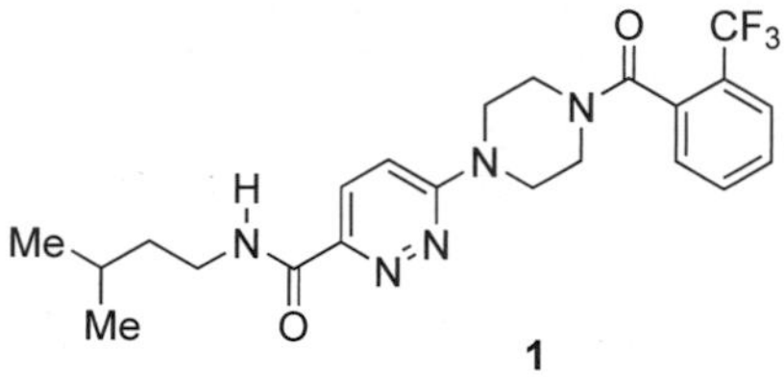

1

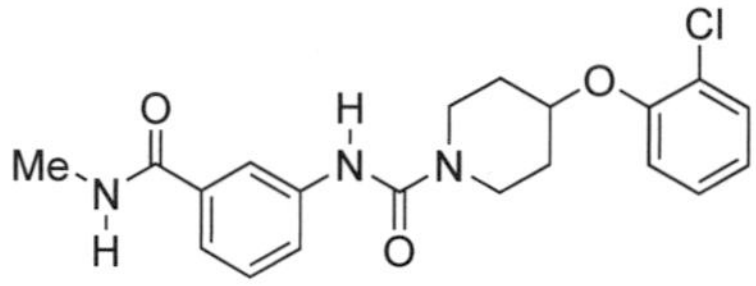

2 (A-939572)

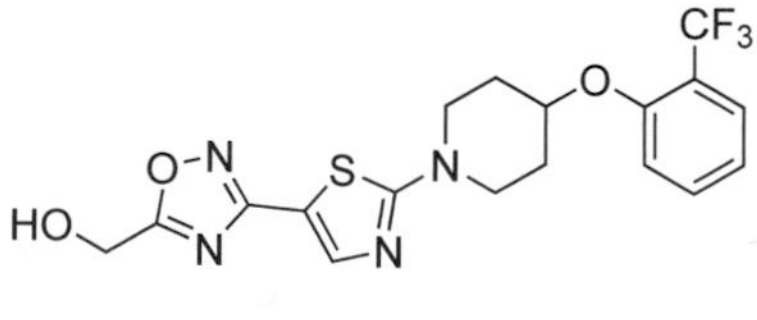

3

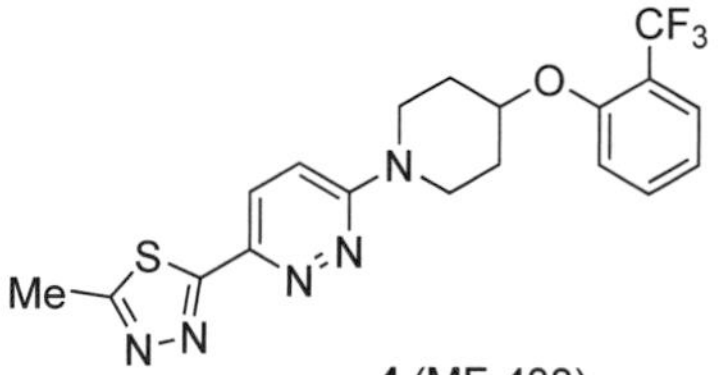

4 (MF-438)

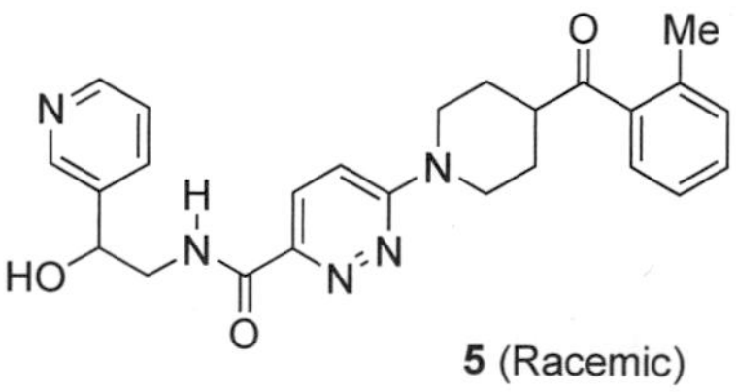

5 (Racemic)

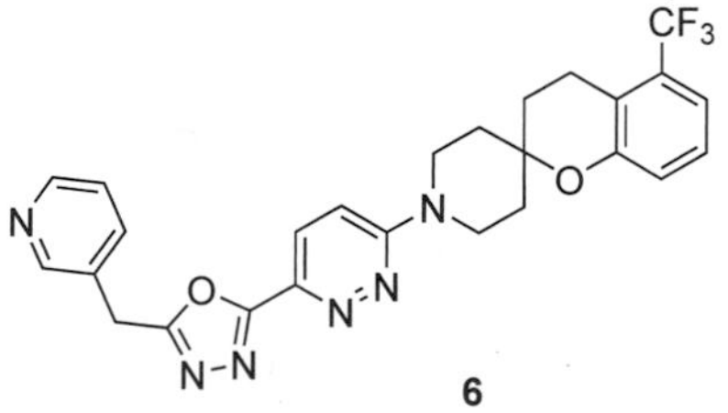

6

Additional leads were identified by independent screening efforts. The quinoxalinone CVT-13036 (7) had an SCD IC_{50} of 50 pM in a human HEPG2 assay, but lacked oral bioavailability [28]. On the other hand, the quinazolinone CVT-11,563 (8) had excellent bioavailability ($F = 90\%$, rat) but only a modest potency for SCD inhibition ($IC_{50} = 268$ nM, rat microsomal assay; $IC_{50} = 68$ nM, human HEPG2 cells) [29]. CVT-12,012 (9) is an analog of 7 that incorporates the hydroxyacetyl group of 8. It had good potency for SCD inhibition and good oral bioavailability ($IC_{50} = 119$ nM, human HEPG2 cells, $F = 78\%$). Compound 9 also showed good *in vivo* efficacy for reduction of the desaturation index and hepatic TGs in high-carbohydrate-fed rats (~50% reduction, 20 mg/kg, po bid, 5 d) [30]. A pyrrolotriazinone analog of 8, CVT-12,805 (10), showed improved potency (SCD $IC_{50} = 10$ nM, rat microsomal assay) and good efficacy for the reduction in desaturation index when dosed orally to high-sucrose-fed mice [31].

Another independent high-throughput screening (HTS) and hit optimization effort resulted in the potent (SCD IC_{50} 2 and 3 nM; mouse and human microsomal assays, respectively) and orally bioavailable ($F = 12\%$) SCD inhibitor 11 [32]. Replacement of the methoxy group of 11 by an ethylamino group afforded 12, which was more potent (SCD $IC_{50} = 40$ pM, human microsomal), had acceptable PK parameters ($Cl_p = 1.26$ L/h/kg, $F = 27\%$, $t_{1/2} = 1.1$ h, mouse) and good *in vivo* reduction of desaturase activity ($ID_{50} = 0.8$ and 2.0 mg/kg at 2–3 and 6–7 h respectively). Thiazole derivative 12 was efficacious in reducing the desaturation index in a dose-dependent fashion when administered qd for 7 days. No eye or skin abnormalities were reported in the study. This apparent success in separating *in vivo* efficacy and untoward skin and eye effects was attributed to the combination of the high potency and short half-life of 12 [33].

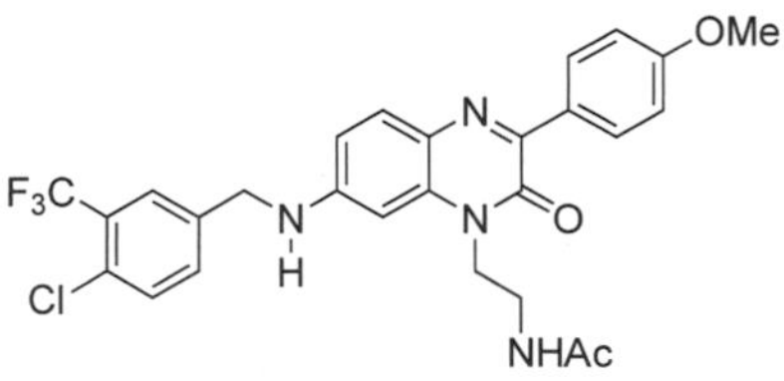

7 (CVT-13,036)

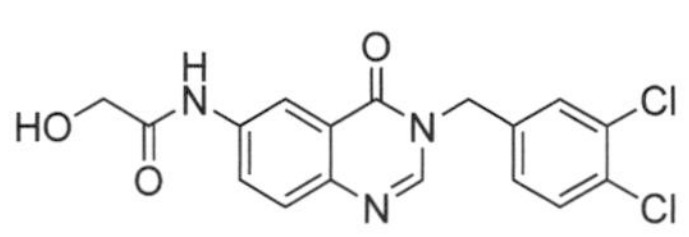

8 (CVT-11,563)

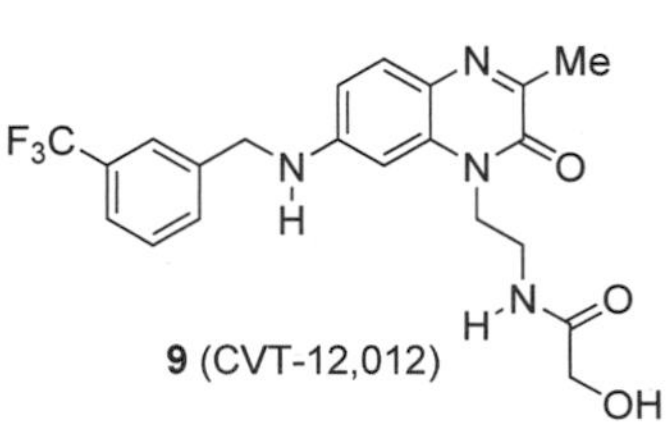

9 (CVT-12,012)

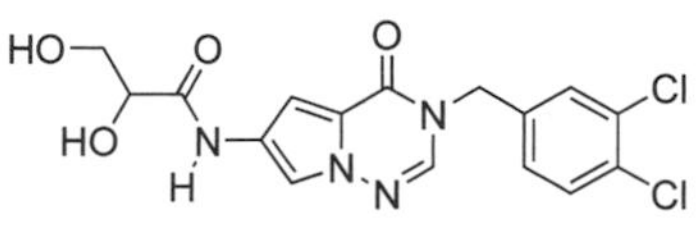

10 (CVT-12,805)
(racemic)

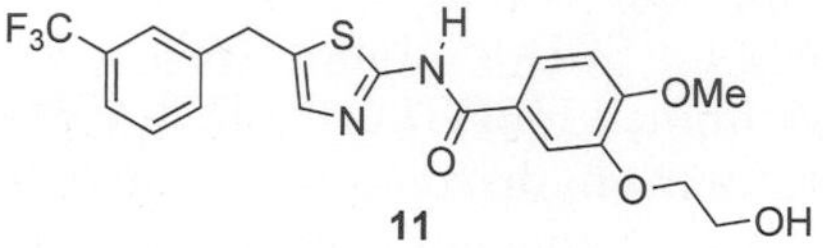

11

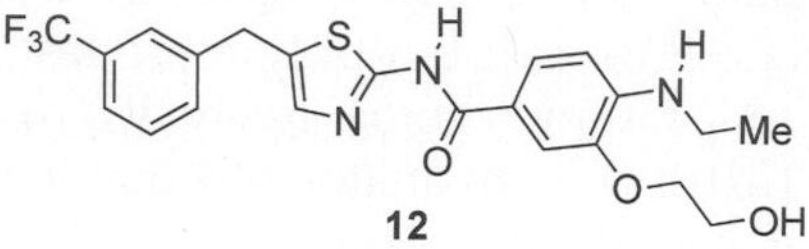

12

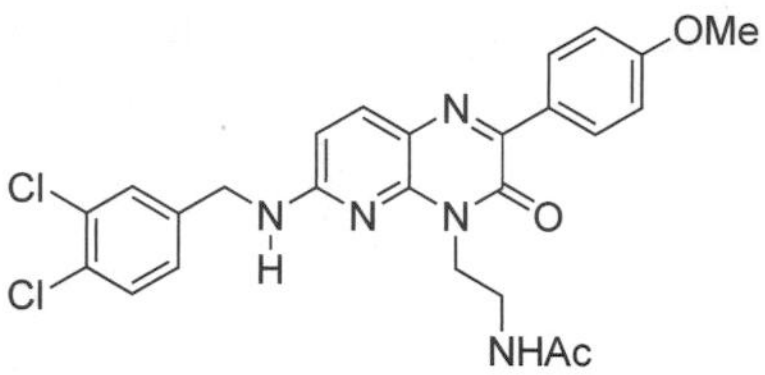

13 (CVT-11,127)

Perhaps in light of the mechanism-based side effects, the therapeutic potential of SCD inhibitors has not been fully evaluated. The possibility exists that at least some of the chronic effects of SCD inhibition might be prevented by coadministration of fish oils or potentially a Toll4 antagonist [20]. However, coformulation of an SCD inhibitor with fish oil has not been reported. Another possibility is that SCD inhibitors with appropriate PK and tissue partitioning properties may limit mechanism-based side effects.

One small-molecule SCD inhibitor is believed to have entered clinical trials as a result of a partnership between Xenon Pharmaceuticals and Novartis. The structure has not been disclosed and the current clinical status is unknown.

A recent development in the field of SCD research is the growing evidence that SCD inhibitors have anticancer and antiproliferative effects. Cancer cells activate the biosynthesis of saturated and monounsaturated FAs in order to sustain an increasing demand for phospholipids with appropriate acyl composition during cell replication [34]. SCD inhibitor CVT-11,127 (13, $IC_{50} = 220$ nM, rat microsomal assay; $IC_{50} = 81$ nM, human HEPG2 cells) [28] reduced cell proliferation *in vitro* in human lung carcinoma cells [34].

3. DIACYLGLYCERIDE *O*-ACYLTRANSFERASES

DGATs [35] along with acyl-CoA:monoacylglycerol acyltrasferase type 3 (MGAT3) [36] are the enzymes responsible for the completion of TG synthesis from an acyl-CoA and a diacylglyceride. Two isoforms of human DGAT are known. DGAT1 is located mainly in enterocytes in the small intestine and is primarily responsible for recombining dietary TGs upon absorption and for their incorporation into nascent chylomicrons for excretion [37,38]. DGAT2 is expressed primarily in liver, adipose tissue, and, to a lesser extent, the small intestine [39]. Both DGAT1 and DGAT2 have been highlighted as potential targets for small-molecule intervention [35]. TG synthesis is not the only role of the DGAT enzymes. DGAT1 is also responsible for retinol esterification, while DGAT2 is involved in the synthesis of wax esters [35].

High-fat diet studies with DGAT1$^{-/-}$ and DGAT1$^{+/-}$ mice demonstrated that the absence of DGAT1 activity may lead to increased insulin sensitivity, leptin sensitivity, protection against diet-induced obesity, and protection against liver steatosis [40–43].

DGAT2$^{-/-}$ mice suffer from severe hypolipidemia and do not survive postnatally [44]. However, this may be only a developmental phenomenon, since adult obese mice treated with DGAT2 antisense oligonucleotides were viable and showed improved hepatic steatosis along with improved insulin sensitivity [45].

Extensive research on small-molecule DGAT1 inhibitors began in the early 2000s. Several patent applications were published in 2004, exemplified by compounds **14** (IC$_{50}$ < 10 nM, recombinant human DGAT1) and **15** [46,47]. These inspired several second-generation designs that retained the characteristic disubstituted cycloalkyl ring featuring a carboxylic acid. One research group reported the use of urea as a bioisostere for the aminobenzothiazole in **15**, affording compound **16** with good potency (IC$_{50}$ = 7 nM, recombinant human DGAT1), excellent rat PK properties (F = 55%, Cl$_p$ = 0.04 L/h/kg, $t_{1/2}$ = 3.9 h), and which reduced plasma and liver TGs in a dose-dependent manner when administered to diet-induced obese (DIO) mice [48]. Another research group reported the use of a bicyclic system in place of *trans*-cyclohexyl ring, resulting in **17**. Compound **17** was potent in a biochemical assay (IC$_{50}$ = 15 nM, recombinant human DGAT1) as well as in a cell-based assay (IC$_{50}$ 3 nM, DGAT1 expressed in HuTu80 cells) and did not inhibit DGAT2. This compound also possessed favorable rat PK properties (F = 72%, Cl$_p$ = 0.09 L/h/kg, $t_{1/2}$ = 7.0 h) and was shown to reduce plasma TGs in a dose-dependent manner in an oral lipid tolerance test in fasted rats and cause body weight reduction when administered to DIO mice [49].

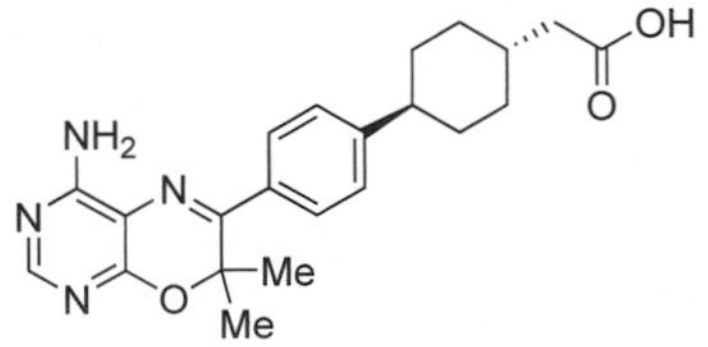

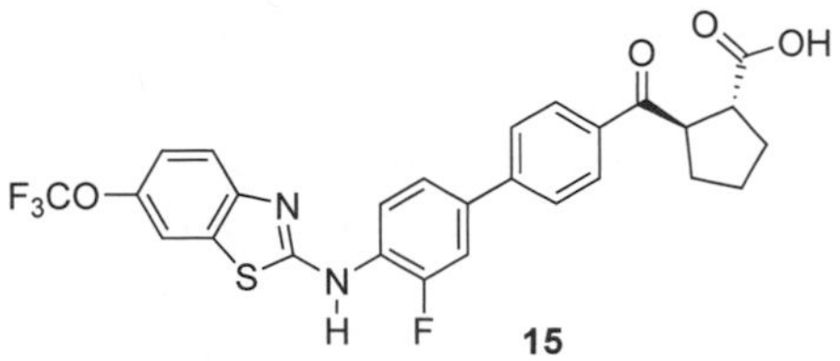

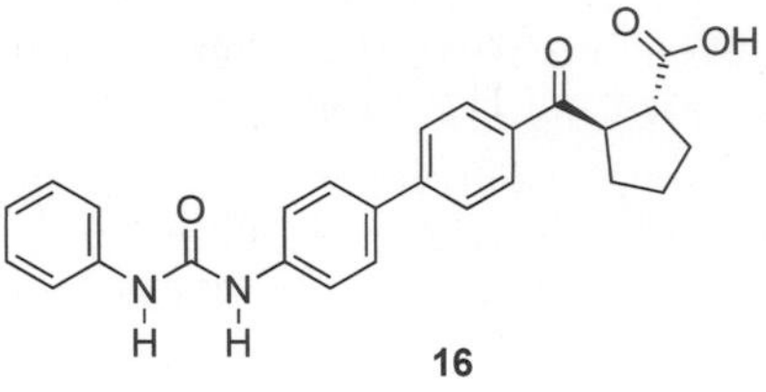

16 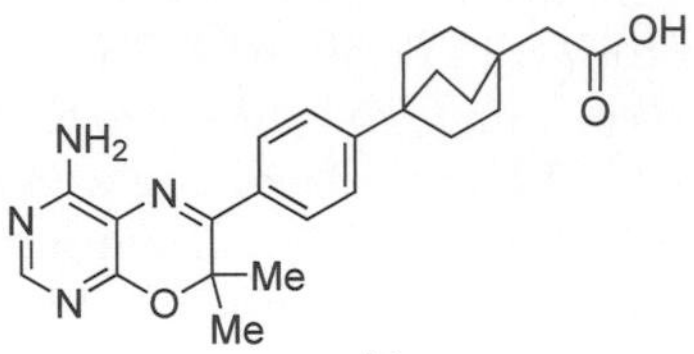   **17**

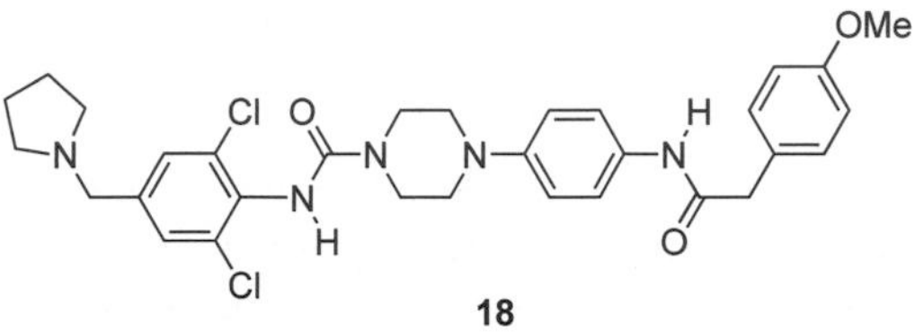

18 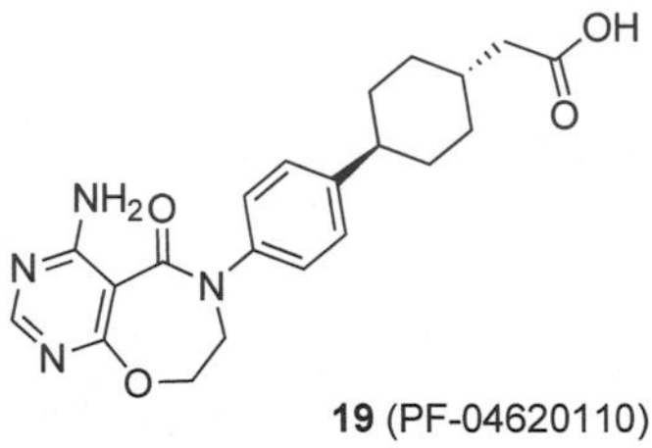   **19** (PF-04620110)

An HTS followed by lead optimization of a resulting hit afforded a novel structural motif represented by ureido piperazine **18** that was potent in a biochemical assay ($IC_{50} = 20$ nM, recombinant human DGAT1) as well as in a cell-based assay ($IC_{50} = 80$ nM, DGAT1 expressed in 3T3-L1 cells) [50]. Compound **18** reduced adiposity and improved insulin sensitivity based on an oral glucose tolerance test after chronic administration to C56Bl6 mice maintained on a high-fat diet. However, untoward side effects of sebaceous gland atrophy and hair loss were observed. In a separate study, compound **18** decreased plasma TG levels and increased circulating GLP-1 levels in dogs after acute oral administration [50].

To date, one small-molecule DGAT inhibitor PF-04620110 (**19**) is reported to be undergoing clinical evaluation for treatment of type 2 diabetes. Compound **19** is a potent and selective DGAT1 inhibitor ($IC_{50} = 19$ nM, recombinant human DGAT1; $IC_{50} = 28$ nM, DGAT1 expressed in HT-29 cells; DGAT2 $IC_{50} > 10 \mu$M) and was well tolerated in Phase I clinical trials [51].

4. MICROSOMAL TRIGLYCERIDE TRANSFER PROTEIN

MTP is responsible for the transfer of TGs and cholesteryl esters from the endoplasmic reticulum (ER) to lipoprotein particles (VLDL in hepatocytes in the liver and chylomicrons in endocytes in the intestine) for secretion [52]. It is a heterodimer consisting of a unique large subunit essential for lipid transfer encoded by the *mttp* gene and a smaller subunit, the ubiquitous ER enzyme protein disulfide isomerase [53].

Abetalipoproteinemia or Bassen–Kornzweig syndrome, a potentially disabling, familial disease characterized by lack of plasma TGs, malabsorption of fat-soluble vitamins, liver steatosis, steatorrhea, and other symptoms, is linked to mutations in the MTP functional subunit [52,53].

This phenotype has been recapitulated with early systemic MTP inhibitors such as CP-346,086 (20). Despite early clinical success in lowering TGs and low-density lipoprotein (LDL) while not affecting HDL levels [54], CP-346,086 was later withdrawn due to safety issues including fat accumulation in the liver. Endocyte-specific MTP inhibitors such as dirlotapide (21, CP-472033) sold under the name Slentrol® and mitratapide (22, R-103,757) sold under the name Yarvitan® have been approved by the US and EU regulatory agencies, respectively, for veterinary use against obesity in dogs [55–58]. A related MTP inhibitor, CP-741952 (structure not disclosed), was withdrawn from human Phase II trials in 2008 because of elevation of liver enzymes and high liver fat content, which are likely to be mechanism based [59].

Currently, there are a number of systemic and intestine-selective MTP inhibitors, including lomitapide (23, BMS-201038, AEGR-733), implitapide (24), JTT-130, SLx-4090, and R-256918 (latter three structures not disclosed) believed to be in active development [60]. In a meta-analysis of three Phase II clinical trials, lomitapide as monotherapy or in combination with ezetimibe, atorvastatin, or fenofibrate significantly reduced LDL cholesterol (up to 35% as monotherapy and 66% in combination with atorvastatin) and was well tolerated with less than 2% discontinuation due to abnormal liver function [61]. Lomitapide has also been granted orphan drug status for the treatment of homozygous familial hypercholesterolemia [59]. Results of a Phase II study of JTT-130 for type 2 diabetes are expected in August 2010 [59,60].

In a recent paper, a series of benzothiazole-based intestine-selective dirlotapide analogs was reported [62]. Benzothiazole 25 is a potent MTP inhibitor with IC_{50} values of 6.2 nM in HEPG2 cells and 11.2 nM in an enzymatic assay. In a 28-day study in DIO mice, 25 reduced food intake and body weight gain, increased circulating levels of the satiety protein PYY, and lowered blood glucose and TG levels. Extensive rat PK studies showed low systemic exposure following oral administration of 25 ($F = 1.3\%$).

There is emerging evidence that VLDL assembly in general and MTP function in particular are intimately tied to assembly of hepatitis C virus (HCV) infectious particles [63]. MTP inhibition and ApoB downregulation in turn inhibit HCV assembly and maturation [64]. This may open new therapeutic opportunities for small-molecule MTP inhibitors in the field of HCV antiviral research.

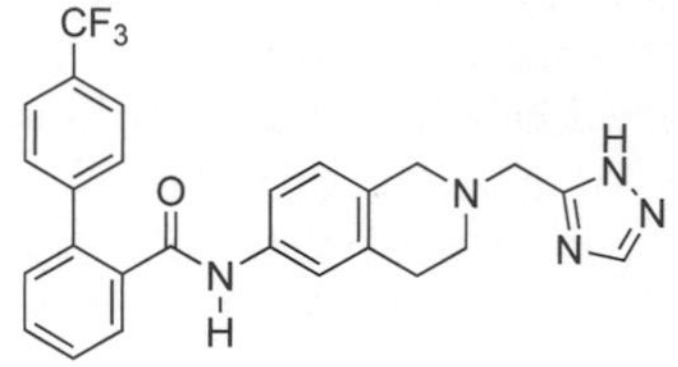

20 (CP-346,086)

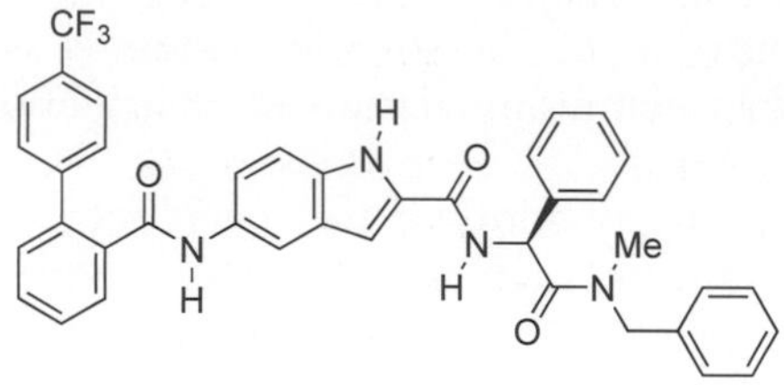

21 (Dirlotapide, CP-742,033)

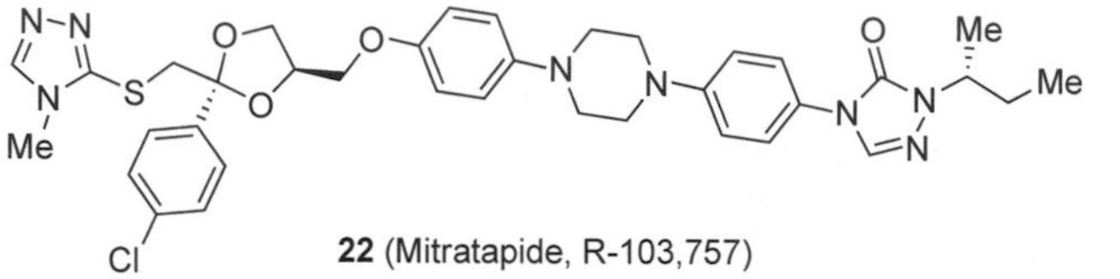

22 (Mitratapide, R-103,757)

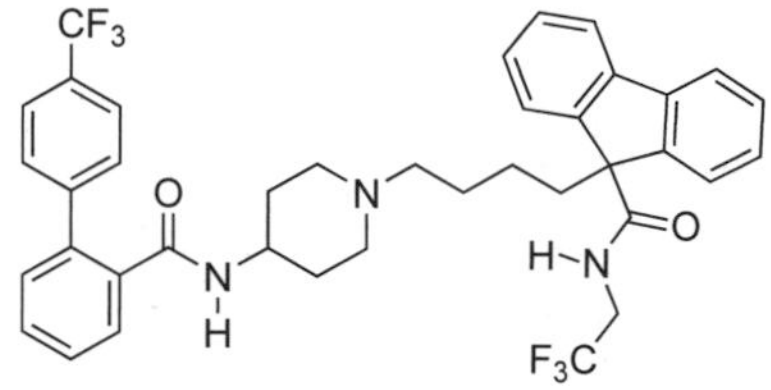

23 (Lomitapide, BMS-201,038, AEGR-733)

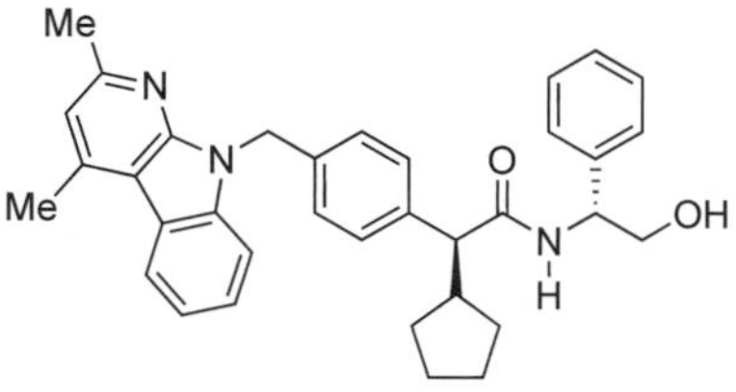

24 (Implitapide, BAY-13-9952)

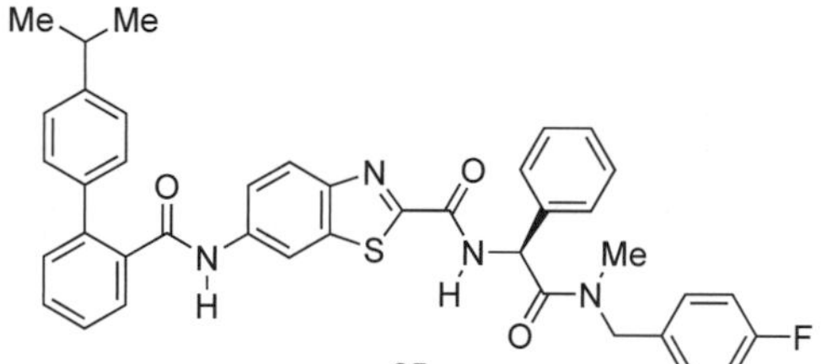

25

5. CONCLUSION

Enzymes involved in TG synthesis continue to represent challenging and intriguing targets for small-molecule intervention against the worldwide epidemic of obesity, metabolic syndrome, and type 2 diabetes, as well as smaller market indications from rare familial diseases to veterinary use. There is also intriguing evidence that inhibition of these enzymes may be beneficial for diseases unrelated to the area of metabolic disorders, such as hepatitis C virus and cancer. The challenges of developing drugs that act on interconnected life-sustaining metabolic pathways are being met with next-generation medicines that are targeted toward individual tissues and organs. Over the next several years, longer term safety and efficacy profiles of intestine-selective MTP inhibitors such as lomitapide (**23**), implitapide (**24**), and JTT-130, as well as the DGAT1 inhibitor PF-04620110 (**19**), and other small-molecule inhibitors of enzymes in the TG synthesis pathway are expected to emerge. These results, along with results from newer enzyme inhibitors in the TG synthesis pathway, will inform the viability of this approach for the treatment of type 2 diabetes, metabolic syndrome, and obesity. Based on information from Xenon Pharmaceuticals, as a result of their collaboration with Novartis, at least one small-molecule SCD inhibitor has been tested in humans, and resulted in a decrease in SCD1 enzyme activity and improvement in lipid profiles. The structure has not been disclosed and the current clinical status is unknown [65].

REFERENCES

[1] World Health Organization: Obesity and Overweight, http://www.who.int/dietphysicalactivity/publications/facts/obesity/en/(accessed July 2, 2010).
[2] A. Tirosh, I. Shai, R. Bitzur, I. Kochba, D. Tekes-Manova, E. Israeli, T. Shochat, and A. Rudich, *Diab. Care*, 2008, **31**, 2032.
[3] D. K. Bloomfield and K. Bloch, *Biochim. Biophys. Acta*, 1958, **30**, 220.
[4] D. K. Bloomfield and K. Bloch, *J. Biol. Chem.*, 1960, **235**, 337.
[5] J. M. Ntambi, S. A. Buhrow, K. H. Kaestner, R. J. Christy, E. Sibley, T. J. Kelly, and M. D. Lane, *J. Biol. Chem.*, 1988, **263**, 17291.
[6] J. Wang, L. Yu, R. E. Schmidt, C. Su, X. Huang, K. Gould, and G. Cao, *Biochem. Biophys. Res. Commun.*, 2005, **332**, 735.
[7] J. M. Ntambi, M. Miyazaki, J. P. Stoehr, H. Lan, C. Kendziorski, B. S. Yandell, Y. Song, P. Cohen, J. M. Friedman, and A. D. Attie, *Proc. Natl. Acad. Sci. USA*, 2002, **99**, 11482.
[8] P. Cohen, M. Miyazaki, N. D. Socci, A. Hagge-Greenberg, W. Liedtke, A. A. Soukas, R. Sharma, L. C. Hudgins, J. M. Ntambi, and J. M. Friedman, *Science*, 2002, **297**, 240.
[9] M. Miyazaki, W. C. Man, and J. M. Ntambi, *J. Nutr.*, 2001, **131**, 2260.
[10] M. Miyazaki, Y. C. Kim, and J. M. Ntambi, *J. Lipid Res.*, 2001, **42**, 1018.
[11] G. Jiang, Z. Li, F. Liu, K. Ellsworth, Q. Las-Yang, M. Wu, J. Ronan, C. Esau, C. Murphy, D. Szalkowski, R. Bergeron, T. Doebber, and B. B. Zhang, *J. Clin. Invest.*, 2005, **115**, 1030.
[12] J. M. Ntambi and M. Miyazaki, *Prog. Lipid Res.*, 2004, **43**, 91.
[13] A. Dobrzyn and J. M. Ntambi, *Prostaglandins Leukot. Essent. Fatty Acids*, 2005, **73**, 35.
[14] A. Dobrzyn and J. M. Ntambi, *Obes. Rev.*, 2005, **6**, 169.
[15] M. Miyazaki, A. Dobrzyn, H. Sampath, S. H. Lee, W. C. Man, K. Chu, J. M. Peters, F. J. Gonzalez, and J. M. Ntambi, *J. Biol. Chem.*, 2004, **279**, 35017.
[16] M. Miyazaki, M. T. Flowers, H. Sampath, K. Chu, C. Otzelberger, X. Liu, and J. M. Ntambi, *Cell Metab.*, 2007, **6**, 484.
[17] J. B. Flowers, M. E. Rabaglia, K. L. Schueler, M. T. Flowers, H. Lan, M. P. Keller, J. M. Ntambi, and A. D. Attie, *Diabetes*, 2007, **56**, 1228.
[18] Y. Zheng, K. J. Eilertsen, L. Ge, L. Zhang, J. P. Sundberg, S. M. Prouty, K. S. Stenn, and S. Parimoo, *Nat. Genet.*, 1999, **23**, 268.

[19] J. M. Brown, S. Chung, J. K. Sawyer, C. Degirolamo, H. M. Alger, T. Nguyen, X. Zhu, M. N. Duong, A. L. Wibley, R. Shah, M. A. Davis, K. Kelley, M. D. Wilson, C. Kent, J. S. Parks, and L. L. Rudel, *Circulation*, 2008, **118**, 1467.

[20] J. M. Brown, S. Chung, J. K. Sawyer, C. Degirolamo, H. M. Alger, T. Nguyen, X. Zhu, M. N. Duong, A. L. Brown, C. Lord, R. Shah, M. A. Davis, K. Kelley, M. D. Wilson, J. Madenspacher, M. B. Fessler, J. S. Parks, and L. L. Rudel, *Atheroscler. Thromb. Vasc. Biol.*, 2010, **30**, 24.

[21] M. Abreo, M. Chafeev, N. Chakka, S. Chowdhury, J. Fu, H. W. Gschwend, M. W. Holladay, D. Hou, R. Kamboj, V. Kodumuru, W. Li, S. Liu, V. Raina, S. Sun, S. Sun, S. Sviridov, C. Tu, M. D. Winther, and Z. Zhang, *Patent Application WO 2005/011655*, 2005.

[22] Z. Xin, H. Zhao, M. D. Serby, B. Liu, M. Liu, B. G. Szczepankiewicz, L. T. J. Nelson, H. T. Smith, T. S. Suhar, R. S. Janis, N. Cao, H. S. Camp, C. A. Collins, H. L. Sham, T. K. Surowy, and G. Liu, *Bioorg. Med. Chem. Lett.*, 2008, **18**, 4298.

[23] G. Liu, Z. Xin, H. Zhao, M. D. Serby, H. T. Smith, N. Cao, T. K. Surowy, A. Adler, A. Mika, T. B. Farb, C. Keegan, K. Landschulz, M. Brune, C. A. Collins, H. L. Sham, and H. S. Camp, Abstracts of Papers, 233rd ACS National Meeting, Chicago, IL, United States, March 25–29, 2007, MEDI 232.

[24] Y. K. Ramtohul, C. Black, C.-C. Chan, S. Crane, J. Guay, S. Guiral, Z. Huang, R. Oballa, L.-J. Xu, L. Zhang, and C. S. Li, *Bioorg. Med. Chem. Lett.*, 2010, **20**, 1593.

[25] S. Léger, W. C. Black, D. Deschenes, S. Dolman, J.-P. Falgueyret, M. Gagnon, S. Guiral, Z. Huang, J. Guay, Y. Leblanc, C. S. Li, F. Massé, R. Oballa, and L. Zhang, *Bioorg. Med. Chem. Lett.*, 2010, **20**, 499.

[26] Y. Uto, T. Ogata, Y. Kiyotsuka, Y. Ueno, Y. Miyawaza, H. Kurata, T. Deguchi, M. Yamada, N. Watanabe, M. Konishi, N. Kurikawa, T. Takagi, S. Wakimoto, and J. Osumi, *Bioorg. Med. Chem. Lett.*, 2010, **20**, 341.

[27] Y. Uto, Y. Kiyotsuka, Y. Ueno, Y. Miyawaza, H. Kurata, T. Ogata, T. Deguchi, N. Watanabe, M. Konishi, R. Okuyama, N. Kurikawa, T. Takagi, S. Wakimoto, K. Kono, and J. Osumi, *Bioorg. Med. Chem. Lett.*, 2010, **20**, 746.

[28] D. O. Koltun, E. Q. Parkhill, N. I. Vasilevich, A. I. Glushkov, T. M. Zilbershtein, A. V. Ivanov, A. G. Cole, I. Henderson, N. A. Zautke, S. A. Brunn, N. Mollova, K. Leung, J. W. Chisholm, and J. Zablocki, *Bioorg. Med. Chem. Lett.*, 2009, **19**, 2048.

[29] D. O. Koltun, N. I. Vasilevich, E. Q. Parkhill, A. I. Glushkov, T. M. Zilbershtein, E. I. Mayboroda, M. A. Boze, A. G. Cole, I. Henderson, N. A. Zautke, S. A. Brunn, N. Chu, J. Hao, N. Mollova, K. Leung, J. W. Chisholm, and J. Zablocki, *Bioorg. Med. Chem. Lett.*, 2009, **19**, 3050.

[30] D. O. Koltun, T. M. Zilbershtein, V. A. Migulin, N. I. Vasilevich, E. Q. Parkhill, A. I. Glushkov, M. J. McGregor, S. A. Brunn, N. Chu, J. Hao, N. Mollova, K. Leung, J. W. Chisholm, and J. Zablocki, *Bioorg. Med. Chem. Lett.*, 2009, **19**, 4070.

[31] E. Q. Parkhill, D. O. Koltun, S. A. Brunn, N. Mollova, K. Leung, J. W. Chisholm, and J. Zablocki, Abstracts of Papers, 238th ACS National Meeting, Washington, DC, United States, August 16–20, 2009, MEDI 147.

[32] Y. Uto, T. Ogata, J. Harada, Y. Kiyotsuka, Y. Ueno, Y. Miyawaza, H. Kurata, T. Deguchi, N. Watanabe, T. Takagi, S. Wakimoto, R. Okuyama, M. Abe, N. Kurikawa, S. Kawamura, M. Yamato, and J. Osumi, *Bioorg. Med. Chem. Lett.*, 2009, **19**, 4151.

[33] Y. Uto, T. Ogata, Y. Kiyotsuka, Y. Miyawaza, Y. Ueno, H. Kurata, T. Deguchi, M. Yamada, N. Watanabe, T. Takagi, S. Wakimoto, R. Okuyama, M. Konishi, N. Kurikawa, K. Kono, and J. Osumi, *Bioorg. Med. Chem. Lett.*, 2009, **19**, 4159.

[34] N. Scaglia, J. W. Chisholm, R. A. Igal, *PLoS ONE*, 2009, **4**, e6812.

[35] V. A. Zammit, L. K. Buckett, A. V. Turnbull, H. Wure, and A. Proven, *Pharmacol. Ther.*, 2008, **188**, 295.

[36] J. Cao, L. Cheng, and Y. Shi, *J. Lipid Res.*, 2007, **48**, 583.

[37] S. Cases, S. J. Smith, Y.-W. Zheng, H. M. Myers, S. R. Lear, E. Sande, S. Novak, C. Collins, C. B. Welch, A. J. Lusis, S. K. Erikson, and R. V. Farese, Jr., *Proc. Natl. Acad. Sci. USA*, 1998, **95**, 13018.

[38] K. K. Buhman, S. J. Smith, S. J. Stone, J. J. Repa, J. S. Wong, F. F. Knapp, Jr., B. J. Burri, R. L. Hamilton, N. A. Abumrad, and R. V. Farese, Jr., *J. Biol. Chem.*, 2002, **277**, 25474.

[39] S. Cases, S. J. Stone, P. Zhou, E. Yen, B. Tow, K. D. Lardizabal, T. Voelker, and R. V. Farese Jr., *J. Biol. Chem.*, 2001, **276**, 38870.

[40] S. J. Smith, S. Cases, D. R. Jensen, H. C. Chen, E. Sande, B. Tow, D. A. Sanan, J. Raber, R. J. Eckel, and R. V. Farese, Jr., *Nat. Genet.*, 2000, **87**, 90.

[41] H. C. Chen, S. J. Smith, Z. Ladha, D. R. Jensen, L. D. Ferreira, L. K. Pulawa, J. G. McGuire, R. E. Pitas, R. J. Eckel, and R. V. Farese, Jr., *J. Clin. Invest.*, 2002, **109**, 1049.

[42] H. C. Chen and R. V. Farese, Jr., *Atheroscler. Thromb. Vasc. Biol.*, 2005, **25**, 482.

[43] H. C. Chen, *Nutr. Metab.*, 2006, **3**, 10.

[44] S. J. Stone, H. M. Myers, S. M. Watkins, B. E. Brown, K. R. Feingold, P. M. Elias, and R. V. Farese, Jr., *J. Biol. Chem.* 2004, **279**, 11767.

[45] X. X. Yu, S. F. Murray, S. K. Pandey, S. L. Booten, D. Bao, X. Z. Song, S. Kelly, S. Chen, R. McKay, B. P. Monia, and S. Bhanot, *Hepatology*, 2005, **42**, 362.

[46] B. M. Fox, N. Furukawa, X. Hao, K. Ho, T. Inaba, S. Jackson, F. Kayser, M. Labelle, K. Li, T. Matsui, D. L. McMinn, N. Ogawa, S. M. Rubenstein, S. Sagawa, K. Sugimoto, M. Suzuki, M. Tanaka, G. Ye, M. Yoshida, and J. Zhang, *Patent Application WO 2004/047755*, 2004.

[47] R. Smith, A.-M. Campbell, P. Coish, M. Dai, S. Jenkins, D. Lower, S. O'Connor, G. Wang, M. Zhang, and L. Z. Zhu, *Patent Application US 2006/7091288*, 2006.

[48] G. Zhao, A. J. Souers, M. Voorbach, H. D. Falls, B. Droz, S. Brodjian, Y. Y. Lau, R. R. Iyengar, J. Gao, A. S. Judd, S. H. Wagaw, M. M. Ravn, K. M. Engstrom, J. K. Lynch, M. M. Mulhern, J. Freeman, B. D. Dayton, X. Wang, N. Grihalde, D. Fry, D. W. A. Beno, K. C. Marsh, Z. Su, G. J. Diaz, C. A. Collins, H. Sham, R. M. Reilly, M. E. Brune, and P. R. Kym, *J. Med. Chem.*, 2008, **51**, 380.

[49] A. M. Birch, S. Birtles, L. K. Buckett, P. D. Kemmitt, G. J. Smith, T. J. D. Smith, A. V. Turnbull, and S. J. Y. Wang, *J. Med. Chem.*, 2009, **52**, 1558.

[50] J. T. M. Linders, P. Roevens, R. Berwaer, S. Boeckx, J.-P. Bongartz, H. Borghys, C. Buyck, E. Coesemans, P. V. Davidenko, R.A.H.J. Gilissen, T. Govaerts, P. Haspeslagh, B. Hrupka, A. Klochkova, K. Kolodziejczyk, G. Van Lommen, J. Peeters, K. De Waepenaert, and G. H. M. Willemsens, Abstracts of Papers, 238th ACS National Meeting, Washington, DC, United States, August 16—20, 2009, MEDI 372.

[51] R. L. Dow, J.-C. Li, L. Patel, C. Perreault, M. J. Munchhof, D. W. Piotrowski, E. M. Gibbs, W. J. Zavadoski, T. B. Manion, J. L. Treadway, and J. L. LaPerle, Abstracts of Papers, 239th ACS National Meeting, San Francisco, CA, United States, March 20–24, 2010, MEDI 315.

[52] C. C. Shoulders, D. J. Brett, J. D. Bayliss, T. M. Narcisi, A. Jarmuz, T. T. Grantham, P. R. Leoni, S. Bhattacharya, R. J. Pease, P. M. Cullen, S. Levi, P. G. H. Byfield, P. Purkiss, and J. Scott, *Hum. Mol. Genet.*, 1993, **2**, 2109.

[53] M. M. Hussain, J. Shi, and P. Dreizen, *J. Lipid Res.*, 2003, **42**, 22.

[54] C. E. Chandler, D. E. Wilder, J. L. Pettini, Y. E. Savoy, S. F. Petras, G. Chang, J. Vincent, and H. J. Harwood, Jr., *J. Lipid Res.*, 2003, **44**, 1887.

[55] Freedom of Information Summary is available from FDA at http://www.fda.gov/downloads/AnimalVeterinary/Products/ApprovedAnimalDrugProducts/FOIA-DrugSummaries/UCM050063.pdf (accessed July 2, 2010).

[56] Public Assessment Report is available from EMEA at http://www.ema.europa.eu/vetdocs/PDFs/EPAR/yarvitan/V-113-en1.pdf (accessed July 2, 2010).

[57] J. Li, B. S. Bronk, J. P. Dirlam, A. E. Blize, P. Bertinato, B. H. Jaynes, A. Hickman, C. Miskell, U. A. Pillai, J. S. Tibbitts, M. L. Haven, N. L. Kolosko, C. J. Barry, and T. B. Manion, *Bioorg. Med. Chem. Lett.*, 2007, **17**, 1996.

[58] B. Dobenecker, M. De Bock, M. Engelen, L. Goossens, A. Scholz, and E. Kienzle, *Vet. Res. Commun.*, 2009, **33**, 839.
[59] Source: FDA, www.clinicaltrials.gov
[60] Source: Prous Integrity, integrity.prous.com
[61] D. J. Rader, W. J. Sasiela, M. Davidson, H. E. Bays, M. Parris, E. Meagher, C. Price, and S. L. Schwartz, 69th Annual Meeting Scientific Session of the American Diabetes Association. (ADA), New Orleans, LA, June 5–9, Abstract 471-P.
[62] C. B. Vu, J. C. Milne, D. P. Carney, J. Song, W. Choy, P. D. Lambert, D. J. Gagne, M. Hirsch, A. Cote, M. Davis, E. Lainez, N. Meade, K. Normington, M. R. Jirousek, and R. B. Perni, *Bioorg. Med. Chem. Lett.*, 2009, **19**, 1416.
[63] C.-I. Popescu and J. Dubuisson, *Biol. Cell.*, 2010, **102**, 63.
[64] P. Gastaminza, G. Cheng, S. Wieland, J. Zhong, W. Liao, and F. V. Chisari, *J. Virology*, 2008, **82**, 2120.
[65] Source: Xenon Pharmaceuticals, http://www.xenon-pharma.com/pages/index_obesity.php (accessed July 19, 2010).

CHAPTER 8

Inhibitors of HIF Prolyl Hydroxylases

Michael H. Rabinowitz, Terrance D. Barrett, Mark D. Rosen and **Hariharan Venkatesan**

1. INTRODUCTION

The ability of animals to detect, absorb, transport, and enzymatically harness the oxidizing potential of molecular oxygen is a crucial factor in their development and evolution, and represents fundamental processes of life. From cellular respiration and electron transport to the anabolism of primary and secondary metabolites and the catabolism of nutrients and drugs, oxidative processes linked to O_2 utilization are some of the most important chemical reactions of life. Because of this, the body's ability to sense and adjust to changing O_2 levels is tightly regulated on the molecular level. The most significant regulator of O_2 sensing and homeostasis is hypoxia inducible factor (HIF) [1], which is a transcription

Johnson & Johnson Pharmaceutical Research and Development, L.L.C., 3210 Merryfield Row, San Diego, CA 92121, USA

Annual Reports in Medicinal Chemistry, Volume 45
ISSN 0065-7743, DOI 10.1016/S0166-526X(10)56008-3

factor discovered in 1992 [2] and subsequently found to be tightly regulated itself via degradation initiated by prolyl hydroxylase (PHD) enzymes [3–6]. HIF is directly responsible for the transcription of hundreds of gene products involved in respiration, metabolism, angiogenesis, erythropoiesis, and many other functions at the cellular and organ levels. HIF-α, the regulatory subunit of the HIF dimer, is constitutively produced with a half-life of approximately 5 minutes, being degraded as it is produced by enzymatic hydroxylation on proline residues leading to recognition of the hydroxyl form by the von Hippel–Lindau tumor suppressor protein (pVHL) and subsequent ubiquitin-mediated proteasomal destruction.

The primary enzymes responsible for the hydroxylation of HIF are the three PHD isoforms PHD1, PHD2, and PHD3 (alternatively named EGLN2, EGLN1, and EGLN3, respectively). PHDs are members of the α-keto acid-dependent nonheme iron-containing family of hydroxylases. These dioxygenases require O_2, iron, and an α-keto carboxylic acid, in this case, 2-oxoglutarate (2-OG) for activity, and mediate the C4 *trans* hydroxylation of HIF-α at Pro402 and Pro564 initiating the path to protein degradation. The K_m values of all HIF PHDs studied to date are in the range of physiological O_2 concentrations (ca. 200 µM). Thus, small decreases in physiological O_2 concentrations lead to large reductions in enzyme reaction rate and consequent buildup or "stabilization" of the constitutively produced HIF-α [7].

Studies of human genetics have demonstrated that PHD enzymes play key roles in regulating erythropoiesis as evidenced by the observation that families with loss of function mutations in PHD2 have elevated blood hemoglobin [8]. It is evident that pharmacological intervention in such a system will have profound implications in physiology and pathophysiology and underpins an important new druggable class of therapeutic protein targets. Moderation of PHD activity with small-molecule inhibitors has already led to the introduction of four compounds into clinical trials for the treatment of various anemias. This review will attempt to broadly characterize the classes of small-molecule PHD inhibitors and summarize their biological profiles.

2. OXYGEN SENSING AND REGULATION VIA PROLYL HYDROXYLASES

2.1 HIF as master regulator of hypoxia responsive gene expression

Acutely, cells respond to low oxygen concentrations with changes in the function of ion channels and enzymes, and, in the case of enzymes involved in electron transport, the generation of reactive oxygen species.

Chronically, cells adapt to hypoxia by changing the regulation of genes that function to make cells more resistant to the extreme conditions of hypoxia and to restore oxygenation to tissues. HIF-α mediates its transcriptional response by forming a heteromeric protein complex that includes HIF-β and p300. The complex binds to the hypoxia response element (HRE) consensus sequences (A/(G)CGTG) in the promoter region of hypoxia responsive genes and regulates gene expression [1,2]. Three HIF-α genes have been identified to date: HIF-1α and HIF-2α have similar but nonredundant gene targets [9]. HIF-3α appears to suppress HIF-1α and HIF-2α gene transcription but much less is known about HIF-3α. It is estimated that up to 1% of the mammalian genome is under the regulation of HIF and stabilization of this protein is believed to represent a physiological response to changes in oxygen concentrations and thus promote a more "orchestrated" improvement in the body's oxygen-carrying capacity than would single agent therapy, that is, erythropoietin.

2.2 Distribution of HIF and PHD subtypes

PHD1, PHD2, and PHD3 differ in their cellular and tissue localization as well as in their physiological role. Fusion of PHDs to fluorescent proteins shows that PHD1 is exclusively located in the nucleus, PHD2 is found mainly in the cytoplasm, and PHD3 is homogeneously distributed in the cytoplasm and nucleus [10]. The physiological role of the PHDs in regulating cellular HIF-α content has been evaluated by various means. It is currently believed that while PHD2 plays a dominant role in controlling the cellular HIF-α levels, PHD1 and PHD3 have more subtle roles [11]. Additional evidence for a dominant role of PHD2 in hematopoiesis is provided by the observation that loss of function mutations in the PHD2 gene in humans is associated with familial polycythemia [8]. Studies in knockout mice also corroborate the central role of PHD2 in determining cellular HIF content as PHD2 knockout mice die during embryonic development due to defects in hematopoiesis and formation of the vasculature [12,13]. In contrast, PHD1 and PHD3 knockout mice are viable and have more subtle phenotypes [14].

3. STRUCTURE OF PHD ENZYMES AND CATALYTIC MECHANISM

The catalytic mechanism of 2-OG-dependent dioxygenases has been thoroughly reviewed elsewhere [15,16]. While full-length PHD enzymes have yet to succumb to crystallization efforts, the x-ray structures of two catalytically active, truncated PHD2 enzymes (PHD2$_{181-426}$ and PHD2$_{181-417}$) in complex with Fe(II) and isoquinoline inhibitor and 2-OG mimetic **1** were reported in 2006 [17]. An additional PHD2 cocrystal structure with isoquinoline inhibitor **2** was deposited in the RCSB protein data bank in 2006 [18]. The three structures were nearly identical with regard to the arrangement of active site residues. Briefly, these structures show the active site to contain a single octahedral Fe atom coordinated by the facial triad of His313, His375, and Asp315 (Figure 1). The iron atom is also ligated by the inhibitor in a bidentate manner through the isoquinoline ring nitrogen atom and the oxygen atom of the amide group, while a water molecule occupies the final apical coordination site. The carboxylate side chain of **1** resides in a nearby pocket where it participates in electrostatic and H-bonding interactions with Arg383 and Tyr329, respectively. Lastly, the phenolic hydroxyl group of **1** and **2** exhibits H-bonding with the Tyr303 residue, which is the only direct interaction observed between these inhibitors and a noncatalytically essential residue. As a corollary to these studies, a recent report provides ^{19}F and ^{13}C NMR spectroscopic evidence that in the case of related azaquinolone PHD2 inhibitors, two distinct binding poses may exist in solution although only a single pose is observed by x-ray crystallography [19].

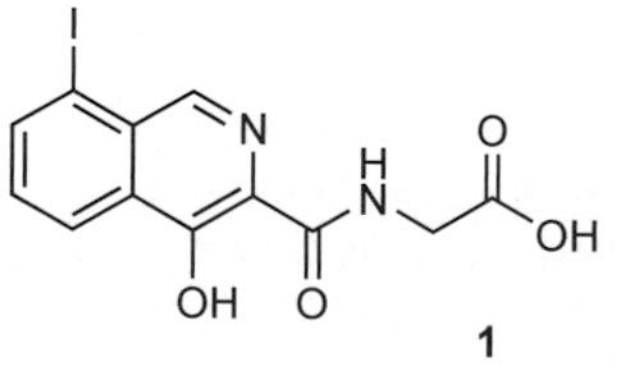

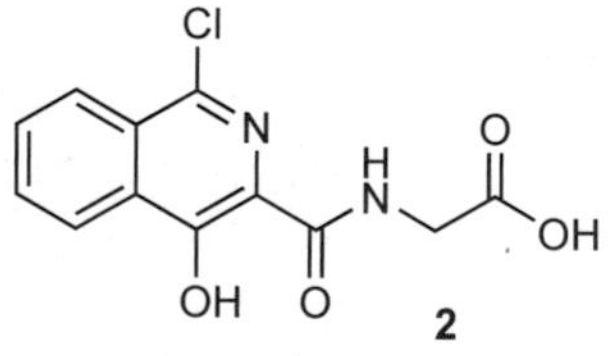

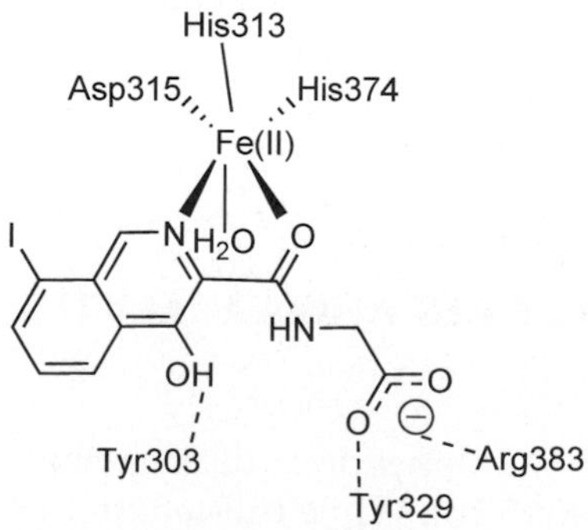

Figure 1 Schematic diagram of key PHD2/**1** interactions based on x-ray cocrystal.

4. INHIBITORS

For the purposes of this review, we will attempt to classify inhibitors of 2-OG dioxygenases into either one of two categories: general iron chelators or structural mimics of 2-OG. While both classes of inhibitors functionally block the activity of these enzymes, typically in a dose-dependent manner, only the latter are viewed as competitive inhibitors possessing enough target protein selectivity to act as relatively nontoxic, potent therapeutics. Furthermore, due to the limited knowledge of the mode of enzyme binding, if any, of the iron chelators, we will attempt to classify compounds as noncompetitive if they contain iron chelation motifs and are not either direct analogs of 2-OG or derived from analogs of 2-OG, although ensuing studies may prove otherwise.

4.1 Noncompetitive inhibitors

Compounds that strongly chelate iron have been known for many years to stabilize HIF-1α as well as upregulate proteins involved in red blood cell production erythropoietin (EPO), angiogenesis, vascular endothelial growth factor (VEGF), and iron transport. Some, but not all, of the pharmacological actions of iron chelators are produced by inhibition of PHD enzymes resulting in elevation of cellular HIF content. The action of selected iron chelators as they relate to PHD inhibition are briefly summarized here.

Deferoxamine (DFO, desferrioxamine), a bacterial siderophore isolated from *Streptomyces* is a potent multidentate iron binder and has been used clinically for the treatment of iron overload and toxicity. Prior to the discovery of the PHD enzymes, DFO was found to increase EPO RNA and HIF-1α levels in Hep3B cells [20]. Additionally, DFO administered intraperitoneally (i.p.) to female mice at a dose of 200 mg/kg showed an increase in EPO mRNA in the kidneys after 22 h, but this effect was transient and lost upon chronic (5 day) dosing. Acclimatization to hypoxia is known to protect the heart from damage due to ischemia/reperfusion (I/R) injury. In rats, DFO has been shown, via HIF-1α stabilization, to simulate hypoxia and protect cardiomyocyte function after I/R [21]. High blood glucose has been shown to produce a reactive metabolite, methylglyoxal, responsible for impairment of HIF-1α binding to the transcription factor p300, thus reducing gene transcription in diabetic models [22]. DFO has been shown to correct the transcription-suppressing effects of high glucose in diabetic mice, restoring neovascularization and efficient wound healing [23]. In addition, in cell culture as well as in Sprague–Dawley rats, iron chelation by DFO was shown to increase glucose uptake and increase insulin receptor activity and signaling in hepatocytes [24].

In a recent patent application, mice treated with a related iron chelator, deferasirox (DFS), showed reduced body weight while on a high fat diet compared to untreated controls. Additionally, DFS was claimed to improve whole body metabolism and energy expenditure as measured by increased O_2 consumption and CO_2 production as well as a reduction in white adipose and visceral fat, despite little difference between food intake in the control and treated animal groups [25].

3,4-Dihydroxybenzoic acid (DHB) is also a commonly used tool to measure the pharmacological effects of HIF-1α stabilization via PHD inhibition. Recently, it was shown that mice pretreated with DHB (100 mg/kg, i.p.) showed a marked resistance to the neurotoxic effects of 1-methyl-4-phenyl-1,2,3,6-tetrahydropyridine (MPTP) via protection of dopaminergic cell loss and striatal denervation. Importantly, this protection was seen to coincide with HIF-1α stabilization, and the prevention of the MPTP-induced loss of ferroportin and striatal iron. Additionally, in these studies, DHB was also observed to block MPTP-induced reduction in mitochondrial pyruvate dehydrogenase, at both the mRNA level and through the measurement of enzyme activity in midbrain substantia nigra [26].

Flavonoids containing putative iron-chelating moieties such as quercetin have also been shown to stabilize intracellular HIF-1α presumably via PHD inhibition. The authors suggest that this property of certain flavonoids is dependent both on their lipophilicity as well as the presence of high-Fe-affinity catechol and/or pyrogallol groups [27]. Interestingly, owing to the polypharmacology seen with many flavonoids, the HIF-1α elevation observed with these compounds in HeLa cells did not produce a correspondingly robust increase in transcriptional activation, and was shown instead to be a result of additional activity of certain flavonoids on kinase inhibition. However, in another publication, quercetin was shown to not only stabilize HIF-1α and HIF-2α, but to also produce a robust, time-dependent increase in VEGF in the human adenocarcinoma prostate cell line, DU145 [28].

It is well known that 1,10-phenanthrolines are highly active iron-chelating agents. The parent compound itself has recently been shown to increase HIF-1α levels in ocular tissue and to suppress O_2-mediated epithelial cell proliferation when administered to mice [29]. A quantitative assay was developed to measure transcriptional potency of certain HIF stabilizers via an HRE-mediated β-lactamase production in which the EC_{50} of 1,10-phenanthroline was measured to be approximately 8 μM. In addition, VEGF was dose-dependently produced in mouse embryonic fibroblasts by 1,10-phenanthroline with an EC_{50} of approximately 10 μM [30]. It is tempting to believe that all iron chelators of this class act solely through noncompetitive enzyme inhibition. However it has been suggested, through a combination of molecular modeling and chemical synthesis, that in the case of the nonheme iron- and 2-OG-dependent asparaginyl hydroxylase FIH, 4-oxo-1,4 dihydro [1,10]phenanthroline 3 carboxylic acid may bind in the enzyme active site through weak carboxylate ligation to Fe [31].

4.2 2-Oxoglutarate analogs

The majority of small-molecule PHD inhibitors in both the recent patent and peer-reviewed literature are structural analogs of 2-OG, many designed with the aid of computational docking techniques employing literature or novel x-ray cocrystal structures. The structure of 2-OG bound to certain dioxygenases is known [15] and more recently, that of 2-OG bound to PHD2 has been solved [32]. Nearly all chemotypes published take advantage of a carboxylate salt bridge interaction with Arg383 in the active site; however, a few analogs that are presumed active site competitive binders have appeared that do not contain the carboxylic acid functional group.

4.2.1 Noncarboxylic acids

A series of 113 quinazolinethione inhibitors of PHD (e.g., **3**) have been disclosed that do not offer the possibility of making a frank salt bridge interaction with Arg383 [33]. Claimed IC_{50} values for compounds in this series range from 1 to 100 nM in an SPA assay, with EC_{50} values for EPO release in Hep3B cells of 1–20 μM. Presumably, strong chelation to the active site Fe by the heterocycle and thione moieties compensates for the loss of the salt bridge binding energy. A pyrazolone [34] inhibitor series has recently been described. The IC_{50} (SPA assay) range for this series of over 180 compounds is given as 0.05–2.8 μM. Limited structure–activity relationship (SAR) data show activity with a variety of simple azoles at the C4 position with N-triazolyl and N-imidazoyl groups evidently preferred in the pyrazolone series (e.g., **4**, IC_{50} = 50 nM). Most of the SAR developed in this series relate to modification of the heteroaryl group at C2, which invariably is a substituted 2-pyridyl or 4-pyrimidyl ring. From this limited SAR it is likely that Fe chelation in this series results from the 2-pyridyl (or 4-pyrimidyl) group and N1 of the pyrazolone ring in its enol tautomeric form. Pyrazolethione (**5**) [35] and triazolones (structure not shown) [36] have been reported from the same group, but with very limited information on the activity of these analogs.

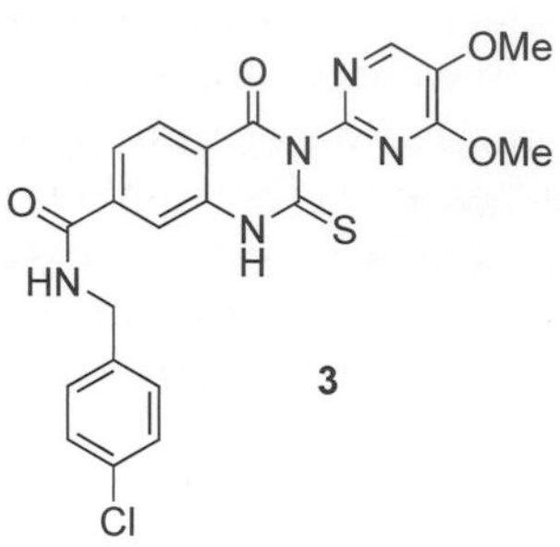

3

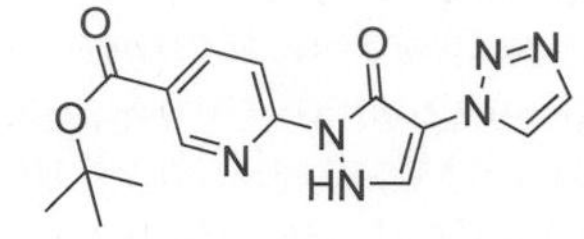

4

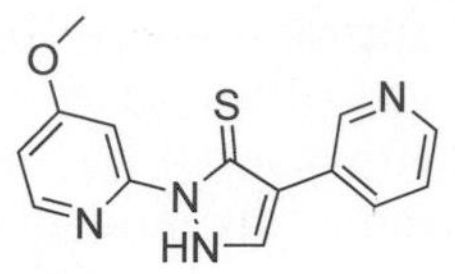

5

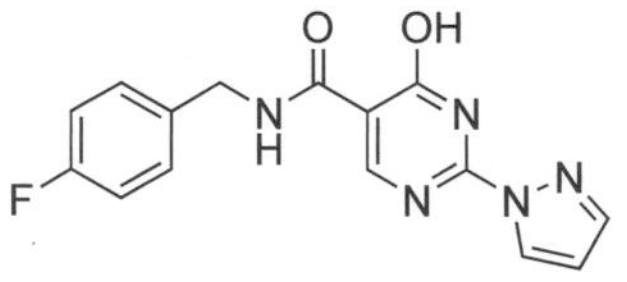

6

2-Pyrazolopyrimidines have also been recently disclosed [37] that take advantage of the Fe chelation by the 2-pyrazoloheterocycle amide isostere but, significantly, achieve potent inhibition ($IC_{50} < 10$ nM for 122 out of 192 examples; e.g., **6**). The authors also found that the pyrazolo group at the pyrimidine C2 position may be replaced with either 2-pyridyl or 3-pyridazinyl groups with retention of high potency in the enzyme assay.

4.2.2 Carboxylic acid glycinamides

Working with collagen PHDs, researchers found that N-oxalyl glycine (NOG, **7**) [38], a noncatalytically competent analog of 2-OG, competitively inhibited collagen PHD with a $K_i = 0.54$ µM. These authors also prepared analogs in which the carboxylic acid group alpha to the amide group was replaced by either a heterocyclic isoquinoline moiety or a cinnoline functionality [39]. Analogs **8** and **9** were shown to be inhibitors of collagen PHD with IC_{50} values of 0.02 and 0.19 µM, respectively.

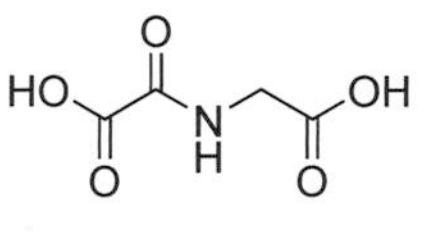

7

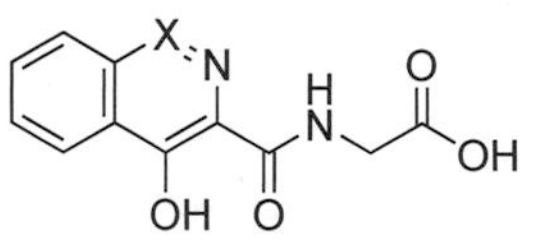

8: X = CH
9: X = N

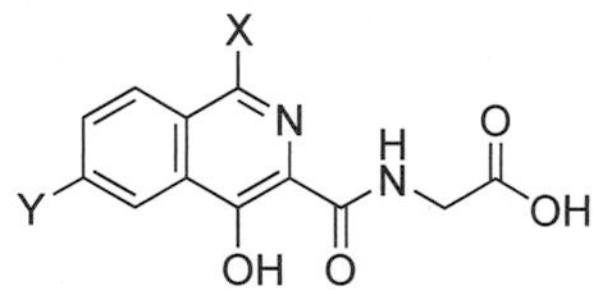

10: X = CN, Y = H
11: X = CN, Y = OPh
12: X = Cl, Y = OPh

Subsequently, in a patent application, isoquinoline glycinamides such as **2** were claimed as collagen PHD inhibitors [40], and an additional patent application claimed isoquinoline glycinamides as HIF PHD inhibitors [41].

It was reported that replacement of the chlorine atom present in 2 with a cyano group provided an improvement in plasma EPO levels in mice after intravenous dosing [42]. Analogs 10 and 11 containing the cyano group at the C-1 position provided fold changes of 4.7 and 449, respectively, in plasma EPO levels relative to their corresponding chloro analogs 2 and 12. Much of the recent research in this area has focused on heterocyclic modifications of the fused phenyl ring present in the iso-quinoline PHD inhibitors with thiophenes [43], thiazoles [44], pyrroles, oxazoles, isoxazoles [45], and isothiazoles [46]. Biological data were either not reported for these analogs or enzymatic activity expressed as percent inhibition at a certain concentration was reported.

Using published x-ray crystal structures [18], pyridine glycinamides were designed, synthesized, and tested in the enzymatic assay for PHD2 [47]. Although the parent molecule 13 was devoid of enzymatic activity, substitution at the 5-position with an aryl ring imparted micromolar activity (14, IC_{50} = 15 μM). Although 15 failed to inhibit PHD2, introduction of a hydroxyl group at the 3-position of the pyridine ring of this compound resulted in 16 (IC_{50} = 0.41 μM), which displayed submicromolar potency [47,48]. Interestingly, replacement of the carboxyl group of 16 with a methyl amide (17) resulted in only a 3-fold reduction in enzymatic potency. This molecule was reported to release VEGF from HEK293 cells with an EC_{50} of 1.4 μM and afforded an increase in plasma EPO levels in mice; however, no additional details were disclosed.

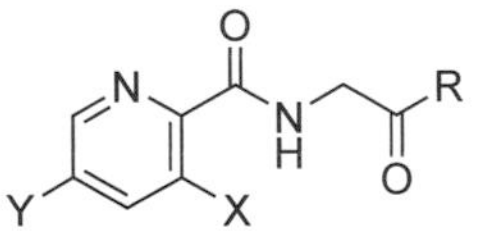

13: X = Y = H; R = OH
14: X = H, Y = p-ClPh, R = OH
15: X = H, Y = m-ClPh, R = OH
16: X = OH, Y = m-ClPh, R = OH
17: X = OH, Y = m-ClPh, R = NHMe

Quinolones and azaquinolones exemplified by general structure 18 have been claimed in multiple patent applications [49–53]. One group reported IC_{50} values in the range of 20 nM to 1 μM (18, Y = CH, X = NH) in a PHD3 assay and EC_{50} values of 1–20 μM in an ELISA EPO assay using Hep3B cells [49]. Compounds wherein the side chain contains either NH or CH_2 displayed IC_{50} values in the range of 11–779 nM in a PHD2 assay [50]. Interestingly, high selectivity was seen for PHD2 over collagen prolyl hydroxylases 1 and 2 (CPH1 and CPH2) by replacing the amide NH with a CH_2 group. For example, while glycinamide analog

19 displayed IC_{50} values of 0.046, 0.351, and 0.111 µM for PHD2, CPH1, and CPH2, respectively, ketone analog **20** displayed IC_{50} values of 0.065, >40, and 13.4 µM, respectively. When **20** (50 mg/kg) was administered orally to rats, approximately a 25-fold increase in plasma EPO levels were observed relative to vehicle control [50].

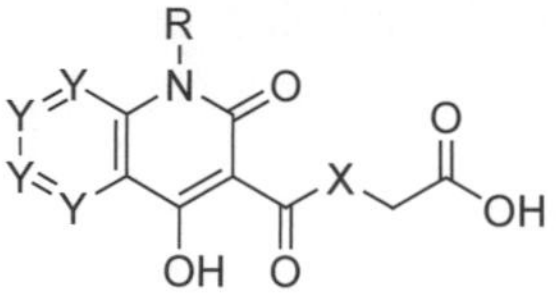

18

X = NH, CH_2
Y = CH, N
R = H, alkyl

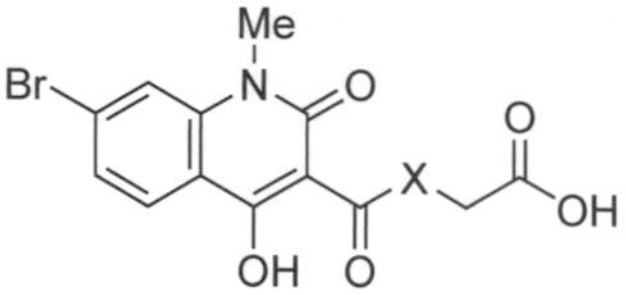

19: X = NH
20: X = CH_2

Hetero-fused pyridone-based analogs were recently disclosed as PHD inhibitors (**18**, "Y" taken together is a 5-membered heterocyclic ring) [54]. IC_{50} values in the range of 1–100 nM for this series of compounds in a PHD3 assay and EC_{50} values of 1–20 µM in an ELISA EPO assay using Hep3B cells were reported. Another group independently disclosed heterocyclic replacement of the fused phenyl ring present in quinolones with thiophene (exemplified by **21**) and thiazole ring systems [55]. In general, improvement in potency is seen in this series of compounds relative to the quinolones, and analogs display IC_{50} values in the range of 3–122 nM in a FRET-based PHD2 assay. Various heterocyclic substitutions at the 2-position of thiophene imparted improvements in potency. Interestingly, replacement of the amide NH with a CH_2 group (analogs **21** and **22**) provided a 25-fold improvement in potency (IC_{50} values of 75 and 3 nM, respectively).

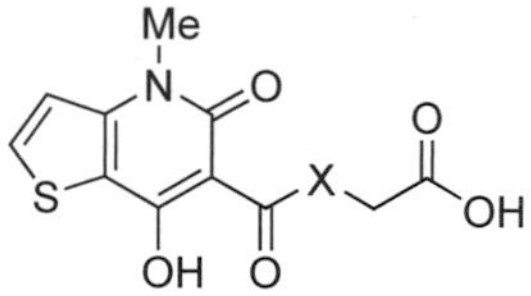

21: X = NH
22: X = CH_2

Pyrimidine- and pyridazine-based analogs have recently emerged as PHD inhibitors and IC_{50} values in the range of 2–175 nM have been reported in a FRET-based PHD2 assay [52]. A number of different substitutions ranging from alkyl, aryl, heteroaryl, and amino groups are accommodated in this series of compounds. While 1-pyrrolidinyl-, **23** and 1-piperidinyl-, **24**, substituted pyrimidine derivatives displayed similar IC_{50} values (7 and 16 nM, respectively), substitution with a morpholine group as in **25** resulted in at least an order of magnitude reduction in potency ($IC_{50} = 240$ nM).

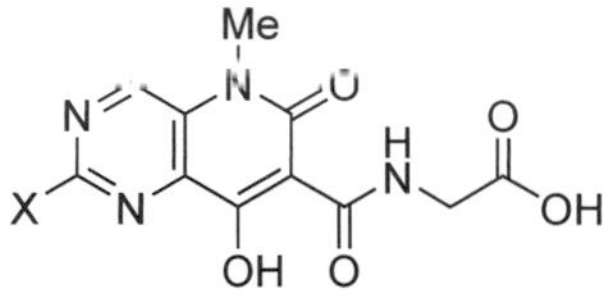

23: X = 1-pyrrolidinyl
24: X = 1-piperidinyl
25: X = 4-morpholinyl

Imidazopyridine analogs exemplified by **26** [56] were designed using the x-ray crystal structure of the known isoquinoline HIF PHD inhibitor **1** [17]. Although the designed chemotype preserves the crucial interaction with Arg383 and iron, the hydroxyl group making an interaction with Tyr303 is absent. However, the presence of an imidazole ring allows for the possibility of coordinating to iron in the active site via two possible binding modes. While coordination to iron is achieved via the pyridine nitrogen atom and the carbonyl group in **27**, the imidazole nitrogen atom and the carbonyl group in structure **28** accomplish similar iron-binding interactions. The designed parent analog **26** displayed potency similar to that of the isoquinoline **1** (PHD2 IC_{50} of 2.8 and 1.4 µM, respectively). Replacement of the imidazole NH of **26** with NPh provided compound **29** showing a 10-fold improvement in potency ($IC_{50} = 0.28$ µM). Introduction of a chlorine substitutent as in **30** or an additional phenyl substituent as in **31** provided improvement in potency, with **31** providing the most active PHD2 inhibitor in this series ($IC_{50} = 3$ nM). Modeling studies suggest that a single binding mode for the chelation to iron in **31** is preferred, and is achieved through the imidazole nitrogen and carbonyl oxygen atoms (i.e., **28**) allowing for van der Waals and hydrophobic interactions of the R^1 and R^3 substituents with Ile256, Met299, Tyr303, and Asp254 residues.

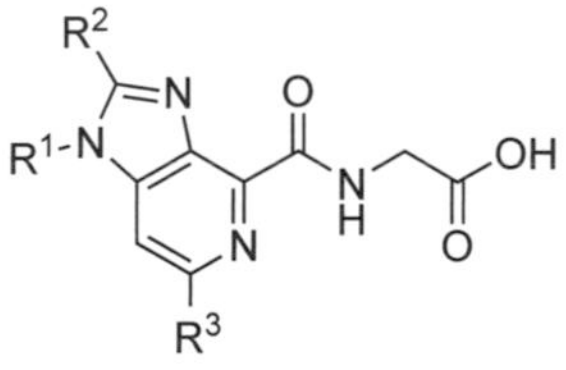

26: $R^1 = R^2 = R^3 = H$
29: $R^1 = Ph$; $R^2 = R^3 = H$
30: $R^1 = Ph$; $R^2 = H$; $R^3 = Cl$
31: $R^1 = Ph$; $R^2 = H$; $R^3 = Ph$

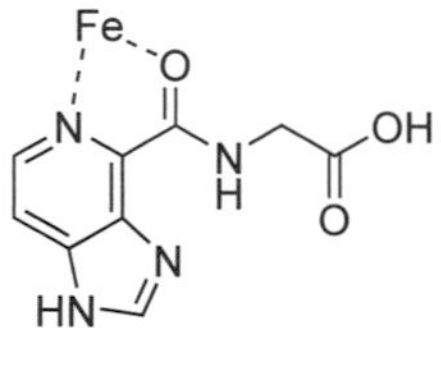

27

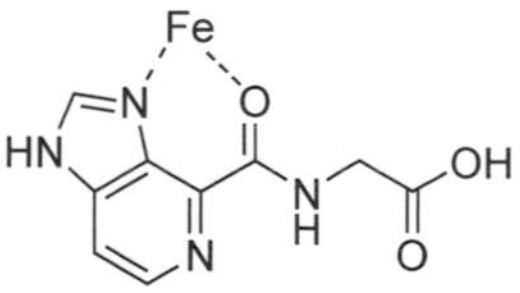

28

A series of naphthyridine analogs exemplified by **32** has been described as PHD inhibitors in a recent patent application with IC_{50} values ranging from 0.4 nM to 1.1 µM in a PHD2 enzymatic assay [51]. Analog **32** (R = H) displayed an IC_{50} value of 2.1 nM in a PHD2 enzymatic assay. In further SAR work, various alpha substituents were incorporated in the side chain. The (S)- and (R)-alanine analogs **33** and **34** displayed remarkably different IC_{50} values (6 nM and 1.1 µM, respectively). Hydroxymethyl substitution or ethyl substitution alpha to the carboxyl group (**35** and **36**) resulted in roughly an 80-fold decrease in potency relative to **32**. Interestingly, aspartic acid-derived analog **37** displayed only a 12-fold decrease in potency relative to **32**.

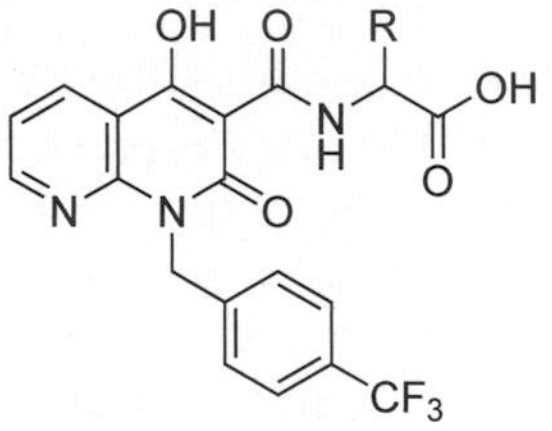

32: R = H
33: R = (S)-CH$_3$
34: R = (R)-CH$_3$
35: R = (S)-CH$_2$OH
36: R = (S)-CH$_2$CH$_3$
37: R = (S)-CH$_2$COOH

Other heterocyclic analogs including those containing the tetrahydrofuro [57], tetrahydropyrrolo [58], and tetrahydrothieno [59] moieties have also been disclosed. These analogs displayed PHD2 IC_{50} values in the range of ≤ 10–100 nM.

Finally, monocyclic pyridazines such as **38** [60] as well as pyrimidine diones (e.g., **39**) [61] and pyrimidine triones (e.g., **40**) [62] have been disclosed as PHD3 inhibitors. Substitution on ring positions led to IC_{50} values in the range of 1 nM to 3.2 µM for the pyridazine diones in a PHD3 assay and EC_{50} values of 0.4–100 µM in an ELISA EPO assay using Hep3B cells were reported. The pyrimidine diones displayed IC_{50} ranges of 0.8 nM to 20 µM in the enzymatic assay and 0.4–100 µM in the cell-based EPO assay.

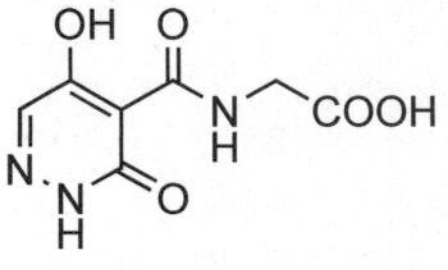

38

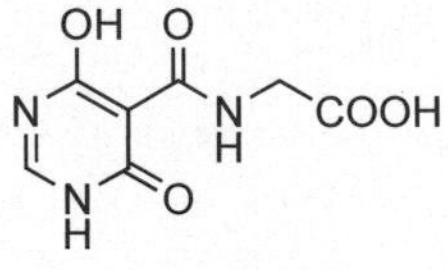

39

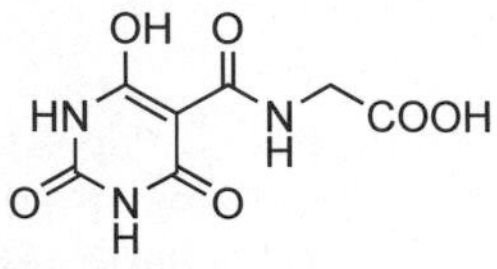

40

4.2.3 Nonglycine 2-oxoglutarate analogs

A series of 2-pyrazolopyridines demonstrate that the pyrazole-4-carboxylic acid moiety may in some cases be an effective glycine isostere in the context of some PHD2 inhibitors. 2-Pyrazolyl-4-arylpyridines were found to possess IC_{50} values in the low micromolar range using a mass spectrometry assay, as exemplified by compound 41 (IC_{50} 1.9 µM) [17]. In a related publication from the same laboratory, pyrazolopyridines with extended side chains in the 4-position of the pyridine ring were found to have improved activities (e.g., 42, $IC_{50} = 0.58$ µM) [63]. Furthermore, several of these compounds were also active in a cell-based VEGF induction assay at concentrations of 1.0 µM or greater.

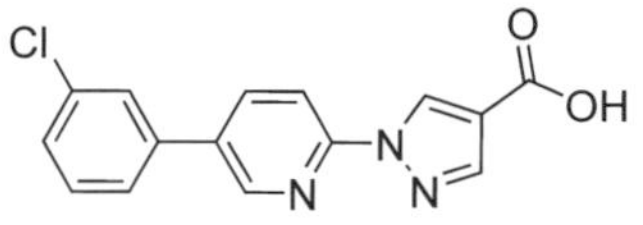

41

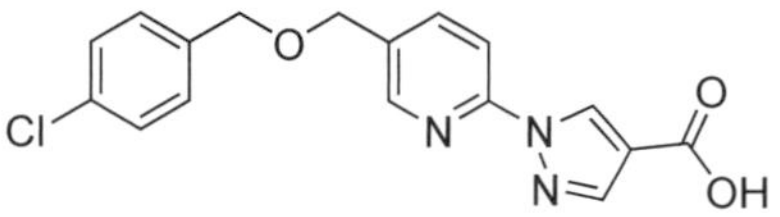

42

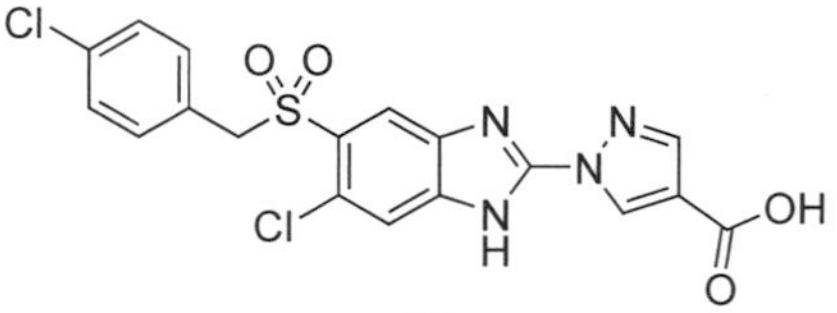

43

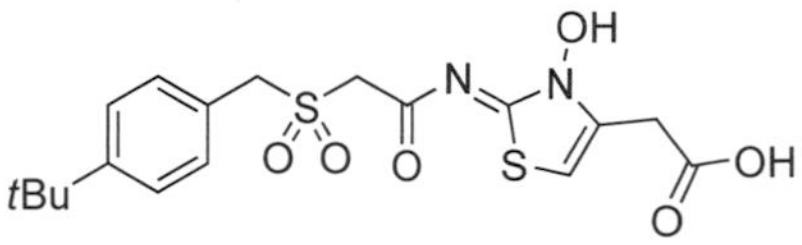

44

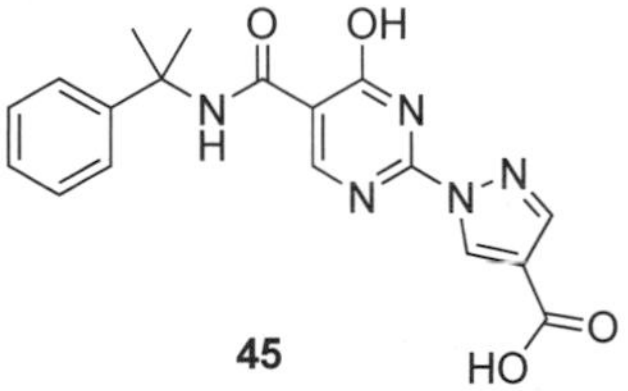

45

A group of benzimidazole-2-pyrazole PHD2 inhibitors appeared in a recent patent application, with IC_{50} values as low as 16 nM for compound 43 in a [14C]-based enzymatic assay [64]. Furthermore, members of this class of compounds were reported to stimulate erythropoietin secretion from Hep3B cells by up to 150% at 100 µM concentrations.

A recently published set of N-hydroxythiazoles is exemplified by compound 44 ($IC_{50} = 3$ nM, TR-FRET assay) [65]. Members of this series were found to be potent inhibitors of PHD in an enzymatic assay, although no cell-based assay data were reported for these compounds. The reported SAR for this series showed the N-hydroxyl group to be crucial for activity, suggesting iron chelation between the N-hydroxyl group and either the amide nitrogen or carbonyl oxygen atom.

In another 2009 patent application, a series of 2-pyrazolopyrimidinones were reported to be PHD inhibitors [37]. Although most of the compounds exemplified in this application lacked a carboxylic acid moiety (*vide supra*), compound **45** was reported as having an IC_{50} value of ≤ 10 nM in a FRET assay, although no cell-based assay data were disclosed.

5. CLINICAL TRIAL STATUS AND FUTURE PROSPECTS

To date, three pharmaceutical companies have entered clinical trials with PHD inhibitors for the treatment of anemia with the most advanced being FG-2216. In clinical studies, compound **2** (likely FG-2216) showed a dose- and time-dependent elevation of plasma erythropoietin after oral administration [66]. Healthy volunteers were orally administered various doses of compound **2** and serum erythropoietin (EPO) concentrations were measured at various times. Compound **2** increased serum EPO levels in a dose-dependent manner and, following administration of the 20 mg/kg dose, a 5-fold increase of EPO levels was observed after 12 h. In the same patent application, the effect of **2** on anemic predialysis patients with no previous rh-EPO exposure was also disclosed. Patients were treated with **2** three times/week for 4 weeks (no dose reported) and the hemoglobin levels were assessed on day 42. The patients who received treatment showed a mean increase in hemoglobin of 1.9 g/dL from baseline values, whereas subjects who received placebo showed a mean decrease of 0.35 g/dL from baseline levels. These data suggest for the first time that an oral PHD inhibitor could be effective for the treatment of anemia.

FG-4592 (structure undisclosed) is also under evaluation in subjects with anemia and chronic kidney disease (CKD) not requiring dialysis. Results have been presented from a single-blind 4-week Phase 2a study employing five dosages administered two to three times weekly for 4 weeks to CKD patients with most of the patients responding with a 1 g/dL increase in Hb levels [67]. Despite their apparent success these clinical programs are not without their challenges. Both FG-2216 and FG-4592 were temporarily put on clinical hold following the occurrence of a single case of fulminant hepatitis in the evaluation of FG-2216 [68].

A study of GSK1278863A (structure undisclosed) in healthy adult subjects for the safety, tolerability, pharmacokinetics, and pharmacodynamics of repeat oral doses up to 300 mg for 14 days has also been completed [69]. Finally, AKB-6548 (structure undisclosed) has completed Phase Ia clinical trials in 48 healthy volunteers for the potential treatment of anemia. Preliminary results claim that doses that significantly increase plasma EPO without raising VEGF levels were identified and that further clinical trials are planned [70].

While the role of PHD inhibitors in the treatment of anemia is now validated, therapeutic validation is less certain in other HIF-associated pathologies such as wound healing, ulcerative colitis, therapeutic angiogenesis, and treatment of acute ischemic events such as myocardial ischemia and stroke. All of these indications are supported by a compelling array of *in vitro* and *in vivo* preclinical studies but their utility in the clinical setting remains to be evaluated and represents exciting possibilities for the future of small-molecule inhibitors of PHD enzymes.

REFERENCES

[1] G. L. Wang, B. H. Jiang, E. A. Rue, and G. L. Semenza, *Proc. Natl. Acad. Sci. USA*, 1995, **92**, 5510.
[2] G. L. Semenza and G. L. Wang. *Mol. Cell. Biol.*, 1992, **12**, 5447.
[3] M. Ivan, K. Kondo, H. Yang, W. Kim, J. Valiando, M. Ohh, A. Salic, J. M. Asara, W. S. Lane, and W. G. Kaelin, Jr., *Science*, 2001, **292**, 464.
[4] P. Jaakkola, D. R. Mole, Y. M. Tian, M. I. Wilson, J. Gielbert, S. J. Gaskell, A. V. Kriegsheim, H. F. Hebestreit, M. Mukherji, S. J. Schofield, P. H. Maxwell, C. W. Pugh, and P. J. Ratcliffe, *Science*, 2001, **292**, 468.
[5] R. K. Bruick and S. L. McKnight, *Science*, 2001, **294**, 1337.
[6] A. C. R. Epstein, J. M. Gleadle, L. A. McNeill, K. S. Hewitson, J. O'Rourke, D. R. Mole, M. Mukherji, E. Metzen, M. I. Wilson, A. Dhanda, Y.-M. Tian, N. Masson, D. L. Hamilton, P. Jaakkola, R. Barstead, J. Hodgkin, P. H. Maxwell, C. W. Pugh, C. J. Schofield, and P. J. Ratcliffe, *Cell*, 2001, **107**, 43.
[7] M. Hirsila, P. Koivunen, V. Gunzler, K. I. Kivirikko, and J. Myllyharju *J. Biol. Chem.*, 2003, **33**, 30772.
[8] M. J. Percy, P. W. Furlow, P. A. Beer, T. R. Lappin, M. F. McMullin, and F. S. Lee *Blood*, 2007, **112**, 2193.
[9] C. J. Hu, A. Sataur, L. Wang, H. Chen, and M. C. Simon. *Mol Biol Cell*, 2007, **18**, 4528.
[10] E. Metzen, U. Berchner-Pfannschmidt, P. Stengel, J. H. Marxsen, I. Stolze, M. Klinger, W. Q. Huang, C. Wotzlaw, T. Hellwig-Bürgel, W. Jelkmann, H. Acker, and J. Fandrey. *J Cell Sci.*, 2003, **116**, 1319.
[11] E. Berra, E. Benizri, A. Ginouvès, V. Volmat, D. Roux, and J. Pouysségur, *EMBO J.*, 2003, **22**, 4082.
[12] K. Takeda, V. C. Ho, H. Takeda, L. J. Duan, A. Nagy, and G. H. Fong. *Mol. Cell. Biol.* 2006, **26**, 8336.
[13] K. Takeda, A. Cowan, and G. H. Fong. *Circulation*, 2007, **116**, 774.
[14] K. Takeda, H. L. Aguila, N. S. Parikh, X. Li, K. Lamothe, L. J. Duan, H. Takeda, F. S. Lee, and G. H. Fong. *Blood*, 2008, **111**, 3229.
[15] V. Purpero and G. R. Morton, *J. Biol. Inorg. Chem.*, 2007, **12**, 587.
[16] P. C. A. Bruijnincx, G. van Koten, and R. J. M. Klein Gebbink, *Chem. Soc. Rev.*, 2008, **37**, 2716.
[17] M.A. McDonough, V. Li, E. Flashman, C. Rasheduzzaman, C. Mohr, B. M. R. Lienard, J. Zondlo, N. J. Oldham, I. J. Clifton, J. Lewis, L. A. McNeill, R. J. M. Kurzeja, K. S. Hewitson, E. Yang, S. Jordan, R. S. Syed, and C. J. Schofield, *Proc. Natl. Acad. Sci. USA*, 2006, **103**, 9814.
[18] A. G. Evodkimov, R. L. Walter, M. Mekel, M. E. Pokross, R. Kawamoto, and A. Boyer, 2006, RCSB protein data bank ID: 2HBT.
[19] L. Poppe, C. M. Tegley, V. Li, J. Lewis, J. Zondlo, E. Yang, R. J. M. Kurzeja, and R. Syed, *J. Am. Chem. Soc.*, 2009, **131**, 16654.

[20] G. L. Wang and G. L. Semenza, *Blood*, 1993, **82**, 3610.
[21] T. Tan, J. A. Luciano, P. M. Scholz, and H. R. Weiss, *Clin. Exp. Pharmacol. Physiol.*, 2009, **36**, 904.
[22] H. Thangarajah, I. N. Vial, R. H. Grogan, D. Yao, Y. Shi, M. Januszyk, R. D. Galiano, E. I. Chang, M. G. Galvez, J. P. Glotzbach, V. W. Wong, M. Brownlee, and G. C. Gurtner, *Cell Cycle*, 2010, **9**, 75.
[23] H. Thangarajah, D. Yao, E. I. Chang, Y. Shi. L. Jazayeri, I. N. Vial, R. D. Galiano, X.-L. Du, R. Grogan, M. G. Galvez, M. Januszyk, M. Brownlee, and G. C. Gurtner, *Proc. Nat. Acad. Sci. USA*, 2009, **106**, 13505.
[24] P. Dongiovanni, L. Valenti, A. L. Fracanzani, S. Gatti, G. Cairo, and S. Fargion, *Am. J. Pathol.*, 2008, **172**, 738.
[25] J. Gunton, *Patent Application WO 2009/086592*, 2009.
[26] D. W. Lee, S. Rajagopalan, A. Siddiq, R. Gwiazda, L. Yang, M. F. Beal, R. R. Ratan, and J. K. Andersen, *J. Biol. Chem.* 2009, **284**, 29065.
[27] A. Ttriantafyllou, I. Mylonis, G. Simos, S. Bonanou, and A. Tsakalof, *Free Radical Biol. Med.*, 2007, **44**, 657.
[28] S.-S. Park, I. Bae, and Y. J. Lee, *J. Cell. Biochem.*, 2008, **103**, 1989.
[29] Y.-B. Shui, J. M. Arbeit, R. S. Johnson, and D. C. Beebe, *Invest. Opthal. Vis. Sci.*, 2008, **49**, 4961.
[30] M. Xia, R. Huang, Y. Sun, G. L. Semenza, S. F. Aldred, K. L. Witt, J. Inglese, R. R. Tice, and C. P. Austin, *Toxicol. Sci.*, 2009, **112**, 153.
[31] B. Banerji, A. Conejo-Garcia, L. A. McNeill, M. A. McDonough, M. R. G. Buck, K. S. Hewitson, N. J. Oldham, and C. J. Schofield, *Chem. Commun.*, 2005, **43**, 5438.
[32] M. D. Rosen, H. Venkatesan, H. M. Peltier, S. D. Bembenek, T. D. Barrett, K. C. Kanelakis, L. X. Zhao, B. Leonard, F. M. Hocutt, X. Wu, H. L. Palomino, T. I. Brondstetter, P. V. Haug, L. Cagnon, W. Yan, L. A. Liotta, A. Young, T. Mirzadegan, N. P. Shankley, and M. H. Rabinowitz, *Med. Chem. Lett.*, 2010 (submitted for publication).
[33] D. B. Gotchev, J. Jin, and Y. Wang, *Patent Application WO 2009/086044-A1*, 2009.
[34] K. Thede, I. Flamme, F. Oehme, J.-K. Ergueden, F. Stoll, J. Schuhmacher, H. Wild, P. Kolkhof, H. Beck, J. Keldenich, M. Akbaba, and M. Jeske, *Patent Application WO 2008/067871-A1*, 2008.
[35] M. Jeske, I. Flamme, F. Stoll, and F. Oehme, *Patent Application DE 102007/048447*, 2009.
[36] M. Jeske, I. Flamme, F. Stoll, F. Oehme, and M. Akbaba, *Patent Application DE 2007/ 102007049157*, 2007.
[37] M. J. Clements, J. S. Debenham, J. J. Hale, C. B. Madsen-Duggan, and T. F. Walsh, *Patent Application WO 2009/117269*, 2009.
[38] C. J. Cunliffe, T. J. Franklin, N. J. Hales, and G. B. Hill, *J. Med. Chem.* 1992, **35**, 2652.
[39] T. J. Franklin, N. J. Hales, D. Johnstone, W. B. Morris, C. J. Cunliffe, A. J. Millest, and G. B. Hill, *Biochem. Soc. Trans.* 1991, **19**, 812.
[40] K. Weidmann, K.-H. Baringhaus, and G. Tschank, U. Werner, *Patent Application DE 19746287-A1*, 1999.
[41] M. P. Arend, L. A. Flippin, V. Guenzler-Pukall, W.-B. Ho, E. D. Turtle, and X. Du, *Patent Application WO 2004/108681-A1*, 2004.
[42] M. P. Arend, L. A. Flippin, M. Wu, E. D. Turtle, W.-B. Ho, and S. Deng, *Patent Application WO 2007/090068-A2*, 2007.
[43] E. D. Turtle, L. A. Flippin, M. P. Arend, and H. Cheng, *Patent Application US 2006/ 199836-A1*, 2006.
[44] S. Deng, M. Wu, E. D. Turtle, W.-B. Ho, M. P. Arend, H. Cheng, and L. A. Flippin, *Patent Application WO 2007/115315-A2*, 2007.
[45] M. P. Arend, H. Cheng, L. A. Flippin, D. Ng, E. Turtle, and M. Wu, *Patent Application WO 2009/073669-A1*, 2009.
[46] X. Zhou, M. P. Arend, M. Wu, and L. A. Flippin, *Patent Application WO 2009/089547-A1*, 2009.

[47] N. C. Warshakoon, S. Wu, A. Boyer, R. Kawamoto, J. Sheville, R. T. Bhatt, S. Renock, K. Xu, M. Pokross, S. Zhou, R. Walter, M. Mekel, A. G. Evdokimov, and S. East, *Bioorg. Med. Chem. Lett.*, 2006, **16**, 5616.

[48] R. M. Kawamoto, *Patent Application US 2007/299086-A1*, 2007.

[49] D. Chai, M. Colon, K. J. Duffy, D. M. Fitch, R. Tedesco, and M. N. Zimmerman, *Patent Application WO 2007/038571-A2*, 2007.

[50] J. R. Allen, R. Burli, M. C. Bryan, G. Q. Cao, S. C. Neira, and A. B. Reed, *Patent Application WO 2008/130600-A2*, 2008.

[51] D. Guiadeen, J. J. Hale, and S. Kothandaraman, *Patent Application WO 2008/130527-A1*, 2008.

[52] J. R. Allen, K. Biswas, R. Burli, J. E. Golden, S. Mercede, and C. M. Tegley, *Patent Application WO 2008/137084-A2*, 2008.

[53] D. M. Fitch and M. Colon, *Patent Application WO 2007/103905-A2*, 2007.

[54] D. M. Fitch and D. Chai, *Patent Application WO 2007/136990-A2*, 2007.

[55] J. R. Allen, R. Burli, M. J. Frohn, R. W. Hungate, S. C. Neira, and A. B. Reed, *Patent Application WO 2008/137060-A1*, 2008.

[56] M. Frohn, V. Viswanadhan, A. J. Pickrell, J. E. Golden, K. M. Muller, R. W. Buerli, G. Biddlecome, S. C. Yoder, N. Rogers, J. H. Dao, R. Hungate, and J. R. Allen, *Bioorg. Med. Chem. Lett.*, 2008, **18**, 5023.

[57] V. J. Colandrea, J. J. Hale, and J. G. McCoy, *Patent Application WO 2009/108496-A1*, 2009.

[58] M. J. Clements, V. J. Colandrea, J. S. Debenham, J. J. Hale, C. B. Madsen-Duggan, and T. F. Walsh, *Patent Application WO 2009/108499-A1*, 2009.

[59] J. S. Debenham, J. J. Hale, C. B. Madsen-Duggan, and T. F. Walsh, *Patent Application WO 2009/108497-A1*, 2009.

[60] A. N. Shaw, K. J. Duffy, W. H. Miller, A. K. Myers, and M. N. Zimmerman, *Patent Application WO 2008/089052-A2*, 2008.

[61] A. N. Shaw, K. J. Duffy, R. Tedesco, and K. Wiggall, *Patent Application US 2008/171756-A1*, 2008.

[62] K. J. Duffy, D. M. Fitch, J. Jin, R. Liu, A. N. Shaw, and K. Wiggall, *Patent Application WO 2007/150011-A2*, 2007.

[63] N. C. Warshakoon, S. Wu, A. Boyer, R. Kawamoto, S. Renock, K. Xu, M. Pokross, A. G. Evdokimov, S. Zhou, C. Winter, R. Walter, and M. Mekel, *Bioorg. Med. Chem. Lett.*, 2006, **16**, 5687.

[64] F. M. Hocutt, B. E. Leonard, H. M. Peltier, V. K. Phuong, M. H. Rabinowitz, M. D. Rosen, K. T. Tarantino, H. Venkatesan, and L. X. Zhao, *Patent Application WO 2009/134750*, 2009.

[65] C. M. Tegley, V. N. Viswanadhan, K. Biswas, M. J. Frohn, T. A. N. Peterkin, C. Chang, R. W. Buerli, J. H. Dao, H. Veith, N. Rogers, S. C. Yoder, G. Biddlecome, P. Tagari, J. R. Allen, and R. W. Hungate, *Bioorg. Med. Chem. Lett.*, 2008, **18**, 3925.

[66] S. J. Klaus and T. B. Neff, *Patent Application US 2006/0276477-A1*, 2006.

[67] P. Frohna, S. Milwee, J. Pinkett, T. Lee, K. Moore-Perry, J. Chou, and R. H. Ellison in *Results from a Randomized, Single-Blind, Placebo-Controlled Trial of FG-4592, a Novel Hypoxia Inducible Factor Prolyl Hydroxylase Inhibitor, in CKD Anemia*, Vol. 2007, American Society of Nephrology Renal Week, San Francisco, November 2–5, 2007, Abstract SU-PO806, as abstracted in N. Agarwal and J. T. Prchal, *Semin. Hematol.*, 2008, **45**, 267.

[68] The FDA Accepts the Complete Response for Clinical Hold of FG-2216/FG-4592 for the Treatment of Anemia, April 2, 2008, http://www.astellas.com/en/corporate/news/pdf/080402_eg.pdf (accessed July 6, 2010).

[69] Repeat Dose Safety Study for Compound to Treat Anemia, November 11, 2009, http://clinicaltrials.gov/show/NCT00840320 (accessed July 6, 2010).

[70] Akebia Announce Positive Results for AKB-6548 Phase 1 Clinical Study, January 5, 2010, http://www.akebia.com/AKB6548_Phase_1a_Completion.pdf (accessed July 6, 2010).

PART III:
Inflammation/Pulmonary/Gastrointestinal Diseases

Editors: Joel C. Barrish and David Weinstein
Bristol-Myers Squibb R&D
Princeton
New Jersey

Opioid Receptor Antagonists for Gastrointestinal Dysfunction

R. William Hipkin and **Roland E. Dolle**

1. INTRODUCTION

1.1 Opioid receptors and GI pharmacology

The G protein-coupled opioid receptors, μ (MOR), δ (DOR), and κ (KOR), are expressed in both the central nervous system (CNS) and a variety of peripheral tissues [1]. Opioid analgesics such as morphine, fentanyl, and oxycodone induce analgesia, miosis, and respiratory depression via activation of MOR in the CNS. However, these drugs also have significant inhibitory effects on GI function, including motility, secretion, absorption, and blood flow via MOR expressed in the enteric nervous system (ENS) in the gut [2]. Indeed, opioid-induced constipation (OIC) is an

Adolor Corporation, 700 Pennsylvania Drive, Exton, PA 19341, USA

Annual Reports in Medicinal Chemistry, Volume 45
ISSN 0065-7743, DOI 10.1016/S0065-7743(10)45009-5

almost universal (and often debilitating) side effect reported by patients receiving opioid therapy [3]. The inhibitory impact of MOR agonists on GI function is so predictable that loperamide, a MOR-selective full agonist with poor oral pharmacokinetics, is used as an antidiarrheal agent via activation of MOR in the ENS [4,5].

Chronic opioid use can result in dose-limiting and debilitating bowel dysfunction, referred to as opioid-induced bowel dysfunction (OBD). OBD is an underrecognized and poorly defined condition, which comprises a constellation of GI symptoms including OIC, incomplete evacuation, bloating, pain, nausea/vomiting, and increased gastric reflux [3]. Several MOR antagonists have been launched recently as treatments to limit the GI side effects of opiate analgesics. For example, TARGIN® combines the MOR analgesic, oxycodone with a low dose of naloxone (1), the nonselective opioid antagonist, in a prolonged release tablet [6]. However, most of the activities in this space have centered on the development of peripherally acting MOR (PAMOR) antagonists, a new class of therapeutic agents that are excluded from the CNS and therefore will not block the central analgesic properties of the MOR agonists while effectively reducing their peripheral actions. Currently, there are two PAMOR antagonists in the market: orally active alvimopan (3; ENTEREG®) for postoperative ileus (POI) [7] and injectable methylnaltrexone (4; RELISTOR®) for the treatment of OIC in palliative care [8].

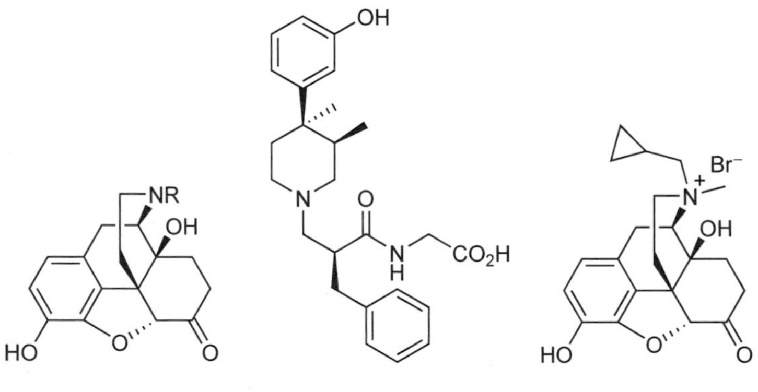

1: R = CH₂CHCH₂
2: R = CH₂C₃H₅

1: $R = CH_2CHCH_2$
2: $R = CH_2C_3H_5$

3

4

1.2 Role of opioid receptors in GI motility

There is a plethora of *in vitro* and *in vivo* pharmacological data with MOR-selective agonist and antagonists implicating MOR in the disruption of GI function by opioid analgesics [2,9]. This pharmacology is consistent with studies demonstrating the expression of MOR mRNA/protein in distinct enteric neurons and intestinal muscle cells in rodents, pigs, and humans [9]. The pharmacological data supporting the essential role of MOR in the inhibition of GI transit by opioids was confirmed genetically using MOR knockout mice [10].

Although the presence of immunoreactive DOR has been reported in the ENS [2,11], the pharmacological evidence is the more compelling data. DOR peptide agonists inhibited contractions of superfused muscle strips from the human sigmoid colon (this response was reduced in the presence of the DOR antagonists, ICI 174864, or naltrindole [12]) and significantly delayed colonic propulsive activity in rats, mice, and cats [13]. The highly selective, nonpeptidic DOR agonist, SNC 80, also significantly decreased GI transit in rodents [14,15]. This effect was blocked by the nonselective opioid antagonist 1 and the DOR-selective antagonist naltrindole.

There is evidence for the expression of KOR and its endogenous ligand, dynorphin, in the myenteric plexus of the ENS [11]. KOR agonists have been shown *in vitro* to inhibit colonic contractility, activity that was blunted with a selective KOR antagonist [16]. However, the KOR-selective agonist U-50,488H did not inhibit intestinal propulsion in either rats or mice when given either i.c.v. or s.c. [17]. Similarly, the KOR agonists U-69,593 and CI-977 did not cause inhibition of gastrointestinal motility in rats [18]. The effect of the KOR agonist asimadoline (or placebo) was tested in 91 healthy volunteers for its effects on visceral sensation, satiation, colonic compliance, tone, perception of colonic distension, and whole gut transit. Asimadoline significantly decreased colonic tone during fasting without affecting postprandial colonic contraction, compliance, or transit [19]. Taken together, preclinical and clinical data suggests that KOR does not have a significant physiological role in controlling GI motility in response to opioid analgesics.

1.3 Opioid-induced GI dysfunction

POI is the impairment of GI motility that follows almost every abdominal surgery and affects all segments of the GI tract. Although the exact mechanism(s) of POI are not well understood, it is clearly complex with inflammatory, neurogenic, and muscular changes resulting in decreased GI motility [20]. The perioperative use of opioid analgesics certainly contributes to this GI dysmotility. Lasting at least 5–6 days, POI delays GI recovery and is the most common cause of delayed hospital discharge after abdominal surgery [20]. The annual national hospital costs arising to the management of POI for both the index hospitalization and any readmissions within 30 days has been estimated at $1.46 billion [21].

OIC is a common and often debilitating side effect of opioid analgesics occurring in 40–95% of patients treated with opioids [22]. Unlike many of the other side effects of opioids (respiratory suppression, sedation, nausea), OIC is unlikely to improve over time [23]. A recent report on the results of a US and European patient survey on the prevalence, severity, and impact of OBD revealed that 81% of 322 patients on oral opioid drugs for pain reported the highest level of "bothersomeness" for constipation, a condition that persisted despite the fact that all the patients were taking laxatives [22]. Another recent study with more than 800 patients suffering from OIC showed that patients with constipation had substantially higher total medical costs arising from significantly higher care requirements (i.e., inpatient, hospice, outpatient, emergency, nursing home care, etc.) compared with control patients [24].

OBD comprises a constellation of GI symptoms including OIC, incomplete evacuation, inhibition of gut peristalsis, bloating, pain, nausea/vomiting, and increased gastric reflux and tone of intestinal sphincters [3]. Approximately 40% of patients taking chronic opioids for nonmalignant pain develop bowel dysfunction [25].

2. MARKETED PAMOR ANTAGONISTS

2.1 Alvimopan (ENTEREG®)

Alvimopan (3) is a potent, orally active PAMOR antagonist with a sub-nanomolar affinity at MOR and approximately 10-fold selectivity versus DOR and KOR [26]. When tested at 1 or 10 μM, 3 had no significant activity in a broad range of nonopioid receptors, ion channels, and enzymes [9]. Relative to full agonists, 3 has no positive intrinsic activity at human MOR and DOR, and guinea pig KOR [26]. In a mouse hot-plate study, subcutaneous dosing of 3 inhibited morphine-induced antinociception ($ID_{50} = 6.0$ mg/kg), with considerably higher potency in small intestinal transit assay ($ID_{50} = 0.03$ mg/kg). Therefore, the peripheral selectivity for 3 in mice (the ratio of ID_{50} values for reversal of analgesia and delayed GI transit) in mice is ~200 [26]. This high degree of peripheral selectivity is thought to arise from its moderately large molecular weight, zwitterionic form, and polarity [27].

In May 2008, the U.S. Food and Drug Administration (FDA) approved 3 for accelerating the time to upper and lower gastrointestinal recovery following partial large or small bowel resection with primary anastomosis.

In healthy subjects, the plasma concentration of **3** peaks ~2 h after ingestion [28]. The steady-state volume of distribution of **3** in humans is estimated to be 30 ± 10 L/kg with a mean terminal half-life ranging from 4.4 to 13.8 h [7]. The drug's oral bioavailability is approximately 6% and it plasma protein binding is 70–80%. Single and multiple oral doses of **3** (up to 18 mg) produced linear pharmacokinetics with C_{min} values exceeding the K_i for opioid receptor binding, supporting a potential for systemically mediated activity of **3**. In patients suffering from chronic constipation, the pharmacokinetics of **3** was similar to that seen in healthy control subjects although drug C_{min} was somewhat elevated. Alvimopan has been studied for both acute (impact on gastrointestinal recovery following major abdominal surgery) and chronic (OBD) indications. In five randomized, placebo-controlled, Phase 3 clinical trials (four North American, one non-US) in bowel resection patients, oral **3** at 12 mg preoperatively and 12 mg b.i.d. for up to 7 days postoperatively, compared with placebo, accelerated the time to GI recovery. In the four North American studies, compared with placebo, treatment with **3** resulted in an earlier time to writing of the hospital discharge order.

Although development activities for long-term use have been discontinued, alvimopan was also evaluated in patients with chronic noncancer pain treated with opioid therapy. Three large, Phase 3, randomized, placebo-controlled, global clinical trials were completed in this population: two comparably designed 3-month efficacy trials and one 12-month safety study with secondary efficacy endpoints. In one of the efficacy trials ($N = 518$, equal randomization) **3**, at a dose of 0.5 mg b.i.d. orally, achieved a statistically significant improvement in the proportion of bowel movement frequency responders (primary endpoint) compared with placebo. Although a numerically higher proportion was observed in the second trial for the 0.5 mg BD dose compared with placebo in the other efficacy trial ($N = 485$, equal randomization), this did not achieve statistical significance (although the placebo responder rate was higher in this study at over 50%) [3].

In the 12-month safety study ($N = 805$, 2:1 (alvimopan:placebo) randomization) **3** at 0.5 mg twice daily was well tolerated and showed evidence of sustained efficacy when taken continuously for 12 months by patients with nonmalignant pain requiring sustained treatment with opioids. However, unexpectedly, there were more reports of myocardial infarctions in patients treated with **3** at 0.5 mg twice daily compared with placebo-treated patients. The majority of myocardial infarctions occurred between 1 and 4 months after initiation of treatment. This imbalance has not been observed in other studies of **3**, including studies in patients undergoing bowel resection surgery who received **3** at 12 mg twice daily for up to 7 days. A causal relationship with **3** has not been established [29].

2.2 Methylnaltrexone bromide (RELISTOR®)

Methylnaltrexone bromide (4) is a derivative of naltrexone (2). Quaternary salt formation increases the polarity and lowers lipid solubility of the parent compound thereby reducing its CNS penetration and ablating its ability to compromise the central analgesic potential of MOR agonists. Methylnaltrexone binds MOR with a $K_i = 10\,nM$, ~25-fold lower affinity than does 3 [26]. Methylnaltrexone is ~3-fold selective versus KOR and approximately 100-fold selective versus DOR [26]. Unlike 3, 4 had a positive intrinsic activity (consistent with partial agonism) both in [^{35}S]GTPγS exchange and guinea pig ileum contraction assays [26].

Methylnaltrexone was launched in 2008 for the treatment of OIC in patients with advanced illness receiving palliative care, when response to laxative therapy has not been sufficient. Due to its poor oral bioavailability, 4 is administered subcutaneously every other day. Approximately 50% of the drug is excreted in the urine and feces with 85% eliminated unchanged [30].

3. NEW PAMOR ANTAGONISTS

3.1 PEGylated naloxol (NKTR-118)

PEGylation of macromolecules is a known strategy for enhancing their pharmacokinetic properties. Small molecules derivatized with low-molecular-weight PEGs can show similar augmented DMPK properties [31]. NKTR-118 is a PEGylated conjugate of naloxol currently in Phase 2 clinical trials as a once daily investigational drug for the treatment of OIC. Although the structure of the NKTR-118 has not been explicitly revealed, structure 5 was reported in a selection invention supported, in part, by clinical data [32]. NKTR-118 has a $K_i = 33.8\,nM$ at MOR, 20-fold lower affinity than 1. NKTR-118 is ~15-fold less brain permeable (rat) than 1 and comparable to atenolol, which has negligible brain penetration [33]. The ED_{50} for NKTR-118 and 1 in rat models of analgesic reversal are 55.4 and 1.14 mg/kg, respectively. The corresponding ED_{50} values for GI transit, antagonizing the peripheral constipative effect of morphine, are reported to be 23.1 mg/kg for NKTR-118 and 0.71 mg/kg for 1. Thus, despite the differences in the absolute potencies of NKTR-118 and 1 in the two *in vivo* assays, the peripheral therapeutic index was ca. 2 for the two agents. This stands in contrast to 3 where this ratio is >200 (Section 2.1) [9]. NKTR-118 is rapidly absorbed across species and demonstrates a dose-dependent increase in oral bioavailability of 7–22% upon single administration of 2 and 20 mg/kg in dog. The NOAEL for NKTR-118 was >200 mg/kg/day in dog. NKTR-118 is metabolically stable against a range of human CYP enzymes and is metabolized *in vitro* only by CYP3A4. It neither significantly inhibits nor causes induction of CYP enzymes *in vitro*. The unbound fraction is >80% in human plasma. A 25-mg dose of NKTR-118 in healthy human subjects established the following PK parameters: $t_{1/2} = 9.4$ h, C_{max} 77 ng/mL, $T_{max} = 1.6$ h, and $AUC_{0-12} = 248$ h*ng/mL. In a placebo-controlled Phase 2 study ($N = 208$) NKTR-118 was administered orally once daily at 5, 25, and 50 mg for 4 weeks. The primary endpoint was the change from baseline in spontaneous bowel movements (SBM) per week during the first week of the double-blind study treatment. The primary endpoint was achieved (statistically significant increase from 1.4 SBM per week baseline to 5 SBM per week) and efficacy demonstrated in OIC patients receiving the 25 and 50 mg dose. The restoration of GI function was not accompanied by any evidence of reversal of analgesia or increase in opioid use. The agent was well tolerated. Dose-dependent GI-related effects were the most commonly reported side effects.

3.2 6-β-Naltrexol (AIKO-150)

6-β-Naltrexol (6; AIKO-150) is the major metabolite of naltrexone (2) [34,35]. 6-β-Naltrexol displays potent MOR affinity ($K_i = 2.12$ nM) and is 100-fold selective versus DOR ($K_i = 213$ nM) and 3.5-fold selective versus KOR ($K_i = 7.42$ nM). 6-β-Naltrexol blocks morphine-induced antinociception in the mouse 55°C tail flick assay with an $ID_{50} = 1$ mg/kg i.p. It is equipotent with naloxone and 4.5-fold less potent than naltrexone. 6-β-Naltrexol antagonizes morphine-induced forward locomotion (mouse) with an $ID_{50} = 0.07$ mg/kg, which is 2.5- and 10-fold less active than 2 or 3. 6-β-Naltrexol precipitates only minimal withdrawal at high doses (100 mg/kg) in an acute morphine dependence model. It is ca. 77- and 30-fold less potent than 2 and 1 in precipitating withdrawal in a chronic dependence model. In drug combination studies, 6 produces a dose-related blockade of hydrocodone-induced antinociception ($ID_{50} = 12$ mg/kg, p.o.; A_{90} dose of hydrocodone, mouse tail flick). This dose is ~10-fold greater than the dose needed to antagonize hydrocodone-induced inhibition of gastrointestinal transit as measured in the charcoal meal assay ($ID_{50} = 1.3$ mg/kg, p.o.). In contrast, 2 showed no such dose separation in the same study design. The duration of action of 6 was longer and it had a ~2-fold slower time to peak effect compared to 2. Collectively, these data clearly demonstrate the CNS action of 6 but indicate a somewhat higher margin for peripheral selectivity relative to 2 in mice. Two Phase 1 clinical evaluations of 6 have been completed. In these trials,

6 reversed OIC at dosages that did not precipitate withdrawal in a highly sensitive (opioid) population and 6 did not interfere with pain relief in healthy subjects [36].

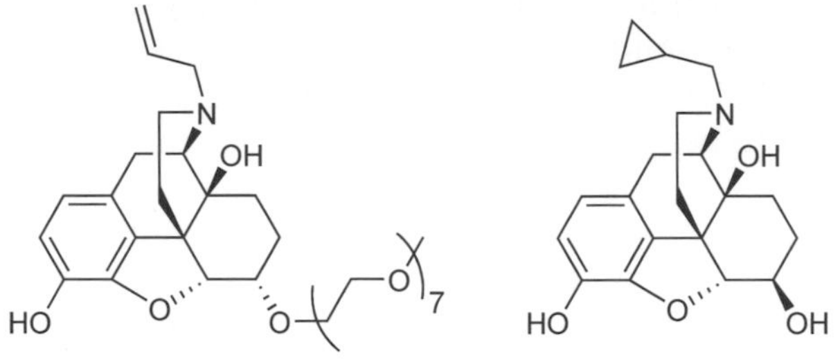

5   6

3.3 Other morphinans and related structures

Numerous patents have appeared describing derivatives of 5 and 6 as potential PAMOR antagonists. With the exception of 5 and 6, there are no peer-reviewed publications on structure–activity relationships, *in vivo* activity, or other preclinical data for the new agents. As viewed from the patent literature, introduction of polar substitutents into the morphinan scaffold is the preferred "peripheralization" strategy.

Morphinanium salts containing a 3-carboxamido substituent were reported [37]. Derivative 7 displays the following opioid receptor-binding affinities: K_i (MOR) = 37 nM, K_i (DOR) >10,000 nM, and K_i (KOR) = 210 nM. The agent is 18.5-fold less potent than 2 at MOR. Opening the 4,5-epoxy ring in 7 yields analog 8 with improved receptor-binding affinity: K_i (MOR) = 1.3 nM, K_i (DOR) = 280 nM, and K_i (KOR) = 7.7 nM. The quaternary salt is a functional antagonist: IC_{50} (MOR) = 52 nM, IC_{50} (KOR) = 7800 nM. At 10 mg/kg p.o., 8 shows 100% reversal of inhibition of PGE_2-induced diarrhea by morphine while the same dose did not affect the response to morphine in the tail flick antinociception test (mouse). The oral bioavailability of 8 in rat was $F = 50\%$ at a 10 mg/kg dose. A 50 mg/kg encapsulated oral dose of ring contracted zwitterion 9, IC_{50} (MOR) = 170 nM, to dogs provided high systemic exposures (C_{max} = 1388 ng/mL, T_{max} = 2 h, $t_{1/2}$ = 2.6 h, and AUC = 6187 h*ng/mL). No *in vivo* activity was reported for the compound [38]. 8-Substituted derivatives of 4 were also reported [39], but again no biological data was provided. A series of morphinan N-oxides was described as represented by naltrexone N-oxide (10), which possesses an IC_{50} = 5.6 nM (K_i < 50 nM) [40].

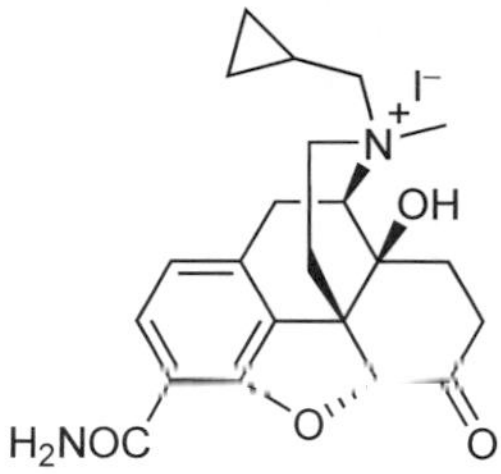

7 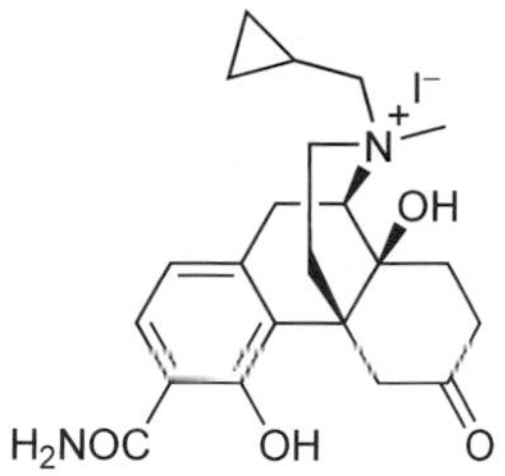   8

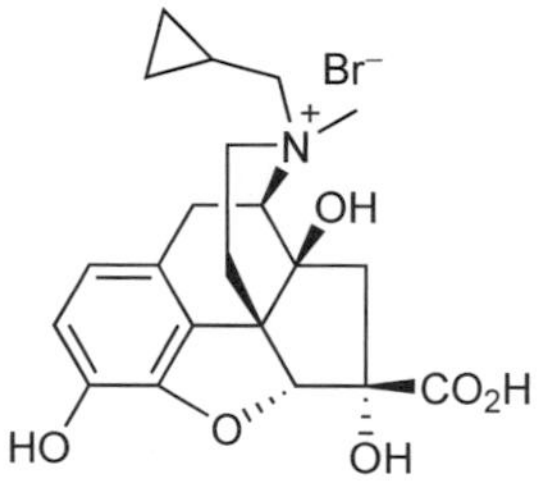

9 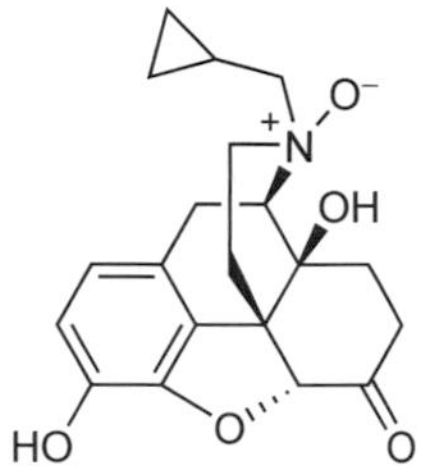   10

Analogs **11–13** are representative of new morphinan-derived zwitter-ions. The orally active 14-substituted ether (**11**) is 10-fold more potent than **2** in reversing morphine-induced decrease in GI motility in rat [41].

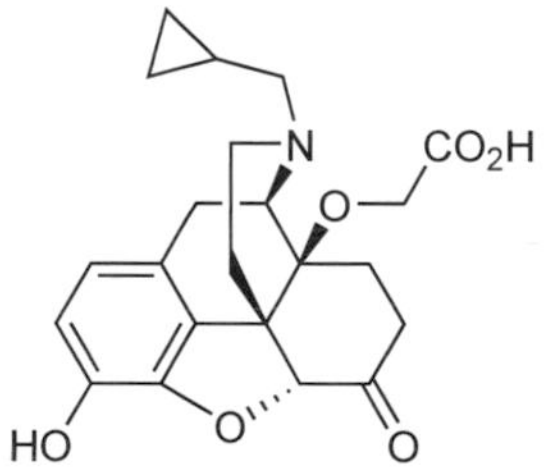

11 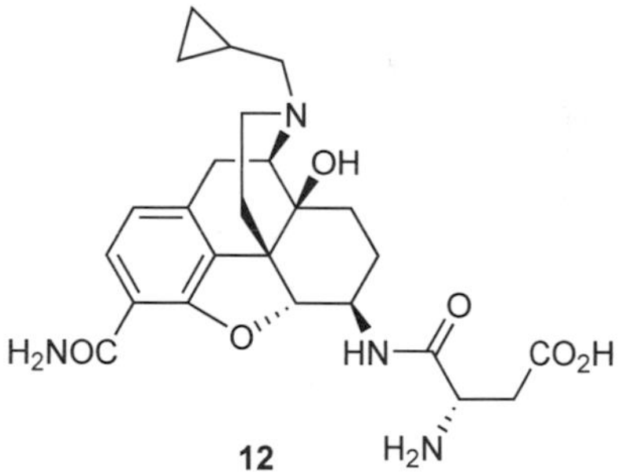   12

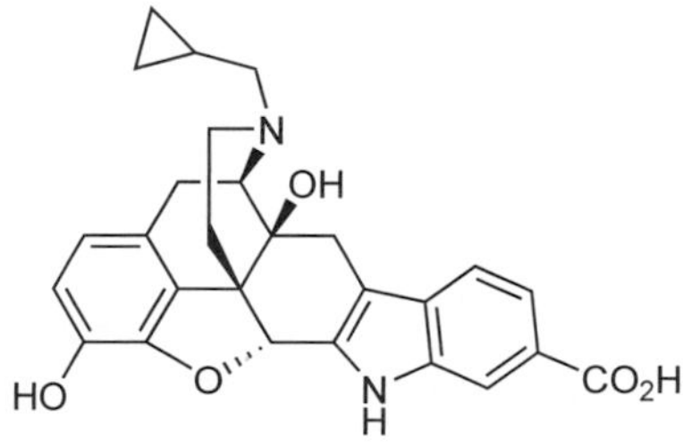

13 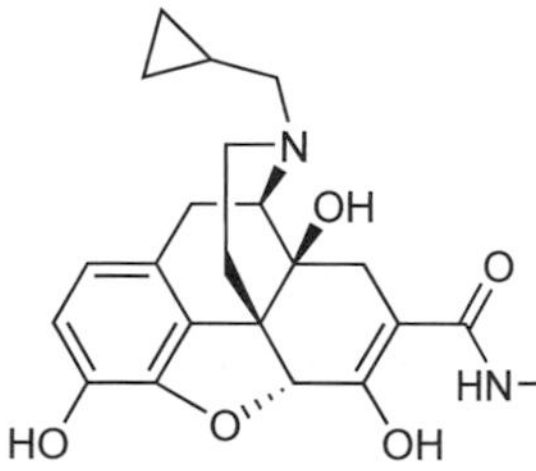   14

No biological data was reported for **12** [42] and **13** [43] or for a series of the 6,7-unsaturated-7-carbamoyl morphinans represented by **14** [44].

3.4 Nonmorphinans

Several patent applications have claimed various *N*-substituted derivatives of 8-azabicyclo[3.2.1]octanes [45–49]. Heteroarylalkyl-substituted **15** is a potent MOR antagonist: K_i (MOR) = 0.95 nM demonstrating full functional antagonism in a GTP–nucleotide exchange assay [46]. In a rat gastric emptying assay, **15** reversed loperamide-induced delayed gastric emptying with an ID_{50} = 0.13 mg/kg, p.o. Crystalline forms of **16** have been claimed [49]. Diol **16** also displays high receptor potency K_i (MOR) 0.1 nM and functional antagonism *in vitro*. In the same rat gastric emptying assay used to evaluate **15**, diol **16** displays an ID_{50} = 0.26 mg/kg, p.o.

2-Aminotetralines, e.g., **17** and **18**, are potent MOR antagonists [50–53]. Compound **17** is orally active in a rat gastric emptying assay with an ID_{50} = 0.48 mg/kg, p.o. [51]. The corresponding diethyl analog (**18**) (K_i (MOR) = 0.094 nM) is more potent than **17** in the rat gastric emptying assay (**18** ID_{50} = 0.11 mg/kg, p.o.). No additional data was provided for these agents.

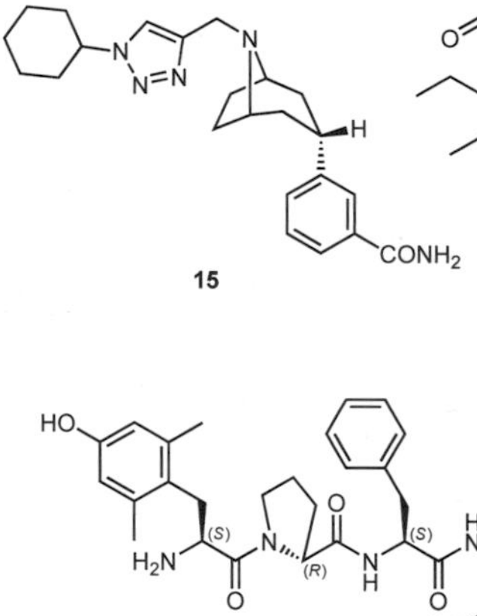

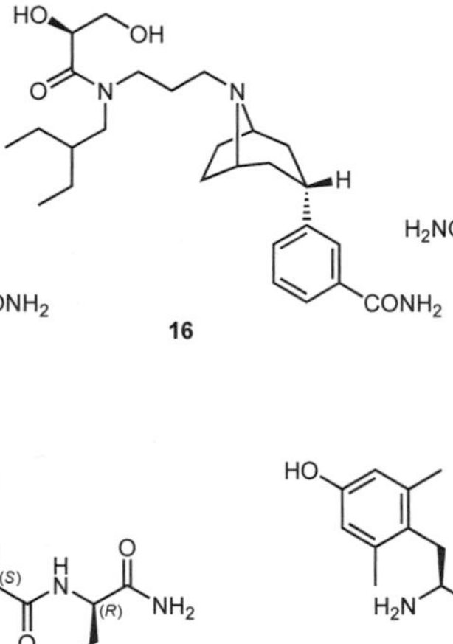

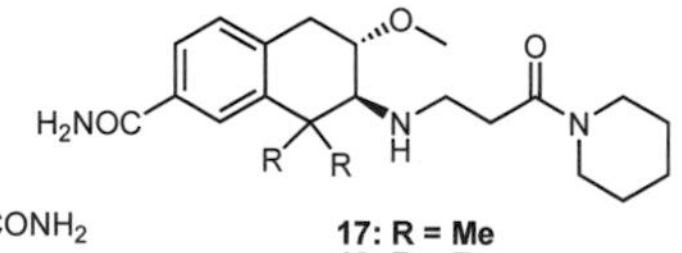

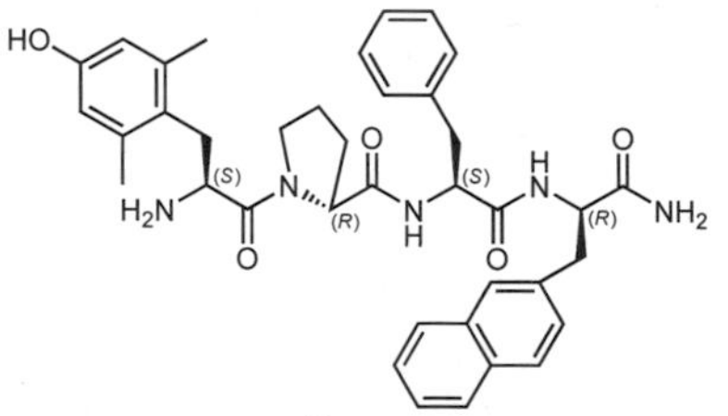

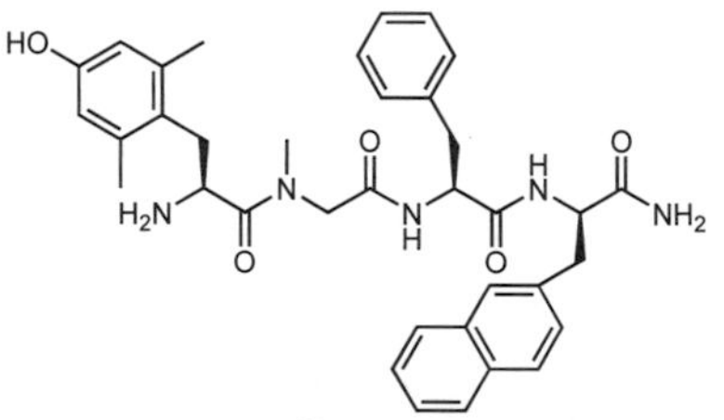

Endomorphin-2 (Tyr–Pro–Phe–NH$_2$), a potent and selective MOR agonist, was used as a template for MOR antagonist design [54,55]. Replacing the N-terminal tyrosine residue with 2,6-dimethyltyrosine and the C-terminal phenylalanine residue with D-(2-naphthyl)alanine afforded a potent and selective MOR antagonist (**19**) (rat brain membrane binding assay $IC_{50} = 1.5$ nM), which is 7-fold selective over DOR. Replacing [D-pro]^2 with sarcosine in **19** generated **20**. Tetrapeptide **20** is an exquisitely potent and selective MOR antagonist: IC_{50} 0.01 nM possessing >10^6-fold selectivity over DOR and KOR. The MOR pA2 values for **19** and **20** are 8.89 and 9.19 in a recombinant mammalian cell line. Tetrapeptides **19** and **20** reverse endomorphin-2-induced smooth muscle contraction in both mouse ileum and colon tissue preparations. Upon i.p. administration (2 mg/kg), **19** and **20** completely normalize morphine (0.1 mg/kg, i.p.)-induced colonic bead expulsion time in mice. Peripheral restriction of **19** and **20** (1 mg/kg, i.p.) was established as the agents did not block endomorphin-2 (3 µg, i.c.v.)-induced increase in jumping latency in mice. In this assay, the CNS permeable control antagonist **1** completely reversed jumping latency. This was further confirmed by demonstrating that **19** and **20** had no effect on antinociception in reversal of endomorphin-2-induced analgesia (hot-plate test). Degradation studies of **19** and **20** were carried out in rat brain homogenate where the compounds were found to have dramatically increased stability ($t_{1/2} = 81$ and 314 min; **19** and **20**, respectively) relative to endomorphin-2 ($t_{1/2} = 11$ min). These data demonstrate the potential utility of peptide-based MOR antagonists in treating OIC.

4. CLINICAL STATUS AND OUTLOOK

ENTEREG® and RELISTOR® are FDA-approved PAMOR antagonists for the treatment of POI and OIC. These agents along with the narcotic/antagonist combination product TARGIN® underscore clinical proof of concept for PAMOR antagonists to treat certain OBDs. There are now seven companies reported to have MOR antagonists in various stages of clinical development. Early research leading to the discovery and development of **3** and **4** focused on generating compounds with >100-fold peripheral selectivity in preclinical models. This was predicated upon the hypothesis that a high degree of peripheral selectivity would be necessary to prevent reversal of opioid analgesia. This hypothesis has been challenged by human clinical data showing that the opioid-conditioned GI tract is exquisitely sensitive to the effects of MOR antagonists and that GI relief occurs well before the detection of attenuated central analgesia. The need for peripheral selectivity notwithstanding, compounds with well-behaved metabolic and pharmacokinetic profiles may ultimately prove to be superior drugs in this class.

REFERENCES

[1] B. N. Dhawan, R. R. Cesselin, T. Reisine, P. B. Bradley, P. S. Portoghese, and M. Hamon, *Pharmacol. Rev.*, 1996, **48**, 567.

[2] A. De Luca and I. M. Coupar, *Pharmacol. Ther.*, 1996, **69**, 103.

[3] J. Moss and C. E. Rosow, *Mayo Clin. Proc.*, 2008, **83**, 1116.

[4] D. L. DeHaven-Hudkins, L. C. Burgos, J. A. Cassel, J. D. Daubert, R. N. DeHaven, E. Mansson, H. Nagasaka, G. Yu, and T. Yaksh, *J. Pharmacol. Exp. Ther.*, 1999, **289**, 494.

[5] F. Baldi, A. Bianco Maria, G. Nardone, A. Pilotto, and E. Zamparo, *World J. Gastroenterol.*, 2009, **15**, 3341.

[6] K. E. Clemens and G. Mikus, *Expert Opin. Pharmacother.*, 2010, **11**, 297.

[7] D. T. Beattie, *Clin. Med. Ther.*, 2009, **1**, 199.

[8] C.-S. Yuan, J. F. Foss, W. A. Williams, and J. Moss, *Drug Dev. Res.*, 2009, **70**, 403.

[9] D. L. DeHaven-Hudkins, R. N. DeHaven, P. J. Little, and L. M. Techner, *Pharmacol. Ther.*, 2008, **117**, 162.

[10] S. Roy, H.-C. Liu, and H. H. Loh, *Mol. Brain Res.*, 1998, **56**, 281.

[11] A. C. Gray, I. M. Coupar, and P. J. White, *Life Sci.*, 2006, **78**, 1610.

[12] P. Chamouard, S. Rohr, C. Meyer, R. Baumann, and F. Angel, *Eur. J. Pharmacol.*, 1994, **262**, 33.

[13] M. Broccardo and G. Improta, *Pharmacol. Res.*, 1992, **25**, 5.

[14] M. Broccardo, G. Improta, and A. Tabacco, *Eur. J. Pharmacol.*, 1998, **342**, 247.

[15] L. Negri, M. Broccardo, R. Lattanzi, and P. Melchiorri, *Br. J. Pharmacol.*, 1999, **128**, 1554.

[16] A. Shahbazian, A. Heinemann, H. Schmidhammer, E. Beubler, U. Holzer-Petsche, and P. Holzer, *Br. J. Pharmacol.*, 2002, **135**, 741.

[17] S. Roy, H.-C. Liu, and H. H. Loh, *Mol. Brain Res.*, 1998, **56**, 281.

[18] A. Tavani, P. Petrillo, A. LaRegina, and M. Sbacchi, *J. Pharmacol. Exp. Ther.*, 1990, **254**, 91.

[19] S. Delgado-Aros, H. J. Chial, M. Camilleri, L. A. Szarka, F. T. Weber, J. Jacob, I. Ferber, S. McKinzie, D. D. Burton, and A. R. Zinsmeister, *Am. J. Physiol.*, 2003, **284**, G558.

[20] K. Holte and H. Kehlet, *Drugs*, 2002, **62**, 2603.

[21] J.L. Goldstein, K.A. Matuszewski, C.P. Delaney, A. Senagore, E.F. Chiao, M. Shah, K. Meye, and T. Bramley, *T.&P.* 2007, **32**, 82.

[22] T. J. Bell, J. P. Sunil, C. Miaskowski, S. C. Bolge, T. Milanova, and R. Williamson, *Pain Med.*, 2009, **10**, 35.

[23] R. Benyamin, M. Trescot Andrea, S. Datta, R. Buenaventura, R. Adlaka, N. Sehgal, E. Glaser Scott, and R. Vallejo, *Pain Physician*, 2008, **11**, S105.

[24] D. Candrilli Sean, L. Davis Keith, and S. Iyer, *J Pain Palliat Care Pharmacother.*, 2009, **23**, 231.

[25] J. Thomas, *J. Pain Symptom Manage.*, 2008, **35**, 103.

[26] D. T. Beattie, M. Cheruvu, N. Mai, M. O'Keefe, S. Johnson-Rabidoux, C. Peterson, E. Kaufman, and R. Vickery, *Naunyn-Schmiedeberg's Arch. Pharmacol.*, 2007, **375**, 205.

[27] D. M. Zimmerman, J. S. Gidda, B. E. Cantrell, D. D. Schoepp, B. G. Johnson, and J. D. Leander, *J. Med. Chem.*, 1994, **37**, 2262.

[28] J. Foss, V. Schmith, B. Wallin, W. Du, and A. Melikian, *J. Clin. Pharmacol. Ther.*, 2005, **77**, P74.

[29] (a) http://us.gsk.com/html/media-news/pressreleases/2007/2007_04_09_GSK1056.htm (b) http://www.entereg.com.

[30] C.-S. Yuan, H. Doshan, M. R. Charney, M. O'Connor, T. Karrison, S. A. Maleckar, R. J. Israel, and J. Moss, *J. Clin. Pharmacol.*, 2005, **45**, 538.

[31] T. Riley and J. Riggs-Sauthier, *Pharm. Technol.*, 2008, **32**, 88. .

[32] K. J. Brodbeck and A. R. Kugler, *Patent Application WO 2009/137086-A1*, 2009.

[33] (a) L. Webster and S. Dhar, 38th Annual American College of Clinical Pharmacology Meeting, San Antonia, TX, September 2009. (b) A. Odinecs, Y. Song, S. Harite, M. G. Lee,

A. R. Kugler, and M. A. Eldon, 38th Annual American College of Clinical Pharmacology Meeting, San Antonia, TX, September 2009. (c) A. Odinecs, R. Gadiraju, J. Sisco, J. Farmer, S. Herzog, A. R. Kugler, and M. A. Eldon, 38th Annual American College of Clinical Pharmacology Meeting, San Antonia, TX, September 2009. (d) T. A. Neumann, H. Van Paachen, A. Marcantonio, D. Song, P. J. Morrison, M. A. Eldon, and A. R. Kugler, American Academy of Pain Management 18th Annual Clinical Meeting, Las Vegas, NV, September 27—30, 2007. (e) M. A. Eldon, D. Song, T. A. Neumann, R. Wolff, L. Cheung, T. X. Viegas, M. D. Bentley, S. Fishburn, and A. R. Kugler, American Academy of Pain Management 18th Annual Clinical Meeting, Las Vegas, NV, September 27—30, 2007. (f) H. van Paaschen, D. Sahner, A. Marcantonio, and M. A. Eldon, American Academy of Pain Management (AAPM) 20th Annual Clinical Meeting, Phoenix, AZ, October, 2009. All of these data are available in pdf documents at http://www.nektar.com/product_-pipeline/cns_ pain_oral _nktr-118and119.html.

[34] K. M. Raehal, J. J. Lowery, C. M. Bhamidipati, R. M. Paolino, J. R. Blair, D. Wang, W. Sadee, and E. J. Bilsky, *J. Pharmacol. Exp. Ther.*, 2005, **313**, 1150.

[35] W. Sadee, E. Bilsky, and J. Yancey-Wrona, *Patent Application WO 2009/051824-A2*, 2009.

[36] http://www.aikobiotech.com/news100325.shtml.

[37] M. Wentland, *Patent Application WO 2009/023567-A1*, 2009.

[38] S. M. Shah, K. A. Ali, F. Kong, T. Zhu, and C. T. Gombar, *Patent Application WO 2008/016704-A1*, 2008.

[39] J. Perez, A. Q. Han, and Y. Rotshteyn, *Patent Application WO 2008/064351-A2*, 2008.

[40] J. Perez, A. Q. Han, G. Kumaran, and Y. Rotshteyn, *Patent Application WO 2008/070462-A2*, 2008.

[41] A. Q. Han, Y. Rotshteyn, and V. Kumar, *Patent Application WO 2009/132313-A2*, 2009.

[42] D. Arnelle, D. Deaver, R. L. Dean III., and M. Todtenkopf, *Patent Application WO 2009/103004-A1*, 2009.

[43] T. Suzuki, T. Sawada, and Y. Ishihara, *Patent Application WO 2006/064780-A1*, 2006.

[44] M. Inagaki, S.-I. Hara, N. Haga, Y. Tamura, Y. Goto, and T. Hasegawa, *Patent Application WO 2006/126637-A1*, 2006.

[45] D. D. Long, D. R. Saito, P. Van Dyke, L. Jiang, T. J. Church, and J. R. Jacobsen, *Patent Application WO 2009/029252-A1*, 2009.

[46] D. D. Long, L. Jiang, D. R. Saito, and P. Van Dyke, *Patent Application WO 2009029253-A1*, 2009.

[47] D. R. Saito, D. D. Long, P. Van Dyke, T. J. Church, L. Jiang, and B. Frieman, *Patent Application WO 2009/029256-A1*, 2009.

[48] D. R. Saito, D. D. Long, and J. R. Jacobsen, *Patent Application WO 2009/029257-A1*, 2009.

[49] S. Daziel, L. M. Prezz, M. Rapta, and P.-J. Colson, *Patent Application WO 2008/106159-A1*, 2008.

[50] J. Perez, A. Q. Han, and Y. Rotshteyn, *Patent Application WO 2009/067275-A1*, 2009.

[51] M. R. Leadbetter, S. G. Trapp, D. D. Long, J. R. Jacobsen, and S. Axt, *Patent Application US 2009/149465-A1*, 2009.

[52] S. G. Trapp, M. R. Leadbetter, D. D. Long, L. Jiang, and S. Axt, *Patent Application WO 2009/124022-A1*, 2009.

[53] S. G. Trapp, M. R. Leadbetter, D. D. Long, L. Jiang, and S. Axt, *Patent Application US 2009/0247627-A1*, 2009.

[54] J. Fichna, J.-C. do-Rego, T. Janecki, R. Staniszewska, J. Poels, J. Vanden Broeck, J. Costentin, P. W. Schiller, and A. Janecka, *Bioorg. Med. Chem. Lett.*, 2008, **18**, 1350.

[55] J. Fichna, J.-C. K. Gach, R. Perlikowska, A. Cravezic, J. J. Bonnet, J.-C. do-Rego, A. Janecki, and M. Storr, *Regul. Pept.*, 2010, **160**, 19.

CHAPTER 10

CFTR Modulators for the Treatment of Cystic Fibrosis

Sabine Hadida, Fredrick Van Goor and **Peter D.J. Grootenhuis**

Vertex Pharmaceuticals Inc., 11010 Torreyana Road, San Diego, CA 92121, USA

Annual Reports in Medicinal Chemistry, Volume 45
ISSN 0065-7743, DOI 10.1016/S0065-7743(10)45010-1

1. INTRODUCTION

Cystic fibrosis (CF) is a lethal genetic disease affecting ~70,000 people worldwide [1,2]. CF is caused by mutations in the gene encoding the CF transmembrane conductance regulator (CFTR), an epithelial chloride- and bicarbonate-selective ion channel activated by cyclic AMP-dependent protein kinase A (PKA) [3–5]. CFTR aids the regulation of epithelial salt and water transport in multiple organs, including the lung, pancreas, liver, and intestinal tract [6]. Although mutations that impair CFTR function have multiple clinical manifestations [7] such as abnormal sweat electrolytes, chronic and progressive respiratory disease, exocrine pancreatic dysfunction, and infertility [8], it is lung disease that is the primary cause of morbidity and mortality. In the lung, the loss of CFTR-mediated Cl^- secretion is believed to cause airway surface dehydration due to both a decrease in CFTR-mediated Cl^- and fluid secretion and a secondary increase in epithelial Na^+ channel (ENaC)-mediated Na^+ and fluid absorption [6,9,10]. This imbalance between Cl^- secretion and Na^+ absorption, and the resulting dehydration of the airway surface, likely contributes to the deleterious cascade of mucus accumulation, infection, inflammation, and destruction that characterizes CF lung disease [11]. The loss of CFTR-mediated HCO_3^- secretion may also contribute to production of the thick, sticky mucus that plugs the passageways in the lung and other organs, such as the pancreas [12].

Current therapies to treat CF lung disease, including mucolytics, antibiotics, and anti-inflammatory agents, target the downstream disease consequences that are secondary to the loss of CFTR function [13]. Since the median predicted survival age is currently about 37 years [14], there is a large medical need for more efficacious therapies that address the underlying defect of CF. To address this need, there has been increased interest in small-molecule therapies that increase CFTR function because such an approach could, in principle, address all consequences of *CFTR* dysfunction as well as slow the progression of the disease. Such therapies are broadly classified as CFTR modulators and include CFTR activators, potentiators, correctors, and antagonists. CFTR activators act on their own to stimulate CFTR-mediated ion transport and include agents that increase cAMP levels, such as β-adrenergic agonists, adenylate cyclase activators, and phosphodiesterase inhibitors [15]. CFTR potentiators act in the presence of endogenous or pharmacological CFTR activators to increase the channel gating activity of cell-surface localized CFTR, resulting in enhanced ion transport [16]. CFTR correctors act by increasing the delivery and amount of functional CFTR protein to the cell surface, resulting in enhanced ion transport [16,17]. Depending on the molecular consequence of the mutation and disease severity, CFTR activators, potentiators, and correctors may be coadministered to maximize clinical efficacy or therapeutic window, if needed [16]. CFTR antagonists act by decreasing CFTR-mediated ion transport and are being developed for the treatment of polycystic kidney disease and cholera-induced secretory diarrhea [18,19]. Although this review focuses on CFTR potentiators and correctors for the treatment of CF, additional approaches to restore salt and fluid transport include ENaC inhibitors and alternative epithelial chloride channel activators, which have been extensively reviewed [15,20,21]. The purpose of this review is to summarize recent progress in the discovery of CFTR potentiators and correctors for the treatment of CF, with focus on compounds that have been most extensively characterized or for which chemical optimization has been reported.

1.1 Molecular consequence of mutations in the CFTR gene

CFTR is a member of the ATP-binding cassette (ABC) superfamily and is localized in the apical membrane of epithelial cells [22]. CFTR is a 1480-amino-acid protein composed of two sets of six helical membrane-spanning domains (MSD1 and MSD2) and two cytoplasmic nucleotide binding domains (NBD1 and NBD2) and is thought to function as a monomer. In addition, it contains a regulatory "R" domain with multiple PKA and protein kinase C (PKC) phosphorylation sites [23,24]. Both the F508del and G551D mutations are located in NBD1.

Following phosphorylation by the cAMP/PKA signaling pathway, CFTR is gated by the binding and hydrolysis of ATP. The open channel passes chloride and bicarbonate to regulate fluid transport across epithelia. The amount of anion flow is determined by the number of channels at the cell surface (density), the fraction of time the channel remains open (open probability, P_o), and its ability to pass anions (single-channel conductance); hence it is defined by the following equation:

Total CFTR current = density * open probability * conductance

The number of channels synthesized, processed, and trafficked to the apical membrane and rate of CFTR internalization at the apical membrane determine the cell surface density. Mutations in CFTR that alter one or more of these parameters decrease total CFTR-mediated anion flux and impair epithelial cell function.

Approximately 90% of CF patients carry a loss-of-function *CFTR* mutation on at least one allele that results in deletion of phenylalanine 508 (F508del) in the first CFTR nucleotide-binding domain [7]. The F508del prevents the proper domain folding and assembly of the multidomain CFTR protein during its biogenesis in the endoplasmic reticulum (ER) [8,25–27]. As a result, virtually all of the misfolded F508del-CFTR is retained and degraded by the ER, so that almost no CFTR is delivered to the cell surface [28–31].

In addition to F508del, more than 1600 *CFTR* mutations have been identified and can be grouped into five classes based on their molecular consequences [7]. Class I, II, and V mutations decrease the cell surface density of CFTR due to complete lack of CFTR protein synthesis (Class I; e.g., G542X, W1282X), impaired processing and trafficking to the membrane (Class II; e.g., F508del, N1303K), or reduced synthesis of otherwise normal protein (Class V; 2789+5G→A, 3849+10kbC→T). Class III (e.g., G551D, G1349D, S1255P, G551S, G1244E) and IV (R117H, R334W) mutations cause CFTR to open less frequently (impaired gating) or reduce its ability to pass Cl⁻ (altered conductance), respectively, but do not decrease the cell surface density of CFTR. In general, the reduction in CFTR-mediated anion secretion is more pronounced for Class I, II, and III mutations compared to Class IV and V mutations. Accordingly, patients with Class IV and V mutations generally have less severe CF disease [7]. Within each class, however, the amount of residual mutant CFTR activity and disease severity can vary and may be influenced by genetic modifiers and environmental factors.

1.2 Structural information on CFTR

Although the domain structure of full-length CFTR is well established, high-resolution three-dimensional structures of native or mutated CFTR remain to be determined. However, X-ray and NMR studies of CFTR subdomains, in particular NBD1 [32,33], and homology-based models for full-length CFTR [34–36] have been published in recent years. Although interpretation of the first NBD1 crystallographic studies may have been influenced by the incorporation of solubilizing mutations that attenuate trafficking and channel effects caused by the F508del mutation [37], a recent comparison of various X-ray structures of isolated native and F508del NBD1 mutants suggests that the effect of the F508 deletion on the overall NBD1 domain structure is rather subtle and localized in the direct vicinity of position 508 [33]. Interestingly, crystallographic analysis and hydrogen/deuterium exchange mass spectrometry indicate increased backbone dynamics and conformational changes in residues 509–511 in F508del-CFTR [33]. Full-length models of native CFTR suggest that this region may interact with the membrane-spanning domains of CFTR [34–36], and may thus confer additional conformational changes leading to misfolded F508del-CFTR. Another possibility is that the altered surface topography alters chaperone interactions leading to protein retention in the ER.

Although attempts to dock compounds in putative binding pockets in a model of the NBD1–NBD2 heterodimer have been published [38], more studies are required to provide the accuracy and resolution needed to guide hit and lead optimization efforts.

1.3 *In vitro* and *in vivo* assays

Fluorescent-based high-throughput screening (HTS) assays designed to detect CFTR potentiators or correctors have been developed to identify chemical starting points from small-molecule libraries and track the structure–activity relationship (SAR). These comprise the use of membrane potential-sensitive fluorescent probes and halide sensors, including genetically targeted fluorescent proteins [15–17]. These cell-based (or phenotypic) assay formats are amenable to automation and the use of high-density plate formats to enable large-scale screening campaigns for the purposes of hit identification. The sensitivity and reproducibility of these HTS assay formats also facilitate SAR evaluation. When using recombinant cell-based models expressing CFTR for monitoring CFTR modulator activity, the choice of CFTR mutation (e.g., F508del vs. G551D) [17], cell background [39], use of human or rodent CFTR [17], and assay format (e.g., membrane potential vs. current measurement) can impact the potency (EC_{50}) and/or efficacy (fold increase in Cl^- transport) of the compound. This can make it difficult to compare the SAR of compounds between different assay formats or laboratories.

To profile compound activity on CFTR-mediated Cl^- transport and epithelial cell function in a physiologically relevant disease model, cultured human bronchial epithelial (HBE) cells derived from CF patients carrying different mutant CFTR forms can be used. Cultured HBE exhibit several of the morphological and functional defects in airway epithelial believed to contribute to the development of CF lung disease. These include the loss of CFTR-mediated Cl^- and fluid secretion, excessive ENaC-mediated Na^+ and fluid absorption, and decreased cilia beating secondary to decreased surface fluid [4–6,40,41]. In addition to cultured HBE, *in vivo* mouse models have been developed by targeted mutations in the CFTR gene. However, the use of these models in the development of CFTR modulators for the treatment of CF is limited because they lack the manifestations typically found in humans with CF, including airway and pancreatic disease [42,43]. In addition, some CFTR modulators have been shown to lack activity on mouse wild-type and mutant CFTR forms [17]. The recent development of the CFTR knockout porcine model of CF indicated that newborn piglets exhibit many of the manifestations that characterize early CF disease in humans, including meconium ileus, pancreatic destruction, early focal biliary cirrhosis, and gall bladder abnormalities [44]. This suggests that the development of genetically engineered pigs carrying CFTR mutations, such as F508del, could provide disease-relevant models of CF to characterize the effects of CFTR modulators on CF disease pathogenesis.

2. CFTR POTENTIATORS

Several excellent reviews have appeared recently covering emerging therapies in CF, including CFTR modulators [45–47] and a collection of CFTR modulators are available through the CFF Consortium for research purposes [48]. Multiple structural classes that potentiate CFTR have been identified by screening compound libraries, natural products, and approved drugs. A subset is described below, which has proved amenable to lead optimization.

2.1 Pyrazoles

High-throughput screen of a diverse collection of small molecules using F508del CFTR-NIH/3T3 cells resulted in the identification of CFTR potentiator **1** [16]. Compound **1** displayed a measured EC_{50} of 2.4 μM. Patch-clamp experiments recorded in temperature-corrected F508del-CFTR in NIH/3T3 cells indicated that 20 μM of **1** increased the open probability of the channel and restored the gating of mutant CFTR to wild-type levels. In addition, **1** potentiated the defective gating of G551D-CFTR mutations expressed in FRT (Fisher rat thyroid) cells with a measured EC_{50} of 20 μM. To confirm the activity of **1** in HBE cells isolated from CF F508del homozygous donor patients, *in vitro* pharmacology experiments were carried out. In CF-HBE, the cells were corrected with **15** (see Section 3) for 48 h prior to the acute addition of **1**. Under these experimental conditions, **1** displayed an EC_{50} of 2.7 μM, which is similar to that obtained in NIH/3T3 cells.

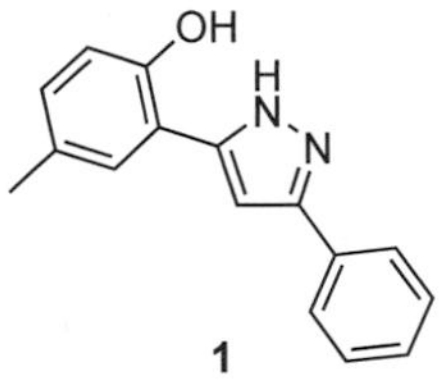

1

2.2 Phenylglycines

Phenylglycine analogs, generically represented by **2**, have been described and characterized as potentiators of various CFTR mutants (wt, F508del, G551D, G1349D) [49]. Compound **3**, the most potent compound described, potentiated F508del-CFTR in FRT cells at concentrations below 100 nM with maximal currents comparable with those of genistein at 50 μM. In FRT cells expressing the G1349D-CFTR mutation, 3 had a measured K_d of ~40 nM. The effect in gating produced by 3 was evaluated by cell-attached patch-clamp experiments using FTR cells. The addition of 3 caused an increase in the open probability of the channel reaching that of the wild-type channel. *In vitro* pharmacology in cultured cells isolated from human nasal epithelium from F508del homozygous patients resulted in significantly increased CFTR-mediated chloride current after addition of nanomolar concentrations of 3. Similar results were obtained in cells from G551D or D1152H/F508del subjects in response to 3. It was postulated that 3 could be a good candidate for inhalation therapy to overcome its poor metabolic stability and short iv half-life.

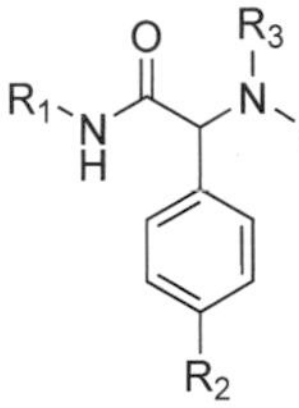

2 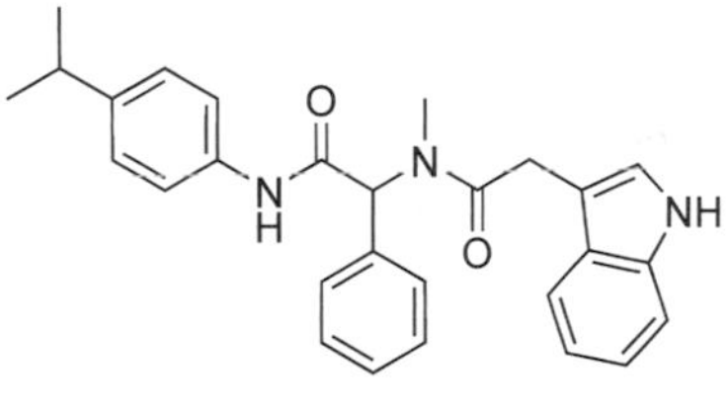   3

2.3 4-Oxoquinoline-3-carboxamides

Another class of CFTR potentiators identified by HTS are the sulfamoyl-4-oxoquinoline-3-carboxamides [49,50]. Compound 4 has been extensively characterized and has been shown to selectively potentiate the gating of F508del-CFTR. Compound 4, at 100 nM, potentiated CFTR-dependent chloride currents to a level similar to that of genistein at 50 μM concentration. In cultured epithelial cells isolated from homozygous F508del patients, temperature corrected by incubation at 27°C for 24 h, compound 4 produced a dose-dependent increase in CFTR-dependent chloride currents at a concentration of 500 nM. To assess the ability of 4 to potentiate different CFTR mutations, compound 4 was tested in FTR cells expressing wt, G551D, and G1349D mutations. Compound 4 was less efficacious than genistein in potentiating wt-CFTR and did not increase chloride current in G551D- and G1349D-CFTR expressing cells. The activity of several analogs was reported, and the most potent compounds described were 5 and 6, with measured K_as of 30 and 20 nM, respectively.

VX-770 (7) has been recently reported [17] and is currently undergoing clinical investigation for the treatment of CF. Compound 7 has been characterized as a potentiator of various CFTR mutants (wt, F508del, G551D) in both recombinant cells and primary HBE cell cultures (G551D/F508del-CFTR, $EC_{50} = 236 \pm 200$ nM; F508del/F508del, $EC_{50} = 22 \pm 10$ nM). In G551D/F508del HBE cell cultures, incubation with compound 7 (10 µM) resulted in an increase in CFTR function followed by an increase in both fluid levels on the apical surface of the cell cultures and cilia beating.

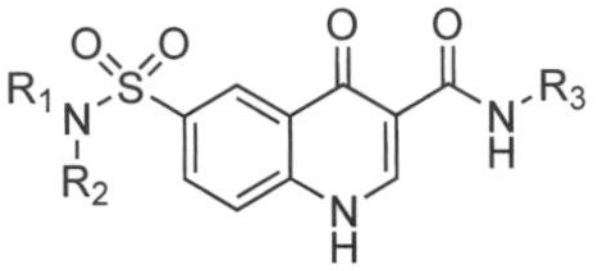

	R_1	R_2	R_3
4	Me	2-EtOPh	allyl
5	Et	Ph	2-OMeBn
6	Et	Ph	cycloheptyl

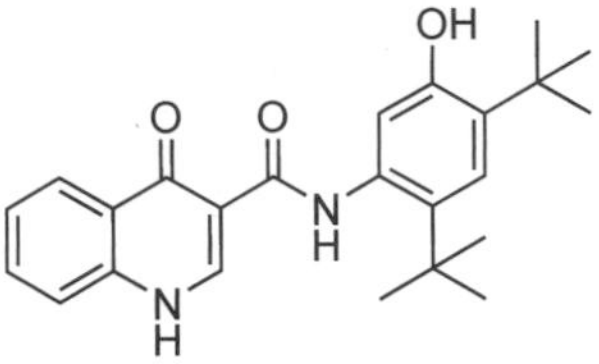

2.4 1,4-Dihydropyridines

1,4-Dihydropyridines (DHPs) are widely used as antihypertensive agents due to their ability to block the L-type calcium channel. DHPs have also been characterized as potentiators of several mutant CFTR channels [51]. The potentiator activity of this class of compounds appears to be mediated by CFTR and not by inhibition of calcium channels. In FRT cells expressing the F508del mutation, felodipine (8) showed a 30-fold potency improvement over genistein ($K_a = 23.6$ µM) with a measured K_a of 0.7 µM. Compound 8 was extensively characterized and was shown to be an effective potentiator of wild-type CFTR ($K_a = 0.5$ µM; genistein, $K_a = 16$ µM) and of the G551D mutation, albeit at higher concentration than those required to potentiate F508del-CFTR ($K_a = 23.3$ µM). Patch-clamp data for 8 using inside-out configuration experiments in FRT cells expressing the G551D were recorded. The data indicated that the addition of 100 µM of 8 resulted in ~10-fold increase in the open probability of the channel. Compound 8 was tested in differentiated primary cultures of airway epithelial cells isolated from CF patients and in nasal epithelial cells. When airway epithelial cells were isolated from G551D patients, a small but significant response was observed when 20 µM of 8 was applied. In temperature-corrected nasal epithelial cells isolated from F508del-CFTR patients, 8 increased the forskolin-stimulated chloride current by 100–150% at concentrations around 5 µM. Several reports have focused on SARs around 8, targeting improvement in the compound selectivity, which led to the identification of DHP-194 (9), a potent

CFTR potentiator (F508del-CFTR K_a= 0.11 μM, G551D-CFTR K_a= 1.2 μM) with ~250-fold selectivity over the L-type voltage-dependent Ca^{2+} channel [52,53].

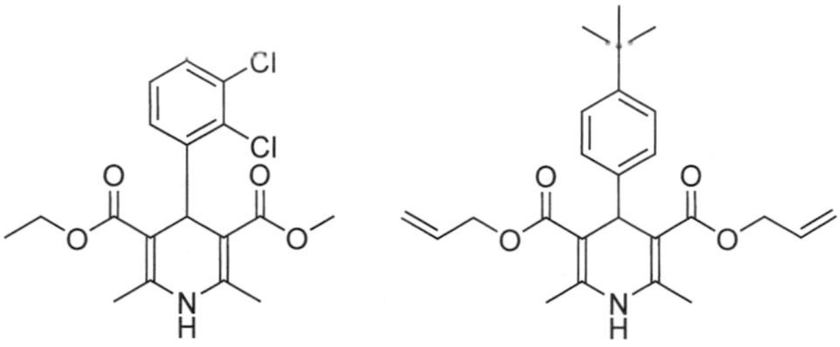

8   9

2.5 Pyrrolo[2,3-*b*]pyrazines

A series of 6-phenylpyrrolo[2,3-*b*]pyrazines, initially described as CDK/GSK-3 inhibitors, were also shown to potentiate CFTR (wt, F508del, and G551D) [54]. Compound **10** was shown to potentiate multiple CFTR mutants with submicromolar affinity (140–152 nM on wt-CFTR (calu-3 and CHO cells), 1.5 nM on G551D (CHO cells), and 111 nM on F508del-CFTR (temperature corrected CF15 cells)) and to stimulate transepithelial ion transport in the proximal colon of mice (wt) under short-circuit conditions with an affinity of ~90 nM.

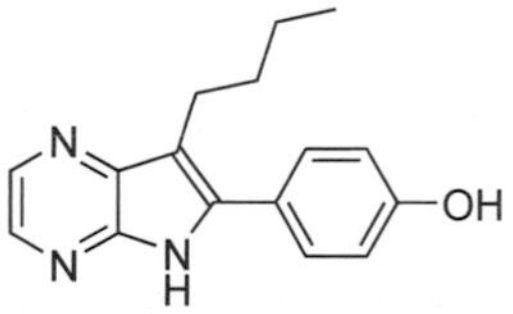

10

2.6 Methylglyoxal adenine derivatives

Water-soluble methylglyoxal adenine derivative **11** has recently been reported as a potentiator of wild-type and F508del-CFTR mutations in several cell lines [55]. *Ex vivo* addition of **11** to *cftr*$^{+/+}$ mice colonic epithelium induced a dose-dependent elevation of chloride secretion with a measured EC_{50} of 175 ± 1.1 μM. *In vivo*, in *cftr*$^{+/+}$ mice, addition of **11** in the presence of isoprenaline induced a dose-dependent salivary secretion with an EC_{50} of 7.1 ± 1.1 μM whereas no effect of isoprenaline and **11** was observed on the salivary secretion of *cftr*$^{-/-}$ mice.

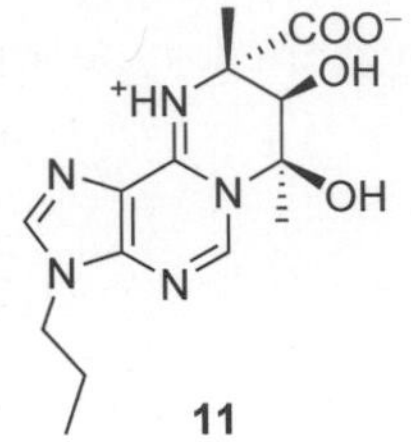

11

3. CFTR CORRECTORS

3.1 Methylbithiazoles

Methylbithiazole derivatives [22] have been reported as CFTR correctors and are exemplified by **12**. It has been reported that **12** corrects the defective folding of F508del-CFTR in FRT cells resulting in approximately ~5.9-fold increase in forskolin/genistein-stimulated chloride currents over DMSO-treated cells. Western blot analysis in BHK (baby hamster kidney) and FRT cells confirmed that incubation with **12** resulted in significant improvement of F508del-CFTR maturation (C-band). The effect of **12** is believed to occur through the improvement of the folding efficiency of F508del-CFTR (2~3-fold), leading to 30–50% delay in ER-associated degradation of the protein and, as a result, an enhanced residence time of the mutant protein at the cell surface. The authors postulate that by improving the folding of mutant CFTR it is less susceptible to the ubiquitin-dependent peripheral quality control mechanism and lysosome degradation than low-temperature-corrected F508del. Incubation of **12** in bronchial epithelial cells isolated from F508del-CFTR homozygous subjects produced a significant increase in CFTR-dependent chloride secretion, equivalent to 8% of measurements in non-CF epithelia.

Extensive lead optimization around **12** ($V_{max} = 122 \pm 9\,\mu M/s$; $K_d = 0.9$ μM) established that peripheral modifications of the methylbithiazole core modulates corrector activity and resulted in **13** ($V_{max} = 127 \pm 8\,\mu M/s$; K_d $0.7\,\mu M$) with improved activity [56]. Further studies focused on the bithiazolyl moiety and concluded that the *s-cis* conformation of the planar dithiazole (as shown) is important for F508del-CFTR correction activity [57].

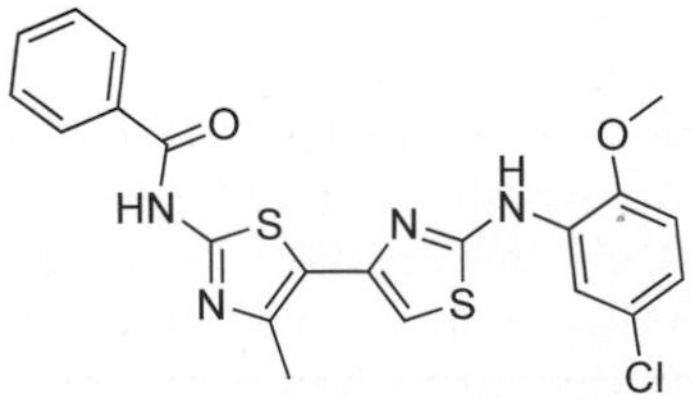

12

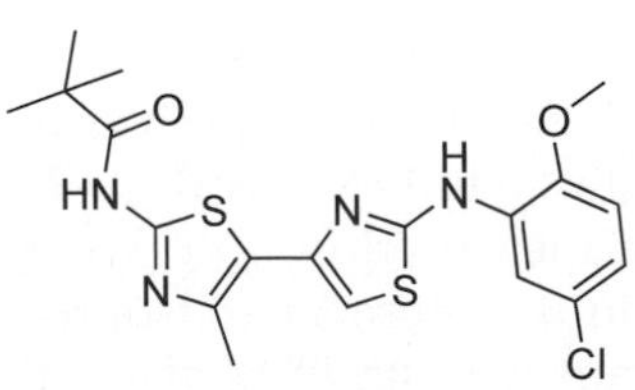

13

3.2 Quinazolines

Compound **14** (EC_{50} of 3.7 µM; efficacy of 47% of 27°C correction) was identified by HTS in NIH/3T3 cells expressing F508del-CFTR [16]. Further optimization resulted in 4-alkoxyquinazoline **15** with significantly improved efficacy (74% of 27°C correction). Western blot techniques in HEK-293 cells showed a dose-dependent increase of the mature form of F508del-CFTR (**14**, $EC_{50} = 4.0$ µM; **15**, $EC_{50} = 0.8$ µM). Whole-cell patch-clamp techniques in F508del-CFTR NIH/3T3 cells were employed to show that incubation with **15** increased the current density stimulated by forskolin and genistein and that the increase in cell surface expression of mature CFTR was leading to increased functional activity.

The corrector activity of **15** was not limited to recombinant systems as it was confirmed in cultured HBE cells isolated from F508del homozygous patients. After 48-h incubation with 6.7 µM **15**, a 2-fold increase in forskolin-stimulated chloride current (compared to DMSO-treated cells) or 14% of the response obtained in non-CF-HBE was observed. It was also reported that **14** and **15** increase the cell surface density of other trafficking-deficient proteins, including G601S-hERG. The authors speculate that this class of compounds may act on the folding pathway shared by both misfolded proteins rather than on CFTR itself [16].

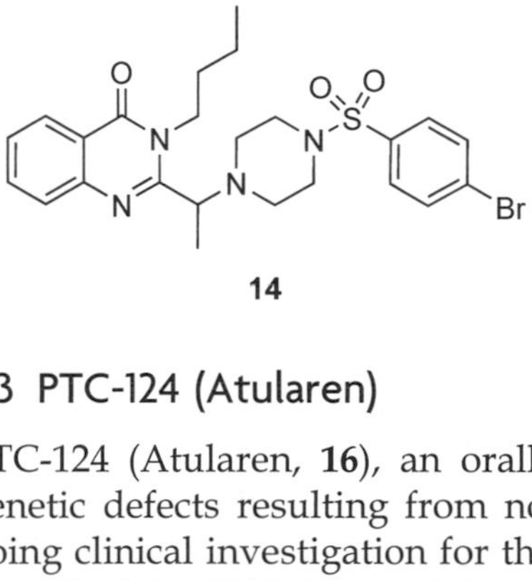

14 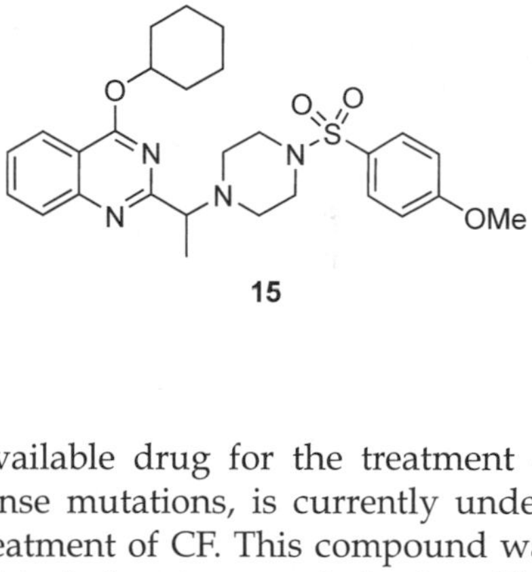   **15**

3.3 PTC-124 (Atularen)

PTC-124 (Atularen, **16**), an orally available drug for the treatment of genetic defects resulting from nonsense mutations, is currently undergoing clinical investigation for the treatment of CF. This compound was identified by HTS followed by medicinal chemistry optimization. This compound allows the cellular machinery to bypass the nonsense mutation and continue the translation process to restore the function of full-length, functional protein. The potential for **16** to treat ~10% of the CF patients that carry an in-frame nonsense mutation that promotes premature termination of translation of the CFTR mRNA was evaluated in a mouse line carrying the knockout of the endogenous CFTR locus ($Cftr^{-/-}$) and expressing the human CFTR (hCFTR) cDNA, containing the most common CFTR-G542X premature stop mutation [58]. Treatment of $Cftr^{-/-}$ hCFTR-G542X mice, with a daily dose of **16** (60 mg/kg, subcutaneous) for 14–21 days, partially restored the expression of functional hCFTR protein and the cAMP-stimulated trans-epithelial chloride currents (29% of wild-type mice).

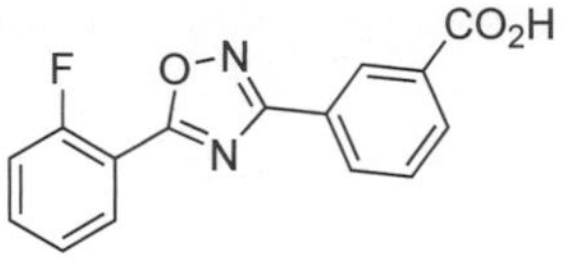

16

3.4 Other chemotypes

One promising approach to identify novel CFTR corrector structural classes is to screen compounds, already developed for other indications, for their ability to correct the trafficking defect of mutant CFTR. Following this approach, glafenine hydrochloride (**17**), miglustat (**18**), and SAHA (**19**) were recently described as promising CFTR corrector leads.

Ester **17**, which is nonsteroidal and anti-inflammatory, partially corrected the defective processing of F508del-CFTR *in vitro* in multiple cell lines (BHK, HEK-293, and CFBE41o– human airway epithelial cells) [59]. Administration of **17** to homozygous F508del-CFTR mice for 48 h, using a micro-osmotic pump delivery system (50 µg/h), increased the salivary secretion to ∼6.6% of normal wild-type control mice. Compound **18**, an inhibitor of the α-1,2 glucosidase, was reported as a CFTR corrector [60]. Daily treatment of a human CF nasal epithelial cell line (JME/CF15) expressing F508del-CFTR for 58 days with 3 µM of **18** resulted in a reversible and sustained correction of F508del-CFTR trafficking, down-regulation of sodium hyperabsorption, and regulation of calcium homeostasis. Histone deacetylase inhibitor SAHA (**19**) was reported to correct the defective processing of F508del-CFTR in human primary airway epithelia to levels that are ∼28% of wild-type CFTR [61].

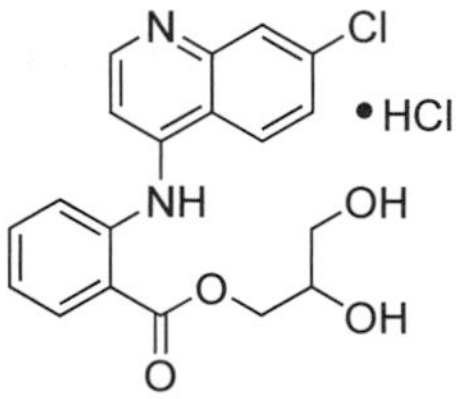

17 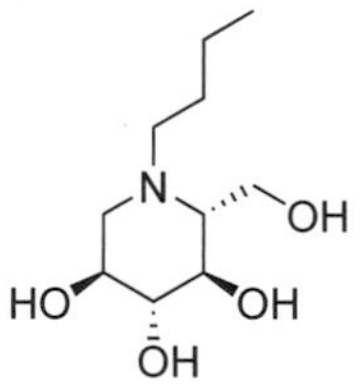   **18** 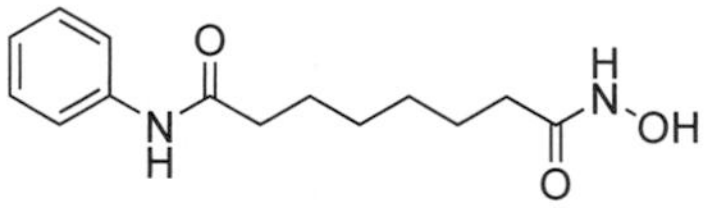   **19**

4. DUAL-ACTING CFTR MODULATORS

Two approaches have been taken to identify CFTR modulators with both CFTR potentiator and CFTR corrector activities in a single molecule. One has been dubbed the "hybrid approach," in which corrector and potentiator compounds are covalently connected through a bio-unstable linker. A recent example is compound **20**, which is composed of a phenylglycine-based potentiator that was coupled with a methylbithiazole corrector using an enzymatically hydrolyzable linker [62]. Compound **20** showed no activity as a CFTR potentiator or corrector but both individual fragments after hydrolysis exhibited strong activity in their corresponding assays. Cleavage of hybrid molecule **20** *in vitro* by carbonic anhydrase, or by incubation with the intestinal contents from mice at 37°C, produced the active potentiator and corrector fragments.

A second approach, recently described in the patent literature, involves a single molecule that displays both potentiator and corrector activities and is exemplified by **21** [63].

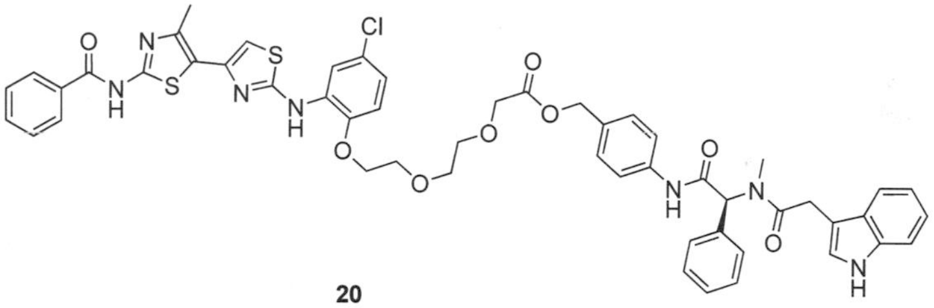

20

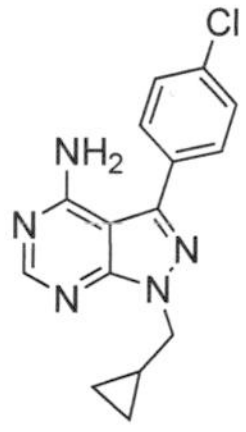

21

5. RECENT CLINICAL DATA

Three orally administered investigational CFTR modulators are currently being evaluated in clinical studies: VX-770 (**7**), VX-809 (structure not currently disclosed), and Atularen (**16**) [64,65]. Compound **16** is being evaluated in a longer–term, multicenter, randomized, double-blind, placebo-controlled Phase III study in patients with nonsense CFTR mutations. In a Phase IIa trial of this investigational drug in pediatric and adult CF patients, it was shown that oral administration of Atularen was well tolerated and resulted in the generation of functional CFTR and statistically significant improvements in CFTR chloride channel function in the nasal airways [66].

Several clinical studies with drug candidates VX-770 (**7**) and VX-809 have been initiated. The more advanced of the two is **7**, which is being evaluated in a registration program that focuses on CF patients with at least one copy of the G551D mutation (Class III). In a Phase IIa study, treatment of 39 patients carrying the G551D mutation on at least one allele with **7** was well tolerated with no treatment discontinuation and resulted in a significant improvement in sweat chloride to near or below the diagnostic threshold for CF as well as statistically significant improvements in CFTR chloride channel function in nasal airways [67]. In addition, improvement in lung function was observed, as measured by spirometry (FEV$_1$) [68]. A 28-day Phase IIa trial of the investigational CFTR corrector compound, VX-809, in CF patients homozygous for the F508del mutation showed that VX-809 was well tolerated and led to a statistically significant improvement in sweat chloride, an important biomarker of CFTR activity. No statistically significant effect on FEV$_1$ was observed [69]. Plans have been disclosed for evaluating 7 in combination with VX-809 in a Phase IIa clinical trial in patients carrying the F508del mutation [69].

6. CONCLUSIONS

In the past 10 years, considerable progress has been made in the field of CFTR modulator drug discovery. The most successful approach so far has been high-throughput phenotypic screening, followed by extensive hit-to-lead optimization. As a result, three compounds have been identified that entered clinical development. The clinical data available to date are limited but encouraging in that they suggest that increasing CFTR function may provide clinical benefits to CF patients with systemic therapies that appear to be well tolerated. These early clinical results indicate that CFTR, as a target for CF drug discovery, has reached the next stage of validation as a drug target. Obviously, more studies will be required to study the effects of long-term use of CFTR modulators in a greater diversity of CF patients in terms of their genetic background, age, and disease stage. With no approved CFTR modulators to date, CFTR drug discovery is still at an early stage.

REFERENCES

[1] H. H. Kazarian Jr., *Hum. Mutat.*, 1994, **4**, 167.
[2] The website http://www.cff.org/ is maintained and updated by the Cystic Fibrosis Foundation. It provides a wealth of data on all aspects of the disease, the patients, science, treatment, and the therapeutic pipeline.
[3] M. R. Knowles, M. J. Stutts, A. Spock, N. Fischer, J. T. Gatzy, and R. C. Boucher, *Science*, 1983, **221**, 1067.
[4] P. M. Quinton, *Nature*, 1983, **301**, 421.
[5] J. R. Riordan, J. M. Rommens, B. Kerem, N. Alon, R. Rozmahel, Z. Grzelczak, J. Zielenski, S. Lok, N. Plavsic, J. L. Chou, M. L. Drumm, M. C. Iannuzzi, F. C. Collins, and L. Tsui, *Science*, 1989, **245**, 1066.
[6] R. C. Boucher, *Trends Mol. Med.*, 2007, **13**, 231.
[7] C. Castellani, H. Cuppens, M. Macek, J. J. Cassiman, E. Kerem, P. Durie, E. Tullis, B. M. Assael, C. Bombieri, A. Brown, T. Casals, M. Claustres, G. R. Cutting, E. Dequeker, J. Dodge, I. Doull, P. Farrell, C. Ferec, E. Girodon, M. Johannesson , B. Kerem, M. Knowles, A. Munck, P. F. Pignatti, D. Radojkovic, P. Rizzotti, M. Schwarz, M. Stuhrmann, M. Tzetis, J. Zielenski, and J. S. Elborn, *J. Cyst. Fibros.*, 2008, **7**, 179.

[8] P. M. Farrell, B. J. Rosenstein, T. B. White, F. J. Accurso, C. Castellani, G. R. Cutting, P. R. Durie, V. A. Legrys, J. Massie,R. B. Parad, M. J. Rock, and P. W. Campbell III, *J. Pediatr.*, 2008, **153**, S4.

[9] N. S. Joo, T. Irokawa, R. C. Robbins, and J. J. Wine, *J. Biol. Chem.*, 2006, **281**, 7392.

[10] H. Matsui, B. R. Grubb, R. Tarran, S. H. Randell, J. T. Gatzy, C.W Davis, and R. C. Boucher, *Cell*, 1998, **95**, 1005.

[11] R. L. Gibson, J. L. Burns, and B. W. Ramsey, *Am. J. Respir. Crit. Care Med.*, 2003, **168**, 918.

[12] P. M. Quinton, *Lancet*, 2008, **372**, 415.

[13] C. Frerichs and A. Smyth, *Expert Opin. Pharmacother.*, 2009, **10**, 1191.

[14] B. P. O'Sullivan and S. D. Freedman, *Lancet*, 2009, **373**, 1891.

[15] A. S. Verkman and L. J. V. Galietta, *Nat. Rev. Drug. Discov.*, 2009, **8**, 153.

[16] F. Van Goor, K. S. Strayley, D. Cao, J. Gonzalez, S. Hadida, A. Hazlewood, J. Joubran, T. Knapp, L. R. Lewis, M. Miller, T. Neuberger, E. Olson, V. Pachenko, J. Rader, A. Singh, J. H. Stack, R. Tung, P. D. J. Grootenhuis, and P. Negulescu, *Am. J. Physiol. Lung Cell Mol. Physiol.*, 2006, **290**, L1117.

[17] F. Van Goor, S. Hadida, P. D. J. Grootenhuis, B. Burton, D. Cao, T. Neuberger, A. Turnbull, A. Singh, J. Joubran, A. Hazlewood, J. Zhou, J. McCartney, V. Arumugam, C. Decker, J. Yang, C. Young, E. R. Olson, J. J. Wine, R. A. Frizzell, M. Ashlock, and P. Negulescu, *Proc. Natl. Acad. Sci. USA*, 2009, **106**, 18825.

[18] A. S. Verkman, P. M. Haggie, and J. L. Galietta, *Rev. Fluores.*, 2004, **1**, 85.

[19] L. Tradtrantip, N. D. Sonawane, W. Namkung, and A. S. Verkman, *J. Med. Chem.*, 2009, **52**, 6447.

[20] P. L. Shah, *Ann. Rep. Med. Chem.*, 2001, **36**, 67.

[21] M.T Clunes and R.C Boucher, *Curr. Opin. Pharmacol.*, 2008, **8**, 292.

[22] N. Pedemonte, G. L. Lukacs, K. Du, E. Caci, O. Zegarra-Moran, L. J. V. Galietta, and A. S. Verkman, *J. Clin. Invest.*, 2005, **115**, 2564.

[23] F. Van Goor, S. Hadida, and P. D. J. Grootenhuis, *Top. Med. Chem.*, 2008, **3**, 29.

[24] D. N. Sheppard and M. J. Welsh, *Physiol. Rev.*, 1999, **79**, S23.

[25] P. G. Noone and M. R. Knowles, *Respir. Res.*, 2001, **2**, 328.

[26] A. M. Cantin, G. Bilodeau, C. Ouellet, J. Liao, and J. W. Hanrahan, *Am. J. Physiol. Cell. Physiol.*, 2006, **290**, C262.

[27] A. M. Cantin, J. W. Hanrahan, G. Bilodeau, L. Ellis, A. Dupuis, J. Liao, J. Zielenski, and P. Durie, *Am. J. Respir. Crit. Care Med.*, 2006, **173**, 1139.

[28] S. H. Cheng, R. J. Gregory, J. Marshall, S. Paul, D. W. Souza, G. A. White, C. R. O'Riordan, and A. E. Smith, *Cell*, 1990, **63**, 827.

[29] G. M. Denning, L. S. Ostedgaard, and M. J. Welsh, *J. Cell. Biol.*, 1992, **118**, 551.

[30] G. L. Lukacs, A. Mohamed, N. Kartner, X. B. Chang, J. R. Riordan, and S. Grinstein, *EMBO J.*, 1994, **13**, 6076.

[31] C. L. Ward and R. R. Kopito, *J. Biol. Chem.*, 1994, **269**, 25710.

[32] H. A. Lewis, X. Zhao, S. G. Buchanan, S. K. Burley, K. Conners, M. Dickey, M. Dorwart, R. Fowler, X. Gao, W. J. Guggino, W. A. Hendrickson, J. F. Hunt, M. C. Kearins, D. Lorimer, P. C. Maloney, K. W. Post, K. R. Rajashankar, S. Shriver, P. H. Thibodeau, P. J. Thomas, M. Zhang, X. Zhao, and S. Emtage, *EMBO J.*, 2004, **23**, 282.

[33] H. A. Lewis, C. Wang, X. Zhao, Y. Hamuro, K. Connors, M. C. Kearins, F. Lu, J. M. Sauder, K. Molnar, S. J. Coales, P. C. Maloney, W. B. Guggino, D. R. Wetmore, P. C. Weber, and J. F. Hunt, *J. Mol. Biol.*, 2010, **396**, 406.

[34] J. L. Mendoza and P. J. Thomas, *J. Bioenerg. Biomembr.*, 2007, **39**, 499.

[35] A. W. R. Serohijos, T. Hegedus, A. A. Aleksandrov, L. He, L. Cui, N. V. Dokholyan, and J. R. Riordan, *Proc. Natl. Acad. Sci. USA*, 2008, **105**, 3256.

[36] J. P. Mornon, P. Lehn, and I. Callebaut, *Cell. Mol. Life Sci.*, 2009, **66**, 3469.

[37] L. S. Pissarra, C. M. Farinha, Z. Xu, A. Schmidt, P. H. Thibodeau, Z. Cai, P. J. Thomas, D. N. Sheppard, and M. D. Amaral, *Chem. Biol.*, 2008, **15**, 62.

[38] O. Moran, L. J. V. Galietta, and O. Zegarra-Moran, *Cell. Mol. Life Sci.*, 2005, **62**, 446.

[39] N. Pedemonte, V. Tomati, E. Sondo, and L. J. V. Galietta, *Am. J. Phys. Cell Physiol.*, 2010, **298**, C866.

[40] M. Knowles, J. Gatzy, and R. Boucher, *J. Clin. Invest.*, 1983, **71**, 1410.

[41] C. Jiang, W. E. Finkbeiner, J. H. Widdicombe, P. B. McCray, and S. S. Miller, *Science*, 1993, **262**, 424.

[42] B. R. Grubb and S. E. Gabriel, *Am. J. Physiol.*, 1997, **273**, G258.

[43] C. Guilbault, Z. Saeed, G. P. Downey, and D. Radzioch, *Am. J. Respir. Cell. Mol. Biol.*, 2007, **36**, 1.

[44] C. S. Rogers, D. A. Stoltz, D. K. Meyerholz, L. S. Ostedgaard, T. Rokhlina, P. J. Taft, M. P. Rogan, A. A. Pezzulo, P. H. Karp, O. A. Itani, A. C. Kabel, C. L. Wohlford-Lenane, G. J. Davis, R. A. Hanfland, T. L. Smith, M. Samuel, D. Wax, C. N. Murphy, A. Rieke, K. Whitworth, A. Uc, T. D. Starner, K. A. Brogden, J. Shilyansky, P. B. McCray Jr., J. Zabner, R. S. Prather, and M. J. Welsh, *Science*, 2008, **321**, 1837.

[45] A. M. Jones and J. M. Helm, *Drugs*, 2009, **69**, 1903.

[46] F. Becq, *Drugs*, 2010, **70**, 241.

[47] M. F. N. Rosser, D. E. Grove, and D. M. Cyr, *Curr. Biol. Chem.*, 2009, **3**, 100.

[48] http://www.cftrfolding.org/files/CFFTCompounds.pdf.

[49] N. Pedemonte, N. D. Sonawane, A. Taddei, J. Hu, O. Zegarra-Moran, Y. F. Suen, L. I. Robins, C. W. Dicus, D. Willenbring, M. H. Nantz, M. J. Kurth, L. J. V. Galietta, and A. S. Verkman, *Mol. Pharm.*, 2005, **67**, 1797.

[50] Y. F. Suen, L. Robins, B. Yang, A. S. Verkman, M. H. Nantz, and M. J. Kurth, *Bioorg. Med. Chem. Lett.*, 2006, **16**, 537.

[51] N. Pedemonte, T. Diena, E. Caci, E. Nieddu, M. Mazzei, R. Ravazzolo, O. Zegarra-Moran, and L. Galietta, *Mol. Pharmacol.*, 2005, **68**, 1736.

[52] N. Pedemonte, D. Boido, O. Moran, M. Giampieri, M. Mazzei, R. Ravazzolo, and L. J. V. Galietta, *Mol. Pharmacol.*, 2007, **72**, 197.

[53] F. Cateni, M. Zacchigna, N. Pedemonte, L. J. V. Galietta, M. T. Mazzei, P. Fossa, M. Giampieri, and M. Mazzei, *Bioorg. Med. Chem. Lett.*, 2009, **17**, 7894.

[54] S. Noel, C. Faveau, C. Norez, C. Rogier, Y. Mettey, and F. Becq, *J. Pharm. Exper. Therap.*, 2006, **319**, 349.

[55] J. Bertrand, B. Boucherle, A. Billet, P. Melin-Heschel, L. Dannhoffer, C. Vandebrouck, C. Jayle, C. Routaboul, M. C. Molina, J. L. Decout, F. Becq, and C. Norez, *Eur. Respir. J.*, 2010, DOI:10.1183/09031936.00122509.

[56] C. L. Yoo, G. J. Yu, B. Yang, L. I. Robins, A. S. Verkman, and M. J. Kurth, *Bioorg. Med. Chem. Lett.*, 2008, **18**, 2610.

[57] G. J. Yu, C. L. Yoo, B. Yang, M. W. Lodewyk, L. Meng, T. T. El-Idreesy, J. C. Fettinger, D. J. Tantillo, A. S. Verkman, and M. J. Kurth, *J. Med. Chem.*, 2008, **51**, 6044.

[58] M. Du, X. Liu, E. M. Welch, S. Hirawat, S. W. Peltz, and D. M. Bedwell, *Proc. Natl. Acad. Sci. USA*, 2008, **105**, 2064.

[59] R. Robert, G. W. Carlile, J. Liao, H. Balghi, P. Lesimple, N. Liu, D. Rotin, M. Wilke, H. R. De Jonge, B. J. Scholte, D. Y. Thomas, and J. W. Hanrahan, *Mol. Pharm.* 2010, **77**, 922.

[60] C. Norez, F. Antigny, S. Noel, C. Vanderbrouck, and F. Becq, *Am. J. Respir. Cell. Mol. Biol.*, 2009, **41**, 217.

[61] D. M. Hutt, D. Herman, A. P. C. Rodrigues, S. Noel, J. M. Pilewski, J. Matteson, B. Hoch, W. Kellner, J. W. Kelly, A. Schmidt, P. J. Thomas, Y. Matsumura, W. R. Skach, M. Gentzsch, J. R. Riordan, E. J. Sorscher, T. Okiyoneda, J. R. Yates III, G. L. Lukacs, R. A. Frizzell, G. Manning, J. M. Gottesfeld, and W. E. Balch, *Nat. Chem. Biol.*, 2009, **275**, 1.

[62] A. D. Mills, C. Yoo, J. D. Butler, B. Yang, A. S. Verkman, and M. J. Kurth, *Bioorg. Med. Chem. Lett.*, 2010, **20**, 87.

[63] C. E. Bulawa, M. Devit, and D. Elbaum, *Patent Application WO 2009/062118-A2*, 2009.

[64] S. Storey and G. Wald, *Nat. Rev. Drug Discov.*, 2008, **7**, 555.
[65] M. A. Ashlock, R. J. Beall, N. M. Hamblett, M. W. Konstan, C. M. Penland, B. W. Ramsey, J. M. Van Dalsen, D. R. Westmore, and P. W. Campbell III, *Sem. Respir. Crit. Care Med.*, 2009, **30**, 611.
[66] E. Kerem, S. Hirawat, S. Armoni, Y. Yaakov, D. Shoseyov, M. Cohen, M. Nissim-Rafinia, H. Blau, J. Rivlin, M. Aviram, G. L. Elfring, V. J. Northcutt, L.L Miller, B. Kerem, and M. Wilschanski, *Lancet*, 2008, **372**, 719.
[67] F. Accurso, S. M. Rowe, P. R. Durie, M. W. Konstan, J. Dunitz, D. Hornick, S. D. Sagel, M. P. Boyle, A. Z. Uluer, R. B. Moss, S. Donaldson, J. Pilewski, R. C. Rubenstein, M. L. Aitken, B. Ramsey, S. D. Freedman, Q. Dong, J. Zha, A. Stone, E. R. Olson, C. Ordonez, P. Campbell, M. Ashlock, and J. P. Clancy, *Pediatr. Pulmonol. Suppl.*, 2009, **32**, 296.
[68] M. Boyle, J. P. Clancy, S. M. Rowe, P. Durie, J. Dunitz, M. W. Konstan, D. Hornick, S. D. Sagel, B. Ramsey, Q. Dong, C. Ordonez, P. Campbell, M. Ashlock, and F. J. Accurso, *Pediatr. Pulmonol. Suppl.*, 2009, **32**, 287
[69] http://www.vrtx.com/.

CHAPTER 11

Targeting B-cells in Inflammatory Disease

Kevin S. Currie

1. INTRODUCTION

1.1 B-cells in autoimmune/inflammatory disease

Autoimmune and inflammatory diseases remain an area of significant unmet medical need. B-cells are critical regulators of immune responses and are implicated in the pathogenesis of conditions such as rheumatoid arthritis (RA), systemic lupus erythematosus (SLE), multiple sclerosis (MS), and immune thrombocytic purpura (ITP) [1–4]. Principal among

Gilead Sciences, Inc., 36 East Industrial Road, Branford, CT 06405, USA

Annual Reports in Medicinal Chemistry, Volume 45
ISSN 0065-2725, DOI 10.1016/S0065-7743(10)45011-3

these conditions is RA. The association of rheumatoid factor (RF) with RA over 30 years ago provided early evidence for B-cell involvement [5,6], and B-cells have recently reemerged as critical players in RA pathogenesis and attractive targets for novel therapies [7,8].

B-cells contribute to the initiation and progression of inflammatory arthritis by three main mechanisms [8,9]. First, B-cells are highly efficient antigen-presenting cells that provide the signals necessary for T-cell activation [10,11]. Once activated by B-cells, T-cells produce proinflammatory cytokines such as tumor necrosis factor α (TNFα) and interleukin (IL)-1 that perpetuate the inflammatory process. Second, autoreactive B-cells produce pathogenic autoantibodies, such as RF and anti-cyclic citrullinated peptide (anti-CCP) antibodies, which are diagnostic and prognostic markers of disease [12,13]. RF immune complexes within the synovium can activate complement [10,14] and macrophages [15]. Activated macrophages in turn secrete proinflammatory cytokines which lead to joint destruction. Finally, B-cells themselves can produce proinflammatory cytokines such as TNFα and IL-6 [11]. This can occur via B-cell receptor (BCR) ligation, binding of the costimulatory ligand on activated T-cells, or interaction with circulating cytokines [16,17]. Thus, B-cells occupy a central position in RA pathogenesis from which they can initiate, amplify, and perpetuate the disease.

1.2 Clinical validation of B-cell therapy

Of the various biologic agents approved for RA, rituximab (Rituxan®) and abatacept (Orencia®) are the first to target B-cells. Rituximab is a monoclonal α-CD20 antibody that causes complete and prolonged depletion of peripheral B-cells, thereby impacting multiple mechanisms through which B-cells contribute to disease progression [4,18,19]. Abatacept is a fusion protein of human cytotoxic T-lymphocyte-associated antigen (CTLA-4) and the Fc portion (the antibody fragment without antigen binding sites) of human immunoglobulin. Abatacept binds CD80/86 on B-cells, thereby blocking the costimulatory interaction with CD28 necessary for T-cell activation. Other biologic agents in clinical development that target B-cell function include belimumab (Benlysta®) and atacicept. Belimumab is an antibody to B-lymphocyte stimulator (BLys) and has met the primary efficacy endpoints in two Phase III studies in SLE [20]. Atacicept binds to BLys and APRIL (a proliferation-inducing ligand) both of which are ligands for TNF family receptors that promote B-cell maturation and proliferation [21]. Atacicept is currently in Phase II trials for RA, MS, and SLE.

The influence of B-cells on disease pathology, and their successful targeting with biologic agents, has stimulated great interest in the discovery of small molecules that modulate B-cell function. Three kinases have emerged as the principal B-cell targets for small-molecule therapeutics: Bruton's tyrosine kinase (Btk), spleen tyrosine kinase (Syk), and phosphoinositide-3 kinase (PI3K).

1.3 Roles of Btk, Syk, and PI3K in B cell biology

Btk, Syk, and PI3K are expressed in B-cells as well as other cell types which may play a role in disease, including monocytes, macrophages, and mast cells. Given the validation of biologic therapies targeting B-cells, this chapter will focus on advances in targeting these kinases to modulate B-cell activity. Btk, Syk, and PI3K are cytoplasmic kinases that sit at the apex of the B-cell signaling cascade and are in position to profoundly impact downstream signaling events and consequent B-cell activation (Figure 1).

Btk is expressed in B-cells, monocyctes, mast cells, and osteoclasts, but not T-cells. Btk is essential for B-cell development and defects in the Btk gene lead to X-linked agammaglobulinemia (XLA) in humans, which is characterized by a lack of circulating B-cells and serum immunoglobulins [22,23]. Btk-deficient mice are resistant to the development of SLE and collagen-induced arthritis (CIA) [24]. Antigen binding to the BCR activates Src family and PI3 kinases, and Btk is recruited to the plasma

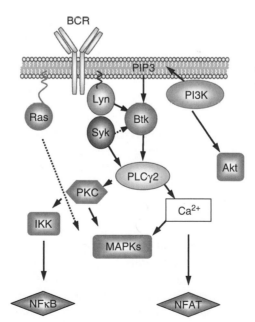

Figure 1   Btk, Syk, and PI3K in B-cell signaling. (See Color Plate 11.1 in the Color Plate Section)

membrane via its PH domain where it is transphosphorylated at Y551 in the activation loop, followed by autophosphorylation at Y223 [25,26]. Fully activated Btk phosphorylates phospholipase-Cγ2 (PLCγ2), leading to increased calcium flux, nuclear factor-kappa B (NFκB) activation, and ultimately B-cell activation and proliferation.

Syk is primarily expressed in hematopoeitic cells, as well as synoviocytes and vascular endothelial cells. Upon ligation of cell-surface immunoreceptors (e.g., BCR, FcR), Src family kinases phosphorylate the cytoplasmic immunoreceptor tyrosine-based activation motifs (ITAMs). Syk binds to phosphorylated ITAMs via its SH2 domain and is subsequently activated. Syk then activates downstream pathways, including PI3K, MAPK, and PLCγ, leading to B-cell activation and cytokine release. Syk is a key element of the BCR signaling cascade, and Syk-deficient mice suffer from an early block in B-cell development [27,28]. Syk is also critical for FcεR-mediated degranulation of mast cells [29,30].

The class 1 PI3 kinases (α, β, γ, δ) all effect the same conversion of phosphatidylinositol bisphosphate (PIP2) to phosphatidylinositol triphosphate (PIP3). The γ and δ isoforms are of most interest for inflammatory indications due to their selective expression in leukocytes. PI3Kδ appears to have a more prominent role in B-cells, while PI3Kγ has a stronger association with T-cell function [31,32]. B-cells isolated from PI3Kδ kinase-inactive knock-in mice do not proliferate in response to BCR stimulation, indicating an essential role for PI3Kδ in B-cell function [33]. PI3Kδ is located downstream of the BCR and CD19 (Figure 1) and plays a key role in establishing a signalosome that also involves Btk and Syk. The biology of PI3K δ and γ has been reviewed recently [34–38].

2. BTK INHIBITORS

Publications describing Btk inhibitors are scarce [39], but a recent upturn in patent activity points to intensifying medicinal chemistry efforts. The field is dominated by two classes: irreversible inhibitors and reversible inhibitors which are based on an imidazopyrazine-amide scaffold (Figure 2).

2.1 Irreversible inhibitors

PCI-32765 1 is the only Btk inhibitor which has been reported to have advanced to clinical trials [40]. Modeling of pyrazolopyrimidine 2 suggested that replacement of the cyclopentyl moiety could position an electrophilic group in proximity to Cys481, and subsequent optimization led to 1 [41]. Compound 1 inhibits Btk with an IC$_{50}$ of 0.8 nM, and covalent binding to Btk was confirmed by mass spectrometry and washout experiments. In the Ramos B-cell line, 1 inhibits BCR-induced calcium

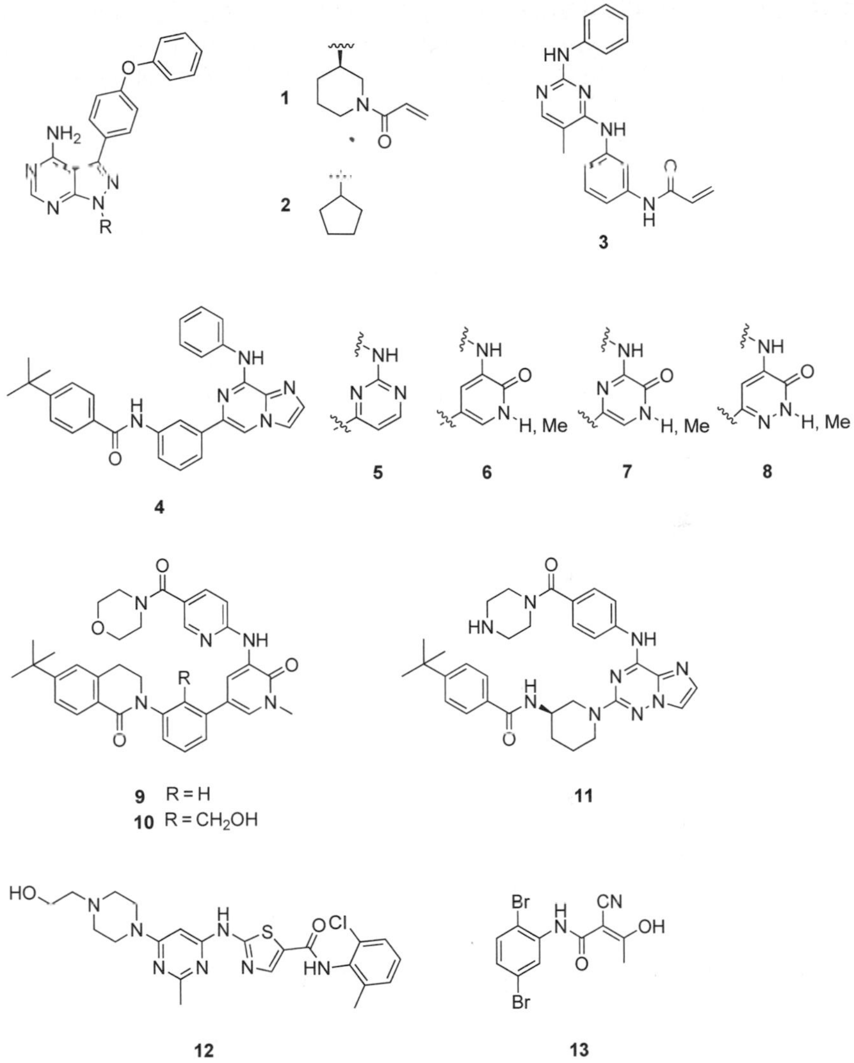

Figure 2 Btk inhibitors.

flux (IC_{50} = 40 nM), but also inhibits Ca flux in T-cells (IC_{50} = 466 nM). Since Btk is not expressed in T-cells, this data suggests functionally relevant off-target activity. Acrylamide **1** was efficacious at 3 mg/kg in murine passive transfer CIA and MRL/lpr lupus models. Not surprisingly, **1** has potent activity (0.1–11 nM) against other kinases containing a cysteine analogous to Btk Cys481, including Blk, Bmx, epidermal growth factor receptor (EGFR), Itk, and Janus kinase 3 (JAK3) [42]. Compound **1** is subject to glutathione conjugation, and analogs with increased substitution on the electrophilic double bond have reduced conjugation rates. The crystal structure of a close analog of **2** bound to Btk shows that the pyrimidine amine interacts with the gatekeeper residue, and the phenoxyphenyl group occupies a hydrophobic pocket beyond the gatekeeper [43].

Preliminary Phase I data for PCI-32765 in refractory B-cell lymphoma indicates a partial response in 5/16 patients [40]. The irreversibility of the binding interaction enables a direct assessment of target occupancy with a fluorescent probe [42]. In the Phase I study, target occupancy was essentially complete and downstream biomarkers (CD63, pERK) were significantly inhibited at 2.5 mg/kg/day.

A diaminopyrimidine scaffold has also been employed to present an electrophile for attack by Cys481 [44]. Compound **3** has an IC_{50} < 10 nM against Btk and inhibits Itk, Bmx, EGFR, and JAK3 with similar potency. Mass spectrometry confirmed that **3** covalently modifies Btk, Tec, and JAK3. Compund **3** inhibited BCR-induced PLCγ2 phosphorylation (IC_{50} = 10 nM) and human primary B-cell proliferation (IC_{50} = 10 nM), and was orally efficacious in rodent arthritis models at 30 mg/kg.

2.2 Imidazopyrazine-amide-based inhibitors

The imidazopyrazine-amide scaffold is exemplified by the early lead CGI560 (**4**), which is an adenosine-5′-triphosphate (ATP)-competitive inhibitor with an IC_{50} of 400 nM against Btk [45–47]. Optimization led to CGI1316 (structure not disclosed) which has significantly improved Btk potency (IC_{50} = 9 nM) and cellular activity (EC_{50} = 15 nM for Btk Y223 autophosphorylation). Notably, CGI1316 was >200-fold selective over a panel of 21 kinases, including Src family members. The imidazopyrazine core has been modified to give imidazopyridine and imidazopyridazine analogs [48] or replaced with monocyclic systems such as pyrimidine **5** [49], pyridone **6** [50,51], pyrazinone **7** [52,53], and pyridazinone **8** [54]. Compounds of this class inhibit transphosphorylation of Btk Y551 in addition to autophosphorylation of Y223, with one example (structure undisclosed) having IC_{50}'s of 112 and 10 nM respectively [55]. Furthermore, these compounds are claimed to inhibit the intramolecular E445/K430 H-bond that is observed in the dasatinib-Btk crystal structure [42], suggestive of binding to an inactive conformation. CGI1746 (structure undisclosed) inhibits Y551 phosphorylation, is highly selective across the kinome, and shows comparable efficacy to TNFα blockade in myeloid cell-dependent CIA models [56].

Recent patent disclosures reveal efforts to optimize within this template. The amide moiety has been cyclized to give dihydroisoquinolone inhibitors such as **9** and **10** [57]. Using various mono- and bicyclic core structure scaffolds, the amide has also been constrained in isoquinolone,

quinolone, and quinazolinone systems [58–61]. The 10-fold greater potency of compound **10** (40 nM) compared to **9** (420 nM) suggests that the hydroxymethyl group contributes an additional binding interaction [57]. Hybridizing an MK2 inhibitor scaffold [62] with the Btk template **4** provided piperidine-containing Btk inhibitors such as **11** (IC_{50} = 13 nM). Analogs within this class inhibit proliferation of mouse splenic and human tonsillar B-cells with EC_{50}'s under 300 nM [63].

Relatively high molecular weight is a feature of the chemotype exemplified by **4–11** and this can require creative formulation techniques. Pharmacokinetic properties of a lead candidate (structure unknown) from the same series that provided **10** were inadequate to provide sufficient exposures at high doses to support preclinical safety studies. However, cocrystal formulations with saccharin or gentisic acid improved water solubility by 50-fold and increased oral exposures up to 10-fold relative to traditional formulations at 20 mg/kg [64].

2.3 Miscellaneous inhibitors

Dasatinib (Sprycel®) **12** inhibits Btk (IC_{50} = 5 nM) as well as the basal and lipopolysaccharide (LPS)-induced secretion of TNFα and IL-6 in U937 cells [65]. However, dasatinib's promiscuity [66] makes it difficult to ascribe these functional effects to direct Btk inhibition. The dasatinib-Btk(Y551E) crystal structure indicates binding to an active conformation, similar to that observed with Abl [43]. LFM-A13, **13**, was first described as a Btk inhibitor more than 10 years ago [67], although its poor Btk activity (IC_{50} = 2.5–17 μM) and selectivity [68] require that caution be taken in interpreting functional effects.

3. SYK INHIBITORS

The clinical efficacy of R406/788 (**14/15**) has heightened interest in Syk inhibition as a therapeutic strategy in RA. There is a clear best-in-class opportunity for Syk inhibitors with improved selectivity and safety profiles (Figure 3).

3.1 R406/R788 (fostamatinib)

R406, **14**, is the active form of the phosphate prodrug R788, **15** [69,70]. Compound **14** has an IC_{50} of 41 nM against Syk and potently inhibits the BCR signaling pathway, as measured by inhibition of CD69 upregulation on primary human B-cells (IC_{50} = 48 nM). In addition, **14** inhibits cytokine release from mast cells, monocytes/macrophages, and neutrophils

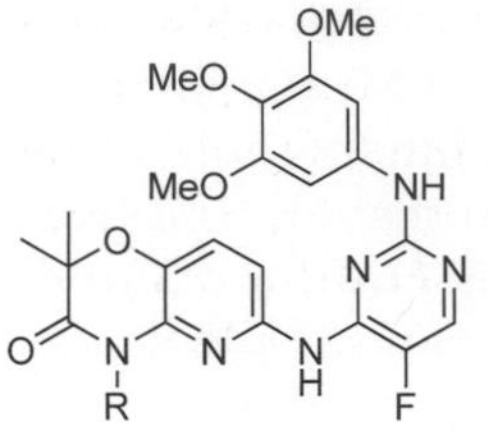

14 R = H
15 R = CH$_2$OP(O)(OH)$_2$

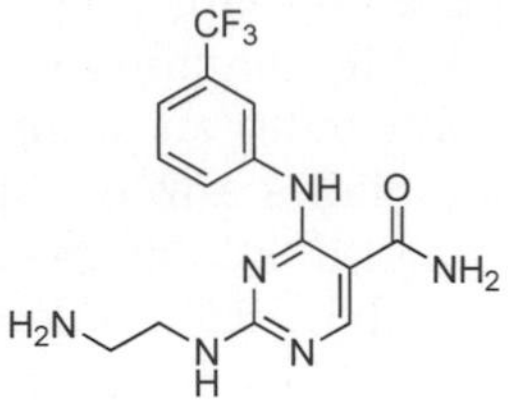

16

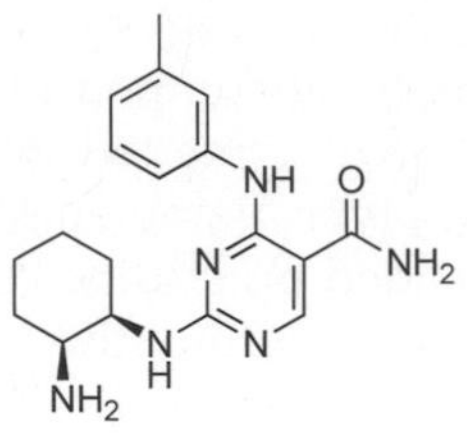

17

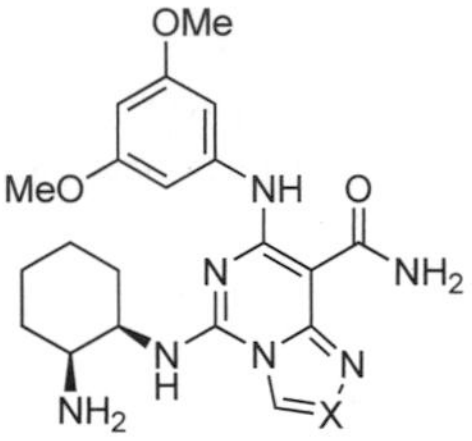

18 X = N
19 X = CH

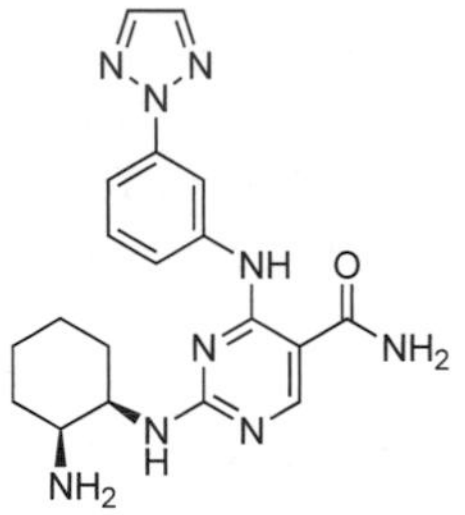

20

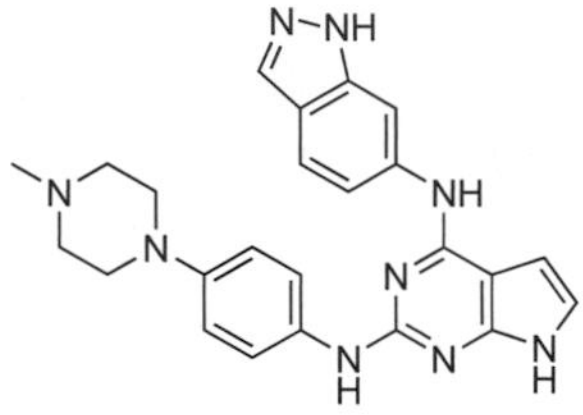

21

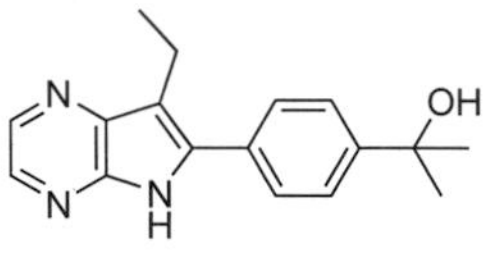

22

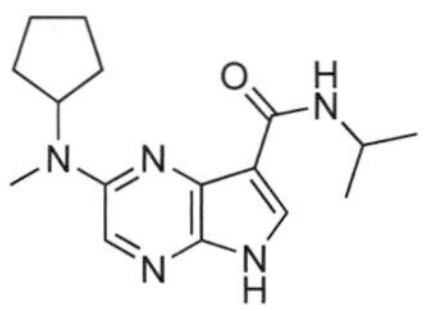

23

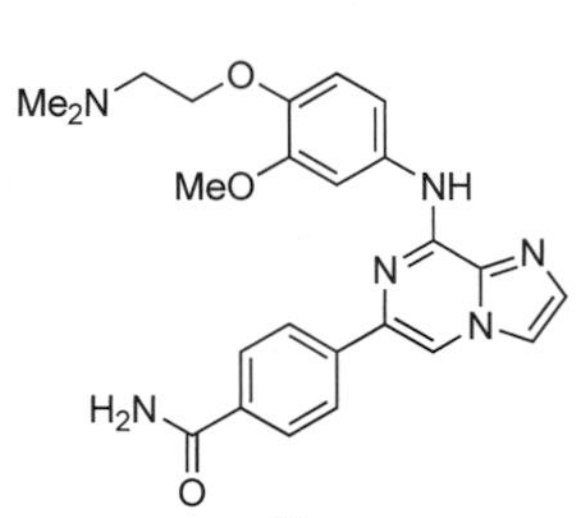

24

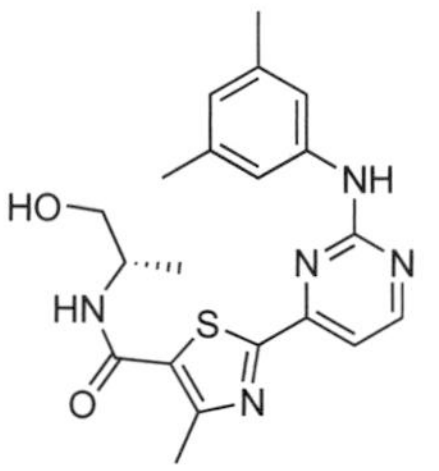

25

Figure 3 Syk inhibitors.

with EC$_{50}$'s ranging from 33 to 158 nM [71]. Crystal structures of the R406-Syk complex show that the aminopyrimidine group forms hydrogen bonds with hinge residue A451 and an aromatic CH–O interaction with E449. The pyrido-oxazinone nitrogens form H-bonds to water molecules and do not interact directly with the protein [71,72].

In an ascending dose Phase IIa study (TASKi1), R788 achieved a significant and dose-related clinical response, with 72% of patients in the 150 mg/bid group achieving ACR20 response compared to 38% on placebo [73]. This efficacy correlated with a reduction in IL-6 and MMP-3 levels. In a Phase IIb trial (TASKi2) in methotrexate failures, R788 showed clinical improvement after 1 week and sustained efficacy thereafter, with 43% of patients achieving ACR50 response (placebo = 19%) after 6 months of dosing at 100 mg/bid [74]. In the subsequent TASKi3 trial, however, ACR scores for R788 were not significantly different from placebo, although MRI scans did show improvements in synovitis and osteitis. The most common adverse effects were diarrhea, hypertension, and neutropenia. In a Phase II trial for refractory ITP, R788 increased and maintained platelet count in 50% of patients. Efficacy increased with dose, as did gastrointestinal adverse events [75]. It is unclear if off-target effects may contribute to efficacy or the reported adverse events, as limited selectivity data has been disclosed [71,76].

3.2 Diaminopyrimidine-based Syk inhibitors

The diaminopyrimidine core present in R406 recurs, with various modifications, in several Syk inhibitor chemotypes. A series of regioisomeric diaminopyrimidines such as **16** ($IC_{50} = 41$ nM) were developed from an HTS hit [77]. SAR studies revealed a requirement for the 5-carboxamide, a preference for *meta*-substitution in the aniline ring, and an important role for the ethylenediamino group in determining Syk potency. This latter motif was incorporated into a diaminocyclohexyl ring to give compound **17**. The X-ray crystal structure of **17** shows that the carboxamide forms two hydrogen bonds with the hinge and the diaminocyclohexane occupies the ribose pocket [72,78]. Unlike R406, **17** forms a favorable H-bond with D512 of the DFG motif, and the aniline NH of **17** does not interact with the protein. The pyrimidine-5-carboxamide motif is found in bicyclic triazolopyrimidine **18** and imidazopyrimidine **19** scaffolds [79,80]. Both show comparable enzyme and cellular potency, but the imidazopyrimidine **18** had superior physicochemical properties and pharmacokinetics.

Cross-reactivity with JAK kinases is a recurring theme with pyrimidine-carboxamides [81,82], but compound **20** exemplifies a Syk-selective series [81]. Pyrimidine amide **20** has an IC_{50} of 6 nM against Syk, and broad screening (at 300 nM concentration) demonstrated significant selectivity for Syk over 270 kinases. In Ramos cells, **20** inhibited BCR-induced phosphorylation of BLNK, a direct substrate of Syk, with an EC_{50} of 500–750 nM and Ca^{2+} flux with an $EC_{50} = 117$ nM. Compound **20** potently inhibited the expression of CD86 and CD69 in primary mouse spleno-cytes, with EC_{50}'s of 50–125 nM. In the mouse collagen antibody-induced arthritis model, **20** reduced clinical score by 44% at 30 mg/kg day, with average peak plasma levels of 7.8 μM.

The diaminopyrimidine has been incorporated into a pyrrolopyrimi-dine scaffold exemplified by **21** [83]. This framework appears in several patent applications [84–87] claiming Syk activity of <1 μM.

3.3 Pyrrolopyrazines

Pyrrolopyrazine **22** has an IC_{50} of 1.7 nM against Syk [88] and signifi-cantly decreased ankle swelling in prophylactic and therapeutic rat CIA models, and micro-CT scans of ankle joints showed significant inhibition of bone erosion. While **22** is clearly a potent Syk inhibitor, no selectivity or *in vitro* functional data has been reported. The pyrrolopyrazine scaffold appears in a series of compounds, exemplified by **23**, that are substituted in the pyrrole and pyrazine rings [89–92]. These compounds are claimed to inhibit Syk and JAK, but the limited data disclosed suggests that JAK is the dominant activity.

3.4 Imidazopyridines and thiazolylpyrimidines

Imidazopyrazines, such as **24**, are one of the few Syk inhibitor chemo-types lacking an embedded *meta*-diamino scaffold [93]. Instead, a phenyl ring at the 6-position and an aniline at the 8-position have been claimed. Preferred compounds have IC_{50}'s < 10 nM against Syk and inhibit B-cell proliferation with EC_{50}'s < 1 μM. As a measure of functional selectivity, the compounds are claimed to be at least fivefold more potent in inhibit-ing proliferation of B-cells versus T cells. The X-ray crystal structure of **25** (Syk $K_i = 9$ nM), bound to the Syk kinase domain, shows that this class of 4-thiazolyl-2-aminopyrimidines form a hydrogen bond from the amide carbonyl to the catalytic lysine (Lys402) [94].

4. PI3K INHIBITORS

Since the discovery of wortmannin and LY294002 as pan-PI3K inhibitors, the search for isoform-selective inhibitors has been a major endeavor. The application of PI3K inhibitors in oncology has been recently reviewed [95,96], as have inhibitors of the δ and γ isoforms [34,97,98]. Given its stronger association with B-cells, the following section will focus on those compounds with some degree of selectivity for PI3Kδ (Figure 4).

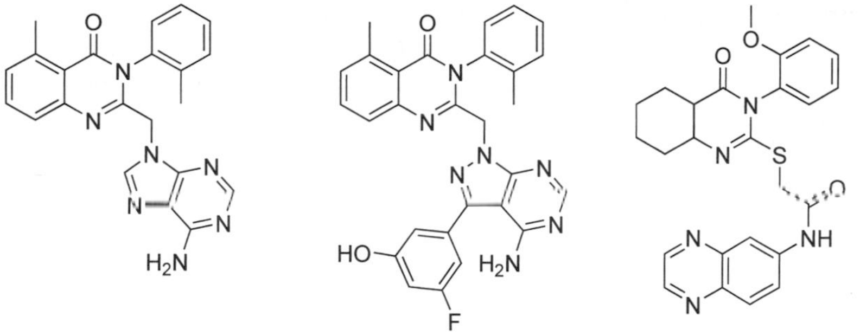

26   **27**   **28**

Figure 4 PI3Kδ inhibitors.

4.1 PI3Kδ-selective inhibitors

Most δ-selective PI3K inhibitors are derived from IC87114 **26**, which has an IC_{50} of 130 nM against PI3Kδ and >100-fold selectivity over the α, β, and γ isoforms [99,100], and showed oral efficacy in a neutrophil-driven model of inflammatory arthritis [101].

Comparing crystal structures of δ-selective versus pan-PI3K inhibitors has revealed a selectivity pocket that is occupied by the *o*-toluyl ring of **26** and related analogs [102]. Selective PI3Kδ inhibitors can adopt a twisted propeller conformation that pan-isoform inhibitors cannot achieve. The fluorophenol group of **27** probes a separate affinity pocket and improves potency against δ while maintaining 50-fold selectivity against α, β, and γ. Compound **28**, with changes to the quiazolinone ring and the hinge binding motif, inhibits δ with an $IC_{50} = 34$ nM and displays >100-fold isoform selectivity [101]. PI3 kinase inhibitors with distinct isoform selectivity profiles (δ, δ/γ, and pan-PI3K) have been used to define the *in vitro* anti-inflammatory signature associated with each profile [103].

CAL-101, which is believed to have been derived from **26**, is a 2.5 nM inhibitor of PI3Kδ with >30-fold isoform selectivity and no activity against 350 other kinases [104]. CAL-101 was the first PI3Kδ-selective inhibitor to enter the clinic and has since shown efficacy in various hematological malignancies [105]. CAL-263 (structure undisclosed) is currently in Phase I trials as a first-in-class PI3Kδ inhibitor for RA [106]. CAL-263 has shown efficacy in animal models of arthritis, including inhibition of bone and cartilage erosion. A decrease in anticollagen antibody production was noted, indicative of inhibition of B-cell activity.

5. CONCLUSION

B-cells play a major role in autoimmune/inflammatory disease, and the clinical success of biologics targeting B-cells has heightened interest in the development of small-molecule modulators of B-cell function. Btk, Syk, and PI3Kδ have emerged as the most compelling B-cell targets to this end. The preclinical evidence implicating them in disease is strong, and their position at the apex of the B-cell signaling cascade offers the potential for inhibitors to exert profound effects on downstream signaling and B-cell activation. Effects in other immune cell types are likely to augment any B-cell-derived efficacy. Medicinal chemistry efforts have intensified significantly in recent years, with selectivity being a prime objective. All three targets are structurally enabled with over 15 small-molecule cocrystal structures deposited in the last 3 years. As this exciting area develops, we can anticipate new compounds entering the clinic for various autoimmune and inflammatory indications.

REFERENCES

[1] G. S. Firestein, *Nature*, 2003, **423**, 356.
[2] D. Franciotta, M. Salvetti, F. Lolli, B. Serafini, and F. Aloisi, *Lancet Neurol.*, 2008, **7**, 852.
[3] F. Martin and A. C. Chan, *Immunity*, 2004, **20**, 517.
[4] P. J. Mease, *J. Rheumatol.*, 2008, **35**, 1245.
[5] T. Dorner, K. Egerer, E. Feist, and G. R. Burmester, *Curr. Opin. Rheumatol.*, 2004, **16**, 246.
[6] T. Hirano, *Nat. Immunol.*, 2002, **3**, 342.
[7] T. Dorner and G. R. Burmester, *Curr. Opin. Rheumatol.*, 2003, **15**, 246.
[8] B. Marston, A. Palanichamy, and J. H. Anolik, *Curr. Opin. Rheuamatol.*, 2010, **23**(3), 307–315.
[9] S. Chaiamnuay and S. L. Bridges, *Pathophysiology*, 2005, **12**, 203.
[10] G. J. Silverman and D. A. Carson, *Arthritis Res. Ther.*, 2003, **5**, S1-S6.
[11] F. E. Lund, B. A. Garvy, T. D. Randall, and D. P. Harris, *Curr. Dir. Autoimmun.*, 2005, **8**, 25.
[12] S. Bas, T. V. Perneger, M. Seitz, J. M. Tiercy, P. Roux-Lombard, and P. A. Guerne, *Rheumatology*, 2002, **41**, 809.
[13] A. Raptopoulou, P. Sidiropoulos, M. Katsouraki, and D. T. Boumpas, *Crit. Rev. Clin. Lab. Sci.*, 2007, **44**, 339.
[14] B. Sutton, A. Corper, V. Bonagura, and M. Taussig, *Immunol. Today*, 2000, **21**, 177.
[15] V. M. Abrahams, G. Cambridge, P. M. Lydyard, and J. C. Edwards, *Arthritis Rheum.*, 2000, **43**, 608.
[16] M. E. Duddy, A. Alter, and A. Bar-Or, *J. Immunol.*, 2004, **172**, 3422.
[17] D. Vishnevetsky, V. A. Kiyanista, and P. J. Gandhi, *Ann. Pharmacother.*, 2004, **38**, 1500.
[18] R. J. Looney, *Drugs*, 2006, **66**, 625.
[19] J. C. Edwards, L. Szczepanski, J. Szechinski, A. Filipowicz-Sosnowska, P. Emery, D. R. Close, R. M. Stevens, and T. Shaw, *N. Engl. J. Med.*, 2004, **350**, 2572.
[20] S. Navarra, R. Guzman, A. Gallacher, R. A. Levy, E. K. Li, M. Thomas, R. Jiminez, M. Leon, S. Hall, J. L. Lan, E. Nasonov, C. Tanasescu, H. -Y. Kim, L. Pineda, Z. J. Zhong, W. Freimuth, and M. A. Petri, Presentation LB1, American College of Rheumatology Annual Meeting, Philadelphia, PA, October 20, 2009.
[21] C. Bracewell, J. D. Isaacs, P. Emery, and W. F. Ng, *Expert Opin. Biol. Ther.*, 2009, **9**, 909.

[22] M. E. Conley, D. Mathias, J. Treadaway, Y. Minegishi, and J. Rohrer, *Am. J. Hum. Genet.*, 1998, **62**, 1034.

[23] F. S. Rosen, M. D. Cooper, and R. J. Wedgewood, *N. Engl. J. Med.*, 1995, **333**, 431.

[24] L. Jansson and R. Holmdahl, *Clin. Exp. Immunol.*, 1993, **94**, 459.

[25] L. Lin, R. Czerwinski, K. Kelleher, M. M. Siegel, P. Wu, R. Kriz, A. Aulabaugh, and M. Stahl, *Biochemistry*, 2009, **48**, 2021.

[26] M. Dinh, D. Grunberger, H. Ho, S. Y. Tsing, D. Shaw, S. Lee, T. Barnett, R. J. Hill, D. C. Swinney, and J. M. Bradshaw, *J. Biol. Chem.*, 2007, **282**, 8768.

[27] A. M. Cheng, B. Rowley, W. Pao, A. Hayday, J. B. Bolen, and T. Pawson, *Nature*, 1995, **378**, 303.

[28] M. Turner, P. J. Mee, P. S. Costello, O. Williams, A. A. Price, L. P. Duddy, M. T. Furlong, R. L. Geahlen, and V. L. Tybulewicz, *Nature*, 1995, **378**, 298.

[29] B. R. Wong, E. B. Grossbard, D. G. Payan, and E. S. Masuda, *Expert Opin. Invest. Drugs*, 2004, **13**, 743.

[30] M Bajpai, P. Chopra, S. G. Dastidar, and A. Ray, *Expert Opin. Invest. Drugs*, 2008, **17**, 641.

[31] A. M. Condliffe, K. Davidson, K. E. Abderson, C. D. Ellson, T. Crabbe, K. Okkenhaug, B. Vanhaesebroeck, M. Turner, L. Webb, M. P. Mymann, E. Hirsch, T. Ruckle, M. Camps, C. Rommel, S. P. Jackson, E. R. Chilvers, L. R. Stephens, and P. T. Hawkins, *Blood*, 2005, **106**, 1432.

[32] L. M. Webb, E. Vigorito, M. P. Wymann, E. Hirsch, and M. Turner, *J. Immunol.*, 2005, **175**, 2783.

[33] K. Okkenhaug, A. Bilancio, G. Farjot, H. Priddle, S. Sanco, E. Peskett, W. Pearce, S. E. Meek, A. Salpekar, M. D. Waterfield, A. J. Smith, and B. Vanhaesebroeck, *Science*, 2002, **297**, 1031.

[34] C. Rommel, M. Camps, and H. Ji, *Nat. Rev. Immunol.*, 2007, **7**, 191.

[35] T. Crabbe, M. J. Welham, and S. G. Ward, *Trends Biochem. Sci.*, 2007, **32**, 450.

[36] K. Okkenhaug and B. Vanhaesebroek, *Biochem Soc. Trans.*, 2003, **31**, 270.

[37] D. A. Fruman, *Biochem Soc. Trans.*, 2004, **32**, 315.

[38] N. T. Ihle and G. Powis, *Curr. Opin. Drug. Discov. Dev.*, 2010, **13**, 41.

[39] Z. Pan, *Drug News Perspect.*, 2008, **21**, 357.

[40] D. A. Pollyea, S. Smith, N. Fowler, T. E. Boyd, A. M. Smith, M. Sirisawad, L. A. Honigberg, A. Hamdy, and R. Advani, Poster 3713, 51st American Society of Hematology Annual Meeting, New Orleans, LA, December 5–8, 2009.

[41] Z. Pan, H. Scheerens, S. J. Li, B. E. Schultz, P. A. Sprengeler, L. C. Burrill, R. V. Mendonca, M. D. Sweeney, K. C. K Scott, P. G. Grothaus, D. A. Jeffery, J. M. Spoerker, L. A. Honigberg, P. R. Young, S. A. Dalrymple, and J. T. Palmer, *ChemMedChem.*, 2006, **2**, 58.

[42] L. Honigberg, E. Verner, J. Buggy, D. Loury, and W. Chen, *Patent Application WO 2010/009342*, 2010.

[43] D. J. Marcotte, Y. -T. Liu, R. M. Arduini, C. A. Hession, K. Miatowski, C. P. Wildes, P. F. Cullen, V. Hong, B. T. Hopkins, E. Mertsching, T. J. Jenkins, M. J. Romanowski, D. P. Baker, and L. F. Silvian, *Protein Sci.*, 2010, **19**, 429.

[44] A. F. Kluge, R. Petter, R. W. Tester, L. Qiao, D. Niu, W. F. Westlin, J. Singh, and H. Mazdiyasni, *Patent Application WO 2009/158571*, 2009.

[45] K. S. Currie, IBC Hit-To-Lead Success Stories, San Diego, CA, January 31, 2005.

[46] K. S. Currie, R. W. DeSimone, S. A. Mitchell, and D. A. I. Pippin, *US Patent 7,393,848*, 2008.

[48] K. S. Currie, R. W. DeSimone, S. A. Mitchell, D. A. Pippin, J. W. Darrow, X. Qian, M. Velleca, and D. Qian, *US Patent 7,405,295*, 2008.

[48] N. J. Dewdney, Y. Lou, and M. Soth, *Patent Application WO 2009/077334*, 2009.

[49] K. S. Currie, S. H. Lee, J. W. Darrow, and P. A. Blomgren, *Patent Application WO 2008/033834*, 2008.

[50] P. A. Blomgren, K. S. Currie, J. E. Kropf, S. H. Lee, J. W. Darrow, S. A. Mitchell, J. Xu, and A. C. Schmitt, *Patent Application WO 2008/033854*, 2008.

[51] P. A. Blomgren, S. H. Lee, S. A. Mitchell, J. Xu, and A. C. Schmitt, *Patent Application WO 2008/033857*, 2008.

[52] D. R. Brittelli, K. S. Currie, J. W. Darrow, J. E. Kropf, S. H. Lee, S. L. Gallion, S. A. Mitchell, D. A. Pippin, P. A. Blomgren, and D. G. Stafford, *Patent Application WO 2006/099075*, 2006.

[53] P. A. Blomgren, K. S. Currie, S. H. Lee, S. A. Mitchell, J. Xu, A. C. Schmitt, Z. Zhao, P. E. Zhichkin, D. G. Stafford, and J. E. Kropf, *Patent Application WO 2009/039397*, 2009.

[54] N. J. Dewdney, T. Gabriel, R. K. Kondru, Y. Lou, and M. Soth, *Patent Application WO 2009/053269*, 2009.

[55] J. A. Whitney, J. DiPaolo, M. A. Velleca, D. R. Brittelli, K. S. Currie, J. W. Darrow, J. E. Kropf, S. H. Lee, S. L. Gallion, S. A. Mitchell, D. A. I. Pippin, P. A. Blomgren, and D. G. Stafford, *Patent Application WO 2008/033858*, 2008.

[56] T. Huang, J. Barbosa, V. Hurez, L. Diehl, W. P. Lee, M. Balazs, K. Currie, J. DiPaolo, and K. Reif, Keystone Symposium: The Macrophage, Banff, AB, February 12–17, 2010.

[57] N. J. Dewdney, J. Kennedy-Smith, R. K. Kondru, B. E. Loe, Y. Lou, J. McIntosh, T. D. Owens, M. Soth, Z. K. Sweeney, and J. P. G. Taygerly, *Patent Application WO 2009/098144*, 2009.

[58] N. J. Dewdney, R. K. Kondru, B. E. Loe, Y. Lou, J. McIntosh, T. D. Owens, and M. Soth, *Patent Application WO 2009/156284*, 2009.

[59] N. J. Dewdney, Y. Lou, E. B. Sjorgen, M. Soth, and Z. K. Sweeney, *Patent Application WO 2010/000633*, 2010.

[60] R. K. Kondru, Y. Lou, E. B. Sjogren, and M. Soth, *Patent Application WO 2010/006970*, 2010.

[61] N. J. Dewdney, Y. Lou, E. B. Sjogren, and M. Soth, *Patent Application WO 2010/006947*, 2010.

[62] W. Vaccaro, Z. Chen, D. S. Dodd, T. N. Huynh, J. Lin, C. Liu, C. P. Mussari, J. Tokarski, D. R. Tortolani, and S. T. Wrobleski, *Patent Application US 2007/0078136*, 2007.

[63] C. Liu, K. Leftheris, and A. J. Tebben, *Patent Application WO 2008/116064*, 2008.

[64] N. Patel, K. Sarma, J. Fretland, Y. Lou, Q. Hu, A. Chow, S. Larrabee, B. Francavilla, D. Stefanidis, M. Brandl, and T. Alfredson, Paper T3401, AAPS Annual Meeting, Los Angeles, CA, November 8–12, 2009.

[65] O. Hantschel, U. Rix, U. Schmidt, T. Burckstümmer, M. Kneidinger, G. Schütze, J. Colinge, K. L. Bennett, W. Ellmeier, P. Valent, and G. Superti-Furga, *Proc. Natl. Acad. Sci. USA*, 2007, **104**, 13283.

[66] M. W. Karamn, S. Herrgard, D. K. Treiber, P. Gallant, C. E. Atteridge, B. T. Campbell, K. W. Chan, P. Ciceri, M. I. Davis, P. Y. Eeded, R. Fataoni, M. Floyd, J. P. Hunt, D. J. Lockhart, Z. V. Milanov, M. J. Morrison, G. Palleres, H. K. Patel, S. Pritchard, L. M. Wodicka, and P. P. Zarrinkar, *Nat. Biotechnol.*, 2008, **26**, 127.

[67] S. Mahajan, S. Ghosh, E. A. Sudbeck, Y. Zheng, S. Downs, M. Hupke, and F. M. Uckun, *J. Biol. Chem.*, 1999, **274**, 9587.

[68] E. van den Akker, T. B. van Dijk, U. Schmidt, L. Felida, H. Beug, B. Lowenberg, and M. von Lindern, *Biol. Chem.*, 2004, **385**, 409.

[69] M. Bajpai, *iDrugs*, 2009, **12**, 174.

[70] R. Singh, S. Bhamidipati, and E. Masuda, *Patent Application WO 2006/078846*, 2006.

[71] S. Braselmann, V. Taylor, H. Zhao, S. Wang, C. Sylvain, M. Baluom, K. Qu, A. Lau, C. Young, B. R. Wong, S. Lovell, T. Sun, G. Park, A. Argade, S. Jucevic, P. Pine, R. Singh, E. B. Grossbard, D. G. Payan, and E. S. Masuda, *J. Pharm. Exp. Ther.*, 2006, **319**, 998.

[72] A. G. Villasenor, R. Kondru, H. Ho, S. Wang, E. Papp, D. Shaw, J. W. Barnett, M. F. Browner, and A. Kuglstatter, *Chem. Biol. Drug Des.*, 2009, **73**, 466.

[73] M. E. Weinblatt, A. Kavanaugh, R. Burgis-Vargas, A. H. Dikranian, G. Medrano-Ramirez, J. L. Motales-Torres, F. T. Murphy, T. K. Musser, N. Straniero, A. V. Vicente-Gonzales, and E. Grossbard, *Arthritis Rheum.*, 2008, **58**, 3309.

[74] TASKi2 & TASKi3 Phase 2b results. Available at www.rigel.com/rigel/rheumatoid_arthritis.

[75] A. Podolanczuk, A. H. Lazarus, A. R. Crow, E. Grossbard, and J. B. Bussel, *Blood*, 2008, **113**, 3154.

[76] G. R. Clemens, R. E. Schroeder, S. H. Magness, E. V. Weaver, J. W. Lech, V. C. Taylor, E. S. Masuda, M. Baluom, and E. B. Grossbard, *Birth Defects Res. A. Clin., Mol. Teratol.*, 2009, **85**, 130.

[77] H. Hisamichi, R. Naito, A. Toyoshima, N. Kawano, A. Ichikawa, A. Orita, M. Orita, N. Hamada, M. Takeuchi, M. Ohta, and S. -I. Tsukamoto, *Bioorg. Med. Chem.*, 2005, **13**, 4936.

[78] E. Papp, J. K. Y. Tse, H. Ho, S. Wang, D. Shaw, S. Lee, J. Barnett, D. C. Swinney, and J. M. Bradshaw, *Biochemistry*, 2007, **46**, 15103.

[79] A. Hirabayashi, H. Mukaiyama, H. Kobayashi, H. Shiohara, S. Nakayama, M. Ozawa, E. Tsuji, K. Miyazawa, H. Ohnota, and M. Isaji, *Bioorg. Med. Chem.* 2008, **16**, 9247.

[80] A. Hirabayashi, H. Mukaiyama, H. Kobayashi, H. Shiohara, S. Nakayama, M. Ozawa, K. Miyazawa, K. Misawa, H. Ohnota, and M. Isaji, *Bioorg. Med. Chem.*, 2008, **16**, 7347.

[81] J. J. Jia, C. Venkataramani, W. Huang, M. Mehrotra, Y. Song, Q. Xu, S. M. Bauer, A. Pandey, and B. Kane, *Patent Application WO 2009/136995*, 2009.

[82] S. M. Bauer, J. J. Jia, Y. Song, Q. Xu, M. Mehrotra, J. W. Rose, W. Huang, C. Venkataramani, and A. Pandey, *Patent Application WO 2009/145856*, 2009.

[83] Y. Song, Q. Xu, and A. Pandey, *Patent Application WO 2009/026107*, 2009.

[84] Y. Song, Q. Xu, S. M. Bauer, J. J. Jia, M. Mehrotra, and A. Pandey, *Patent Application WO 2009/131687*, 2009.

[85] P. M. Gore, V. K. Patel, and A. L. Walker, *Patent Application US 2008/0004295*, 2008.

[86] R. A. Ancliff, F. L. Atkinson, M. D. Barker, P. C. Box, C. Daniel, P. M. Gore, S. B. Guntrip, M. Hasegawa, G. A. Inglis, K. Kano, Y. Miyazaki, V. K. Patel, T. J. Ritchie, S. Swanson, A. L. Walker, C. R. Wellaway, and M. Woodrow, *Patent Application WO 2007/042299*, 2007.

[87] P. M. Gore, V. K. Patel, A. L. Walker, and M. Woodrow, *Patent Application WO 2007/042298*, 2007.

[88] T. A. Gillespy, P. Eynott, E. M. Allen, K. T. Yu, and A. Zilberstein, *Patent Application WO 2008/033798*, 2008.

[89] T. R. Elworthy, R. T. Hendricks, R. K. Kondru, Y. Lou, T. D. Owens, M. Soth, and H. Yang, *Patent Application US 2009/0215788*, 2009.

[90] D. J. DuBois, R. T. Hendricks, J. C. Hermann, R. K. Kondru, Y. Lou, T. D. Owens, and C. W. Yee, *Patent Application US 2009/0215724*, 2009.

[91] J. T. Bamberg, M. Bartlett, D. J. DuBois, T. R. Elworthy, R. T. Hendricks, J. C. Hermann, R. K. Kondru, R. Lemoine, Y. Lou, T. D. Owens, J. Park, D. B. Smith, M. Soth, H. Yang, and C. Q. Yee, *Patent Application US 2009/0215750*, 2009.

[92] D. J. DuBois, T. R. Elworthy, R. T. Hendricks, J. C. Hermann, R. K. Kondru, Y. Lou, T. D. Owens, and D. B. Smith, *Patent Application US 2009/0215785*, 2009.

[93] S. A. Mitchell, K. S. Currie, P. A. Blomgren, D. M. Armistead, and J. Raker, *Patent Application WO 2009/000919*, 2009.

[94] L. J. Farmer, G. Bemis, S. D. Britt, J. Cochran, M. Connors, E. M. Harrington, T. Hoock, W. Markland, S. Nanthakumar, P. Taslimi, E. Ter Harr, J. Wang, D. Zhaveri, and F. G. Salituro, *Bioorg. Med. Chem. Lett.*, 2008, **18**, 6231.

[95] J. M. Nuss, A. L. Tsuhako, and N. K. Anand, *Ann. Rep. Med. Chem.*, 2009, **44**, 339.

[96] R. Marone, V. Cmiljanovic, B. Giese, and M. P. Wymann, *Biochim. Biophys. Acta*, 2008, **1784**, 159.

[97] M. K. Ameriks and J. D. Venable, *Curr. Top. Med. Chem.*, 2009, **9**, 738.

[98] T. Ruckle, M. K. Schwarz, and C. Rommel, *Nat. Rev. Drug Disc.*, 2006, **5**, 903.

[99] C. Sadhu, B. Masinovsky, K. Dick, C. G. Sowell, and D. E. Staunton, *J. Immunol.*, 2003, **170**, 2647.

[100] Z. A. Knight, B. Gonzalez, M. E. Feldmann, E. R. Zunder, D. D. Goldenberg, O. Williams, R. Loewith, D. Stokoe, A. Balla, B. Toth, T. Balla, A. W. Weiss, R. L. Williams, and K. M. Shokat, *Cell*, 2006, **125**, 733.

[101] T. M. Randis, K. D. Puri, H. Zhuo, and T. G. Diacovo, *Eur. J. Immunol.*, 2008, **38**, 1215.

[102] A. Berndt, S. Miller, O. Williams, D. D. Lee, B. T. Houseman, J. I. Pacold, F. Gorrec, W. C. Hon, Y. Liu, C. Rommel, P. Gaillard, T. Ruckle, M. K. Scwarz, K. M. Shokat, J. P. Shaw, and R. L. Williams, *Nat. Chem. Biol.*, 2010, **6**, 244.

[103] O. Williams, B. T. Houseman, E. J. Kunkel, B. Aizenstein, R. Hoffman, Z. A. Knight, and K. M. Shokat, *Chem. Biol.*, 2010, **17**, 123.

[104] B. J. Lannutti, S. A. Meadows, A. Kashishian, B. Steiner, G. Pogosov, O. Sala-Torra, A. J. Johnson, J. C. Byrd, J. Radich, and N. A. Giese, Paper 286, American Society of Hematology Annual Meeting, New Orleans, LA, December 5–9, 2009.

[105] I. W. Flinn, J. C. Byrd, R. R. Furman, J. R. Brown, D. M. Benson, S. E. Coutre, B. S. Kahl, B. D. Smith, N. D. Wagner-Johnston, S. E. Spurgeon, N. A. Giese, and A. S. Yu, Paper 2922, American Society of Hematology Annual Meeting, New Orleans, LA, December 5–9, 2009.

[106] www.calistogapharma.com/products.php.

CHAPTER 12

Modulators of Toll-Like Receptor (TLR) Signaling

Mark Spyvee, Lynn D. Hawkins and **Sally T. Ishizaka**

Eisai, Inc., Frontier Product Creation Unit, 4 Corporate Drive, Andover, MA 01810, USA

Annual Reports in Medicinal Chemistry, Volume 45
ISSN 0065-7743, DOI 10.1016/S0065-7743(10)45012-5

1. INTRODUCTION

The role of the toll-like receptor (TLR) family in immune defense was first appreciated in *Drosophila*, where knockout of the *Toll* gene resulted in florid fungal infection [1]. The mammalian homologs of *Toll* are receptors for molecules like endotoxin, whose immunostimulatory properties have been known for nearly a century [2,3]. The 10 TLRs expressed in humans belong to a system of innate sensors that bind ligands associated with viral, bacterial, or parasitic pathogens, or tissue damage [4,5]. Within hours, TLR signaling elicits antipathogen mediators such as defensins, interferons, nitric oxide, and chemokines that attract immune cells to amplify the response. Through effects on antigen-presenting dendritic cells (DCs), TLRs have an important role in initiating and shaping the antigen-specific T and B cell responses that protect the host in the longer term [6].

Consistent with their role as immune receptors, each human TLR is expressed by at least one subset of myeloid cells (MCs) or lymphocytes [7,8]. TLRs are also present on stromal elements like endothelium particularly after local inflammatory stimulus [9–11]. These distribution patterns can determine the physiological consequences of stimulation or antagonism, and affect the balance of toxicity versus therapeutic effect. Another consideration for medicinal chemistry is subcellular localization of TLRs. While most are expressed on the cell surface, some (TLRs 3, 7, 8, and 9) can localize to endosomes where they survey ingested material for ligands, so drug access to this compartment can be crucial when targeting these TLRs [12].

1.1 Structural features of the TLRs

TLRs are type 1 transmembrane glycoproteins with a single transmembrane domain and a conserved intracellular domain [13]. The extracellular domain has 16–28 leucine-rich repeating (LRR) units containing conserved "LxxLxLxxN" motifs [14]. Ligand binding induces homo- or heterodimerization of the receptor, leading to recruitment of adaptor proteins such as MyD88, MAL (TIRAP), TRIF, and TRAM to the intracellular toll/interleukin-1 receptor (TIR) [15].

High-resolution X-ray crystal structures have been reported for TLR1/2 [16], TLR2/6 [17], TLR3 [18], TLR4/MD2 with *Escherichia coli* lipopolysaccharide (LPS) [19], MD2 with Lipid IVa (a precursor to Lipid A) [20], and the TLR4/MD2 complex with the antagonist E5564 [21]. Homology modeling of human TLRs 7, 8, and 9 with TLR3 predicts that they will have the general horseshoe-shaped monomer unit for the ligand-binding domain and possibly similar dimer structures upon interaction with their ligand as found for TLR3 [22]. Upon dimerization the horseshoe-like monomeric structures join into "m"-shaped hetero- (TLR2 with TLR1 or TLR6) or homodimers

(TLR3 and TLR4/MD2 complexes) [15]. Although this dimer conformation is common between TLRs, the binding of the individual ligands is quite different from one TLR complex to the next and from species to species [23,24]. The lipid portions of the ligands for the TLR2/TLR1, TLR2/TLR6, and TLR4/MD2 complexes bind in the hydrophobic regions of the receptor domains, while the double stranded (ds) RNAs for TLR3 bind in the hydrophilic portion where the sugar–phosphate backbone is crucial for this interaction. The TLR–ligand forms a 2:1 complex that draws the C-terminal region in the ectodomain of the two receptors sufficiently close to allow protein–protein interactions. The ligand binding and this protein–protein interaction provide the stabilized dimeric complex. It is believed that this interaction also facilitates the dimerization of the cytoplasmic TIR domains and thus initiates the internal signaling. Besides the ligand-binding domain of the complex, one may project that targeting the protein–protein interface of the dimer at the C-terminal region of the ectodomain may provide an additional opportunity for the design of agonists or antagonists [18].

Since the discovery of the TLRs, a number of exogenous and endogenous ligands have been reported for each of the receptors [25]. Other work done over the past 15 years has focused on the effects of stimulating or ablating TLR signaling in models of disease. Agonists were investigated long before the receptors were known, and this work continues with efforts to use TLR ligands as vaccine adjuvants, as prophylactics against pathogen exposure in the area of biodefense [26], or as immunostimulators alone or with other drugs [4], particularly in oncology. Most recently, there is an appreciation that TLRs also respond to endogenous ligands, as in the case of TLR7 and TLR9 activation by autologous single-stranded (ss) RNA (TLR7) and DNA (TLR9) in systemic lupus erythematosus (SLE) [27]. This raises the possibility that antagonists to these two TLRs could be used for this indication. The complex signaling pathways downstream of TLRs have also been investigated and some essential components have been identified. Among these, IRAK-4 and MyD88 have attracted attention as therapeutic targets [28].

2. TLR1, TLR2, AND TLR6

TLR2 forms heterodimers with either TLR1 or TLR6 at the cell surface to recognize a wide range of pathogens and associated ligands.

2.1 Agonists

Bacterial lipopeptides such as Pam3CysSK4 (**1**) and MALP-2 (**2**) are highly potent agonists of the TLR2–TLR1 and TLR2–TLR6 heterodimers, respectively, and have therapeutic potential as vaccine adjuvants [29], for prophylaxis of sepsis [30], and wound healing [31]. Structure–activity relationship (SAR) investigations of related lipopeptides showed that the thioether bridge and palmityl esters are essential for activity, while the cysteinyl stereochemistry and the palmityl amide are not [32].

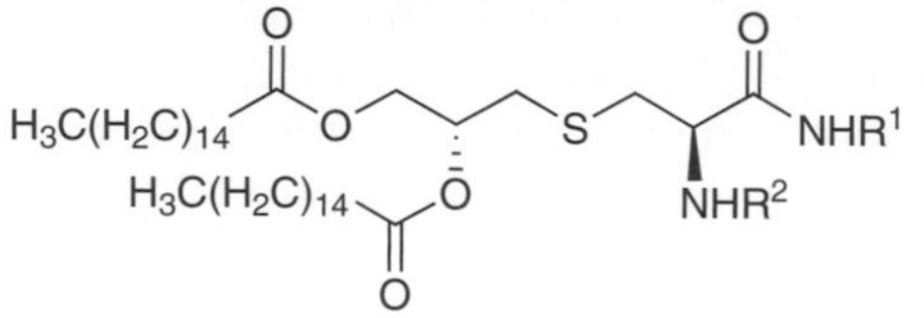

1 (R^1=SKKKK, R^2=C(O)(CH$_2$)$_{14}$CH$_3$)
2 (R^1=GNNDESNISFKEK, R^2=H)

2.2 Antagonists

The few reported synthetic TLR2 antagonists appear to mimic bacterial lipopeptide TLR2 agonists. Most recently, SAR within a class of lipo-lanthionine TLR2 antagonists (e.g., 3) were reported [33]. Compound 3 not only bears close structural similarity to agonist 1 but also to some previously disclosed TLR2 antagonists (e.g., 4–7) [34].

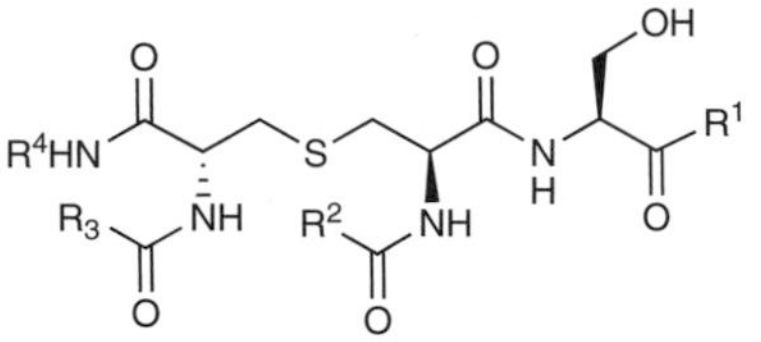

3
(R^1=KKKK, R^2=R^3=(CH$_2$)$_{12}$CH$_3$, R^4=(CH$_2$)$_7$CH$_3$)

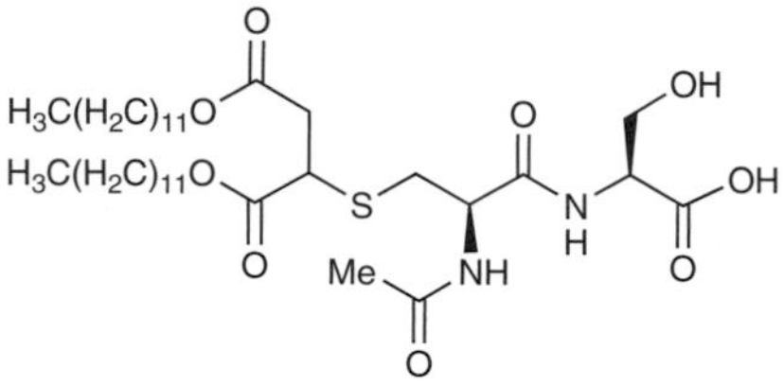

4

Within this class of lipopeptide mimics the SAR suggests that lipid chain length is a factor determining antagonistic potency, with structural core changes being less important [34–36]. Antagonists 5–7 demonstrated 100-fold selectivity for the TLR2–TLR1 heterodimer over the TLR2–TLR6 heterodimer [36].

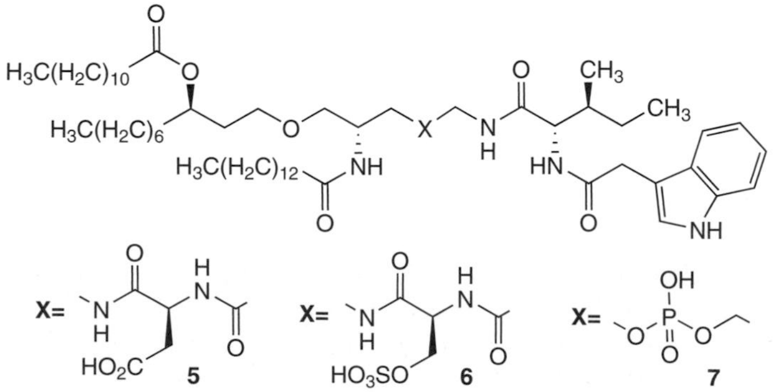

Compounds **3–8** are dual TLR2/4 antagonists, having varying degrees of selectivity, with **8** being the most selective (26-fold) for antagonism of TLR2 over TLR4 [36].

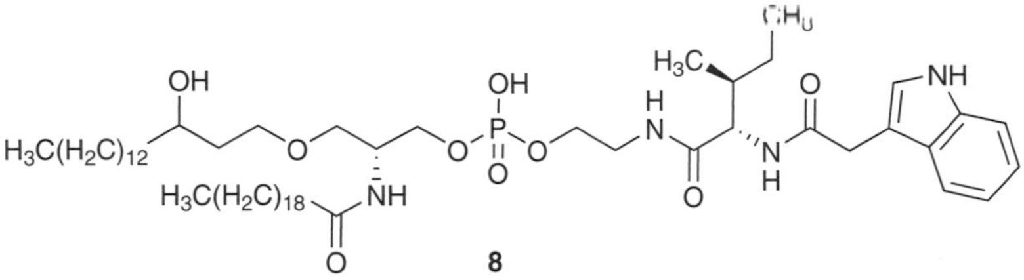

8

Bacterial lipid A structures typically act as TLR2 agonists; however, there are some cases where such compounds act as antagonists (e.g., *Bordetella pertussis* lipid A, **9**), which antagonize the TLR2–TLR6 hetero-dimer [37].

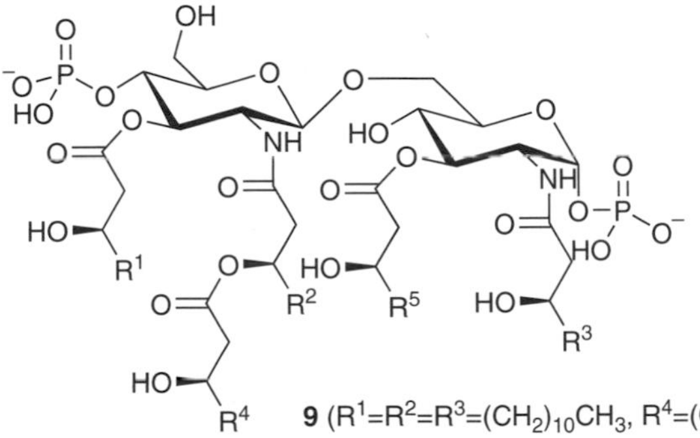

9 ($R^1=R^2=R^3=(CH_2)_{10}CH_3$, $R^4=(CH_2)_8CH_3$, $R^5=(CH_2)_6CH_3$)

3. TLR3

TLR3 is activated by dsRNA, which is typically associated with intra-cellular viral infection [38]. The first recognized ligand was poly-ino-sine-cytosine, an RNA sequence that self-assembles into hairpin structures. Two synthetic agonists are close structural analogs, the ear-liest dating to the 1980s. One is a poly(I:C12U), with uridine mismatches incorporated for more rapid degradation and lower toxicity [39,40]. A patent has also been filed on poly(AU) to target TLR3 [41]. There is no published research on antagonists of TLR3, and little concerning their potential as therapeutics other than for treatment of West Nile virus infection [4].

4. TLR4

TLR4 was the first human toll-like receptor found to have homology with *Drosophila melanogaster* toll [42]. The crystal structure of TLR4 complexed with MD2 and its natural ligand *E. coli* LPS confirmed that the lipid A portion is the key component for receptor complex binding, consistent with previous biological studies [43].

4.1 Agonists

Numerous natural and synthetic TLR4 ligands have been reported. *E. coli* LPS and its toxicophore, the lipid A moiety, represent some of the most potent forms [44]. TLR4 agonists are primarily used as immunomodulators or as vaccine adjuvants. *E. coli* lipid A's toxicity prevented its clinical development, but less toxic semisynthetic analogs are known; for example, MPL (monophosphoryl lipid A) is used in marketed vaccine formulations [45]. Other synthetic analogs being pursued as adjuvants include RC-529, **10** [46], E6020, **11** [47], and OM-174 (**12**) [48].

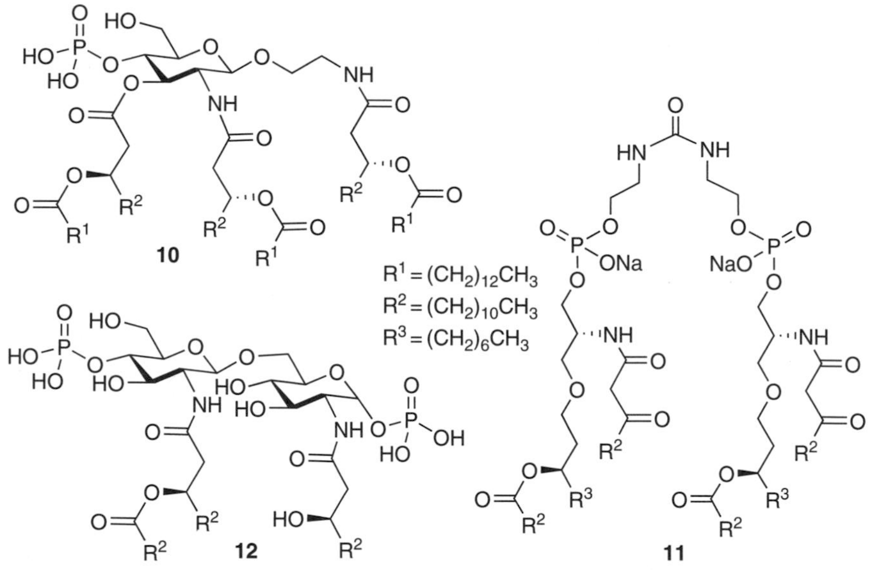

$R^1 = (CH_2)_{12}CH_3$
$R^2 = (CH_2)_{10}CH_3$
$R^3 = (CH_2)_6CH_3$

4.2 Antagonists

Several inflammatory conditions including sepsis, atherosclerosis, and Alzheimer's disease are implicated as targets for therapy with TLR4 antagonists based on the presence of pathogen-related or endogenous ligands [4,49,50]. Gram-negative sepsis is of key interest [51]. The TLR4 antagonist E5564 (13) demonstrated a nonstatistically significant trend toward a lower mortality rate in a Phase II trial in patients diagnosed with severe sepsis [52]. An intracellular TLR4 inhibitor TAK-242 (14) was withdrawn from a Phase III sepsis trial [53]. Additional TLR4 antagonists in preclinical development include simplified lipid A mimics such as CRX-526, 15 [54], and the benzylammonium lipid (16) [55]. High-resolution X-ray structural information on TLR4 reveals potential structural information that may be of use for the design of new antagonists. This approach was demonstrated by the discovery of a short-chain peptide (17 amino acids) from the TLR4-binding region of MD2, which noncompetitively inhibited LPS binding to TLR4. Further modification of this peptide was reported to be in progress [50].

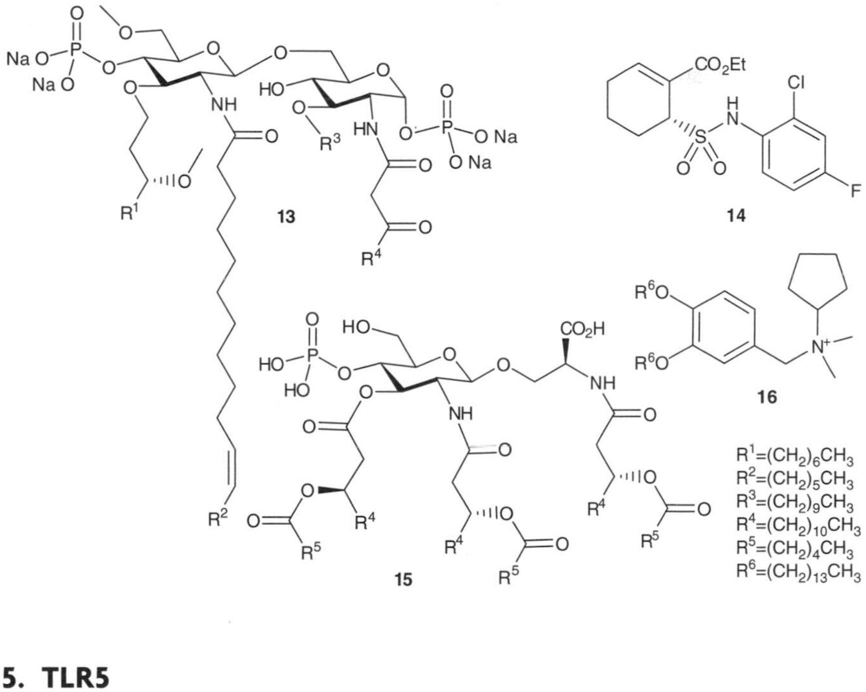

$R^1=(CH_2)_6CH_3$
$R^2=(CH_2)_5CH_3$
$R^3=(CH_2)_9CH_3$
$R^4=(CH_2)_{10}CH_3$
$R^5=(CH_2)_4CH_3$
$R^6=(CH_2)_{13}CH_3$

5. TLR5

TLR5 recognizes the flagellin of many bacteria and has been implicated in the pathology of Crohn's disease [4,56]. Although a flagellin recognition site has been located on the TLR5 extracellular region [57], the precise structural aspects have yet to be established. Flagellins have been investigated as vaccine adjuvants [58–60], and as radioprotectants [61]. Thus far, no antagonists or low-molecular-weight agonists are reported.

6. TLR7 AND TLR8

A natural ligand for TLR7 and TLR8 is viral ssRNA containing GU-rich or poly-U sequences. TLR7 is mainly expressed on plasmacytoid DCs (pDCs) while TLR8 is expressed on myeloid DCs (mDCs), neutrophils, and mono-cytes [62]. pDCs stimulated by ssRNAs express type I interferons (IFNα and β), which play a crucial role in blocking viral infection and replication [63]. As for other TLRs, there is evidence of endogenous ligands that may play an important role in "sterile" inflammation and autoimmunity [64].

6.1 Agonists

Members of the imidazoloquinoline family of synthetic, low-molecular-weight, immunostimulatory compounds can be either dual TLR7/TLR8 agonists (e.g., imiquimod, **17**) or TLR7-selective agonists (e.g., resiqui-mod, **18**). Imiquimod was approved for topical treatment of papilloma-virus infection along with other viral and oncology-based therapies including genital warts, precancerous actinic keratosis, and basal cell carcinomas [35]. Other antiviral and vaccine adjuvant uses are under investigation [65]. More soluble imidazoloquinoline agonists are being pursued as chemotherapeutics (e.g., 852A, **19**) [66].

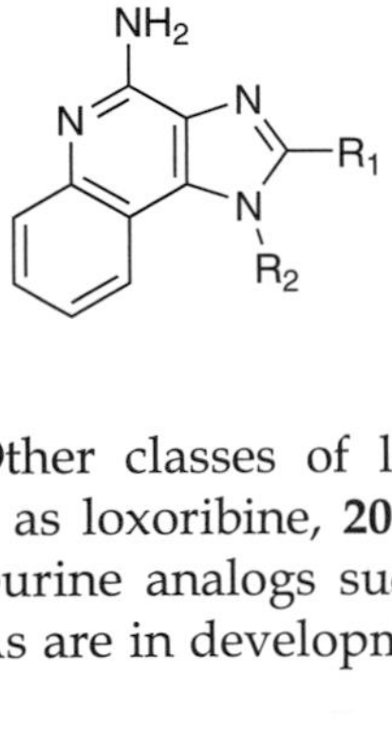

17	$R_1 = H$	$R_2 = CH_2CH(CH_3)_2$
18	$R_1 = CH_2OCH_2CH_3$	$R_2 = CH_2C(OH)(CH_3)_2$
19	$R_1 = CH_2CH_3$	$R_2 = (CH_2)_4NHS(O)_2CH_3$

Other classes of low-molecular-weight TLR7-selective compounds such as loxoribine, **20** [67], a prodrug of isatoribine, ANA-975, **21** [68], oxopurine analogs such as **22** [69], and stabilized immunomodulatory RNAs are in development [70].

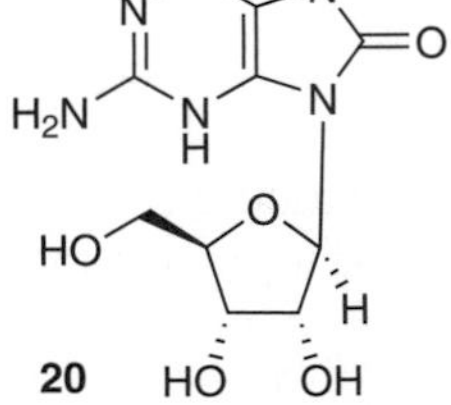

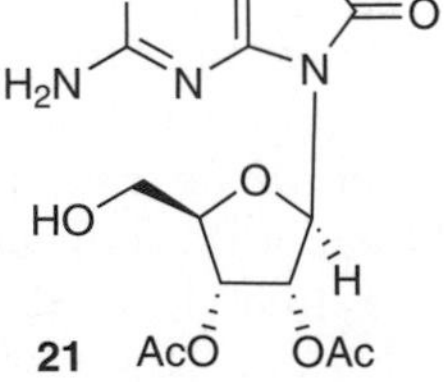

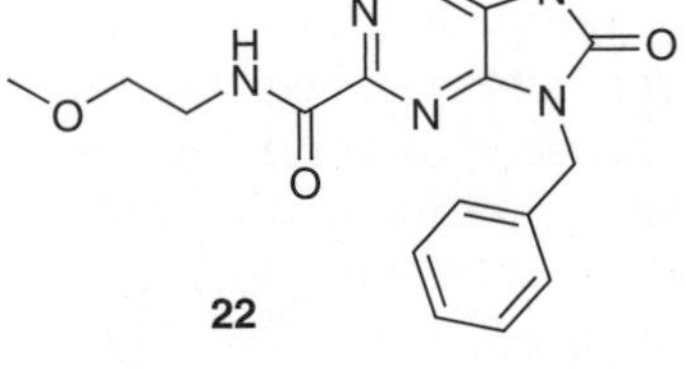

6.2 Antagonists

The most compelling evidence for TLR activation associated with self-nucleic acids is with the overproduction of type I IFN in patients with SLE, scleroderma, and Sjögren's syndrome. Antibody against autologous DNA is seen in these diseases, and studies using purified patient-derived IgG mixed with apoptotic or necrotic cell debris show activation of pDCs [71]. These cell populations are active in production of interferon and autoantibody, which are elevated in this disease. Activation is decreased upon treatment of these complexes with DNase or RNAase. Antimalarial drugs such as chloroquine (23) and hydroxychloroquine (24) are effective in SLE [72,73], and are antagonists of TLR7, TLR8, and TLR9 receptors in the submicromolar range *in vitro* [74–76].

Preclinical research strategies toward obtaining selective TLR7, TLR8, or combined TLR7/8 antagonists for the potential treatment of inflammation and various autoimmune diseases can be categorized into two classes: low-molecular-weight oral compounds and immunoregulatory DNA sequences (IRS) [77]. The first low-molecular-weight compounds are derived from the modification of the imidazoloquinoline series (17–19) where the substitution pattern has been reversed and modified to provide compounds with single-digit micromolar inhibitory potency against TLR7 (e.g., compound 25) [78]. The most advanced nonantimalarial antagonist is the orally available TLR7/8/9 antagonist CPG 52364 (26), which has demonstrated single-digit nanomolar inhibitory potency in hTLR9-transfected HEK293 cells [79]. IRSs are TLR7, TLR7/8, or TLR7/9 antagonists that have demonstrated efficacy in mouse models for SLE, rheumatoid arthritis, and psoriasis; they include DV1079 (IRS954 5′-TGC TCC TGG AGG GGT TGT-3′) [77].

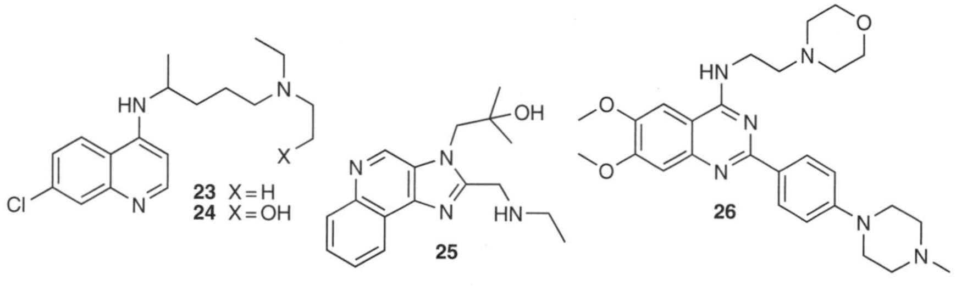

7. TLR9

ssDNA is the ligand for TLR9 [80]. TLR9 ligands were first envisioned as vaccine adjuvants, and sequences were optimized based on cell activation and cytokine responses. Interest has since expanded to oncologic uses predicated on activation of antitumor immunity, or initiation of apoptosis in TLR9-expressing tumor cells [4]. Antagonists of TLR9 are of interest because endogenous DNA can stimulate this receptor, particularly when introduced into responding cells by a protein or peptide carrier such as autoantibody or a peptide carrier like the cathelicidin LL37, which acts as a chaperone for nucleic acids [27,81]. This raises the possibility of deleterious results from endogenous DNA stimulation (e.g., in SLE or psoriasis).

7.1 Agonists

Synthetic polynucleotide agonists typically include phosphorothioate linkages in their backbone to limit DNAase-mediated metabolism and in this context activate most efficiently when the sequence incorporates an unmethylated CpG motif [82]. There is extensive patent and peer-reviewed literature on the agonist activity of specific short oligonucleotide sequences of 20–30 bases [83]. Sequences flanking the CpG motif determine species specificity [84] and affect the rate of transit through different endosomal compartments, thereby modulating the production of class I interferon, which is dependent on signaling from a transferrin-positive compartment [85,86]. The sensitivity of stimulatory activity to modifications of the stimulatory cytosine and guanine residues has also been examined [83,87–89]. Synthetic sequences in clinical trials include agatolimod and EMD1201081. Nonnucleotide TLR9 agonists have not been disclosed.

7.2 Antagonists

Both nucleotide-based and small-molecule antagonists of TLR9 are known, and due to the similarity in receptor and ligand structures are generally effective against both TLR9 and TLR7. Inhibitory oligonucleotide sequences are typically guanine-rich competitive antagonists that bind receptor without causing activation [90]. Two series of small-molecule inhibitors have been described. The first series, exemplified by compound **27**, was optimized on the substituted quinoline structure of chloroquine and has nanomolar potency in a CpG oligo-stimulated viability assay [91,92], and in a transfected TLR9 reporter assay [93]. A more recently described series of diamino-substituted phenylbenzoxazoles (**28–30**) also show nanomolar potency in cell-based assays [94,95].

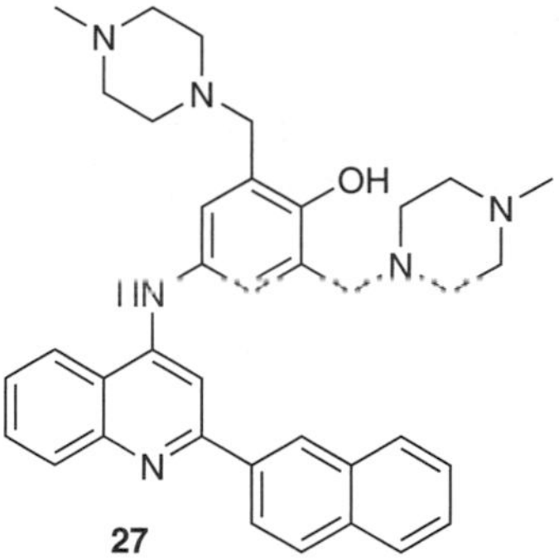

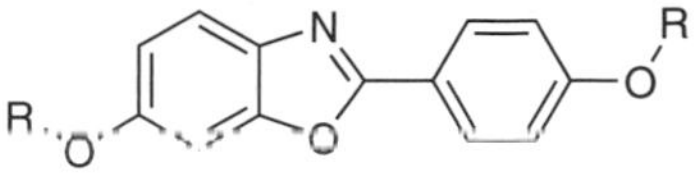

28 R = -3-pyrrolidine
29 R = -(CH₂)₃N(CH₃)₂
30 R = -(CH₂)₃-N-pyrrolidine

27

8. MYD88

Myeloid differentiation primary response protein (MyD88) is a cytosolic adapter protein essential for TLR signal transduction from all the TLRs except TLR3, and is a key component for IL-1 and IL-18 signaling [96]. Blocking MyD88-mediated interactions is a promising therapeutic strategy for various inflammatory diseases since this would attenuate both initiation and amplification of inflammatory responses. MyD88, like TLRs and the IL-1 receptor, contains a TIR domain, which is critical in the formation of (1) MyD88–TLR heterodimers, (2) a MyD88–IL-1R heterodimer, and (3) a MyD88–MyD88 homodimer. Homodimerization of MyD88 is a common critical step in the TLR and IL-1 signaling pathways since it triggers recruitment of essential downstream signal transducers such as IRAK-4 and IRAK-1 [97].

8.1 Inhibitors

Given the functional importance of the MyD88 TIR domain, some groups have focused on the rational design of inhibitors of MyD88-mediated protein–protein interactions. Compound **31** was designed to mimic a tripeptide region of the MyD88 TIR domain and shown to block association with the IL-1 receptor, and attenuate IL-1β-induced fever in mice at 200 mg/kg when dosed i.p. [98]. Interestingly, **31** did not affect interactions between MyD88 and TLR4, suggesting differential specificity for TIR domains of MyD88 and the IL-1 receptor versus TLR4. Consistent with a trend toward bifunctional binders for blocking protein–protein interactions, TIR domain mimics such as compound **32** have also been engineered [96]. More recently, the peptidomimetic approach has been extended to a heptapeptide region of the MyD88 TIR domain. ST2825 (**33**) inhibits the MyD88–MyD88 homodimerization process [97,99,100]. ST2825 inhibited IL-6 production in response to TLR9 or IL-1 receptor stimulation and showed oral activity in an acute *in vivo* mouse inflammation model at 100 mg/kg [97]. *In vitro*, ST2825 blocked proliferation of human B cells activated through TLR9, suggesting promise for autoimmune diseases such as SLE.

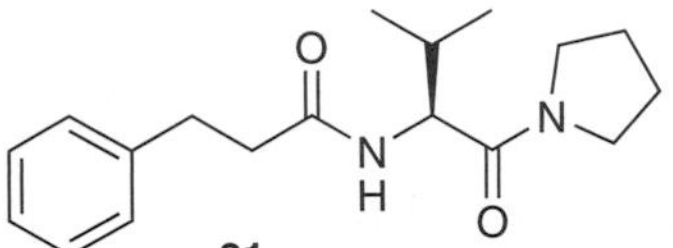

9. IRAK-4

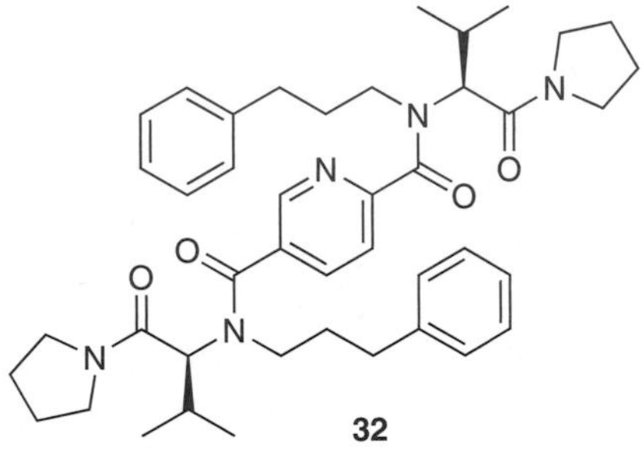

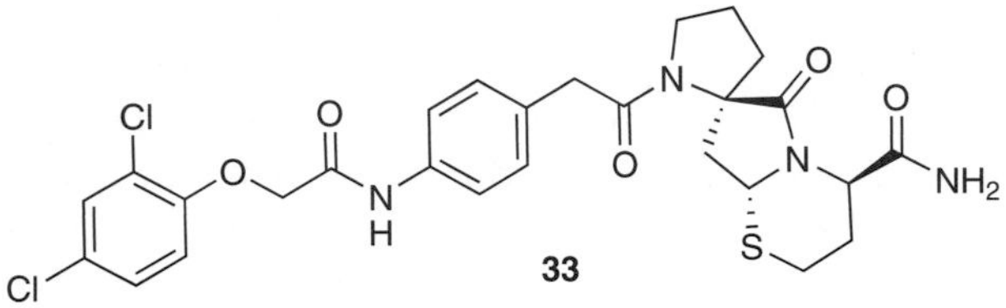

Many protein kinases are involved in TLR signaling pathways but IRAK-4 is particularly important because it is the first kinase downstream of the TLRs and is absolutely required for TLR signal transduction [28]. IRAK-4 is also essential for IL-1 signaling and thus plays a key role in both inflammatory initiation and propagation. Cells from IRAK-4-deficient mice and humans have impaired responses to TLR ligands and IL-1, suggesting this kinase to be a potentially valuable target for treating inflammatory diseases including sepsis and autoimmunity [101].

9.1 Inhibitors

Potent IRAK-4 inhibitors have been reported by several groups including some structurally related benzimidazoles (e.g., **34–36**) as well as alternative fused heterocycles such as the imidazopyridine (**37**) and imidazopyridazine (**38**) [102–107]. Such compounds are reported to have IRAK-1 and IRAK-4 activity with varying degrees of selectivity. Although it is debatable whether selectivity would be advantageous [108], some evidence suggests that IRAK-4-selective compounds could be more valuable [109]. An X-ray cocrystal structure of IRAK-4 with benzimidazole (35) revealed a uniquely shaped ATP-binding pocket that may provide useful insights into the design of optimized inhibitors [110].

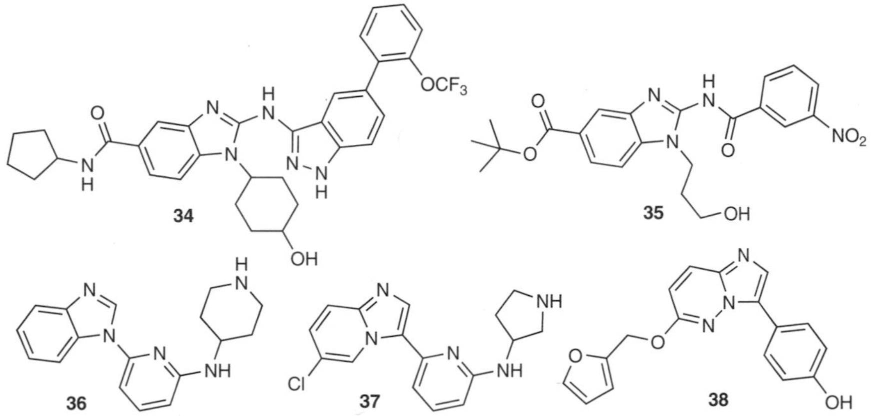

10. CONCLUSIONS

The rate of TLR-related publications has risen consistently over the past decade, and shows no signs of slowing down with over 2400 papers published in 2009 alone. Despite this increasing wealth of information, relatively few TLR modulators have been described. Among the pharmaceutically relevant articles published over the past year or so, TLR4, TLR7, and TLR9 have featured most prominently. In terms of development, the most advanced compound is TLR4 antagonist E5564, which is in Phase III sepsis trials [111]. Also of interest is the TLR7/8/9 antagonist CPG 52364, which is in early clinical development for SLE [112]. TLR2 (as a heterodimer with TLR1 or TLR6), MyD88, and IRAK-4 may also prove to be attractive therapeutic targets for various inflammatory diseases; however, few agents have been reported and these have not yet advanced beyond early discovery. This situation may change as recent publications have disclosed valuable structural information for TLR1/2 [16], TLR2/6 [17], TLR3 [18], MyD88 [113], and IRAK-4 [110], which may help stimulate additional structure-based drug design efforts in these areas.

REFERENCES

[1] B. Lemaitre, E. Nicolas, L. Michaut, J. M. Reichhart, and J. A. Hoffmann, *Cell*, 1996, **86**, 973.

[2] E. T. Rietschel and O. Westphal in *Endotoxin in Health and Disease*, (ed. H. Brade), Marcel Dekker, Inc., New York, NY, p. 1, 1999.

[3] E. J. Hennessy, A. E. Parker, and L. A. J. O'Neill, *Nat. Rev. Drug Discov.*, 2010, **9**, 293.

[4] L. A. O'Neill, C. E. Bryant, and S. L. Doyle, *Pharmacol. Rev.*, 2009, **61**, 177.

[5] K. J. Ishii, S. Koyama, A. Nakagawa, C. Coban, and S. Akira, *Cell Host Microbe*, 2008, **3**, 352.

[6] T. Kaisho and S. Akira, *Trends Immunol.*, 2001, **22**, 78.

[7] N. Kadowaki, S. Ho, S. Antonenko, R. W. Malefyt, R. A. Kastelein, F. Bazan, and Y. J. Liu, *J. Exp. Med.*, 2001, **194**, 863.

[8] M. Muzio, D. Bosisio, N. Polentarutti, G. D'Amico, A. Stoppacciaro, R. Mancinelli, C. van't Veer, G. Penton-Rol, L. P. Ruco, P. Allavena, and A. Mantovani, *J. Immunol.*, 2000, **164**, 5998.

[9] M. C. Banas, B. Banas, K. L. Hudkins, T. A. Wietecha, M. Iyoda, E. Bock, P. Hauser, J. W. Pippin, S. J. Shankland, K. D. Smith, B. Stoelcker, G. Liu, H. J. Grone, B. K. Kramer, and C. E. Alpers, *J. Am. Soc. Nephrol.*, 2008, **19**, 704.

[10] A. Schaffler, J. Scholmerich, and B. Salzberger, *Trends Immunol.*, 2007, **28**, 393.

[11] J. R. Ward, S. E. Francis, L. Marsden, T. Suddason, G. M. Lord, S. K. Dower, D. C. Crossman, and I. Sabroe, *Immunology*, 2009, **128**, 58.

[12] M. M. Brinkmann, E. Spooner, K. Hoebe, B. Beutler, H. L. Ploegh, and Y. M. Kim, *J. Cell. Biol.*, 2007, **177**, 265.

[13] N. J. Gay and M. Gangloff, *Ann. Rev. Biochem.*, 2007, **76**, 141.

[14] N. Matsushima, T. Tanaka, P. Enkhbayar, T. Mikami, M. Taga, K. Yamada, and Y. Kuroki, *BMC Genomics*, 2007, **8**, 124.

[15] M. S. Jin and J. O. Lee, *Immunity*, 2008, **29**, 182.

[16] M. S. Jin, S. E. Kim, J. Y. Heo, M. E. Lee, H. M. Kim, S. G. Paik, H. Lee, and J. O. Lee, *Cell*, 2007, **130**, 1071.

[17] J. Y. Kang, X. Nan, M. S. Jin, S.-J. Youn, Y. H. Ryu, S. Mah, S. H. Han, H. Lee, S. -G. Paik, and J. -O. Lee, *Immunity*, 2009, **31**, 873.

[18] L. Liu, I. Botos, Y. Wang, J. N. Leonard, J. Shiloach, D. M. Segal, and D. R. Davies, *Science*, 2008, **320**, 379.

[19] B. S. Park, D. H. Song, H. M. Kim, B. S. Choi, H. Lee, and J. O. Lee, *Nature*, 2009, **458**, 1191.

[20] U. Ohto, K. Fukase, K. Miyake, and Y. Satow, *Science*, 2007, **316**, 1632.

[21] H. M. Kim, B. S. Park, J. I. Kim, S. E. Kim, J. Lee, S. C. Oh, P. Enkhbayar, N. Matsushima, H. Lee, O. J. Yoo, and J. O. Lee, *Cell*, 2007, **130**, 906.

[22] T. Wei, J. Gong, F. Jamitzky, W. M. Heckl, R. W. Stark, and S. C. Rossle, *Protein Sci.*, 2009, **18**, 1684.

[23] N. J. Gay, M. Gangloff, and A. N. Weber, *Nat. Rev. Immunol.*, 2006, **6**, 693.

[24] D. Werling, O. C. Jann, V. Offord, E. J. Glass, and T. J. Coffey, *Trends Immunol.*, 2009, **30**, 124.

[25] G. Srikrishna and H. H. Freeze, *Neoplasia*, 2009, **11**, 615.

[26] C. Amlie-Lefond, D. A. Paz, M. P. Connelly, G. B. Huffnagle, N. T. Whelan, and H. T. Whelan, *J. Allergy Clin. Immunol.*, 2005, **116**, 1334.

[27] M. W. Boulé, C. Broughton, F. Mackay, S. Akira, A. Marshak-Rothstein, and I. R. Rifkin, *J. Exp. Med.*, 2004, **199**, 1631.

[28] Z. Wang, H. Wesche, T. Stevens, N. Walker, and W. C. Yeh, *Curr. Top. Med. Chem.*, 2009, **9**, 724.

[29] C. Olive, K. Schulze, H. K. Sun, T. Ebensen, A. Horvath, I. Toth, and C. A. Guzman, *Vaccine*, 2007, **25**, 1789.

[30] C. Zeckey, T. Tschernig, F. Hildebrand, M. Frink, C. Fromke, M. Dorsch, C. Krettek, and T. Barkhausen, *Shock*, 2010, **33**, 614.

[31] M. Niebuhr, U. Mai, A. Kapp, and T. Werfel, *Exp. Dermatol.*, 2008, **17**, 953.

[32] W. Wu, R. Li, S. S. Malladi, H. J. Warshakoon, M. R. Kimbrell, M. W. Amolins, R. Ukani, A. Datta, and S. A. David, *J. Med. Chem.*, 2010, **53**, 3198.

[33] T. Seyberth, S. Voss, R. Brock, K. -H. Wiesmuller, and G. Jung, *J. Med. Chem.*, 2006, **49**, 1754.

[34] M. R. Spyvee, H. Zhang, L. D. Hawkins, and J. C. Chow, *Bioorg. Med. Chem. Lett.*, 2005, **15**, 5494.

[35] M. Czarniecki, *J. Med. Chem.*, 2008, **51**, 6621.

[36] J. Chow, F. Gusovsky, L. Hawkins, and M. Spyvee, *Patent Application US 2004/35447*, 2004.

[37] R. Girard, T. Pedron, S. Uematsu, V. Balloy, M. Chignard, S. Akira, and R. Chaby, *J. Cell Sci.*, 2003, **116**, 293.

[38] L. Alexopoulou, A. C. Holt, R. Medzhitov, and R. A. Flavell, *Nature*, 2001, **413**, 732.

[39] H. R. Hubbell, K. Kvalnes-Krick, W. A. Carter, and D. R. Strayer, *Cancer Res.*, 1985, **45**, 2481.

[40] B. Jasani, H. Navabi, and M. Adams, *Vaccine*, 2009, **27**, 3401.

[41] E. Aubin, C. Belmant, L. Gauthier, Y. Morel, C. Paturel, and D. Bregeon, *Patent Application WO 2009/130616*, 2009.

[42] R. Medzhitov, P. Preston-Hurlburt, and C. A. Janeway Jr., *Nature*, 1997, **388**, 394.

[43] O. Westphal and O. Luderitz, *Angew. Chem.*, 1954, **66**, 407.

[44] E. T. Rietschel, T. Kirikae, F. U. Schade, U. Mamat, G. Schmidt, H. Loppnow, A. J. Ulmer, U. Zahringer, U. Seydel, and F. Di Padova, *FASEB J.*, 1994, **8**, 217.

[45] N. Qureshi, K. Takayama, and E. Ribi, *J. Biol. Chem.*, 1982, **257**, 11808.

[46] H. G. Bazin, L. S. Bess, M. T. Livesay, K. T. Ryter, C. L. Johnson, J. S. Arnold, and D. A. Johnson, *Tetrahedron Lett.*, 2006, **47**, 2087.

[47] S. T. Ishizaka and L. D. Hawkins, *Expert Rev. Vaccines*, 2007, **6**, 773.

[48] F. Savoy, D. M. Nicolle, D. Rivier, C. Chiavaroli, B. Ryffel, and V. F. Quesniaux, *Immunobiology*, 2006, **211**, 767.

[49] S. K. Drexler and B. M. Foxwell, *Int. J. Biochem. Cell Biol.*, 2009, **42**, 506.

[50] J. Krishnan, G. Lee, and S. Choi, *Arch. Pharm. Res.*, 2009, **32**, 1485.

[51] D. P. Rossignol and M. Lynn, *Curr. Opin. Invest. Drugs*, 2005, **6**, 496.

[52] M. Tidswell, W. Tillis, S. P. Larosa, M. Lynn, A. E. Wittek, R. Kao, J. Wheeler, J. Gogate, and S. M. Opal, *Crit. Care Med.*, 2009, **38**, 72.

[53] http://clinicaltrials.gov/ct2/show/NCT00633477?term=NCT00633477&rank=1.

[54] H. G. Bazin, T. J. Murray, W. S. Bowen, A. Mozaffarian, S. P. Fling, L. S. Bess, M. T. Livesay, J. S. Arnold, C. L. Johnson, K. T. Ryter, C. W. Cluff, J. T. Evans, and D. A. Johnson, *Bioorg. Med. Chem. Lett.*, 2008, **18**, 5350.

[55] M. Piazza, L. Yu, A. Teghanemt, T. Gioannini, J. Weiss, and F. Peri, *Biochemistry*, 2009, **48**, 12337.

[56] C. Lunardi, C. Bason, M. Dolcino, R. Navone, R. Simone, D. Saverino, L. Frulloni, E. Tinazzi, D. Peterlana, R. Corrocher, and A. Puccetti, *J. Intern. Med.*, 2009, **265**, 250.

[57] E. Andersen-Nissen, K. D. Smith, R. Bonneau, R. K. Strong, and A. Aderem, *J. Exp. Med.*, 2007, **204**, 393.

[58] W. McDonald, T. J. Powell, A. E. Price, and R. S. Becker, *Patent Application WO 2009/128949-A1*, 2009.

[59] A. E. Price, A. R. Shaw, T. J. Powell, V. Nakaar, L. Song, L. Zhang, M. D. Yeager, R. S. Becker, and L. G. Tussey, *Patent Application WO 2009/128951-A2*, 2009.

[60] A. R. Shaw, L. Song, and L. G. Tussey, *Patent Application WO 2009/128952-A1*, 2009.

[61] L . G. Burdelya, V. I. Krivokrysenko, T. C. Tallant, E. Strom, A. S. Gleiberman, D. Gupta, O. V. Kurnasov, F. L. Fort, A. L. Osterman, J. A. Didonato, E. Feinstein, and A. V. Gudkov, *Science*, 2008, **320**, 226.

[62] K. B. Gorden, K. S. Gorski, S. J. Gibson, R. M. Kedl, W. C. Kieper, X. Qiu, M. A. Tomai, S. S. Alkan, and J. P. Vasilakos, *J. Immunol.*, 2005, **174**, 1259.

[63] C. Guiducci, R. L. Coffman, and F. J. Barrat, *J. Intern. Med.*, 2009, **265**, 43.

[64] A. Marshak-Rothstein, *Nat. Rev. Immunol.*, 2006, **6**, 823.

[65] http://clinicaltrials.gov/ct2/show/NCT00810758?id=NCT+00810758&rank=1.

[66] http://clinicaltrials.gov/ct2/show/NCT00189332?term=NCT00189332&rank=1.

[67] S. S. Agarwala, J. M. Kirkwood, and J. Bryant, *Cytokines Cell. Mol. Ther.*, 2000, **6**, 171.

[68] S. Fletcher, K. Steffy, and D. Averett, *Curr. Opin. Invest. Drugs*, 2006, **7**, 702.

[69] P. Jones and D. C. Pryde, *Patent Application WO 2006/227670*, 2006.

[70] T. Lan, L. Bhagat, D. Wang, M. Dai, E. R. Kandimalla, and S. Agrawal, *Bioorg. Med. Chem. Lett.*, 2009, **19**, 2044.

[71] T. K. Means, E. Latz, F. Hayashi, M. R. Murali, D. T. Golenbock, and A. D. Luster, *J. Clin. Invest.*, 2005, **115**, 407.

[72] G. Ruiz-Irastorza, M. Ramos-Casals, P. Brito-Zeron, and M. A. Khamashta, *Ann. Rheum. Dis.*, 2010, **69**, 20.

[73] S. Shinjo, E. Bonfa, D. Wojdyla, E. Borba, L. Ramirez, H. Scherbarth, J. Brenol, R. Chacon-Diaz, O. Neira, G. Berbotto, I. De La Torre, E. Acevedo-Vazquez, L. Massardo, L. Barile-Fabris, F. Caeiro, L. Silveira, E. Sato, S. Buliubasich, G. Alarcon, and B. BPons-Este, *Arthritis and Rheum.*, 2010, **62**, 855.

[74] S. Sun, N. L. Rao, J. Venable, R. Thurmond, and L. Karlsson, *Inflamm. Allergy Drug Targets*, 2007, **6**, 223.

[75] J. Lee, T. H. Chuang, V. Redecke, L. She, P. M. Pitha, D. A. Carson, E. Raz, and H. B. Cottam, *Proc. Natl. Acad. Sci. USA*, 2003, **100**, 6646.

[76] S. Kalia and J. P. Dutz, *Dermatol. Ther.*, 2007, **20**, 160.

[77] D. Wang, L. Bhagat, D. Yu, F. G. Zhu, J. X. Tang, E. R. Kandimalla, and S. Agrawal, *J. Med. Chem.*, 2009, **52**, 551.

[78] N. M. Shukla, M. R. Kimbrell, S. S. Malladi, and S. A. David, *Bioorg. Med. Chem. Lett.*, 2009, **19**, 2211.

[79] G. B. Lipford, A. Forsbach, and C. M. Zepp, *Patent Application WO 2005/007672-A1*, 2005.

[80] H. Hemmi, O. Takeuchi, T. Kawai, T. Kaisho, S. Sato, H. Sanjo, M. Matsumoto, K. Hoshino, H. Wagner, K. Takeda, and S. Akira, *Nature*, 2000, **408**, 740.

[81] R. Lande, J. Gregorio, V. Facchinetti, B. Chatterjee, Y. H. Wang, B. Homey, W. Cao, B. Su, F. O. Nestle, T. Zal, I. Mellman, J. M. Schroder, Y. J. Liu, and M. Gilliet, *Nature*, 2007, **449**, 564.

[82] A. M. Krieg, A. K. Yi, S. Matson, T. J. Waldschmidt, G. A. Bishop, R. Teasdale, G. A. Koretzky, and D. M. Klinman, *Nature*, 1995, **374**, 546.

[83] J. Kindrachuk, J. Potter, H. L. Wilson, P. Griebel, L. A. Babiuk, and S. Napper, *Mini Rev. Med. Chem.*, 2008, **8**, 590.

[84] S. Bauer, C. J. Kirschning, H. Hacker, V. Redecke, S. Hausmann, S. Akira, H. Wagner, and G. B. Lipford, *Proc. Natl. Acad. Sci. USA*, 2001, **98**, 9237.

[85] C. Guiducci, G. Ott, J. H. Chan, E. Damon, C. Calacsan, T. Matray, K. D. Lee, R. L. Coffman, and F. J. Barrat, *J. Exp. Med.*, 2006, **203**, 1999.

[86] M. Kerkmann, L. T. Costa, C. Richter, S. Rothenfusser, J. Battiany, V. Hornung, J. Johnson, S. Englert, T. Ketterer, W. Heckl, S. Thalhammer, S. Endres, and G. Hartmann, *J. Biol. Chem.*, 2005, **280**, 8086.

[87] M. Jurk, A. Kritzler, H. Debelak, J. Vollmer, A. M. Krieg, and E. Uhlmann, *ChemMedChem.*, 2006, **1**, 1007.

[88] E. R. Kandimalla, L. Bhagat, Y. Li, D. Yu, D. Wang, Y. P. Cong, S. S. Song, J. X. Tang, T. Sullivan, and S. Agrawal, *Proc. Natl. Acad. Sci. USA*, 2005, **102**, 6925.

[89] E. R. Kandimalla, D. Yu, Q. Zhao, and S. Agrawal, *Bioorg. Med. Chem.*, 2001, **9**, 807.

[90] T. Haas, J. Metzger, F. Schmitz, A. Heit, T. Muller, E. Latz, and H. Wagner, *Immunity*, 2008, **28**, 315.

[91] L. Manzel, L. Strekowski, F. M. Ismail, J. C. Smith, and D. E. Macfarlane, *J. Pharmacol. Exp. Ther.*, 1999, **291**, 1337.

[92] L. Strekowski, O. Zegrocka, M. Henary, M. Say, M. J. Mokrosz, B. M. Kotecka, L. Manzel, and D. E. Macfarlane, *Bioorg. Med. Chem. Lett.*, 1999, **9**, 1819.

[93] G. D. Liplord, A. Forbach, and C. Zepp, *US Patent 10/872,196*, 2005.

[94] W. Zheng, M. Spyvee, F. Gusovsky, and S. T. Ishizaka, *Patent Applications WO 2010/036905-A1 and WO 2010/036908-A1*, 2010.

[95] M. Lamphier, E. Latz, W. Zheng, H. Hansen, M. Spyvee, J. Rose, M. Genest, H. Yang, C. Shaffer, Y. Shen, C. Liu, D. Liu, C. Rowbottom, J. Chow, F. Gusovsky, and S. Ishizaka, *unpublished results*.

[96] C. N. Davis, E. Mann, M. M. Behrens, S. Gaidarova, M. Rebek, J. Rebek Jr., and T. Bartfai, *Proc. Natl. Acad. Sci. USA*, 2006, **103**, 2953.

[97] M. Loiarro, F. Capolunghi, N. Fanto, G. Gallo, S. Campo, B. Arseni, R. Carsetti, P. Carminati, R. De Santis, V. Ruggiero, and C. Sette, *J. Leukoc. Biol.*, 2007, **82**, 801.

[98] T. Bartfai, M. M. Behrens, S. Gaidarova, J. Pemberton, A. Shivanyuk, and J. Rebek, Jr., *Proc. Natl. Acad. Sci. USA*, 2003, **100**, 7971.

[99] C. Sette, *J. Leukoc. Biol.*, 2007, **82**, 811.

[100] N. Fanto, G. Gallo, A. Ciacci, M. Semproni, D. Vignola, M. Quaglia, V. Bombardi, D. Mastroianni, M. P. Zibella, G. Basile, M. Sassano, V. Ruggiero, R. De Santis, and P. Carminati, *J. Med. Chem.*, 2008, **51**, 1189.

[101] C. Wietek and L. A. O'Neill, *Mol. Interv.*, 2002, **2**, 212.

[102] G. M. Buckley, T. A. Ceska, J. L. Fraser, L. Gowers, C. R. Groom, A. P. Higueruelo, K. Jenkins, S. R. Mack, T. Morgan, D. M. Parry, W. R. Pitt, O. Rausch, M. D. Richard, and V. Sabin, *Bioorg. Med. Chem. Lett.*, 2008, **18**, 3291.

[103] G. M. Buckley, R. Fosbeary, J. L. Fraser, L. Gowers, A. P. Higueruelo, L. A. James, K. Jenkins, S. R. Mack, T. Morgan, D. M. Parry, W. R. Pitt, O. Rausch, M. D. Richard, and V. Sabin, *Bioorg. Med. Chem. Lett.*, 2008, **18**, 3656.

[104] G. M. Buckley, L. Gowers, A. P. Higueruelo, K. Jenkins, S. R. Mack, T. Morgan, D. M. Parry, W. R. Pitt, O. Rausch, M. D. Richard, V. Sabin, and J. L. Fraser, *Bioorg. Med. Chem. Lett.*, 2008, **18**, 3211.

[105] J. P. Powers, S. Li, J. C. Jaen, J. Liu, N. P. Walker, Z. Wang, and H. Wesche, *Bioorg. Med. Chem. Lett.*, 2006, **16**, 2842.

[106] T. Durand-Reville, C. Jewell, C. Hammond, and D. Chin, *Patent Application WO 2008/030579-A2*, 2008.

[107] K. Guckian, C. Jewell, P. Conlan, E. Y. -S. Lin, and T. Chan, *Patent Application WO 2008/030584-A2*, 2008.

[108] K. W. Song, F. X. Talamas, R. T. Suttmann, P. S. Olson, J. W. Barnett, S. W. Lee, K. D. Thompson, S. Jin, M. Hekmat-Nejad, T. Z. Cai, A. M. Manning, R. J. Hill, and B. R. Wong, *Mol. Immunol.*, 2009, **46**, 1458.

[109] T. -T. Yamin and D. K. Miller, *J. Biol. Chem.*, 1997, **272**, 21540.

[110] Z. Wang, J. Liu, A. Sudom, M. Ayres, S. Li, H. Wesche, J. P. Powers, and N. P. Walker, *Structure*, 2006, **14**, 1835.

[111] http://www.eisai.com/view_press_release.asp?ID=147&press=253.

[112] http://clinicaltrials.gov/ct2/show/NCT00547014.

[113] H. Ohnishi, H. Tochio, Z. Kato, K. E. Orii, A. Li, T. Kimura, H. Hiroaki, N. Kondo, and M. Shirakawa, *Proc. Natl. Acad. Sci. USA*, 2009, **106**, 10260.

PART IV:
Oncology

Editor: Shelli R. McAlpine
Department of Chemistry and Biochemistry
San Diego State University
San Diego
California

CHAPTER 13

Janus Kinase 2 (JAK2) Inhibitors for the Treatment of Myeloproliferative Neoplasm (MPN)

Ashok V. Purandare, Matthew V. Lorenzi and **Louis J. Lombardo**

1. INTRODUCTION

JAK2 is a member of the Janus kinase (JAK) family of nonreceptor tyrosine kinases comprising of four family members (JAK1, JAK2, JAK3, and TYK2). Each JAK is composed of two kinase domains in its carboxy terminal half, JH1 and JH2, with the JH2 domain representing a pseudokinase domain that negatively regulates enzymatic activity. The amino terminal half of the protein contains a FERM domain that associates with cytokine receptors. JAKs are essential components to cytokine receptor signaling; upon ligand binding they activate and subsequently phosphorylate a STAT (signal transducers and activators of transcription). The STAT is then translocated to the

Bristol-Myers Squibb, Research and Development, Princeton, NJ 08543, USA

Annual Reports in Medicinal Chemistry, Volume 45
ISSN 0065-7743, DOI 10.1016/S0065-7743(10)45013-7

nucleus, initiating gene expression. Imbalances in the JAK–STAT pathway have been implicated in various cancer and inflammatory diseases.

1.1 Myeloproliferative neoplasms: disease background

Myeloproliferative neoplasms (MPNs) are a subset of myeloid malignancies that are characterized by the expansion of a multipotent hematopoietic progenitor stem cell. Chronic MPNs can be classified into two categories: those harboring the BCR-ABL oncogene and those that do not [1]. This latter category of neoplasms encompasses polycythemia vera (PV), essential thrombocythemia (ET), and primary myelofibrosis (PMF). Recently, PV, ET, and PMF have been shown to be associated with somatic mutations that constitutively activate the JAK2 enzyme [2,3]. In normal hematopoiesis, JAK family signaling is regulated by the binding of a cytokine to its cognate receptor, which promotes the maturation and/or proliferation of a particular hematopoietic lineage. In MPNs, the acquisition of a JAK2 pathway mutation results in cytokine-independent activation of the pathway leading to uncontrolled growth of erythrocytes, platelets, and granulocyte/monocytes being the predominant lineages expanded in ET, PV, and PMF, respectively. The uncontrolled growth of these cell lineages in MPN results in pathologies including splenomegaly, hemorrhage, thrombosis, bone marrow fibrosis, and transformation to acute myeloid leukemia [4]. The overall survival rate for patients afflicted with advanced disease is estimated at 3–5 years and hence represents a high unmet medical need.

1.2 JAK2 pathway activation in MPNs

Interest in JAK2 inhibitors for the treatment of MPN was bolstered by the discovery of a point mutation in JAK2 at position 617 (valine to phenylalanine, V617F) in the majority of patients with MPN [2,3]. This mutation occurs in the JH2 pseudokinase domain and relieves an autoinhibitory function of this domain, constitutively activating the catalytic function of the kinase via a cytokine-independent mechanism. Although this is the predominant mutation in the disease, additional mutations in the enzyme, its receptor, or within other levels of the network have also been identified that result in JAK2–STAT pathway activation [5–9] (Figure 1).

1.3 Rationale for selective targeting of JAK2 for MPNs

JAK2 is critical for thrombopoietin (TPO) and erythropoietin (EPO) receptor signaling that controls the growth and differentiation of platelets and erythrocytes, respectively [10,11]. Transgenic reconstitution of the activated

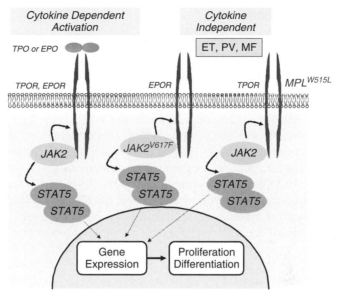

Figure 1 JAK2-STAT Pathway Activation in MPN. (See Color Plate 13.1 in the Color Plate Section)

JAK2^{V617F} gene into hematopoietic progenitor cells in mice results in the hallmark features of human MPN including elevated blood counts, bone marrow fibrosis, and splenomegaly [12,13]. Similarly, introduction of the JAK2^{V617F} gene into the human hematopoietic stem cell compartment *ex vivo* results in EPO-independent erythroid colony formation [14]. Importantly, both of these features are inhibited by administration of small-molecule JAK2 inhibitors [15]. Small-molecule JAK2 inhibitors are effective at clinically reducing some of MPN's key features (described below). Collectively, these data substantiate the causal nature of the JAK2 signaling pathway in the development of MPNs and highlight the disease-modifying potential of JAK2 inhibitors. Given that other JAK family members are involved in the regulation of immune function, it is important to maintain selectivity for JAK2 over these family members in order to mitigate the risks associated with undesired immunosuppression.

2. CLINICAL TRIALS AND SUPPORTING PRECLINICAL DATA

Several JAK2 inhibitors with varying JAK family selectivity are in early to midstage clinical trials. The primary indications for this disease are myelofibrosis, or post-PV, or ET myelofibrosis. As candidates show promising responses to treatment, indicated by reducing their symptoms and the drug demonstrates long-term safety, it is anticipated that the drug's use on patients with less aggressive forms of MPNs will be pursued.

2.1 Clinical candidates

INCB18424 (**1**, ruxolitinib) is the most advanced ATP competitive JAK2 inhibitor and has completed Phase (Ph) II clinical trials [16]. This compound is a potent inhibitor of both JAK1 and JAK2 with selectivity versus JAK3 and TYK2 (IC_{50} values of 2.7, 4.5, 332, and 19 nM, respectively, in biochemical assays) [17]. INCB18424 demonstrated potent *in vitro* cellular activity (100–130 nM) in an engineered Ba/F3 cell line harboring the JAK2^{V617F} mutation and EPO-independent erythroid colony growth of precursor cells from PV patients (IC_{50} of 67 nM) while showing lesser effect on normal colony formation from healthy donors ($IC_{50} > 400$ nM). Oral administration of INCB18424 to mice bearing BaF/3-JAK2^{V617F} xenografts resulted in significant reduction of associated splenomegaly [17]. In addition, elimination of neoplastic cells from the spleen, liver, and bone marrow was observed, accompanied by normalization of the histology of affected organs and marked improved survival benefit.

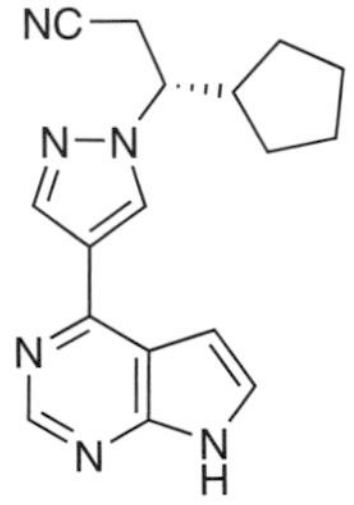

1

Clinically, INCB18424 is administered twice a day with doses ranging from 10 to 25 mg, depending on patient tolerability [18a, 18b]. The molecule was well tolerated for continuous dosing up to 15 months with few Grade III–IV toxicities. The compound showed durable effects in reducing spleen size and relieving constitutional symptoms associated with PMF (e.g., reduction of night sweats and increased appetite). Consistent with potent JAK1 activity, INCB18424 also reduced levels of circulating inflammatory cytokines. However, only a modest decrease in JAK2^{V617F} allele burden was noticed in the treated patients. In Ph II clinical trials with advanced PV and ET patients resistant to hydroxyurea, INCB18424 showed normalization of hematological cell parameters and a reduction in phlebotomy frequency [19]. Given the positive outcomes from Ph II clinical trials, INCB18424 is recruiting for Ph III studies.

TG101348 (**2**) is reported to be a potent and selective ATP competitive inhibitor of JAK2 with high JAK family selectivity (IC_{50} values for JAK1, JAK2, JAK3, and TYK2 in the biochemical enzymatic assay are 105, 3,

1000, and 405 nM, respectively) [15]. TG101348 inhibited growth of HEL, Ba/F3-MPL^{W515L}, and Ba/F3-JAK2^{V617F} cell lines bearing JAK2 pathway mutations with IC$_{50}$ values of 300, 620, and 580 nM, respectively [15]. It preferentially inhibited colony growth of erythroid precursors from PV patients (IC$_{50}$ of ~600 nM) compared to those in healthy volunteers. In a mouse PV model, oral administration of TG101348 reversed polycythemia, leukocytosis, extramedullary hematopoiesis, and splenomegaly in a dose-dependent manner. TG101348 was also shown to inhibit JAK2-driven tumor growth and cell engraftment *in vivo* [20]. In Ph I/II clinical trials, TG101348 was dosed 680–800 mg, once a day where it was generally well tolerated, with manageable Grade I–II gastrointestinal toxicities at higher doses [21]. Two-thirds of MF patients treated with the drug achieved ≥50% reduction in palpable splenomegaly, while ~30% had a complete response. TG101348 showed antimyeloproliferative activity with virtually all treated patients with complete resolution of leukocytosis and thrombocytosis. In contrast to INCB18424, the compound showed minimal effect on serum levels of proinflammatory cytokines and reportedly decreased JAK2^{V617F} allele burden in a substantial proportion of treated patients, thus showing promise as a future treatment [22].

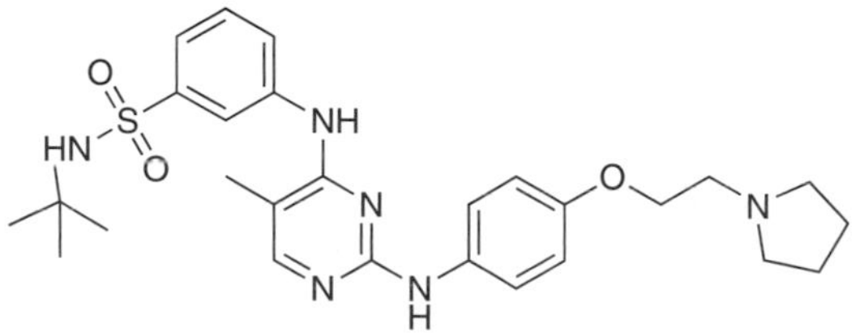

2

A highly JAK family-selective inhibitor, XL019 (structure not disclosed), has been reported, and has enzyme inhibition IC$_{50}$ values of 130, 2, 250, and 340 nM, for JAK1, JAK2, JAK3, and TYK2, respectively [23]. The molecule demonstrated equivalent selectivity in functional assays conducted in primary human cells stimulated with respective JAK family cytokines. XL019 inhibited JAK2 signaling in HEL erythroleukemia cells *in vitro* and *in vivo* and delayed tumor growth up to 70% in a mouse HEL xenograft model [24]. In Ph I/II clinical trials, XL019 demonstrated sustained spleen size reduction (>50%) in >50% of patients at a dose of 100 mg/day. A long half-life in humans (>30 h), coupled with reversible neuropathy in some patients, necessitated dose adjustment and schedule to 25 mg on alternate days. Using this dosing schedule, partial to complete resolution of splenomegaly and constitutional symptoms were noticed in short duration (3.5 months) [25].

SB1518 (structure not disclosed) is a JAK2/FLT3 inhibitor (biochemical IC_{50} values for JAK1, JAK2, JAK3, TYK2, and FLT3 are 1280, 23, 520, 50, and 22 nM, respectively) that is being evaluated in Ph I/II for MPN [26a-b] and lymphoma indications [27]. The compound blocked STAT signaling pathways in JAK2 and FLT3 mutated tumor cell lines, as well as STAT5 signaling in blood cells from normal mice [28]. In humans, SB1518 has a mean elimination half-life of 2–3 days and T_{max} ranging from 3–5 h. The compound demonstrated early (4–6 h post dose) pharmacodynamic effects by inhibiting p-STAT3 and p-STAT5 formation in lysed peripheral blood mononuclear cell (PBMC) at all dose levels. Overall, the drug was well tolerated with manageable GI toxicities. Early signs of clinical activity in PMF were observed during the first dosing cycle (28 days).

A JAK1/JAK2 inhibitor CYT387 (3) (biochemical IC_{50} values for JAK1, JAK2, and JAK3 are 18, 11, and 155 nM, respectively) was discovered through an extensive SAR investigation of phenylaminopyrimidines [29]. In a broad kinase screen, only 8 out of 150 kinases were inhibited with IC_{50} values < 100 nM. CYT387 inhibited *in vitro* growth of JAK2^{V617}-bearing cell lines (SET2 IC_{50} 232 nM) and erythroid colonies from PV patients at low micromolar concentrations [30]. Based upon a favorable preclinical profile, Ph I/II dose escalation studies have been initiated with a 100–150 mg dose.

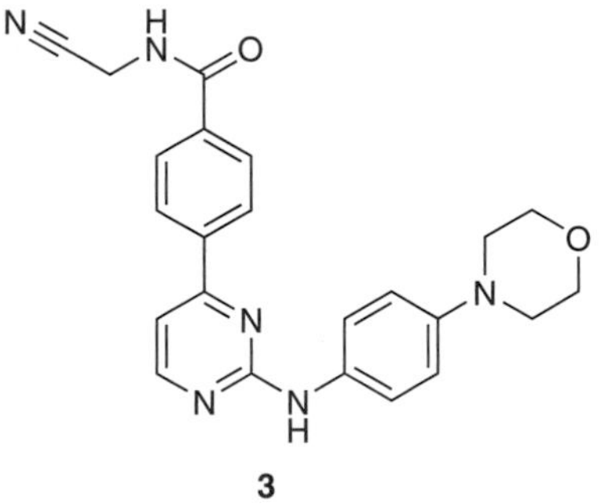

3

AZD1480 (4) is a pyrazolyl pyrimidine analog currently in Ph I clinical trials. The compound is a JAK1/JAK2 ATP competitive inhibitor (IC_{50} values for JAK1, JAK2, and JAK3 are 1.3, <0.4, and 3.9 nM, at respective ATP K_m values for JAK1: 55 μM, JAK2: 15 μM, and JAK3: 3 μM). However, when the assay was carried out at 5 mM of ATP (high end of ATP concentration in cells), significant JAK2 selectivity over JAK3 ($\sim$30 ×) was observed. The compound had GI_{50} values for Ba/F3 TEL–JAK2 and JAK3 cells of 60 and 2100 nM, respectively, consistent with biochemical selectivity. AZD1480 is the first JAK2 inhibitor that suppressed growth of multiple solid tumors with constitutive STAT3 activation, thereby suggesting applicability beyond MPN indications for a JAK2 inhibitor [31]. Dose escalation studies have been initiated.

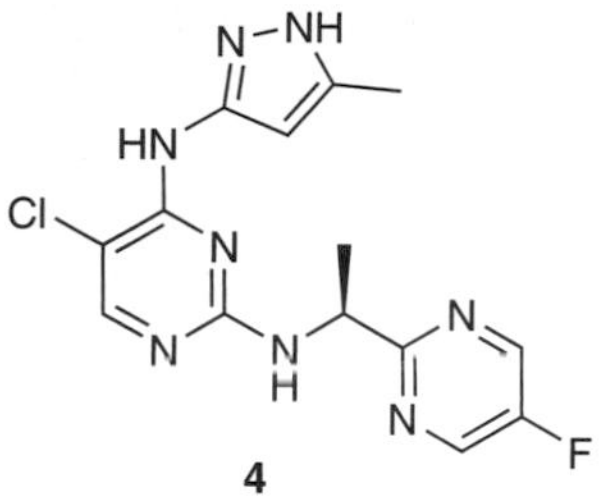

4

As a result of the complex signaling network associated with kinases, it is speculated that multitargeted kinase inhibitors may offer advantages over selective inhibitors in the treatment of MPNs. Hence, in addition to the selective JAK2 inhibitors, candidates with "opportunistic" JAK2 activity have been examined in the clinic.

The indolocarbazole analog, CEP701 (5, lestaurtinib), is a multikinase ATP competitive inhibitor with potent activity against JAK2, FLT3, and TrkA (IC$_{50}$ values of 0.9, 3, and 4 nM, respectively). The compound strongly inhibited phosphorylation of JAK2 and its downstream targets (p-STAT3 and p-STAT5) in HEL cells with an IC$_{50}$ of 10–30 nM, while inhibiting growth with an IC$_{50}$ of 100–300 nM. CEP701 also preferentially inhibited the growth of erythroid precursors from subjects with MPNs in a dose-dependent manner (IC$_{50}$ ~ 100 nM) while showing minimal effects on samples from healthy donors [32]. Unfortunately, in a Ph II clinical trial, CEP701 orally administered 80 mg twice a day resulted in only modest efficacy (6% response with reduction in splenomegaly or transfusion independence) while inducing mild and frequent gastrointestinal toxicity [33,34].

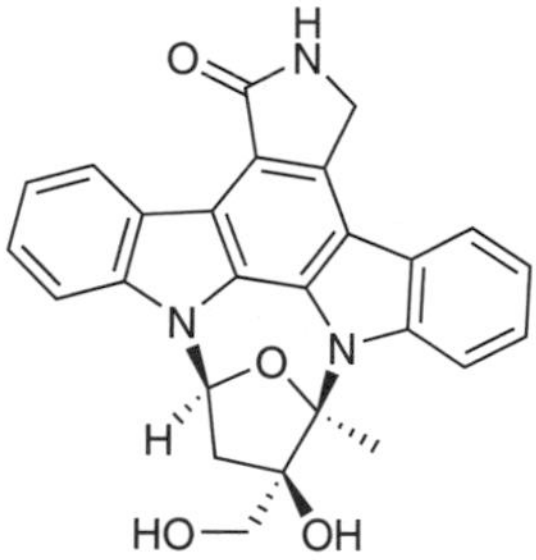

5

Similarly, AT9283 (6), a multitargeted kinase inhibitor, is being examined in a Ph I/II clinical trial. The compound is a potent inhibitor of JAK2, JAK3, Aurora A, Aurora B, and Abl (T315I) with IC$_{50}$ values less than 5 nM [35]. In HEL cells, AT9283 inhibited proliferation and phosphorylation of JAK2 pathway markers (e.g., p-STAT5) at an IC$_{50}$ of 100 nM. Also, the molecule demonstrated antiproliferative activity in the primary cells taken from patients diagnosed with MPNs [36].

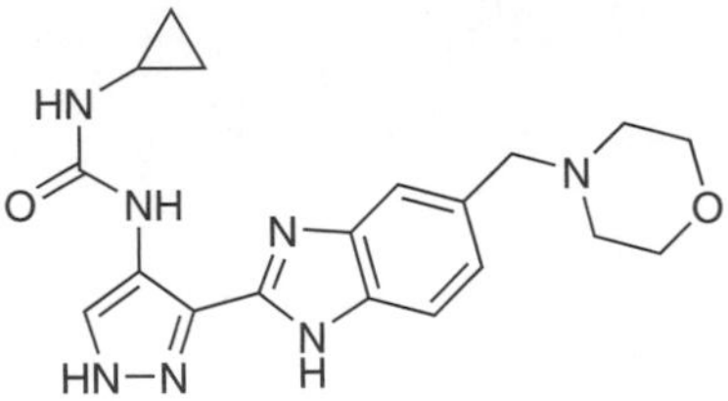

6

3. PRECLINICAL JAK2 INHIBITORS

Interest in the identification of JAK2 inhibitors is reinforced by the extensive number of preclinical reports of lead compounds in the primary and patent literature. Multiple industrial and academic groups have published data supporting the use of such agents for MPN indications. Recent compounds possess a range of JAK family and kinome activity, suggesting that research groups do not embrace a common strategy for dealing with kinase selectivity. However, it must be assumed that second-generation agents advancing into clinical trials will offer the potential for greater efficacy and safety relative to more advanced lead compounds.

The ATP-competitive JAK3 inhibitor CP690550 (**7**, tasocitinib), presently in advanced rheumatoid arthritis clinical trials, was recently reported to more potently suppress the proliferation of murine factor-dependent cell Patersen-erythropoietin receptor (FDCP-EpoR) cells, which harbor the $JAK2^{V617F}$ mutation, over the $JAK2^{wt}$ cells (IC_{50} values of 0.25 and 2.1 µM, respectively). This antiproliferative activity was correlated with the inhibition of downstream STAT3 and STAT5 phosphorylation. Additionally, the compound demonstrated similar sensitivity to erythroid progenitor cells isolated from $JAK2^{V617F}$-positive PV patients relative to healthy controls in proliferation and apoptosis assays [37].

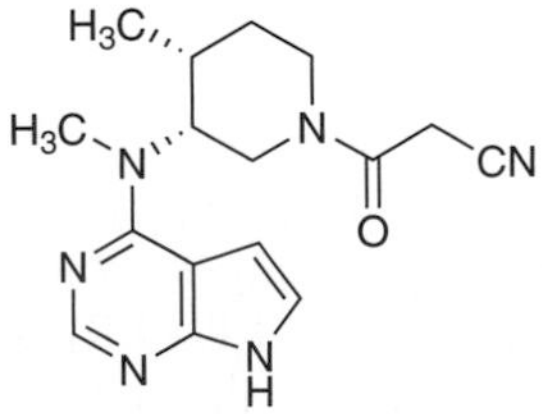

7

Similar activity has been reported for WP1066 (**8**), a tyrphostin analog optimized from AG490. The biochemical potency and kinase selectivity profile of this compound has not been reported, but compound **8** blocked the proliferation of $JAK2^{V617F}$-bearing HEL cells with an IC_{50} of 2.3 μM and had significant *ex vivo* activity in a clonogenic assay utilizing PBMCs isolated from $JAK2^{V617F}$-positive PV patients [38].

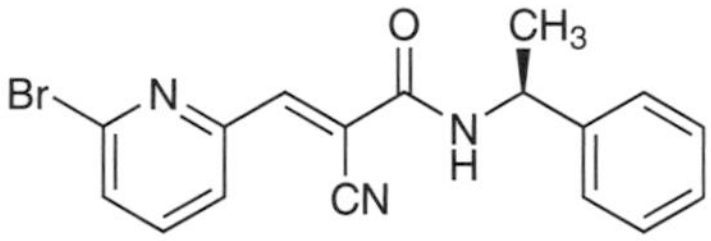

8

Butanone derivative Z3 (**9**) was identified from a library of 20,000 compounds by DOCK, a program used to predict the ability of these compounds to interact with a structural pocket adjacent to the ATP-binding site in murine JAK2. This compound inhibited $JAK2^{V617F}$ and $JAK2^{wt}$ autophosphorylation in a concentration-dependent manner, and displayed selectivity for JAK2 relative to Tyk2 and c-Src kinase function. Using the HEL 92.1.7 cell line, Z3's antiproliferative activity was correlated with reduced JAK2 and STAT3 phosphorylation as well as cell cycle arrest. Z3 also inhibited the growth of hematopoietic progenitor cells isolated from the bone marrow of both $JAK2^{V617F}$-positive essential thrombocythemia (ET) and $JAK2^{F537I}$-positive PV patients in *ex vivo* colony-forming assays [39].

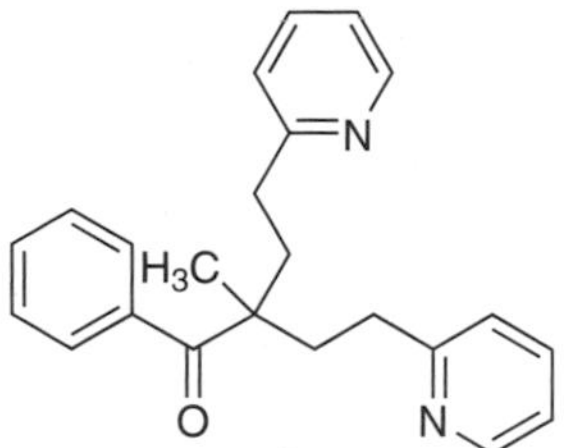

9

Structure-based virtual screening employing a JAK2 homology model based on available JAK3 crystal structures identified stilbene analog, G6 (**10**). G6 is a potent inhibitor of $JAK2^{V617F}$ activity, with a reported IC_{50} of 60 nM, and selectivity versus Tyk2 and c-Src kinases. In addition to selectively blocking HEL cell viability relative to three other cell lines bearing non-JAK2 activating mutations, at 2.5 μM the compound inhibited the *ex vivo* growth of mononuclear cells derived from a PV patient bearing the JAK2^{V617F} mutation by greater than 50%. G6 is proposed to bind to the ATP site in JAK2, forming critical H-bond interactions with the backbone NH group of Leu932 and the backbone C=O group of Glu930 and a salt bridge interaction with Asp994 of the DFG-motif [40].

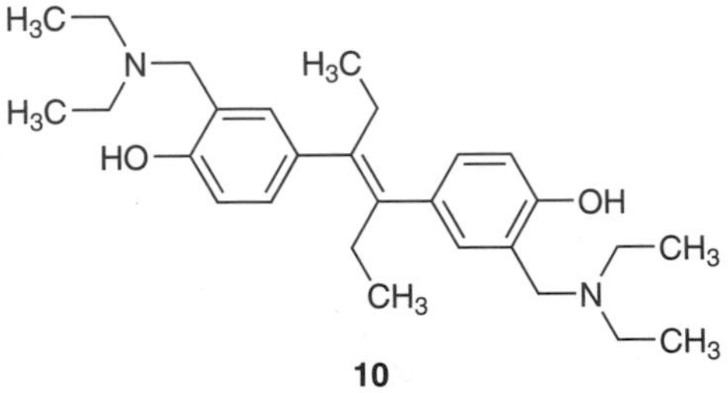

10

A combination of high-throughput screening and rational drug design led to the discovery of polycyclic azaindole ATP-competitive JAK2 inhibitors exemplified by **11**. Compound **11** exists at equilibrium as a 2:1 mixture of atropisomers, and is a potent inhibitor of both JAK2 and JAK3 (IC$_{50}$ values of 1 and 6 nM, respectively). The compound demonstrated moderate clearance, high oral bioavailability, and a long half-life in rodent PK studies. It also showed less than 1 μM activity in an EPO-independent erythroid colony formation assay, which used bone marrow isolates cultured from a PV patient bearing the JAK2^{V617F} mutation, as well as *in vivo* activity in a PV mouse survival assay at a dose of 15 mg/kg BID for 32 days [41].

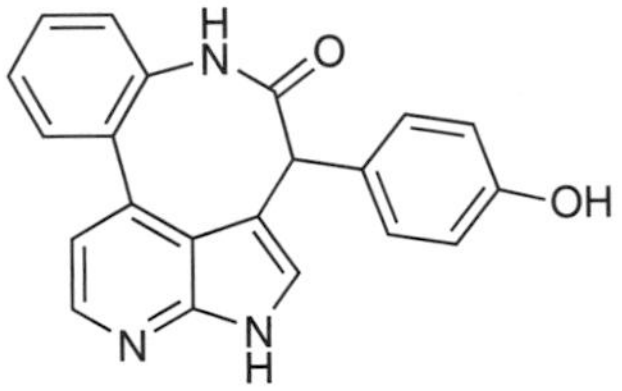

11

Pyrazolopyrimidine JAK2 inhibitors were identified through high-throughput screening. Compounds **12** and **13** represent optimized examples that bind at the ATP site with subnanomolar kinase inhibitory activity, modest selectivity for JAK3 relative to JAK2, and submicromolar activity in a JAK2-dependent granulocyte macrophage colony stimulating factor

(GMCSF)-stimulated STAT5 phosphorylation assay in TF1 cells. The kinase selectivity of these compounds was reproduced functionally in a JAK3-driven, IL-2-stimulated STAT5 phosphorylation assay in HT2 cells. Both examples demonstrated favorable iv pharmacokinetics in Sprague–Dawley rats [42].

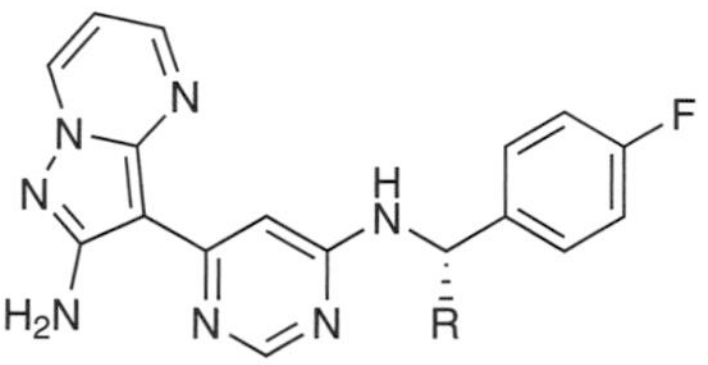

12; R = n-Pr

13; R = i-Pr

INCB16562 (**14**) is an ATP-competitive JAK1/2 inhibitor (IC_{50} values for JAK1, JAK2, and JAK3 are 9.1, 2.1, and 1895 nM, respectively) that demonstrated potent pharmacodynamic effects in the Ba/F3 JAK2^{V617F} cell line by blocking JAK2 autophosphorylation and STAT5 phosphorylation (IC_{50} values of 50 and 128 nM, respectively). It also exhibited good antiproliferative activity against human HEL and SET2 leukemia cell lines that bear mutant JAK2^{V617F}. In a mouse disease model of PV, **14** reduced JAK2^{V617F} cell burden, and splenomegaly, prevented weight loss, and prolonged survival when dosed twice daily at 25 and 75 mg/kg. Additionally, drug-treated animals demonstrated reduced inflammatory cytokine levels relative to vehicle-treated controls, an effect attributed to the JAK1 activity of the compound [43].

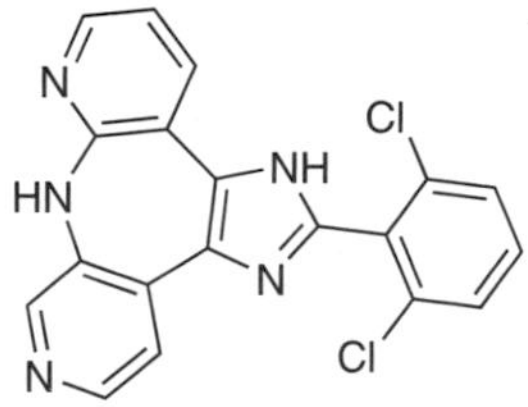

14

The structurally related pan-JAK kinase inhibitor pyridone P1 (**15**) has been evaluated in a mouse model of PV induced through the chronic administration of EPO. Administration of EPO (10 U/g) every other day for 5 days resulted in progressive polycythemia and splenomegaly. Administration of **15** at 100 mg/kg/day prevented EPO-induced increases in hematocrit and blocked splenomegaly relative to untreated control animals. Reductions in p-STAT5 levels in peripheral blood were also observed, consistent with JAK2 inhibition [44].

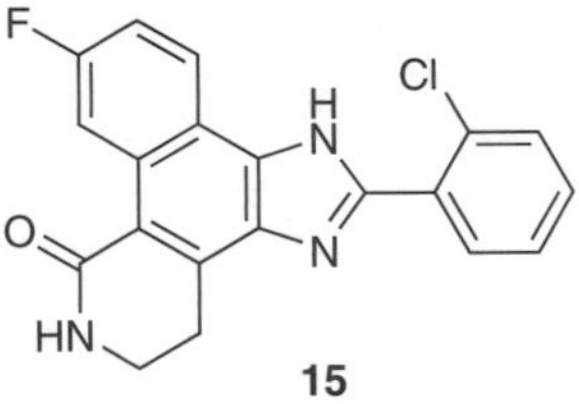

15

AZ960 (**16**), a pyrazolo-nicotinonitrile analog, was reported to be a tight-binding, ATP-competitive JAK2 inhibitor with a K_i of 0.4 nM. The compound demonstrated greater than 3-fold selectivity for JAK2 over JAK3 when tested in biochemical assays. Functional selectivity of the compound for JAK family members was established in an isogenic cell line panel expressing constitutively active TEL–JAK fusion proteins, where the compound blocked p-STAT5 formation with 15–30-fold selectivity for TEL–JAK2 signaling relative to other JAK kinase family members. AZ960 also induced apoptosis in SET2 cells, which bear a constitutively active JAK2 pathway [45].

16

Modification of the AZ960 structure resulted in the identification of pyrazolylaminopyrazines, exemplified by compound **17**. In biochemical assays, **17** had an IC_{50} of <3 nM when tested at K_m concentrations of ATP, and demonstrated >300-fold selectivity for JAK2 over JAK3 when evaluated at 5 mM ATP. The cellular selectivity of the compound was determined using the TEL–JAK cellular assay above, and confirmed the compound's selectivity for JAK2 over JAK3 (GI_{50} values of 0.22 and 7.4 µM, respectively). In a mouse pharmacodynamic study, a single oral dose of **17** (25 mg/kg) demonstrated greater than 95% inhibition of STAT5 phosphorylation in splenic infiltrates of TEL–JAK2 transfected Ba/F3 cells [46].

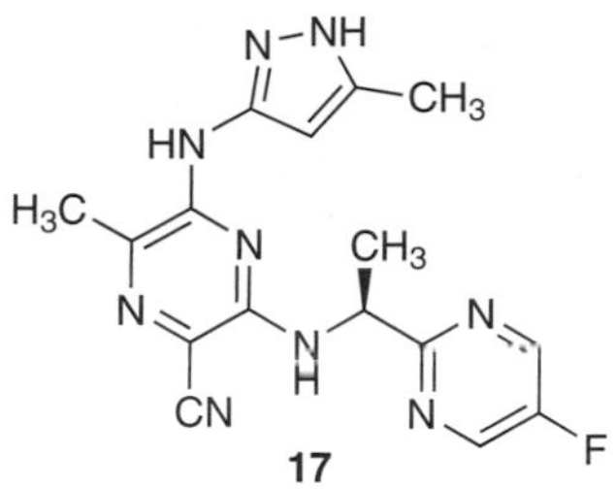

17

Replacement of the pyrazole ring in **17** with a thiazole, followed by scaffold optimization, afforded compounds with low nanomolar JAK2 inhibitory activity, exemplified by compound **18**. However, **18** also showed reduced potency relative to the pyrazoles in the high ATP concentration assay. Promisingly, when evaluated in the TEL–JAK2 cellular assay, **18** demonstrated potent and durable effects (ca. 50% inhibition of STAT5 phosphorylation for 6 h) following oral administration at 10 mg/kg [47].

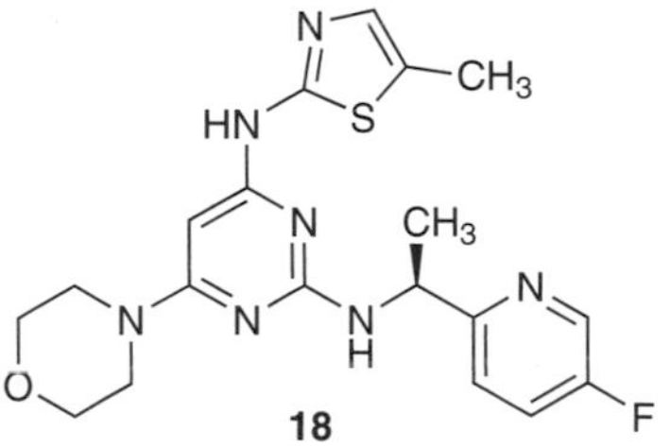

18

The ATP-competitive focal adhesion kinase (FAK) inhibitor **19** served as a starting point for the design of quinoxaline and benzoxazole JAK kinase inhibitors [48]. Compound **20** is a selective JAK2 inhibitor with 7-fold selectivity for JAK1, 44-fold selectivity for JAK3, and 20-fold selectivity for Tyk2 family members (IC$_{50}$ values of 5, 35, 220, and 97 nM, respectively). Compound **22** was identified from **21** using "scaffold morphing strategy". In a rat pharmacokinetic study, **22** demonstrated high oral bioavailability, while iv administration showed low clearance and a low volume of distribution [49].

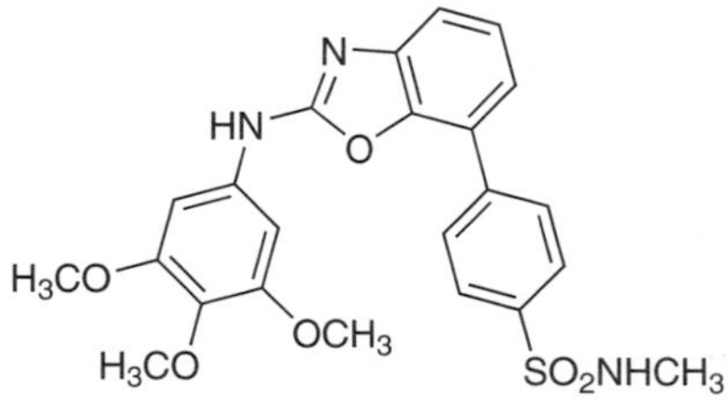

19

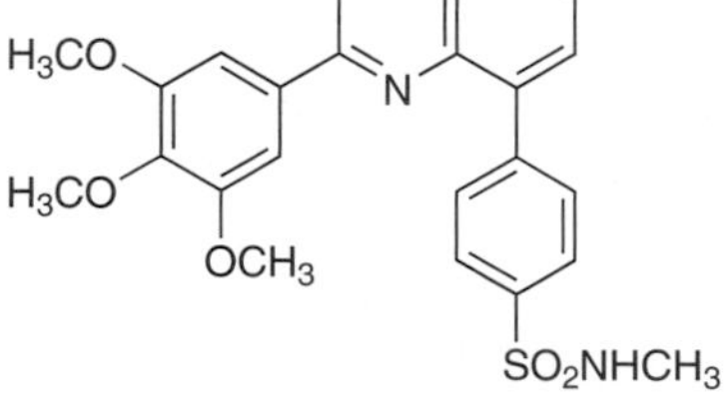

20

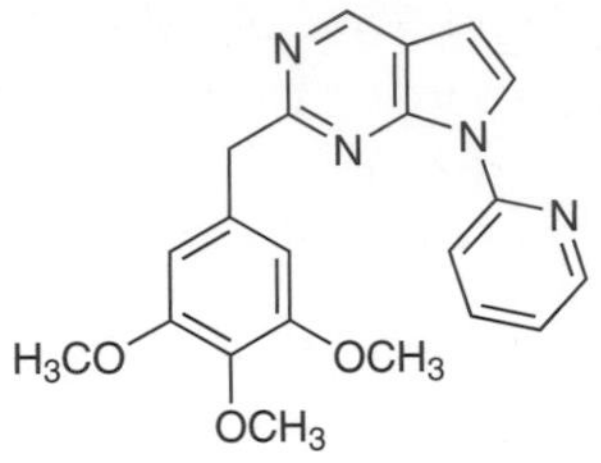

21

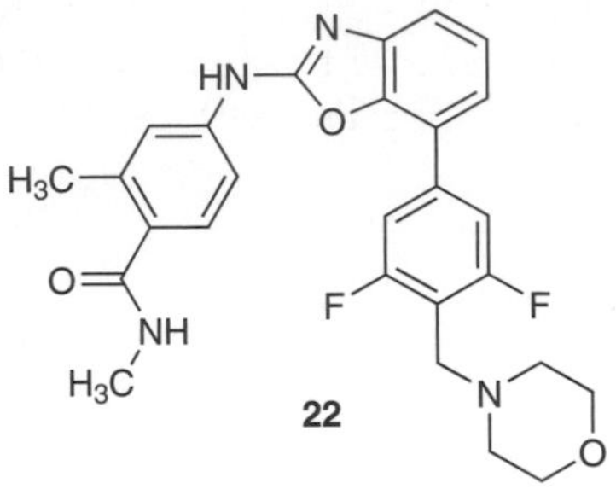

22

Fragment-based hit identification coupled with structure-based drug design enabled identification of **23** (JAK2 IC_{50} 78 nM) with greater than 35-fold selectivity for JAK2 over JAK3. Compound **23** was identified following two iterations of optimization of the initial fragment hit **24** (JAK2 IC_{50} 41 μM), and represents a greater than 500-fold improvement in target potency [50].

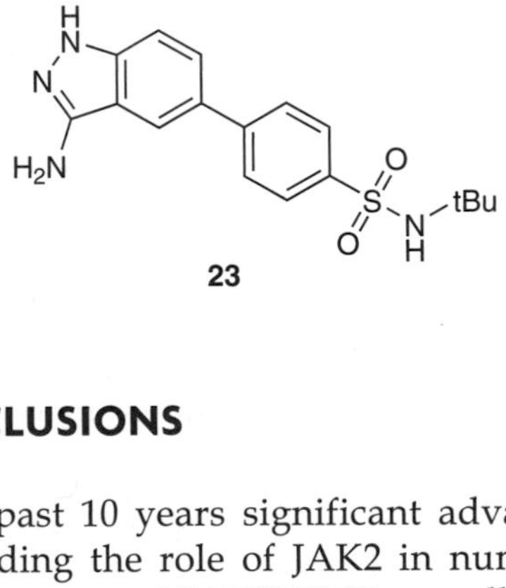

23

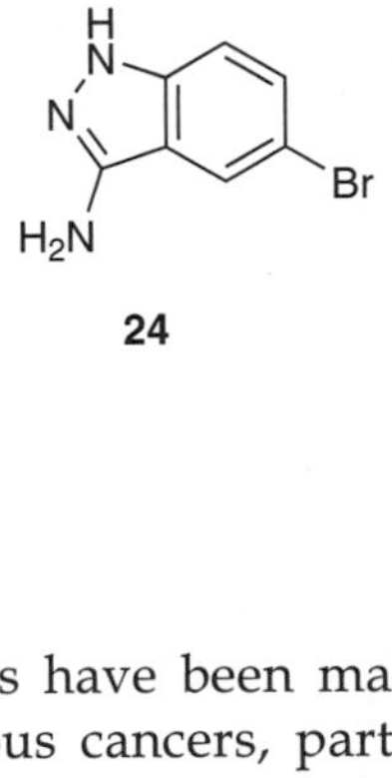

24

4. CONCLUSIONS

Over the past 10 years significant advances have been made toward understanding the role of JAK2 in numerous cancers, particularly in MPNs. Discovery of JAK2V617F as well as JAK2–STAT pathway mutations and their relevance to MPNs have fuelled research collaborations between academia and the pharmaceutical industry. These collaborations have resulted in the advancement of several candidates into early and midstage clinical trials. Early clinical data has shown promise of JAK2 inhibitors to relieve constitutional symptoms and improve hematological parameters and overall quality of life for moderately aggressive MPNs, particularly myelofibrosis; however, only modest effects on the disease pathology or molecular responses are observed. Success with kinase-targeted therapies, such as imatinib for the BCR-ABL mutation-driven CML, has raised hopes for using this same approach for JAK2 inhibitors. The ultimate success of a JAK2 inhibitor for the treatment of MPN would depend upon optimal kinase selectivity, long-term safety and tolerability of the agent, and its effect on the pathophysiology of the disease.

REFERENCES

[1] C. Kumar, A. V. Purandare, F. Y. Lee, and M. V. Lorenzi, *Oncogene*, 2009, **28**, 2305.

[2] R. L. Levine, M. Wadleigh, J. Cools, B. L. Ebert, G. Wernig, B. J. Huntly, T. J. Boggon, I. Wlodarska, J. J. Clark, S. Moore, J. Adelsperger, S. Koo, J. C. Lee, S. Gabriel, T. Mercher, A. D'Andrea, S. Fröhling, K. Döhner, P. Marynen, P. Vandenberghe, R. A. Mesa, A. Tefferi, J. D. Griffin, M. J. Eck, W. R. Sellers, M. Meyerson, T. R. Golub, S. J. Lee, and D. G. Gilliland, *Cancer Cell*, 2005, **7**, 387.

[3] E. J. Baxter, L. M. Scott, P. J. Campbell, C. East, N. Fourouclas, S. Swanton, G. S. Vassiliou, A. J. Bench, E. M. Boyd, N. Curtin, M. A. Scott, W. N. Erber, and A. R. Green, *Lancet*, 2005, **365**, 1054.

[4] P. J. Campbell and A. R. Green, *N. Engl. J. Med.*, 2006, **355**, 2452.

[5] L. M. Scott, W. Tong, R. L. Levine, M. A. Scott, P. A. Beer, M. R. Stratton, P. A. Futreal, W. N. Erber, M. F. McMullin, C. N. Harrison, A. J. Warren, D. G. Gilliland, H. F. Lodish, and A. R. Green, *N. Engl. J. Med.*, 2007, **356**, 459.

[6] E. Jost, O. N. do, E. Dahl, C. E. Maintz, P. Jousten, L. Habets, S. Wilop, J. G. Herman, R. Osieka, and O. Galm, *Leukemia*, 2007, **21**, 505.

[7] Y. Pikman, B. H. Lee, T. Mercher, E. McDowell, B. L. Ebert, M. Gozo, A. Cuker, G. Wernig, S. Moore, I. Galinsky, D. J. DeAngelo, J. J. Clark, S. J. Lee, T. R. Golub M. Wadleigh, D. G. Gilliland, and R. L. Levine, *PLoS Med.*, 2006, 3 1140.

[8] C. G. Mulligan, J. R. Collins-Underwood, L. A. Phillips, M. G. Loudin, W. Liu, J. Zhang, J. Ma, E. Coustan-Smith, R. C. Harvey, C. L. Willman, F. M. Mikhail, J. Meyer, A. J. Carroll, R. T. Williams, J. Cheng, N. A. Heerema, G. Basso, A. Pession, C. H. Pui, S. C. Raimondi, S. P. Hunger, J. R. Downing, W. L. Carroll, and K. R. Rabin, *Nat. Genet.*, 2009, **41**, 1243.

[9] O. Galm, H. Yoshikawa, M. Esteller, R. Osieka, and J. G. Herman, *Blood*, 2003, **101**, 2784.

[10] P. J. Murray, *J. Immuno.*, 2007, **178**, 2623.

[11] H. Neubauer, A. Cumano, M. Müller, H. Wu, U. Huffstadt, and K. Pfeffer, *Cell*, 1998, **93**, 397.

[12] T. G. Bumm, C. Elsea, A. S. Corbin, M. Loriaux, D. Sherbenou, L. Wood, J. Deininger, R. T. Silver, B. J. Druker, and M. W. Deininger, *Cancer Res.*, 2006, **66**, 11156.

[13] G. Wernig, T. Mercher, R. Okabe, R. L. Levine, B. H. Lee, and D. G. Gilliland, *Blood*, 2006, **107**, 4274.

[14] I. Geron, A. E. Abrahamsson, C. F. Barroga, E. Kavalerchik, J. Gotlib, J. D. Hood, J. Durocher, C. C. Mak, G. Noronha, R. M. Soll, A. Tefferi, K. Kaushansky, and C. H. Jamieson, *Cancer Cell*, 2008, **13**, 321–330.

[15] G. Wernig, M. G. Kharas, R. Okabe, S. A. Moore, D. S. Leeman, D. E. Cullen, M. Gozo, E. P. McDowell, R. L. Levine, J. Doukas, C. C. Mak, G. Noronha, M. Martin, Y. D. Ko, B. H. Lee, R. M. Soll, A. Tefferi, J. D. Hood, and D. G. Gilliland, *Cancer Cell*, 2008, **13**, 311.

[16] S. Verstovsek, H. Kantarjian, A. Pardanani, D. Thomas, J. Cortes, R. Mesa, W. Hogan, J. Redman, R. Levy, K. Vaddi, J. Fridman, and A. Tefferi, *Haematol. Haematol. J.*, 2008, **93**, (suppl 1), 179.

[17] J. Fridman, R. Nussenzveig, P. Liu, J. Rodgers, T. Burn, P. Haley, P. Scherle, R. Newton, G. Hollis, S. Friedman, S. Verstovsek, and K. Vaddi, *Blood (ASH Annual Meeting Abstracts)* 2007, **110**, Abstract 3538.

[18a] S. Verstovsek, H. M. Kantarjian, A. D. Pardanani, D. Thomas, J. Cortes, R. A. Mesa, W. J. Hogan, J. R. Redman, S. Erickson-Viitanen, R. Levy, K. Vaddi, E. Bradley, J. Fridman, and A. Tefferi, *Blood (ASH Annual Meeting Abstracts,)* 2008, **112**, Abstract 1762.

[18b] S. Verstovsek, H. Kantarjian, R. A. Mesa, J. Cortes-Franco, A. D. Pardanani, D. A. Thomas, Z. Estrov, E. C Bradley, S. Erickson-Viitanen, K. Vaddi, R. Levy, and A. Tefferi, *Blood (ASH Annual Meeting Abstracts)* 2009, **112**, Abstract 756.

[19] S. Verstovsek, F. Passamonti, A. Rambaldi, G. Barosi, P. Rosen, R. Levy, E. Bradley, L. Schacter, W. Garrett, K. Vaddi, N. Contel, E. Rumi, E. Gattoni, M. Cazzola, H. Kantarjian, T. Barbui, and A. Vannucchi, *Blood (ASH Annual Meeting Abstracts)*, 2009, **114**, Abstract 311.

[20] T. L. Lasho, A. Tefferi, J. D. Hood, S. Verstovsek, D. G. Gilliland, A. Pardanani, *Leukemia*, 2008, **22**, 1790–1792.

[21] A. D. Pardanani, J. Gotlib, C. Jamieson, J. Cortes, M. Talpaz, R. M. Stone, M. H Silverman, J. Shorr, D. G. Gilliland, and A. Tefferi, *Blood (ASH Annual Meeting Abstracts)*, 2008, **112**, Abstract 97.

[22] A. D. Pardanani, J. R Gotlib, C. Jamieson, J. Cortes, M. Talpaz, R. Stone, M. H. Silverman, J. Shorr, D. G. Gilliland, and A. Tefferi, *Blood (ASH Annual Meeting Abstracts)*, 2009, **114**, Abstract 755.

[23] S. Verstovsek, A. D. Pardanani, N. P. Shah, L. Sokol, M. Wadleigh, D. G. Gilliland Alan F. List, A. Tefferi, H. M. Kantarjian, L. A. Bui, and D. O. Clary, *Blood (ASH Annual Meeting Abstracts)*, 2007, **110**, Abstract 553.

[24] R. L. Paquette, L. Sokol, N. P. Shah, R. Silver, A. F. List, J. -F. Martini, D. O. Clary, L. A. Bui, and M. Talpaz, *Blood (ASH Annual Meeting Abstracts)*, 2008, **112**, Abstract 2810.

[25] N. P. Shah, M. Wadleigh, L. Sokol, A. F. List, J. Cortes, H. M. Kantarjian, D. G. Gilliland, P. Olszynski, C. Smith, D. O. Clary, L. A. Bui, and S. Verstovsek, *Blood (ASH Annual Meeting Abstracts)*, 2008, **112**, Abstract 98.

[26a] K. C. Goh, W. C. Ong, C. Hu, H. Hentze, A. L. Liang, W. Stunkel1, Y. C. Tan, K. Sangthongpitag, S. K. Goh, S. Sieh, Z. Q. Bonday, A. D. William, A. Lee, S. Blanchard, T. C. Liu, M. Sattler, J. D. Griffin, B. W. Dymock, E. Kantharaj, and J. M. Wood, *Blood (ASH Annual Meeting Abstracts)*, 2007, **110**, Abstract 538.

[26b] K. C. Goh, S. Hart, Y. C. Tan, A. Chithra, K. H. Ong, and J. Wood, *Blood (ASH Annual Meeting Abstracts)*, 2009, **114**, Abstract 2913.

[27] S. Verstovsek, O. Odenike, B. Scott, Z. Estrov, J. Cortes, D. A. Thomas, J. Wood, K. Ethirajulu, A. Lowe, H. J. Zhu, H. Kantarjian, and H. J. Deeg, *Blood (ASH Annual Meeting Abstracts)*, 2009, **114**, Abstract 3905.

[28] A. Younes, M. Fanale, P. McLaughlin, A.R Copeland, S. de Castro Faria, J. Wood, K. Ethirajulu, and H. J. Zhu, *Blood (ASH Annual Meeting Abstracts)*, 2009 **114**, Abstract 588.

[29] C. J. Burns, D. G. Bourke, L. Andrau, X. Bu, S. A. Charman, A. C. Donohue, E. Fantino, M. Farrugia, J. T. Feutrill, M. Joffe, M. R. Kling, M. Kurek, T. L. Nero, T. Nguyen, J. T. Palmer, I. Phillips, D. M. Shackleford, H. Sikanyika, M. Styles, S. Su, H. Treutlein, J. Zeng, and A. F. Wilks, *Bioorg. Med. Chem. Lett.*, 2009, **19**, 5887.

[30] A. Pardanani, T. Lasho, G. Smith, C. J. Burns, E. Fantino, and A. Tefferi, *Leukemia*, 2009, **23**, 1441–1445.

[31] M. Hedvat, D. Huszar, A. Herrmann, J. M. Gozgit, A. Schroeder, A. Sheehy, R. Buettner, D. Proia, C. M. Kowolik, H. Xin, B. Armstrong, G. Bebernitz, S. Weng, L. Wang, M. Ye, K. McEachern, H. Chen, D. Morosini, K. Bell, M. Alimzhanov, S. Ioannidis, P. McCoon, Z. A. Cao, H. Yu, R. Jove, and M. Zinda, *Cancer Cell*, 2009, **16**, 487.

[32] E. O. Hexner, C. Serdikoff, M. Jan, C. R. Swider, C. Robinson, S. Yang, T. Angeles, S. G. Emerson, M. Carroll, B. Ruggeri, and P. Dobrzanski, *Blood*, 2008, **111**, 5663.

[33] S. Verstovsek, A. Tefferi, S. Kornblau, D. Thomas, J. Cortes, F. Ravandi-Kashani, G. Garcia-Manero, and H. Kantarjian, *Blood (ASH Annual Meeting Abstracts)*; **2007, 110,** Abstract 3543.

[34] E. Hexner, J. D. Goldberg, J. T. Prchal, E.P Demakos, S. Swierczek, R. Singer Weinberg, J. Tripodi, V. Najfeld, M. Carroll, R. Marchioli, L. R. Silverman, and R. Hoffman, *Blood (ASH Annual Meeting Abstracts)*, 2009, **114,** Abstract 754.

[35] S. Howard, V. Berdini, J. A. Boulstridge, M. G. Carr, D. M. Cross, J. Curry, L. A. Devine, T. R. Early, L. Fazal, A. L. Gill, M. Heathcote, S. Maman, J. E. Matthews, R. L. McMenamin, E. F. Navarro, M. A. O'Brien, M. O'Reilly, D. C. Rees, M. Reule, D. Tisi, G. Williams, M. Vinkovic, and P. G. Wyatt, *J. Med. Chem.*, 2009, **52,** 379.

[36] M. S. Squires, J. E. Curry, M. A. Dawson, M. A. Scott, K. Barber, M. Reule, T. Smyth, N. Wallis, M. Yule1, N. T. Thompson, J. F. Lyons, and A. R. Green, *Blood (ASH Annual Meeting Abstracts)*, 2007, **110,** Abstract 3537.

[37] T. Manshouri, A. Quintas-Cardama, R. H. Nussunzveig, A. Gaikwad, Z. Estrov, J. Prchal, J. E. Cortes, H. M. Kantarjian, and S. Verstovsek, *Cancer Sci.*, 2008, **99,** 1265.

[38] S. Verstovsek, T. Manshouri, A. Quintas-Cardama, D. Harris, J. Cortes, F. J. Giles, H. M. Kantarjian, W. Priebe, and Z. Estrov, *Clin. Cancer Res.*, 2008, **14,** 788.

[39] J. Sayyah, A. Magis, D. A. Ostrov, R. W. Allan, R. C. Braylan, and P. P. Sayeski, *Mol. Cancer Ther.*, 2008, **7,** 2308.

[40] R. Kiss, T. Polgar, A. Kirabo, J. Sayyah, N. C. Figueroa, A. F. List, L. Sokol, K. S. Zuckerman, M. Gali, K. S. Bisht, P. P. Sayeski, and G. M. Keseru, *Bioorg. Med. Chem. Lett.*, 2009, **19,** 3598.

[41] T. Wang, J. P. Duffy, J. Wang, S. Halas, F. G. Salituro, A. C. Pierce, H. J. Zuccola, J. R. Black, J. K. Hogan, S. Jepson, D. Shlyakter, S. Mahajan, Y. Gu, T. Hoock, W. Wood, B. F. Furey, J. D. Frantz, L. M. Dauffenbach, U. A. Germann, B. Fan, M. Namchuk, Y. L. Bennani, and M. W. Ledeboer, *J. Med. Chem.*, 2009, **52,** 7938.

[42] M. W. Ledeboer, A. C. Pierce, J. P. Duffy, H. Gao, D. Messersmith, F. G. Salituro, S. Nanthakumar, J. Come, H. J. Zuccola, L. Swenson, D. Shlyakter, S. Mahajan, T. Hoock, B. Fan, W. -J. Tsai, E. Kolaczkowski, S. Carrier, J. K. Hogan, R. Zessis, S. Pazhanisamy, and Y. L. Bennani, *Bioorg. Med. Chem. Lett.*, 2009, **19,** 6529.

[43] P.C.C. Liu, E. Caulder, J. Li, P. Waeltz, A. Margulis, R. Wynn, M. Becker-Pasha, Y. Li, E. Crowgey, G. Hollis, P. Haley, R. B. Sparks, A. P. Combs, J. D. Rodgers, T. C. Burn, K. Vaddi, and J. S. Fridman, *Clin. Cancer Res.*, 2009, **15,** 6891.

[44] A. Mathur, J.-R. Mo, M. Kraus, E. O'Hare, P. Sinclair, J. Young, S. Zhao, Y. Wang, J. Kopinja, X. Qu, J. Reilly, D. Walker, L. Xu, D. Aleksandrowicz, G. Marshall, M. L. Scott, N. E. Kohl, and E. Bachman, *Biochem. Pharmacol.*, 2009, **78,** 382.

[45] J. M. Gozgit, G. Bebernitz, P. Patil, M. Ye, J. Parmentier, J. Wu, N. Su, T. Wang, S. Ioannidis, A. Davies, D. Huszar, and M. Zinda, *J. Biol. Chem.*, 2008, **283,** 32334.

[46] S. Ioannidis, M. L. Lamb, A. M. Davies, L. Almeida, M. Su, G. Bebernitz, M. Ye, K. Bell, M. Alimzhanov, and M. Zinda, *Bioorg. Med. Chem. Lett.*, 2009, **19,** 6524.

[47] S. Ioannidis, M. L. Lamb, L. Almeida, H. Guan, B. Peng, G. Bebernitz, K. Bell, M. Alimzhanov, and M. Zinda, *Bioorg. Med. Chem. Lett.*, 2010, **20,** 1669.

[48] P. Furet, M. Gerspacher, and C. Pissot-Soldermann, *Bioorg. Med. Chem. Lett.*, 2010, **20,** 1858.

[49] M. Gerspacher, P. Furet, C. Pissot-Soldermann, C. Gaul, P. Holzer, E. Vangrevelinghe, M. Lang, D. Erdmann, T. Radimerski, C. H. Regnier, P. Chene, A. De Pover, F. Hofmann, F. Baffert, T. Buhl, R. Aichholz, F. Blasco, R. Endres, J. Trappe, and P. Drueckes, *Bioorg. Med. Chem. Lett.*, 2010, **20,** 1724.

[50] S. Antonysamy, G. Hirst, F. Park, P. Sprengeler, F. Stappenbeck, R. Steensma, M. Wilson, and M. Wong, *Bioorg. Med. Chem. Lett.*, 2009, **19,** 279.

CHAPTER 14

Development of Poly (ADP-Ribose)Polymerase (PARP) Inhibitors in Oncology

Philip Jones

Merck Research Labs Boston, Avenue Louis Pasteur 33, Boston, MA 02115-5727, USA

Annual Reports in Medicinal Chemistry, Volume 45
ISSN 0065-7743, DOI 10.1016/S0065-7743(10)45014-9

1. INTRODUCTION

Nearly 40 years have passed since the discovery of the nuclear enzymatic activity that is responsible for synthesizing an adenine-containing RNA polymer [1]. Subsequent research led to the identification of poly(ADP-ribose)polymerase (PARP)-1, the founding member of poly(ADP-ribosyl) ating family of proteins [2,3]. PARP-1 has been implicated in DNA repair, resulting in PARP inhibitors (PARPi) being investigated clinically as chemo- and radio-sensitizers. Recent publications reporting that particular cancer cells are highly sensitive to PARP inhibition has rejuvenated research in this area, resulting in novel PARPi advancing into clinical trials.

2. PARP FAMILY

2.1 Role of PARP-1 and PARP-2 in DNA repair

Several PARPs share a catalytic domain capable of catalyzing the transfer of ADP-ribose units using nicotinamide adenine dinucleotide (NAD^+) to form long and branched poly(ADP)ribose (PAR) chains [2]. PARP-1 and the closely related PARP-2 are nuclear proteins and possess N-terminal DNA-binding domains [4]. These domains serve to localize and bind PARP to sites of DNA single- and double-strand breaks (SSB and DSB). A central automodification domain permits these enzymes to poly(ADP-ribosyl)ate themselves, while the C-terminus contains the catalytic active site, which is responsible for polymerizing the ADP-ribose. Upon DNA binding, the catalytic activity of these enzymes is stimulated more than 500-fold, resulting in the addition of PAR polymer chains to itself and to associated DNA repair proteins [5]. This leads to subsequent chromatin relaxation and fast recruitment of DNA repair factors, resulting in DNA break repair through a process known as base excision repair.

The knockout of PARP-1 or PARP-2 significantly reduces the cells' ability to repair damaged DNA following their exposure to radiation or cytotoxic insult [6,7]. Poly(ADP-ribose) polymer synthesis consumes substantial amounts of NAD^+, and hyperactivation of PARP results in depletion of cellular NAD^+ pools and energy stores, leading to cell death by necrosis.

2.2 Biology of other PARP family members

Genomic sequence analysis has identified 17 structurally related proteins containing a PARP catalytic domain [3]; however, closer inspection suggests that only PARPs 1–3, PARP-4 (vault PARP), and tankyrases 1 and 2 may truly function as poly(ADP-ribosyl)ating proteins. The remaining family members may be mono-ADP-ribosyltransferases, or catalytically inactive [8], as they have the key glutamic acid, which is essential for polymerase activity, replaced with leucine, isoleucine, or tyrosine. Vault PARP is a catalytic component of vault particles, which are ribonucleo-protein complexes involved in multidrug resistance of tumors [9], and may be a tumor suppressor. Tankyrases 1 and 2 are known to regulate telomere homeostasis and play a key role during mitotic segregation [10].

3. SYNTHETIC LETHALITY OF INHIBITING PARP-1 AND PARP-2 IN BRCA MUTANT TUMORS

Recently, data has emerged suggesting that targeting more than one DNA repair pathway in tumor cells could induce "synthetic lethality," which is a phenomenon whereby deletion of a single gene or inhibition of the gene product is compatible with viability, but two simultaneous deletions, mutations, and/or inhibition are lethal [11–13]. Specifically, PARPi selectively kill breast cancer type 1 susceptibility protein (BRCA)-1- or BRCA-2-deficient tumor cells while, in contrast, PARPi have minimal effects on normal cells with intact BRCA pathways. BRCA-1 and BRCA-2 are known tumor suppressors, and are key components involved in the repair of DNA DSB by the homologous recombination (HR) pathway. Mutations in these genes predispose individuals to hereditary breast and ovarian cancer. This selectivity is attributed to the fact that PARP-1 and PARP-2 usually play key roles in DNA repair. In the presence of a PARPi, DNA damage results in persistent SSB and these subsequently develop into DSB. In BRCA-1- and BRCA-2-deficient cells, these lesions are unable to be repaired by the HR pathway, resulting in gross genomic instability and apoptosis.

Two groups demonstrated that BRCA-1- and BRCA-2-deficient cells are acutely sensitive to PARPi [11,12]. Potent inhibitors like KU0058684 (5), KU0058948 (6), and AG14361 (26) were cytotoxic at nanomolar concentrations in HR-defective cells, and displayed excellent selectivity for BRCA-1- and BRCA-2-deficient cells over wild-type cells. After 24 h of exposure, 5 elicited G2 or M phase cell cycle arrest and a tetraploid DNA content. The applicability of this discovery was revealed when BRCA-2-deficient and BRCA-2-proficient cells were injected into mice and tumors were allowed to develop. Daily treatment with 5 or 26 had no effect on the BRCA-2 wild-type cells; however, when BRCA-2-deficient cells were treated with PARPi, no tumors developed.

To evaluate PARP inhibition in a realistic setting, olaparib (7) was tested in a $Brca1^{-/-};p53^{-/-}$ mouse breast cancer model. Treatment with olaparib caused tumor growth inhibition without generating signs of toxicity [14]. Interestingly, upon cessation of treatment and tumor regrowth, a second course of olaparib failed to elicit responses due to increased expression of drug efflux transporters.

4. RECENT DEVELOPMENTS IN PARP INHIBITOR DESIGN

Most PARPi are based upon nicotinamide (1), the by-product of the PARylation reaction, which itself is a micromolar PARPi. PARPi have been developed using three design approaches: constraining the amide using an internal hydrogen bond, forming a lactam, and incorporation of the amide into a heterocycle [3,15].

4.1 Discovery of clinical candidates

A series of phthalazinone PARPi has been developed from micromolar screening hits [16]. Initial optimization identified 4-benzyl derivative 2, which was quickly optimized to *meta*-derived analog 3 (PARP-1 $IC_{50} = 770$ and 20 nM, respectively). Constraint, as with imide 4, improved activity ($IC_{50} = 12$ nM), although 4 displayed only modest cellular activity ($PF_{50} = 1.74$, where PF_{50} is a measure of the potentiation of methyl methanesulfonate (MMS) defined as $GI_{50}(MMS)/GI_{50}(MMS+PARPi)$). Introduction of a fluorine group at the *ortho* position enhanced cellular activity, giving 5 ($PF_{50} = 5.62$). Further structure activity relationship (SAR) exploration demonstrated that the amide could be reversed, and a wide range of amines could be tolerated [17]. Homopiperazine 6 proved to be a potent PARPi ($IC_{50} = 7$ nM, $PF_{50} = 12.6$) that also demonstrated potent and specific killing of BRCA-defective cells, with $SF_{50} = 15$ nM for BRCA-2-deficient cells compared to $2 \mu M$ for wild type.

In an effort to develop orally bioavailable derivatives, compounds were prioritized on the basis of activity and mouse PK, leading to the identification of cyclopropylamide 7, which resulted in a significant enhancement of cellular activity ($IC_{50} = 5$ nM, $PF_{50} = 25.8$) [18]. This compound is a dual PARP-1 and PARP-2 inhibitor, displaying PARP-2 $IC_{50} = 1$ nM, and significantly weaker tankyrase activity $IC_{50} = 1.5 \mu M$. In monotherapy, 7 was shown to selectively inhibit the growth of BRCA-1-deficient MDA-MB-436 and HCC1937 cells. In a SW620 xenograft, the combination of 7 with methylating agent temozolomide (TMZ) significantly enhanced tumor growth inhibition over TMZ treatment alone, resulting in a 59% increase in tumor doubling time. On the basis of these data, 7 (olaparib) was selected for clinical development.

More recently, an alternative series of 2-alkoxybenzamides was designed where an intramolecular hydrogen bond mimics the phthalazinone scaffold [19]. Gratifyingly 8 exhibited good PARP inhibition, $IC_{50} = 18$ nM.

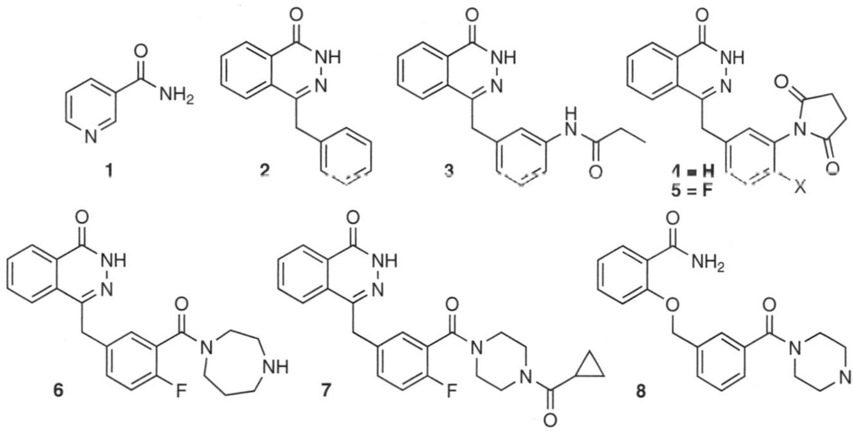

An entirely different screening approach was used during the optimization of a series of indazole PARPi, which were screened for antiproliferation activity in BRCA-1 silenced HeLa cells and the corresponding isogenic line [20]. The benchmark **9** displayed $IC_{50} = 24$ nM, and *para*-substitution improved activity to 3.8 nM, with **10** inhibiting PARP activity in cells, $EC_{50} = 68$ nM and the proliferation of BRCA-deficient cells with $CC_{50} = 460$ nM, displaying a 12-fold selectivity over BRCA-proficient cells. Excellent levels of PARP inhibition were necessary to inhibit proliferation of BRCA-deficient cells. Poor rat PK was improved by the introduction of a 3-piperidinyl group, resulting in the identification of MK-4827 (**11**). The compound inhibited PARP-1 with an $IC_{50} = 3.2$ nM and caused growth inhibition in BRCA-deficient cells with $CC_{50} = 33$ nM, displaying 25-fold selectivity. This indazole inhibited the proliferation of the BRCA-1 and BRCA-2 mutant cell lines MDA-MB-436 and CAPAN-1, with $CC_{50} = 18$ and 90 nM, respectively. *In vivo*, **11** was shown to cause tumor regression in a MDA-MB-436 xenograft and growth inhibition in a BRCA-2 mutant CAPAN-1 xenograft.

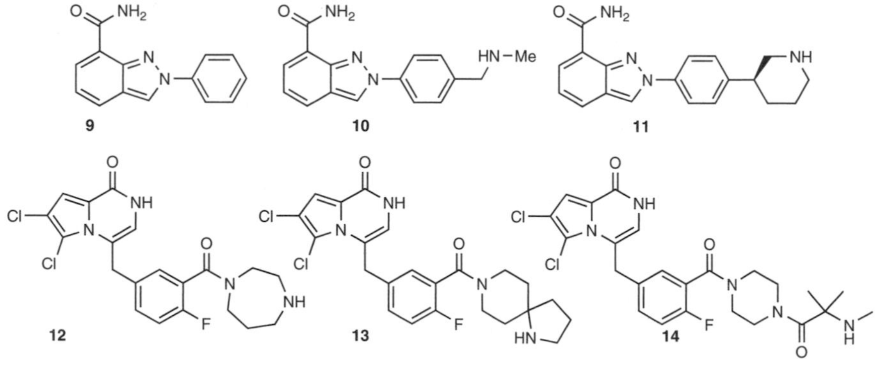

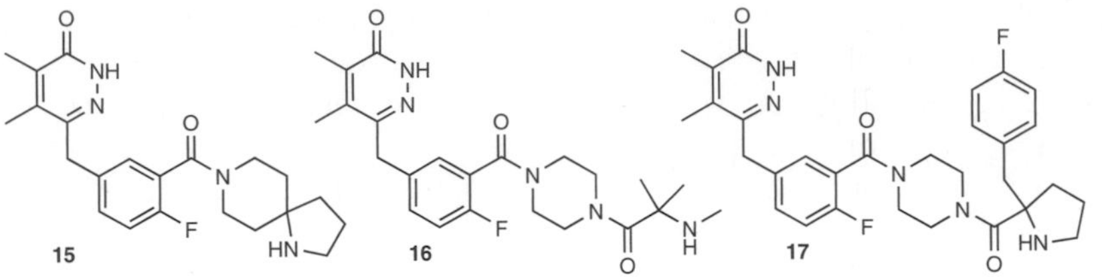

Rational design led to a series of pyrrolo[1,2-*a*]pyrazin-1(2*H*)-one PARPi, including **12**, which displayed an $IC_{50} = 2.6$ nM and inhibited the proliferation of BRCA-1-deficient cells with a $CC_{50} = 170$ nM [21]. SAR was conducted to identify the optimal substituent for cellular activity, which led to the discovery of **13** and **14** ($CC_{50} = 48$ and 53 nM, respectively). Modification to the analogous 4,5-dimethylpyridazin-3(2*H*)-one scaffold gave compounds **15** and **16**, which displayed comparable activity in BRCA-1-deficient cells relative to that of **13** and **14**. They also showed improved PK properties, and superior off-target profiles [22]. Further optimization resulted in the identification of an α-substituted proline derivative **17**, which displayed nanomolar antiproliferative activity against BRCA-1-deficient cells, and 300-fold selectivity for BRCA-1-deficient cells over BRCA-1-proficient cells.

Benzimidazole PARPi have been reported [23], with **18** displaying high intrinsic potency against PARP-1, $K_i = 240$ nM. Introduction of phenyl groups yielded compounds with good enzymatic but poor cellular activity. Incorporation of basic amines resulted in a marked improvement in cellular activity, and contemporarily enhanced PK properties. SAR studies identified **19** ($K_i = 8$ nM against PARP-1), which inhibited PARP activity in cells following H_2O_2-induced DNA damage ($EC_{50} = 3$ nM). Pyrrolidine **20** displayed $K_i = 30$ nM and $EC_{50} = 16$ nM yet, interestingly, methylated analog **21** gave $K_i = 5$ nM and $EC_{50} = 3$ nM [23], indicating that introduction of a methyl-substituted quaternary carbon significantly improved enzymatic and/or cellular potency. The *R*-enantiomer ABT-888 (**22**) demonstrated similar potency to **21** ($K_i = 5.2$ and 2.9 nM against PARP-1 and PARP-2, respectively, and $EC_{50} = 2$ nM cellular activity), while displaying superior oral bioavailability. Thus, **22** was selected for further development. It shows good *in vivo* activity in a melanoma model in combination with TMZ, with 64% TGI compared to the TMZ control, and in a breast cancer model potentiated both carboplatin and cyclophosphamide, with 91 and 61% TGI compared to the respective monotherapy control group. A new generation of related PARPi has been developed, diversifying through the incorporation of biaryl motifs [24]. The biphenyl **23**, while displaying good potency, $K_i = 2\,nM$, showed weak cellular activity. Introduction of heterocyclic motifs, like oxadiazole **24**, improved cellular activity, $EC_{50} = 3.7\,nM$. An innovative series of 3-oxoisoindoline-4-carboxamides, wherein the amide is locked in a seven-membered intramolecular hydrogen bond [25], produced compound **25** with $K_i = 33\,nM$.

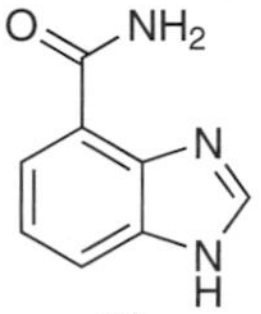

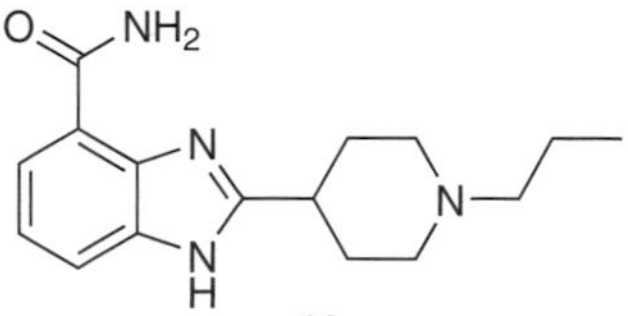

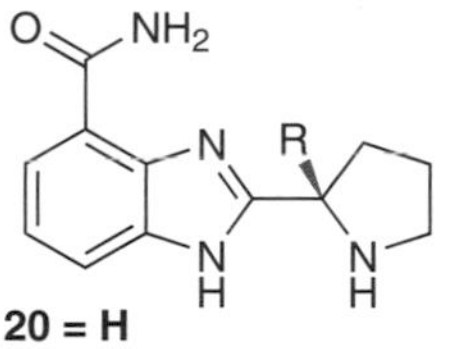

PF-01367338 (**27**) was identified as a clinical chemo- and radio-sensitizer [26] from several series of previously reported bicyclic and tricyclic PARPi. Compounds were evaluated based on their ability to potentiate the activity of TMZ and topotecan in cells, in comparison to control AG14361 (**26**). Further triage *in vivo* identified **27** displaying a $K_i = 1.4\,nM$ and $PF_{50} = 8.1$ in cells.

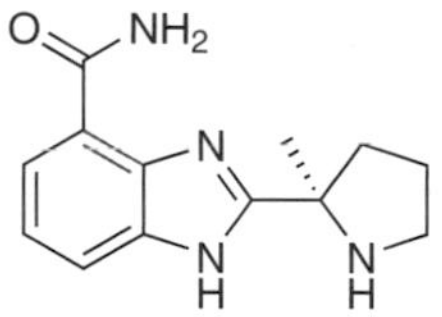

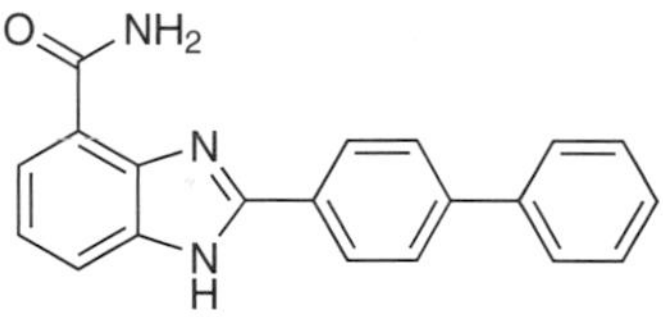

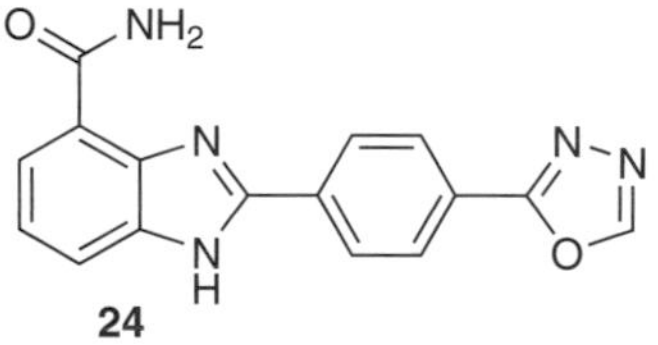

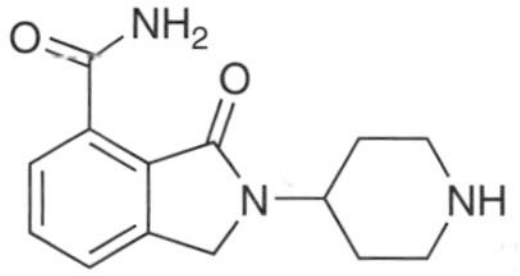

A series of indeno[1,2-*c*]isoquinolines have been described [27] and INO-1001 has been investigated in the clinic. Although the structure of INO-1001 has not been reported, sulfonamide **28** is a representative structure ($IC_{50} = 1\,nM$, $EC_{50} = 10\,nM$). CEP-9722 is in Phase I clinical trials, while its exact structure has not been disclosed, it is a prodrug of CEP-8983 (**29**) and patent applications highlight **30**. Preclinical efficacy in combination with TMZ and camptothecin have been reported, with 32–60 and 51–83% TGI compared to the respective monotherapy controls [28,29].

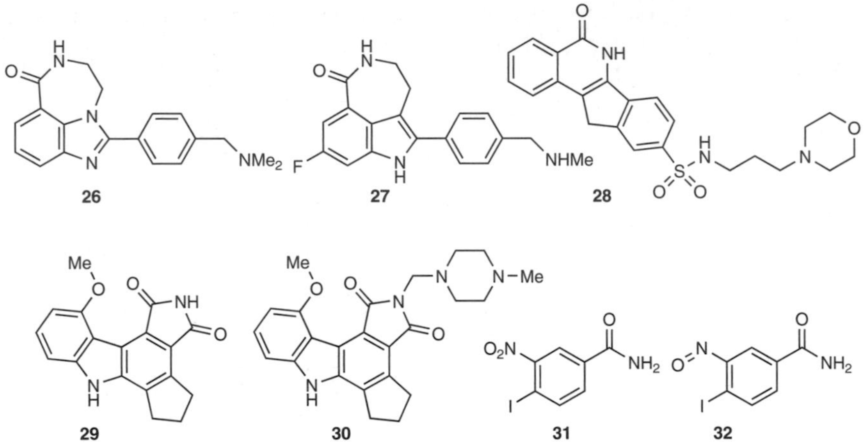

26   **27**   **28**

29   **30**   **31**   **32**

BSI-201 is being developed as a potential treatment for triple-negative breast cancer (TNBC) and other solid tumors, and is currently in late-stage clinical trials. On the basis of extensive patent applications, the structure of BSI-201 is suspected to be **31**, 4-iodo-3-nitrobenzamide [30,31]. This compound undergoes metabolic reduction to the 3-nitroso derivative (**32**), an irreversible inhibitor of PARP [32]. Recently, preclinical studies have discovered multiple metabolites whose biological activity has been evaluated [30]. It has also been reported that BSI-201 inhibits other processes critical for cancer cell proliferation, including fatty acid synthesis [31]. The activity of BSI-201 in combination with topotecan in ovarian xenografts has been reported [33]. The combination produced significant antitumor activity and increased the percentage of complete tumor regression compared with topotecan alone.

4.2 Other inhibitors

The majority of PARPi bear close resemblance to nicotinamide (**1**), although notable exceptions include imidazoloquinolinones **33** and imidazolopyridine **34**. Despite the absence of the classical amide group, these derivatives are dual PARP1/2 inhibitors with pIC_{50} 8.36, and 6.40 for PARP-1, respectively. Both compounds are reversible NAD^+-competitive inhibitors [34].

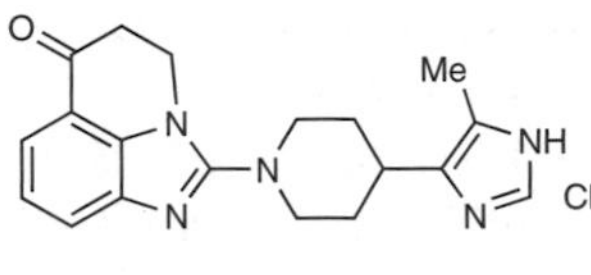

4.3 Subtype selective inhibitors

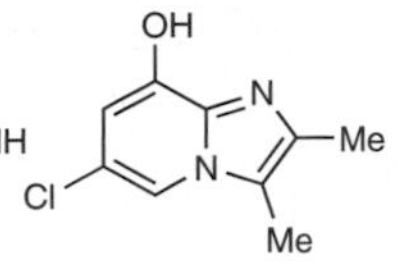

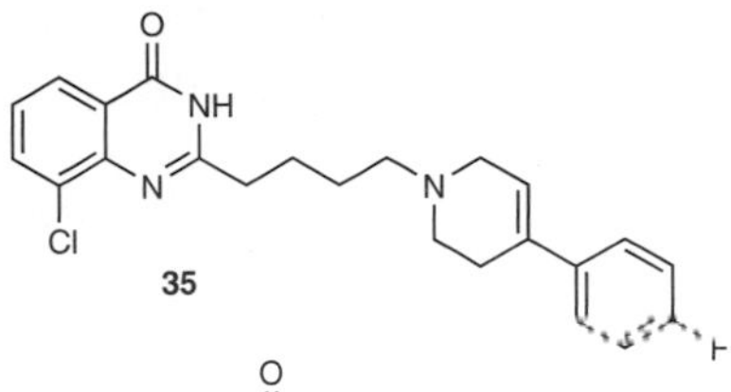

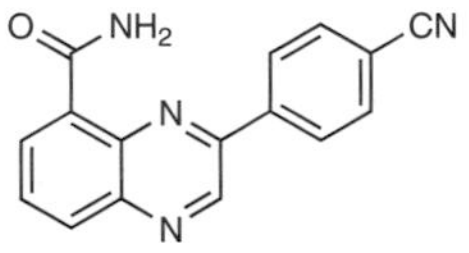

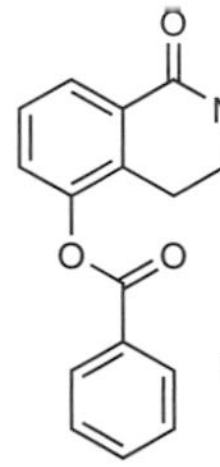

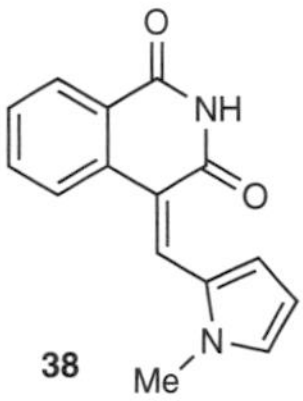

Very few compounds discriminate between PARP-1 and PARP-2, which is unsurprising given their highly conserved active sites. Exceptions to this are a series of quinazolinones that display relatively high selectivity for PARP-1, and quinoxalines that exhibit superior PARP-2 activity [35]. Quinazolinone **35** displayed high affinity for PARP-1, $IC_{50} = 13$ nM. The 40-fold selectivity for PARP-1 over PARP-2 is rationalized by the ability of the 4-phenyltetra-hydropyridine motif to bind tightly to the adenine-ribose pocket in PARP-1 as it has a favorable interaction with Leu769; this interaction is significantly less favorable in the PARP-2 pocket because Leu769 is replaced by a glycine (Gly314). In contrast, the quinoxaline **36** was found to be 10-fold more active for PARP-2, with $IC_{50} = 8$ nM, where the phenyl group extends into a groove containing Glu311 and Gln308 on PARP-2, generating a highly effective binding interaction that is not available in PARP-1.

Another chemotype displaying appreciable isoform selectivity is the 5-benzoyloxy-3,4-dihydroisoquinolin-1(2H)-one (**37**), displaying $IC_{50} = 0.8$ and 13 µM against PARP-2 and PARP-1, respectively [36]. Interestingly, **38** is a selective PARP-1 inhibitor, $pIC_{50} = 7.35$, displaying around 100-fold selectivity for PARP-1 over PARP-2 [34].

5. PRECLINICAL STUDIES WITH PARP INHIBITORS

The combination of PARPi olaparib (**7**) with cisplatin and carboplatin was evaluated in the $Brca1^{-/-};p53^{-/-}$ model where olaparib potentiated the effects of these platinum drugs [14]. In a $Brca2^{-/-};p53^{-/-}$ model, olaparib monotherapy caused regression or tumor growth inhibition in 46 of 52 tumors, while vehicle-treated mice showed rapid tumor progression [37]. However, upon extended treatment with olaparib, the tumors always regressed, primarily due to efflux transporter upregulation. The activity of olaparib on the selective inhibition of BRCA-2-deficient murine mammary tumors has also been investigated in comparison to a panel of cytotoxic agents [38].

ABT-888 (22) shows limited activity as monotherapy; however, it strongly potentiates the activity of multiple DNA-damaging agents in preclinical models [39]. ABT-888 potentiated TMZ in a glioma model in a dose-dependent manner, with maximal efficacy achieved at 50 mg/kg, which reduced tumor volume by 63%, 44% better than TMZ alone. In the MX-1 breast xenograft model, ABT-888, at 5 mg/kg/day in combination with cisplatin, caused sustained regressions in 8/9 mice compared to 3/9 for cisplatin monotherapy.

6. RECENT CLINICAL RESULTS

For many years, clinical development of PARPi has focused on evaluating their potential as chemo- and radio-potentiating agents. However, clinical trials have recently been initiated to investigate single agent activity in specific subpopulations [40,41].

6.1 Monotherapy in BRCA mutant tumors

Olaparib is the first PARPi to show antitumor activity when administered as monotherapy, confirming the preclinical findings of synthetic lethality in BRCA-1 and BRCA-2 mutant tumors. The results of a Phase I study have recently been reported [42], in patients carrying BRCA-1 and BRCA-2 mutations where clinical benefit was observed in 63% of patients following doses up to 600 mg bid. Olaparib was administered at a dose of 400 mg bid in two Phase II studies involving BRCA-1 and BRCA-2 mutation carriers in breast and ovarian cancer. In the breast cancer study, 41% of patients achieved an objective response [43], while in the ovarian cancer study 33% of patients treated achieved a response (measured by response evaluation criteria in solid tumors (RECIST)) [44]. The response rate climbed to 61% if responses measured by CA125 were included. Side effects were mild, including fatigue, nausea, and vomiting.

In 2008, a Phase I clinical trial using MK-4827 was initiated, with a cohort expansion planned in BRCA-1- and BRCA-2-mutant ovarian cancer patients [45].

6.2 Combination therapy with DNA-damaging agents

The intravenous PARPi PF-1367338 has been investigated in the clinic for several years, and recently a Phase II study coupling PF-1367338 with cisplatin was initiated in TNBC patients [46]. A Phase II study with combination therapy of PF-1367338 and TMZ in metastatic malignant melanoma patients has been reported [47]. In this melanoma clinical trial, there was an increase in both the partial response rate and disease stabilization compared to the use of TMZ monotherapy alone. This improved response was accompanied by an increased frequency of myelosuppression.

Arguably, the most clinically advanced PARPi is BSI-201 and this agent is currently being investigated in Phase III studies in TNBC as a combination with both gemcitabine and carboplatin [48]. The results of a randomized Phase II studies have been reported, where TNBC patients were treated with both gemcitabine and carboplatin and BSI-201 was given to one group of patients [49]. Interim results showed clinical benefit in 62% of patients on the BSI-201 combination, compared to 21% on chemotherapy alone. Improvements in survival were also noted, with median overall survival of 9.2 months compared with 5.7 months, respectively. The BSI-201 combination was well tolerated; common severe side effects included neutropenia, thrombocytopenia, and anemia.

Combination studies of olaparib are being explored clinically with a number of agents, including irinotecan, dacarbazine, carboplatin, gemcitabine, doxorubicin, cisplatin, and topotecan.

A Phase 0 clinical trial was conducted using ABT-888 [50] to determine a dose range and time course that would inhibit PARP activity. The compound was rapidly absorbed, with the target C_{max} exceeded at a 10 mg dose. Statistically significant PARP inhibition was observed in PBMC and tumor biopsies following single 25 and 50 mg doses. Currently, ABT-888 is being used in combination with several chemotherapeutics, and results of a Phase I study of ABT-888 combined with TMZ in patients with nonhematological malignancies and metastatic melanoma have been reported [51]. Of 20 patients evaluated, one had a partial response, and 10 had stable disease. Finally, Phase I clinical trials with the PARPi CEP-9722 were initiated in 2009 [52].

7. APPLICATION TO TUMORS WITH OTHER DNA DAMAGE REPAIR DEFICIENCIES

Besides BRCA-deficient cells being sensitive to PARPi, efforts have been undertaken to extend the utility of PARPi to cancers containing other genetic defects. Given the role of BRCA-1 and BRCA-2 in the DNA repair by HR, the effects of deficiencies in other proteins involved in HR were investigated [13]. Deficiencies of RAD51, ATR, ATM, CHK1, FANCD2, FANCA, and others were shown to cause a similar sensitivity to PARPi. Interestingly, mutations in the ATM gene have been found in patients with leukemia, breast cancer, and mantel cell lymphoma, and

ATM-deficient cell lines were found to be sensitive to olaparib [53]. In xenograft studies using ATM-deficient Granta-519 cells, olaparib was shown to decrease tumor growth and prolong overall survival by 42%.

Similarly, mutations in the PTEN gene and loss of PTEN expression have both been associated with a range of cancers and it has recently been established that PTEN-deficient cells are exquisitely sensitive to PARPi [54]. Indeed, HCT116 PTEN$^{-/-}$ cells are 25 times more sensitive to olaparib than isogenically matched wild-type cells. In a PTEN$^{-/-}$ xenograft, olaparib (15 mg/kg, i.p., 5 qdw) was shown to cause statistically significant tumor growth inhibition.

8. TANKYRASE INHIBITION AS A STRATEGY FOR Wnt/β-CATENIN PATHWAY MODULATION

The Wnt/β-catenin pathway is frequently inappropriately activated in many tumors and, consequently, has long been a target for therapeutic intervention. Inhibiting tankyrase disrupts the Wnt/β-catenin pathway, suggesting that this may be a new approach for targeting the pathway. Two independent groups identified small-molecule inhibitors using cellular Wnt/β-catenin pathway reporter assays [55,56]. Compounds **39** and **40** inhibited Wnt signaling with IC$_{50}$ = 180 and 350 nM, respectively, and **39** inhibited proliferation in colon DLD-1 and prostate DU145 cancer cells, both of which have an aberrant Wnt pathway.

The second group identified **41**, which inhibited Wnt pathway activity and proliferation of cancer cells with an aberrant Wnt pathway at micromolar concentrations; in contrast, the structurally related analog **42** had no effect. Immobilization of **41** and affinity capture followed by proteomics identified several potential targets, including TANK-1 and TANK-2, as well as PARP-1 and PARP-2. Codepletion of TANK-1 and TANK-2 by RNA interference phenocopied the effects seen with **41**. Subsequent biochemical studies showed that both **39** and **41** are potent dual TANK-1/-2 inhibitors, with **39** displaying $IC_{50} = 131/56$ nM and **41** displaying 11/4 nM, respectively, and have weaker affinity for PARP-1.

9. SUMMARY

Through the initial success of inhibitors in the clinic, PARP has been firmly established as a therapeutic target for the treatment of BRCA-1 and BRCA-2 mutant cancers. Preclinical studies suggest that this class of inhibitor may have further clinical utility in treating cancer patients specifically when used against tumors that have other defects in the proteins involved in the HR pathway. Finally, the recent finding that selective tankyrase inhibitors are capable of modulating the Wnt pathway could provide additional avenues given the frequent disregulation of the Wnt pathway in many tumors.

REFERENCES

[1] P. Chambon, J. D. Weill, and P. Mandel, *Biochem. Biophys. Res. Commun.*, 1963, **11**, 39.
[2] P. Jagtap and C. Szabó, *Nat. Rev. Drug Discov.*, 2005, **4**, 421.
[3] D. V. Ferraris, *J. Med. Chem.*, 2010, **53**, 4561.
[4] A. Huber, P. Bai, J. M. de Murcia, and G. de Murcia, *DNA Repair*, 2004, **3**, 1103.
[5] B. W. Durkacz, O. Omidiji, D. A. Gray, and S. Shall, *Nature*, 1980, **283**, 593.
[6] J. M. de Murcia, C. Niedergang, C. Trucco, M. Ricoul, B. Dutrillaux, M. Mark, F. J. Oliver, M. Masson, A. Dierich, M. LeMeur, C. Walztinger, P. Chambon, and G. de Murcia, *Proc. Natl. Acad. Sci. USA*, 1997, **94**, 7303.
[7] J. M. de Murcia, M. Ricoul, L. Tartier, C. Niedergang, A. Huber, F. Dantzer, V. Schreiber, J. C. Amé, A. Dierich, M. LeMeur, L. Sabatier, P. Chambon, and G. de Murcia, *EMBO J.*, 2003, **22**, 2255.
[8] H. Kleine, E. Poreba, K. Lesniewicz, P. O. Hassa, M. O. Hottiger, D. W. Litchfield, B. H. Shilton, and B. Lüscher, *Mol. Cell*, 2008, **32**, 57.
[9] S. Raval-Fernandes, V. A. Kickhoefer, C. Kitchen, and L. H. Rome, *Cancer Res.*, 2005, **65**, 8846.
[10] S. J. Hsiao and S. Smith, *Biochimie*, 2008, **90**, 83.
[11] H. E. Bryant, N. Schultz, H. D. Thomas, K. M. Parker, D. Flower, E. Lopez, S. Kyle, M. Meuth, N. J. Curtin, and T. Helleday, *Nature*, 2005, **434**, 913.
[12] H. Farmer, N. McCabe, C. J. Lord, A. N. J. Tutt, D. A. Johnson, T. B. Richardson, M. Santarosa, K. J. Dillon, I. Hickson, C. Knights, N.M.B. Martin, S. P. Jackson, G. C. M. Smith, and A. Ashworth, *Nature*, 2005, **434**, 917.
[13] N. McCabe, N. C. Turner, C. J. Lord, K. Kluzek, A. Bialkowska, S. Swift, S. Giavara, M. J. O'Connor, A. N. Tutt, M. Zdzienicka, G.C.M. Smith, and A. Ashworth, *Cancer Res.*, 2006, **66**, 8109.
[14] S. Rottenberg, J. E. Jaspers, A. Kersbergen, E. van der Burg, A.O.H. Nygren, S. A. L. Zander, P. W. B. Derksen, M. de Bruin, J. Zevenhoven, A. Lau, R. Boulter, A. Cranston, M. J. O'Connor, N.M.B. Martin, P. Borst, and J. Jonkers, *Proc. Natl. Acad. Sci. USA*, 2008, **105**, 17079.

[15] S. Peukert and U. Schwahn, *Expert Opin. Ther. Pat.*, 2004, **14**, 1531.

[16] V. M. Loh, X. Cockcroft, K. J. Dillon, L. Dixon, J. Drzewiecki, P. J. Eversley, S. Gomez, J. Hoare, F. Kerrigan, I.T.W. Matthews, K. A. Menear, N.M.B. Martin, R. F. Newton, J. Paul, G.C.M. Smith, J. Vile, and A. J. Whittle, *Bioorg. Med. Chem. Lett.*, 2005, **15**, 2235.

[17] X. Cockcroft, K. J. Dillon, L. Dixon, J. Drzewiecki, F. Kerrigan, V. M. Loh, N.M.B. Martin, K. A. Menear, and G.C.M. Smith, *Bioorg. Med. Chem. Lett.*, 2006, **16**, 1040.

[18] K. A. Menear, C. Adcock, R. Boulter, X. Cockcroft, L. Copsey, A. Cranston, K. J. Dillon, J. Drzewiecki, S. Garman, S. Gomez, H. Javaid, F. Kerrigan, C. Knights, A. Lau, V. M. Loh, I.T.W. Matthews, S. Moore, M. J. O'Connor, G. C. M. Smith, and N. M. B. Martin, *J. Med. Chem.*, 2008, **51**, 6581.

[19] K. A. Menear, C. Adcock, F. C. Alonso, K. Blackburn, L. Copsey, J. Drzewiecki, A. Fundo, A. Le Gall, S. Gomez, H. Javaid, C. F. Lence, N. M. B. Martin, C. Mydlowski, and G. C. M. Smith, *Bioorg. Med. Chem. Lett.*, 2008, **18**, 3942.

[20] P. Jones, S. Altamura, J. Boueres, F. Ferrigno, M. Fonsi, C. Giomini, S. Lamartina, E. Monteagudo, J. M. Ontoria, M. V. Orsale, M. C. Palumbi, S. Pesci, G. Roscilli, R. Scarpelli, C. Schultz-Fademrecht, C. Toniatti, and M. Rowley, *J. Med. Chem.*, 2009, **52**, 7170.

[21] G. Pescatore, D. Branca, F. Fiore, O. Kinzel, L. Llauger Bufi, E. Muraglia, F. Orvieto, M. Rowley, C. Toniatti, C. Torrisi, and P. Jones, *Bioorg. Med. Chem. Lett.*, 2010, **20**, 1094.

[22] F. Ferrigno, D. Branca, O. Kinzel, S. Lillini, L. Llauger Bufi, E. Monteagudo, E. Muraglia, M. Rowley, C. Schultz-Fademrecht, C. Toniatti, C. Torrisi, and P. Jones, *Bioorg. Med. Chem. Lett.*, 2010, **20**, 1100.

[23] T. D. Penning, G.-D. Zhu, V. B. Gandhi, J. Gong, X. Liu, Y. Shi, V. Klinghofer, E. F. Johnson, C. K. Donawho, D. J. Frost, V. Bontcheva-Diaz, J. J. Bouska, D. J. Osterling, A. M. Olson, K. C. Marsh, Y. Luo, and V. L. Giranda, *J. Med. Chem.*, 2009, **52**, 514.

[24] Y. Tong, J. J. Bouska, P. A. Ellis, E. F. Johnson, J. Leverson, X. Liu, P. A. Marcotte, A. M. Olson, D. J. Osterling, M. Przytulinska, L. E. Rodriguez, Y. Shi, N. Soni, J. Stavropoulos, S. Thomas, C. K. Donawho, D. J. Frost, Y. Luo, V. L. Giranda, and T. D. Penning, *J. Med. Chem.*, 2009, **52**, 6803.

[25] V. J. Gandhi, Y. Luo, X. Liu, Y. Shi, V. Klinghofer, E. F. Johnson, C. Park, V. L. Giranda, T. D. Penning, and G. -D. Zhu, *Bioorg. Med. Chem. Lett.*, 2010, **20**, 1023.

[26] H. D. Thomas, C. R. Calabrese, M. A. Batey, S. Canan, Z. Hostomsky, S. Kyle, K. A. Maegley, D. R. Newell, D. Skalitzky, L. -Z. Wang, and S. E. Webber, *Mol. Cancer Ther.*, 2007, **6**, 945.

[27] P. G. Jagtap, E. Baloglu, G. J. Southan, J. G. Mabley, H. Li, J. Zhou, J. van Duzer, A. L. Salzman, and C. Szabó, *J. Med. Chem.*, 2005, **48**, 5100.

[28] S. Miknyoczki, H. Chang, J. Grobelny, S. Pritchard, C. Worrell, N. McGann, M. Ator, J. Husten, J. Deibold, R. Hudkins, A. Zulli, R. Parchment, and B. Ruggeri, *Mol. Cancer Ther.*, 2007, **6**, 2290.

[29] J. L. Diebold, R. L. Hudkins, S. J. Miknyoczki, and B. Ruggeri, *Patent Application WO 2008/063644-A1*, 2008.

[30] J. Moore, B. Keyt, J. Burnier, B. Sherman, M. Totrov, and V. Ossovskaya, *Patent Application WO 2008/030883-A2*, 2008.

[31] V. Ossovskaya and B. Sherman, *Patent Application WO 2008/030891-A2*, 2008.

[32] J. Mendeleyev, E. Kirsten, A. Hakam, K. G. Buki, and E. Kun, *Biochem. Pharmacol.*, 1995, **50**, 705.

[33] V. Ossovskaya, L. Li, C. Bradley, and B. Sherman, Abstract 2311, 99th AACR Annual Meeting, San Diego, CA, 2008.

[34] T. Eltze, R. Boer, T. Wagner, S. Weinbrenner, M. C. McDonald, C. Thiemermann, A. Buerkle, and T. Klein, *Mol. Pharmacol.*, 2008, **74**, 1587.

[35] J. Ishida, H. Yamamoto, Y. Kido, K. Kamijo, K. Murano, H. Miyake, M. Ohkubo, T. Kinoshita, M. Warizaya, A. Iwashita, K. Mihara, N. Matsuoka, and K. Hattori, *Bioorg. Med. Chem.*, 2006, **14**, 1378.

[36] R. Pellicciari, E. Camaioni, G. Costantino, L. Formentini, P. Sabbatini, F. Venturoni, G. Eren, D. Bellocchi, A. Chiarugi, and F. Moroni, *ChemMedChem*, 2008, **3**, 914.

[37] T. Hay, J. R. Matthews, L. Pietzka, A. Lau, A. Cranston, A. O. H. Nygren, A. Douglas-Jones, G. C. M. Smith, N. M. B. Martin, M. O'Connor, and A. R. Clarke, *Cancer Res.*, 2009, **69**, 3850.

[38] B. Evers, R. Drost, E. Schut, M. de Bruin, E. van der Burg, P.W.B. Derksen, H. Holstege, X. Liu, E. van Drunen, H. Beverloo, G. C. M. Smith, N.M.B. Martin, A. Lau, M. J. O'Connor, and J. Jonkers, *Clin. Cancer Res.*, 2008, **14**, 3916.

[39] C. K. Donawho, Y. Luo, Y. Luo, T. D. Penning, J. L. Bauch, J. J. Bouska, V. D. Bontcheva-Diaz, B. F. Cox, T. L. DeWeese, L. E. Dillehay, D. C. Ferguson, N. S. Ghoreishi-Haack, D. R. Grimm, R. Guan, E. K. Han, R. R. Holley-Shanks, B. Hristov, K. B. Idler, K. Jarvis, E. F. Johnson, L. R. Kleinberg, V. Klinghofer, L. M. Lasko, X. Liu, K. C. Marsh, T. P. McGonigal, J. A. Meulbroek, A. M. Olson, J. P. Palma, L. E. Rodriguez, Y. Shi, J. A. Stavropoulos, A. C. Tsurutani, G. -D. Zhu, S. H. Rosenberg, V. L. Giranda, and D. J. Frost, *Clin. Cancer Res.*, 2007, **13**, 2728.

[40] S. K. Sandhu, T. A. Yap, and J. S. de Bono, *Eur. J. Cancer*, 2010, **46**, 9.

[41] http://www.clinicaltrials.gov/ct2/results?term=PARP.

[42] P. C. Fong, D. S. Boss, T. A. Yap, A. Tutt, P. Wu, M. Mergui-Roelvink, P. Mortimer, H. Swaisland, A. Lau, M. J. O'Connor, A. Ashworth, J. Carmichael, S. B. Kaye, J. H. M. Schellens, and J. S. de Bono, *N. Engl. J. Med.*, 2009, **361**, 123.

[43] A. Tutt, M. Robson, J. E. Garber, S. Domchek, M. W. Audeh, J. N. Weitzel, M. Friedlander, and J. Carmichael, *J. Clin. Oncol.*, 2009, **27** (suppl 18), CRA501.

[44] M. W. Audeh, R. T. Penson, M. Friedlander, B. Powell, K. M. Bell-McGuinn, C. Scott, J. N. Weitzel, J. Carmichael, and A. Tutt, *J. Clin. Oncol.*, 2009, **27** (suppl 15), 5500.

[45] http://clinicaltrials.gov/ct2/show/NCT00749502.

[46] http://clinicaltrials.gov/ct2/show/NCT01074970.

[47] R. Plummer, P. Lorigan, J. Evans, N. Steven, M. Middleton, R. Wilson, K. Snow, R. Dewji, and H. Calvert, *J. Clin. Oncol.*, 2006, **24** (suppl 18), 8013.

[48] http://www.clinicaltrials.gov/ct2/show/NCT00938652.

[49] J. O'Shaughnessy, C. Osborne, J. Pippen, M. Yoffe, D. Patt, G. Monaghan, C. Rocha, V. Ossovskaya, B. Sherman, and C. Bradley, *J. Clin. Oncol.*, 2009, **27** (suppl 18), 3.

[50] S. Kummar, R. Kinders, M. E. Gutierrez, L. Rubinstein, R. E. Parchment, L. R. Phillips, J. Ji, A. Monks, J. A. Low, A. Chen, A. J. Murgo, J. Collins, S. M. Steinberg, H. Eliopoulos, V. L. Giranda, G. Gordon, L. Helman, R. Wiltrout, J. E. Tomaszewski, and J. H. Doroshow, *J. Clin. Oncol.*, 2009, **27**, 2705.

[51] J. Molina, C. Erlichman, D. W. Northfelt, J. Lensing, and V. Giranda, Abstract 3602, 100th AACR Annual Meeting, Denver, CO, 2009.

[52] http://clinicaltrials.gov/ct2/show/NCT00920595.

[53] C. T. Williamson, H. Muzik, A. G. Turhan, A. Zamo, M. J. O'Connor, D. G. Bebb, and S. P. Lees-Miller, *Mol. Cancer Ther.*, 2010, **9**, 347.

[54] A. M. Mendes-Pereira, S. A. Martin, R. Brough, A. McCarthy, J. R. Taylor, J. -S. Kim, T. Waldman, C. J. Lord, and A. Ashworth, *EMBO Mol. Med.*, 2009, **1**, 315.

[55] B. Chen, M. E. Dodge, W. Tang, J. Lu, Z. Ma, C. -W. Fan, S. Wei, W. Hao, J. Kilgore, N. S. Williams, M. G. Roth, J. F. Amatruda, C. Chen, and L. Lum, *Nat. Chem. Biol.*, 2009, **5**, 100.

[56] S. -M. A. Huang, Y. M. Mishina, S. Liu, A. Cheung, F. Stegmeier, G. A. Michaud, O. Charlat, E. Wiellette, Y. Zhang, S. Wiessner, M. Hild, X. Shi, C. J. Wilson, C. Mickanin, V. Myer, A. Fazal, R. Tomlinson, F. Serluca, W. Shao, H. Cheng, M. Shultz, C. Rau, M. Schirle, J. Schlegl, S. Ghidelli, S. Fawell, C. Lu, D. Curtis, M. W. Kirschner, C. Lengauer, P. M. Finan, J. A. Tallarico, T. Bouwmeester, J. A. Porter, A. Bauer, and F. Cong, *Nature*, 2009, **461**, 614.

CHAPTER 15

Epigenetic Targets and Cancer Drug Discovery

Patrick M. Woster

Contents

1. INTRODUCTION

Chromatin architecture is a key determinant in regulating gene expression and is strongly influenced by posttranslational modifications of histones. Histones, which are the chief protein component of chromatin, are complex macromolecules comprised of core protein subunits termed H_{2a}, H_{2b}, H_3, and H_4. These core histone subunits differ slightly in primary sequence and three-dimensional structure. Histones occur as octamers that consist of one H_3–H_4 tetramer and two H_{2A}–H_{2B} dimers [1] and are directed to DNA by their predominating positive charges [2]. Approximately 146 base pairs of DNA are wrapped around each histone octamer, forming a unit known as a nucleosome. Multiple nucleosomes form on the DNA strand and ultimately condense to form chromatin [3]. Lysine-containing histone tails, consisting of up to 40 amino acid residues, protrude through the DNA strand; and lysines and arginines on this tail act as sites for posttranslational modification. These modifications allow alteration of higher order nucleosome structure and thus contribute to structural changes in chromatin [4]. Methylation, phosphorylation, ubiquitination, and acetylation of histone tails are known

Pharmaceutical Sciences, Wayne State University, 3132 Applebaum Hall, 259 Mack Ave, Detroit, MI 48202, USA

Annual Reports in Medicinal Chemistry, Volume 45
ISSN 0065-7743, DOI 10.1016/S0065-7743(10)45015-0

posttranslational modifications, with acetylation being the best characterized process [5]. Posttranslational modifications of histone tails produce epigenetic changes in gene expression via a mechanism that does not involve changes to the primary DNA sequence. Changes that are created at the epigenetic level persist for the life of the cell and can be passed on for multiple generations. Drug discovery efforts have mainly concentrated on histone acetylation/deacetylation, but more recently histone methylation/demethylation enzymes have become validated targets for drug discovery. This chapter deals primarily with the discovery of novel agents that modulate epigenetic control mechanisms, such as those described below. For a more detailed discussion of the biological aspects of epigenetic control of gene expression, the reader is referred to Chapter 20.

2. HISTONE ACETYLATION/DEACETYLATION

Histone acetylation status is controlled by a balance between histone acetyltransferases (HATs), which promote the open chromatin conformation and are associated with active gene expression, and histone deacetylases (HDACs), which promote closed chromatin (heterochromatin) and repression of gene transcription [1,5]. HATs have received far less attention than HDACs as potential therapeutic targets, mainly because their relationship to disease has not been firmly established. HATs can be divided into three families—the Gen-5-related N-acetyltransferse (GNAT) family, the p300 cAMP-responsive element-binding protein-binding protein (p300/CBP) family, and the MOZ/YBF2/SAS2/TIP60 (MYST) family [6–8]—and their effects are either gene-specific or global, depending on structure and whether they are part of a multisubunit protein complex. The p300/CBP and GNAT family HATs have been implicated in cancer, and elevated HAT activity has been associated with asthma and chronic obstructive pulmonary disease (COPD)[9]. Only a few HAT inhibitors have been studied *in vitro*, among them the natural products curcumin, garcinol, and anacardic acid [9]. In addition, isothiazolone derivatives [10] have been shown to inhibit the GNAT family acetyltransferase p300/CBP-associated factor (PCAF, while α-methylene-γ-butyrolactones while have been shown to inhibit GNC5, a GNAT family HAT [11].

Specific DNA methylation at cytosine - phosphate – guanine (CpG) (CpG) islands (a region with at least 200 base pairs and with a GC percentage that is greater than 50%) leads to recruitment of HDACs to gene promoters and epigenetic silencing of specific genes [12,13]. Mammalian cells exhibit an exquisite level of control of chromatin architecture through balancing HAT and HDAC activity [14]. There are 11 known zinc-dependent HDAC isoforms [15–17] that belong to four structural classes: class I (isoforms 1, 2, 3, and 8) [18], class IIa (isoforms 4, 5, 7, and 9) [15], class IIb (isoforms 6 and 10) [15], and class IV (HDAC 11) [15].

In addition to the zinc-dependent HDACs, class III HDACs include sirtuins 1 though 7, which have homology to yeast Sir2 and require NAD+ [19]. In some tumor cell types, hypoacetylation of histones by one or more of the zinc-dependent HDACs results in aberrant gene silencing, which leads to the underexpression of crucial growth regulatory proteins, such as the cyclin dependent kinase inhibitor p21^{Waf1}[1,5]. This underexpression of regulatory proteins creates cellular conditions that facilitate tumorigenesis. Despite extensive study, it is still uncertain which HDAC isoforms play the most important roles in human cancer. Increases in histone acetylation caused by HDAC inhibitors (Figure 1), such as trichostatin A (TSA, 1), N-(2-aminophenyl)-4-[N-(pyridin-3-ylmethoxycarbonyl)-aminomethyl]benzamide (MS-275, 2), and suberoylanilide hydroxamic acid (SAHA, 3), can cause growth arrest in transformed cells and can inhibit the growth of human tumor xenografts [5,18,20–22], especially when used in combination with DNA methyltransferase inhibitors

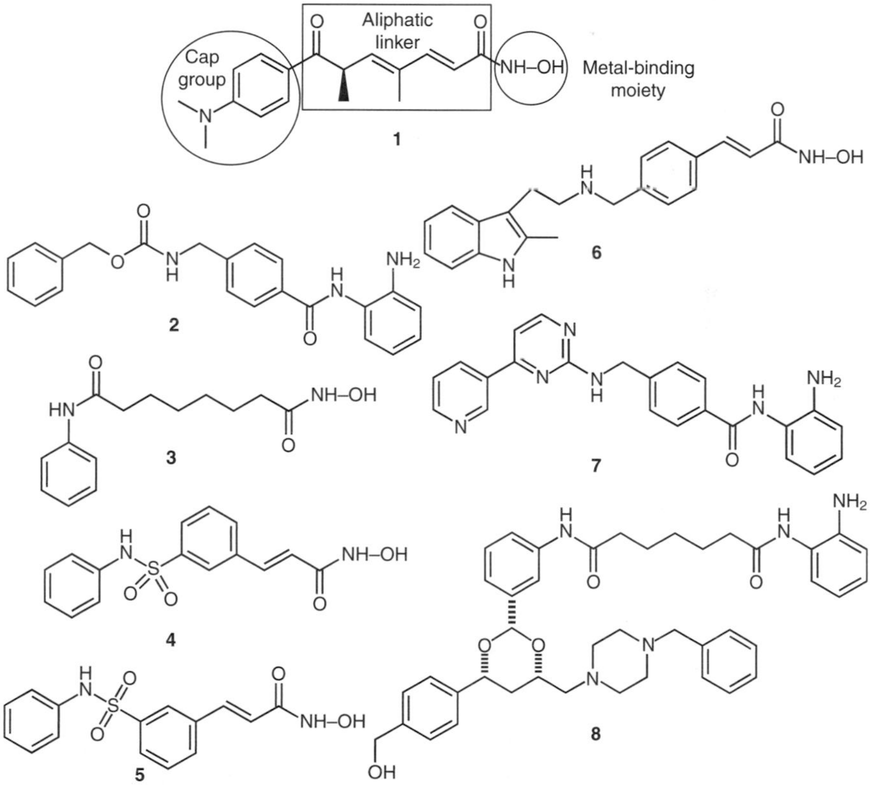

Figure 1 Structures of traditional HDAC inhibitors **1–3** and HDAC inhibitors now in clinical trials (**4–8**) .

[23–26]. Known HDAC inhibitors fall into several structural classes, the most common of which are short-chain carboxylic acids, hydroxamic acids (e.g., 1 and 3), and benzamides (e.g., 2) [21]. Cyclic peptide inhibitors, such as apicidin, depsipeptide, trapoxin, and CHAPs, have also shown promising activity *in vivo*. Although several HDAC inhibitors are in clinical trials and two agents have been marketed, dose-limiting toxicity remains a problem [27]. Some toxicity issues could be eliminated by identifying isoform-specific HDAC inhibitors to minimize off-target effects. These inhibitors would be of great value in elucidating the roles of HDAC isoforms in both normal and cancer cells. Recent structure/activity studies involving analogs of 1–3 have focused largely on modifications to the aromatic "cap group" moiety and the aliphatic linker region present in these molecules (see the structure of 1 above) [21], but very few isoform-selective inhibitors have been identified [28]. Drug discovery in this area has been accelerated by the commercial availability of all 11 HDAC isoforms and the availability of X-ray crystal structures for some HDAC isoforms [29–31]. The discovery of histone deacetylase inhibitors and subsequent clinical trials have been recently reviewed [32–37], and as such this topic is not discussed in depth in this chapter. A number of clinical candidates based on known pharmacophores have emerged, including the recently marketed cyclic peptide romedepsin [38,39], hydroxamic acid derivatives PXD101 (4), belinostat (5), panobinostat (6), and benzamide derivatives: MGCD0103 (7) and histacin (8) (Figure 1), among many others. More recent studies have identified HDAC inhibitors with more diverse zinc-binding moieties, including thiols [40], sulfamides [41], polyamine-containing hydroxamic acids and benzamides [42,43], fluoroalkenylmercaptoacetoamides [44], boronic acids [45], and selenium analogs of SAHA [46], to name a few. Although HDAC inhibitors have traditionally been developed as anticancer agents, reexpression of aberrantly silenced genes by HDAC inhibitors can impact other diseases, such as parasitic infection [47,48], diabetes [49–51], and neurological disorders [52 –54]

3. HISTONE METHYLATION/DEMETHYLATION

In cancer, hypermethylation at specific DNA promoters, combined with other chromatin modifications that decrease activating marks and increase repressive marks on histones 3 and 4, is associated with silencing of tumor suppressor genes [55]. The important role of promoter CpG island methylation, and its relationship to covalent histone modifications, has recently been reviewed [56]. N-terminal histone tails undergo numerous posttranslational modifications, including methylation at specific histone lysine chromatin marks [5,57]. To date, 17 lysine residues and 7 arginine residues on histone proteins have been shown to undergo methylation [58]. It has been shown that aberrant methylation of histone arginines and lysines is linked to carcinogenesis [59,60] and possibly to coronary disease [61]. There are three classes of histone-methylating enzymes: Su(var), enhancer of zeste, trithorax (SET) domain lysine methyltransferases (KMTs), non-SET domain KMTs, and arginine methyltransferases (AMTs). All three classes use S-adenosylmethionine (SAM) as a methyl group donor. These methylations are catalyzed by ten KMTs [62,63] and nine AMTs [57,64]. Histone KMTs generally target a particular lysine residue, and can promote activation or repression of gene transcription, depending on the specific lysine residue involved [63,65,66]. Additional information on protein KMTs and methyllysine-binding proteins can be found elsewhere in this volume (Chapter 20).

Efforts to identify specific inhibitors of histone methyltransferases are in their early stages, and only two specific KMT inhibitors have been identified. The fungal mycotoxin chaetocin (9, Figure 2) inhibits the KMT Su(var)3–9 from *Drosophila* with an IC_{50} of 0.8 µM and produces a

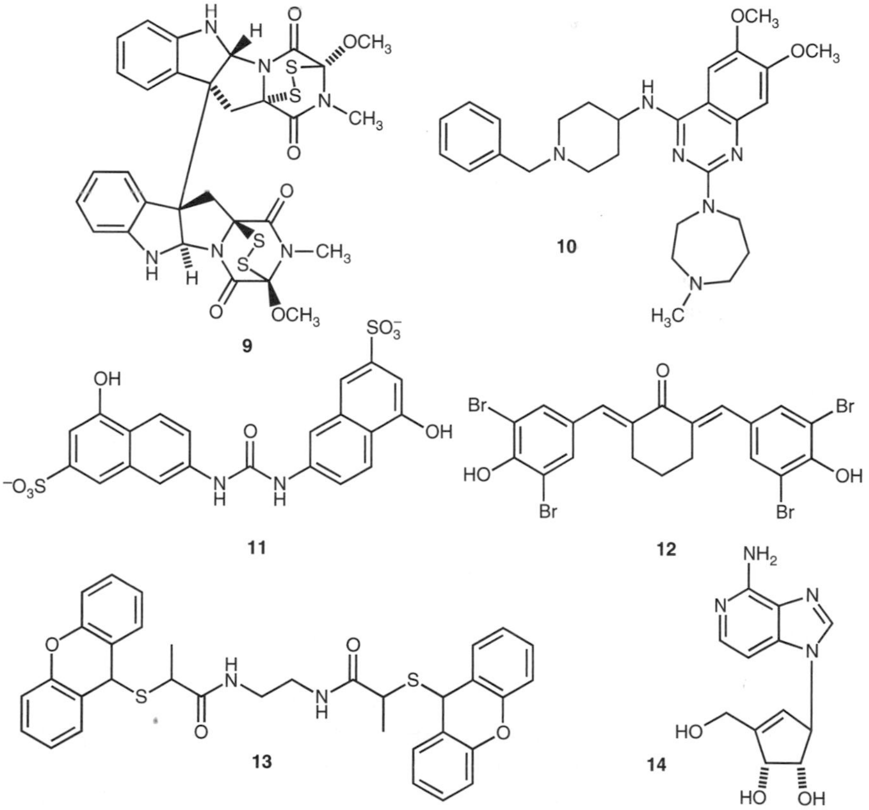

Figure 2   Known inhibitors of histone methylation by KMTs and AMTs.

cytotoxic effect *in vivo*[67]. The compound BIX-01294 (10) inhibits the G9a KMT and produces a reduction in histone H3 lysine 9 (H3K9) dimethylation in several cell lines [68]. Importantly, chromatin immunoprecipitation (ChIP) revealed that the mono- or trimethyl H3K9 was unaffected, and that other lysine methylation sites, like H3K27 or H4K20, were not altered. These results demonstrate the tightly controlled substrate specificity for KMTs and demethylases, and the complexity of the enzyme systems that control gene expression through modifications at specific lysine residues. Some progress has also been made in the search for selective AMT inhibitors. A random compound library screen revealed nine AMT inhibitors with structures related to AMI-1 (11, Figure 2). Compound 11 was selective for AMT over KMT and produced an *in vivo* reduction in arginine methylation [69]. Analogs of 11 (Figure 2), bearing 2-bromo- and 2,6-dibromophenol moieties linked through unsaturated cyclic spacers, such as 12, equally inhibited protein AMTs, KMTs, HATs, and the sirtuins, thus behaving as multiple ligands for enzymes involved in the epigenetic control of transcription [70]. Compound 13 (RM-65) inhibited the AMTs known as RmtA and PRMT1 with equal strength. *In silico* docking showed that 13 is a bisubstrate mimic, occupying both the arginine- and SAM-binding pockets [71]. More recently, the KMT inhibitor 3-aza-neplanocin A (14), in combination with panobinostat (6, Figure 1), enhanced antitumor activity against cultured acute myeloid leukemia cells [72] .

Histone methylation, once thought to be an irreversible modification, has been shown to be a dynamic process regulated by histone methyltransferases and histone demethylases. Methyl groups are selectively removed from mono- and dimethyllysines by lysine-specific demethylase 1 (LSD1, also called BHC110 or KDM1) [73–75] and from mono-, di-, and trimethyllysines by specific Jumonji C (JmjC) domain-containing demethylases [75–77]. A more detailed description of the JmjC demethylases can be found in Chapter 20. Recent results suggest that LSD1 also plays a role in demethylation of nonhistone proteins [78,79]. LSD1 has been shown to demethylate K370me1 and K370me2 on the p53 tumor suppressor protein [78]. Through this process, LSD1 can be targeted to chromatin by p53, likely in a gene-specific manner [80]. These findings define a molecular mechanism by which p53 mediates transcriptional repression during differentiation *in vivo*. Additional evidence suggests that LSD1 is required for maintenance of global DNA methylation [81], indicating that LSD1 mediates a general mechanism for transcriptional control.

LSD1 catalyzes the oxidative demethylation of histone 3 methyllysine 4 (H3K4me1) and histone 3 dimethyllysine 4 (H3K4me2) (Figure 3), resulting in transcriptional repression [74,82]. The LSD1 reaction is flavin adenine dinucleotide (FAD)-dependent oxidative demethylation (Figure 3) that proceeds through a protonated imine intermediate [83]. The K_m for H3K4me2 is

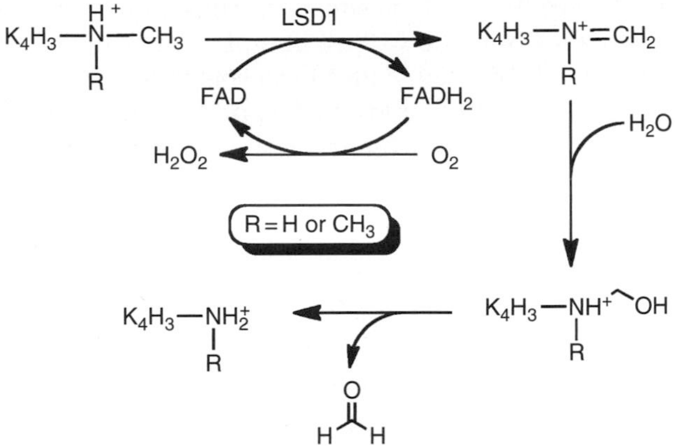

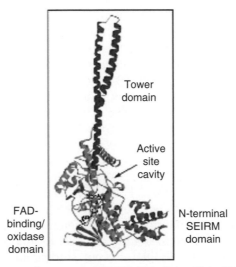

Figure 3   The catalytic mechanism for oxidative demethylation by LSD1.

30 µM [74], which is comparable to substrate K_m values for other histone-modifying enzymes. H3K4me2 is a transcription-activating chromatin mark at gene promoters; and demethylation of this mark prevents expression of tumor suppressor genes important in human cancer [73]. Thus, LSD1 is emerging as an important new target for the development of specific inhibitors that could become a new class of antitumor agents [84]. The X-ray structure and catalytic mechanism of LSD1 have recently been reported [85]. Nascent LSD1 (MW 110 kDa) folds so as to produce three distinct structural domains (Figure 4). The tower domain, which is

Figure 4   The X-ray crystal structure of LSD1. (See Color Plate 15.4 in the Color Plate Section)

directly connected to the catalytic site, is thought to be a platform for the binding of factors such as the CoRest complex [86,87]. CoRest enhances the ability of LSD1 to demethylate nucleosomal lysine residues on the H3K4 mark [88]. LSD1 also binds to the protein factor BHC80, which inhibits the demethylation process, as well as to the androgen receptor (AR). The AR is thought to act as a transcriptional activator and is also believed to alter LSD1 substrate specificity by promoting demethylation of the deactivating marks mono- and dimethyl histone H3K9 [89]. The LSD1 SWi3p/Rsc8p/Moira (SWIRM) domain [90] is typical of analogous domains found in other chromatin-remodeling proteins [91] and is potentially the site of DNA binding. The oxidase domain has two functional lobes: one for binding the FAD cofactor and another for substrate binding. The FAD-binding/oxidase domain is bound to the SWIRM domain through a series of hydrophobic interactions and point mutations that disrupt these interactions greatly reduce catalytic activity. The LSD1 active site is large enough to bind the substrate lysine residue, as well as several adjacent amino acids in the histone tail. The enzyme demethylates H3K4me1 and H3K4me2, but chemically (rather than sterically) excludes the trimethylated lysine H3K4me3 [77].

Multiple studies suggest that LSD1 hyperactivity plays an important role in tumorigenesis by promoting aberrant silencing of tumor suppressor genes. LSD1 colocalizes with the AR in normal human prostate and in prostate tumors [89], where it interacts with AR *in vitro* and *in vivo*, and stimulates AR-dependent transcription. Conversely, knockdown of LSD1 protein levels in these tumors abrogates androgen-induced transcriptional activation and cell proliferation [89]. It has been suggested that LSD1 is a prognostic marker in prostate cancer [92]. High levels of LSD1, nuclear expression of the FHL2 coactivator (a protein factor that transmits signals from the cell membrane to the nucleus), high Gleason score and grade, and very strong staining of nuclear p53 correlate significantly with relapse of prostate carcinoma during follow-up. Thus, LSD1 and nuclear FHL2 may serve as novel biomarkers predictive for prostate cancer with aggressive biology and point to a role of LSD1 and FHL2 in constitutive activation of AR-mediated growth signals [93]. In neuroblastoma, siRNA –mediated knockdown of LSD1 decreased cellular growth, induced expression of differentiation-associated genes, and increased target gene-specific H3K4 methylation. LSD1 inhibition, using monoamine oxidase inhibitors, increases global H3K4 methylation, inhibits neuroblastoma cell growth, and reduces neuroblastoma xenograft growth *in vivo*. LSD1 is involved in maintaining the undifferentiated, malignant phenotype in neuroblastoma and inhibition of LSD1 reprograms the transcriptome of neuroblastoma cells [94]. Recent studies suggest that LSD1 is highly expressed in ER-negative breast cancer, providing a predictive marker for aggressive biology and a novel attractive therapeutic target for treatment [95]. These data provide a clear indication that LSD1 is a valid target for the design of antitumor agents.

The active-site structure of LSD1 has considerable sequence homology to monoamine oxidases A and B (MAO A and B) and to N^1-acetylpolyamine oxidase (APAO) and spermine oxidase (SMO) [74,82,96]. It was therefore postulated that MAOs and other classical amine oxidase inhibitors would also inhibit LSD1. The MAO inhibitors nialamide, clorgyline, deprenyl, and pargyline were devoid of any inhibitory activity, while phenelzine 15 and tranylcypromine 16 (Figure 5) effectively inhibited LSD1/CoRest at 1 μM [82]. Tranylcypromine 15 was titrated to reveal an IC_{50} of <2 μM, and it increased global levels of H3K4me2 in the P19 cell line. LSD1 inhibition by tranylcypromine was subsequently shown to be irreversible and mechanism based [96].

The substrate analogs aziridinyl-K4H3$_{1-21}$ (17), propargyl-K4H3$_{1-21}$ (18), and N-methylpropargyl-K4H3 $_{1-21}$ (19) (Figure 5) were synthesized in an effort to identify novel active-site-directed inhibitors of LSD1. Compound 17 reversibly inhibited LSD1 with an IC_{50} of 15.6 μM, while 18 produced time-dependent inactivation with a K_i of 16.6 μM [83] through formation of a covalent adduct with the enzyme-bound flavin cofactor [96,97]. Because these compounds are peptides with 21 amino acid residues, it is unlikely that they will become useful therapeutic agents. In a study comparing small-molecule inhibitors of LSD1 to 21-mer analogs of the peptide substrate, the peptide-based inhibitors

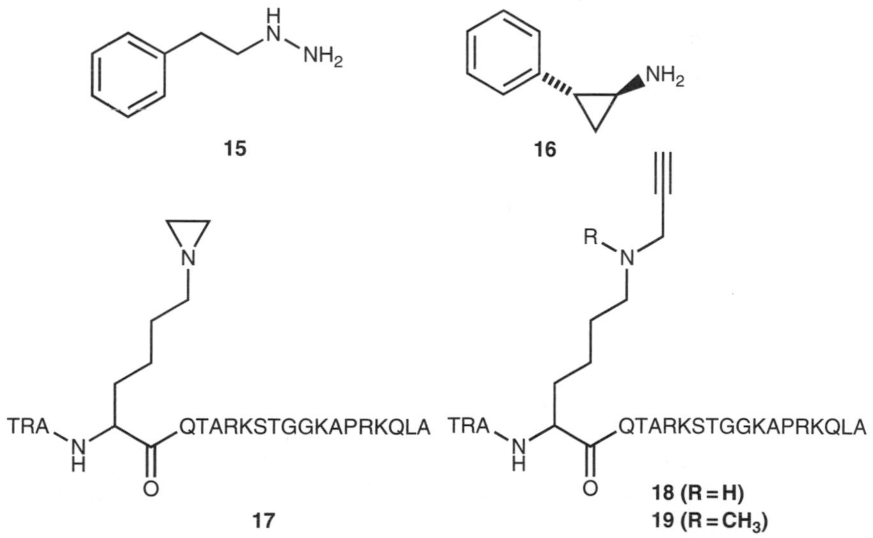

Figure 5 Structures of phenylzine 15, tranylcypromine 16, and the peptide substrate mimics 17–19.

were more potent, but **15** possessed more potent activity than originally reported and increased global H3K4 methylation [98].

McCafferty et al. recently described the synthesis of a series of *trans*-2-arylcyclopropylamine analogs that inhibit LSD1 with K_i values between 188 and 566 μM [99]. However, these analogs were 1–2 orders of magnitude more potent against MAO A and MOA B. Compound **20** (Figure 6) inhibited LSD1 with an IC_{50} value of 188 μM, but the cellular effects were not determined. Similar analogs, identified by structure-based design, were synthesized and evaluated as LSD1 inhibitors, the most potent of which was **21** ($IC_{50} = 1.9$ μM) [100]. It was determined by mass spectral analysis that compound **21** forms a covalent adduct with

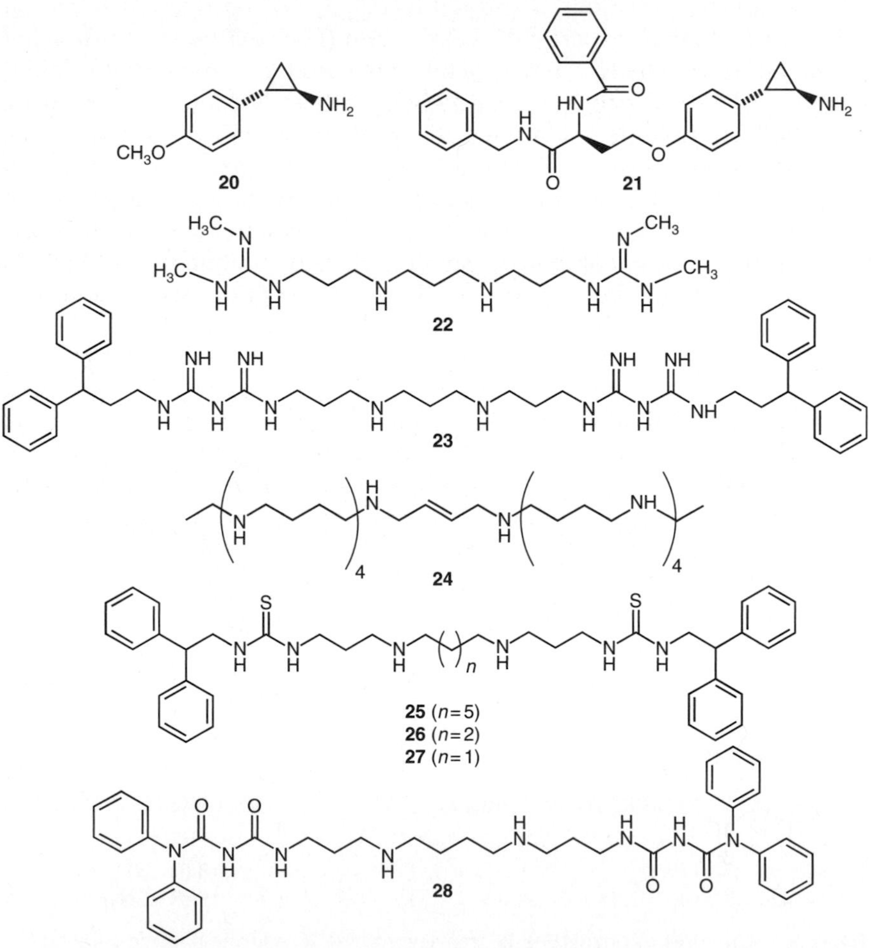

Figure 6   Structures of novel small-molecule LSD1 inhibitors **20–28**.

the FAD cofactor that is analogous to the adduct formed by **18** (Figure 5). In addition, **21** caused a dose-dependent increase in global H3K4 methylation in HEK293 cells *in vitro*.

The C-terminal portion of the LSD1 protein chain shares significant sequence homology with the amine oxidases APAO and SMO [74,101]. Several groups have identified amines, guanidines, or similar analogs that act as selective modulators of these two amine oxidases. The synthesis of a novel series of (bis)guanidine and (bis)biguanide antitrypanosomal agents with structures related to **22** and **23** (Figure 6) has been reported [102]. These analogs were modest inhibitors of APAO and SMO, and exhibited some selectivity for either APAO or SMO. Based on the similarity between LSD1 and the amine oxidases APAO and SMO, these (bis)guanidines and (bis)biguanides were examined for the ability to inhibit purified recombinant LSD1. Nine of the 13 compounds tested inhibited LSD1 by >50% at 1 μM [73]. Compounds **22** and **23** promoted increases in cellular levels of secreted frizzle-related proteins (SFRPs) 1–5, which are soluble modulators of Wnt signaling, and the zinc-finger transcription factor GATA5, in the HCT116 human colon carcinoma line [73]. Similar effects were observed with the oligamine PG-11144 (**24** Figure 6), and both **23** and **24** were synergistic with 5-azacytidine in limiting the growth of HCT116 xenografts *in vivo*[103]. Isosteric derivatives of **22** and **23**, such as the substituted (bis)thioureas **25** –**27** and carbamoylurea **28**, retain the activity of the parent compound. Compounds **25–28** (Figure 6) inhibit purified recombinant LSD1 by 80.5%, 82.9%, 75.2%, and 73.9% at a 10 μM concentration, respectively. Although the inhibitory potencies of **25–27** remain constant, they have dramatically different effects on global H3K4 methylation, producing 17.4-, 3.5-, and 2.4-fold elevation of cellular H3K4me2 in the HCT116 cell line. In addition, **25–27** promoted significant dose-dependent increases in the SFRP2 signaling protein (1.3- to 4.8-fold), but not in the GATA4 transcription factor, in the Calu-6 human lung carcinoma cell line following 24-h treatment.

The number of target enzymes involved in epigenetic regulation of transcription will continue to increase as arginine demethylases and additional JmjC demethylases continue to be identified, and as new cellular roles for LSD1 are elucidated [104–106]. Interestingly, it has recently been shown that histone 3 arginine 2 acts as an additional epigenetic control point at the 3′ end of moderately transcribed genes, but to date a specific arginine demethylase has not been identified [107]. A second isoform of lysine-specific demethylase, LSD2, has recently been described [108]. LSD2 appears to be specific for H3K4me1 and H3K4me2, but does not bind to CoRest [108]. LSD2 contains a SWi3p/Rsc8p/Moira (SWIRM)-type zinc-finger motif with zinc-binding sites not present in LSD1. These data indicate that LSD2 participates in chromatin remodeling that is distinct from

LSD1. A recent study identified LSD2 inhibitors that exhibited a 10–55-fold selectivity for LSD2 over LSD1, but that also potently inhibited MAO A and MAO B [109]. As specific inhibitors for individual demethylases and their isoforms are identified, it will be possible to elucidate the role that each enzyme plays in the epigenetic control of gene expression.

4. FUTURE DIRECTIONS AND CONCLUSION

The success of HDAC inhibitors in the treatment of cancer underscores the importance of epigenetic changes in tumorigenesis. These agents cause the reexpression of aberrantly silenced gene products that restore a "normal" cell cycle in tumor cells and make them more susceptible to apoptosis. For this reason, HDAC inhibitors are used in the clinic in combination with traditional DNA methyltransferase inhibitors, such as 5-azacytidine. The presence of multiple HDAC isoforms and the interrelationship between HDACs and CpG island methylation at specific promoters contribute to difficulties in understanding the complexity of gene expression control mechanisms. Recent drug discovery efforts in this area have focused on the identification of isoform-specific inhibitors, but an incomplete understanding of the role of the individual HDAC isoforms in tumorigenesis complicates the discovery process. Continued research will result in the discovery of isoform-specific HDAC inhibitors and could result in clinical indications for HDAC inhibitors in other disease states. The discovery of agents that modulate histone methylation and demethylation is in its infancy. These processes appear to be ubiquitous epigenetic control mechanisms; and histone methylases and demethylases will soon become drug targets of equal significance to HDACs. A number of groups have recognized this potential, and lead compounds have already begun to appear in the literature. Epigenetic control, via histone methylation/demethylation, adds a significant degree of complexity to the process of gene expression, but also provides numerous targets for the modulation of specific genes that are important in cancer, and perhaps other diseases. As such, systematic drug discovery will likely result in the generation of important new therapeutic agents.

REFERENCES

[1] P. A. Marks, V.M. Richon, R. Breslow, and R. A. Rifkind, *Curr. Opin. Oncol.*, 2001, **13**, 477.
[2] K. Luger, A.W. Mader, R. K. Richmond, D. F. Sargent, and T. J. Richmond, *Nature*, 1997, **389**, 251.
[3] J. Zlatanova and S. H. Leuba, *J. Mol. Biol.*, 2003, **331**, 1.
[4] T. Jenuwein and C.D. Allis, *Science*, 2001, **293**, 1074.
[5] R. W. Johnstone, *Nat. Rev. Drug Discov.*, 2002, **1**, 287.

[6] T. Kouzarides, *Cell*, 2007, **128**, 693.
[7] S. C. Hodawadekar and R. Marmorstein, *Oncogene*, 2007, **26**, 5528.
[8] R. C. Trievel, J. R. Rojas, D. E. Sterner, R. N. Venkataramani, L. Wang, J. Zhou, C. D. Allis, S. L. Berger, and R. Marmorstein, *Proc. Natl. Acad. Sci. USA*, 1999, **96**, 8931.
[9] F. J. Dekker and H.J. Haisma, *Drug Discov. Today*, 2009, **14**, 942.
[10] L. Stimson, M. G. Rowlands, Y. M. Newbatt, N. F. Smith, F. I. Raynaud, P. Rogers, V. Bavetsias, S. Gorsuch, M. Jarman, A. Bannister, T. Kouzarides, E. McDonald, P. Workman, and G.W. Aherne, *Mol. Cancer Ther.*, 2005, **4**, 1521.
[11] M. Biel, A. Kretsovali, E. Karatzali, J. Papamatheakis, and A. Giannis, *Angew. Chem. Int. Ed. Engl.*, 2004, **43**, 3974.
[12] J. G. Herman and S. B. Baylin, *N. Engl. J. Med.*, 2003, **349**, 2042.
[13] K. D. Robertson, *Oncogene*, 2001, **20**, 3139.
[14] M. Shogren-Knaak, H. Ishii, J. -M. Sun, M.J. Pazin, J. R. Davie, and C. L. Peterson, *Science*, 2006, **311**, 844.
[15] M. Haberland, R.L. Montgomery, and E.N. Olson, *Nat. Rev. Genet.*, 2009, **10**, 32.
[16] S. G. Gray and T. J. Ekstrom, *Exp. Cell Res.*, 2001, **262**, 75.
[17] J. Taunton, C. A. Hassig, and S. L. Schreiber, *Science*, 1996, **272**, 408.
[18] C. M. Grozinger and S. L. Schreiber, *Chem. Biol.*, 2002, **9**, 3.
[19] B. J. North and E. Verdin, *Genome Biol.*, 2004, **5**, 224.
[20] P. Marks, R. A. Rifkind, V. M. Richon, R. Breslow, T. Miller, and W. K. Kelly, *Nat. Rev. Cancer*, 2001, **1**, 194.
[21] H. Weinmann and E. Ottow, *Ann. Rep. Med. Chem.*, 2004, **39**, 185.
[22] G. Kouraklis and S. Theocharis, *Oncol. Rep.*, 2006, **15**, 489.
[23] E. E. Cameron, K. E. Bachman, S. Myohanen, J. G. Herman, and S. B. Baylin, *Nat. Genet.*, 1999, **21**, 103.
[24] J. G. Herman, C.I. Civin, J. P. Issa, M. I. Collector, S. J. Sharkis, and S. B. Baylin, *Cancer Res.*, 1997, **57**, 837.
[25] J. G. Herman, J. Jen, A. Merlo, and S. B. Baylin, *Cancer Res.*, 1996, **56**, 722.
[26] P. G. Corn, B. D. Smith, E. S. Ruckdeschel, D. Douglas, S. B. Baylin, and J. G. Herman, *Clin. Cancer Res.*, 2000, **6**, 4243.
[27] Q. C. Ryan, D. Headlee, M. Acharya, A. Sparreboom, J. B. Trepel, J. Ye, W. D. Figg, K. Hwang, E. J. Chung, A. Murgo, G. Melillo, Y. Elsayed, M. Monga, M. Kalnitskiy, J. Zwiebel, and E. A. Sausville, *J. Clin. Oncol.*, 2005, **23**, 3912.
[28] S. Balasubramanian, E. Verner, and J. J. Buggy, *Cancer Lett.*, 2009, **280**, 211.
[29] E. Hu, Z. Chen, T. Fredrickson, Y. Zhu, R. Kirkpatrick, G. -F. Zhang, K. Johanson, C. -M. Sung, R. Liu, and J. Winkler, *J. Biol. Chem.*, 2000, **275**, 15254.
[30] E. Hu, E. Dul, C. -M. Sung, Z. Chen, R. Kirkpatrick, G. -F. Zhang, K. Johanson, R. Liu, A. Lago, G. Hofmann, R. Macarron, M. De Los Frailes, P. Perez, J. Krawiec, J. Winkler, and M. Jaye, *J. Pharmacol. Exp. Ther.*, 2003, **307**, 720.
[31] R. Ficner, *Curr. Top. Med. Chem.*, 2009, **9**, 235.
[32] H. M. Prince, M. J. Bishton, and S. J. Harrison, *Clin. Cancer Res.*, 2009, **15**, 3958.
[33] K. T. Smith and J. L. Workman, *Int. J. Biochem. Cell Biol.*, 2009, **41**, 21.
[34] J. Tan, S. Cang, Y. Ma, R. L. Petrillo, and D. Liu, *J. Hematol. Oncol.*, 2010, **3**, doi:10.1186/1756-8722-3-5.
[35] H. Wang and B. W. Dymock, *Expert Opin. Ther. Patent*, 2009, **19**, 1727.
[36] E. Pontiki and D. Hadjipavlou-Litina, *Med. Res. Rev.*, 2010, DOI 10.1002/med.20200.
[37] D. Brittain, H. Weinmann, and E. Ottow, *Ann. Rep. Med. Chem.*, 2007, **42**, 337.
[38] L. Paoluzzi, L. Scotto, E. Marchi, J. Zain, V.E. Seshan, and O.A. O–Connor, *Clin. Cancer Res.*, 2010, **16**, 554.
[39] W. Zhang, M. Peyton, Y. Xie, J. Soh, J. D. Minna, A. F. Gazdar, and E. P. Frenkel, *J. Thorac. Oncol.*, 2009, **4**, 161.
[40] Y. Chen, R. He, M. A. D'Annibale, B. Langley, and A. P. Kozikowski, *ChemMedChem*, 2009, **4**, 842.

[41] S. Manku, M. Allan, N. Nguyen, A. Ajamian, J. Rodrigue, E. Therrien, J. Wang, T. Guo, J. Rahil, A .J. Petschner, A. Nicolescu, S. Lefebvre, Z. Li, M. Fournel, J.M. Besterman, R. Deziel, and A. Wahhab, *Bioorg. Med. Chem. Lett.*, 2009, **19**, 1866.

[42] S. Varghese, D. Gupta, T. Baran, A. Jiemjit, S.D. Gore, R. A. Casero, Jr., and P. M. Woster, *J. Med. Chem.*, 2005, **48**, 6350.

[43] S. Varghese, T. Senanayake, T. Murray-Stewart, K. Doering, A. Fraser, R. A. Casero, and P. M. Woster, *J. Med. Chem.*, 2008, **51**, 2447.

[44] S. Osada, S. Sano, M. Ueyama, Y. Chuman, H. Kodama, and K. Sakaguchi, *Bioorg. Med. Chem.*, 2010, **18**, 605.

[45] N. Suzuki, T. Suzuki, Y. Ota, T. Nakano, M. Kurihara, H. Okuda, T. Yamori, H. Tsumoto, H. Nakagawa, and N. Miyata, *J. Med. Chem.*, 2009, **52**, 2909.

[46] D. Desai, U. Salli, K. E. Vrana, and S. Amin, *Bioorg. Med. Chem. Lett.*, 2010, **20**, 2044.

[47] V. Patil, W. Guerrant, P. C. Chen, B. Gryder, D.B. Benicewicz, S. I. Khan, B. L. Tekwani, and A. K. Oyelere, *Bioorg. Med. Chem.*, 2010, **18**, 415.

[48] S. Agbor-Enoh, C. Seudieu, E. Davidson, A. Dritschilo, and M. Jung, *Antimicrob. Agents Chemother.*, 2009, **53**, 1727.

[49] L. Susick, T. Senanayake, R. Veluthakal, P. M. Woster, and A. Kowluru, *J. Cell Mol. Med.*, 2009, **13**, 1877.

[50] S. G. Gray and P. De Meyts, *Diabetes Metab. Res. Rev.*, 2005, **21**, 416.

[51] S. E. Wardell, O.R. Ilkayeva, H. L. Wieman, D. E. Frigo, J. C. Rathmell, C. B. Newgard, and D. P. McDonnell, *Mol. Endocrinol.*, 2009, **23**, 388.

[52] N. A. Shein, N. Grigoriadis, A. G. Alexandrovich, C. Simeonidou, A. Lourbopoulos, E. Polyzoidou, V. Trembovler, P. Mascagni, C. A. Dinarello, and E. Shohami, *FASEB J.*, 2009, **23**, 4266.

[53] J. M. Gottesfeld and M. Pandolfo, *Future Neurol.*, 2009, **4**, 775.

[54] H. E. Covington, 3rd, I. Maze, Q.C. LaPlant, V. F. Vialou, Y. N. Ohnishi, O. Berton, D. M. Fass, W. Renthal, A.J. Rush, 3rd, E.Y. Wu, S. Ghose, V. Krishnan, S. J. Russo, C. Tamminga, S.J. Haggarty, and E.J. Nestler, *J. Neurosci.*, 2009, **29**, 11451.

[55] S. B. Baylin and J. E. Ohm, *Nat. Rev. Cancer*, 2006, **6**, 107.

[56] P. A. Jones and S. B. Baylin, *Cell*, 2007, **128**, 683.

[57] B. C. Smith and J. M. Denu, *Biochim. Biophys. Acta (BBA)Gene Regul. Mech.*, 2009, **1789**, 45.

[58] A. J. Bannister and T. Kouzarides, *Nature*, 2005, **436**, 1103.

[59] R. Schneider, A. J. Bannister, and T. Kouzarides, *Trends Biochem. Sci.*, 2002, **27**, 396.

[60] A. Spannhoff, A. T. Hauser, R. Heinke, W. Sippl, and M. Jung, *ChemMedChem*, 2009, **4**, 1568.

[61] X. Chen, F. Niroomand, Z. Liu, A. Zankl, H.A. Katus, L. Jahn, and C. P. Tiefenbacher, *Basic Res. Cardiol.*, 2006, **101**, 346.

[62] Q. Feng, H. Wang, H. H. Ng, H. Erdjument-Bromage, P. Tempst, K. Struhl, and Y. Zhang, *Curr. Biol.*, 2002, **12**, 1052.

[63] Y. Zhang and D. Reinberg, *Genes Dev.*, 2001, **15**, 2343.

[64] C. D. Krause, Z. -H. Yang, Y.-S. Kim, J. -H. Lee, J. R. Cook, and S. Pestka, *Pharmacol. Ther.*, 2007, **113**, 50.

[65] T. Kouzarides, *Curr. Opin. Genet. Dev.*, 2002, **12**, 198.

[66] C. Martin and Y. Zhang, *Nat. Rev. Mol. Cell Biol.*, 2005, **6**, 838.

[67] D. Greiner, T. Bonaldi, R. Eskeland, E. Roemer, and A. Imhof, *Nat. Chem. Biol.*, 2005, **1**, 143.

[68] S. Kubicek, R. J. O'Sullivan, E. M. August, E. R. Hickey, Q. Zhang, M. L. Teodoro, S. Rea, K. Mechtler, J. A. Kowalski, C. A. Homon, T. A. Kelly, and T. Jenuwein, *Mol. Cell*, 2007, **25**, 473.

[69] D. Cheng, N. Yadav, R.W . King, M. S. Swanson, E. J. Weinstein, and M. T. Bedford, *J. Biol. Chem.*, 2004, **279**, 23892.

[70] A. Mai, D. Cheng, M. T. Bedford, S. Valente, A. Nebbioso, A. Perrone, G. Brosch, G. Sbardella, F. De Bellis, M. Miceli, and L. Altucci, *J. Med. Chem.*, 2008, **51**, 2279.

[71] A. Spannhoff, R. Machmur, R. Heinke, P. Trojer, I. Bauer, G. Brosch, R. Schule, W. Hanefeld, W. Sippl, and M. Jung, *Bioorg. Med. Chem. Lett.*, 2007, **17**, 4150.

[72] W. Fiskus, Y. Wang, A. Sreekumar, K.M. Buckley, H. Shi, A. Jillella, C. Ustun, R. Rao, P. Fernandez, J. Chen, R. Balusu, S. Koul, P. Atadja, V.E. Marquez, and K.N. Bhalla, *Blood*, 2009, **114**, 2733.

[73] Y. Huang, E. Greene, T. Murray Stewart, A.C. Goodwin, S. B. Baylin, P. M. Woster, and R. A. Casero, Jr, *Proc. Natl. Acad. Sci. USA*, 2007, **104**, 8023.

[74] Y. Shi, F. Lan, C. Matson, P. Mulligan, J. R. Whetstine, P. A. Cole, R. A. Casero, and Y. Shi, *Cell* 2004, **119**, 941.

[75] Y. Tsukada and Y. Zhang, *Methods*, 2006, **40**, 318.

[76] M. Huarte, F. Lan, T. Kim, M.W. Vaughn, M. Zaratiegui, R.A. Martienssen, S. Buratowski, and Y. Shi, *J. Biol. Chem.*, 2007, **282**, 21662.

[77] J. R. Whetstine, A. Nottke, F. Lan, M. Huarte, S. Smolikov, Z. Chen, E. Spooner, E. Li, G. Zhang, M. Colaiacovo, and Y. Shi, *Cell*, 2006, **125**, 467.

[78] J. Huang, R. Sengupta, A. B. Espejo, M. G. Lee, J. A. Dorsey, M. Richter, S. Opravil, R. Shiekhattar, M. T. Bedford, T. Jenuwein, and S.L. Berger, *Nature*, 2007, **449**, 105.

[79] T. B. Nicholson and T. Chen, *Epigenetics*, 2009, **4**, 129.

[80] W. W. Tsai, T. T. Nguyen, Y. Shi, and M. C. Barton, *Mol. Cell Biol.*, 2008, **28**, 5139.

[81] J. Wang, S. Hevi, J.K. Kurash, H. Lei, F. Gay, J. Bajko, H. Su, W. Sun, H. Chang, G. Xu, F. Gaudet, E. Li, and T. Chen, *Nat. Genet.*, 2009, **41**, 125.

[82] M. G. Lee, C. Wynder, D.M. Schmidt, D.G. McCafferty, and R. Shiekhattar, *Chem. Biol.*, 2006, **13**, 563.

[83] J. C. Culhane, L.M. Szewczuk, X. Liu, G. Da, R. Marmorstein, and P. A. Cole, *J. Am. Chem. Soc.*, 2006, **128**, 4536.

[84] P. Stavropoulos and A. Hoelz, *Expert Opin. Ther. Targets*, 2007, **11**, 809.

[85] P. Stavropoulos, G. Blobel, and A. Hoelz, *Nat. Struct. Mol. Biol.*, 2006, **13**, 626.

[86] M. G. Lee, C. Wynder, N. Cooch, and R. Shiekhattar, *Nature*, 2005, **437**, 432.

[87] F. Forneris, C. Binda, A. Adamo, E. Battaglioli, and A. Mattevi, *J. Biol. Chem.*, 2007, **282**, 20070.

[88] X. Tian and J. Fang, *Acta Biochim. Biophys. Sin (Shanghai)*, 2007, **39**, 81.

[89] E. Metzger, M. Wissmann, N. Yin, J. M. Muller, R. Schneider, A. H. Peters, T. Gunther, R. Buettner, and R. Schule, *Nature*, 2005, **437**, 436.

[90] N. Tochio, T. Umehara, S. Koshiba, M. Inoue, T. Yabuki, M. Aoki, E. Seki, S. Watanabe, Y. Tomo, M. Hanada, M. Ikari, M. Sato, T. Terada, T. Nagase, O. Ohara, M. Shirouzu, A. Tanaka, T. Kigawa, and S. Yokoyama, *Structure*, 2006, **14**, 457.

[91] G. Da, J. Lenkart, K. Zhao, R. Shiekhattar, B. R. Cairns, and R. Marmorstein, *Proc. Natl. Acad. Sci. USA*, 2006, **103**, 2057.

[92] E. Metzger, M. Wissmann, and R. Schule, *Curr. Opin. Genet. Dev.*, 2006, **16**, 513–517.

[93] P. Kahl, L. Gullotti, L.C. Heukamp, S. Wolf, N. Friedrichs, R. Vorreuther, G. Solleder, P.J. Bastian, J. Ellinger, E. Metzger, R. Schule, and R. Buettner, *Cancer Res.*, 2006, **66**, 11341.

[94] J. H. Schulte, S. Lim, A. Schramm, N. Friedrichs, J. Koster, R. Versteeg, I. Ora, K. Pajtler, L. Klein-Hitpass, S. Kuhfittig-Kulle, E. Metzger, R. Schule, A. Eggert, R. Buettner, and J. Kirfel, *Cancer Res.*, 2009, **69**, 2065.

[95] S. Lim, A. Janzer, A. Becker, A. Zimmer, R. Schule, R. Buettner, and J. Kirfel, *Carcinogenesis*, 2010, **31**, 512.

[96] D. M. Schmidt and D. G. McCafferty, *Biochemistry*, 2007, **46**, 4408.

[97] L. M. Szewczuk, J. C. Culhane, M. Yang, A. Majumdar, H. Yu, and P. A. Cole, *Biochemistry*, 2007, **46**, 6892.

[98] J. C. Culhane, D. Wang, P. M. Yen, and P. A. Cole, *J. Am. Chem. Soc.*, 2010, **132**, 3164.

[99] D. M. Gooden, D. M. Schmidt, J. A. Pollock, A. M. Kabadi, and D. G. McCafferty, *Bioorg. Med. Chem. Lett.*, 2008, **18**, 3047.

[100] R. Ueda, T. Suzuki, K. Mino, H. Tsumoto, H. Nakagawa, M. Hasegawa, R. Sasaki, T. Mizukami, and N. Miyata, *J. Am. Chem. Soc.*, 2009, **131**, 17536.

[101] Y. Wang, T. Murray-Stewart, W. Devereux, A. Hacker, B. Frydman, P.M. Woster, and R. A. Casero, Jr, *Biochem. Biophys. Res. Commun.*, 2003, **304**, 605.

[102] X. Bi, C. Lopez, C. J. Bacchi, D. Rattendi, and P. M. Woster, *Bioorg. Med. Chem. Lett.*, 2006, **16**, 3229.

[103] Y. Huang, T. M. Stewart, Y. Wu, S. B. Baylin, L. J. Marton, B. Perkins, R. J. Jones, P. M. Woster, and R. A. Casero, Jr., *Clin. Cancer Res.*, 2009, **15**, 7217.

[104] G. Liang, R. J. Klose, K.E. Gardner, and Y. Zhang, *Nat. Struct. Mol. Biol.*, 2007, **14**, 243.

[105] J. Secombe, L. Li, L. Carlos, and R. N. Eisenman, *Genes Dev.*, 2007, **21**, 537.

[106] L. Yu, Y. Wang, S. Huang, J. Wang, Z. Deng, Q. Zhang, W. Wu, X. Zhang, Z. Liu, W. Gong, and Z. Chen, *Cell Res.*, 2010, **20**, 166.

[107] A. Kirmizis, H. Santos-Rosa, C. J. Penkett, M. A. Singer, M. Vermeulen, M. Mann, J. Bahler, R. D. Green, and T. Kouzarides, *Nature*, 2007, **449**, 928.

[108] A. Karytinos, F. Forneris, A. Profumo, G. Ciossani, E. Battaglioli, C. Binda, and A. Mattevi, *J. Biol. Chem.*, 2009, **284**, 17775.

[109] C. Binda, S. Valente, M. Romanenghi, S. Pilotto, R. Cirilli, A. Karytinos, G. Ciossani, O. A. Botrugno, F. Forneris, M. Tardugno, D.E. Edmondson, S. Minucci, A. Mattevi, and A. Mai, *J. Am. Chem. Soc.*, 2010, **132**, 6827.

PART V:
Infectious Diseases

Editor: John Primeau
Astra Zeneca
Waltham
Massachusetts

CHAPTER 16

HIV-1 Integrase Strand Transfer Inhibitors

Brian A. Johns

1. INTRODUCTION

Human immunodeficiency virus (HIV)-1 is the etiological agent responsible for acquired immunodeficiency syndrome (AIDS). The number of individuals infected worldwide has grown to 33 million according to the most recent Joint United Nations Programme on HIV/AIDS (UNAIDS) estimates, including 1 million infections in the United States [1]. These statistics are staggering and, unfortunately, solidify HIV/AIDS as a major threat to public health globally. Additionally, concerns are emerging about the transmission of resistant viruses that are limiting patient treatment options upon initial diagnosis and are capable of causing self-sustaining epidemics [2]. Hence, there remains a need for novel agents that are orthogonal in their mechanism of action to existing antiretroviral agents. These would increase treatment regimen options during highly active antiretroviral therapy (HAART) to combat the evolving epidemic [3]. This chapter is focused on a new class of antiretroviral agents called integrase (IN) strand transfer

GlaxoSmithKline, Five Moore Drive, Research Triangle Park, NC 27709, USA

Annual Reports in Medicinal Chemistry, Volume 45
ISSN 0065-7743, DOI 10.1016/S0065-7743(10)45016-2

inhibitors that exert their pharmacological effect by preventing the incorporation of viral DNA into host chromatin. The body of this chapter is broken into three primary areas. The first section presents a brief synopsis on the biological role of DNA integration during the retroviral viral life cycle with a focus on the bond-making and -breaking biochemical steps and why these are amenable to inhibition. The second section presents the basis for a two-metal chelation pharmacophore and the key design aspects used to discover several first generation clinical candidates, including raltegravir and elvitegravir. The final section discusses the progress and supporting data for several compounds that have been designated as next generation IN strand transfer inhibitors including MK-2048, MK-0536, and S/GSK1349572.

2. HIV-1 INTEGRASE AND RETROVIRAL INTEGRATION

HIV IN is a 32-kDa protein that is responsible for the integration of reverse-transcribed viral double-stranded DNA (dsDNA) into host chromosomal DNA during the retroviral replication cycle. The IN enzyme consists of a 288 amino acid primary sequence divided into N-terminal, C-terminal, and catalytic core domains. The catalytic core domain consisting of residues 51–212 contains a highly conserved triad of carboxylate residues at positions Asp64, Asp116, and Glu152 known as a DD(35)E. motif. The DD(35)E. catalytic triad residues are essential to enzyme function and coordinate two magnesium ions which catalyze DNA phosphodiester bond breaking and formation, in two discrete steps called 3'-processing and strand transfer, respectively, during the retroviral integration sequence [4]. It is this second step that has been most amenable to inhibition thus far and the target of the molecules discussed below [5]. An excellent molecular view of the enzymatic catalysis of this process has been described by Kawasuji and coworkers [6].

3. TWO-METAL-BINDING PHARMACOPHORE AND FIRST GENERATION STRAND TRANSFER INHIBITORS

It took nearly a decade from the time the first purified IN protein was reported to the elucidation of the basic pharmacophore through the discovery of diketo acids (DKAs) and acid isostere analogs which were nearly simultaneously reported by independent groups [7]. Examples of these early generation IN inhibitors (INIs) are shown in Figure 1.

From the SAR data acquired during the DKA investigations, a model of the basic inhibitor pharmacophore consisting of a planar, metal-coordinating region and tethered hydrophobic appendage was established (Figure 2). The field quickly expanded through systematic modifications

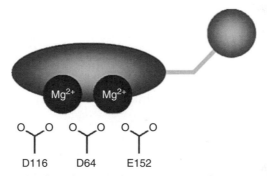

Figure 1 Diketo acid (DKA) and acid isostere analogs.

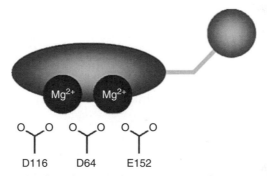

Figure 2 Two-metal-binding pharmacophore. (See Color Plate 16.2 in the Color Plate Section)

of the DKA structure using this model as a template. Much of this work has been reviewed thoroughly elsewhere [8] and will not be discussed in this chapter.

While the DKAs and their modified counterparts described above greatly advanced the field as the first INIs and provided encouraging preclinical data, they typically lacked sufficient potency when protein-binding and pharmacokinetic properties were taken into account to make good drug candidates. It was the advent of the heterocyclic DKA mimetics that maintained the key elements of the two-metal-binding pharmacophore but were more "druglike" in their structure that propelled the field from tool molecules into clinically interesting drugs.

The naphthyridine carboxamides of L-870,810 and L-870,812 (6 and 7) (Figure 3) are particularly noteworthy since they were the first INIs to

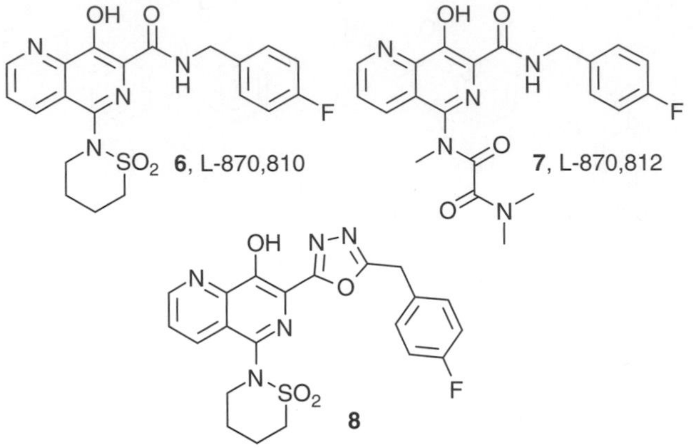

Figure 3   Naphthyridine DKA mimetics.

demonstrate efficacy in human trials and a rhesus macaques simian-human immunodeficiency virus (SHIV) model, respectively [9,10]. The sultam-containing analog L-870,810 was advanced into phase IIa studies and demonstrated a 1.7-$\log_{10}$ reduction in viral load in HIV-1-infected subjects when dosed orally up to 400 mg BID. Development of L-870,810 was discontinued due to hepatotoxicity observed in long-term safety studies. Importantly, however, clinical validation of INIs as efficacious antivirals had been established. An alternative metal chelation motif was demonstrated using five-membered heterocycles, as shown in compound **8** [11,12]. It was found that the use of either a triazole or isomeric oxadiazoles imparted potent enzyme and antiviral activity demonstrating the amide isostere's ability to participate in the two-metal-binding pharmacophore.

The naphthyridinone core has been used in combination with a benzyl carboxamide substituent to provide the pharmacophore elements of a planar chelation motif and aromatic group, as depicted in **9** (Figure 4) [13]. Presumably, the oxo group in the naphthyridinone core serves to form a hydrogen bond with the amide NH and firm up the planarity of the metal-coordinating elements. The core templates discussed thus far are generally planar and the phenolate-like central coordinating group is flanked on each side by Lewis basic groups. This subtle pseudo-C2 symmetry of the chelation moiety allows for a reversed binding mode as depicted in Figure 5 using the naphthyridinone template as an example. Slight tilting of the scaffold is necessary in order to align the requisite heteroatoms and if the benzyl carboxamide utilitized in binding mode I is repositioned as an arylmethyl substituent in the 3-position of the pyridine ring, as shown in binding mode II, there is a remarkably good overlap of the basic pharmacophore [14].

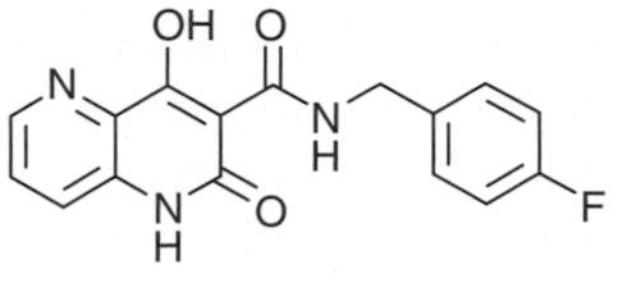

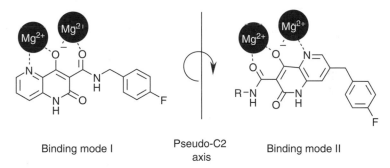

Figure 4 Naphthyridinone and tricyclic integrase inhibitor scaffolds.

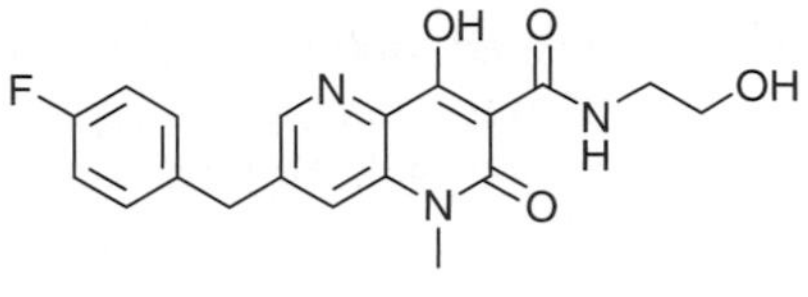

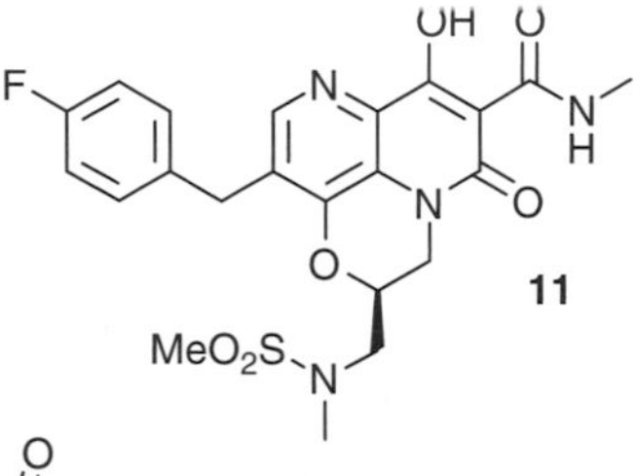

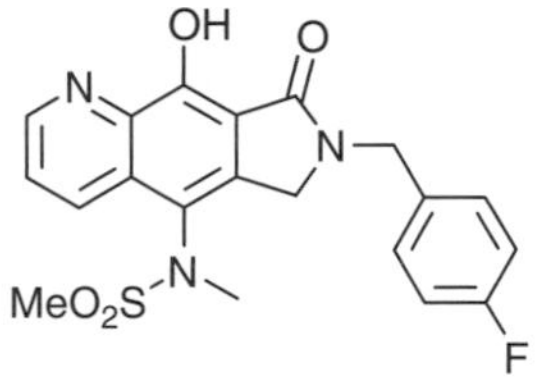

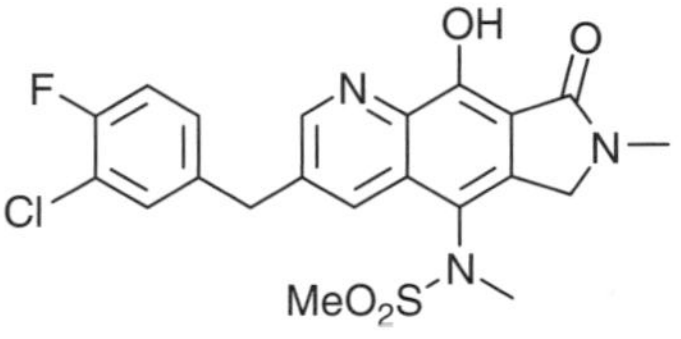

Binding mode I   Pseudo-C2 axis   Binding mode II

Figure 5 Symmetry of the planar two-metal binding motif.

This symmetry has been taken advantage of in a series of naphthyridinone analogs that utilize binding mode II. The most notable example of which is S/GSK364735 (**10**) which progressed into clinical trials (Figure 4) [15,16]. This compound is a potent antiviral (PBMC $IC_{50} = 2.1\,nM$) demonstrating a 2.2-$\log_{10}$ reduction of HIV-1 RNA levels at the highest dose examined. S/GSK364735 (**10**) was terminated shortly after the phase IIa results were obtained as a result of hepatotoxicity in long-term safety studies. Further elaboration of this naphthyridinone series has recently appeared, with structure **11** representing a very potent antiviral [17].

GS-9160 (**12**) takes advantage of a lactam ring to hold the metal-chelating groups in a coplanar orientation [18]. This tricyclic analog has low nanomolar potency in several antiviral assay systems. GS-9160 was advanced into phase I clinical trials but was terminated due to insufficient exposure to facilitate QD dosing [19]. Symmetry-reversed analogs such as **13** have also been shown to be potent antivirals ($IC_{50} = 1.1$ nM) with good pharmacokinetic properties (Clp 0.12 L/h/kg, 60% F) in dogs [20].

Pyrimidinone derivative **14** (raltegravir, MK-0518, RAL) (Figure 6) was the first INI to receive Food and Drug Administration (FDA) approval. The discovery story for RAL has been nicely documented by Rowley [21] and the focus here will be on the clinical program results to date. In a pair of identical phase III clinical studies (BENCHMRK 1 and 2), approximately 700 highly treatment-experienced patients were stratified between optimized background therapy (OBT) alone (plus placebo pill) or OBT plus 400 mg of RAL twice daily [22,23]. Subjects had documented resistance to at least one drug in each of the nucleoside reverse transcriptase inhibitor (NRTI), non-nucleoside reverse transcriptase inhibitor (NNRTI), and protease inhibitor (PI) classes of antiretrovirals. In both studies, the RAL combination arms had significantly greater ($p < 0.001$) virologic responses than placebo combination arms. At week 16, 77.5% of patients on the RAL arm had achieved a reduction in viral load to <400 copies/mL of HIV-1 RNA versus 41.9% of placebo patients. Responses of 61.8% versus 34.7% in the RAL and placebo arms, respectively, were reported to have achieved <50 copies/mL during the 16-week assessment. Similar results were maintained through week 48. A treatment-naïve study (STARTMRK) of 563 patients added 400 mg RAL BID or the NNRTI efavirenz (EFV) at a dose of 600 mg QD to a background therapy of tenofovir disoproxil and emtricitabine [24]. The primary endpoint in this study was HIV-1 RNA of <50 copies/mL. At 48 weeks, analysis showed 86.1% and 81.9% of patients had achieved the

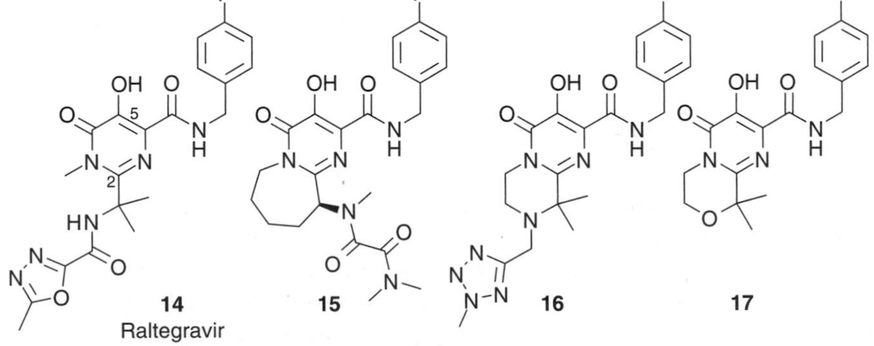

Figure 6   Raltegravir and related pyrimidinone integrase inhibitors.

primary endpoint for the RAL and EFV arms, respectively. It is note-worthy that in the RAL arm, a statistically significant shorter time to achieve viral suppression was observed along with significantly fewer adverse events than the EFV arm (124 drug-related adverse events for the RAL arm vs. 217 for the EFV arm). Serious drug-related clinical adverse events were <2% of patients in both arms. Further studies (SWITCHMRK 1 and 2) designed to compare RAL with a boosted protease inhibitor (lopi-navir/ritonavir, LPV/RTV) gave unexpected results [25]. The original goal was to determine if switching patients from a boosted PI-containing regi-men (containing two NRTIs) would sustain the suppression of viral repli-cation. The potential benefits would result from being able to remove patients from a PI regimen which can have significant lipid change side effects. Patients were randomly removed from LPV/RTV and placed on 400 mg BID RAL and at week 12 improved lipid profiles were clearly observed with a lowering of total cholesterol (−12.6% vs. 1.0%) and non-high-density lipoprotein (non-HDL) cholesterol (−15.0% vs. 2.6%) for the RAL versus LPV/RTV arms. However, at week 24 only 84.4% of patients on the RAL arm compared to 90.6% on the LPV/RTV arm achieved the virological endpoint of HIV-1 RNA <50 copies/mL. This was outside of the statistically significant noninferiority range and the study was termi-nated as a result. It has been suggested that these results can be explained by stratifying the patients into two groups to account for previous ther-apeutic failures. Patients with previous treatment failures had similar virological success rates on both arms of the study (89% and 90% in the switch and continued arms, respectively). Importantly, these results sug-gest that RAL may lose effectiveness in the setting of one major viral mutation while the PI requires more changes. At the core, this is possibly suggestive of a lower genetic barrier to resistance for the INI RAL com-pared to the higher barrier for the PI [26]. Genetic barrier to resistance in its simplest definition is the fold change in activity against successive viral mutations. Small fold changes and the need for several successive muta-tions before viral breakthrough are the key to a compound having a "high" genetic barrier to resistance [27,28]. This concept is discussed further in the context of next generation compounds at the end of the chapter.

The primary route of metabolism for RAL is via glucuronidation, primarily by UDP-glucuronosyltransferase 1A1 (UGT1A1). Little interac-tion with cytochrome P450s (CYP450s) is observed and RAL is not an inhibitor of the UGT1A1 so drug–drug interactions are not a big problem. The free fraction in rat, dog, rhesus monkey, and human are 27%, 29%, 15%, and 17%, respectively, showing that a significant amount of drug is available for target interaction. It appears that this higher free fraction is also part of the reason for a modest half-life in humans and results in a clinical twice-daily dose of 400 mg [29–33]. Extensive SAR investigations on RAL and related series have been published [34]. These primarily focus on various C2 substitutions that are distal to the metal -binding region of the molecule. There has also been an extensive investigation of additional rings between the N1 and C2 positions over the past few years, and this work continues to appear in the current literature. The azepine derivative **15** has been reported to be in clinical development; however, results from these studies have not been disclosed [35,36]. A similar modification is evident in structure **16** resulting in a very potent inhibitor even in the presence of 50% human serum ($IC_{95} = 31$ nM) [37]. A different group has reported a similar compound containing modifications of the extra ring to deliver cyclic ether **17** and related analogs. To date, this analog and its cogeners have only appeared in the patent literature but an apparent focus on crystal forms and process development suggests evidence of extensive preclinical development efforts [38].

Elvitegravir (**18**, GS-9137, JTK-303, EVG) is the second IN strand transfer inhibitor to advance into phase III clinical trials (Figure 7). EVG was derived from the quinolone antibiotics which do not show IN activity [39–41]. Through careful optimization, this work resulted in EVG displaying enzyme and antiviral activity of 7.2 and 0.9 nM, respectively. EVG has moderate bioavailability in preclinical species (29 and 34%), low clearance (Clp 0.5 and 1.0 L/h/kg) and a moderate half-life of 2.3 and 5.2 h in rats and dogs, respectively [42]. It is primarily metabolized via CYP450 oxidation and shows a marked increase in human exposure with RTV boosting.

This was confirmed in phase I and IIa clinical studies where EVG was shown to be boosted by the CYP450 inhibitor RTV. The results from a 10-day monotherapy trial in HIV-1-infected patients showed mean $\log_{10}$ viral load reductions of 1.48 (200 mg BID), 1.94 (400 mg BID), 1.91 (800 mg BID), 0.98 (800 mg QD), and 1.99 (50 mg QD plus 100 mg RTV QD) [43]. Area under the curve (AUC) and trough concentrations corresponded to the viral load changes showing clear evidence of the RTV boosting directly affecting efficacy. A phase IIb dose range-finding study designed to compare RTV-boosted EVG versus a RTV-boosted

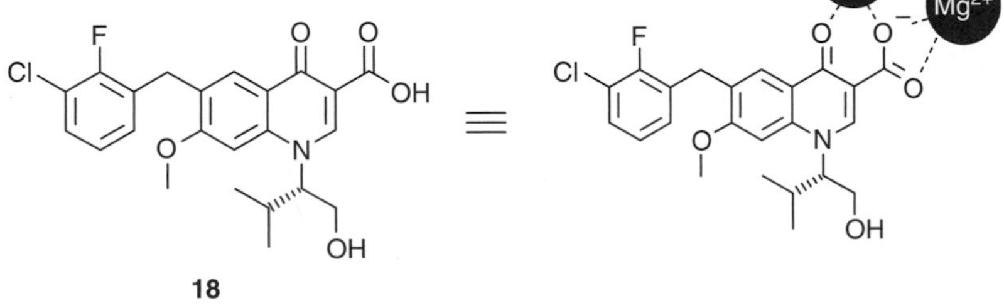

18
Elvitegravir

Figure 7   Elvitegravir.

PI for noninferiority was conducted at 20, 50, and 125 mg once-daily EVG with 100 mg RTV boosting across all doses in addition to OBT in treatment-experienced patients for 24 weeks. The 50 and 125 mg QD EVG/RTV achieved a >2 $\log_{10}$ copies/mL reduction in HIV-1 RNA in 69 and 76% of patients, respectively. The 125 mg EVG/100 mg RTV cohort was significantly different from the boosted PI arm. Phase III studies are currently underway comparing EVG/RTV-dosed QD to 400 mg RAL-dosed BID in a 700-patient antiretroviral-experienced patient population [44]. EVG is also being studied as part of a three-active drug regimen combined with a boosting agent in a fixed dose combination phase IIb study. Rather than using RTV, which is itself an early generation HIV-1 PI that has a suboptimal antiviral profile, a novel boosting agent, GS-9350 that is devoid of antiviral potency, is being developed as part of this four-agent (QUAD)-containing regimen [45]. Results from an ongoing phase IIb dose range-finding study that compared the QUAD regimen (150 mg EVG/150 mg GS-9350/tenofovir disoproxil/emtricitabine) to the triple combination regimen of Atripla (EFV/tenofovir disoproxil/emtricitabine) in treatment-naïve patients have recently been reported [46]. The 24-week primary endpoint in this study showed that 90 and 83% of patients achieved a viral load of <50 copies of HIV-1 RNA/mL for the QUAD and Atripla cohorts, respectively.

At first glance, EVG would appear to not contain a two-metal chelation motif; however, it seems plausible that the carboxylate in the deprotonated form can both mimic the phenolate-like group present in many of the previously discussed structures and play the role of the outside coordinating group and effectively coordinate as a 1,3-bidentate ligand (Figure 7). This binding mode would be consistent with the two-metal chelation pharmacophore model. It has been shown that EVG is mechanistically identical to the other classes of strand transfer inhibitors and shows cross-resistance with DKA- and RAL-derived mutations [47–50]. In addition, passage studies show that similar mutations to those reported for RAL arise from EVG-selective pressure, consistent with the two-metal-binding motif. In the clinical setting, three pathways for resistance have emerged (N155, Q148, and Y143) and appear to be similar for both EVG and RAL. These pathways toward viral escape will be important considerations for next generation inhibitors discussed below.

4. NEXT GENERATION INTEGRASE INHIBITORS

The field is now poised to move into a new phase that will address opportunities for improvement on the profiles of RAL and EVG. These "next generation" drugs will need to deliver notable advantages in resistance profile along with having a high genetic barrier to resistance [51,52].

In addition, they will almost certainly need to have low dose, once-daily pharmacokinetic profiles that do not require boosting agents. While a high genetic barrier to resistance is key to the genetic durability goal, both tolerability and patient compliance with dosing regimens are similarly important factors in managing the emergence of resistance.

Three potential next generation IN strand transfer inhibitors have recently appeared in the literature (Figure 8). Data has been presented for the fused pyrrole derivative **19** (MK-2048) that shows significant improvement in retained potency against a small panel of IN mutant viruses [53]. These include some important clinical isolates containing single or double mutants involving the Q148 and N155 residues. Some of the Q148 double and triple mutants appear to have a 10–20-fold decrease in activity compared to wild-type virus. To date, no preclinical pharmacokinetic or clinical exposure data has been presented that would establish *in vivo* efficacy or pharmacokinetic properties. A second series, exemplified by **20** (MK-0536), has also shown a very good retention of antiviral activity against the same panel of mutants [54,55]. It was reported that **20** had clearance values of 2.3 and 2.2 mL/min/kg and intravenous (IV) half-life durations of 5.7 and 4.5 h in rats and dogs, respectively. Oral bioavailability ranged from 29% in rats to 60% in dogs. As with **19**, it has not been disclosed whether **20** has been taken into clinical trials.

The recent disclosure of **21** (S/GSK1349572) as a next generation INI has been supported by preclinical data showing this compound to be a potent antiviral (PBMC $IC_{50} = 0.15$ ng/mL) with a low clearance and good oral bioavailability Cl, (%F) across species (rat 0.2 mL/min/kg, (35%), dog 2.2, (35%), cyno 2.1, (25%)) [56]. The protein-binding-adjusted

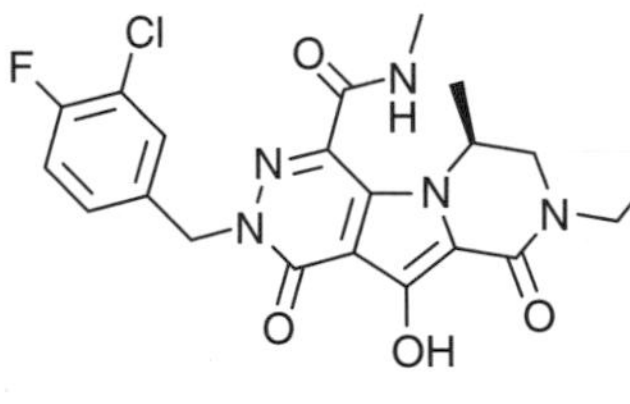

19, MK-2048

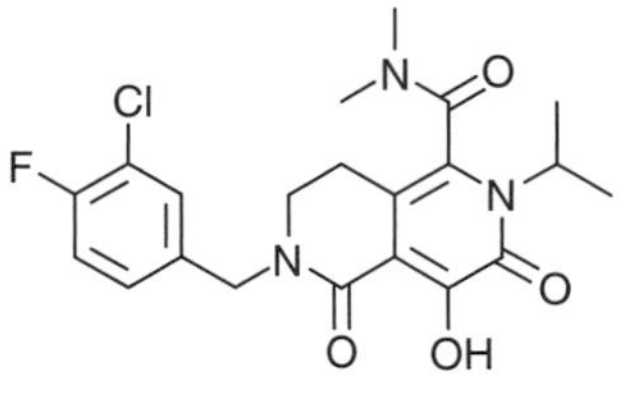

20, MK-0536

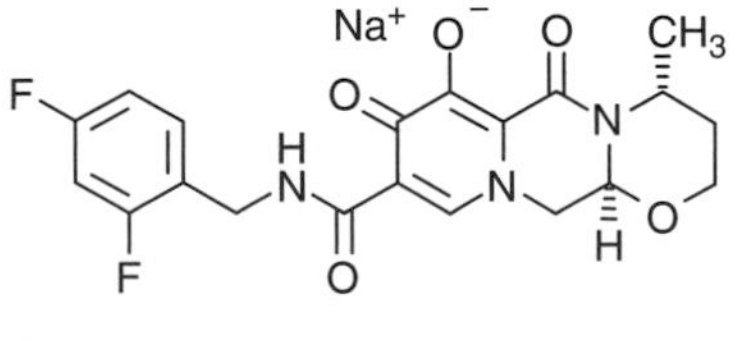

21, S/GSK1349572

Figure 8   Next generation integrase inhibitors.

IC_{90} clinical trough target concentration was reported to be 64 ng/mL. In addition, **21** showed a low fold change (<5) against a panel of 21 single mutation site-directed mutants and 18/21 double mutants [57]. Notably, it is reported that **21** has a resistance profile distinct from that of RAL and EVG resulting from *in vitro* passaging studies. These combined data demonstrate a potential for S/GSK1349572 to have a higher genetic barrier to resistance. S/GSK1349572 was reported to be well tolerated in early clinical investigations studying unboosted, once-daily single and repeat doses which resulted in a half-life of approximately 15 h [58] and exposures above the clinical target well beyond 24 h. In a phase IIa 10-day monotherapy trial, S/GSK1349572 was dosed at 2, 10, and 50 mg QD versus placebo resulting in reductions in HIV-1 RNA levels up to $2.46 \log_{10}$ copies/mL. Importantly, low variability and a predictable exposure–response relationship were observed, further demonstrating a well-behaved human pharmacokinetic profile [59]. S/GSK1349572 is currently being studied in phase IIb dose range-finding studies.

Progress has recently been reported in the area of structure elucidation of an IN/DNA complex [60]. A full-length homolog of HIV-1 IN from the prototype foamy virus (PFV) was crystallized with viral donor DNA. In addition to the binary enzyme–DNA structure, the authors were able to soak both RAL and EVG into the crystals. The details presented by the new data give a glimpse of the active site architecture and clarify some of the SAR understandings that had been empirically derived previously. Probably, the most striking detail is the formation of an inhibitor pocket through a combination of residues from the protein along with both the 3'-recessed and 5' ends of the DNA.

5. CONCLUSIONS

The evolution of HIV-1 IN strand transfer inhibitors has been one of the most significant advances in the field of HIV/AIDS chemotherapy over the past decade. The progression of scaffold designs from early DKAs to heterocyclic replacements and eventually into clinical candidates that are beginning to deliver benefit for patients is a tribute to the many researchers that appear in the accompanying citations. While the accomplishment of the first INIs has been monumental, it would be naïve to think that the field is complete and no more can be done to improve patient's lives. The emergence of IN resistance has already begun and the need for convenient dosing regimens and tolerable next generation inhibitors cannot be overstated. It appears that the field is again poised to raise the bar to a new level with exciting data for investigational drugs such as S/GSK1349572 and possibly others yet to come.

REFERENCES

[1] United Nations program on HIV/AIDS (UNAIDS), AIDS epidemic update 2008, December, 2008, available from www.UNAIDS.org.

[2] R. J. Smith, J. T. Okano, J. S. Cahn, E. N. Bodine, and S. Blower, *Science*, 2010, **327**, 697.

[3] D. Hazuda, M. Iwamoto, and L. Wenning, *Annu. Rev. Pharmacol. Toxicol.*, 2009, **49**, 377.

[4] J. A. Grobler, K. Stillmock, B. Hu, M. Witmer, P. Felock, A. S. Espeseth, A. Wolfe, M. Egbertson, M. Bourgeois, J. Melamed, J. S. Wai, S. Young, J. Vacca, and D. J. Hazuda, *Proc. Natl. Acad. Sci. USA*, 2002, **99**, 6661.

[5] M. De Vivo, M. Dal Peraro, and M. L. Klein, *J. Am. Chem. Soc.*, 2008, **130**, 10955.

[6] T. Kawasuji, M. Fuji, T. Yoshinaga, A. Sato, T. Fujiwara, and R. Kiyama, *Bioorg. Med. Chem.*, 2006, **14**, 8420.

[7] A. Savarino, *Expert Opin. Invest. Drugs*, 2006, **15**, 1507.

[8] B. A. Johns and A. C. Svolto, *Expert Opin. Ther. Patents*, 2008, **18**, 1225.

[9] D. J. Hazuda, S. D. Young, J. P. Guare, N. J. Anthony, R. P. Gomez, J. S. Wai, J. P. Vacca, L. Handt, S. L. Motzel, H. J. Klein, G. Dornadula, R. M. Danovich, M. V. Witmer, K. A. A. Wilson, L. Tussey, W. A. Schleif, L. S. Gabryelski, L. Jin, M. D. Miller, D. R. Casimiro, E. A. Emini, and J. W. Shiver, *Science*, 2004, **305**, 528.

[10] S. Little, G. Drusano, R. Schooley, D. Haas, P. Kumar, S. Hammer, D. McMahon, K. Squires, R. Asfour, D. Richman, J. Chen, A. Saah, R. Leavitt, D. Hazuda, and B. Y. Nguyen, Abstract 161, 12th Conference of Retroviruses and Opportunistic Infections, Boston, MA, February, 2005.

[11] B. A. Johns, J. G. Weatherhead, S. H. Allen, J. B. Thompson, E. P. Garvey, S. A. Foster, J. L. Jeffrey, and W. H. Miller, *Bioorg. Med. Chem. Lett.*, 2009, **19**, 1802.

[12] B. A. Johns, J. G. Weatherhead, S. H. Allen, J. B. Thompson, E. P. Garvey, S. A. Foster, J. L. Jeffrey, and W. H. Miller, *Bioorg. Med. Chem. Lett.*, 2009, **19**, 1807.

[13] M. Egbertson, J. Y. Melamed, H. M. Langford, and S. D. Young, *Patent application WO 2003/062204*, 2003.

[14] E. E. Boros, C. E. Edwards, S. A. Foster, M. Fuji, T. Fujiwara, E. P. Garvey, P. L. Golden, R. J. Hazen, J. L. Jeffrey, B. A. Johns, T. Kawasuji, R. Kiyama, C. S. Koble, N. Kurose, W. H. Miller, A. L. Mote, H. Murai, A. Sato, J. B. Thompson, M. C. Woodward, and T. Yoshinaga, *J. Med. Chem.*, 2009, **52**, 2754.

[15] E. P. Garvey, B. A. Johns, M. Gartland, S. A. Foster, W. H. Miller, R. G. Ferris, R. J. Hazen, M. R. Underwood, E. E. Boros, J. B. Thompson, J. G. Weatherhead, C.S. Koble, S. H. Allen, L. T. Schaller, R. G. Sherrill, T. Yoshinaga, M. Kobayashi, C. Wakasa-Morimoto, S. Miki, K. Nakahara, T. Noshi, A. Sato, and T. Fujiwara, *Antimicrob. Agents Chemother.*, 2008, **52**, 901.

[16] Y. S. Reddy, S. S. Min, J. Borland, I. Song, J. Lin, S. Palleja, and W. T. Symonds, *Antimicrob. Agents Chemother.*, 2007, **51**, 4284.

[17] E. Aktoudi-anakis, A. A. Celebi, Z. Du, S. Y. Jabri, H. Jin, C.U. Kim, J. Li, S. E. Metobo, M. Mish, B. W. Phillips, J. H. Saugier, Z. -Y. Yang, and C. S. Zonte, *Patent application WO 2010/011959-A1*, 2010.

[18] H. Jin, M. Wright, R. Pastor, M. Mish, S. Metobo, S. Jabri, R. Lansdown, R. Cai, P. Pyun, M. Tsiang, X. Chen, and C. U. Kim, *Bioorg. Med. Chem. Lett.*, 2008, **18**, 1388.

[19] G. S. Jones, F. Yu, A. Zeynalzadegan, J. Hesselgesser, X. Chen, J. Chen, H. Jin, C. U. Kim, M. Wright, R. Geleziunas, and M. Tsiang, *Antimicrob. Agents Chemother.*, 2009, **53**, 1194.

[20] S. Metobo, M. Mish, H. Jin, S. Jabri, R. Lansdown, X. Chen, M. Tsiang, M. Wright, and C. U. Kim, *Bioorg. Med. Chem. Lett.*, 2009, **19**, 1187.

[21] M. Rowley, *Prog. Med. Chem.*, 2008, **46**, 1.

[22] R. A. Steigigel, D. A. Cooper, P. N. Kumar, J. E. Enron, M. Schecter, M. Markowitz, M. R. Loutfy, J. L. Lennox, J. M. Gatell, J. K. Rockstroh, C. Katlama, P. Yenio, A. Lazzarin, B. Clotet, J. Zhao, J. Chen, D. M. Ryan, R. R. Rhodes, J. A. Killar, L. R. Gilde,

K. M. Strohmaier, A. R. Meibohm, M. D. Miller, D. J. Hazuda, M. L. Nessly, M. J. DiNubile, R. D. Isaacs, B. -Y. Nguyen, and H. Teppler, *New Engl. J. Med.*, 2008, **359**, 339.

[23] D. A. Cooper, R. T. Steigbigel, J. M. Gatell, J. K. Rockstroh, C. Katlama, P. Yeni, A. Lazzarin, B. Clotet, P. N. Kumar, J. E. Enron, M. Schechter, M. Markowitz, M. R. Loutfy, J. L. Lennox, J. Zhao, J. Chen., D. M. Ryan, R. R. Rhodes, J. A. Killar, L. R. Gilde, K. M. Strohmaier, A. R. Meibohm, M. D. Miller, D. J. Hazuda, M.L. Nessly, M. J. DiNubile, R. D. Isaacs, H. Teppler, and B. -Y. Nguyen, *New Engl. J. Med.*, 2008, **359**, 355.

[24] J. L. Lennox, E. Dejesus, A. Lazzaron, R. B. Pollard, J. V. R. Madruga, D. S. Berger, J. Zhao, X. Xu, A. Williams-Diaz, A. J. Rogers, R. J. O. Barnard, M. D. Miller, M. J. DiNubile, B. -Y Nguyen, R. Leavitt, and P. Sklar, *Lancet*, 2009, **374**, 796.

[25] J. J. Enron, B. Young, D.A. Cooper, M. Youle, E. DeJesus, J. Andrade-Villanueva, C. Workman, R. Zajdenverg, G. Fatkenheuer, D. S. Berger, P. N. Kumar, A. J. Rodgers, M. A. Shaughnessy, M. L. Walker, R. J. O. Barnard, M. D. Miller, M. J. DiNubile, B. -Y. Nguyen, R. Leavitt, X. Xu, and P. Sklar, *Lancet*, 2010, **375**, 396.

[26] J. M. Kilby, *Lancet*, 2010, **375**, 353.

[27] J. L. Martinez-Cajas and M. A. Wainberg, *Antiviral Res.*, 2007, **76**, 203.

[28] S. Dandache, C. A. Coburn, M. Oliveira, T. J. Allison, M. K. Holloway, J. J. Wu, B.R. Stranix, C. Panchal, M. A. Wainberg, and J. P. Vacca, *J. Med. Virol.*, 2008, **80**, 2053.

[29] M. Iwamoto, L. A. Wenning, A. S. Petry, M. Laethem, M. De Smet, J. T. Kost, S. A. Merschman, K. M. Strohmaier, S. Ramael, K. C. Lasseter, J. A. Stone, K. M. Gottesdiener, and J. A. Wagner, *Clin. Pharm. Ther.*, 2008, **83**, 293.

[30] J. D. Croxtall, K. A. Lyseng-Williamson, and C. M. Perry, *Drugs*, 2008, **68**, 131.

[31] M. Markowitz, J. O. Morales-Ramirez, B. -Y. Nguyen, C. M. Kovacs, R. T. Steigbigel, D. A. Cooper, R. Liporace, R. Schwartz, R. Isaacs, L. R. Gilde, L. Wenning, J. Zhao, and H. Teppler, *J. Acquir. Immune Defic. Syndr.*, 2006, **43**, 509.

[32] S. G. Deeks, S. Kar, S. I. Gubernickj, and P. Kirkpatrick, *Nat. Rev.*, 2008, **7**, 117.

[33] M. Anker and R. B. Corales, *Expert Opin. Invest. Drugs*, 2008, **17**, 97.

[34] P. Pace and M. Rowley, *Curr. Opin. Drug Discov. Dev.*, 2008, **11**, 471.

[35] Y. -L. Zhong, S. W. Krska, H. Zhou, R. A. Reamer, J. Lee, Y. Sun, and D. Askin, *Org. Lett.*, 2009, **11**, 369.

[36] M. del Rosario Rico Ferreira, G. Cecere, P. Pace, and V. Summa, *Tetrahedron Lett.*, 2009, **50**, 148.

[37] A. Petrocchi, P. Jones, M. Rowley, F. Fiore, and V. Summa, *Bioorg. Med. Chem. Lett.*, 2009, **19**, 4245.

[38] D. Digiugno and B. N. Narasimhulu *Patent application WO 2007/064502-A1*, 2007.

[39] M. Sato, T. Motomura, H. Aramaki, T. Matsuda, M. Yamashita, Y. Ito, H. Kawakami, Y. Matsuzaki, W. Watanabe, K. Yamataka, S. Ikeda, E. Kodama, M. Matsuoka, and H. Shinkai, *J. Med. Chem.*, 2006, **49**, 1506.

[40] T. Kirschberg and J. Parrish, *Curr. Opin. Drug Discov. Dev.*, 2007, **10**, 460.

[41] M. Sato, H. Kawakami, T. Motomura, H. Aramaki, T. Matsuda, M. Yamashita, Y. Ito, Y. Matsuzaki, K. Yamataka, S. Ikeda, and H. Shinkai, *J. Med. Chem.*, 2009, **52**, 4869.

[42] L. A. Sorbera and N. Serradell, *Drugs Future*, 2006, **31**, 310.

[43] E. DeJesus, D. Berger, M. Markowitz, C. Cohen, T. Hawkins, P. Ruance, R. Elion, C. Farthing, L. Zhong, A. K. Cheng, D. McColl, and B. P. Kearney, *J. Acquir. Immune Defic. Syndr.*, 2006, **43**, 1.

[44] D. J. McColl and X. Chen, *Antiviral Res.*, 2010, **85**, 101.

[45] A. A. Mathias, P. German, B. P. Murray, L. Wei, A. Jain, S. West, D. Warren, J. Hui, and B. P Kearney, *Clin. Pharm. Ther.*, 2010, **87**, 322.

[46] C. Cohen, D. Shamblaw, P. Ruane, R. Elion, E. DeJesus, H. Liu, L. Zhong, D. Warren, B. Kearney, and S. Chuck, Abstract 58-LB, 17th Conference on Retroviruses and Opportunistic Infections, San Francisco, CA. February 16–19, 2010.

[47] O. Goethals, R. Clayton, M. Van Ginderen, I. Vereycken, E. Wagemans, P. Geluykens, K. Dock, R. Strijbos, V. Smits, A. Vos, G. Meersseman, D. Jochmans, K. Vermeire, D. Schols, S. Hallenberger, and K. Hertogs, *J. Virol.*, 2008, **82**, 10366.

[48] K. Shimura, E. Kodama, Y. Sakagami, Y. Matsuzaki, W. Watanabe, K. Yamataka, Y. Watanabe, Y. Ohata, S. Doi, M. Sato, M. Kano, S. Ikeda, and M. Matsuoka, *J. Virol.*, 2008, **82**, 764.

[49] J. Marinello, C. Marchand, B. T. Mott, A. Bain, C. J. Thomas, and Y. Pommier, *Biochemistry*, 2008, **47**, 9345.

[50] A. Hombrouck, A. Voet, B. Van Remoortel, C. Desadeleer, M. De Maeyer, Z. Debyser, and M. Witvrouw, *Antimicrob. Agents Chemother.*, 2008, **52**, 2069.

[51] E. Serrao, S. Odde, K. Ramkumar, and N. Neamati, *Retrovirology*, 2009, **6**, 25.

[52] A. R. D. Voet, M. De Maeyer, Z. Debyser, and F. Christ, *Future Med. Chem.*, 2009, **1**, 1259.

[53] J. Wai, T. Fisher, M. Embrey, M. Egbertson, J. Vacca, D. Hazuda, M. Miller, M. Witmer, L. Gabryelski, and T. Lyle, Abstract 87, 14th Conference on Retroviruses and Opportunistic Infections, Los Angeles, CA. February 25–28, 2007.

[54] W. Han, M. Egbertson, J. Wai, D. S. Perlow, L. S. Payne, J. Vacca, D. Hazuda, M. Miller, P. J. Felock, K. Stillmock, M.V. Witmer, L. J. Gabryelski, W. A. Schleif, J. Ellis, M. R. Anari, and T. A. Lyle, MEDI-005, 235th National Meeting of the Americal Chemical Society, New Orleans, LA, April 6–10, 2008.

[55] D. S. Perlow, M. Egbertson, J. Wai, L. S. Payne, W. Han, C. D. Martyr, V. E. Obligado, K. Hoffman, J. Vacca, D. Hazuda, P. J. Felock, K. A. Stillmock, W. A. Schleif, L. Gabryelski, M. R. Anari, J. Ellis, M. V. Witmer, M. Miller, N. N. Tsou, M. Biba, C. J. Welch, and T. A. Lyle, MEDI-055, 235th National Meeting of the Americal Chemical Society, New Orleans, LA, April 6–10, 2008

[56] B. Johns, T. Kawasuji, T. Taishi, H. Yoshida, E. Garvey, W. Spreen, M. Underwood, A. Sato, T. Yoshinaga, and T. Fujiwara, Abstract 55, 17th Conference on Retroviruses and Opportunistic Infections, San Francisco, CA. February 16–19, 2010.

[57] T. Seki, M. Kobayashi, C. Wakasa-Morimoto, T. Yoshinaga, A. Sato, T. Fujiwara, M. Underwood, E. Garvey, and B. Johns, Abstract 555, 17th Conference on Retroviruses and Opportunistic Infections, San Francisco, CA. February 16–19, 2010.

[58] S. Min, I. Song, J. Borland, S. Chen, Y. Lou, T. Fujiwara, and S.C. Piscitelli, *Antimicrob. Agents Chemother.*, 2010, **54**, 254.

[59] J. Lalezari, L. Sloan, E. DeJesus, T. Hawkins, L. McCurdy, I. Song, J. Borland, R. Stroder, S. Chen, Y. Lou, M. Underwood, T. Fujiwara, S. Piscitelli, and S. Min, Abstract TUAB105, 5th International Aids Society Conference on HIV Pathogenesis and Treatment, Cape Town, South Africa, July, 19–22, 2009.

[60] S. Hare, S. S. Gupta, E. Valkov, A. Engelman, and P. Cherepanov, *Nature*, 2010, **464**, 232.

CHAPTER 17

Recent Advances in Drug Discovery for Neglected Tropical Diseases Caused by Infective Kinetoplastid Parasites

Robert T. Jacobs[*] and Charles Ding[**]

[*] SCYNEXIS, Inc., PO Box 12878, Research Triangle Park, NC 27709-2878, USA

[**] Anacor Pharmaceuticals Inc., East Meadow Circle, Palo Alto, CA 94303, USA

Annual Reports in Medicinal Chemistry, Volume 45
ISSN 0065-7743, DOI 10.1016/S0065-7743(10)45017-4

1. INTRODUCTION

Diseases of the developing world, caused by infection with kinetoplastid parasites, affect millions of people, not only those infected, but also the communities in which they live, contributing to a perpetual cycle of poverty [1]. In sub-Saharan Africa, human African trypanosomiasis (HAT), or African sleeping sickness, is caused by *Trypanosoma brucei* spp. This disease is fatal if left untreated and is estimated to affect between 50,000 and 70,000 individuals [2]. Leishmaniases, including visceral (VL), cutaneous (CL), and mucocutaneous (MCL) forms, caused by *Leishmania* spp., are endemic across Africa, the Middle East, South America, and Asia; and affects over 2 million people annually [3]. Finally, Chagas disease, caused by *Trypanosoma cruzi*, is endemic across Central and South America and is estimated to affect nearly 10 million individuals in this region [4,5]. Due to limited resources available to be spent on drug treatments for these diseases, a dearth of research into new, safe, and effective treatments was evident throughout the end of the 20th century. Over the past decade, however, there has been an increased awareness of the impact of these diseases, and a renewed research commitment has emerged within the medicinal chemistry community, supported by a number of philanthropic organizations and public–private partnerships.

As the causative parasites for these diseases are closely related, and the genome of each has been sequenced, it is not surprising that a number of common biochemical pathways have been targeted by drug discovery efforts [6]. In addition, screening for parasiticidal activity of compounds has frequently been performed simultaneously against representative strains of the three major parasites, hence many of the chemotypes described in this chapter exhibit activity against all three.

2. HUMAN AFRICAN TRYPANOSOMIASIS

2.1 Disease background and current treatments

African sleeping sickness is endemic across sub-Saharan Africa and is fatal if untreated. The disease is transmitted through bites from infected tsetse flies, with introduction of *T. brucei* to the bloodstream. The hemolymphatic infection, or stage 1 disease, can persist for weeks or months, but parasites ultimately cross the blood–brain barrier, resulting in stage 2 (CNS) disease. While stage 1 disease can be treated with pentamidine (**1**), treatment of stage 2 disease is particularly problematic, as the only drugs that are effective are the highly toxic melarsoprol (**2**) and eflornithine (**3**), which are difficult to administer [7]. A review of drug discovery for HAT has recently been published [8].

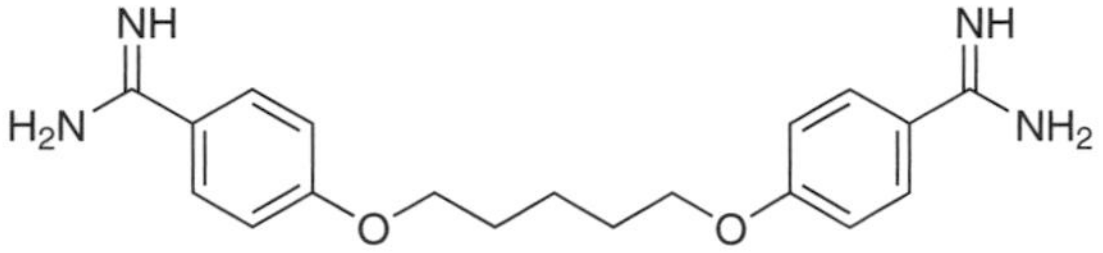

2.2 Diamidines

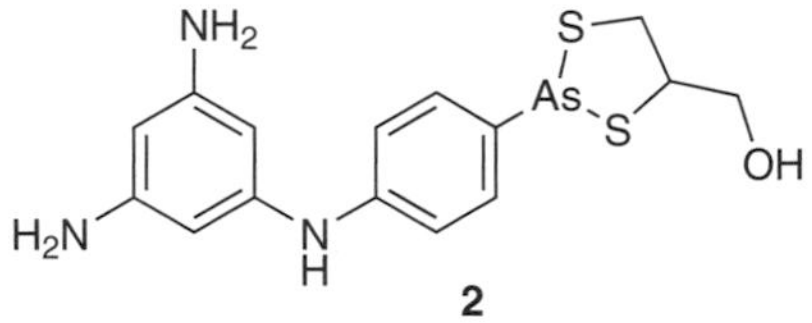

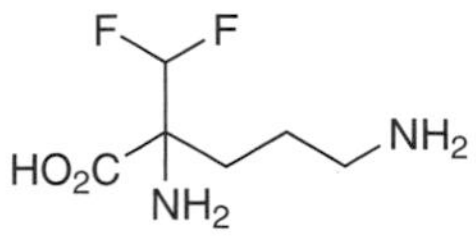

One of the most widely explored classes of compounds with robust activity against *T. brucei* spp. is the diamidine class, based on pentamidine, a compound able to kill *T. b. rhodesiense* parasites *in vitro* with an $IC_{50} = 3$ nM. A program focused on conformationally constrained analogs of pentamidine delivered DB75 (**4**, *T.b.r.* $IC_{50} = 4.5$ nM) and the related bisalkoxyamidine prodrugs, DB289 (**5**, *T.b.r.* $IC_{50} > 14$ µM) and DB377 (**6**, *T.b.r.* $IC_{50} = 645$ nM) [9]. DB289 was progressed to clinical trials on the basis of good activity in murine [10] and vervet monkey models of stage 1 HAT, where cures were observed following oral administration for 4 days at 100 mg/kg in mice and following 10 mg/kg, p.o. × 3 days in monkeys [11]. DB289 was withdrawn from clinical trials in 2009 due to liver and renal toxicity [12]. Efforts to identify additional clinical candidates from this series have delivered compounds with azaterphenyl (**7–9**, *T.b.r.* $IC_{50} = 1–10$ nM), aryloxyxylene (**10**, *T.b.r.* $IC_{50} = 46$ nM), and isoxazole (**11**, *T.b.r.* $IC_{50} = 5$ nM) linkers [13–15]. A series of diamide-linked compounds (**12**, **13**, and **14**, *T.b.r.* $IC_{50} = 2200, 96, 2$ nM, respectively) demonstrate the relationship of spatial distance between the two amidines and the activity [16].

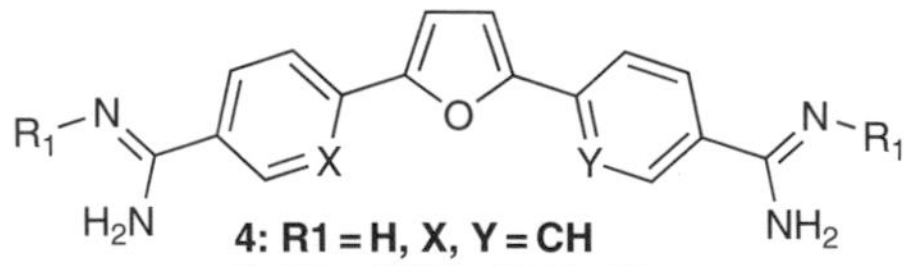

4: R1 = H, X, Y = CH
5: R1 = OCH₃, X, Y = H
6: R1 = OEt, X, Y = H
15: R1 = H, X, Y = N
16: R1 = OCH₃, X = N, Y = CH
17: R1 = OCH₃, X, Y = N

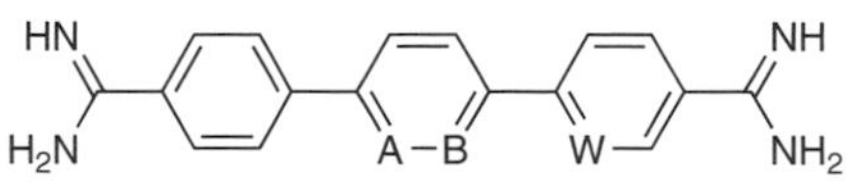

7: A, B = N; W = CH
8: W = N; A, B = CH
9: A, W = N; B = CH

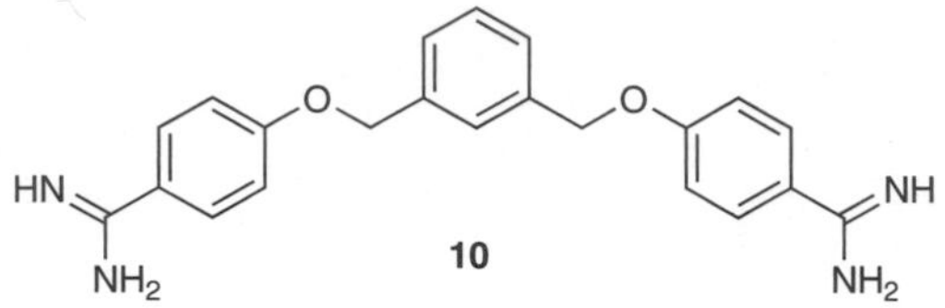

In addition to high *in vitro* trypanocidal potency, several of these series have demonstrated activity in animal models of stage 1 HAT following intraperitoneal administration. For example, compound **9** cured 4/4 mice infected with the STIB900 strain of *T. b. rhodesiense* when dosed at 5 mg/kg, i.p. × 4 days [13]. Activity in stage 2 HAT models has been more difficult to achieve, presumably due to the highly cationic nature of the diamidines, which limits their access to the CNS. This challenge has been addressed by a series of aza analogs, exemplified by DB829 (**15**), DB844 (**16**), and DB868 (**17**) [17]. These analogs exhibited activity in a murine model of stage 2 HAT, with DB844 and DB868 effecting cures in 5/5 mice treated for 5 days at a dose of 100 mg/kg, po.

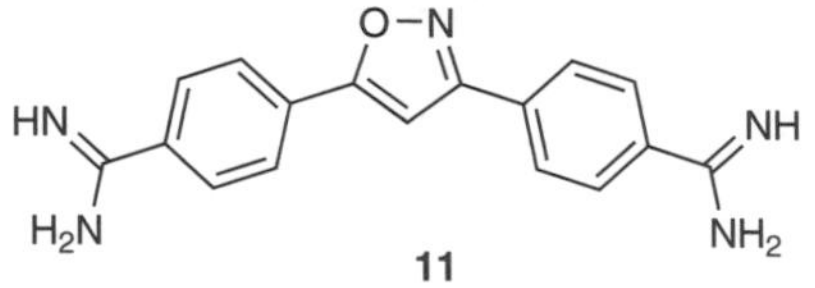

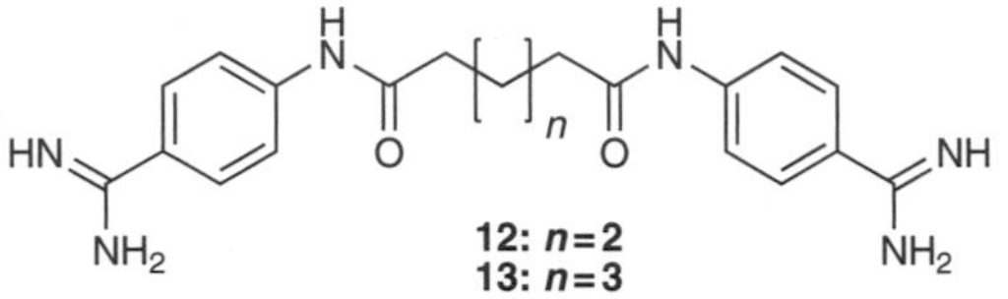

2.3 Polyamine biosynthesis inhibitors

Biosynthesis of polyamines is essential for growth and multiplication of *T. brucei*, hence discovery of drug candidates that inhibit enzymes in the polyamine biosynthesis pathway represent an attractive approach to development of trypanocides. The consequences of gene knockout of ornithine decarboxylase (ODC), the target of eflornithine (**3**), have been further characterized and suggest that new inhibitors of this enzyme may be particularly effective [18].

Analogously, inhibition of trypanothione reductase (TR) has been pursued due to the genetic validation by RNAi of this target in both *T. brucei* and *T. cruzi* [19]. The homology between the two TR enzymes (83% identity, 90% similarity) is quite high, hence it has been suggested that compounds active against either TR should exhibit parasitical activity in both species. A screen of commercially available compounds against

T. brucei TR, followed by evaluation of actives against *T.b. rhodesiense*, identified a series of tricyclic piperazines, such as **18** (TR IC$_{50}$ = 7.5 μM, *T.b.r.* IC$_{50}$ = 4.5 μM) which was optimized to **19** (TR IC$_{50}$ = 0.75 μM, *T.b.r.* IC$_{50}$ = 0.78 μM) by modification of both the tricyclic ring and the piperazine substituent [20]. Also identified in this screen were substituted piperidines, such as **20** (TR IC$_{50}$ = 3.3 μM, *T.b.r.* IC$_{50}$ = 10 μM), which was converted to **21** (TR IC$_{50}$ = 12 μM, *T.b.r.* IC$_{50}$ = 4.3 μM) by modification of the carbocyclic framework to incorporate a substituted piperidine [21]. In a similar high-throughput screen against the *T. cruzi* TR, a number of compounds, including the substituted piperidine **22** (*T.c.* TR IC$_{50}$ = 33 μM, *T.b.r.* IC$_{50}$ = 3 μM), were identified [22]. Similarly, quinazoline **23** was identified as an inhibitor of *T. cruzi* TR (K_i = 7.5 μM) which was a quite potent (IC$_{50}$ = 0.12 μM) inhibitor of *T. b. rhodesiense* growth [23].

A third enzyme in the polyamine biosynthesis pathway, trypanothione synthetase-amidase (TRYS), has been shown to be essential by both genetic and chemical methods [24]. Specifically, the indazole analog **24** has been identified in a TRYS screen (IC$_{50}$ = 140 nM) and shown to inhibit growth of wild-type (IC$_{50}$ = 5.1 μM) and TRYS-dKO (IC$_{50}$ = 0.46 mM) *T. brucei* parasites in culture [25].

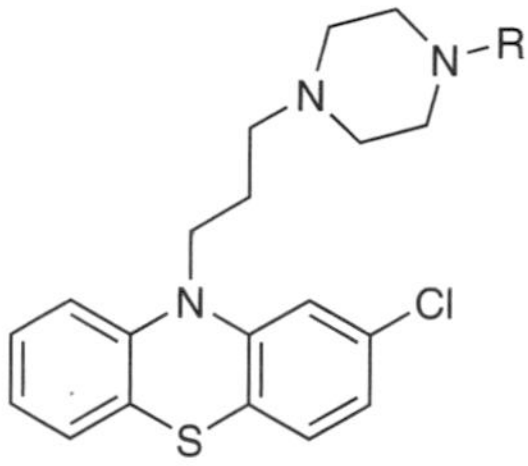

18: R = H
19: R = −(CH$_2$)$_3$Ph

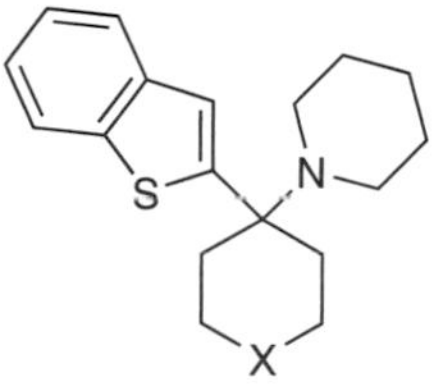

20: X = CH$_2$
21: X = NCH$_2$Ph

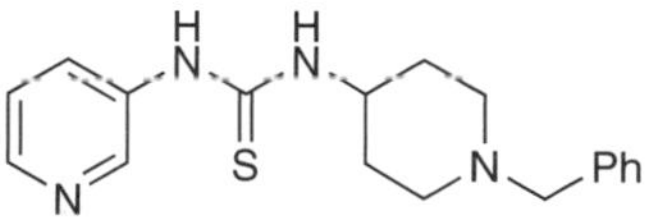

22

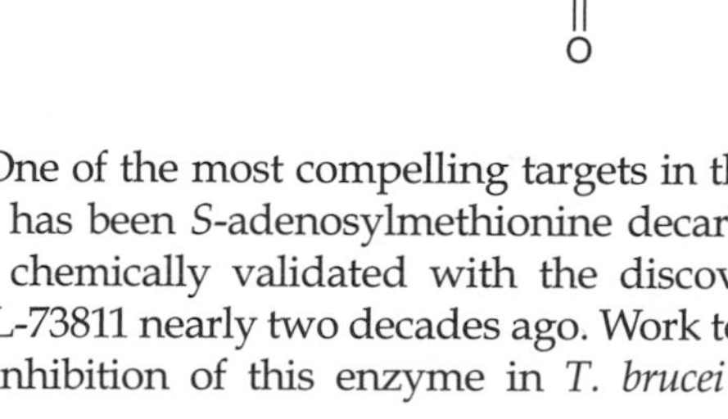

23

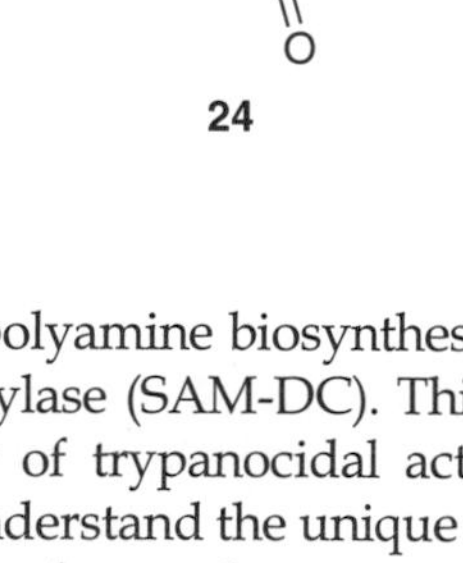

24

One of the most compelling targets in the polyamine biosynthesis pathway has been S-adenosylmethionine decarboxylase (SAM-DC). This target was chemically validated with the discovery of trypanocidal activity of MDL-73811 nearly two decades ago. Work to understand the unique kinetics for inhibition of this enzyme in *T. brucei* has shown that a catalytically inactive "prozyme" is required for functional activity of SAM-DC [26,27]. Recent efforts have delivered a series of modified purines that demonstrate improved potency and pharmacokinetics over MDL-73811 [28]. The ability to inhibit SAM-DC for these compounds, expressed as the kinetic enzyme inhibition constant k_{inact}/K_iapp, varies depending on substitution pattern (**25**: $1.50\,M^{-1}\,min^{-1}$; **26**: $0.02\,M^{-1}\,min^{-1}$; **27**: $7.78\,M^{-1}\,min^{-1}$; **28**: $0.19\,M^{-1}\,min^{-1}$) and correlates well with activity in a *T. b. rhodesiense* growth inhibition assay (*T.b.r.* IC_{50}: **25**: 0.011 μM; **26**: 1.7 μM; **27**: 0.001 μM; **28**: 1.4 μM). The 8-methyl analog **27** was evaluated for *in vivo* pharmacokinetic properties and for activity in a murine model of stage 1 HAT, where it exhibited the ability to clear parasites from the blood using a 50 mg/kg, i.p., q.d. × 7 day dosing paradigm [29]. More extensive evaluation of the *in vivo* activity of **27** demonstrated robust activity in the stage 1 model at doses as low as 2 mg/kg (i.p., q.d. × 7 days), but was only partial active in a murine model of stage 2 HAT (100 mg/kg, i.p. × 14 days) [30].

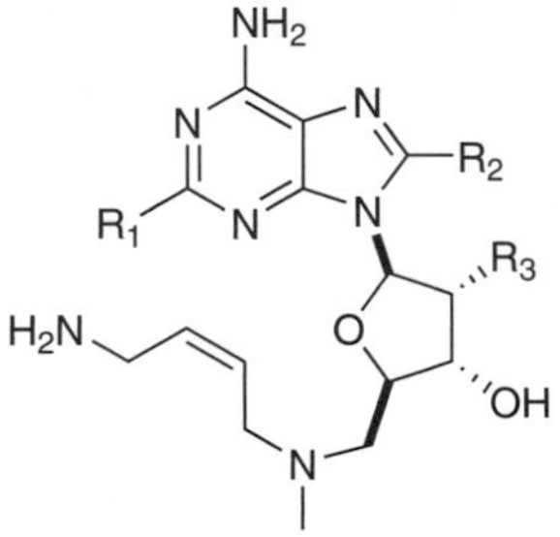

25: R^1, R^2 = H; R^3 = OH
26: R^1 = Cl; R^2 = H; R^3 = OH
27: R^1 = H; R^2 = CH$_3$; R^3 = OH
28: R^1 = H; R^2 = H; R^3 = F

2.4 Energy metabolism—glycolysis inhibition

Inhibitors of several enzymes in the glycolytic pathway, upon which survival of *T. brucei* is dependent, have been described. Lonidamine (**29**) has been shown to inhibit *T. brucei* hexokinase (IC_{50} = 850 μM) and be toxic to the parasite (*T.b.* LD_{50} = 50 μM) in culture [31]. A series of mannitol derivatives have been discovered, which inhibit *T. brucei* phosphofructokinase (TbPFK) [32]. The most potent compound (**30**) within this series exhibits an IC_{50} = 23 μM in a recombinant enzyme assay and inhibits parasite growth *in vitro* (IC_{50} = 30 μM).

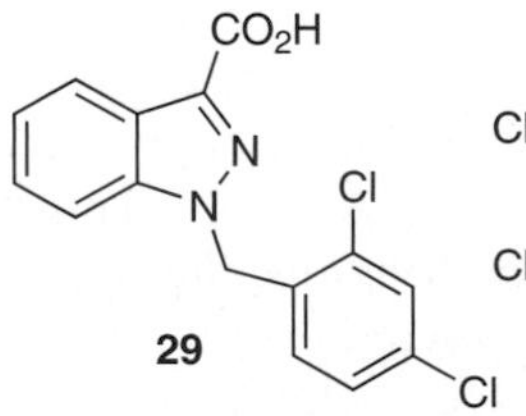

29

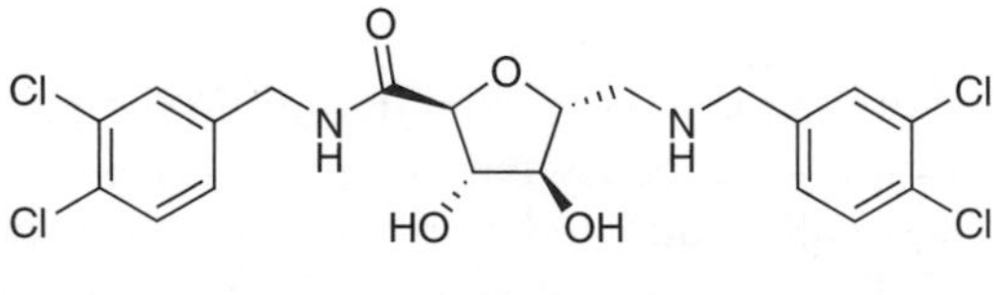

30

2.5 Signal transduction pathways—kinase inhibition

As in all eukaryotic cells, protein kinases play an important role in the life cycle of the kinetoplastids and, as such, are attractive targets. Recent efforts, predominantly through genetic (RNAi) means, have validated a number of kinases as essential for survival of *T. brucei*, but few have been explored with chemical probes [33–35].

The essentiality of glycogen synthase kinase 3 (GSK-3) to survival of *T. brucei* has been demonstrated by both RNAi and chemical means. A series of pyrimidines, represented by **31**, have been shown to be potent inhibitors of both *T. brucei* GSK-3 (IC_{50} = 4 nM) and parasite growth (*T.b.b.* IC_{50} = 50 nM) [36]. The *T. brucei* Aurora kinase (TbAUK1) has also been chemically validated as a potential target through use of human AUK inhibitors. The AUK inhibitor, VX-680 (**32**), inhibits TbAUK1 (IC_{50} = 190 nM) and disrupts cell cycle progression in the parasite [37]. Similarly, hesperadin (**33**) has been shown to inhibit TbAUK1 (IC_{50} = 40 nM), inhibit parasite growth (*T.b.b.* IC_{50} = 50 nM), and halt nuclear division [38].

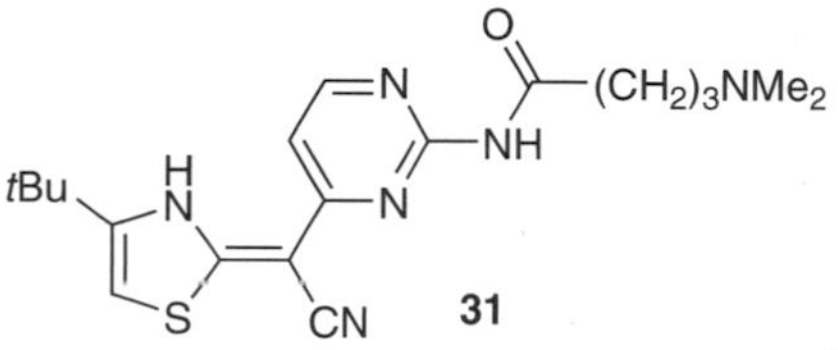

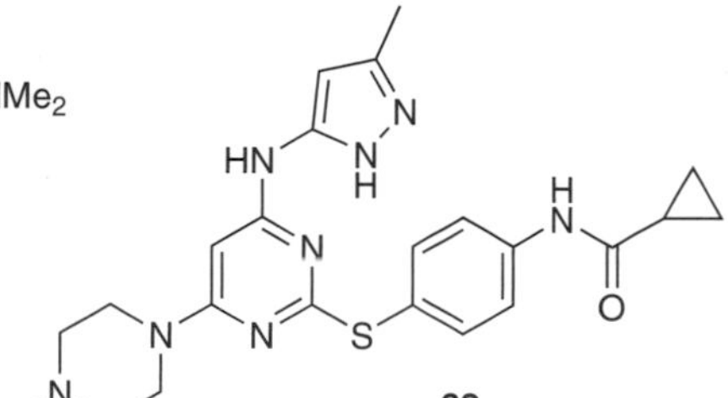

2.6 Nucleosides—purine uptake and metabolism

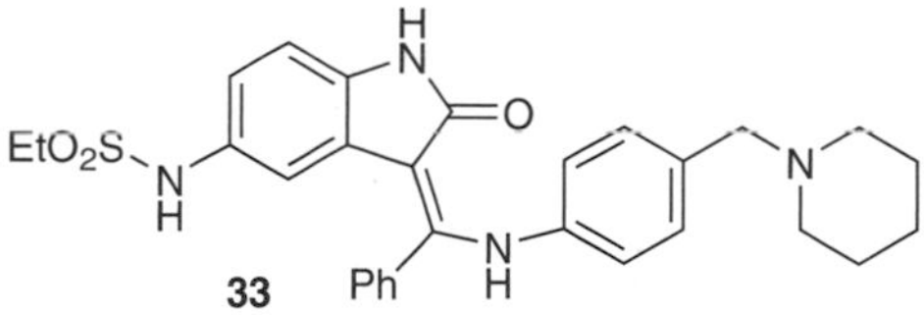

T. brucei is unable to synthesize purines *de novo* and, as such, is dependent upon salvage mechanisms from the host. A number of transporters and enzymes are used by *T. brucei* to accomplish this task, and inhibition of these targets offers promise for development of trypanocides [39]. This strategy has been validated by demonstration that cordycepin (**34**), a substrate for *T. brucei* adenosine kinase (TbAK), which terminates RNA synthesis and parasite growth, can cure stage 2 HAT infections in mice when coadministered with deoxycoformycin (**35**), an adenosine deaminase inhibitor [40].

Further support for the importance of TbAK has been obtained through exploration of a series of pyrazoles, reported to exhibit activity against *T. b. rhodesiense, Leishmania donovani,* and *T. cruzi.* The morpholine derivative **36** (*T.b.r.* IC50 = 1.0 μM) has been identified as a lead based on inhibition of *T. b. rhodesiense* growth [41]. Immobilization of the 4-amino analog **37** on a solid matrix, followed by incubation with a *T. b. rhodesiense* cell lysate, gel electrophoresis, and identification of bound proteins by liquid chromatography/mass spectrometry (LC/MS) of a trypsin digest, identified *T. b. r.* adenosine kinase as the primary target of this chemotype [42]. A complementary approach to inhibit uptake of purine bases has prompted exploration of N(6)-substituted purines as inhibitors of the P1 or P2 transporters of *T. brucei* [43]. Lipophilic amides such as **38** (*T.b.r.* IC$_{50}$ = 3.4 μM) and **39** (*T.b.r.* IC$_{50}$ = 4.0 μM) are particularly active in this regard.

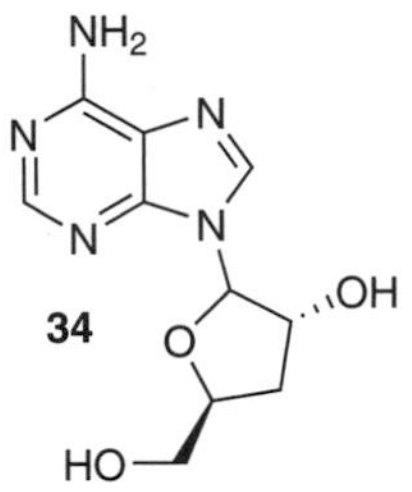

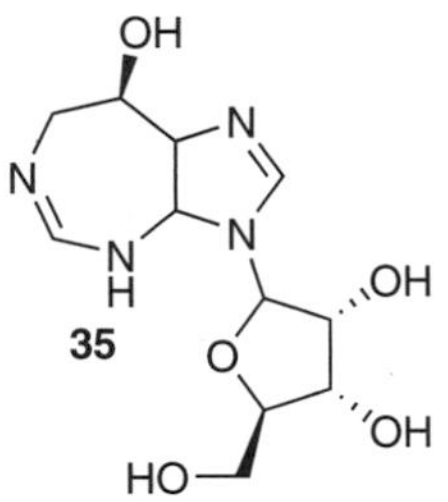

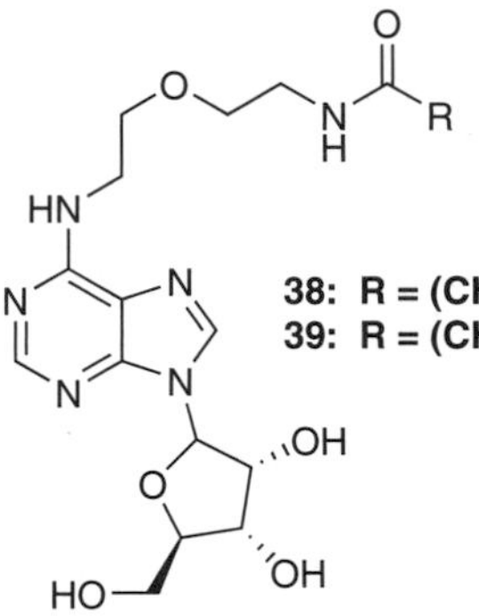

3. LEISHMANIASES

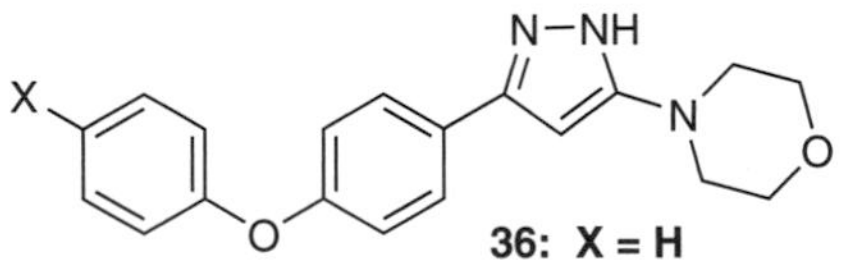

3.1 Disease background and current treatments

Leishmaniases are caused by an intracellular protozoan parasite of the genus *Leishmania* spp. Depending on the geographic area, either *L. donovani* or *Leishmania infantum* is the causative agent. The first line of defense is pentavalent antimony, administered as either meglumine antimoniate (**40**) or sodium stibogluconate (**41**). These compounds were discovered 50 years ago and have undesirable side effects and resistance problems. The second line of treatment is pentamidine (**1**) and amphotericine (**42**), though the latter suffers from nephrotoxicity and high cost of therapy. Recent additional treatment options are miltefosine (**43**) and paromomycin (**44**). Although miltefosine can be given orally, it is teratogenic and has severe gastrointestinal side effects. Sitamaquine (**45**) is still in clinical trials where it has shown acceptable efficacy, but its side-effect profile is unknown. Research for new antileishmanial drugs is progressing on multiple fronts, and chemistry patents in this area have recently been summarized [44].

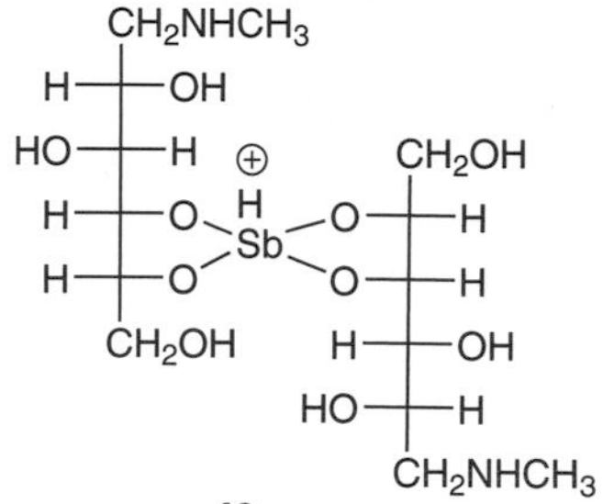

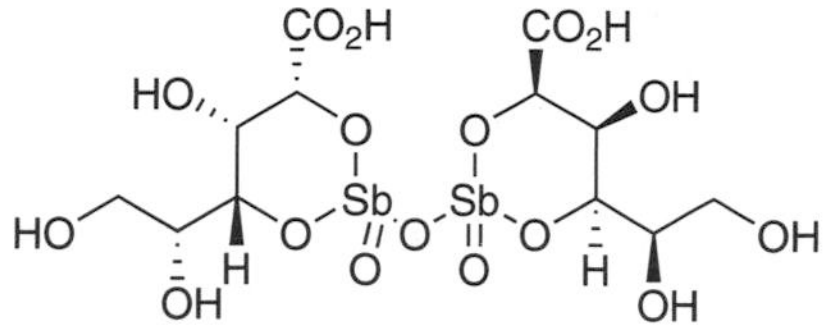

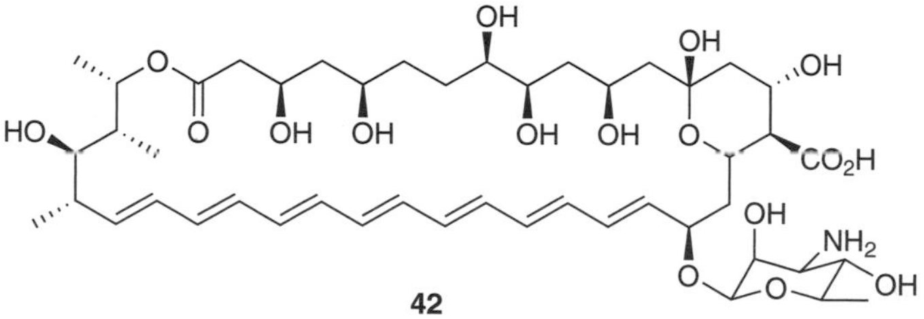

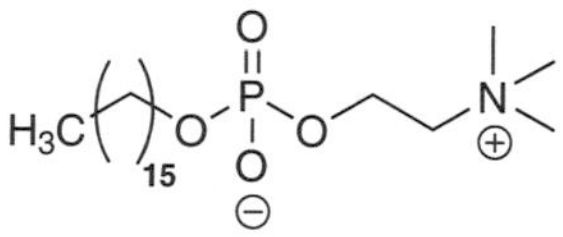

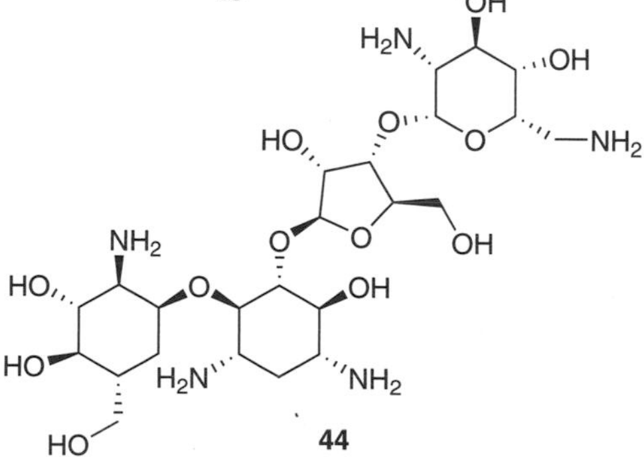

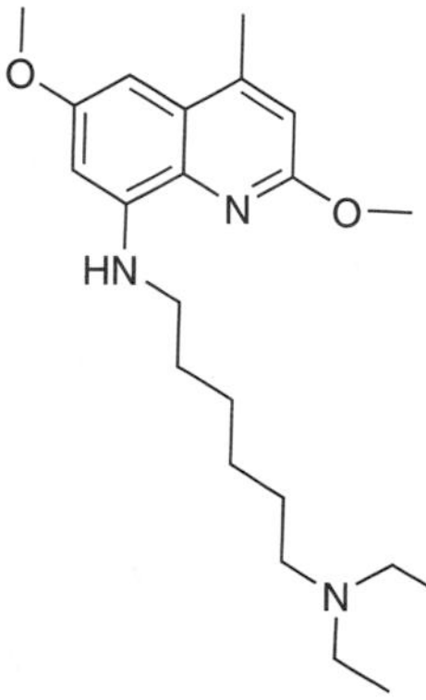

3.2 Sterol biosynthesis pathway inhibitors

Protozoan parasites *Leishmania*, like fungi, require the synthesis of 24-substituted sterols, such as ergosterol, whereas mammals just need cholesterol. The azole antifungals that work on the same pathway possess good antileishmanial activity. A series of aryloxyalkyl imidazoles, exemplified by 46 (IC_{50} =0.47 µg/mL), were characterized for activity in an *in vitro* anti-amastigote assay [45]. This compound showed 60% inhibition of parasite growth *in vivo* (dose: 50 mg/kg, i.p. × 10 days). In the same assay, amphotericin B (dose: 8 mg/kg, i.v. × 5 days) exhibited 92% inhibition of parasite growth. An SAR study of new examples of azasterols containing N-25 amine and amide substituents, such as 47 (IC_{50} = 1.6 µM), was reported as active in an *in vitro* L. donovani amastigote assay [46]. These compounds are based on the 22,26-azasterol 48, an inhibitor (IC_{50} = 24 nM) of 24-sterol methyltransferase (24-SMT), which has been shown to be active (IC_{50} = 8.9 µM) in an anti-amastigote *L. donovani* assay [47].

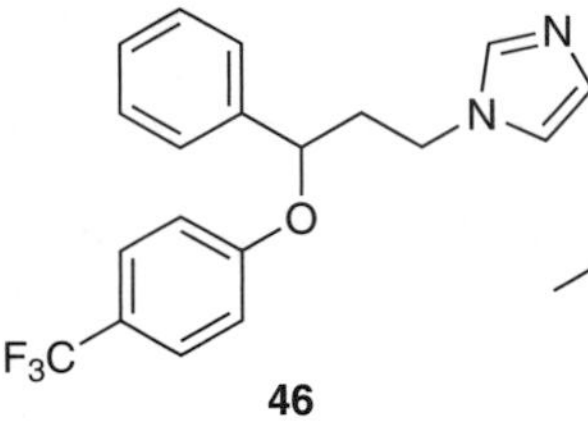

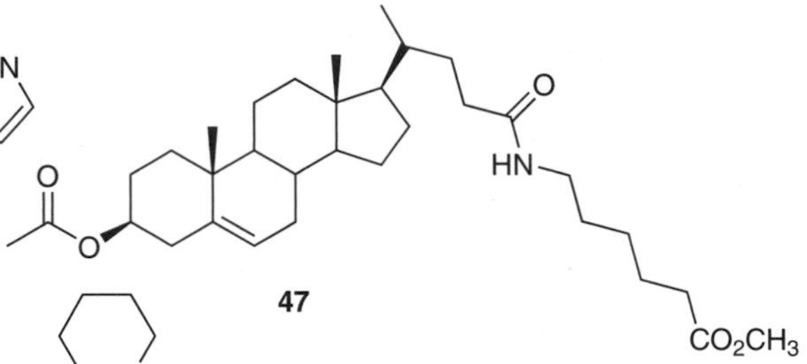

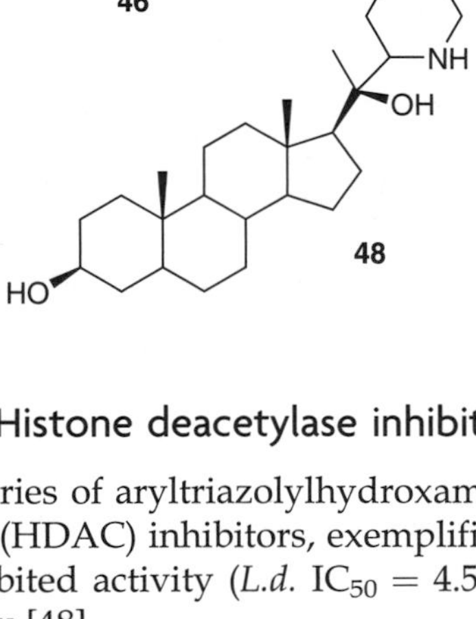

3.3 Histone deacetylase inhibitors

A series of aryltriazolylhydroxamates were reported as histone deacetylase (HDAC) inhibitors, exemplified by 49 (HDAC IC_{50} = 9.6 nM) which exhibited activity (*L.d.* IC_{50} = 4.5 µg/mL) in an *in vitro* antileishmanial assay [48].

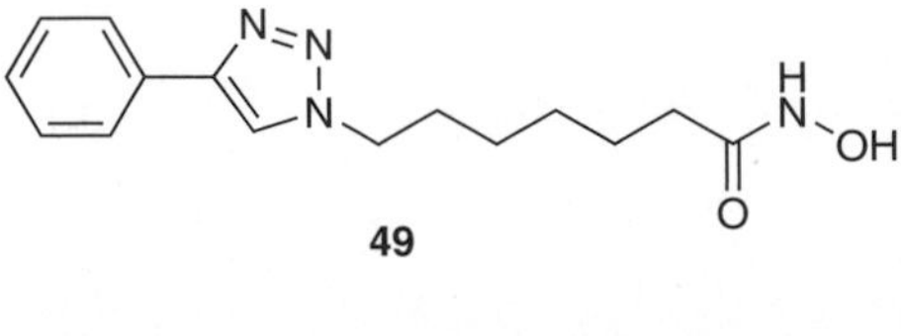

3.4 Chalcones

Licochalcone (**50**) is a natural product that is isolated from the roots of Chinese liquorice and is reported to have antileishmanial activity [49]. A series of chromene-substituted chalcones related to licochalcone have been reported to have antileishmanial activity [50]. Compound **51** was reported to have an IC_{50} of 1.2 μM against *Leishmania major* promastigotes versus meglumine antimoniate (IC_{50} =30 μM). Various compounds related to **51** have potent antileishmanial activity (IC_{50} < 3 μM) with potency similar to **51**, but they did not show cytotoxicity.

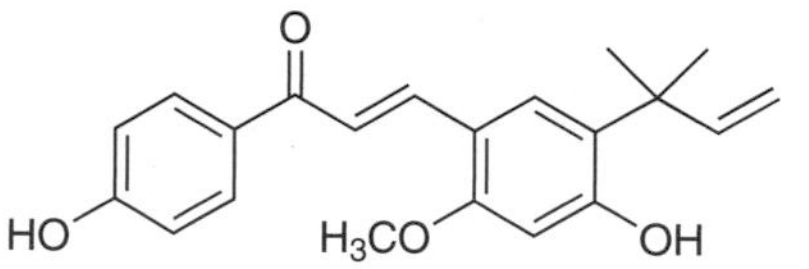

50 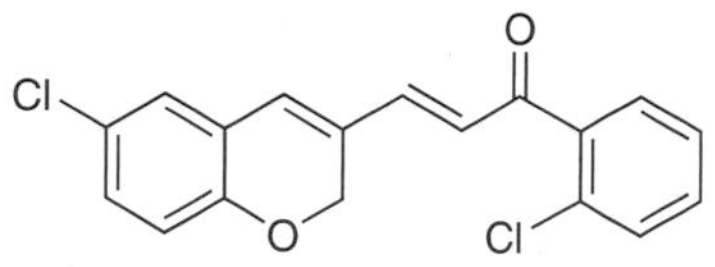   **51**

3.5 Nitroheteroarenes

In order to adapt to the low oxygen conditions present in host environments, most parasites do not use the oxygen from the host, rather employ their own system's anaerobic metabolic pathways to generate ATP. Nitrobenzylidene hydrazides, such as nifuroxazide (**52**), were reported to exhibit promising activity when screened against *L. donovani* [51]. Analogs of trifluralin (**53**) were reported as antileishmanial agents, for example, compounds **54** and **55**. Compound **54** showed *in vitro* potency against promastigotes of both *L. infantum* (IC_{50} = 2.2 μM) and *L. donovani* (IC_{50} = 0.60 μM). The compound is also potent in an intracellular assay (IC_{50} = 1.8 μM). However, compound **55** is not active against promastigotes, but showed activity in the intracellular assay. Trifluralin (**53**) was effective for cutaneous form of leishmaniasis as an ointment.

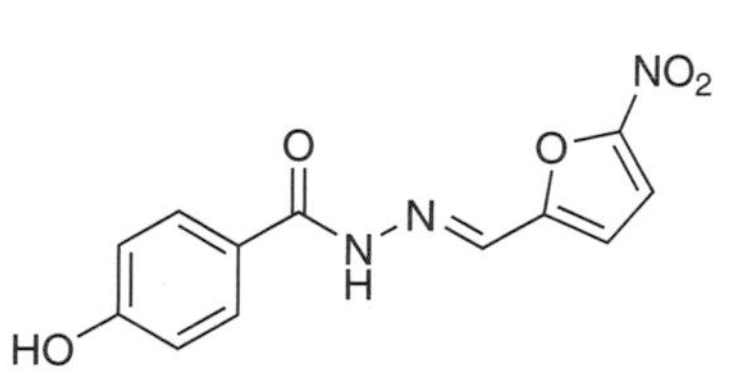

52 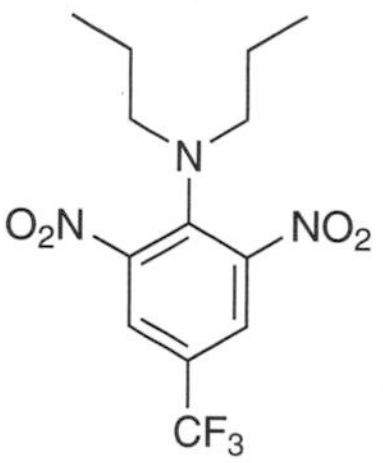   **53**

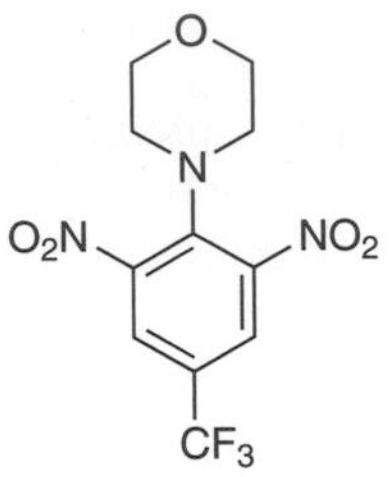

55

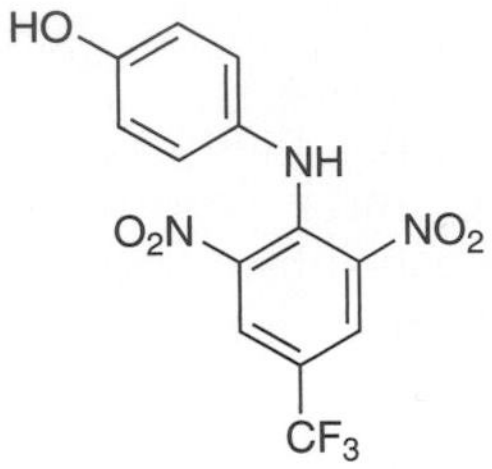

54

3.6 Folate biosynthesis inhibitors

Dihydrofolate reductase-thymidylate synthase (DHFR-TS) enzymatic activity has been shown to be fundamental to parasite survival in *L. major*. A DHFR-TS knockout mutant does not survive *in vivo*, which further demonstrates the essentiality of this target [52]. A series of quinazoline derivatives were designed as antileishmanial agents, targeting the DHFR-TS pathway [53]. Compound **56** showed *in vitro* activity of 99.9% inhibition against *L. donovani* promastigote and an IC_{50} of 3.2 µg/mL against an amastigote luciferase system.

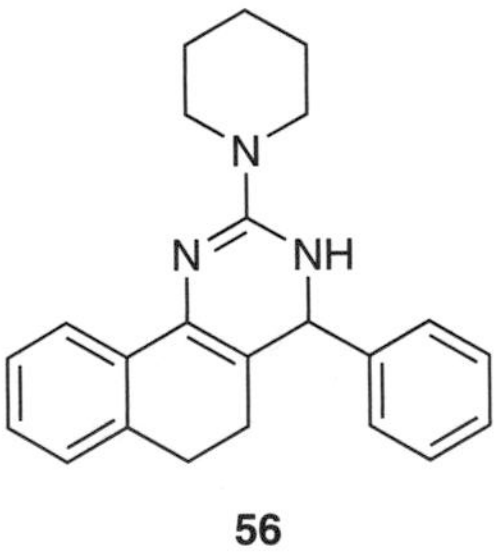

56

4. CHAGAS DISEASE

4.1 Disease background and current treatments

The parasite *T. cruzi* is the causative agent for Chagas disease, which affects about 12–14 million people in Central and South America. The affected patients present with symptoms of hypertrophy, dilatation of the heart, the esophagus, and the large intestine. Earlier drug discovery efforts have been reviewed [54–56]. Two nitroheteroarene drugs were discovered in the late 1960s and the early 1970s, nifurtimox (**57**) and benznidazole (**58**), respectively. These compounds are the only currently available drugs for the specific treatment of Chagas disease. Both have significant efficacy in the acute phase of the disease, but poor efficacy in the chronic phase, which is more prevalent among Chagas patients. Nifurtimox treatment is accompanied by nausea and vomiting, and benznidazole treatment is accompanied by allergic dermopathy.

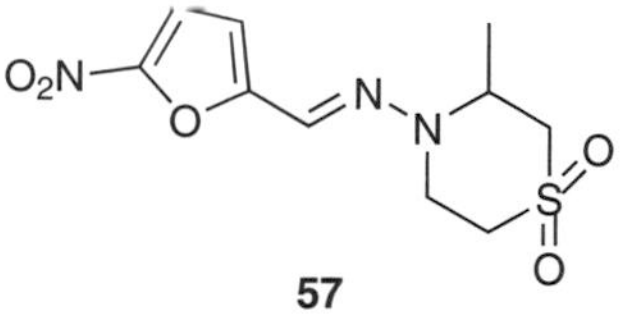

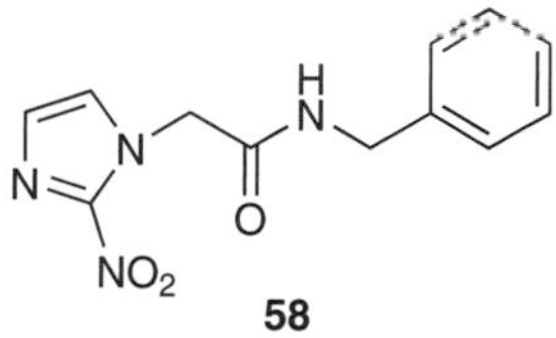

4.2 Sterol biosynthesis pathway inhibitors

One promising area of recent research on Chagas disease is sterol biosynthesis inhibition [5]. Like *Leishmania, T. cruzi* requires specific sterols for cell viability and proliferation in all stages of its life cycle. Consequently, several enzymes in the ergosterol biosynthesis pathway have been validated as targets. Posaconazole (**59**), a selective cytochrome P450-dependent C14α sterol demethylase (CYP51) inhibitor, has shown parasitological cure in murine models of both acute and chronic stages of Chagas disease [57].

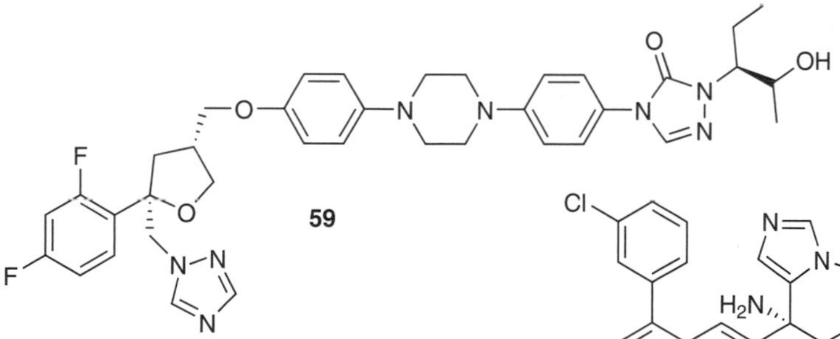

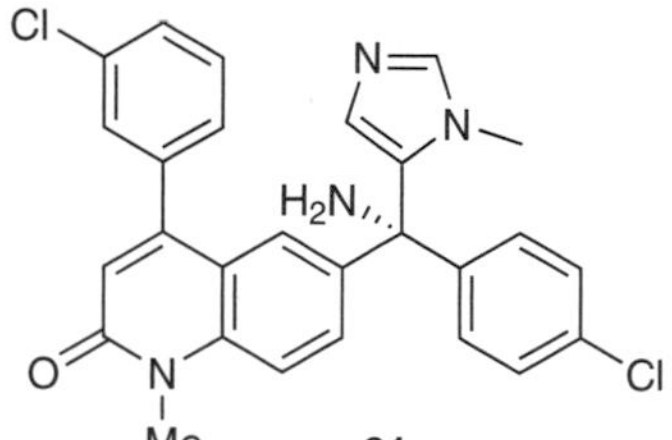

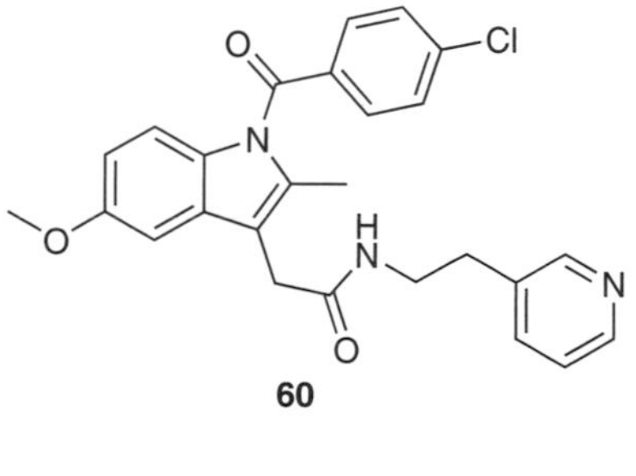

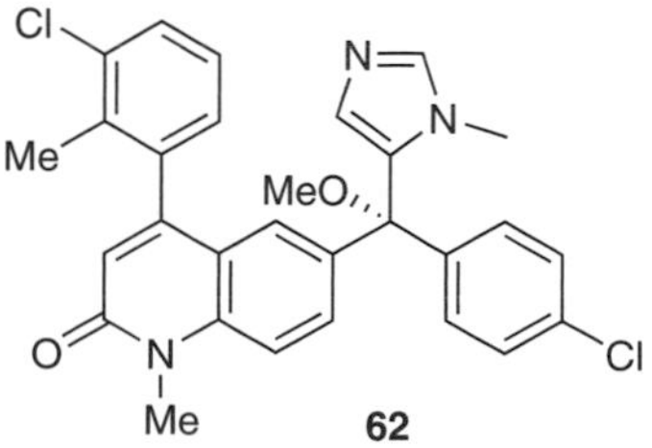

Indomethacin amides were also disclosed as a novel series of sterol 14α-demethylase inhibitors leading to antiparasitic effects in cultured *T. cruzi* cells [58]. It was shown that specific inhibition of *T. cruzi* 14α-demethylase (TCCYP51) is effective in killing both extracellular and intracellular human stages of *T. cruzi* and may have the advantage of decreased probability of resistance because of this specificity. Treatment of *T. cruzi* trypomastigotes with compound **60** showed clear dose-dependent antiparasitic effects in the human form of *T. cruzi*; and at 20 μM concentration, the parasite was practically eliminated from cardiomyocytes *in vitro*. Good inhibitory activity of **60**, compared to other analogs within this class, was attributed to better compound permeability and metabolic stability.

Imidazole-containing farnesyltransferase inhibitors, like tipifarnib (**61**), originally developed as anticancer agents, kill *T. cruzi* parasites *in vitro* and *in vivo* [59]. The mechanism of action is inhibition of lanosterol 14α-demethylase, which is required for synthesis of ergosterol, a necessary and irreplaceable component of the parasite's membranes. In a search for a parasite-specific inhibitor, tipifarnib analogs were found to have better potency and selectivity [60]. Compound **62** was designed based on an X-ray structural understanding of the interaction of tipifarnib with mammalian protein farnesyltransferase and docking of **61** to a homology model of *T. cruzi* lanosterol 14α-demethylase. Compound **62** has decreased activity against the protein farnesyl transferase ($IC_{50} > 5000$ nM), but exhibits 10-fold better potency in killing *T. cruzi* amastigotes ($EC_{50} = 0.6$ nM) than tipifarnib.

4.3 Trypanosomal cysteine protease (cruzain) inhibitors

The cysteine protease cruzain is essential for *T. cruzi* survival. A peptidic inhibitor of cruzain, K777 (**63**), has demonstrated activity in an *in vitro* parasiticidal assay, an *in vivo* model of Chagas disease, and has been shown to prevent cardiac damage from the infection in dog [61,62]. Efforts to discover nonpeptidic inhibitors of cruzain have been reported, with a new series of triazine nitriles identified from a high-throughput screen [63]. Chemical modifications of the lead series provided the purine nitrile **64** as a very potent enzyme inhibitor ($IC_{50} = 10$ nM), but activity in an *in vitro* parasiticidal assay was not observed, potentially due to poor cellular permeability.

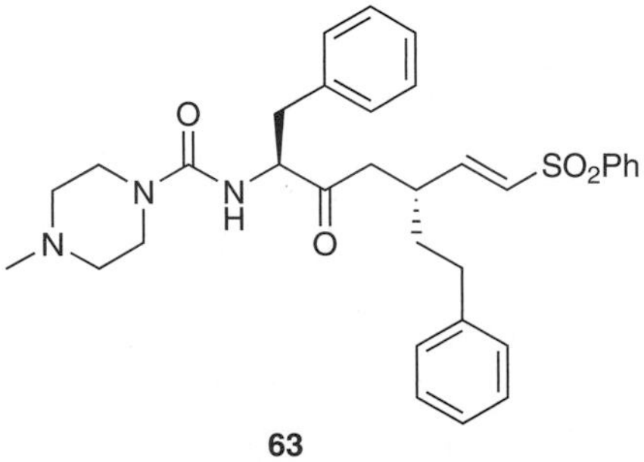

63

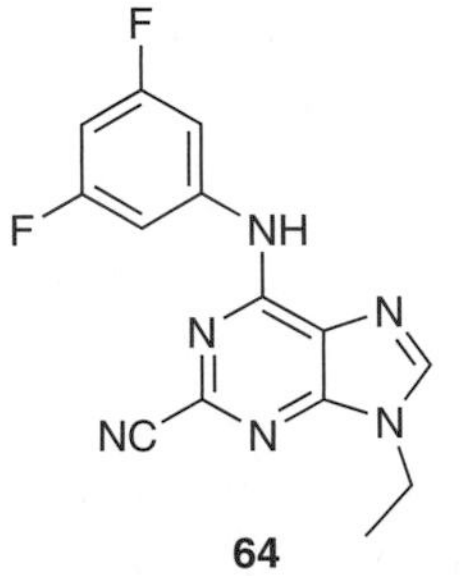

64

4.4 Miscellaneous *T. cruzi* compounds

Nitroheteroarenes continue to attract research activity. A series of 5-nitro-2-furyl derivatives were evaluated for *in vivo* efficacy against *T cruzi* [64] and an improved toxicity profile. Compound **65** showed a good efficacy profile *in vivo* and better acute toxicity profile when compared to nifurtimox. A series of 5-nitroindazoles, represented by **66** (IC_{50} = 7.4 µM), were reported as having activity similar to nifurtimox (IC_{50} = 3.4 µM) in a growth inhibition assay [65].

2,3-Diphenyl-1,4-naphthoquinone (**67**) was described as a potential chemotherapeutic agent against *T. cruzi* [66]. The compound is highly toxic (LD_{50} = 2.5 µM) in *T. cruzi* epimastigote, tripomastigote, and intracellular amastigote assays, but less toxic against mammalian cells (LD_{50} = 130 µM). In a mouse *in vivo* assay (dose = 10 mg/kg i.p. × 3 days), the compound suppressed parasitemia and prolonged survival 70 days post-infection. No apparent drug toxicity effects were observed in an uninfected control group treated with **67**.

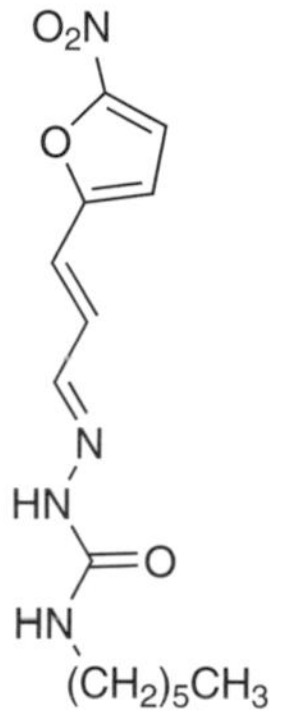

65 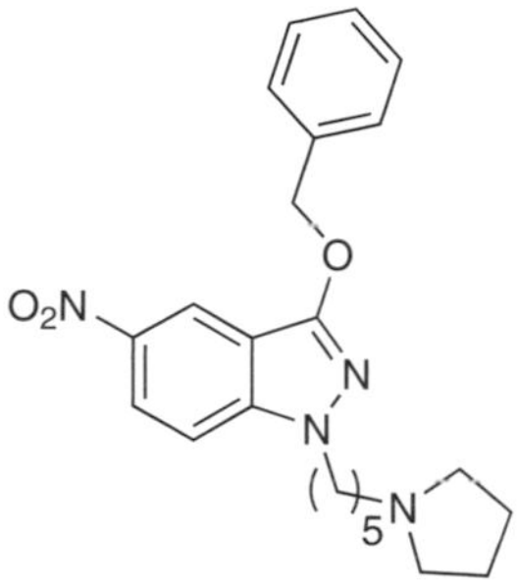   66 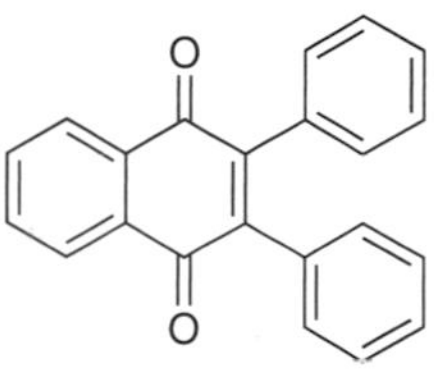   67

5. CONCLUSIONS

There has been a resurgence in interest in the discovery and development of new treatments for neglected diseases in the past several years. In particular, diseases caused by kinetoplastid parasites have received considerable attention. Most efforts continue to rely on initial evaluation of compounds in whole-cell parasiticidal assays, but approaches based on specific biochemical targets and pathways have increasingly been reported due to the availability of the complete genome of these parasites and the subsequent understanding afforded by this knowledge. Several new clinical and preclinical candidates have been progressed that target each parasite, but continued efforts are required to have significant impact on these devastating diseases.

REFERENCES

[1] K. Stuart, R. Brun, S. Croft, A. Fairlamb, R. E. Gurtler, J. McKerrow, S. Reed, and R. Tarleton, *J. Clin. Invest.*, 2008, **118**, 1301.
[2] P. J. Hotez and A. Kamath, *PLoS Negl. Trop. Dis.*, 2009, **3**, e412.
[3] S. L. Croft and V. Yardley, *Curr. Pharm. Des.*, 2002, **8**, 319.
[4] F. Sanchez-Sancho, N. E. Campillo, and J. A. Paez, *Curr. Med. Chem.*, 2010, **17**, 423.
[5] J. A. Urbina, *Acta Trop.*, 2009, **115**, 55.
[6] N. M. El-Sayed, P. J. Myler, G. Blandin, M. Berriman, J. Crabtree, G. Aggarwal, E. Caler, H. Renauld, E. A. Worthey, C. Hertz-Fowler, E. Ghedin, C. Peacock, D. C. Bartholomeu, B. J. Haas, A. N. Tran, J. R. Wortman, U. C. Alsmark, S. Angiuoli, A. Anupama, J. Badger, F. Bringaud, E. Cadag, J.M. Carlton, G.C. Cerqueira, T. Creasy, A. L. Delcher, A. Djikeng, T.M. Embley, C. Hauser, A.C. Ivens, S.K. Kummerfeld, J.B. Pereira-Leal, D. Nilsson, J. Peterson, S.L. Salzberg, J. Shallom, J.C. Silva, J. Sundaram, S. Westenberger, O. White, S. E. Melville, J. E. Donelson, B. Andersson, K. D. Stuart and N. Hall, *Science*, 2005, **309**, 404.
[7] C. J. Bacchi, J. Garofalo, D. Mockenhaupt, P. P. Mccann, K.A. Diekema, A. E. Pegg, H. C. Nathan, E. A. Mullaney, L. Chunosoff, A. Sjoerdsma, and S. H. Hutner, *Mol. Biochem. Parasitol.*, 1983, **7**, 209.
[8] R. T. Jacobs, B. Nare, and M. A. Phillips, *Curr. Top. Med. Chem.*, 2010, in press.
[9] J. H. Ansede, M. Anbazhagan, R. Brun, J. D. Easterbrook, J. E. Hall, and D. W. Boykin, *J. Med. Chem.*, 2004, **47**, 4335.
[10] J. K. Thuita, S. M. Karanja, T. Wenzler, R. E. Mdachi, J. M. Ngotho, J. M. Kagira, R. Tidwell, and R. Brun, *Acta Trop.*, 2008, **108**, 6.
[11] R. E. Mdachi, J. K. Thuita, J. M. Kagira, J. M. Ngotho, G. A. Murilla, J. M. Ndung'u, R. R. Tidwell, J. E. Hall, and R. Brun, *Antimicrob. Agents Chemother.*, 2009, **53**, 953.
[12] G. Pohlig, S. Bernhard, J. Blum, C. Burri, A. M. Kabeya, J. -P .F. Lubaki, A. M. Mpoto, B. F. Munungu, G. K. M. Deo, P. N. Mutantu, F. M. Kuikumbi, A. F. Mintwo, A. K. Munungi, A. Dala, S. Macharia, C. M. M. Bilenge, V. K. B. K. Mesu, J. R. Franco, N. D. Dituvanga, and C. Olson., Abstract 542, 57th Meeting of the American Society of Tropical Medicine and Hygiene, New Orleans, LA, December 7–11, 2008. www.astmh.org/Meeting-Archives.htm
[13] L. Hu, R. K. Arafa, M. A. Ismail, A. Patel, M. Munde, W.D. Wilson, T. Wenzler, R. Brun, and D.W. Boykin, *Bioorg. Med. Chem.*, 2009, **17**, 6651.
[14] D.A. Patrick, S.A. Bakunov, S.M. Bakunova, E.V.K. Suresh Kumar, H. Chen, S.K. Jones, T. Wenzler, T. Barzcz, K.A. Werbovetz, R. Brun, and R.R. Tidwell, *Eur. J. Med. Chem.*, 2009, **44**, 3543.
[15] D. A. Patrick, S.A. Bakunov, S.M. Bakunova, E. V. Kumar, R . J. Lombardy, S. K. Jones, A.S. Bridges, O. Zhirnov, J. E. Hall, T. Wenzler, R. Brun, and R. R. Tidwell, *J. Med. Chem.*, 2007, **50**, 2468.
[16] T. L. Huang, J. J. Vanden Eynde, A. Mayence, M. S. Collins, M. T. Cushion, D. Rattendi, I. Londono, L. Mazumder, C. J. Bacchi, and N. Yarlett, *Bioorg. Med. Chem. Lett.*, 2009, **19**, 5884.
[17] T. Wenzler, D. W. Boykin, M. A. Ismail, J.E. Hall, R. R. Tidwell, and R. Brun, *Antimicrob. Agents Chemother.*, 2009, **53**, 4185.
[18] Y. Xiao, D. E. Mccloskey, and M. A. Phillips, *Eukaryot. Cell*, 2009, **8**, 747.
[19] S. Krieger, W. Schwarz, M. R. Ariyanayagam, A. H. Fairlamb, R. L. Krauth-Siegel, and C. Clayton, *Mol. Microbiol.*, 2000, **35**, 542.
[20] J. L. Richardson, I.R. Nett, D.C. Jones, M.H. Abdille, I.H. Gilbert, and A.H. Fairlamb, *ChemMedChem*, 2009, **4**, 1333.
[21] S. Patterson, D. C. Jones, E.J. Shanks, J. A. Frearson, I. H. Gilbert, P. G. Wyatt, and A. H. Fairlamb, *ChemMedChem*, 2009, **4**, 1341.

[22] G. A. Holloway, W. N. Charman, A. H. Fairlamb, R. Brun, M. Kaiser, E. Kostewicz, P. M. Novello, J. P. Parisot, J. Richardson, I. P. Street, K. G. Watson, and J. B. Baell, *Antimicrob. Agents Chemother.*, 2009, **53**, 2824.

[23] A. Cavalli, F. Lizzi, S. Bongarzone, R. Brun, R. Luise Krauth-Siegel, and M.L. Bolognesi, *Bioorg. Med. Chem. Lett.*, 2009, **19**, 3031.

[24] S. Wyllie, S. L. Oza, S. Patterson, D. Spinks, S. Thompson, and A. H. Fairlamb, *Mol. Microbiol.*, 2009, **74**, 529.

[25] L. S. Torrie, S. Wyllie, D. Spinks, S. L. Oza, S. Thompson, J. R. Harrison, I. H. Gilbert, P. G. Wyatt, A. H. Fairlamb, and J. A. Frearson, *J. Biol. Chem.*, 2009, **284**, 36137.

[26] C. J. Bacchi, H. C. Nathan, N. Yarlett, B. Goldberg, P. P. Mccann, A. J. Bitonti, and A. Sjoerdsma, *Antimicrob. Agents Chemother.*, 1992, **36**, 2736.

[27] E. K. Willert, R. Fitzpatrick, and M. A. Phillips, *Proc. Natl. Acad. Sci. USA*, 2007, **104**, 8275.

[28] B. Hirth, R. H. Barker, Jr., C. A. Celatka, J. D. Klinger, H. Liu, B. Nare, A. Nijjar, M. A. Phillips, E. Sybertz, E.K. Willert, and Y. Xiang, *Bioorg. Med. Chem. Lett.*, 2009, **19**, 2916.

[29] R. H. Barker, Jr., H. Liu, B. Hirth, C. A. Celatka, R. Fitzpatrick, Y. Xiang, E.K. Willert, M. A. Phillips, M. Kaiser, C. J. Bacchi, A. Rodriguez, N. Yarlett, J. D. Klinger, and E. Sybertz, *Antimicrob. Agents Chemother.*, 2009, **53**, 2052.

[30] C. J. Bacchi, R. H. Barker, Jr., A. Rodriguez, B. Hirth, D. Rattendi, N. Yarlett, C. L. Hendrick, and E. Sybertz, *Antimicrob. Agents Chemother.*, 2009, **53**, 3269.

[31] J. W. Chambers, M. L. Fowler, M. T. Morris, and J. C. Morris, *Mol. Biochem. Parasitol.*, 2008, **158**, 202.

[32] M. W. Nowicki, L. B. Tulloch, L. Worralll, I. W. Mcnae, V. Hannaert, P. A. Michels, L. A. Fothergill-Gilmore, M. D. Walkinshaw, and N. J. Turner, *Bioorg. Med. Chem.*, 2008, **16**, 5050.

[33] M. D. Urbaniak, *Mol. Biochem. Parasitol.*, 2009, **166**, 183.

[34] S. Monnerat, C. Clucas, E. Brown, J. C. Mottram, and T. C. Hammarton, *BMC Res. Notes*, 2009, **2**, 46.

[35] C. L. De Graffenried, H. H. Ho, and G. Warren, *J. Cell Biol.*, 2008, **181**, 431.

[36] K. K. Ojo, J. R. Gillespie, A.J. Riechers, A. J. Napuli, C. L. Verlinde, F. S. Buckner, M.H. Gelb, M. M. Domostoj, S. J. Wells, A. Scheer, T. N. Wells, and W. C. Van Voorhis, *Antimicrob. Agents Chemother.*, 2008, **52**, 3710.

[37] Z. Li, T. Umeyama, and C.C. Wang, *PLoS Pathog.*, 2009, **5**, e1000575.

[38] N. Jetton, K.G. Rothberg, J. G. Hubbard, J. Wise, Y. Li, H. L. Ball, and L. Ruben, *Mol. Microbiol.*, 2009, **72**, 442.

[39] A. Luscher, H.P. De Koning, and P. Maser, *Curr. Pharm. Des.*, 2007, **13**, 555.

[40] S. K. Vodnala, M. Ferella, H. Lunden-Miguel, E. Betha, N. Van Reet, D. N. Amin, B. Oberg, B. Andersson, K. Kristensson, H. Wigzell, and M. E. Rottenberg, *PLoS Negl. Trop. Dis.*, 2009, **3**, e495.

[41] S. Kuettel, A. Zambon, M. Kaiser, R. Brun, L. Scapozza, and R. Perozzo, *J. Med. Chem.*, 2007, **50**, 5833.

[42] S. Kuettel, M. Mosimann, P. Mäser, M. Kaiser, R. Brun, L. Scapozza, and R. Perozzo, *PLoS Negl. Trop. Dis.*, 2009, **3**, e506.

[43] A. Link, P. Heidler, M. Kaiser, and R. Brun, *Eur. J. Med. Chem.*, 2009, **44**, 3665.

[44] H. C. Maltezou, *Recent Patents on Anti Infect. Drug Discov.*, 2008, **3**, 192.

[45] K. Bhandari, N. Srinivas, V. K. Marrapu, A. Verma, S. Srivastava, and S. Gupta, *Bioorg. Med. Chem. Lett.*, 2010, **20**, 291.

[46] F. Gigante, M. Kaiser, R. Brun, and I. H. Gilbert, *Bioorg. Med. Chem.*, 2009, **17**, 5950.

[47] L. Gros, S. O. Lorente, C. J. Jimenez, V. Yardley, L. Rattray, H. Wharton, S. Little, S. L. Croft, L.M. Ruiz-Perez, D. Gonzalez-Pacanowska, and I. H. Gilbert, *J. Med. Chem.*, 2006, **49**, 6094.

[48] V. G. Patil, W. Guerrant, P. C. Chen, B. Gryder, D. B. Benicewicz, S. I. Khan, B. L. Tekwani, and A. K. Oyelere, *Bioorg. Med. Chem.*, 2009, **20**, 291.

[49] M. Chen, L. Zhai, S. B. Christensen, T. G. Theander, and A. Kharazmi, *Antimicrob. Agents Chemother.*, 2001, **45**, 2023.

[50] Z. Nazarian, S. Emami, S. Heydari, S.K. Ardestani, M. Nakhjiri, F. Poorrajab, A. Shafiee, and A. Foroumadi, *Eur. J. Med. Chem.*, 2010, **45**, 1424.

[51] D. G. Rando, M. A . Avery, B. L. Tekwani, S. I. Khan, and E. I. Ferriera, *Bioorg. Med. Chem.*, 2008, **19**, 6724.

[52] A. Cruz, C. M. Coburn, and S. M. Beverley, *Proc. Natl. Acad. Sci. USA*, 1991, **88**, 7170.

[53] K. C. Agarwal, V. Sharma, N. Shakya, and S. Gupta, *Bioorg. Med. Chem. Lett.*, 2009, **19**, 5474.

[54] Z. Brener, *Adv. Pharmacol. Chemother.*, 1975, **13**, 1.

[55] J. R. Coura and S.L. De Castro, *Mem. Inst. Oswaldo Cruz*, 2002, **97**, 3.

[56] J. C. Pinto Dias, *Ann. Int. Med.*, 2006, **144**, 772.

[57] J. A. Urbina and J. L. Concepcion, *Antimicrob. Agents Chemother.*, 2004, **48**, 2379.

[58] M. E. Konkle, T. Y. Hargrove, Y. Y. Kleshchenko, J. P. von Kries, W. Ridenour, M. J. Uddin, R. M. Caprioli, L. J. Marnett, W. D. Nes, F. Villalta, M. R. Waterman, and G. I. Lepesheva, *J. Med. Chem.*, 2009, **52**, 2846.

[59] F. Buckner, K. Yokoyama, J. Lockman, K. Aikenhead, J. Ohkanda, M. Sadilek, S. Sebti, W. van Voorhis, A. Hamilton, and M.H. Gelb, *Proc. Natl. Acad. Sci. USA*, 2003, **100**, 15149.

[60] J. M. Kraus, C. L. Verlinde, M. Karimi, G. I. Lepesheva, M. H. Gelb, and F.S. Buckner, *J. Med. Chem.*, 2009, **52**, 1639.

[61] S. C. Carr, K.L. Warner, B.G. Kornreic, J. Piscitelli, A. Wolfe, and L. Benet, *Antimicro. Agents Chemother.*, 2005, **49**, 5160.

[62] P. S. Doyle, Y. Zhou, J. C. Engel, and J. H. McKerrow, *Antimicro. Agents Chemother.*, 2007, **51**, 3932.

[63] B. T. Mott, R. S. Ferreira, A. Simeonov, A. Jadhav, K. K. Ang, W. Leister, M. Shen, J.T. Silveira, P.S. Doyle, M. Arkin, J.H. McKerrow, J. Inglese, C.P. Austin, C.J. Thomas, B.K. Shoichet, and D.J. Maloney, *J. Med. Chem.*, 2010, **53**, 52.

[64] E. M. Cabrera, G. M. Murguiondo, M. G. Arias, C. Arredondo, C. Pintos, G. Aguirre, M. Fernandez, Y. Basmadjian, R. Rosa, J. P. Pacheco, S. Raymondo, R. D. Maio, M. Gonzalez, and H. Cercetto, *Eur. J. Med. Chem.*, 2009, **44**, 3909.

[65] J. A. Rodriguez, V. J. Aran, L. Boiani, C. Olea-Azar, M. L. Lavaggi, M. Gonzalez, H. Cerecetto, J.D. Maya, C. Carrasco-Pozo, and H.S. Cosoy, *Bioorg. Med.Chem.*, 2009, **17**, 8186.

[66] E. I. Ramos, K. M. Garza, R.L. Krauth-Siegel, J. Bader, L. E. Martinez, and R. A. Maldonado, *J. Parasitol.*, 2009, **95**, 461.

CHAPTER 18

Recent Advances in the Inhibition of Bacterial Fatty Acid Biosynthesis

Vincent Gerusz

Mutabilis, 102 Avenue Gaston Roussel, Romainville F-93230, France

Annual Reports in Medicinal Chemistry, Volume 45
ISSN 0065-7743, DOI 10.1016/S0065-7743(10)45018-6

1. INTRODUCTION

The bacterial fatty acid biosynthesis pathway has recently generated much interest for the development of novel classes of antibacterial agents [1,2]. The organization of the bacterial fatty acid synthase type II system (FASII system, Figure 1) based on individual enzymes is fundamentally different from the multifunctional fatty acid synthase type I system found in eukaryotes, therefore providing good prospects for selective inhibition. Moreover, all the representative targets of the FASII system have already been characterized by X-ray crystallography or nuclear magnetic resonance (NMR), which should facilitate the rational design of inhibitors. Some of them also share a high degree of conservation among bacterial species, raising hopes of developing broad-spectrum agents. A few FASII enzymes have already been successfully validated as antibacterial targets. For instance, InhA is inhibited by isoniazid as a first-line medication against tuberculosis [3], while the widely used consumer goods preservative triclosan has been found to inhibit FabI [4].

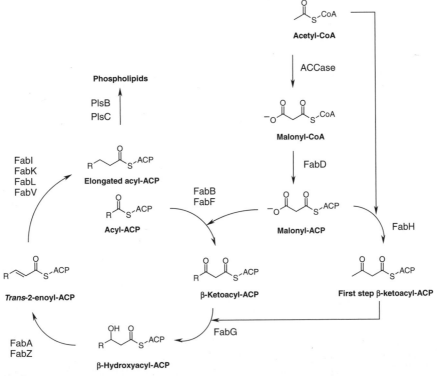

Figure 1   General scheme for FASII system.

However, a recent communication expressed reason for caution by reporting that FASII is not a suitable antibiotic target for Gram-positive pathogens [5]. While relevant *in vivo* experiments were described in favor of the virulent potential of *fabF* and *fabM* mutants of *Streptococcus agalactiae*, the evidence provided for other strains was based solely on their reduced *in vitro* susceptibility for cerulenin and triclosan in the presence of added fatty acids such as Tween 80 or human serum. Although not discussed in the communication, such Minimum inhibitory concentration (MIC) shifts might be explained by either the binding of these highly lipophilic molecules to serum albumin or the sequestration by Tween 80 micelles. To further substantiate their claims, the same authors have also described exogenous fatty acid incorporation in *Staphylococcus aureus* using a medium containing C18:2 fatty acid [6]. In a counter-experiment, knocking out *fabI* by antisense RNA led to growth-compromised *S. aureus* strains after preincubation in 50% human serum, bringing some evidence to the essentiality of this pathway for this pathogen [7].

Significant work continues by both the academic groups and the pharmaceutical industry to validate the FASII targets and discover relevant inhibitors. A selection of some efforts to inhibit bacterial fatty acid biosynthesis at various stages is presented in this chapter.

2. ACCase INHIBITORS

Acetyl-CoA carboxylase (ACCase) carboxylates acetyl-CoA into malonyl-CoA and therefore represents the first committed step in fatty acid biosynthesis. ACCase is a multimer essential for cell growth whose components are highly conserved among bacteria, making it a promising broad-spectrum target [8].

2.1 Natural product lead

Moiramide B (**1**) was identified as a nanomolar inhibitor of AccAD, the carboxyltransferase subunit of ACCase. First elements of structure–activity relationship (SAR) revealed a high degree of tolerance for substitution or replacement of the fatty acid side chain, while very few changes were permitted on the pyrrolidinedione group. The best derivatives proved to be substrates for *Escherichia coli* efflux mechanism and displayed single-digit MICs against *E. coli* and *S. aureus* and double-digit MICs against *Streptococcus pneumoniae* [9].

2.2 High-throughput screening synthetic leads

Screening a large pharmaceutical library derived from eukaryotic drug discovery programs afforded pyridopyrimidine 2, a lead targeting the adenosine-5'-triphosphate (ATP)-binding site of the biotin carboxylase portion of ACCase. Although this site shares structural similarity to known human targets, **2** displays nanomolar potency and excellent selectivity for its bacterial target over a range of eukaryotic protein kinases. However, it is sensitive to efflux in *E. coli* and is inactive against major Gram-positive pathogens [10].

Library screening by affinity selection–mass spectrometry yielded benzimidazole hits, which were optimized through structure-based drug design to imidazo[4,5-*b*]pyridine **3**. This compound is selective for bacterial biotin carboxylase with an IC_{50} of 20 nM against the *E. coli* enzyme. Sensitivity to efflux might here again be an issue since MICs are only presented for a pump-deleted *E. coli* strain [11].

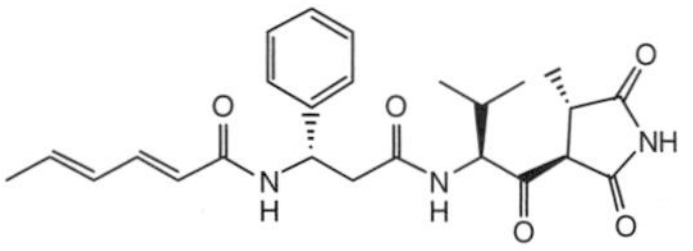

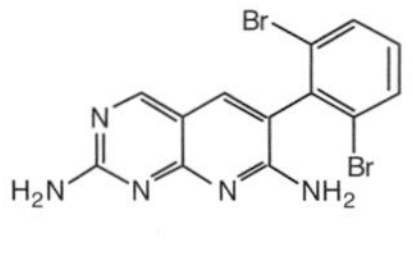

3. FabD INHIBITORS

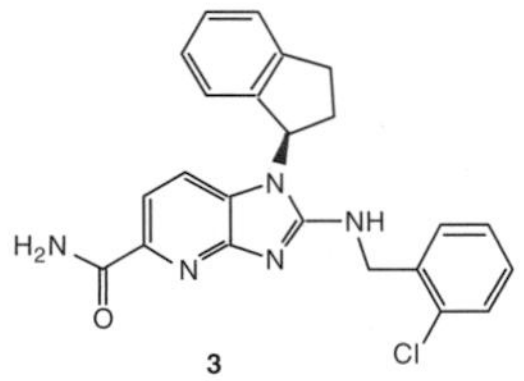

Malonyl-CoA:Acyl carrier protein transacylase (MCAT or FabD) catalyzes the transfer of a malonyl group between coenzyme A and acyl carrier proteins that are the privileged transporters of the FASII system. Although essential, this enzyme is present in excess and does not have a regulatory role in the FASII pathway, which might explain the relative paucity of reported inhibitors [1].

3.1 Virtual screening leads

A range of thiolactones was obtained by virtual screening with the rationale to be covalently cleaved by FabDSer92. Among them, **4** was described as a potent *E. coli* FabD inhibitor with antibiotic properties although no further biological data was provided [12].

3.2 High-throughput screening synthetic leads

A large screening campaign on *Enterococcus faecium* FabD provided the two hits A000326719 (**5**) and A000124358 (**6**). Carboxypyrrole **5** shows an IC_{50} of 55 µM, while 4-hydroxyquinoline **6** is reported as a low micromolar inhibitor. The latter compound was cocrystallized in *E. faecium* FabD, allowing structure-based design. Although inactive against *E. coli* and *Pseudomonas aeruginosa*, **6** displays broad-spectrum Gram-positive antibacterial activity with sub-µg/mL MICs [13].

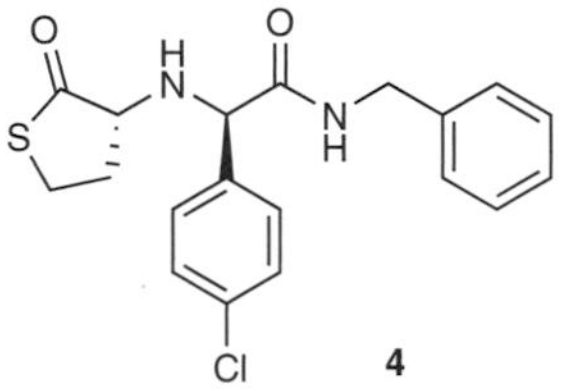

4. FabH INHIBITORS

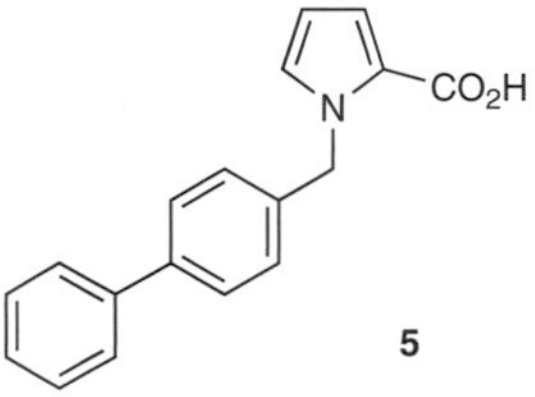

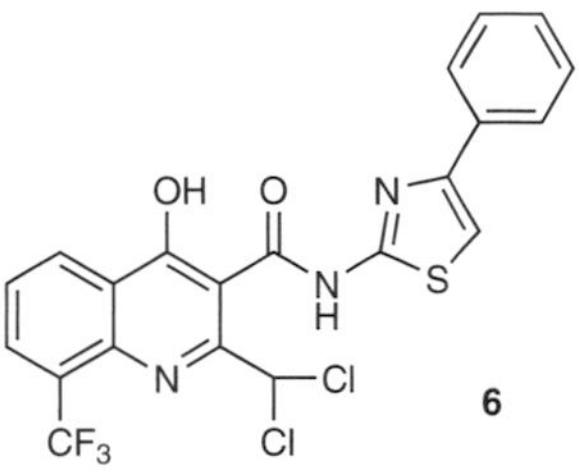

β-Ketoacyl-ACP-synthase III (FabH) catalyzes the transfer of an acetyl group from acetyl-CoA to malonyl-ACP through decarboxylative Claisen condensation. It is an essential and regulatory enzyme that initiates the elongation cycle of bacterial fatty acid synthesis [14]. Methyl-CoA disulfide, a mechanism-based inhibitor designed to irreversibly inactivate the catalytic cysteine responsible for the acyl transfer shed light on important structural changes that happen with *E. coli* FabH [15].

4.1 Virtual screening leads

Mining the three-dimensional (3D) National Cancer Institute database for analogs of thiolactomycin (a natural product binding the malonyl-ACP pocket, see FabB/F Inhibitors) led to the discovery of the three following leads. 1,2-Dithiole-3-one **7** shows low micromolar inhibition of FabH with an MIC of 1 µg/mL against *E. coli* K12. However, its structure and enzymatic studies suggest possible irreversible multitarget binding [16]. Thiazoline-2-one-1,1-dioxide **8** displays a similar potency [17], while sulfonyl-naphthalene-1,4-diol **9** is a low nanomolar inhibitor of FabH with an MIC of 14 µg/mL against an *E. coli* TolC mutant [18]. Their antibacterial activities were also shown to involve other targets.

A virtual screening on pharmacophores derived from available structures of FabH delivered YKAs3003 (**10**). This Schiff base displays a K_D of 20 nM against *E. coli* FabH but has no significant antibacterial activity [19]. Other derivatives based on similar templates have been subsequently reported, such as **11** [20] and **12** [21]. They show low micromolar IC_{50}'s against *E. coli* FabH with broad-spectrum antibacterial activity

(MICs 0.39–3.13 µg/mL) for the former and interesting Gram-negative activity for the latter (0.78 µg/mL on *E. coli* and 1.56 µg/mL on *P. aeruginosa*). No biological evidence linking the antibacterial activity to target inhibition has been disclosed yet for these compounds.

Another virtual screen based on FabH-specific queries yielded a benzoylaminobenzoic acid lead. Using structure-based drug design, it was optimized to afford potent inhibitors of *Enterococcus faecalis* FabH such as compound **13** with an IC_{50} of 4 nM [22]. However, the best representatives of this series display double-digit MICs against a range of Gram-positive species, indicating poor permeability and/or efflux problems. Disappointingly, the most active compounds from this series against the *E. faecalis* and *Streptococcus pyogenes* enzymes lost at least 2-logs of potency against the *S. aureus* and *Haemophilus influenzae* enzymes despite apparent active site conservation. Recent publication of crystal structures of FabH from different species by the same team highlighted important differences in the active sites, suggesting that the quest of a broad-spectrum agent could be more challenging than previously thought for this target.

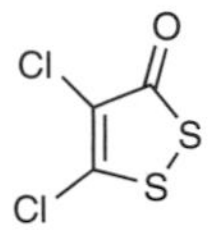

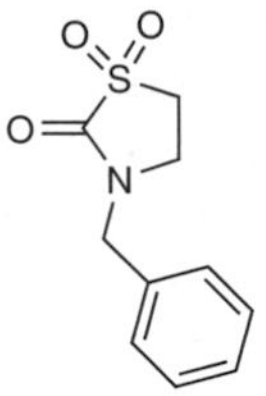

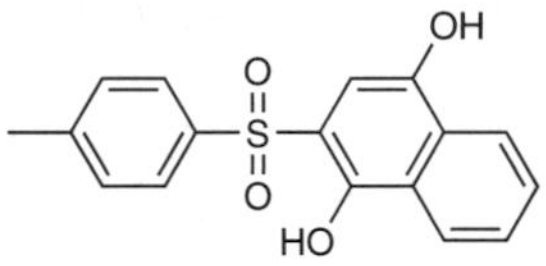

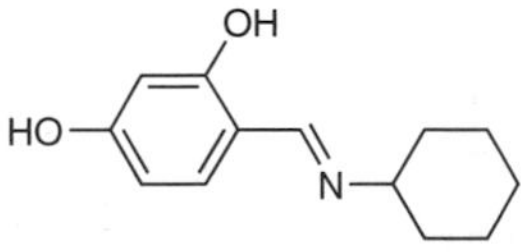

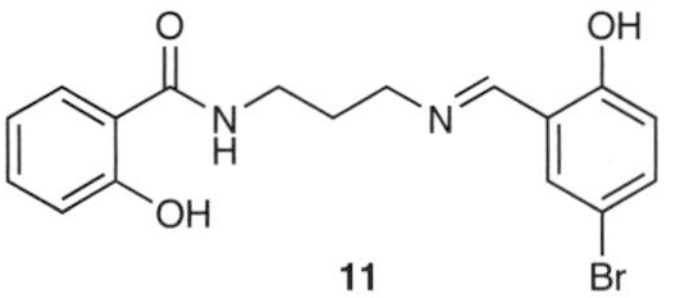

4.2 High-throughput screening synthetic leads

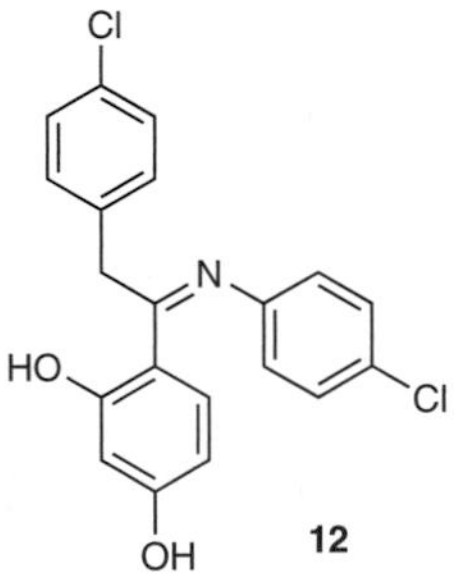

A high-throughput screening (HTS) campaign identified indolyl derivatives as *S. pneumoniae* FabH inhibitors. Since crystallizing this enzyme proved unsuccessful, homology modeling was used to design a more soluble analog that was cocrystallized in the *E. coli* homolog to facilitate rational design [23]. One of the best compounds in this series (SB418011, **14**) displays IC_{50}'s of 16 nM for *S. pneumoniae*, 590 nM for *H. influenzae*, and 1.20 µM for *E. coli* enzymes without any activity on the human enzyme. However, no MICs were disclosed, and the report of the *S. aureus* structure by the same team discusses important differences with the *E. coli* structure, indicating again active site architecture variability [24].

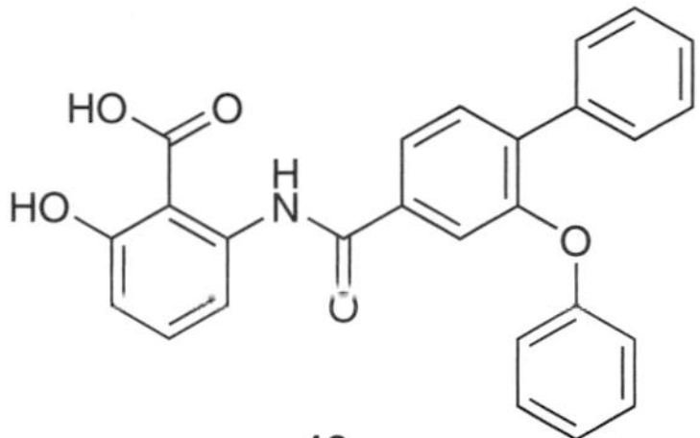

13 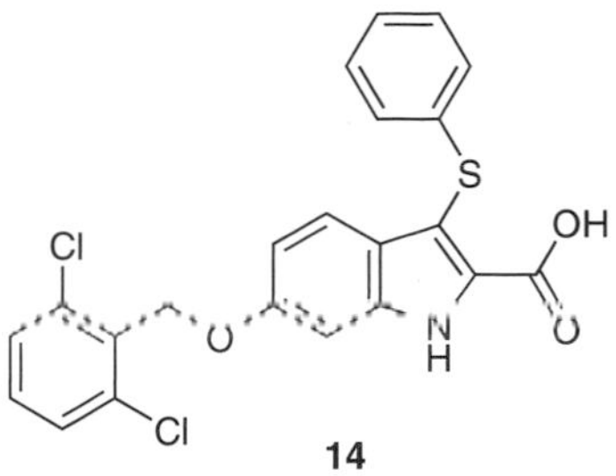   14

5. FabB/F INHIBITORS

β-Ketoacyl-ACP-synthase I (FabB) and β-ketoacyl-ACP-synthase II (FabF) share around 40% identity in their primary sequences and are, like FabH, condensing enzymes. They take a rate-controlling part in the elongation cycle, contributing to the addition of two carbons to the growing fatty acid chain. The first step involves transfer of the growing acyl-ACP to an active cysteine, which is followed by decarboxylative Claisen condensation on malonyl-ACP. FabB and FabF both occur in Gram-negative proteobacteria, FabB being essential to fatty acid synthesis, whereas Gram-positive pathogens contain only FabF [25]. The differences between the malonyl-ACP and acyl-ACP sites offer multiple opportunities for inhibition, and the fact that natural products are known to specifically target them has reinforced the attention devoted to these enzymes. Nevertheless, good synthetic leads have not yet been reported for these targets.

The first disclosed natural product was cerulenin (15), an irreversible inhibitor of FabB. This hydrophobic epoxide locates itself in the hydrophobic groove of the acyl site and reacts covalently with the active site cysteine [26]. However, 15 was also found to inhibit eukaryotic fatty acid synthase.

Thiolactomycin (16) is another natural product that reversibly inhibits E. coli FabF, FabB, and FabH with respective IC_{50}'s of 6, 25 and 110 μM. Unlike cerulenin, it binds the malonyl-ACP site of the enzyme [27]. Despite modest double-digit MICs on E. coli, S. aureus, Serratia marcescens, and Mycobacterium tuberculosis, 16 has generated quite some interest due to its good in vivo protection against an oral or intramuscular S. marcescens urinary tract infection model where it displayed rapid tissue distribution [28]. Despite several medicinal chemistry efforts, thiolactomycin has proven difficult to optimize due to some strict functional group requirements for its SAR [29].

An ingenious whole-cell mechanism-based assay using an antisense fabF RNA was developed to increase the detection threshold to FabH/F inhibitors in S. aureus. A range of new natural products was thus discovered, including phomallenic acid C (**17**), platensimycin (**18**), and platencin (**19**). Compound **17** shows an MIC of 3.9 μg/mL against *S. aureus* and *H. influenzae*, without any activity against wild-type or *tolC*-deleted *E. coli*. Fatty acid product analyses revealed both FabH and FabF as its targets in *S. aureus* [30].

The *Spirulina platensis* secondary metabolite **18** is a potent FabF inhibitor (IC$_{50}$'s of 48 and 160 nM for the *S. aureus* and *E. coli* enzymes, respectively) and only a weak FabH inhibitor (IC$_{50}$ of 67 μM on the *S. aureus* enzyme). The mechanism involves direct binding to the acyl-enzyme intermediate. Platensimycin has attracted much attention since it displays excellent broad-spectrum antibacterial activity with MICs ranging from 0.1 to 1 μg/mL on major Gram-positive pathogens without any observed toxicity [31]. However, it is inactive on *E. coli* because of efflux and had to be administered at high doses via continuous flow pump during an efficacy experiment (mice infected with *S. aureus*). These presumed pharmacokinetic problems may be due to fast clearance and/or plasma instability and thus might preclude clinical experiments on the natural product itself. A few analogs obtained by total synthesis showed that while some modifications were permitted in the lipophilic polycyclic region, small changes on the benzoic acid area resulted in a drastic loss of activity. In any case, potency could not be improved over the natural product [32].

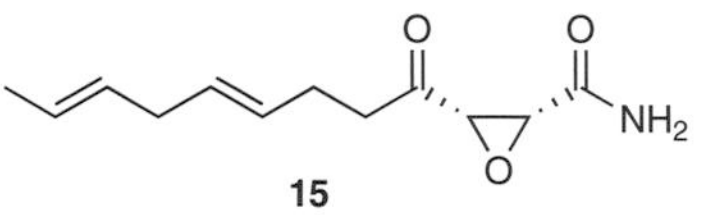

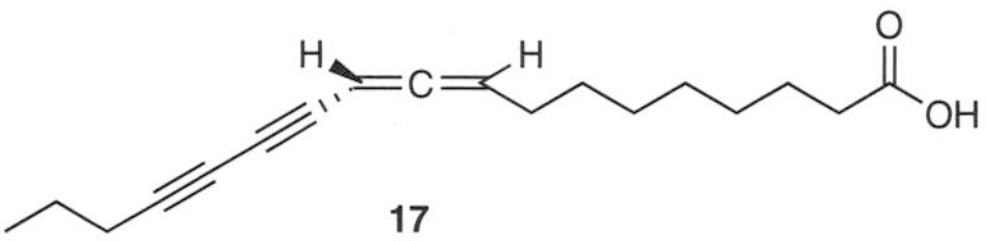

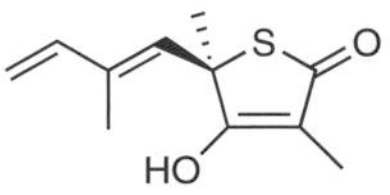

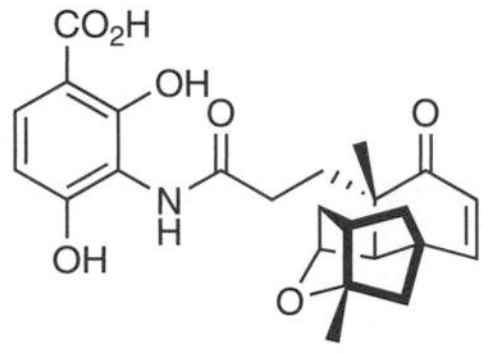

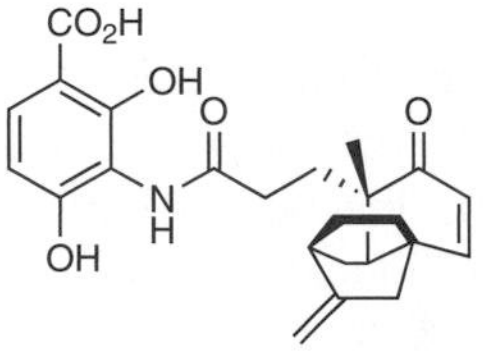

Another isolate from *S. platensis*, **19** exhibits a different lipophilic polycyclic region than **18**. Unlike platensimycin, platencin is a poorer FabF inhibitor but a better FabH one (*S. aureus* IC$_{50}$'s of 113 and 16.2 μM, respectively). Antibacterial activities are generally comparable to **18**, with better potencies against vancomycin-resistant *Enterococcus* (MIC < 0.06 μg/mL). Compared to **18**, IC$_{50}$ values of **19** for both enzymes were >10-fold higher than that obtained from the whole-cell labeling assay, suggesting a synergistic effect in *S. aureus* by dual targeting. By virtue of this dual inhibition, platencin also raises hopes of drugs less likely to develop resistance since the chances of both targets mutating simultaneously is extremely low [33]. The same group has recently identified natural hydroxylated analogs of **18** and **19**. Although less potent than their initial congeners, they are contributing to a better understanding of the SAR of this series.

6. FabG INHIBITORS

β-Ketoacyl-ACP-reductase (FabG) catalyzes the Nicotinamide adenine dinucleotide phosphate (NADPH)-dependent reduction of β-ketoacyl-ACP into β-hydroxyacyl-ACP as the first reductase step of the elongation cycle. Although not rate controlling, this enzyme presents all the characteristics of an appealing target since it is essential, widely expressed across the bacterial genome, and highly conserved since no other isoforms are known [34]. The crystal structure of *E. coli* FabG has also been determined. This tetrameric protein undergoes significant conformational changes upon cofactor binding that might prove important for the rational design of slow binding inhibitors [35]. Yet, this target remains largely unexploited since only the non-specific polyphenol epigallocatechin gallate has so far been reported as an *E. coli* FabG mixed-type inhibitor with an IC_{50} of 5 µM [36].

7. FabA/Z INHIBITORS

β-Hydroxydecanoyl-ACP dehydratase/isomerase (FabA), an enzyme involved in the formation of unsaturated fatty acids in Gram-negative bacteria, is irreversibly inhibited by the mechanism-based suicide inhibitor 3-decynoyl-*N*-acetylcysteamine (**20**) [37]. However, the lack of FabA in Gram-positive pathogens and the growth of *E. coli* FabA mutants in exogenous fatty acid supplementation provide little incentives to pursue this target.

β-Hydroxyacyl-ACP dehydratase (FabZ) catalyzes the formation of *trans*-2-enoyl-ACP. It appears more desirable than FabA since it is essential, ubiquitous, and well conserved among bacteria, with structures available. However, the fact that FabZ is not a rate-determining step and that the equilibrium lies more on the side of the β-hydroxyacyl-ACP has tempered interest in this target [38]. Among the few inhibitors of FabZ, hydrazide **21** is the result of a structure-based optimization of a Biacore HTS hit. Although submicromolar on the *Helicobacter pylori* enzyme, it did not show any significant antibacterial activity [39].

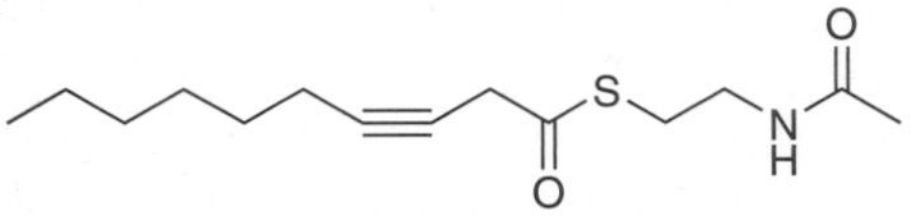

20 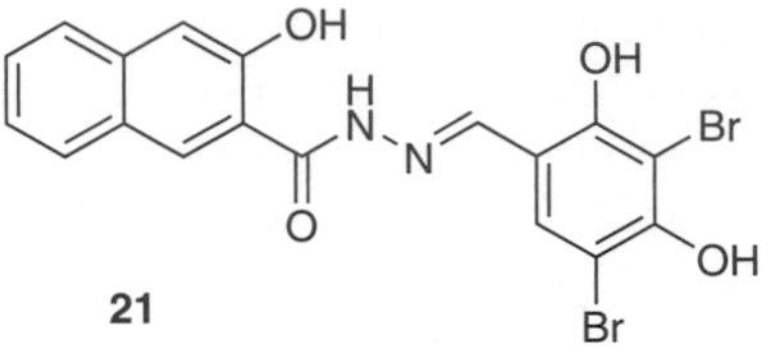   21

8. FabI INHIBITORS

Enoyl-ACP reductase (ENR) uses the cofactor NAD(P)H to reduce the double bond in the *trans*-2-enoyl-ACP to the corresponding acyl-ACP product in the ultimate and rate-limiting step of the elongation cycle [38]. The FabI isoform is an essential target in pathogens such as *E. coli*, *S. aureus*, *Acinetobacter baumannii*, *Bacillus anthracis*, *H. pylori*, *Klebsiella pneumoniae*, *Neisseria meningitidis*, *Salmonella typhi*, and *M. tuberculosis* (also called InhA for the latter). Clinical success mentioned in the introduction has validated this target as one of the most attractive of the FASII pathway. However, other isoforms have also been isolated such as FabK from *S. pneumoniae* [40], FabL from *Bacillus subtilis*, and FabV from *Vibrio cholerae* and *P. aeruginosa*. Although FabI is structurally and mechanistically unrelated to FabK, its relative similarity with FabL, FabV, and PfENR (*Plasmodium falciparum*) still offers opportunities of interesting activity spectra.

8.1 Triclosan and analogs

In addition to its broad-spectrum biocidal activity, triclosan (**22**) displays reversible inhibition of *E. coli* FabI with a picomolar K_i for binding the enzyme–cofactor complex [4]. Triclosan entry results in the reordering of a loop of amino acids close to the active site, making it a slow, tight-binding inhibitor [41].

NB2001 (**23**), a prodrug of triclosan, has been developed based on the enzyme-catalyzed therapeutic activation (ECTA) concept. Evidence supporting ring opening of the cephalosporin moiety by penicillin-binding proteins and/or β-lactamases to release triclosan at the bacterium site, as well as sub-µg/mL MIC on β-lactamase-positive strains, demonstrates the potential of this approach [42].

Several academic and pharmaceutical groups have conducted structure-based design and SAR studies around triclosan aiming at harvesting its potency without its biocidal component. Among the results of these efforts, CPP (**24**) was found to be sevenfold more tightly bound to *E. coli* FabI than triclosan with an MIC value of 0.07 µg/mL, fourfold lower than that of triclosan [43]. Phenoxyphenol **25** is extremely potent against *Francisella tularensis* with an MIC_{90} of 0.00018 μg/mL, and the 5-hexyl-2-phenoxyphenol analog (SBPT04) displays *in vivo* activity against the same bacteria [44]. Both MUT037307 (**26**) [45] and MUT056399 (**27**) [46] are low nanomolar inhibitors of *S. aureus* FabI that are remarkably potent on an extended range of *Staphylococci* with MIC_{90} of ≤0.06 μg/mL (MIC_{90} 2 μg/mL on coagulase negative strains). Animal pharmacokinetic data show an elimination half-life in the range of 2 to 5.8 h for **27** with good antibacterial efficacies via subcutaneous administration and a clean safety pharmacological pro-file. MUT056399 has been renamed FAB-001 and Phase I studies were begun in September 2009. CG400549 (**28**) has been designed to replace the triclosan phenolic hydroxyl by a metabolically more stable carbonyl of a 2-pyridone, which was confirmed by crystallographic studies. Compound **28** displays an MIC_{90} of ≤0.5 μg/mL against *S. aureus* and MIC_{90}'s of 4–8 μg/mL against coagulase-negative strains. Good sub-cutaneous and oral antibacterial efficacy has been reported in animal models, and toxicity studies helped set up the no-observed-effect level for a Phase I study due in late 2009 [47]. A series of 2-pyridones shows high nanomolar activities against *Bacillus anthracis* FabI with modest double-digit MICs [48].

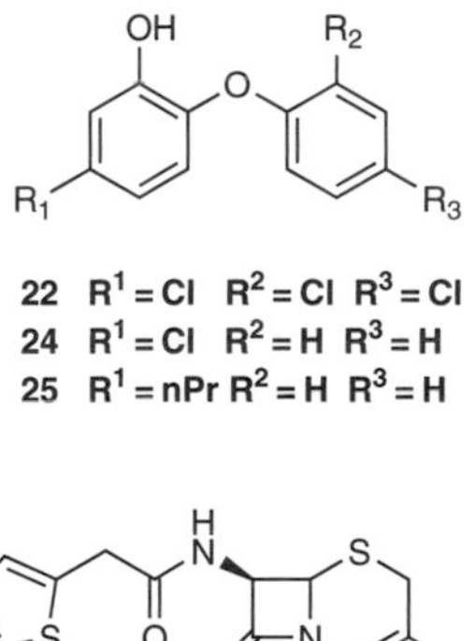

22 R^1 = Cl R^2 = Cl R^3 = Cl
24 R^1 = Cl R^2 = H R^3 = H
25 R^1 = nPr R^2 = H R^3 = H

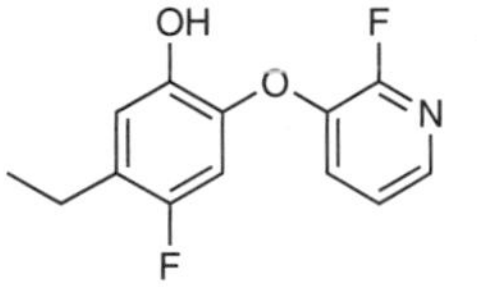

26

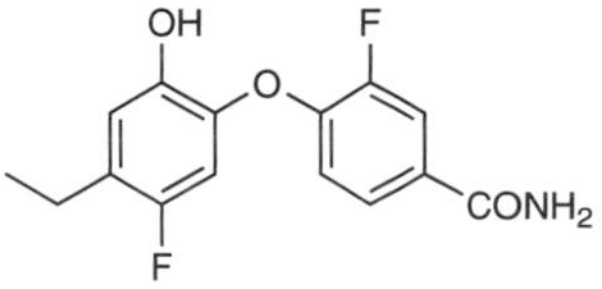

27

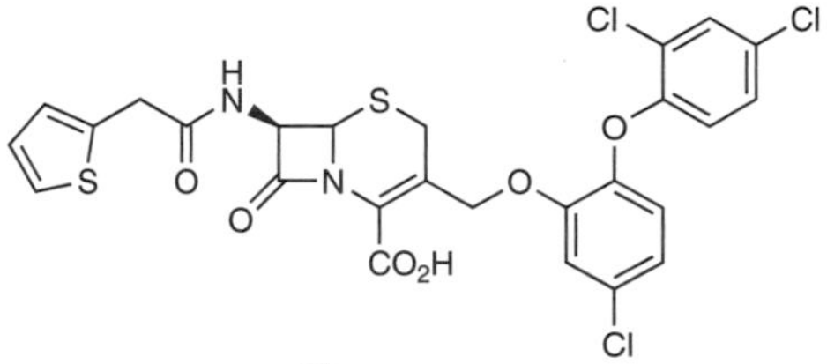

23

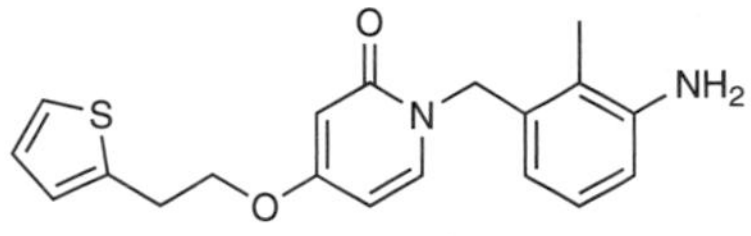

28

8.2 HTS synthetic leads

4-Pyridone HTS hits have been optimized to derivative **29**, which has an MIC_{90} of 0.5 μg/mL against *S. aureus* [49]. Retention of activity against a triclosan-resistant strain implies that these compounds are binding differently than the preceding 2-pyridones.

A different screening campaign has yielded several low micromolar hits against *S. aureus* FabI. The most intensively pursued has been optimized through several rounds of structure-based design in acrylamide **30** (API-1252/AFN-1252) [50]. Although some analogs displayed additional FabK activity, **30** is a pure FabI inhibitor with a *Staphyloccocci*-restricted spectrum. It displays excellent *in vitro* activities (MIC$_{90}$ ≤0.008 µg/mL against *S. aureus* and 0.03 µg/mL against *Staphylococcus epidermidis*) and is also active on triclosan-resistant strains. Good oral efficacy has been demonstrated in animal models, and oral Phase I studies indicated a human half-life range from 8 to 12 h [51]. Poor solubility of **30** is the critical factor limiting absorption, and more soluble analogs such as **31** and **32** (CG400462) have been developed [52,53].

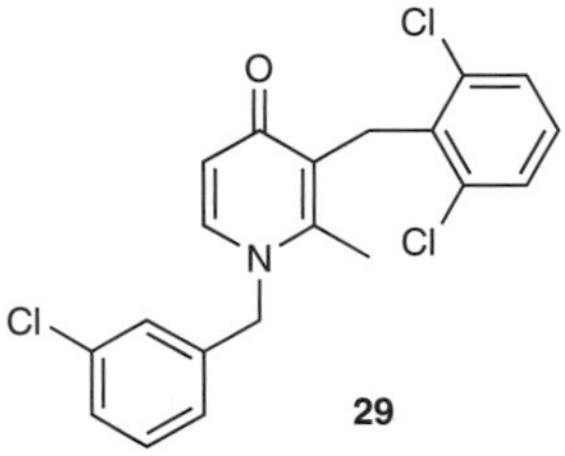

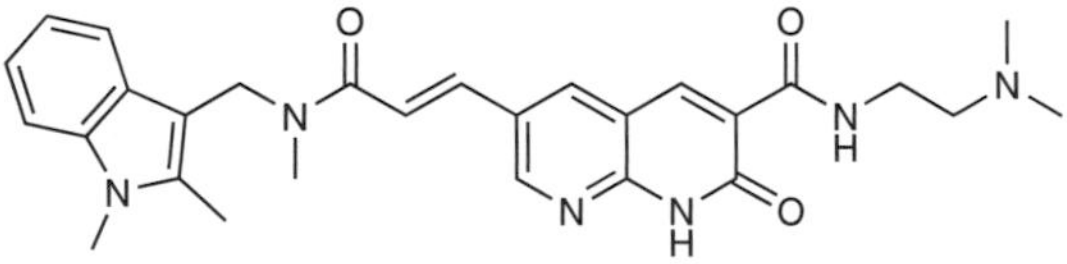

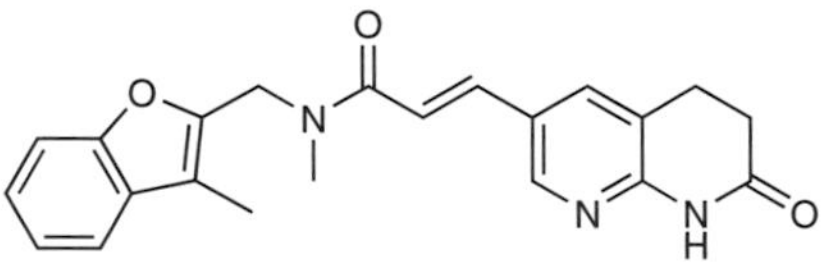

8.3 Natural product leads

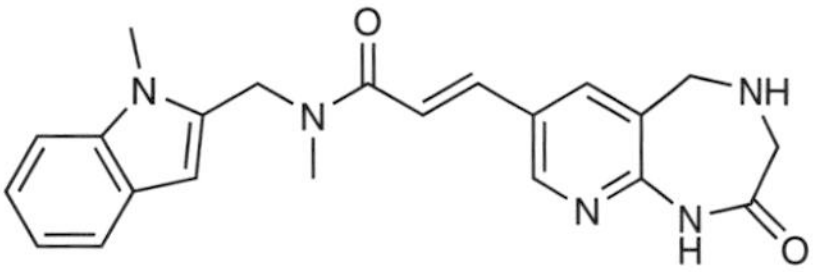

A few natural products, mostly polyphenols and fatty acids, are low micromolar inhibitors of FabI. Based on a different template, (*E*)-oroidin (**33**) displays an uncompetitive binding mechanism, similar to the one observed for triclosan [54].

8.4 InhA inhibitors

The prodrug isoniazid (**34**) targets *M. tuberculosis* InhA [3] after activation by a mycobacterial catalase–peroxidase by reacting irreversibly with the cofactor nicotinamide adenine dinucleotide (NAD). This covalent adduct is a slow tight binder of the enzyme. Related clinically used thioamides such as ethionamide (**35**) are flavin monooxygenase-activated prodrugs that form similar adducts [55].

Direct InhA inhibitors have also been sought to avoid isoniazid resistance mediated by catalase–peroxidase mutation. Lipophilic analogs of triclosan such as **36** show a nanomolar K_i on the enzyme with an MIC of 1–2 μg/mL on isoniazid-resistant strains [56]. Structure-based optimization of two separate HTS leads afforded **37** and **38**, both submicromolar inhibitors of InhA but devoid of any significant antibacterial activity [57,58].

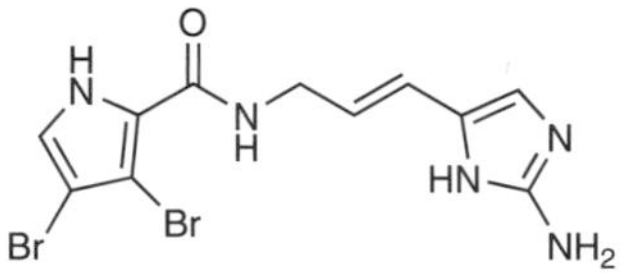

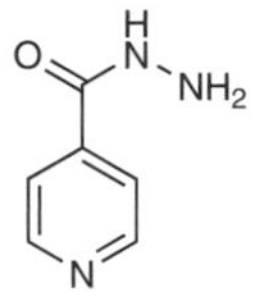

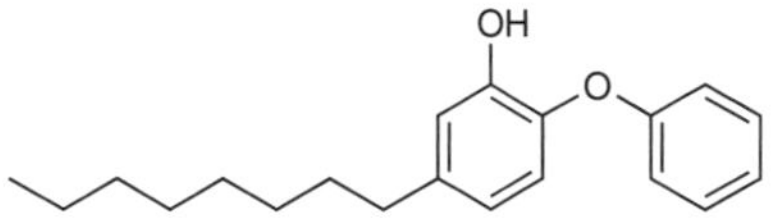

9. FabK INHIBITORS

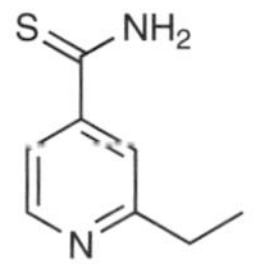

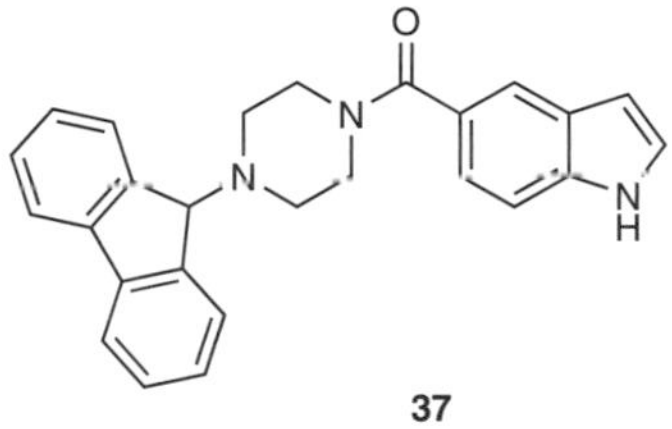

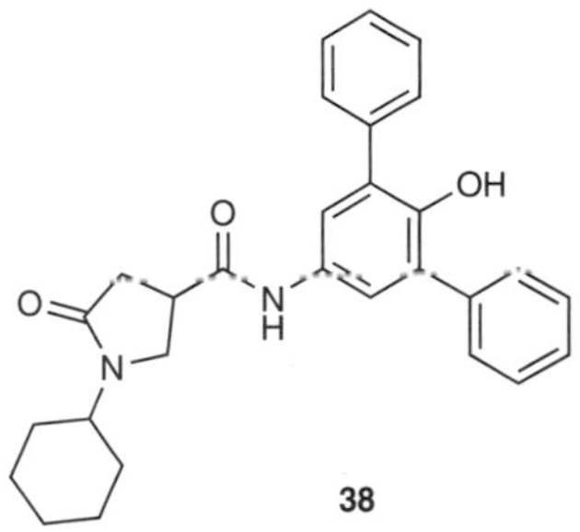

Inhibition of ENR FabK is appropriate for either a narrow spectrum against *Streptococci* and *Clostridium difficile* since it is an essential target for these species, or a broader spectrum in combination with a FabI inhibitor since some bacteria such as *E. faecalis* share both isoforms.

Among the few reported inhibitors, the optimized HTS lead phenylimidazole **39** is an interesting candidate against *S. pneumoniae* with nanomolar FabK inhibition and an MIC_{90} of 4 μg/mL [59]. The same team hybridized **29**, a FabI pyridine inhibitor with **39** to provide **40**, in an attempt to realize a compound with a dual mode of action. Although inhibition of both enzymes as well as sub-μg/mL MIC against *S. pneumoniae* is retained, the FabI-related antibacterial activity against *S. aureus* is lost [60].

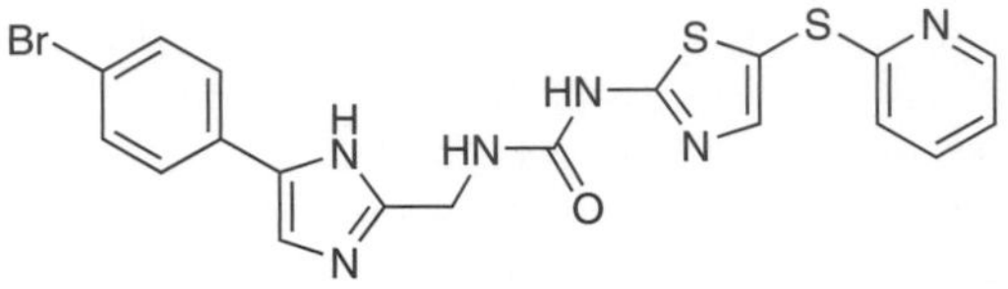

39

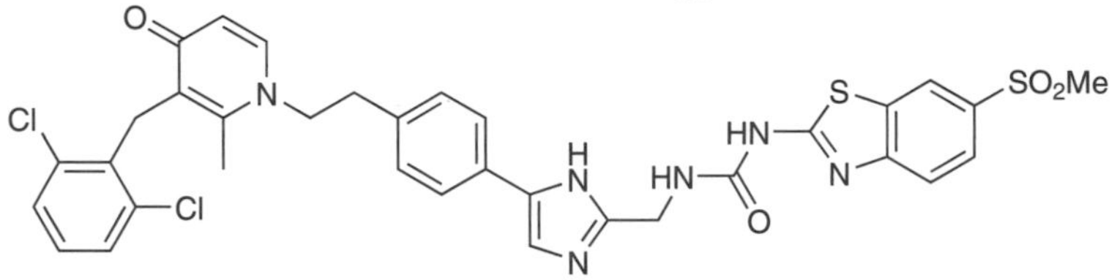

40

10. CONCLUSION

Despite the clinical introduction of isoniazid in 1952, the elucidation of its mode of action along with the complete description of the FASII system has only occurred within the past 15 years. During this time, advances in genomics and structural biology coupled to screening and medicinal chemistry efforts have yielded a wealth of information on this pathway and some of its inhibitors. While the essentiality of some enzymes is still being debated, convincing *in vivo* data have already been disclosed with inhibitors of FabB/F and FabI. The latter enzyme with three inhibitors in early clinical trials might well constitute the target of the next marketed FASII drugs. Their synthetic origin is a testimony of the success of synthetic library screening and structure-based methodologies. Although some reported natural products have also been exciting, they have generally been less amenable to the realization of good human pharmacokinetic characteristics. In addition to the aforementioned enzymes, targeting ACP, CoA and protein–protein interactions as well as integrating computational chemistry in metabolic network analysis [61] provide many opportunities to develop broad- or narrow-spectrum agents and to capitalize on their potential synergistic inhibition. FASII targeting represents therefore an important emerging research area to address resistance issues as well as the ongoing paucity of antibacterial agents with a novel mode of action.

REFERENCES

[1] Y. M. Zhang, S. W. White, and C. O. Rock, *J. Biol. Chem.*, 2006, **281**, 17541.
[2] H. T. Wright and K. A. Reynolds, *Curr. Opin. Microbiol.*, 2007, **10**, 447.
[3] A. Quémard, J. C. Sacchettini, A. Dessen, C. Vilcheze, R. Bittman, W. R. Jacobs, Jr., and J. S. Blanchard, *Biochemistry*, 1995, **34**, 8235.

[4] W. H. Ward, G. A. Holdgate, S. Rowsell, E. G. McLean, R. A. Pauptit, E. Clayton, W. W. Nichols, J. G. Colls, C. A. Minshull, D. A. Jude, A. Mistry, D. Timms, R. Camble, N. J. Hales, C. J. Britton, and I. W. Taylor, *Biochemistry*, 1999, **38**, 12514.

[5] S. Brinster, G. Lamberet, B. Staels, P. Trieu-Cuot, A. Gruss, and C. Poyart, *Nature*, 2009, **458**, 83.

[6] S. Brinster, G. Lamberet, B. Staels, P. Trieu-Cuot, A. Gruss, and C. Poyart, *Nature*, 2010, **463**, E4.

[7] W. Balemans, N. Lounis, R. Gilissen, J. Guillemont. K. Simmen, K. Andries, and A. Koul, *Nature*, 2010, **463**, E3; discussion E4.

[8] J. E. Cronan and G. L. Waldorf, *Prog. Lipid Res.*, 2002, **41**, 407.

[9] J. Pohlmann, T. Lampe, M. Shimada, P. G. Nell, J. Pernerstorfer, N. Svenstrup, N. A. Brunner, G. Schiffer, and C. Freiberg, *Bioorg. Med. Chem. Lett.*, 2005, **15**, 1189.

[10] J. R. Miller, S. Dunham, I. Mochalkin, C. Banotai, M. Bowman, S. Buist, B. Dunkle, D. Hanna, H. J. Harwood, M. D. Huband, A. Karnovsky, M. Kuhn, C. Limberakis, J. Y. Liu, S. Mehrens, W. T. Mueller, L. Narasimhan, A. Ogden, J. Ohren, J. V. Prasad, J. A. Shelly, L. Skerlos, M. Sulavik, V. H. Thomas, S. VanderRoest, L. Wang, Z. Wang, A. Whitton, T. Zhu, and C.K. Stover, *Proc. Natl. Acad. Sci. USA*, 2009, **106**, 1737.

[11] C. C. Cheng, G. W. Shipps, Jr., Z. Yang, B. Sun, N. Kawahata, K. A. Soucy, A. Soriano, P. Orth, L. Xiao, P. Mann, and T. Black, *Bioorg. Med. Chem. Lett.*, 2009, **19**, 6507.

[12] A. Doemling, *Patent Application US 2010/0022520-A1*, 2010.

[13] J. M. Michel, C. Miossec, I. Pozzato, P. Broto, J. Houtmann, A. Bonnefoy, A. Parent, L. Durant, and C. Dini,, K-003, 104th ASM General Meeting, New Orleans, LA, 2004.

[14] C. Y. Lai and J. E. Cronan, *J. Biol. Chem.*, 2003, **278**, 51494.

[15] M. M. Alhamadsheh, F. Musayev, A. A. Komissarov, S. Sachdeva, H. T. Wright, N. Scarsdale, G. Florova, and K. A. Reynolds, *Chem. Biol.*, 2007, **14**, 513.

[16] X. He, A. M. Reeve, U. R. Desai, G. E. Kellogg, and K. A. Reynolds, *Antimicrob. Agents Chemother.*, 2004, **48**, 3093.

[17] M. M. Alhamadsheh, N. C. Waters, D. P. Huddler, M. Kreishman-Deitrick, G. Florova, and K. A. Reynolds, *Bioorg. Med. Chem. Lett.*, 2007, **17**, 879.

[18] M. M. Alhamadsheh, N. C. Waters, S. Sachdeva, P. Lee, and K. A. Reynolds, *Bioorg. Med. Chem. Lett.*, 2008, **18**, 6402.

[19] J. Y. Lee, K. W. Jeong, J. U. Lee, D. I. Kang, and Y. Kim, *Bioorg. Med. Chem.*, 2009, **17**, 1506.

[20] K. Cheng, Q. Z. Zheng, Y. Qian, L. Shi, J. Zhao, and H. L. Zhu, *Bioorg. Med. Chem.*, 2009, **17**, 7861.

[21] H. Q. Li, Y. Luo, P. C. Lv, L. Shi, C. H. Liu, and H. L. Zhu, *Bioorg. Med. Chem. Lett.*, 2010, **20**, 2025.

[22] Z. Nie, C. Perretta, J. Lu, Y. Su, S. Margosiak, K. S. Gajiwala, J. Cortez, V. Nikulin, K. M. Yager, K. Appelt, and S. Chu, *J. Med. Chem.*, 2005, **48**, 1596.

[23] R. A. Daines, I. Pendrak, K. Sham, G. S. Van Aller, A. K. Konstantinidis, J. T. Lonsdale, C. A. Janson, X. Qiu, M. Brandt, S. S. Khandekar, C. Silverman, and M. S. Head, *J. Med. Chem.*, 2003, **46**, 5.

[24] X. Qiu, A. E. Choudhry, C. A. Janson, M. Grooms, R. A. Daines, J. T. Lonsdale, and S. S. Khandekar, *Protein Sci.*, 2005, **14**, 2087.

[25] D. J. Payne, P. V. Warren, D. J. Holmes, Y. Ji, and J. T. Lonsdale, *Drug Discov. Today*, 2001, **6**, 537.

[26] S. Kauppinen, M. Siggaard-Andersen, and P. von Wettstein-Knowles, *Carlsberg Res Commun.*, 1988, **53**, 357.

[27] A. C. Price, K. H. Choi, R. J. Heath, Z. Li, S. W. White, and C. O. Rock, *J. Biol. Chem.*, 2001, **276**, 6551.

[28] S. Miyakawa, K. Suzuki, T. Noto, Y. Harada, and H. Okazaki, *J. Antibiot. (Tokyo)*, 1982, **35**, 411.

[29] P. Kim, Y. M. Zhang, G. Shenoy, Q. A. Nguyen, H. I. Boshoff, U. H. Manjunatha, M. B. Goodwin, J. Lonsdale, A. C. Price, D. J. Miller, K. Duncan, S. W. White, C. O. Rock, C. E. Barry, 3rd, and C. S. Dowd, *J. Med. Chem.*, 2006, **49**, 159.

[30] K. Young, H. Jayasuriya, J. G. Ondeyka, K. Herath, C. Zhang, S. Kodali, A. Galgoci, R. Painter, V. Brown-Driver, R. Yamamoto, L. L. Silver, Y. Zheng, J. I. Ventura, J. Sigmund, S. Ha, A. Basilio, F. Vicente, J. R. Tormo, F. Pelaez, P. Youngman, D. Cully, J. F. Barrett, D. Schmatz, S. B. Singh, and J. Wang, *Antimicrob. Agents Chemother.*, 2006, **50**, 519.

[31] J. Wang, S. M. Soisson, K. Young, W. Shoop, S. Kodali, A. Galgoci, R. Painter, G. Parthasarathy, Y. S. Tang, R. Cummings, S. Ha, K. Dorso, M. Motyl, H. Jayasuriya, J. Ondeyka, K. Herath, C. Zhang, L. Hernandez, J. Allocco, A. Basilio, J. R. Tormo, O. Genilloud, F. Vicente, F. Pelaez, L. Colwell, S. H. Lee, B. Michael, T. Felcetto, C. Gill, L. L. Silver, J. D. Hermes, K. Bartizal, J. Barrett, D. Schmatz, J. W. Becker, D. Cully, and S. B. Singh, *Nature*, 2006, **441**, 358.

[32] K. C. Nicolaou, A. F. Stepan, T. Lister, A. Li, A. Montero, G. S. Tria, C. I. Turner, Y. Tang, J. Wang, R. M. Denton, and D. J. Edmonds, *J. Am. Chem. Soc.*, 2008, **130**, 13110.

[33] J. Wang, S. Kodali, S.H. Lee, A. Galgoci, R. Painter, K. Dorso, F. Racine, M. Motyl, L. Hernandez, E. Tinney, S. L. Colletti, K. Herath, R. Cummings, O. Salazar, I. González, A. Basilio, F. Vicente, O. Genilloud, F. Pelaez, H. Jayasuriya, K. Young, D. F. Cully, and S. B. Singh, *Proc. Natl. Acad. Sci. USA*, 2007, **104**, 7612.

[34] Y. Zhang and J. E. Cronan, Jr., *J. Bacteriol.*, 1998, **180**, 3295.

[35] A. C. Price, Y.M. Zhang, C. O. Rock, and S. W. White, *Structure*, 2004, **12**, 417.

[36] Y. M. Zhang and C. O. Rock, *J. Biol. Chem.*, 2004, **279**, 30994.

[37] K. Endo, G. M. Helmkamp, Jr., and K. Bloch, *J. Biol. Chem.*, 1970, **245**, 4293.

[38] R. J. Heath and C. O. Rock, *J. Biol. Chem.*, 1995, **270**, 26538.

[39] L. He, L. Zhang, X. Liu, X. Li, M. Zheng, H. Li, K. Yu, K. Chen, X. Shen, H. Jiang, and H. Liu, *J. Med. Chem.*, 2009, **52**, 2465.

[40] R. J. Heath and C. O. Rock, *Nature*, 2000, **406**, 145.

[41] M. J. Stewart, S. Parikh, G. Xiao, P. J. Tonge, and C. Kisker, *J. Mol. Biol.*, 1999, **290**, 859.

[42] Q. Li, J. Y. Lee, R. Castillo, M. S. Hixon, C. Pujol, V. R. Doppalapudi, H. M. Shepard, G. M. Wahl, T. J. Lobl, and M. F. Chan, *Antimicrob. Agents Chemother.*, 2002, **46**, 1262.

[43] S. Sivaraman, T. J. Sullivan, F. Johnson, P. Novichenok, G. Cui, C. Simmerling, and P. J. Tonge, *J. Med. Chem.*, 2004, **47**, 509.

[44] K. England, C. am Ende, H. Lu, T. J. Sullivan, N. L. Marlenee, R. A. Bowen, S. E. Knudson, D. L. Knudson, P. J. Tonge, and R. A. Slayden, *J. Antimicrob. Chemother.*, 2009, **64**, 1052.

[45] S. Escaich, S. Fischer, C. Soulama, E. Malacain, J. M. Genevard, F. Faivre, Y. Bonvin, V. Gerusz, M. Oxoby, and F. Moreau, F1-331, 48th Annual ICAAC Meeting, Washington, DC, 2008.

[46] S. Escaich, L. Prouvensier, M. Saccomani, L. Durant, E. Malacain, S. Floquet, A. Walton, V. Sam-Sambo, F. Faivre, M. Oxoby, Y. Bonvin, V. Gerusz, V. Vongsouthi, F. Moreau, and C. Soulama, F1-2012, 49th Annual ICAAC Meeting, San Francisco, CA, 2009.

[47] S. Ro, K. H. Son, Y. E. Kim, H. J. Chang, S. B. Park, J. R. Choi, and J. M. Cho, F1-2007, 49th Annual ICAAC Meeting, San Francisco, CA, 2009.

[48] S. K. Tipparaju, S. Joyasawal, S. Forrester, D. C. Mulhearn, S. Pegan, M. E. Johnson, A. D. Mesecar, and A.P. Kozikowski, *Bioorg. Med. Chem. Lett.*, 2008, **18**, 3565.

[49] H. Kitagawa, K. Kumura, S. Takahata, M. Iida, and K. Atsumi, *Bioorg. Med. Chem.*, 2007, **15**, 1106.

[50] M. A. Seefeld, W. H. Miller, K. A. Newlander, W. J. Burgess, W. E. DeWolf, Jr., P. A. Elkins, M. S. Head, D. R. Jakas, C. A. Janson, P. M. Keller, P. J. Manley, T. D. Moore, D. J. Payne, S. Pearson, B. J. Polizzi, X. Qiu, S. F. Rittenhouse, I. N. Uzinskas, N. G. Wallis, and W. F. Huffman, *J. Med. Chem.*, 2003, **46**, 1627.

[51] N. Kaplan, H. Flanner, and B. Hafkin, F1-2006, 49th Annual ICAAC Meeting, San Francisco, CA, 2009.

[52] J. Ramnauth, M. D. Surman, P. B. Sampson, B. Forrest, J. Wilson, E. Freeman, D. D. Manning, F. Martin, A. Toro, M. Domagala, D. E. Awrey, E. Bardouniotis, N. Kaplan, J. Berman, and H. W. Pauls, *Bioorg. Med. Chem. Lett.*, 2009, **19**, 5359.

[53] H. S. Park, Y. M. Yoon, S. J. Jung, I. N. Yun, C. M. Kim, J. M. Kim, and J. H. Kwak, *Int. J. Antimicrob. Agents*, 2007, **30**, 446.

[54] D. Tasdemir, B. Topaloglu, R. Perozzo, R. Brun, R. O'Neill, N. M. Carballeira, X. Zhang, P. J. Tonge, A. Linden, and P. Rüedi, *Bioorg. Med. Chem.*, 2007, **15**, 6834.

[55] F. Wang, R. Langley, G. Gulten, L. G. Dover, G. S. Besra, W. R. Jacobs, Jr., and J. C. Sacchettini, *J. Exp. Med.*, 2007, **204**, 73.

[56] T. J. Sullivan, J. J. Truglio, M. E. Boyne, P. Novichenok, X. Zhang, C. F. Stratton, H. J. Li, T. Kaur, A. Amin, F. Johnson, R. A. Slayden, C. Kisker, and P. J. Tonge, *ACS Chem. Biol.*, 2006, **1**, 43.

[57] M. R. Kuo, H. R. Morbidoni, D. Alland, S. F. Sneddon, B. B. Gourlie, M. M. Staveski, M. Leonard, J.S. Gregory, A. D. Janjigian, C. Yee, J. M. Musser, B. Kreiswirth, H. Iwamoto, R. Perozzo, W. R. Jacobs, Jr., J. C. Sacchettini, and D. A. Fidock, *J. Biol. Chem.*, 2003, **278**, 20851.

[58] X. He, A. Alian, R. Stroud, and P. R. Ortiz de Montellano, *J. Med. Chem.*, 2006, **49**, 6308.

[59] T. Ozawa, H. Kitagawa, Y. Yamamoto, S. Takahata, M. Iida, Y. Osaki, and K. Yamada, *Bioorg. Med. Chem.*, 2007, **15**, 7325.

[60] H. Kitagawa, T. Ozawa, S. Takahata, M. Iida, J. Saito, and M. Yamada, *J. Med. Chem.*, 2007, **50**, 4710.

[61] Y. Shen, J. Liu, G. Estiu, B. Isin, Y. Y. Ahn, D. S. Lee, A. L. Barabási, V. Kapatral, O. Wiest, and Z. N. Oltvai, *Proc. Natl. Acad. Sci. USA*, 2010, **107**, 1082.

PART VI:
Topics in Biology

Editor: John Lowe
JL3Pharma LLC
Stonington
Connecticut

CHAPTER 19

Back to the Plaque: Emerging Studies That Refocus Attention on the Neuritic Plaque in Alzheimer's Disease

Robert B. Nelson

Lundbeck Research USA, 215 College Drive, Paramus, NJ 07652, USA

Annual Reports in Medicinal Chemistry, Volume 45
ISSN 0065-7743, DOI 10.1016/S0065-7743(10)45019-8

ABBREVIATIONS

AD   Alzheimer's disease
NP   Neuritic plaque
PiB   Pittsburgh compound B

1. INTRODUCTION

Alzheimer's disease (AD) is the most common neurological disorder of advanced aging, affecting nearly one in ten individuals over the age of 65 years. An additional 10% of the population will develop AD with each additional decade of life. AD leads to profound impairments in memory and executive function, and places a tremendous burden on society with regard to patient care costs, lost productivity, and high burden on caregivers. With the demographic aging of the population, the prevalence of AD in the United States is projected to grow to over 14 million by the year 2050 [1].

The most definitive diagnosis of AD is a postmortem examination of the brain for the presence of two characteristic lesions: the neuritic plaque (NP) and the neurofibrillary tangle. Both structures were originally described in 1906 by Alois Alzheimer using silver-based histological stains. The discovery of NPs was hailed as a watershed moment in the history of neurological disease as it helped shift society's perception of age-related dementia from social stigma to physical disease [2].

Given this historical perspective, it is somewhat ironic that the role of the NP in the pathogenesis of AD has been deeply questioned in recent years, with opinion shifting from an assumption that NPs play a causal role in disease to the opposing hypothesis that NPs are either benign or protective in AD. This chapter begins with a description of the cardinal features of the NP and the changing context in which these features have been viewed relative to disease pathogenesis. Key observations in the literature that have served to deemphasize the pathologic role of NPs in AD are revisited in parallel with more recent findings that have begun to refocus attention back toward the NP as a potential causal factor in the development of AD. In the final section, a proposed sequence of events leading to the formation of NPs and ultimately to clinical AD is put forth, and the therapeutic implications of focusing on targets related to NPs are discussed.

2. CHANGING PERSPECTIVES ON CARDINAL FEATURES OF THE NP

2.1 Key features of the NP

The NP is composed first and foremost of a dense core of fibrillar material (termed amyloid) that in 1984 was identified as a 40–42-amino acid-long peptide termed the Aβ peptide [3]. The altered conformation of Aβ peptide into folded "stacks" having an antiparallel β-pleated sheet structure defines amyloid. A second key feature of the NP is an almost invariable association of the amyloid core with a cluster of tightly bound macrophages [4]. At the ultrastructural level, the amyloid core resembles a sea urchin, with the "spikes" from the plaque core extending in tortuous membrane-lined invaginations that penetrate far into the cytoplasm of the macrophages [5]. A third defining characteristic of the plaques is a sphere of reactive astrocytes encircling the macrophage-bound plaque core [4]. Finally, a fourth characteristic of the NP from whence its name is derived is a "halo" of dystrophic neuritic processes that surround the plaque core, stain positive for silver, and balloon in size wherever they come in direct contact with the plaque core.

A sizeable number of transgenic mouse models have now been generated that overexpress the Aβ peptide precursor protein with or without other transgenes related to either increased liberation of the Aβ-peptide or production of the longer and more deposition-prone version of the peptide ($A\beta_{1-42}$) [6]. These transgenic models have been disappointing in some regards in that they fail to produce a full spectrum of disease pathology, most notably lacking neurofibrillary tangle formation and associated hyperphosphorylation of the tau protein. On the other hand, they faithfully recapitulate the four cardinal features of an NP noted above and offer the opportunity to study this hallmark feature of AD pathology in isolation from full disease expression.

2.2 The amyloid core

A better understanding of the biochemical nature of the amyloid core of the NP was only gained decades after its initial description. In 1984, it was discovered that boiling the plaque cores in formic acid could dissociate them, revealing that they were proteinaceous in nature and composed primarily of the Aβ peptide [3].

Antibodies raised against the Aβ peptide were subsequently used to identify more diffuse types of Aβ-containing deposits in AD brain by immunohistochemistry. Diffuse Aβ peptide deposits lack the characteristic fibrillar β-sheet structure that defines "amyloid" in the NPs, as well as the associated macrophage, astrocyte, and neurite pathology [7]. Thus NPs, but not diffuse Aβ deposits, appear to disrupt the neuropil in which they are found and to be selectively associated with an innate immune response.

NPs have largely been presumed to form late in the disease and to gradually increase in size. Both of these presumptions are in question following the emergence of new technologies for imaging NP growth *in vivo*. The clinical imaging agent ^{11}C Pittsburgh compound B (PiB), which recognizes the fibrillar structure of Aβ, has revealed that NP cores form in cognitively normal or mildly impaired humans over a fairly brief period [8]. Recent longitudinal studies suggest that the occurrence of PiB labeling in human brain is associated with a subsequent higher rate of cognitive decline and/or conversion to AD [9]. The rate of formation of individual amyloid cores has also been studied using two-photon microcopy in Aβ-depositing transgenics, where it was found that amyloid cores form and reach their terminal size within 24–48 h [10]. After forming, these plaque cores are remarkably stable and in fact serve as landmarks for detecting the formation of additional NPs.

2.3 Macrophage role in NPs

Macrophages are an almost invariant feature of NPs, and were first identified in association with this plaque type in the late 1980s [4]. The conventional view of NP-associated macrophages is that they arrive after the plaque core has formed and represent an unsuccessful attempt to phagocytose the amyloid core. However, investigators examining the relationship between plaques and macrophages at the electron microscope level have questioned this presumption. Macrophages surrounding the amyloid core do not have the characteristic morphology of cells engaged in phagocytosis, including a conspicuous lack of phagolysosomes that would indicate attempted amyloid ingestion and degradation [11]. Instead, these macrophages are characterized by a labyrinth of channels that are filled with bundles of fibrillar Aβ and that remain contiguous with the outside of the cell. This unusual relationship between amyloid and macrophage instead suggested that the macrophages might actually be the site of fibrillar Aβ generation.

Cultured macrophages are able to rapidly refold Aβ peptide into its amyloidogenic conformation, which also supports the hypothesis that macrophages play a causal role in the generation of the plaque core [12,13]. Macrophage-facilitated refolding of amyloidogenic peptides may be a phenomenon common to different amyloidoses since it has also been observed for serum amyloid A-derived amyloidogenic peptide following addition to peripheral macrophage cultures. In both cultured macrophages and the brain, Aβ-derived amyloid is found in long invaginations of the cell that terminate in clathrin-coated "heads," suggesting that clathrin-coated pits may provide an environment conducive to amyloid fibril formation [12,14].

Macrophages associated with NPs have historically been presumed to be of brain origin (i.e., microglia). Once again, it was close examination of electron micrographs that first led to a questioning of this assumption [11]. Serial reconstruction of electron micrographs revealed that (1) virtually all examined NPs are closely associated with cerebral microvessels and (2) cells of monocyte lineage are seen infiltrating at the interface between blood vessels and plaques. Subsequent studies used green fluorescent protein-labeled bone marrow cells grafted into an Aβ-depositing transgenic mouse to demonstrate recruitment of blood monocytes into brain and around NPs [15].

More recent papers have attempted to establish a role for macrophages in either plaque generation or plaque containment [16,17]. The prevailing strategy has been to cross genetically a mouse that expresses thymidine kinase via a monocyte/macrophage-specific promoter with an Aβ-depositing transgenic mouse, and then induce death of the dividing monocytes/macrophages through exposure to exogenously administered gancyclovir. Systemic administration of gancyclovir to these transgenic crosses proved too toxic to allow organism-wide depletion of macrophages, so local infusion of gancyclovir into brain was utilized instead.

The first paper to employ this strategy found an increase in "amyloid load" after central gancyclovir administration and concluded that recruited blood monocytes play a role in limiting NP development [16]. However, the authors made some assumptions to reach this conclusion, namely, that centrally delivered gancyclovir would kill invading monocytes, which were presumed to be dividing, and spare resident microglia, which were presumed to be terminally differentiated. A recent study employed the same strategy and examined the pattern of surviving monocytes/macrophages in brain. This second study revealed that resident microglia were almost completely ablated by exposure to central gancyclovir, casting doubt on the conclusions of the earlier paper [17]. The second study also found that "plaque load" was unaltered by centrally delivered gancyclovir despite a dramatic overall lowering of central macrophage number. The authors concluded that macrophages play no role in plaque formation.

The conclusion from Grothwahl et al. [17] may also have some caveats. Since evidence from *in vivo* longitudinal two-photon microscopy studies indicates that full-size plaque cores can form within 24 h [10], while depletion of central macrophage populations by gancyclovir requires 2 weeks [17], the experimental design used is unable to rule out the possibility that monocytes are recruited into brain, generate plaque cores within 24 h, and then slowly succumb to local gancyclovir exposure before degrading and being resorbed from the plaque cores. Higher resolution figures in this paper reveal macrophages clustered around a subset of plaque cores after weeks of central gancyclovir exposure. This is consistent with the possibility that continuous blood monocyte recruitment and plaque generation occur in this experimental design. Additional studies are needed to resolve the role of macrophages in NP generation or clearance.

2.4 Neuritic pathology associated with NPs

A defining feature of the NP is the "halo" of dystrophic neuritic processes that surround the plaque core. These structures were identified at the same time that plaque cores were first described in the early 20th century. While the morphology of these dystrophic neurites clearly points to a pathological state, more recent studies have expanded the characterization of this pathology. Classic silver-based histopathological stains only reveal changes in neurite structure close to the plaque, but anterograde tracing studies that inject phytohemagglutinin in the entorhinal cortex reveal that dystrophic neurites are part of an ectopic sprouting event that culminates in a marked increase in labeled fibers associated with plaque cores, and a corresponding denervation of perforant path fibers in the middle molecular layer of the hippocampal formation [18]. This redirection of fibers is suggestive of chemotactic gradients that peak at the plaque core and attract extending neurites (most likely axons). The plaque-bound macrophages and surrounding astrocytes are a logical candidate source for such factors, and it is of interest that gradients of brain-derived neurotrophic factor identified by laser capture dissection of tissue immediately surrounding NPs originate from both macrophages and astrocytes [19]. The NP is in many ways reminiscent of a "wound response" phenotype seen when macrophages migrate to the edge of a stab wound in the brain and promote reinnervation of the area. Notably different is that "wound responses" resolve over time, whereas NPs appear to be relatively permanent structures.

Dystrophic neurites show decreased spine density and decreased spine stabilization with increasing proximity to the plaque core [20]. Ca^{2+}-sensitive dyes reveal a profound elevation in Ca^{2+} in neurites that are closer to the plaque core [21]. The cause of this Ca^{2+} dysregulation is unknown, but it is interesting that host defense mechanisms associated with NPs, including activation of the complement cascade and potentially oligomerization of the Aβ peptide, are in turn associated with the formation in membranes of Ca^{2+}-permeable pores intended to lyse microorganisms [22,23]. "Bystander lysis" of neurons associated with NPs is a long-standing hypothesis for neuronal dysfunction in NP-associated neurites.

2.5 Microvascular association of NPs

A selective association of NPs with the microvasculature has been proposed for many years, but this observation failed to gain widespread acceptance due to arguments that microvessels are so plentiful that any specific association with them is by chance. Recent studies have challenged that view. Serial reconstruction of electron micrographs reveals a perivascular localization of all NPs examined [11]. The "random association" argument was tested by creating overlays of blood vessels and NPs from AD brain, then scoring the association for the correct orientation versus the association found with multiple random shifts of the overlays [24]. Two main conclusions emerged from the article: (1) the association rate between plaque and vessel was far higher for the correct overlay orientation than for the random overlays and (2) each NP appeared to be the site of a microinfarct based on blood proteins found to be associated with the plaque. Interestingly, this same NP association with blood vessels has been reported in several Aβ-depositing transgenic mouse models as well, suggesting that both the process and the localization of NP formation in human disease is largely recapitulated in the transgenic models [25].

An association of amyloid cores with small blood vessels and microcapillaries is perhaps not surprising given the well-known association of amyloid with larger bore blood vessels that gives rise to congophilic amyloid angiopathy [26]. Microvascular origin of NPs also complements the emerging consensus that blood monocytes migrate across the blood–brain barrier to become associated with and likely generate NP cores. Finally, the microhemorrhage and vascular edema liabilities associated with Aβ immunotherapy [27] can be considered in light of the possibility that removal of plaque material may compromise a physical barrier formed between blood and brain in the area of a microinfarct.

3. REVISITING THE IMPORTANCE OF NPs IN ALZHEIMER'S DISEASE

At least four lines of evidence have emerged that have been used to argue that NPs have a small or negligible role in the pathogenesis of AD. In this section, these lines of evidence are revisited in light of emerging data that reexamine this view.

3.1 Lack of correlation between NP load and dementia

One of the early findings used to argue against a pathological role for NPs in the pathogenesis of AD was the lack of correlation between extent of Aβ plaque deposition and degree of dementia measured using standard clinical scales [28]. One source of confusion in many of these studies was a failure to distinguish different types of Aβ immunoreactivity. Plaque deposition as defined by Aβ immunoreactivity occurs as large diffuse areas of staining that show minimal to no perturbation of neuropil associated with it.

While NPs are also labeled by Aβ antibodies, the total area of this labeling is a small fraction of the total area of diffuse Aβ immunoreactivity. Thus, these studies sought to correlate dementia with a form of Aβ that seems to have little effect on associated neuropil. These studies, not surprisingly, were for the most part negative. Other studies have restricted their analysis to fibrillar forms of Aβ that are relevant to NPs and have been more equivocal in their reporting of a link between plaque load and dementia [29].

Studies that have focused on neuritic response to the NP rather than to presence of the plaque core itself have been more successful in finding a relationship between plaque-associated dystrophic neurites and dementia [30]. These studies suggest that plaque cores per se may only be an initiating event and that the gradual effects of the plaque core on adjacent neuronal pathway architecture may actuate the ultimate disruption of function. The timeframe over which the latter occurs, and the nature of how plaque-induced changes in synaptic connectivity ultimately affect cognition, is not known.

An understanding of the relationship between NP formation and disease progression in AD has been greatly boosted in recent years with the advent of PiB imaging in humans. A number of recent studies have employed longitudinal PiB scans on cognitively normal individuals or individuals with amnestic mild cognitive impairment and found that the presence of PiB labeling is predictive of more rapid cognitive decline [9]. This finding is consistent with the interpretation that plaque core formation may be necessary, but is not sufficient for cognitive decline to occur. Rather it may be the protracted response to the presence of plaque cores in the brain that leads to cognitive decline and onset of dementia.

3.2 Subtle behavioral deficits in Aβ plaque-depositing transgenic mice

The availability of transgenic mice that faithfully recapitulate NP generation but fail to progress to neurofibrillary tangle formation, show little or no neuronal loss, and demonstrate modest at best cognitive deficits became a key finding leading to the prevalent view that NPs are an epiphenomenon of AD and do not cause cognitive decline [6]. Given the disruption of normal synaptic connectivity that is evident in NP-bearing mice [18], it is remarkable that these mice do not show more profound behavioral deficits in cognition. However, the most widely used cognitive assessments in rodents were developed by and large to detect global deficits in cognition induced by scopolamine or complete lesions of areas critical for memory consolidation. These behavioral indices may not be sensitive enough to detect the more subtle deficits in pathway connectivity associated with NPs.

While Aβ-depositing transgenic mice clearly recapitulate many features of the NP also present in AD brain, no additional pathological features associated with AD appear to develop. Without additional genetic manipulations, mice do not go on to develop hyperphosphorylation of tau, neurofibrillary tangles, and neuronal loss, which are all cardinal features of human disease. One striking difference between human NPs and transgenic mouse NPs is the presence of intense immunolabeling for hyperphosphorylated tau epitopes in neurites surrounding the former but not the latter [31]. Surprisingly, little work has been done to understand the signal transduction pathways that might tie NP pathology to tau-related pathology. Arresting AD progression at the stage of NP formation by blocking downstream pathological signaling events leading to tau hyperphosphorylation and neurofibrillary tangle formation could be an attractive therapeutic option. Humans who have only NPs in their brains (i.e., a high PiB load) with little tau-related pathology appear to have only subtle impairments in cognition [8] and in this sense are reminiscent of the subtle cognitive changes seen in Aβ-depositing transgenic mice.

3.3 Oligomeric Aβ as the toxic species in Alzheimer's disease

The "amyloid hypothesis" has been a guiding principle in AD research for many years [32]. Based largely on genetic links to Aβ from familial AD, this hypothesis posits that the Aβ peptide is an obligatory contributor to the neurodegeneration that occurs in AD. The de-emphasis of NPs as a focus for the amyloid hypothesis gave rise to an alternative version of the amyloid hypothesis, namely, that soluble oligomeric species of the Aβ peptide rather than fibrillar forms of Aβ were responsible for "toxicity" associated with Aβ [33]. While a comprehensive review of oligomeric toxicity hypothesis is outside the scope of this chapter, it is of note that this literature was largely built around studies conducted in cultured neurons. One difficulty for this hypothesis has been the observation that perturbation of neuronal processes in AD brain is focused around NPs. This localization of pathology is difficult to reconcile with the concept of a soluble toxin that should theoretically have a fairly uniform distribution pattern of toxicity. A recent paper offers a potential reconciliation for this discrepancy. It reveals that NPs are a potential reservoir of oligomeric Aβ species, such that concentration gradients of oligomeric Aβ peak near the plaque core [34]. Oligomeric Aβ immunoreactivity is reported to colocalize with postsynaptic densities located near plaque cores and to be associated with spine collapse. While confirmation is needed, this finding potentially reconciles the competing schools of thought that posit "plaque" versus "oligomeric Aβ" as the synaptotoxic species in the brain of AD patients. In the amended framework, both are important to the manifestation of spatially restricted neurotoxicity.

3.4 *Post hoc* findings from Aβ immunotherapy clinical trials

Immunization with either Aβ peptide or monoclonal antibodies raised against the Aβ peptide is reported to block NP formation or to have a modest effect on reducing existing NP load in Aβ-depositing transgenic mice [35]. Clinical trials involving active immunization with the Aβ peptide were halted due to toxicity [36]. However, former subjects enrolled in the trial have continued to come to autopsy. Case studies suggested that some of these subjects may have experienced a lowering of amyloid load, yet showed a continuing decline in cognition [37]. These results have been used to argue that Aβ deposition is not important in cognition. There are several caveats to this conclusion. First, these are uncontrolled *post hoc* analyses from an aborted clinical trial. Second, if there was in fact a clearance of amyloid load induced by an elevated titer of Aβ antibodies, the timeframe over which this antibody titer formed and the clearance occurred is unknown. Third, it is not known in what temporal window Aβ immunotherapy needs to be administered to be clinically efficacious. If amyloid deposition is a triggering event for subsequent pathological events, we now know that the mild-to-moderate AD patient population is well past the point at which NPs form in brain. In sum, post hoc analysis of the original Aβ immunotherapy trial is not able to test the "amyloid hypothesis" [32].

4. SUMMARY: AN INTEGRATED HYPOTHESIS OF NP FORMATION AND PATHOLOGY

While the role of NPs in AD is far from resolved at this time, the emerging literature reviewed here reopens the case for considering NPs as a triggering pathology in the development of AD. An integration of this literature suggests the following sequence of events that lead to NP pathology and subsequent dementia (Figure 1). The goal in constructing

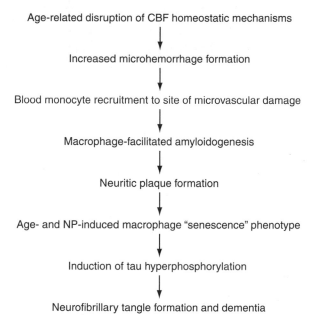

Age-related disruption of CBF homeostatic mechanisms

↓

Increased microhemorrhage formation

↓

Blood monocyte recruitment to site of microvascular damage

↓

Macrophage-facilitated amyloidogenesis

↓

Neuritic plaque formation

↓

Age- and NP-induced macrophage "senescence" phenotype

↓

Induction of tau hyperphosphorylation

↓

Neurofibrillary tangle formation and dementia

Figure 1   Hypothetical sequence of events in the NP-mediated progression of AD.

this hypothetical sequence of events is to provide testable hypotheses and potential new therapeutic avenues for AD.

The first hypothesized pathological step in AD is a disruption of cerebral blood flow autoregulation, brought about in part by the effects of vascular Aβ on the microvasculature [38]. This loss of autoregulation would lead to intermittent elevation of blood pressure at brain microcapillaries at times when mean arterial pressure is elevated. The normal role that Aβ peptide likely plays in cerebrovascular function is likely to be regulation of functional hyperemia, a response in which blood flow is increased to those brain regions involved in greater neuronal activity with higher metabolic need [39]. A chronic pathological loss of cerebrovascular autoregulation might be predicted to result in a local spiking of blood pressure in microcapillaries, which in turn leads to damage of microvessels. The local damage to the microvasculature in turn induces monocytes to migrate to the site of damage, resulting in monocyte recruitment across the blood–brain barrier [15].

Recruited blood monocytes encountering local elevated levels of Aβ peptide would then refold the Aβ into a fibrillar conformation, thus forming the core of an NP adjacent to a blood vessel [24]. These plaque-bound monocytes differentiate into macrophages and mount an aberrant "wound response" through the release of neurotrophic and chemotactic agents, inducing an inappropriate ectopic sprouting response toward the plaque core [18]. Those recruited neurites coming in contact with the plaque core swell and become dystrophic. The ectopic sprouting response causes denervation of normal terminal fields that these neurites would normally migrate toward and innervate.

While the disruption of neuropil surrounding the NP is extensive, the effects of this disruption on cognition are fairly subtle [8]. However, macrophage "senescence" appears to be accelerated by the presence of NPs, and it has been postulated that the appearance of dystrophic macrophages in AD brain is accompanied by loss of trophic support to neurons [40]. Dystrophic macrophages appear in AD brain in concert with phosphotau-positive dystrophic neurites [40]. Multiple phosphotau epitopes identified in AD brain are consistent with activity by glycogen synthase kinase 3β (GSK-3β), a key metabolic regulatory enzyme. Elevated GSK-3β-mediated protein phosphorylation in brain accompanies hypometabolic states such as hibernation and pentobarbital-induced hypothermia [41,42]. It is tempting to speculate that a hypometabolic state brought about by loss of macrophage trophic support in later stages of AD is a triggering event that moves AD pathology from an NP-centered event to a phosphotau-centered event, culminating in neurofibrillary tangle formation, widespread neuronal loss, and manifestation of clinical AD.

This sequence of events suggests several therapeutic strategies. Early in the disease process, strategies aimed at blocking recruitment of blood monocytes or shifting the phenotype of these cells away from macrophage-facilitated amyloidogenesis toward successful clearance would be desirable. Later in the disease process, strategies that block signaling events downstream of NPs would be favored, including those that modulate phenotype of macrophages physically associated with NPs. If macrophage senescence and loss of trophic support to neurons is a trigger for tau pathology, then novel strategies to support macrophage function may be called for. Success might be monitored by the ability to suppress GSK-3β-mediated tau hyperphosphorylation in brain.

In conclusion, emerging evidence indicates that the role of NPs in AD pathology may have been prematurely discounted. The formation of NPs is likely to be a dynamic process that can be altered through many strategies not directly targeting generation of the Aβ peptide. The targeting of NPs in the search for new disease-modifying strategies offers a complementary approach to Abeta oligomer-directed strategies, and is well situated to take advantage of emerging imaging agents that visualize NPs in the living brain.

REFERENCES

[1] L. E. Hebert, P. A. Scherr, J. L. Bienias, D. A. Bennett, and D. A. Evans, *Arch. Neurol.*, 2003, **60**, 1119.

[2] G. E. Berrios In *Dementia* (ed. J. O'Brien, D. Ames, and A. Burns), Arnold, London, 2000, p. 3.

[3] G. Glenner and C. Wong, *Biochem. Biophys. Res. Comm.*, 1984, **120**, 885.

[4] S. Itagaki, P. McGeer, H. Akiyama, S. Zhu, and D. Selkoe, *J. Neuroimmunol.*, 1989, **24**, 173.

[5] J. Wegiel and H.M. Wisniewski, *Acta Neuropathol.*, 1990, **81**, 116.

[6] D. Games, M. Buttini, D. Kobayashi, D. Schenk, and P. Seubert, *J. Alzheimers Dis.*, 2006, **9**, 133.

[7] L. C. Cork, C. Masters, and K. Beyreuther, *Am. J. Pathol.*, 1996, **137**, 1383.

[8] C. R. Jack, V. J. Lowe, S. D. Weigand, H. J. Wiste, M. L. Senjem, D. S. Knopman, M. M. Shiung, J. L. Gunter, B. F. Boeve, B. J. Kemp, M. Weiner, and R. C. Petersen, *Brain*, 2009, **132**, 1355.

[9] J. C. Morris, C. M. Roe, E. A. Grant, D. Head, M. Storandt, A. M. Goate, A. M. Fagan, D. M. Holtzman, and M. A. Mintun, *Arch. Neurol.*, 2009, **66**, 1469.

[10] M. Meyer-Luehmann, T. L. Spire-Jones, C. Prada, M. Garcia-Alloza, A. de Calignon, A. Rozkalne, J. Koenigsknecht-Taiboo, D. M. Holtzmann, B. J. Backskai, and B. T. Hyman, *Nature*, 2008, **451**, 720.

[11] J. Wegiel, H. Imaki, K. -C. Wang, J. Wegiel, A. Wronska, M. Osuchowski, and R. Rubenstein, *Acta Neuropathol.*, 2003, **105**, 393.

[12] G. P. Gellerman, K. Ullrich, A. Tannert, C. Unger, G. Habicht, S. R. N. Sauter, P. Hortschansky, U. Horn, U. Mollmann, M. Decker, J. Lehmann, and M. Fandrich, *J. Mol. Biol.*, 2006, **360**, 251.

[13] J. E. Finley, R. B. Nelson, and C. E. Nolan, *Patent Application WO 0606091637*, 2006.

[14] J. Wegiel, K. -C. Wang, H. Imaki, R. Rubenstein, A. Wronska, M. Osuchowski, W. J. Lipinski, L. C. Walker, and H. LeVine, *Neurobiol. Aging*, 2001, **22**, 41.

[15] A. K. Stalder, F. Ermini, L. Bondolfi, W. Krenger, G. J. Burbach, T. Deller, J. Coomaraswamy, M. Staufenbiel, R. Landmann, and M. Jucker, *J. Neurosci.*, 2005, **25**, 11125.

[16] A. R. Simard, D. Soulet, G. Gowing, J. P. Julien, and S. Rivest, *Neuron*, 2006, **49**, 489.

[17] S. A. Grothwohl, R. E. Kalin, T. Bolmont, S. Prokop, G. Winkelmann, S. A. Kaeser, J. Odenthal, R. Radde, T Eldh, S. Gandy, A. Aguzzi, M. Staufenbiel, P. M. Mathews, H. Wolburg, F. L. Heppner, and M. Jucker, *Nat. Neurosci.*, 2009, **12**, 1361.

[18] A. L. Phinney, T. Deller, M. Stalder, M. E. Calhoun, M. Frotscher, B. Sommer, M. Staufenbiel, and M. Jucker, *J. Neurosci.*, 1999, **19**, 8552.

[19] G. J. Burbach, R. Hellweg, C. A. Haas, D. Del Turco, U. Deicke, D. Abramowski, M. Jucker, M. Staufenbiel, and T. Deller, *J. Neurosci.*, 2004, **24**, 2421.

[20] T. L. Spires-Jones, M. Meyer-Luehmann, J. D. Osetek, P. B. Jones, B. J. Backskai, and B. T. Hyman, *Am. J. Pathol.*, 2007, **171**, 1304.

[21] K. V. Kuchibhotla, S. T. Goldman, C. R. Lattarulo, H. Y. Wu, B. T. Hyman, and B. J. Bacskai, *Neuron*, 2008, **59**, 214.

[22] D. H. Small, R. Gasperini, A. J. Vincent, A. J. Hung, and L. Foa, *J. Alzheimers Dis.*, 2009, **16**, 225.

[23] B. P. Morgan, J. P. Luzio, and A. K. Campbell, *Cell Calcium*, 1986, **7**, 399.

[24] K. M. Cullen, Z. Kocsi, and J. Stone, *Neurobiol. Aging*, 2006, **27**, 1786.

[25] S. Kumar-Singh, D. Pirici, E. McGowan, S. Serneels, C. Ceuterick, J. Hardy, K. Duff, D. Dickson, and C. Van Broeckhoven, *Am. J. Pathol.*, 2005, **167**, 527.

[26] R. O. Weller, D. Boche, and J. A. Nicoll, *Acta Neuropathol.*, 2009, **118**, 87.

[27] D. M. Wilcock, P. T. Jantzen, Q. Li, D. Morgan, and M. N. Gordon, *Neuroscience*, 2007, **144**, 950.

[28] P. A. Arrigada, J. H. Growdon, and E. T. Hedley-White, *Neurology*, 1992, **42**, 631.

[29] K. Hsiao, P. Chapman, and S. Nilsen, *Science*, 1996, **274**, 99.

[30] R. B. Knowles, T. Gomez-Isla, and B. T. Hyman, *J. Neuropathol. Exp. Neurol.*, 1998, **57**, 1122.

[31] O. Yasuhara, T. Kawamata, Y. Aimi, E. G. McGeer, and P. L. McGeer, *Neurosci. Lett.*, 1994, **171**, 73.

[32] J. Hardy, *J. Neurochem.*, 2009, **110**, 1129.

[33] W. L. Klein, W. B. Stine, and D. B. Teplow, *Neurobiol. Aging*, 2004, **25**, 569.
[34] R. M. Koffie, M. Meyer-Luehmann, T. Hashimoto, K. W. Adams, M. L. Miclkc, M. Garcia-Alloza, K. D. Micheva, S. J. Smith, M. L. Kim, V. M. Lee, B. T. Hyman, and T. L. Spires-Jones, *Proc. Natl. Acad. Sci. USA*, 2009, **106**, 4012.
[35] T. Bussière, F. Bard, R. Barbour, H. Grajeda, T. Guido, K. Khan, D. Schenk, D. Games, P. Seubert, and M. Buttini, *Am. J. Pathol.*, 2004, **165**, 987.
[36] K. Marder, *Curr. Neurol. Neurosci. Rep.*, 2003, **3**, 371.
[37] C. Holmes, D. Boche, D. Wilkinson, G. Yadegarfar, V. Hopkins, A. Bayer, R. W. Jones, R. Bullock, S. Love, J. W. Neal, E. Zotova, and J. A. Nicoll, *Lancet*, 2008, **372**, 216.
[38] K. Niwa, K. Kazama, L. Younkin, S. G. Younkin, G. A. Carlson, and C. Iadecola, *Am. J. Physiol. Heart Circ. Physiol.*, 2002, **283**, H315.
[39] K. Niwa, L. Younkin, C. Ebeling, S. K. Turner, D. Westaway, S. Younkin, K. H. Ashe, G. A. Carlson, and C. Iadecola, *Proc. Natl. Acad. Sci. USA*, 2000, **97**, 9735.
[40] W. J. Streit, H. Braak, Q. S. Xue, and I. Bechmann, *Acta Neuropathol.*, 2009, **118**, 475.
[41] T. Arendt, J. Stieler, A. M. Strijkstra, R. A. Hut, J. Rüdiger, E. A. Van der Zee, T. Harkany, M. Holzer, and W. Härtig, *J. Neurosci.*, 2003, **23**, 6972.
[42] E. Planel, K. E Richter, C. E Nolan, J. E. Finley, L. Liu, Y. Wen, P. Krishnamurthy, M. Herman, L. Wang, J. B. Schachter, R. B Nelson, L. F. Lau, and K. E. Duff, *J. Neurosci*, 2007, **27**, 3090.

CHAPTER 20

Targeting Methyl Lysine

Stephen V. Frye[*], **Tom Heightman**[**] and **Jian Jin**[*]

ABBREVIATIONS

2-OG	2-Oxoglutarate
HTS	High-throughput screening
ITC	Isothermal titration calorimetry
KDM	Lysine demethylase

[*] The Center for Integrative Chemical Biology and Drug Discovery, Eshelman School of Pharmacy, The University of North Carolina at Chapel Hill, Genetics Medicine Building, Campus Box 7363, Chapel Hill, NC 27599-7363, USA

[**] Structural Genomics Consortium, Oxford University, Oxford OX3 7DQ, UK

Annual Reports in Medicinal Chemistry, Volume 45
ISSN 0065-7743, DOI 10.1016/S0065-7743(10)45020-4

KMe   Lysine methylation marks
MAOI   Monoamine oxidase inhibitor
PHD   Plant homeodomain
PTM   Posttranslational modification
PMT   Protein lysine methyl
             transferase
SAH   S-adenosyl-L-homocysteine
SAM   S-adenosyl-L-methionine
SET domain   Su(var)3-9 (suppressor of
             variegation 3-9), E(z)
             (enhancer of zeste), and
             trithorax domain

1. INTRODUCTION: EPIGENETICS

Multicellular organisms have evolved elaborate mechanisms to enable differential and cell-type-specific expression of genes. Epigenetics refers to the heritable changes in how the genome is accessed in different cell types and during development and differentiation [1]. This capability permits specialization of function between cells even though each cell contains essentially the same genetic code. Over the past decade, the cellular machinery that creates these heritable changes has been the subject of intense scientific investigation as there is no area of biology or indeed, human health, where epigenetics may not play a fundamental role [2].

The template upon which the epigenome is written is chromatin—the complex of histone proteins, RNA, and DNA that efficiently package the genome in an appropriately accessible form within each cell. The basic building block of chromatin structure is the nucleosome—an octamer of histone proteins (associated dimers of H3 and H4 capped with dimers of H2A and H2B) around which approximately 150 base pairs of DNA are wound. The amino-terminal tails of histone proteins project from the nucleosome structure and are subject to more than 100 posttranslational modifications (PTMs) [2]. The state of chromatin, and therefore access to the genetic code, is largely regulated by specific PTMs to histone proteins and DNA, and the recognition of these marks by other proteins and protein complexes [3,4]. The enzymes that produce these modifications (the "writers"), the proteins that recognize them (the "readers"), and the enzymes that remove them (the "erasers," Figure 1) are critical targets for manipulation in order to further understand the histone code and its role in biology and human disease [5,6]. Indeed, small-molecule inhibitors of histone deacetylases have already proven useful in the treatment of cancer [7,8] and the role of lysine acetylation is rivaling that of

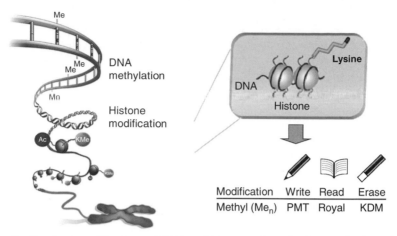

Figure 1 Covalent modifications of DNA and histones play a fundamental role in the regulation of differentiation and development. The writers, readers, and erasers of this dynamic code are potentially amenable to modulation with small molecules. Lysine methylation is a critical posttranslational modification influencing chromatin function (PMT = protein lysine methyltransferase, royal family proteins bind KMe, KDM = lysine demethylase).

phosphorylation in importance as a PTM of protein function [9,10] (See Chapter 15 in this volume for a discussion of histone acetylation.)

Among the chemical marks that make up the histone code, the methylation of lysine residues plays a central role in development, differentiation, and cellular response to the environment through its influence on activation and repression of gene expression [11,12]. This review will focus on the writers (protein lysine methyl transferases, PMTs), readers (methyl lysine-binding proteins), and erasers (lysine demethylases, KDMs) of methyl lysine marks and the emerging chemical approaches to modulating the function of these protein families. General prospects for drug discovery in this arena will also be discussed. A more thorough presentation of diverse epigenetic targets implicated in oncology drug discovery appears in Chapter 15 in this volume.

2. LYSINE METHYLATION AS A DYNAMIC REGULATOR OF CHROMATIN FUNCTION

The dynamic and differential methylation of lysine residues in histone tails plays a central role in the creation of the histone code and the regulation of chromatin structure and function that the code implies [12]. For example, different lysine methylation marks (KMe) are associated with active (histone 3, lysine 4 trimethylation, i.e., H3K4Me3) and repressed transcriptional states (H3K9Me3). Additionally, global differences in the patterns of lysine methylation correlate with cellular differentiation state and are markedly different in stem cells versus terminally differentiated cells [13]. In addition to the site-specific function of methyl lysine marks, the existence of three degrees of methylation (KMe1–3) has also been shown to regulate the state of chromatin via recruitment of different readers based on the degree of modification [14]. While active methylation of specific lysine residues by lysine methyl transferases has been appreciated for many years, the resulting methyl lysine mark was generally considered to be a stable PTM until the discovery of the amine oxidase LSD1, which can demethylate KMe1 and KMe2 [15] and the subsequent discovery of demethylases acting on KMe3—the Jumonji domain-containing histone demethylases [16].

The protein families that maintain this cellular literacy for KMe marks are therefore central to the creation and interpretation of the histone code and are fascinating potential targets for chemical biology and drug discovery [16–21]. In addition to the attraction of their potential therapeutic relevance, many of the functional domains of the readers, writers, and erasers have had their three-dimensional structure defined by X-ray crystallography [5] and therefore present a rather unique opportunity to explore the chemical biology and medicinal chemistry of a set of relatively uncharted protein families with high-quality structural information in hand. While this review will largely focus on the role of histone lysine methylation in chromatin function, it is clear that other proteins are subject to this modification [22] and a complete understanding of the activity of chemical probes in this area must also consider their effect on nonhistone proteins [23].

3. PROTEIN LYSINE METHYL TRANSFERASES

3.1 Protein family

PMTs catalyze mono-, di-, or trimethylation of lysine residues of various proteins including histones. With the exception of DOT1L, all PMTs contain the evolutionarily conserved SET domain, named after *Drosophila* Su(var)3-9 (suppressor of variegation 3-9), E(z) (enhancer of zeste), and trithorax [24]. Over the past decade, more than 50 human PMTs have been identified (Figure 2) [11,12]. The SET domain-containing PMTs are classified into five subfamilies based on SET domain homology and are named after their founding members: SUV39, SET1, SET2, RIZ, and SMYD3.

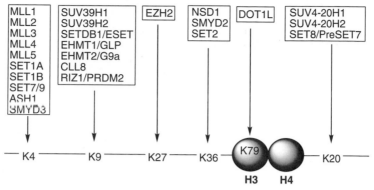

Figure 2 Selected human protein lysine methyltransferases and their target sites on histones [11].

Three-dimensional X-ray crystal structures of the SET domains of >10 PMTs and the catalytic domain of DOT1L have been reported to date [25–27]. These structures, either in the apo-state or when bound to the cofactor product S-adenosyl-L-homocysteine (SAH), a histone peptide, or an inhibitor, yield key structural insights into enzyme/substrate/cofactor/inhibitor interactions and inform approaches to further inhibitor design.

3.2 Chemical approaches

To date, three selective small-molecule PMT inhibitors have been reported [25,28,29]. Chaetocin, a fungal mycotoxin that belongs to the class of 3,6-epidithio-diketopiperazines, was identified as the first selective small-molecule inhibitor of *Drosophila melanogaster* SU(VAR)3-9 (IC_{50} = 0.6 μM) via screening of 2,967 compounds using a standard radioactive filter-binding assay [28]. Chaetocin also inhibited SUV39H1 (IC_{50} = 0.8 μM), the human ortholog of SU(VAR)3-9 that is an H3K9 PMT. Although chaetocin inhibited other H3K9 PMTs such as *Neurospora crassa* DIM5 (IC_{50} = 2.5 μM) and mouse G9a (IC_{50} = 3.0 μM), it was selective for H3K9 PMTs over PMTs that do not target H3K9. Mechanistically, chaetocin was characterized as a S-adenosyl-L-methionine (SAM) competitive inhibitor and surprisingly its inhibitory activity was not decreased by the presence of dithiothreitol, which reduces the disulfide bonds of chaetocin. Although cytotoxic under some conditions, 0.5 μM chaetocin showed marked cellular reduction of dimethylation and trimethylation of H3K9 and no changes in the methylation state of H3K27, H3K36, and H3K79.

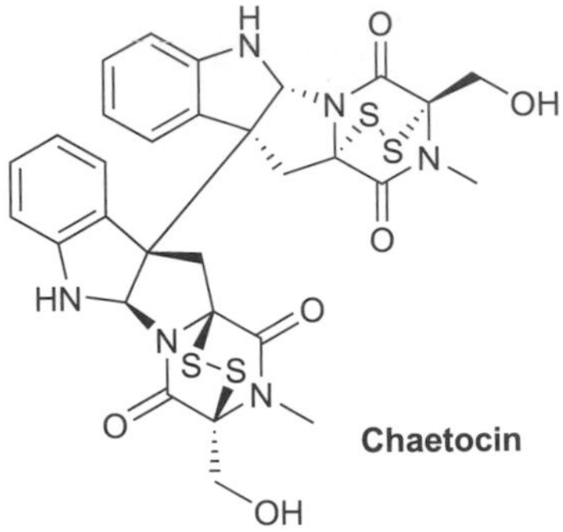

Chaetocin

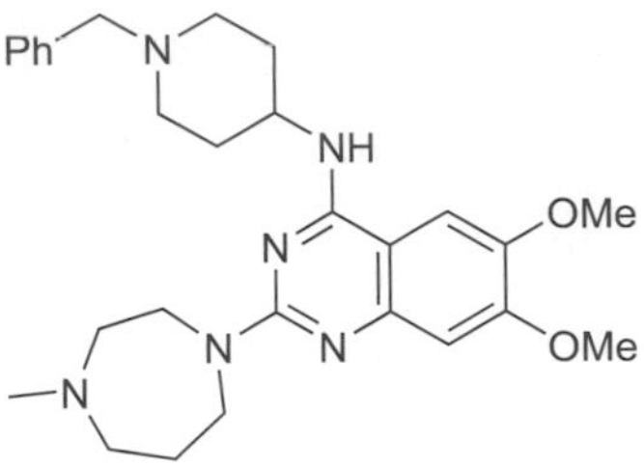

BIX01294

The first selective small-molecule G9a and GLP inhibitor, BIX01294, was discovered via screening of a library of 125,000 synthetic compounds [29]. BIX01294 has excellent selectivity for G9a ($IC_{50} = 1.9$ μM) and GLP ($IC_{50} = 0.7$ μM) over several H3K9 PMTs including SUV39H1 ($IC_{50} > 45$ μM) and ESET ($IC_{50} > 45$ μM), other PMTs such as SET7/9 ($IC_{50} > 45$ μM), and the arginine methyltransferase PRTM1 ($IC_{50} > 45$ μM). In contrast to chaetocin, BIX01294 did not compete with the SAM cofactor [26,29]. The X-ray crystal structure of GLP-BIX01294 confirmed that BIX01294 bound to the histone peptide-binding pocket and did not interact with the lysine-binding channel or cofactor-binding site [26]. In cellular assays, BIX01294 at 4.1 μM reduced the H3K9Me2 level of bulk histones; however, it was toxic at higher concentrations [29].

Structure–activity relationships (SAR) of the quinazoline scaffold of BIX01294 were recently investigated based on structural insights from the GLP–BIX01294 complex [25,26]. Exploring the lysine-binding channel via the 7-methoxy moiety of the quinazoline template resulted in the discovery of UNC0224 as a more potent and selective G9a inhibitor [25]. In the G9a ThioGlo assay, UNC0224 (IC_{50} of 15 nM) was seven times more potent than BIX01294 ($IC_{50} = 106$ nM). The high potency of UNC0224 was confirmed in isothermal titration calorimetry (ITC) experiments ($K_d = 23$ nM vs. 130 nM for BIX01294). Although UNC0224 also potently inhibited GLP with an IC_{50} of 20 nM in the ThioGlo assay, it was more than 1,000-fold selective for G9a over SET7/9 and SET8/PreSET7. A high-resolution (1.7 Å) X-ray crystal structure of the G9a–UNC0224 complex was also obtained. This was the first cocrystal structure of a G9a–small-molecule inhibitor complex, and it confirmed that the 7-dimethylamino propoxy side chain of UNC0224 occupied the lysine-binding channel of G9a—thus validating the design hypothesis for UNC0224 and providing an explanation for its higher potency [25,26]. The combination of high potency and good selectivity makes UNC0224 a potentially useful tool compound for the biomedical research community to further investigate the biology of G9a and its role in chromatin remodeling.

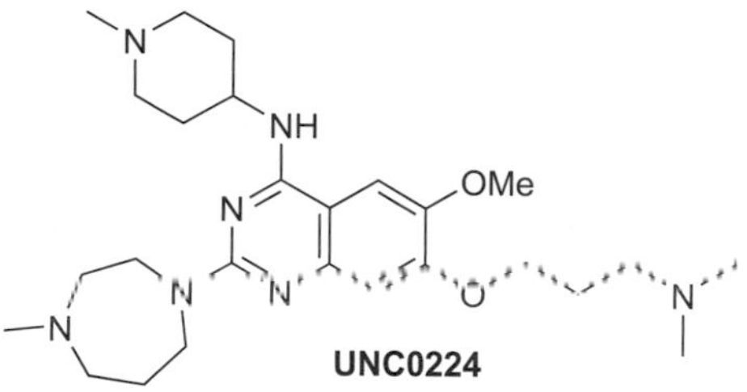

UNC0224

3.3 Drug discovery opportunities

Growing evidence suggests that PMTs play critical roles in the development of various human diseases including cancer [24,30], inflammation [31], and drug addiction [32]. For example:

- EZH2 is overexpressed in various human cancers and its high expression correlates with poor prognosis in prostate and breast cancers [33,34].
- G9a, also known as EHMT2, is overexpressed in human cancers and knockdown of G9a inhibits cancer cell growth [35]. In addition, G9a inactivates the tumor suppressor p53 via dimethylation of lysine 373 [22].
- SET7/9 modulates p53 activity in human cancer cells *in vitro* and *in vivo* [22] and also regulates NF-κB-dependent inflammatory genes [31].
- BIX01294, which inhibits G9a [29], is efficacious as a replacement for oct3/4, one of the four original genetic factors used for reprogramming of mammalian somatic cells into induced pluripotent stem cells [36].

With this evidence of disease relevance, discovering and developing high-quality chemical probes [23] of PMTs has been gaining momentum in both the academic research community and the pharmaceutical industry. Progress in this area is keenly awaited.

4. METHYL LYSINE-BINDING PROTEINS

4.1 Protein families

The specific domains that recognize KMe have been described within several protein families: the so-called "royal family" of Tudor, Agenet, chromo, PWWP, and MBT domains; the plant homeodomain (PHD); the WD40 repeat protein—WDR5; and ankyrin repeats [37,38]. Given the rapid rate of progress in this area, the relatively low affinity of these interactions (K_d values typically 10–100 µM), the wide variety of

Table 1 Readers of methyl lysine marks and representative binding partners

Reader module	PTM mark
Chromodomain	H3K9Me2/3, H3K27Me2/3
Double chromodomain	H3K4Me1/2/3
Chromo barrel	H3K36Me2/3
Tudor/tandem Tudor	H3K4Me3, H4K20Me3, H4K20Me1/2
MBT	H4K20Me1/2, H1K26Me1/2, H3K4Me1, H3K9Me1/2
PHD finger	H3K4Me2/3, H3K4Me0, H3K9Me3, H3K36Me3
WD40 repeat	H3K4Me2
Ankyrin repeats	H3K9Me1/2

techniques applied to determine binding partners, and increasing evidence of cross-talk between modifications [39–42], the summary provided in Table 1 must be considered indicative of the biologically significant interactions and certainly not a comprehensive or completely accurate contextual summary. Indeed, current estimates of the number of KMe-binding modules in the human proteome is >150 and there is no doubt that this estimate will grow over time.

A recent review of the available high-resolution X-ray crystal structures of these domains summarizes the key recognition features of various KMe states as an electron-rich aromatic cage interacting with the lysine cation with additional charge neutralization and H-bonding from 0 to 2 acidic functionalities—depending upon the methylation state of lysine [43]. This π-cation recognition motif is well known in structural biology [44] and has been investigated in peptide model systems and computationally to further explore the physical basis for the energetics driving this interaction [45,46].

4.2 Chemical approaches

Successful ligand discovery is highly dependent upon the development of appropriate assays. Early assays for methyl lysine binders relied upon ITC [47] or fluorescent anisotropy measurements [48] and only methylated histone peptides were examined [49]. These methods have the disadvantage of either being low-throughput or requiring concentrations of protein in excess of 70 μM, making high-throughput screening (HTS) unfeasible. This limitation can be attributed to the binding affinities of methyl lysine proteins for their cognate peptides, with K_d values in the micromolar range [40]. However, recently, HTS quality assays based on

Amplified Luminescence Proximity Homogeneous Assay (AlphaScreen™) technology have been reported [50,51]. With an HTS methodology in hand and orthogonal assays previously developed [47–49], nonpeptide antagonists of methyl lysine binding should appear from diversity screening approaches in the near future.

Given the amount of structural information available for the readers of KMe, strategies other than diversity screening for ligand discovery are also possible and a recent paper [52] outlines a method to glean chemical leads from the protein databank (PDB) by searching for binding pockets analogous to KMe-binding sites with bound, small-molecule ligands. Not surprisingly, quaternary ammonium species and tertiary amines were well represented as ligands for π-cation recognition sites but additional less basic ligands were also identified. The 150 compounds selected in this study could form the basis for a focused screening effort to disrupt KMe-driven protein–protein interactions.

4.3 Drug discovery opportunities

Among the methyl lysine readers, writers, and erasers (see Figure 1), the readers of KMe marks are perhaps the most challenging targets for chemical probe development because they are a large and diverse family, have relatively weak interactions with histone peptides, and lack enzymatic activity—a function of proteins that is traditionally targeted by medicinal chemists. Notwithstanding these potential limitations for ligand discovery, the biological roles of the readers are varied and significant enough to inspire efforts toward probe discovery. For example:

- PHD dysregulation is associated with hematopoietic cancers [18].
- In *Drosophila*, the two-MBT-repeat protein Scm is a core component in the PRC1 complex, and is essential for the repression of Hox genes [53].
- *Drosophila* dl(3)MBT has been shown to recruit a lysine deacetylase to chromatin bearing the H4K20Me1 mark leading to sequential maturation of the posttranslational state of the epigenome.
- In medulloblastoma tumors in humans, misregulation of H3K9 methylation is implicated as a carcinogenic event with L3MBTL3 possibly serving as a tumor suppressor as evidenced by its deletion in some medulloblastoma cell lines [21].

These studies suggest that the biological consequences of antagonism of KMe recognition may include dedifferentiation, reexpression of silenced genes, and possibly contributions to cellular reprogramming [36] with legitimate concern for selectivity versus tumor-suppressive effects of some readers. The utility of small-molecule probes in exploring these areas of biology are manifest and while the challenges are substantial, selective probe discovery may lead to significant drug discovery opportunities in the future.

5. PROTEIN LYSINE DEMETHYLASES

5.1 Protein family

Lysine demethylases fall into two major classes defined by their mechanism: (1) The LSD family are homologues of the flavin-containing monoamine oxidases that use the cofactor FAD to oxidize methylated lysines to the corresponding imine intermediate that undergoes hydrolysis to demethylated lysine and formaldehyde [54]. LSDs are incapable of demethylating trimethyl lysine residues, because the quaternary ammonium group cannot form the requisite imine intermediate. (2) Jumonji domain-containing demethylases belong to the family of 2-oxoglutarate (2-OG)-containing oxygenases, which also includes HIF-prolyl hydroxylase. These enzymes use Fe(II) together with 2-OG to oxygenate methyl groups on methylated lysines and generate the corresponding hydroxymethyl amines that undergo the same fate as in the LSD1 mechanism [15,16]. This mechanism allows for demethylation of all three possible methylation states of lysine residues, but individual enzymes show methylation state selectivity that is apparently driven by steric accommodation, with trimethyl demethylases (e.g., JMJD2A) having larger methyl lysine-binding pockets than dimethyl demethylases (e.g., FBXL11, PHF8) [55,56]. These mechanistic insights are informed by high-resolution crystal structures for JMJD2A, FBXL11, PHF8, and KIA1718 [55,56] and further advances in the structural biology of demethylases are anticipated.

The sequence selectivity of demethylation within histones has been established for many of the demethylases (Figure 3) [5]. Demethylase

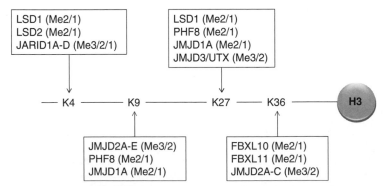

Figure 3 Human protein lysine demethylases and their target sites on histones [5].

catalytic domains have intrinsic sequence selectivity, but this can be modulated by complex formation. Hence, LSD1 has been shown to repress gene expression through the demethylation of H3K4Me1/2, while its association with the androgen receptor leads to enhanced transcription by demethylation of H3K9Me1/2 [57].

In some cases, the sequence selectivity of demethylation is partly controlled by other domains within the enzymes, for example, PHF8 contains a PHD finger that binds to H3K4Me3, directing the catalytic domain toward H3K9Me2 and thereby increasing its activity and selectivity; while for KIAA1718, PHD finger binding to H3K4Me3 directs the catalytic domain to preferentially demethylate H3K27Me2 [56]. The extent to which similar binding domain control occurs in the substrate selectivity of other demethylase subfamilies (for example, JMJD2 enzymes that contain methyl lysine-binding Tudor domains) remains to be explored.

The susceptibility of histone lysines to demethylation is also influenced by posttranslational modifications to neighboring residues, for example, phosphorylation of H3T11 by the kinase PRK1 has been shown to accelerate demethylation of H3K9Me3 by the demethylase JMJD2C [58]; in contrast, phosphorylation of H3S10 prevents demethylation of H3K9 by the JMJD2 demethylases [55].

5.2 Chemical approaches

Tractability of KDMs for inhibitor discovery has been demonstrated by the development of high-throughput screens for JMJD2 isoforms [59] and several small-molecule LSD1 inhibitors have been reported; these include known nonspecific monoamine oxidase inhibitors (MAOIs), including tranylcypromine [60] and pargyline [57], and thalidomide derivatives pomalidomide and lenalidomide [61]. These inhibitors show effects on histone H3K9 and H3K4 methylation levels in cells that are mediated by inhibition of LSD1-catalyzed demethylation. The inhibition by tranylcypromine has been shown to occur by formation of a covalent adduct with the FAD cofactor. (See Chapter 15 in this volume for a more detailed discussion of LSD1.)

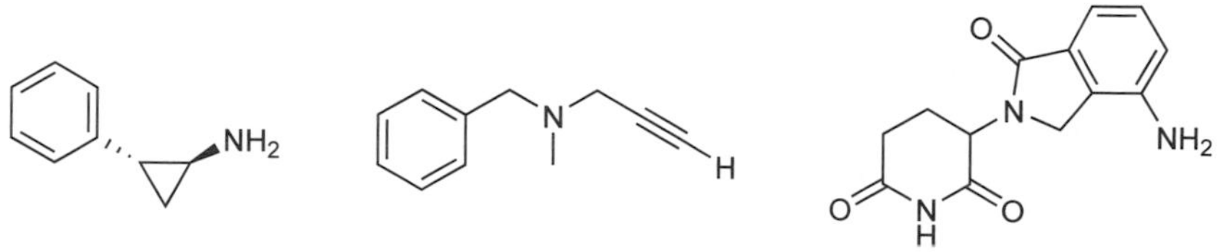

Tranylcypromine   **Pargyline**   **Pomalidomide**

JMJD2 demethylases are inhibited by analogues of the cofactor 2-OG that include *N*-oxalylamino acids, pyridine dicarboxylates, and related bipyridyl derivatives. Other chemotypes that are also presumed to bind to the active-site Fe(II) include catechols, hydroxamic acids (including the clinically used HDAC inhibitor SAHA/Vorinostat), and tricarboxylic acid cycle intermediates, such as succinate and fumarate [59,62].

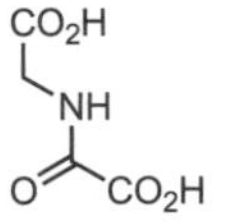

N-Oxalylglycine 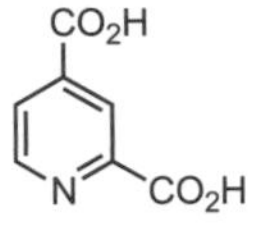   **2,4-PDCA** 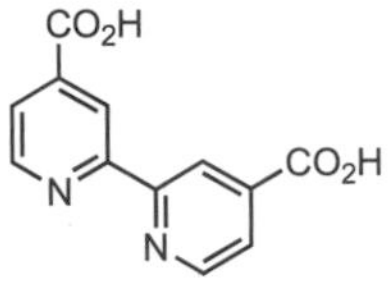   **Bipyridyl-bis-carboxylic acid** 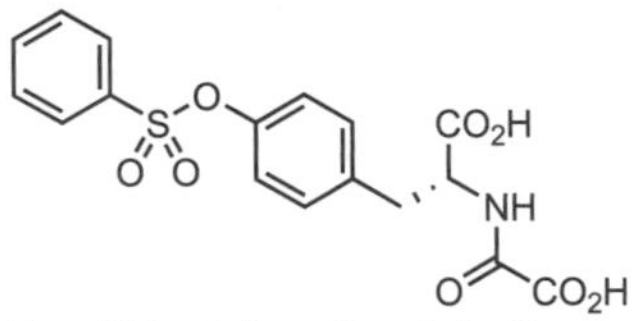   **N-Oxalyltyrosine derivative**

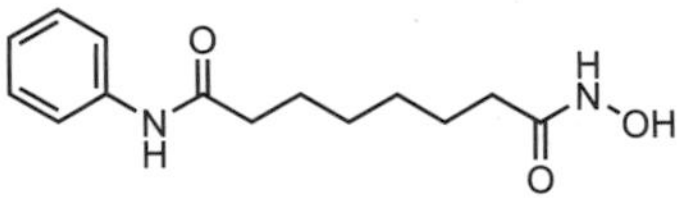

SAHA (Vorinostat) 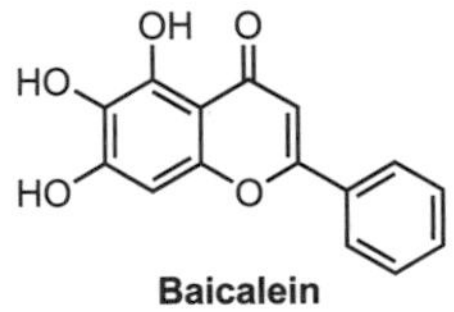   **Baicalein**

Targeting the catalytic Fe(II) brings with it the challenge of achieving selectivity since the cofactor and Fe(II)-binding sites of 2-OG oxygenases are generally very similar. Recently, Schofield and coworkers used the protein crystal structure of JMJD2E with a dynamic combinatorial chemistry approach to derive a series of substituted oxalyl tyrosines that exploit a subpocket of this enzyme to achieve selectivity over PHD2 [63].

5.3 Drug discovery opportunities

Generally, an imbalance between histone methylation and demethylation appears to be correlated with tumorigenesis, and several demethylases are implicated in the pathogenesis of cancer. Both LSD1 and JARID1B are overexpressed in prostate cancer, and LSD1 expression correlates with tumor recurrence during therapy [64]. LSD1 also demethylates p53, repressing p53-mediated transcriptional activation and inhibiting the role of p53 in promoting apoptosis [22]. JARID1B and JMJD2C are over-expressed in breast and testis cancer and esophageal squamous carci-noma [65]. RNAi knockdown of JMJD2C inhibited cell proliferation and highlights this isoform as a potential therapeutic target [16].

A growing body of evidence links demethylase dysfunction to other disease areas. JMJD3 expression is induced in macrophages by the inflammatory transcription factor NFκB in response to microbial stimuli, and 70% of lipopolysaccharide-inducible genes have been shown to be JMJD3 targets [66]. Invading viral pathogens that depend upon the host cell's transcriptional machinery are also subject to the regulatory impact of histone modifications, and this has been specifically demonstrated for LSD1 where depletion or inhibition of its activity with MAOIs resulted in blockade of herpes simplex virus and varicella zoster virus gene expression [67].

While tractability for screening and ligand discovery has been demonstrated for these enzymes, significant challenges remain, notably in identifying chemotypes that show potent and selective inhibition of isoforms of interest while retaining physicochemical properties suitable for the intracellular site of action.

6. CONCLUSIONS

With the enormous influence of protein kinase drug discovery efforts, both therapeutically and scientifically, the modulation of the PTM state of proteins has fully come of age as a paradigm for the systematic exploration of the druggable genome [68,69]. Targeting KMe as a key PTM to which this strategy can be applied holds great promise for both exciting chemical biology and significant drug discovery.

REFERENCES

[1] S. L. Berger, T. Kouzarides, R. Shiekhattar, and A. Shilatifard, *Genes Dev.*, 2009, **23**, 781.

[2] B. E. Bernstein, A. Meissner, and E. S. Lander, *Cell*, 2007, **128**, 669.

[3] K. A. Gelato and W. Fischle, *Biol. Chem.*, 2008, **389**, 353.

[4] A. J. Ruthenburg, H. Li, D. J. Patel, and C. David Allis, *Nat. Rev. Mol. Cell Biol.*, 2007, **8**, 983.

[5] R. Marmorstein and R. C. Trievel, *Biochim. Biophys. Acta.*, 2009, **1789**, 58.

[6] B. T. Seet, I. Dikic, M. M. Zhou, and T. Pawson, *Nat. Rev. Mol. Cell Biol.*, 2006, **7**, 473.

[7] M. Esteller, *N. Engl. J. Med.*, 2008, **358**, 1148.

[8] P. A. Marks, V. M. Richon, T. Miller, and W. K. Kelly, *Adv. Cancer Res.*, 2004, **91**, 137.

[9] A. Norvell and S. B. McMahon, *Science*, 2010, **327**, 964.

[10] M. Haberland, R. L. Montgomery, and E. N. Olson, *Nat. Rev. Genet.*, 2009, **10**, 32.

[11] C. Martin and Y. Zhang, *Nat. Rev. Mol. Cell Biol.*, 2005, **6**, 838.

[12] N. P. Blackledge and R. J. Klose, *Epigenomics*, 2010, **2**, 151.

[13] T. S. Mikkelsen, M. Ku, D. B. Jaffe, B. Issac, E. Lieberman, G. Giannoukos, P. Alvarez, W. Brockman, T. K. Kim, R. P. Koche, W. Lee, E. Mendenhall, A. O'Donovan, A. Presser, C. Russ, X. Xie, A. Meissner, M. Wernig, R. Jaenisch, C. Nusbaum, E. S. Lander, and B. E. Bernstein, *Nature*, 2007, **448**, 553.

[14] J. Kim, J. Daniel, A. Espejo, A. Lake, M. Krishna, L. Xia, Y. Zhang, and M. T. Bedford, *EMBO Rep.*, 2006, **7**, 397.

[15] A. J. Bannister and T. Kouzarides, *Nature*, 2005, **436**, 1103.

[16] P. A. Cloos, J. Christensen, K. Agger, and K. Helin, *Genes Dev.*, 2008, **22**, 1115.

[17] R. A. Copeland, M. E. Solomon, and V. M. Richon, *Nat. Rev. Drug. Discov.*, 2009, **8**, 724.
[18] G. G. Wang, J. Song, Z. Wang, H. L. Dormann, F. Casadio, H. Li, J. L. Luo, D. J. Patel, and C. D. Allis, *Nature*, 2009, **459**, 847.
[19] S. Karberg, *Cell*, 2009, **139**, 1029.
[20] S. Akbarian and H.-S. Huang, *Biol. Psychiatry*, 2009, **65**, 198.
[21] P. A. Northcott, Y. Nakahara, X. Wu, L. Feuk, D. W. Ellison, S. Croul, S. Mack, P. N. Kongkham, J. Peacock, A. Dubuc, Y. -S. Ra, K. Zilberberg, J. McLeod, S. W. Scherer, J. Sunil Rao, C. G. Eberhart, W. Grajkowska, Y. Gillespie, B. Lach, R. Grundy, I. F. Pollack, R. L. Hamilton, T. Van Meter, C. G. Carlotti, F. Boop, D. Bigner, R. J. Gilbertson, J. T. Rutka, and M. D. Taylor, *Nat. Genet.*, 2009, **41**, 465.
[22] J. Huang and S. L. Berger, *Curr. Opin. Genet. & Dev.*, 2008, **18**, 152.
[23] S. V. Frye, *Nat. Chem. Biol.*, 2010, **6**, 159.
[24] C. K. Fog, K. T. Jensen, and A. H. Lund, *APMIS*, 2007, **115**, 1060.
[25] F. Liu, X. Chen, A. Allali-Hassani, A. M. Quinn, G. A. Wasney, A. Dong, D. Barsyte, I. Kozieradzki, G. Senisterra, I. Chau, A. Siarheyeva, D. B. Kireev, A. Jadhav, J. M. Herold, S. V. Frye, C. H. Arrowsmith, P. J. Brown, A. Simeonov, M. Vedadi, and J. Jin, *J. Med. Chem.*, 2009, **52**, 7950.
[26] Y. Chang, X. Zhang, J. R. Horton, A. K. Upadhyay, A. Spannhoff, J. Liu, J. P. Snyder, M. T. Bedford, and X. Cheng, *Nat. Struct. Mol. Biol.*, 2009, **16**, 312.
[27] C. Qian and M. M. Zhou, *Cell. Mol. Life Sci.*, 2006, **63**, 2755.
[28] D. Greiner, T. Bonaldi, R. Eskeland, E. Roemer, and A. Imhof, *Nat. Chem. Biol.*, 2005, **1**, 143.
[29] S. Kubicek, R. J. O'Sullivan, E. M. August, E. R. Hickey, Q. Zhang, M. L. Teodoro, S. Rea, K. Mechtler, J. A. Kowalski, C. A. Homon, T. A. Kelly, and T. Jenuwein, *Mol. Cell*, 2007, **25**, 473.
[30] A. Spannhoff, W. Sippl, and M. Jung, *Int. J. Biochem. Cell Biol.*, 2009, **41**, 4.
[31] Y. Li, M. A. Reddy, F. Miao, N. Shanmugam, J. -K. Yee, D. Hawkins, B. Ren, and R. Natarajan, *J. Biol. Chem.*, 2008, **283**, 26771.
[32] I. Maze, H. E. Covington, 3rd, D. M. Dietz, Q. LaPlant, W. Renthal, S. J. Russo, M. Mechanic, E. Mouzon, R. L. Neve, S. J. Haggarty, Y. Ren, S. C. Sampath, Y. L. Hurd, P. Greengard, A. Tarakhovsky, A. Schaefer, and E. J. Nestler, *Science*, 2010, **327**, 213.
[33] K. Collett, G. E. Eide, J. Arnes, I. M. Stefansson, J. Eide, A. Braaten, T. Aas, A. P. Otte, and L. A. Akslen, *Clin. Cancer. Res.*, 2006, **12**, 1168.
[34] R. J. Bryant, N. A. Cross, C. L. Eaton, F. C. Hamdy, and V. T. Cunliffe, *Prostate*, 2007, **67**, 547.
[35] Y. Kondo, L. Shen, S. Ahmed, Y. Boumber, Y. Sekido, B. R. Haddad, and J. P. Issa, *PLoS One*, 2008, **3**, e2037.
[36] Y. Shi, J. T. Do, C. Desponts, H. S. Hahm, H. R. Scholer, and S. Ding, *Cell Stem Cell*, 2008, **2**, 525.
[37] S. D. Taverna, H. Li, A. J. Ruthenburg, C. D. Allis, and D. J. Patel, *Nat. Struct. Mol. Biol.*, 2007, **14**, 1025.
[38] R. E. Collins, J. P. Northrop, J. R. Horton, D. Y. Lee, X. Zhang, M. R. Stallcup, and X. Cheng, *Nat. Struct. Mol. Biol.*, 2008, **15**, 245.
[39] D. J. Bua, A. J. Kuo, P. Cheung, C. L. Liu, V. Migliori, A. Espejo, F. Casadio, C. Bassi, B. Amati, M. T. Bedford, E. Guccione, and O. Gozani, *PLoS One*, 2009, **4**, e6789.
[40] N. Nady, J. Min, M. S. Kareta, F. Chedin, and C. H. Arrowsmith, *Trends Biochem. Sci.*, 2008, **33**, 305.
[41] E. D. Martinez, A. B. Dull, J. A. Beutler, G. L. Hager, and I. James in J. Inglese (Ed.), *Methods in Enzymology*, Academic Press, New York, 2006, **414**, p. 21.
[42] J. A. Latham and S. Y. Dent, *Nat. Struct. Mol. Biol.*, 2007, **14**, 1017.
[43] M. A. Adams-Cioaba and J. Min, *Biochem. Cell. Biol.*, 2009, **87**, 93.

[44] N. Zacharias and D. A. Dougherty, *Trends Pharmacol. Sci.*, 2002, **23**, 281.

[45] R. M. Hughes, K. R. Wiggins, S. Khorasanizadeh, and M. L. Waters, *Proc. Natl. Acad. Sci. USA*, 2007, **104**, 11184.

[46] Z. Lu, J. Lai, and Y. Zhang, *J. Am. Chem. Soc.*, 2009, **131**, 14928.

[47] Y. Guo, N. Nady, C. Qi, A. Allali-Hassani, H. Zhu, P. Pan, M. A. Adams-Cioaba, M. F. Amaya, A. Dong, M. Vedadi, M. Schapira, R. J. Read, C. H. Arrowsmith, and J. Min, *Nucleic Acids Res.*, 2009, **37**, 2204.

[48] W. Fischle, Y. Wang, S. A. Jacobs, Y. Kim, C. D. Allis, and S. Khorasanizadeh, *Genes Dev.*, 2003, **17**, 1870.

[49] T. Klymenko, B. Papp, W. Fischle, T. Kocher, M. Schelder, C. Fritsch, B. Wild, M. Wilm, and J. Muller, *Genes Dev.*, 2006, **20**, 1110.

[50] T. J. Wigle, J. M. Herold, G. A. Senisterra, M. Vedadi, D. B. Kireev, C. H. Arrowsmith, S. V. Frye, and W. P. Janzen, *J. Biomol. Screen*, 2010, **15**, 62.

[51] A. M. Quinn, M. T. Bedford, A. Espejo, A. Spannhoff, C. P. Austin, U. Oppermann, and A. Simeonov, *Nucleic Acids Res.*, 2009, **38**, e11.

[52] V. Campagna-Slater and M. Schapira, *Mol. Inform.*, 2010, **29**, 322.

[53] D. Bornemann, E. Miller, and J. Simon, *Genetics*, 1998, **150**, 675.

[54] M. Yang, J. C. Culhane, L. M. Szewczuk, P. Jalili, H. L. Ball, M. Machius, P. A. Cole, and H. Yu, *Biochemistry*, 2007, **46**, 8058.

[55] S. S. Ng, K. L. Kavanagh, M. A. McDonough, D. Butler, E. S. Pilka, B. M. Lienard, J. E. Bray, P. Savitsky, O. Gileadi, F. von Delft, N. R. Rose, J. Offer, J. C. Scheinost, T. Borowski, M. Sundstrom, C. J. Schofield, and U. Oppermann, *Nature*, 2007, **448**, 87.

[56] J. R. Horton, A. K. Upadhyay, H. H. Qi, X. Zhang, Y. Shi, and X. Cheng, *Nat. Struct. Mol. Biol.*, 2009, **17**, 38.

[57] E. Metzger, M. Wissmann, N. Yin, J. M. Muller, R. Schneider, A. H. Peters, T. Gunther, R. Buettner, and R. Schule, *Nature*, 2005, **437**, 436.

[58] E. Metzger, N. Yin, M. Wissmann, N. Kunowska, K. Fischer, N. Friedrichs, D. Patnaik, J. M. Higgins, N. Potier, K. H. Scheidtmann, R. Buettner, and R. Schule, *Nat Cell. Biol.*, 2008, **10**, 53.

[59] M. Sakurai, N. R. Rose, L. Schultz, A. M. Quinn, A. Jadhav, S. S. Ng, U. Oppermann, C. J. Schofield, and A. Simeonov, *Mol. Biosyst.*, 2010, **6**, 357.

[60] J. C. Culhane, D. Wang, P. M. Yen, and P. A. Cole, *J. Am. Chem. Soc.*, 2010, **132**, 3164.

[61] L. Escoubet-Lozach, I. L. Lin, K. Jensen-Pergakes, H. A. Brady, A. K. Gandhi, P. H. Schafer, G. W. Muller, P. J. Worland, K. W. Chan, and D. Verhelle, *Cancer Res.*, 2009, **69**, 7347.

[62] N. R. Rose, S. S. Ng, J. Mecinovic, B. M. Lienard, S. H. Bello, Z. Sun, M. A. McDonough, U. Oppermann, and C. J. Schofield, *J. Med. Chem.*, 2008, **51**, 7053.

[63] N. R. Rose, E. C. Woon, G. L. Kingham, O. N. King, J. Mecinovic, I. J. Clifton, S. S. Ng, J. Talib-Hardy, U. Oppermann, M. A. McDonough, and C. J. Schofield, *J. Med. Chem.*, 2010, **53**, 1810.

[64] P. Kahl, L. Gullotti, L. C. Heukamp, S. Wolf, N. Friedrichs, R. Vorreuther, G. Solleder, P. J. Bastian, J. Ellinger, E. Metzger, R. Schule, and R. Buettner, *Cancer Res.*, 2006, **66**, 11341.

[65] K. Yamane, K. Tateishi, R. J. Klose, J. Fang, L. A. Fabrizio, H. Erdjument-Bromage, J. Taylor-Papadimitriou, P. Tempst, and Y. Zhang, *Mol. Cell.*, 2007, **25**, 801.

[66] F. De Santa, V. Narang, Z. H. Yap, B. K. Tusi, T. Burgold, L. Austenaa, G. Bucci, M. Caganova, S. Notarbartolo, S. Casola, G. Testa, W. K. Sung, C. L. Wei, and G. Natoli, *EMBO J.*, 2009, **28**, 3341.

[67] Y. Liang, J. L. Vogel, A. Narayanan, H. Peng, and T. M. Kristie, *Nat. Med.*, 2009, **15**, 1312.

[68] S. V. Frye, *Chem. Biol.*, 1999, **6**, R3.

[69] A. L. Hopkins and C. R. Groom, *Nat. Rev. Drug Discov.*, 2002, **1**, 727.

CHAPTER 21

Chemical Proteomic Technologies for Drug Target Identification

Kieran F. Geoghegan and Douglas S. Johnson

Contents

ABBREVIATIONS

LC–MS	liquid chromatography–mass spectrometry
ABPP	activity-based protein profiling
ABP	activity-based probe
CC	click chemistry
MAGL	monoacylglycerol lipase
FAAH	fatty acid amide hydrolase
SILAC	stable isotope labeling with amino acids in cell culture

PharmaTherapeutics Research, Pfizer Inc., Groton, CT 06340, USA

Annual Reports in Medicinal Chemistry, Volume 45
ISSN 0065-7743, DOI 10.1016/S0065-7743(10)45021-6

1. INTRODUCTION

Rapid progress in DNA sequencing and recombinant DNA technologies made target-focused drug discovery a dominant paradigm in the pharmaceutical industry of the 1990s. In time, however, increasing doubt has come to surround its adequacy [1,2]. Despite its technical coherence, the express-and-screen approach has returned insufficient rewards to justify its exclusive selection as a platform for discovery. Even some of its spectacular successes, such as selective kinase inhibitors for cancer treatment, were found to be less than completely specific [3]. The surprise in that case was beneficial, but the emergence of unpredicted effects still highlights our inability to anticipate the full scope of drug activities.

A solitary focus on target-specific screening leaves reserves of value untapped. Thoughts are turning again to discovery conducted at the cellular [4] and whole-organism [5] levels, where complex systems exist intact. This is in spite of the fact that the molecular action of lead compounds will probably need to be explained at a later stage. Such approaches will produce new uncertainties, but the target-based approach generates its own share of enigmas such as the origins of off-target activities.

Proteomics—more precisely, the technologies that underpin it—offers assistance on this front. Cross-referencing experimental data from peptide mass spectrometry with protein sequence databases allows proteins to be identified rapidly, reliably, and with high sensitivity [6]. New experimental strategies exist in which proteins that bind selectively to immobilized drugs or react selectively with chemically or photochemically reactive drug analogs are identified by these methods. The objective is to rationalize functional properties of drugs by finding out which proteins they bind to.

The challenge being addressed by newer methods was successfully solved in the past for a range of drugs using Edman degradation, affinity chromatography, and photoaffinity labeling [7]. These included FK506 [8], trichostatin [9], and p38 kinase inhibitors [10]. Therefore, proteomics is refreshing an approach that has long been attractive to chemists but challenging to implement.

2. FUNDAMENTALS OF PROTEOMICS

Proteomics is a child of disparate parents: the revolutionary advent of genomic sequencing and the evolutionary extension of mass spectrometry to permit the analysis of peptides. The fusion of these advances initially created a vision of full inventories of proteins in a biological unit, such as a cell, a subcellular fraction, or a physiological fluid. Good progress has been made toward this in some cases [11], but the focus is shifting from encyclopedic surveys toward an emphasis on quantitative detection of clinically or scientifically valued targets [12]. Meanwhile, proteomic technologies are being recruited to serve other purposes, as we discuss in this chapter.

Methods based on liquid chromatography–mass spectrometry (LC–MS) and universally accepted search algorithms permit reliable identifications of low levels of proteins at high sensitivity [6]. Even semispecialized protein chemistry labs can readily identify proteins at the level of a few picomoles (10 pmol of a 50-kDa protein is 500 ng). Specialized groups with access to the latest advances in HPLC and mass spectrometry routinely work with subpicomolar quantities. Chemical proteomics as discussed here requires the more advanced equipment.

Proteins need not be purified to be identified by these methods. Mixtures of proteins are first converted into collections of fragment peptides using more or less selective proteases (trypsin is favored), and the peptides are subjected to tandem (two-stage) mass spectrometry. The intact mass of each peptide is recorded, after which the masses of the many products generated by a physical or chemical fragmentation process performed inside the mass spectrometer are also acquired. In the search phase, peptides are independently identified by comparing their experimental properties with the corresponding calculated properties of candidate peptides predicted from a sequence database. Detecting multiple peptides from any protein makes it nearly certain that the parent protein was present in the sample. The quality of identifications is calibrated against the probability that an assigned identity is due to chance. An increasing wealth of genomic resources provides a vast opportunity to explore the details of drug action in species ranging from human to bacteria.

3. PRINCIPAL STRATEGIES

In the methods surveyed here, a specific interaction between target and ligand is created and then preserved so that the target can be identified. Two of the three approaches reviewed depend on forming a covalent bond between the molecular partners, but the first does not.

3.1 Affinity capture and identification of drug targets

Before the era of artificially recombinant DNA, affinity chromatography emerged as a potentially highly selective approach to protein purification [13]. It revived a laborious art based on selective precipitations and a limited range of chromatographic media.

Successful affinity chromatography requires that the protein interact with an immobilized ligand tightly enough to be captured from solution, but weakly enough for binding to be reversed without denaturing the protein. Retention of the protein's native structure is usually a high priority, and elution with soluble ligand or a pH shift is often used with this in mind.

Priorities shift in bead-based chemical proteomics, a modified form of affinity chromatography in which an immobilized ligand is used to fish for an unknown protein partner in a complex mixture [14]. The main difference is that the native protein structure of the target need not be retained, so that elution by denaturation is acceptable. Ideally, proteins lacking affinity for the probe are washed away, while a specifically bound protein dissociates slowly enough during washing that its interaction with the ligand is preserved. It can then be eluted with no need to preserve the native structure, often by preparing the protein for sodium dodecyl sulfate–polyacrylamide gel electrophoresis.

Nonspecific protein binding to the solid phase complicates the method and is a selective pressure driving its evolution. The adaptive response has been the development of intrinsically comparative methods in which specific binding to an immobilized ligand is blocked in one out of two otherwise identical samples. When the respective protein components of the samples are compared, specifically bound proteins are present in one but severely depleted in the other. To allow relative quantitation, the two samples can be made isotopically distinct by a chemical or metabolic process and then mixed for an analytical step that avoids intersample variability [15].

Recently, a bead-based proteomic approach was used to identify proteins that selectively bound a PDE5 inhibitor (its "interactome") [16]. PF-4540124 (1) inhibits PDE5 with a low-nanomolar IC_{50}, and was designed for immobilization through a 6-carbon linker ending in a primary amine. Recombinant PDE5 bound to agarose beads linked to the compound (2) but not to control beads.

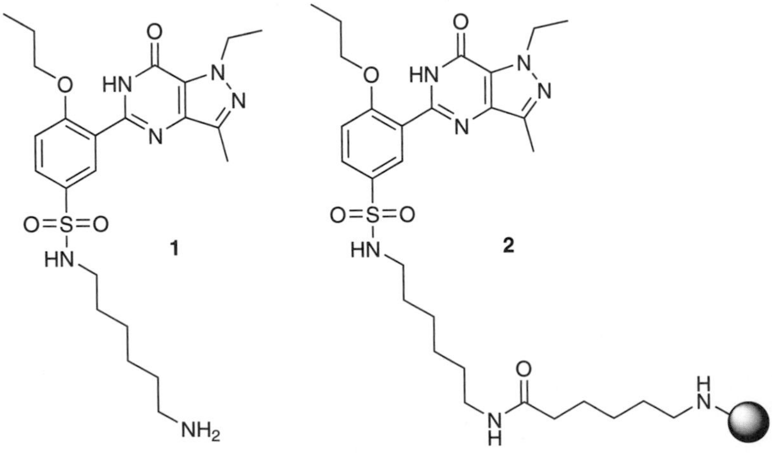

Bead-bound PF-4540124 (**2**) was incubated with the soluble fraction of a mouse lung homogenate in the presence and absence of **1** (100 μM). The type of protein captured was restricted by having 1 mM ADP and GDP in both samples. The beads were pelleted and washed before being treated (both samples) with **1** (100 μM) to elute specifically bound proteins.

Gel electrophoresis showed that each sample contained only a few proteins. Separate tryptic digests of each sample were subjected to chemically uniform but isotopically distinct reductive dimethylations of their amino groups. Sodium cyanoborohydride with [1]H-formaldehyde was used for the control, but the reducing agent was added with deuterated formaldehyde to the drug-eluted sample. The digests were then combined and proteomic analysis was used to identify proteins and extract their relative abundance in the two samples. The respective relative abundances measured for multiple peptides were combined to give estimates for their parent proteins, and prenyl-binding protein was detected as a previously unknown interactor of the inhibitor.

A related 2009 method from the groups of Schreiber and Carr has attracted wide interest [17]. It resembles the one described above, except that the two protein samples required for bead-binding trials are distinguished by metabolic labeling with stable isotopes, avoiding the need to chemically derivatize peptides midway through the experiment. Apart from the presence of ^{13}C and/or ^{15}N in selected protein amino acids, the samples are otherwise identical to begin with. Protein ratios are measured by observing heavy-to-light isotope ratios in the peptides.

The approach recruited to chemical proteomics in Reference [17] is called SILAC (stable isotope labeling with amino acids in cell culture) and is important in comparative proteomics (Figure 1). SILAC works well with cultured mammalian cells, but prokaryotes defeat it by metabolizing the label (usually supplied in lysine and arginine) into other amino acids. For applications beyond cultured eukaryotic cells, the reductive methylation route to differential labeling [18] is among the alternatives [15].

An important factor in all these experiments is the choice of bead used to immobilize the probe. Biochemists have considered cross-linked agarose beads to be exceptionally hydrophilic with a low tendency to bind proteins nonspecifically, and these beads have the further attraction of being commercially available in activated forms (succinimidyl esters, epoxides, and maleimides, for example). However, early trials of bead-based chemical proteomics have shown that many proteins in mammalian cell lysates bind tenaciously to agarose beads. This was unimportant in many studies in which protein–protein interactions were detected by coimmunoprecipitation with immunochemical

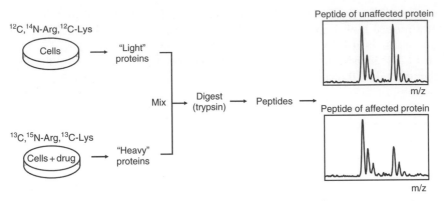

Figure 1 The SILAC method for comparative proteomics. A real experiment requires collection of thousands of mass spectra corresponding to different peptides. In the case portrayed, the drug has lowered the abundance of the affected protein in the treated cells compared to the control cells. Multiple peptide fragments of the affected protein would show this effect, supporting the conclusion that the protein was affected.

readouts, but it is significant in proteomic studies. This set of proteins has been dubbed the "agarose bead proteome" [19].

Alternative materials may become increasingly important in chemical proteomics. Agarose beads for biochemistry typically have diameters of at least 20 μm, but Handa's group (Tokyo) has reported the use of methacrylate-based 200 nm nanobeads ("SG beads") in an effort to reduce nonspecific binding [20]. A magnetic form of these beads is called FG beads [21,22]; magnetic capture of functionalized beads can replace centrifugation and filtration [23]. Results with these materials make them a potential alternative to agarose, and they will be more widely evaluated during the next couple of years as they are now being produced commercially.

3.2 Activity-based detection of enzymes

Activity-based protein profiling (ABPP) is a chemical proteomic strategy in which active-site-directed covalent probes are used to profile the functional states of enzymes in complex proteomes. Activity-based probes (ABPs) can distinguish active enzymes from their inactive zymogens or inhibitor-bound forms. They contain a reactive group intended to modify enzyme active sites covalently and a reporter group (typically rhodamine or biotin) that assists in detection and identification of protein targets.

Gel-based ABPP enables visualization of labeling events using gel electrophoresis accompanied by in-gel fluorescence (rhodamine) or avidin blotting (biotin). Mass spectrometry-based platforms permit the enrichment and identification of probe-labeled proteins from a complex proteome. ABPs have been developed for enzyme classes that include serine hydrolases (e.g., **3**), cysteine proteases (e.g., **4**), protein phosphatases, glycosidases, ubiquitin-conjugating/hydrolyzing enzymes, proteasomes, oxidoreductases, ATP-binding enzymes (e.g., kinases), and cytochrome P450s [24–26].

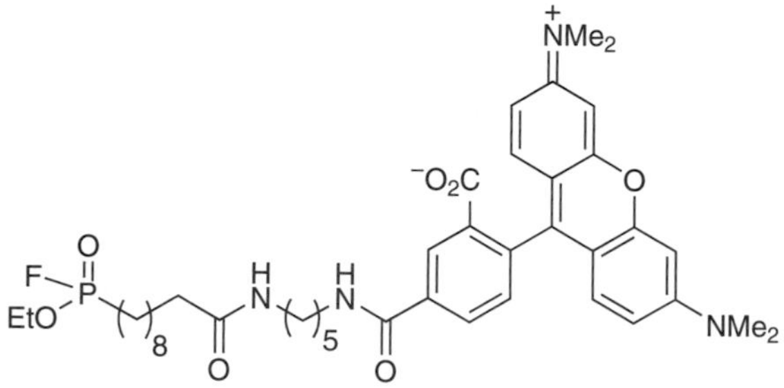

3

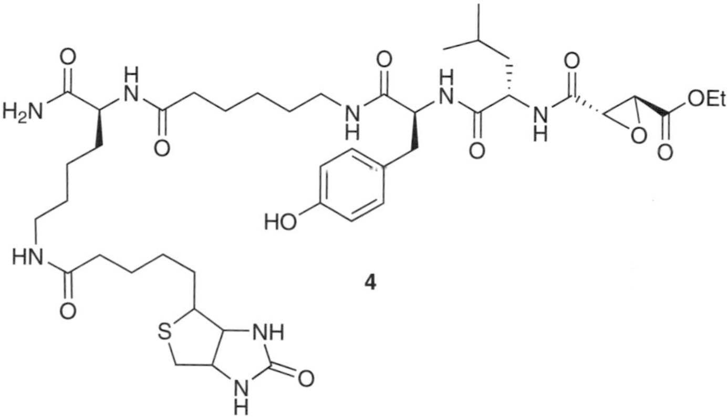

4

A competitive version of ABPP identifies the target(s) and assesses the selectivity of an enzyme inhibitor in biological systems by gauging how well the inhibitor slows the enzyme's reaction with an ABP. For example, fluorophosphonate ABP **3** was used to profile the selectivity of fatty acid amide hydrolase (FAAH) inhibitors within the serine hydrolase superfamily [27] (FAAH hydrolyzes endocannabinoids such as anandamide). Serine hydrolases that exhibited reduced labeling by the probe in the presence of inhibitor were scored as targets of the inhibitor. Urea FAAH inhibitors exemplified by PF-3845 (**5**) that covalently modify the active-site serine nucleophile of FAAH were found to be exquisitely selective for FAAH in brain and liver proteomes [28,29]. In contrast, carbamates such as URB597 (7) that labeled FAAH also hit multiple serine hydrolases in liver cells, including carboxylesterases.

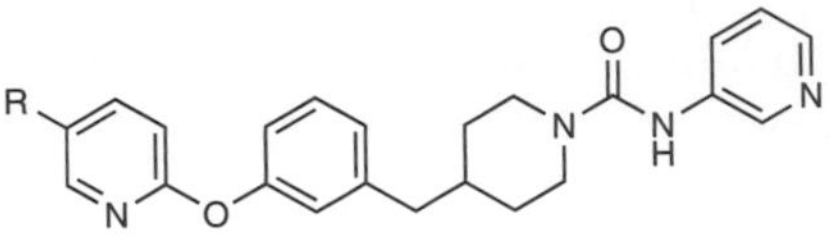

5: R=CF$_3$
6: R= ≡

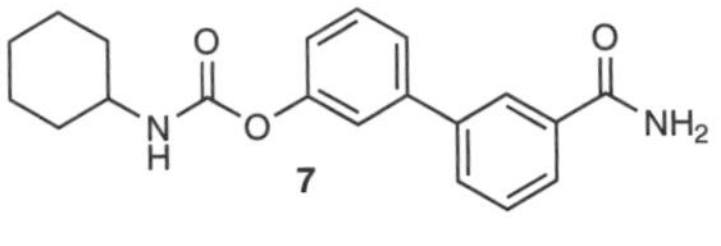

7

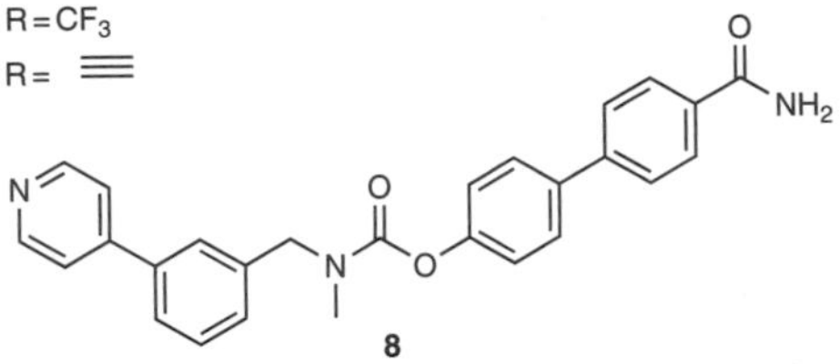

8

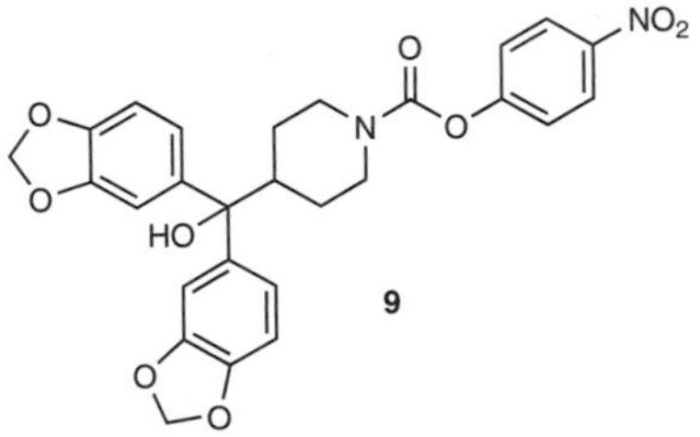

9

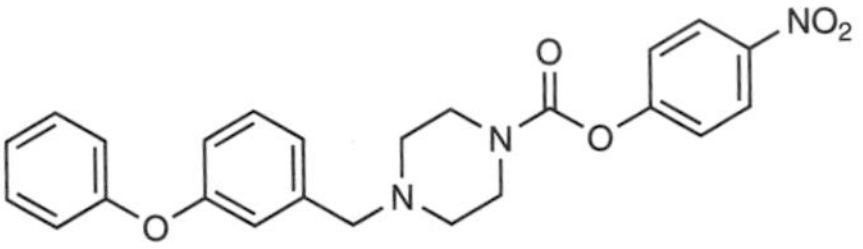

10

ABPP allows screening for inhibitors of members of an enzyme family. Recently, competitive ABPP screening of a carbamate library led to the development of a carbamate inhibitor of the uncharacterized serine hydrolase, alpha/beta-hydrolase-6 (ABHD6) [30]. This inhibitor, WWL70 (8), exhibited an IC$_{50}$ value of 70 nM in brain membranes and was highly selective for ABHD6 over 27 other serine hydrolases in these proteomes. Similar methods were used to identify a selective carbamate inhibitor (9, JZL184) against monoacylglycerol lipase (MAGL). JZL184 is based on a piperidine–carbamate scaffold and demonstrates high *in vivo* potency in mice resulting in near-complete blockage of MAGL activity at 4 mg/kg ip with minimal effects on other brain serine hydrolases, including FAAH [31]. The structural similarity between the piperidine–carbamate MAGL inhibitors (e.g., 9) and piperazine/piperidine–urea FAAH inhibitors (e.g., 5) also inspired the use of competitive ABPP to develop dual FAAH–MAGL inhibitors such as JZL195 (10) [32,33]. ABPP facilitates the development of potent covalent inhibitors of uncharacterized enzymes, a valuable tool for elucidating their biological functions.

Another ABPP platform integrates click chemistry (CC), where an alkyne or azide group in the ABP acts as a latent attachment point for a reporter tag [34,35]. Replacing bulky reporter groups with an alkyne extends ABP design to include probes that function in live cells and organisms (alkynes are more likely to be cell-permeable). After cell lysis, CC is used to append an azide-functionalized reporter tag to labeled proteins to help identify target proteins. For example, the *in vivo* selectivity of a clickable covalent FAAH inhibitor was profiled using CC-ABPP. Following confirmation that the urea, PF-3845 (5), covalently modified the serine nucleophile of FAAH, the alkynyl analog, PF-3845-yne (6), was synthesized and the protein targets were analyzed *in vivo* by CC-ABPP in FAAH(+/+) and (−/−) mice. PF-3845-yne selectively reacted with FAAH in both the nervous system and peripheral tissue [29].

β-Lactam antibiotics, such as cephalosporins, and penicillins, such as ampicillin (11) and aztreonam, covalently modify their protein targets. Alkyne-functionalized versions of these antibiotics, for example, AmpN (12), were used to probe various penicillin-binding proteins *in vitro* and *in vivo* using CC-ABPP [36,37].

Yao's group applied CC-ABPP to seek cellular targets of orlistat (13), an FDA-approved antiobesity drug that covalently modifies active-site serines of gastric and pancreatic lipases. The alkyne-modified orlistat analog (14) confirmed targeting of the previously reported fatty acid synthase (FAS), but also modified eight new targets including GAPDH and β-tubulin at their active-site cysteine residues [38].

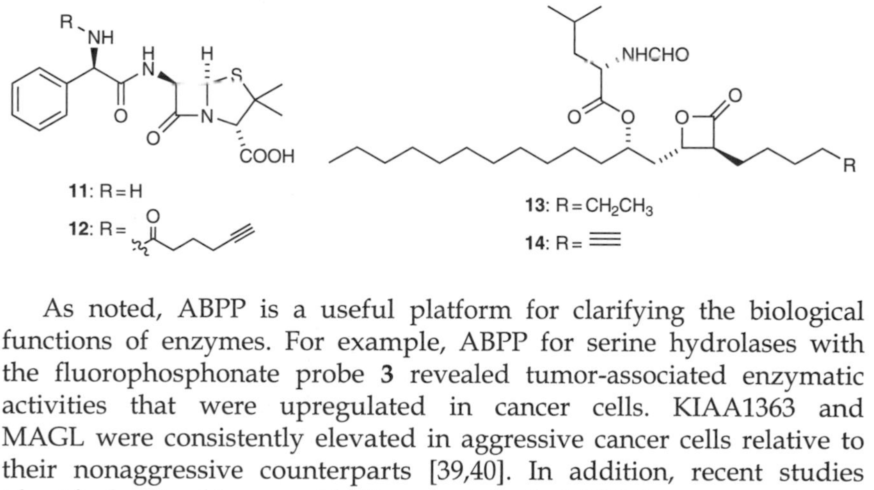

11: R=H
12: R=
13: R=CH$_2$CH$_3$
14: R= ≡

As noted, ABPP is a useful platform for clarifying the biological functions of enzymes. For example, ABPP for serine hydrolases with the fluorophosphonate probe 3 revealed tumor-associated enzymatic activities that were upregulated in cancer cells. KIAA1363 and MAGL were consistently elevated in aggressive cancer cells relative to their nonaggressive counterparts [39,40]. In addition, recent studies identified retinoblastoma-binding protein 9 (RBBP9) as a previously uncharacterized tumor-associated serine hydrolase with elevated activity in pancreatic carcinomas [41].

Bogyo's group developed several ABPs for cysteine proteases with epoxide, vinyl sulfone, and acyloxymethyl ketone reactive groups [42,43]. Particularly noteworthy is DCG04 (4), a biotinylated papain-family protease inhibitor with an epoxide reactive group that targets active-site cysteines. When used to profile the regulation of different cysteine proteases throughout the *Plasmodium falciparum* life cycle, DCG04 identified selective inhibitors for DPAP1 and DPAP3, proteases involved in parasite egress. A role for calpains in host cell rupture was also detected [44,45]. As an extension of Bogyo's ABPs, a Rockefeller University group recently developed a cell-permeable ABP with an acyloxymethyl ketone reactive group that selectively labels the cysteine protease cathepsin B in the cytosol of apoptotic cells [46]. This probe revealed a role for cytosolic cathepsin B as a positive regulator of a cell death feed-forward loop.

These examples highlight the utility of competitive ABPP and CC-ABPP to identify the targets and selectivity of existing small molecules, to identify novel covalent enzyme inhibitors, and to characterize enzymes of unknown function.

3.3 Photoaffinity probes

Photoaffinity reagents combine a structure that binds to a target biomacromolecule with a "warhead" that remains inert until photoactivated. The strategy was introduced by Westheimer's group in the 1960s [47,48], and has been newly invigorated by proteomic methods for the identification of labeled targets. It has traditionally been used with the goal of identifying proteins labeled by a probe, but has increasingly been applied to locating specific sites within proteins. This transformation is undoubtedly due to advances in proteomic methods.

In photoaffinity labeling, the probe is added to the biological milieu that contains its target and is presumed to bind to the target at equilibrium. Photolysis then generates a radical, carbene, or nitrene reactive intermediate that reacts with a nearby group on the target. No *chemical* complementarity between probe and target is required. Binding to the target is biospecific, but probe-to-target coupling is relatively indiscriminate, as explained by an elegant early review by Knowles [49]. Difficulties associated with the production, storage, and use of the prototypical arylazide photoprobes have been partly overcome by alternatives such as benzophenones [50,51] and substituted 3-aryl-3-(trifluoromethyl)diazirines [52,53].

ABPP is only applicable to targets that possess a nucleophilic active-site residue (Ser, Cys, Lys) susceptible to covalent labeling by an electrophile. When this is lacking, an alternative is to add a photoaffinity group to an inhibitor scaffold so that a covalent adduct with the target can be created by exposure to UV light.

If the probe must also possess a reporter group for detection and/or enrichment of labeled targets, it can be functionalized with an alkyne to allow CC-mediated conjugation with azide-reporter tags (e.g., rhodamine-azide or biotin-azide) after photocross-linking to target proteins. This strategy was employed to create photoreactive ABPP probes for metalloproteases (e.g., **15**) [54–56], histone deacetylases (e.g., **16**) [57], the nicotinic acetylcholine receptor (e.g., **17**) [58], and Abl kinase (e.g., **18**) [59].

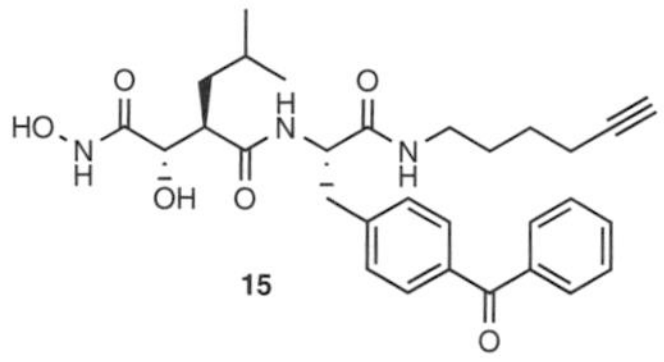

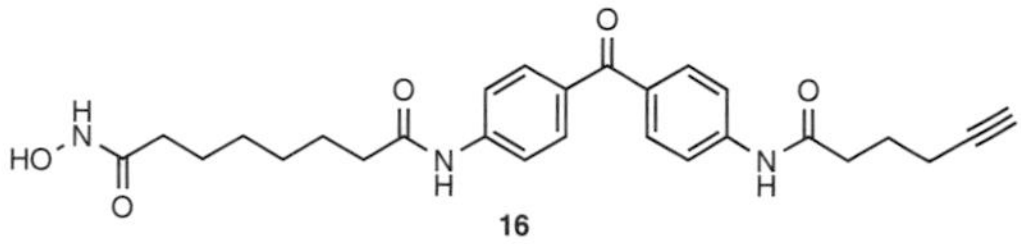

The sustained attractiveness of photolabeling is apparent from its prominence in studies of γ-secretase, an intramembrane protease that contributes to forming amyloid-β peptides and is a major target in Alzheimer's disease [60–62]. γ-Secretase is a complex of at least four different polypeptides, and is difficult to engage with high-resolution structural methods. However, in a case of this kind that involves a known target, immunodetection of proteins can often specify the target of γ-secretase inhibitor photoaffinity probes such as **19**, and proteomic mass spectrometry is not needed.

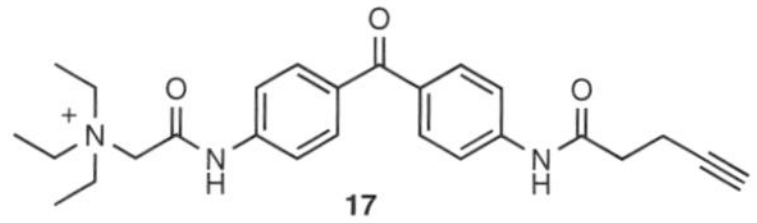

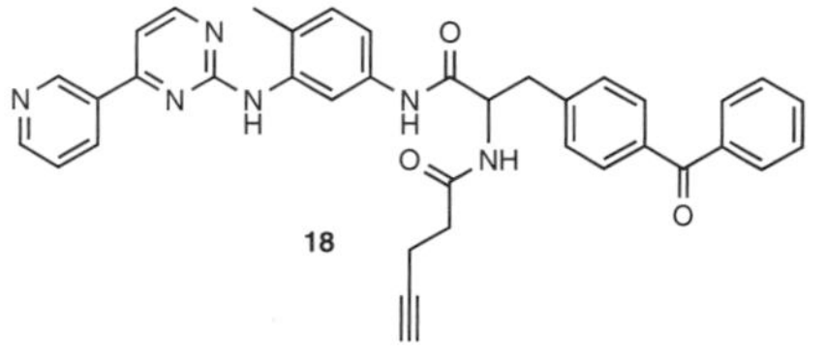

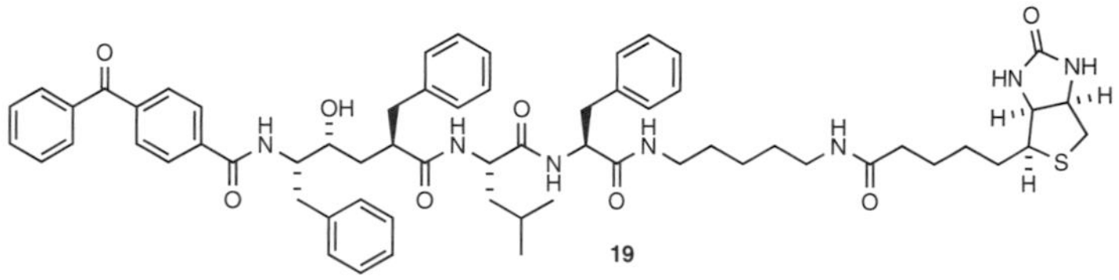

Advancing technology for modified biopolymer synthesis is being used to make biomolecular photoprobes that can be used to elucidate details of drug action. For example, a benzophenone photoprobe based on cisplatin-modified DNA was used to identify proteins that interact with cisplatin interstrand cross-links in the genetic material (Figure 2A) [63]. Also noteworthy, although not directly connected to drug function, were double-stranded deoxynucleotide photoprobes intended to be recognized by bacterial DNA repair proteins. These included a fluorescein probe and a biotin handle for recovery as well as two diazirine

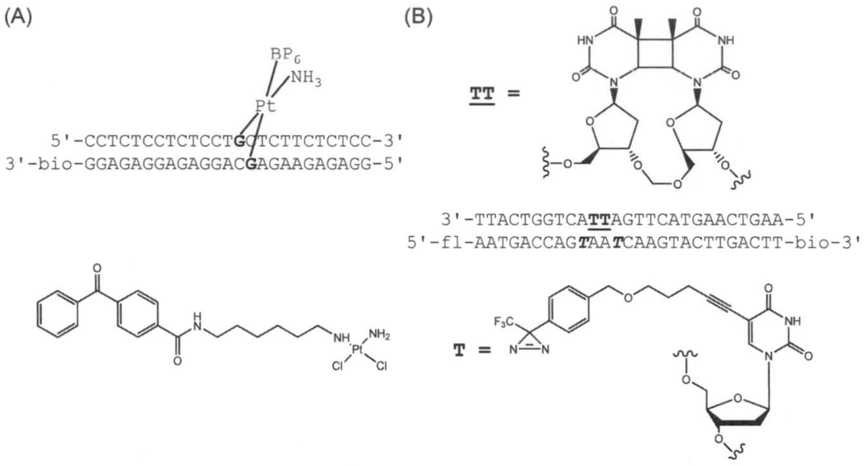

(A)

(B)

BP$_6$
/NH$_3$
Pt

5'-CCTCTCCTCTCCTG*C*TCTTCTCTCC-3'
3'-bio-GGAGAGGAGAGGAC*G*AGAAGAGAGG-5'

TT =

3'-TTACTGGTCA**TT**AGTTCATGAACTGAA-5'
5'-fl-AATGACCAG*TAA T*CAAGTACTTGACTT-bio-3'

T =

Figure 2 Double-stranded oligonucleotide photoprobes that simulate modified DNA and intended to cross-link to DNA-binding proteins. (A) Probe modeling interstrand cross-linking by cisplatin *Source:* From Ref. [63], with permission from the American Chemical Society via the Rightslink service (license number 2458870278307 granted June 30, 2010). The benzophenone probe prior to reaction with DNA is shown in the lower part of the panel. (B) Photoaffinity probe for bacterial DNA repair proteins. **TT** is a simulated thymine dimer intended to be recognized as a site of damage in DNA, and **T** (two instances) is the diazirine thymine derivative T *Source:* From Ref. [64], with permission from Wiley.

groups for coupling to proteins with affinity for bait regions simulating sites of DNA damage (Figure 2B) [64].

Peptidic photoprobes can be based on the photoreactive amino acid *p*-benzoyl-L-phenylalanine inserted into a peptide in place of a natural aromatic residue by peptide synthesis [65] or by manipulation of the genetic code [66]. The use of *p*-benzoyl-L-phenylalanine for this purpose is not new, but the nature of peptide probes naturally offers opportunities for the location of linkage sites by proteomic analysis [67].

Photoaffinity probes have also been used to identify the location of drug-binding sites. Protein digestion and LC–MS were used to deduce that the chloroquine (**20**)-binding site of the chloroquine resistance transporter of *P. falciparum* was near a particular exposed loop of the transporter following labeling with a perfluorophenylazido chloroquine analog (**21**) [68]. A diazirine derivative **22** of cinnamoyl inhibitors of liver transglutaminase was used to probe the binding mode of these compounds, an issue for which direct structural data were not previously available [69]. This study also required proteomic analysis.

Finally, GTP is bound by many proteins, and diazirine-bearing thiol-cleavable GTP analogs (e.g., **23**) have been proposed as probes to detect such proteins. By this method, H-Ras was successfully labeled in a nuclear extract of cultured human cells [70].

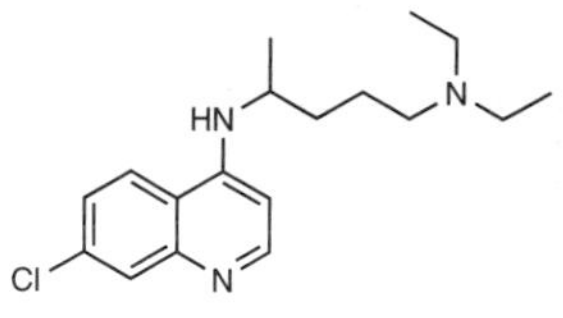

With apologies to authors whose relevant work was excluded because of space constraints, we conclude that the use of photoaffinity/proteomic strategies in connection with drug function has gained significant momentum from the availability of proteomic analytical methods.

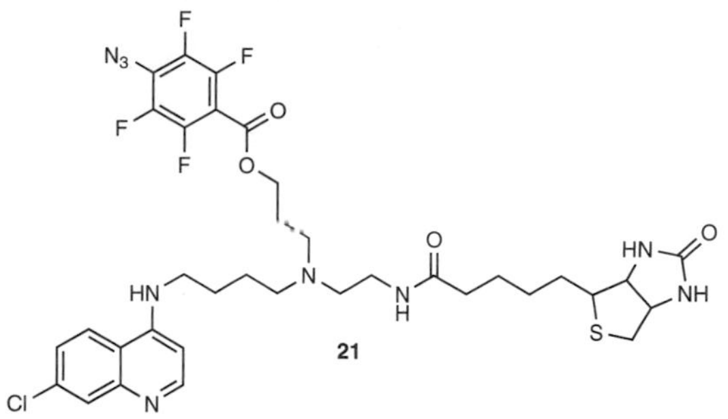

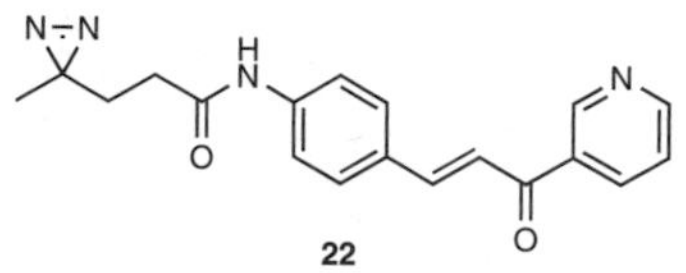

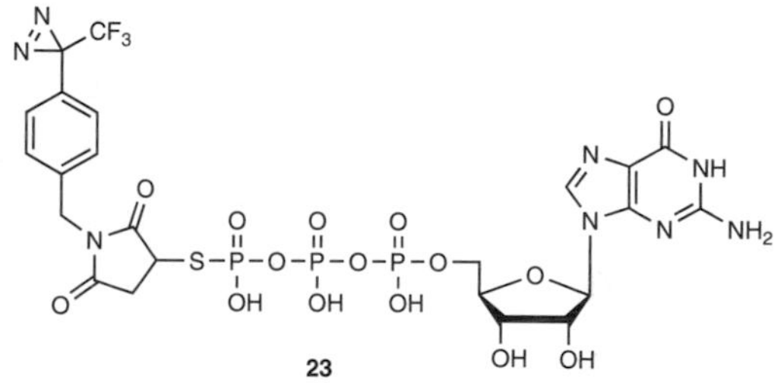

4. SUMMARY AND COMMENTS

Of the front-end methods that we have described in chemical proteomics, affinity capture and photoaffinity labeling have histories of nearly 50 years in biochemistry, while ABPP is a newcomer. The recent literature makes it clear that proteomic analytical methods have been utilized effectively with these methods for target capture and labeling, and that each of these approaches makes a natural interface with protein identification based on mass spectrometry and sequence databases. The results are that the prospects for "reverse pharmacology" have been greatly improved, reviving the feasibility of completing the process of drug discovery when starting with a complex, intact biological system.

REFERENCES

[1] F. Sams-Dodd, *Drug Discov. Today*, 2006, **11**, 465.
[2] S. Carney, *Drug Discov. Today*, 2005, **10**, 1011.

[3] R. Capdeville, E. Buchdunger, J. Zimmermann, and A. Matter, *Nat. Rev. Drug Discov.*, 2002, **1**, 493.

[4] Y. Feng, T. J. Mitchison, A. Bender, D. W. Young, and J. A. Tallarico, *Nat. Rev. Drug Discov.*, 2009, **8**, 567.

[5] M. Austen and C. Dohrmann, *Drug Discov. Today*, 2005, **10**, 275.

[6] J. R. Yates, C. I. Ruse, and A. Nakorchevsky, *Annu. Rev. Biomed. Eng.*, 2009, **11**, 49.

[7] C. P. Hart, *Drug Discov. Today*, 2005, **10**, 513.

[8] M. W. Harding, A. Galat, D. E. Uehling, and S. L. Schreiber, *Nature*, 1989, **341**, 758.

[9] M. Yoshida, M. Kijima, M. Akita, and T. Beppu, *J. Biol. Chem.*, 1990, **265**, 17174.

[10] J. C. Lee, J. T. Laydon, P. C. McDonnell, T. F. Gallagher, S. Kumar, D. Green, D. McNulty, M. J. Blumenthal, J. R. Keys, S. W. Land Vatter, J. E. Strickler, M. M. McLaughlin, I. R. Siemens, S. M. Fisher, G. P. Livi, J. R. White, J. L. Adams, and P. R. Young, *Nature*, 1994, **372**, 739.

[11] P. Denny, F. K. Hagen, M. Hardt, L. Liao, W. Yan, M. Arellanno, S. Bassilian, G. S. Bedi, P. Boontheung, D. Cociorva, C. M. Delahunty, T. Denny, J. Dunsmore, K. F. Faull, J. Gilligan, M. Gonzalez-Begne, F. Halgand, S. C. Hall, X. Han, B. Henson, J. Hewel, S. Hu, S. Jeffrey, J. Jiang, J. A. Loo, R. R. Ogorzalek Loo, D. Malamud, J. E. Melvin, O. Miroshnychenko, M. Navazesh, R. Niles, S. K. Park, A. Prakobphol, P. Ramachandran, M. Richert, S. Robinson, M. Sondej, P. Souda, M. A. Sullivan, J. Takashima, S. Than, J. Wang, J. P. Whitelegge, H. E. Witkowska, L. Wolinsky, Y. Xie, T. Xu, W. Yu, J. Ytterberg, D. T. Wong, J. R. Yates, and S. J. Fisher, *J. Proteome Res.*, 2008, **7**, 1994.

[12] N. L. Anderson, N. G. Anderson, T. W. Pearson, C. H. Borchers, A. G. Paulovich, S. D. Patterson, M. Gillette, R. Aebersold, and S. A. Carr, *Mol. Cell. Proteomics*, 2009, **8**, 883.

[13] P. Cuatrecasas, *Advan. Enzymol. Relat. Areas Mol. Biol.*, 1972, **36**, 29.

[14] C. Saxena, R. E. Higgs, E. Zhen, and J. E. Hale, *Expert Opin. Drug Discov.*, 2009, **4**, 701.

[15] S.-E. Ong and M. Mann, *Nat. Chem. Biol.*, 2005, **1**, 252.

[16] P. Dadvar, M. O'Flaherty, A. Scholten, K. Rumpel, and A. J. R. Heck, *Mol. BioSyst.*, 2009, **5**, 472.

[17] S.-E. Ong, M. Schenone, A. A. Margolin, X. Li, K. Do, M. K. Doud, D. R. Mani, L. Kuai, X. Wang, J. L. Wood, N. J. Tolliday, A. N. Koehler, L. A. Marcaurelle, T. R. Golub, R. J. Gould, S. L. Schreiber, and S. A. Carr, *Proc. Natl. Acad. Sci. USA*, 2009, **106**, 4617.

[18] P. J. Boersema, R. Raijmakers, S. Lemeer, S. Mohammed, and A. J. R. Heck, *Nat. Protoc.*, 2009, **4**, 484.

[19] L. Trinkle-Mulcahy, S. Boulon, Y. W. Lam, R. Urcia, F.-M. Boisvert, F. Vandermoere, N. A. Morrice, S. Swift, U. Rothbauer, H. Leonhardt, and A. Lamond, *J. Cell Biol.*, 2008, **183**, 223.

[20] H. Uga, C. Kuramori, A. Ohta, Y. Tsuboi, H. Tanaka, M. Hatakeyama, Y. Yamaguchi, T. Takahashi, M. Kizaki, and H. Handa, *Mol. Pharmacol.*, 2006, **70**, 1832.

[21] C. Kuramori, M. Azuma, K. Kume, Y. Kaneko, A. Inoue, Y. Yamaguchi, Y. Kabe, T. Hosoya, M. Kizaki, M. Suematsu, and H. Handa, *Biochem. Biophys. Res. Commun.*, 2009, **379**, 519.

[22] K. Nishio, Y. Masaike, M. Ikeda, H. Narimatsu, N. Gokon, S. Tsubouchi, M. Hatakeyama, S. Sakamoto, N. Hanyu, A. Sandhu, H. Kawaguchi, M. Abe, and H. Handa, *Colloids Surf.*, B 2008, **64**, 162.

[23] J. F. Peter and A. M. Otto, *Proteomics*, 2010, **10**, 628.

[24] M. J. Evans and B. F. Cravatt, *Chem. Rev.*, 2006, **106**, 3279.

[25] B. F. Cravatt, A. T. Wright, and J. W. Kozarich, *Annu. Rev. Biochem.*, 2008, **77**, 383.

[26] G. M. Simon and B. F. Cravatt, *J. Biol. Chem.*, 2010, **285**, 11051.

[27] Y. Liu, M. P. Patricelli, and B. F. Cravatt, *Proc. Natl. Acad. Sci. USA*, 1999, **96**, 14694.

[28] K. Ahn, D. S. Johnson, L. R. Fitzgerald, M. Liimatta, A. Arendse, T. Stevenson, E. T. Lund, R. A. Nugent, T. K. Nomanbhoy, J. P. Alexander, and B. F. Cravatt, *Biochemistry*, 2007, **46**, 13019.

[29] K. Ahn, D. S. Johnson, M. Mileni, D. Beidler, J. Z. Long, M. K. McKinney, E. Weerapana, N. Sadagopan, M. Liimatta, S. E. Smith, S. Lazerwith, C. Stiff, S. Kamtekar, K. Bhattacharya, Y. Zhang, S. Swaney, K. Van Becelaere, R. C. Stevens, and B. F. Cravatt, *Chem. Biol.*, 2009, **16**, 411.

[30] W. Li, J. L. Blankman, and B. F. Cravatt, *J. Am. Chem. Soc.*, 2007, **129**, 9594.

[31] J. Z. Long, W. Li, L. Booker, J. J. Burston, S. G. Kinsey, J. E. Schlosburg, F. J. Pavon, A. M. Serrano, D. E. Selley, L. H. Parsons, A. H. Lichtman, and B. F. Cravatt, *Nat. Chem. Biol.*, 2009, **5**, 37.

[32] J. Z. Long, X. Jin, A. Adibekian, W. Li, and B. F. Cravatt, *J. Med. Chem.*, 2010, **53**, 1830.

[33] J. Z. Long, D. K. Nomura, R. E. Vann, D. M. Walentiny, L. Booker, X. Jin, J. J. Burston, L. J. Sim-Selley, A. H. Lichtman, J. L. Wiley, and B. F. Cravatt, *Proc. Natl. Acad. Sci. USA*, 2009, **106**, 20270.

[34] A. E. Speers, G. C. Adam, and B. F. Cravatt, *J. Am. Chem. Soc.*, 2003, **125**, 4686.

[35] A. E. Speers and B. F. Cravatt, *Chem. Biol.*, 2004, **11**, 535.

[36] I. Staub and S. A. Sieber, *J. Am. Chem. Soc.*, 2008, **130**, 13400.

[37] I. Staub and S. A. Sieber, *J. Am. Chem. Soc.*, 2009, **131**, 6271.

[38] P. Y. Yang, K. Liu, M. H. Ngai, M. J. Lear, M. R. Wenk, and S. Q. Yao, *J. Am. Chem. Soc.*, 2010, **132**, 656.

[39] K. P. Chiang, S. Niessen, A. Saghatelian, and B. F. Cravatt, *Chem. Biol.*, 2006, **13**, 1041.

[40] D. K. Nomura, J. Z. Long, S. Niessen, H. S. Hoover, S. W. Ng, and B. F. Cravatt, *Cell*, 2010, **140**, 49.

[41] D. J. Shields, S. Niessen, E. A. Murphy, A. Mielgo, J. S. Desgrosellier, S. K. Lau, L. A. Barnes, J. Lesperance, M. Bouvet, D. Tarin, B. F. Cravatt, and D. A. Cheresh, *Proc. Natl. Acad. Sci. USA*, 2010, **107**, 2189.

[42] D. Kato, K. M. Boatright, A. B. Berger, T. Nazif, G. Blum, C. Ryan, K.A.H. Chehade, G. S. Salvesen, and M. Bogyo, *Nat. Chem. Biol.*, 2005, **1**, 33.

[43] A. W. Puri and M. Bogyo, *ACS Chem. Biol.*, 2009, **4**, 603.

[44] S. Arastu-Kapur, E. L. Ponder, U. P. Fonovic, S. Yeoh, F. Yuan, M. Fonovic, M. Grainger, C. I. Phillips, J. C. Powers, and M. Bogyo, *Nat. Chem. Biol.*, 2008, **4**, 203.

[45] R. Chandramohanadas, P. H. Davis, D. P. Beiting, M. B. Harbut, C. Darling, G. Velmourougane, M. Y. Lee, P. A. Greer, D. S. Roos, and D. C. Greenbaum, *Science*, 2009, **324**, 794.

[46] M. R. Pratt, M. D. Sekedat, K. P. Chiang, and T. W. Muir, *Chem. Biol.*, 2009, **16**, 1001.

[47] J. Shafer, P. Barnowsky, R. Laursen, F. Finn, and F. H. Westheimer, *J. Biol. Chem.*, 1966, **241**, 421.

[48] A. Singh, E. R. Thornton, and F. H. Westheimer, *J. Biol. Chem.*, 1962, **237**, PC3006.

[49] J. R. Knowles, *Acc. Chem. Res.*, 1972, **5**, 155.

[50] J. D. Olszewski, G. Dorman, J. T. Elliott, Y. Hong, D. G. Ahern, and G. D. Prestwich, *Bioconjugate Chem.*, 1995, **6**, 395.

[51] G. Dorman and G. D. Prestwich, *Biochemistry*, 1994, **33**, 5661.

[52] M. Hashimoto and Y. Hatanaka, *Eur. J. Org. Chem.*, 2008, **2008**, 2513.

[53] A. Blencowe and W. Hayes, *Soft Matter*, 2005, **1**, 178.

[54] S. A. Sieber, S. Niessen, H. S. Hoover, and B. F. Cravatt, *Nat. Chem. Biol.*, 2006, **2**, 274.

[55] A. Saghatelian, N. Jessani, A. Joseph, M. Humphrey, and B. F. Cravatt, *Proc. Natl. Acad. Sci. USA*, 2004, **101**, 10000.

[56] E. W. Chan, S. Chattopadhaya, R. C. Panicker, X. Huang, and S. Q. Yao, *J. Am. Chem. Soc.*, 2004, **126**, 14435.

[57] C. M. Salisbury and B. F. Cravatt, *J. Am. Chem. Soc.*, 2008, **130**, 2184.

[58] M. Tantama, W. C. Lin, and S. Licht, *J. Am. Chem. Soc.*, 2008, **130**, 15766.

[59] K. A. Kalesh, D. S. Sim, J. Wang, K. Liu, Q. Lin, and S. Q. Yao, *Chem. Commun. (Cambridge, U. K.)*, 2010, **46**, 1118.

[60] Y.-M. Li, M. Xu, M.-T. Lai, Q. Huang, J. L. Castro, J. DiMuzio-Mower, T. Harrison, C. Lellis, A. Nadin, J. G. Neduvelil, R. B. Register, M. K. Sardana, M. S. Shearman, A. L. Smith, X.-P. Shi, K.-C. Yin, J. A. Shafer, and S. J. Gardell, *Nature*, 2000, **405**, 689.

[61] C. C. Shelton, L. Zhu, D. Chau, L. Yang, R. Wang, H. Djaballah, H. Zheng, and Y.-M. Li, *Proc. Natl. Acad. Sci. USA*, 2009, **106**, 20228.

[62] H. Fuwa, Y. Takahashi, Y. Konno, N. Watanabe, H. Miyashita, M. Sasaki, H. Natsugari, T. Kan, T. Fukuyama, T. Tomita, and T. Iwatsubo, *ACS Chem. Biol.*, 2007, **2**, 408.

[63] G. Zhu and S. J. Lippard, *Biochemistry*, 2009, **48**, 4916.

[64] M. Winnacker, S. Breeger, R. Strasser, and T. Carell, *ChemBioChem*, 2009, **10**, 109.

[65] J. C. Kauer, S. Erickson-Viitanen, H. R. Wolfe, and W. F. DeGrado, *J. Biol. Chem.*, 1986, **261**, 10695.

[66] J. W. Chin, A. B. Martin, D. S. King, L. Wang, and P. G. Schultz, *Proc. Natl. Acad. Sci. USA*, 2002, **99**, 11020.

[67] K. Dimova, S. Kalkhof, I. Pottratz, C. Ihling, F. Rodriguez-Castaneda, T. Liepold, C. Griesinger, N. Brose, A. Sinz, and O. Jahn, *Biochemistry*, 2009, **48**, 5908.

[68] J. K. Lekostaj, J. K. Natarajan, M. F. Paguio, C. Wolf, and P. D. Roepe, *Biochemistry*, 2008, **47**, 10394.

[69] C. Pardin, I. Roy, A. Chica Roberto, E. Bonneil, P. Thibault, W. D. Lubell, J. N. Pelletier, and J. W. Keillor, *Biochemistry*, 2009, **48**, 3346.

[70] M. Kaneda, S. Masuda, T. Tomohiro, and Y. Hatanaka, *ChemBioChem*, 2007, **8**, 595.

CHAPTER 22

NIH Translational Programs for Assisting Pre-Clinical Drug Discovery and Development

Raj N. Misra

National Institutes of Health, National Cancer Institute, Developmental Therapeutics Program, 6130 Executive Blvd., EPN Suite 8032, Bethesda, MD 20892, USA

Annual Reports in Medicinal Chemistry, Volume 45
ISSN 0065-7743, DOI 10.1016/S0065-7743(10)45022-8

ABBREVIATIONS

ADME	absorption, distribution, metabolism, elimination
API	active pharmaceutical ingredient
ASP	Anticonvulsant Screening Program
ATDP	Addiction Treatment Discovery Program
CBC	Chemical Biology Consortium
CCR	Center for Cancer Research
CMC	Chemistry and Manufacturing Controls
CNS	central nervous system
CPB	Chemistry and Pharmaceutics Branch
CSR	Center for Scientific Review
CTCL	cutaneous T-cell lymphoma
DCTD	Division of Cancer Treatment and Diagnosis
DDG	Drug Development Group
DMID	Division of Microbiology and Infectious Diseases
DPMCDA	Division of Pharmacotherapies and Medical Consequences of Drug Abuse
DTP	Developmental Therapeutics Program
FDA	Food and Drug Administration
GLP	Good Laboratory Practices
GMP	Good Manufacturing Practices
HDAC	histone deacetylase
HTS	high-throughput screening
IDE	Investigational Device Exemption
IND	Investigational New Drug
INDA	Investigational New Drug Application
IP	intellectual property
MDD	Medications Development Division
MDP	Medications Development Program
MDTP	Medications Discovery and Toxicology Branch
miRNA	microRNA
MLPCN	Molecular Libraries Probe Production Centers Network
MLSMR	Molecular Libraries Small Molecule Repository
NCGC	NIH Chemical Genomics Center

NCI	National Cancer Institute
NCI60	NCI 60 Cancer Cell Line Screen
NExT	NCI Experimental Therapeutics
NHGRI	National Human Genome Research Institute
NHLBI	National Heart, Lung, and Blood Institute
NIAID	National Institute of Allergy and Infectious Diseases
NIDA	National Institute on Drug Abuse
NIDDK	National Institute of Diabetes and Digestive and Kidney Diseases
NIH	National Institutes of Health
NINDS	National Institute of Neurological Disorders and Stroke
NME	new molecular entity
ORDR	Office of Rare Diseases Research
PARP	poly(ADP-ribose) polymerase
PBMC	peripheral blood mononuclear cell
PD	pharmacodynamics
PK	pharmacokinetics
R&D	research and development
RAID	Rapid Access to Interventional Development
RAND	Rapid Access to NCI Discovery Services
RNA	ribonucleic acid
SBIR	Small Business Innovation Research
STTR	Small Business Technology Transfer
siRNA	small inhibitory RNA
shRNA	small hairpin RNA
SMA	Spinal Muscular Atrophy
T1D	Type 1 Diabetes
T1D-PTP	Type 1 Diabetes Pre-Clinical Testing Program
TK	toxicokinetics
TRND	Therapeutics for Rare and Neglected Diseases

1. INTRODUCTION

The National Institutes of Health (NIH) is the largest government-funded medical research institution in the world. Its origins trace back to the Laboratory of Hygiene at the Marine Hospital in Staten Island, NY. The laboratory was created by Congress in 1887 to study cholera, yellow fever, and other infectious diseases of returning U.S. sailors [1,2]. In 1891 the Laboratory moved to Washington, D.C., and in 1930 was expanded and designated by the Ransdell Act of Congress as the National Institute of Health. In 1948 the designation was changed to the plural, National Institutes of Health, as research moved into the additional areas of mental health, dental disease, and heart disease. The NIH has since grown to 18,000 employees spread over 27 semi-autonomous Institutes and Centers (ICs) that are focused on basic medical research and organized primarily along specific disease areas. The NIH is located on a rangy 300-acre campus in Bethesda, MD, on the outskirts of Washington, D.C. The vast majority of the current $28 billion NIH budget is invested in supporting medical research at institutions throughout the United States through the distribution and administration of competitive extramural research grants. About 10% of the NIH budget is directed toward the support of intramural programs carried out by its own 6000 scientists. In addition to supporting basic medical research, the NIH also supports and had re-emphasized, as part of the NIH Roadmap for Medical Research (http://nihroadmap.nih.gov), a number of translational medicine initiatives, including the pre-clinical development of agents for the diagnosis and treatment of disease. The goals of these programs are to resurrect fallow pipelines and stimulate the discovery of new agents by providing critical resources that can overcome translational roadblocks on the route to clinical evaluation. These programs are often the only bridge forward for investigators, especially in the costly late pre-clinical stages of the development process.

This review is an overview of current and major NIH translational programs that support pre-clinical drug discovery and development. This review is not comprehensive in its scope, for example, focused discovery-related services (e.g., *in vitro* screening) provided by many ICs outside of full-development programs are generally not included. Additional program information can be found through the NIH website (http://www.nih.gov) or individual IC websites. Support to investigators from the cited programs is generally provided as pre-clinical services at no charge, rather than as direct funding grants. SBIR/STTR (Small Business Innovation Research and Small Business Technology Transfer) programs are available, which can provide direct funding. It is suggested that the referenced websites be reviewed for updated information since programs evolve in response to changing needs and budgetary conditions.

2. NIH MOLECULAR LIBRARIES PROBE PRODUCTION CENTERS NETWORK

Purpose: The Molecular Libraries Roadmap offers public sector biomedical researchers access to the large-scale screening resources through the Molecular Libraries Probe Production Centers Network (MLPCN). The MLPCN has established a collection of ~250,000 chemically diverse small molecules in the Molecular Libraries Small Molecule Repository (MLSMR) for use by the MLPCN. The collection contains molecules with known biological activities, while others have the potential to modulate novel biological functions. The resources can be used to identify small-molecule chemical probes to study the functions of genes, cells, and biochemical pathways. These probes may also be suitable as starting points for drug discovery programs.

Website: http://mli.nih.gov/mli/
Eligibility: Academic and nonprofit institutions
Background: For additional information, a detailed review recently appeared in this series [3].

3. NIH-RAID PROGRAM

Purpose: The NIH-RAID (Rapid Access to Interventional Development) program was developed to provide core pre-clinical drug development services as part of the NIH Roadmap "bench to bedside" translational medicine initiative. NIH-RAID provides a path forward to clinical trials for academics, non-profits, and small businesses, and is meant to serve as a bridge to future funding. The program potentially encompasses all therapeutic areas, and is currently administered through the National Institute of Neurological Disorders and Stroke (NINDS) and the National Heart, Lung and Blood Institute (NHLBI) with scientific oversight of tasks by National Cancer Institute (NCI) scientists.

Website: http://commonfund.nih.gov/raid/
Eligibility: Academic and government investigators (U.S. and foreign), non-profit organizations, SBIR-eligible small businesses
Intellectual Property: In general, the investigator retains ownership and development rights to their invention, consistent with Bayh–Dole provisions [4]. The investigator is expected to submit an Investigational New Drug Application (INDA) with the FDA under their own sponsorship. Specific information can be found at the NIH-RAID website.

Background: The NIH Roadmap for Medical Research was launched in September 2004, by former NIH Director Dr. Elias A. Zerhouni. The program was proposed in order to address roadblocks to research, and to transform the manner in which publically funded biomedical research is conducted by overcoming specific hurdles or filling defined knowledge gaps. As part of the Roadmap theme of Re-engineering the Clinical Research Enterprise, there has been a re-invigorated effort to bring new medicines from the basic research laboratory to the patient. Although the role that NIH should play in drug development has been controversial, there is growing support for this effort. The NIH-RAID program is one of the key initiatives in this area. NIH-RAID is a drug development program designed to aid investigators in obtaining the data and material necessary to initiate a Phase 1 clinical trial.

The emphasis of the program is to support projects that complement, rather than compete with the private sector. Some examples are projects related to high-risk and novel targets, orphan diseases, and areas in which the monetary return may not be immediately clear.

Logistics: NIH-RAID projects are operated as a collaborative effort with administrative responsibilities lying within NINDS, and scientific responsibilities with the Institute/Center (IC) staff. Applications to the NIH-RAID program are accepted three times a year and undergo Center for Scientific Review (CSR)-peer review and scoring. Selected proposals are discussed in a collaborative meeting with the investigator and NIH-RAID and IC staff with the goal of establishing firm development tasks, milestones, and project plan. Once these are determined, a funding decision occurs. If funded, the mutually agreed development tasks are executed with scientific oversight by NCI and NHLBI staff. Regular updates between NIH-RAID, IC staff, and the investigator are scheduled to monitor progress. Once complete, the data and materials are transferred to the investigator.

Program Capabilities: The NIH-RAID program is capable of performing all the necessary development tasks to support the submission of an INDA. A summary of the services is below. A complete listing can be found at the NIH-RAID website.

For small molecules, natural products, peptides, oligonucleotides, gene vectors, recombinant proteins and monoclonal antibodies:

- Synthesis and scale-up production of active pharmaceutical ingredient (API)
- Development of analytical methods
- Development, manufacture, and stability of clinical formulation
- Isolation and purification of natural products
- Pharmacokinetic/ADME (absorption, distribution, metabolism, elimination) studies including bioanalytical method development
- Range-finding and IND-directed toxicology
- Product development planning and advice in IND preparation

3.1 Case Study 1: Safety Pharmacology Studies for an IND for Beta Thalassemia: Susan Perrine, M.D., Professor of Pediatrics, Medicine, Pharmacology, and Experimental Therapeutics, Boston University School of Medicine

Summary: Thalassemias as a group are the most common genetic diseases in the world. Beta thalassemia is an inherited blood disorder in which the body produces an abnormal form of hemoglobin. The disorder results in excessive destruction of red blood cells, which in a severe form manifests as life-shortening anemia shortly after birth. Short-chain fatty acids had previously been shown to be useful in the treatment of beta thalassemia [5]. As part of a drug development program, sodium ST20, a short-chain fatty acid, was identified as exhibiting beneficial actions in thalassemia models. ST20 had shown favorable oral PK in baboons and an acceptable safety profile and was selected for clinical development by the investigators prior to applying to NIH-RAID. Good Manufacturing Practices (GMP) synthesis and pre-clinical IND-enabling toxicology studies in two species had also been performed using several funding mechanisms; however, additional safety studies were requested in discussions between the investigator and FDA reviewers before a first-in-human clinical trial would be allowed to proceed. Subsequent to discussions with the FDA, the investigator submitted a proposal and requested Good Laboratory Practices (GLP) central nervous system (CNS) and cardiovascular safety studies, as well as a genotoxicity study from NIH-RAID. The program successfully completed all required safety studies, developed a clinically acceptable formulation, and manufactured early clinical supplies for the project. An INDA was submitted by the investigator and allowed by the FDA in early 2008. The clinical development was subsequently partnered with a biotech company, HemaQuest. The agent (re-designated as HQK-1001) has demonstrated safety and tolerability in healthy volunteers and encouraging clinical trials are ongoing in the United States, Thailand, Lebanon, and the United Kingdom [6].

3.2 Case Study 2: Pre-Clinical Development of CDD-0102A for the Treatment of Alzheimer's Disease: William S. Messer, Ph.D., Department of Pharmacology, The University of Toledo

Summary: Muscarinic agonists have the potential to treat memory and cognitive deficits associated with Alzheimer's disease, and the potential to slow or stop the disease process. CDD-0102A, a small molecule, was discovered at The University of Toledo, and was characterized as a selective muscarinic agonist. It displayed promising biochemical activity, functional selectivity, *in vivo* efficacy, and was well tolerated at doses that improve memory function in animal models. NIH-RAID contributed to the development of CDD-0102A by preparation of GMP bulk drug, development of a clinical formulation, and manufacture of Phase 1 clinical supplies. These Chemistry and Manufacturing Controls (CMC) tasks were part of a larger full-development program carried out by the investigator and was supported through several funding mechanisms. An INDA was submitted for CDD-0102A and allowed by the FDA in 2009. The clinical development has been partnered with a biotech company, Mithridion, Inc. The agent (re-designated as MCD-386) is currently in clinical trials [7].

4. NATIONAL CANCER INSTITUTE: NEXT PROGRAM

Purpose: The NCI Experimental Therapeutics (NExT) program is designed to assist investigators in "bench to bedside" translation of novel anticancer therapeutic interventions, synthetic, natural product, or biologic, arising from academic, industrial, or government entities. The program provides the resources for selected discovery tasks, comprehensive pre-clinical IND-enabling tasks, and biomarker development for Phase 0 clinical studies. The tasks are completed by NCI staff and contractors, rather than through direct investigator grants. The program goal is to provide NCI with an integrated pre-clinical pipeline of novel anticancer agents.

Website: http://next.cancer.gov/

Eligibility: Academic investigators, non-profit organizations, SBIR-eligible small businesses

Intellectual Property: A review of intellectual property, relevant to the project, is requested as part of the application. The guidelines for ownership of intellectual property generated will vary with each project and generally depend on the stage of the project.

Background: The NCI has traditionally had a strong pre-clinical drug development program, dating back over 50 years. These programs include Drug Development Network, Rapid Access to NCI Discovery Services (RAND), RAID, and Drug Development Group (DDG). The NexT program is a consolidation of these programs into a single integrated pipeline. The purpose of an integrated pipeline is to ensure that the available NCI resources are allocated to the most meritorious programs. The pipeline is managed by multiple collaborative governance committees using a milestone and stage-gate approach.

Logistics: Applications to the NExT program are submitted electronically (https://proposalcentral.altum.com/default.asp?GMID=76) and reviewed by a panel of experts. Application deadlines are on a quarterly cycle.

Program Capabilities: The NExT program has the capability of performing both discovery and development tasks including target identification, lead small-molecule optimization, and early toxicology and pharmacokinetic analysis. The NexT program will also support all the necessary development tasks for the submission of an INDA for both small molecules and biological products. A summary of the services is given below.

Discovery: Projects classified as discovery would include, but are not limited to, the following tasks:

- Identification of targets (genes, pathways, molecules, biologics, etc.)
- Biological function of targets (pathway dissection, miRNA/siRNA/shRNA studies, model building *in vitro* and *in vivo*)
- Exploratory screen development and high-throughput screening (HTS) optimization

- Novel leads for medicinal chemistry optimization
- Exploratory toxicology studies and pharmacokinetic evaluation

Development: Projects classified as development would include, but are not limited to, the following activities to support INDA submission:

- *In vivo* efficacy studies
- Bulk API synthesis (GMP or non-GMP) including analytical methods development
- Formulation development, production, and stability testing of clinical dosage form
- Development of pharmacology assays
- Pharmacokinetic (PK) and pharmacodynamic (PD) studies
- Range-finding and IND-enabling toxicity studies
- Planning of clinical trials
- INDA submission consultation

4.1 Case Study: Development of Istodax® (romidepsin, NSC 630176, depsipeptide), Fujisawa Pharmaceutical Co.

Summary: Romidepsin is a novel natural product histone deacetylase (HDAC) inhibitor that was developed in collaboration with Fujisawa Pharmaceuticals (now Astellas Pharma). Fujisawa originally submitted romidepsin as NSC 630176 in 1990 for NCI60 anti-proliferative activity evaluation [8]. Based partly on its unique, bicyclic peptide structure, and pattern of activity in the NCI60, the DTP within NCI subsequently performed animal PK and efficacy evaluation, as well as *in vitro* and *in vivo* safety studies under the oversight of the Decision Network committee, a forerunner to NExT. Importantly, DTP studies defined a specific dosing schedule that ameliorated profound cardiotoxicity, normally associated with the drug, and had derailed earlier development. NCI-sponsored Phase 1 studies began in 1997 and romidepsin for injection was approved by the FDA in 2009 for treatment of cutaneous T-cell lymphoma (CTCL). Romidepsin is marketed under the brand name Istodax® by the Celgene Corporation and clinical trials are continuing for other indications [9].

5. NATIONAL CANCER INSTITUTE: CBC PROGRAM

Purpose: The NCI Chemical Biology Consortium (CBC) is a developing biotech-like venture designed to assist investigators in the development of novel cancer therapeutics. The program goal is to increase the number of early-stage drug candidates entering the NCI development pipeline by establishing a drug discovery consortium on the scale of a small biotech company. The focus of the group will be on high-risk, underrepresented areas, and in significantly advancing the discovery of novel agents against specific molecular and genetic cancer targets. The program will provide access to cutting-edge tools for iterative drug discovery, optimization, and development, including, when necessary, Phase 0 clinical trials. Project tasks are generally to be completed in collaboration with CBC investigators, NCI's Division of Cancer Treatment and Diagnosis (DCTD), the Center for Cancer Research (CCR) staff, and NCI/DTP staff and contractors, rather than through directed investigator grants. The program differs significantly from other NCI drug development programs in that eligibility for resources requires information sharing by members in a collaborative consortium (as noted under Intellectual Property below).

Website: http://dctd.cancer.gov/CurrentResearch/ChemicalBioConsortium.htm

Eligibility: Government, academia, and industry

Intellectual Property: All consortium members will be required to sign a good-faith Intellectual Property Agreement. This agreement details the management structure of the CBC, and provides a mechanism for cooperation and information sharing among the participants. The agreement outlines confidentiality requirements, resource and data sharing, and describes intellectual property protections under the existing statutory Bayh–Dole statutory framework. NCI will have the option to clinically develop successful compounds (new molecular entities (NMEs)) created by the CBC.

Background: The NCI has historically supported a vigorous and successful pre-clinical drug development program for identified drug candidates. The effort has focused on late-stage development leading to first-in-man studies. Within this effort, the NCI has supported, among other tasks, animal toxicology and the manufacture of both bulk drug and clinical dosage form. These specific tasks have traditionally presented a formidable monetary roadblock to many investigators and NCI's role has been to provide a bridge for academic investigators to move their agents from the lab into clinical trials. Lesser emphasis has been placed on the early discovery tasks, such as target identification and validation, lead identification, and optimization. A major goal of the CBC will be to re-invigorate NCI's early discovery effort, and establish the framework for state-of-the-art iterative drug discovery. Resource allocation for projects within the CBC will be managed as part of the integrated NExT pipeline.

Logistics: Applications to the CBC will be managed through the NExT program application (https://dctd.cancer.gov/nextapp/setUp.do), and are to be submitted electronically and will be reviewed by a panel of experts. Questions should be directed to individuals listed on the website.

Program Capabilities: The CBC program has the capability of performing a full range of standard drug discovery and development tasks, in addition to the more specialized tasks listed below.

- HTS of compound libraries, including natural products
- Access to a network of highly experienced medicinal chemists
- Molecular and small-animal imaging
- Animal modeling of targeted therapies, including pre-clinical PK and PD
- Development and validation of PD assays to confirm drug effect on molecular target in pre-clinical studies and clinical trials conducted under an exploratory INDA, or in a traditional Phase I/II setting

5.1 Case Study: Phase 0 Clinical Trial of ABT-888 in Patients with Advanced Malignancies, Abbott Laboratories

Summary: This example demonstrates the early clinical development capabilities of the CBC program, although the reported study was performed outside of the formal confines of the CBC program. The NCI, in collaboration with Abbott Laboratories, conducted the first oncology clinical trial under the FDA Exploratory IND Guidance with novel PARP inhibitor, ABT-888. ABT-888 was administered as a single oral dose to determine the time course in which it inhibits PARP activity in tumor biopsies and peripheral blood mononuclear cells (PBMCs). The study determined that ABT-888 inhibited PARP activity in tumor biopsies and PBMCs at clinically achievable concentrations. PARP inhibition was observed in tumor biopsies and PBMCs at 25- and 50-mg doses. Initial biochemical data was available within 5 months of initiation, and showed the value of a Phase 0 trial in accelerating proof-of-principle studies in early clinical development. In this case, pivotal data was rapidly obtained and employed in the design of subsequent Phase 1 trials [10].

6. NATIONAL INSTITUTE OF ALLERGY AND INFECTIOUS DISEASES

Purpose: The Division of Microbiology and Infectious Diseases (DMID) at the National Institute for Allergy and Infectious Disease (NIAID) supports drug discovery and development services for researchers. The purpose is to facilitate the translation of ideas, generated through basic research, into safe and effective drugs, vaccines, and diagnostics to control and prevent infectious diseases. Pre-clinical drug development services are focused only on therapeutics, and must fill a gap in the drug development pathway.

Website: http://www3.niaid.nih.gov/LabsAndResources/resources/dmid

Eligibility: Academia, not-for-profit organizations, industry, and government worldwide

Intellectual Property: In general the investigator retains ownership and development rights to their invention, consistent with Bayh–Dole provisions

Logistics: Applications are written to include the information outlined on the website (http://www3.niaid.nih.gov/LabsAndResources/resources/dmid/pretheraagents/preclinapp.htm), and submitted to the relevant Program Officer. The application should include the information below, in addition to, other specific information listed.

A description of the overall product development plan

- Background and significance (including why the service requested fits into the overall development plan for the product)
- Preliminary data that supports the request to advance the product through provision of contract services
- A list of other support from NIAID or other Federal agencies, including funded and pending grant applications, and a brief description of what activities are covered
- Description of the overall plan for advancing the product beyond completion of the services requested

Program Capabilities: The NIAID program has the capability of performing all the necessary development tasks to support the submission of an INDA for small molecules and biological products, including, but not limited to, the services below:

- Chemical screening
- Synthesis of chemical analogues including limited lead optimization
- Use *in silico* systems to predict ADME and toxic properties
- *In vitro* microbiological characterization
- GMP bulk drug synthesis, including stability studies
- Formulation development
- IND-enabling toxicity studies
- Pharmacokinetic/toxicokinetic (PK/TK)
- Bioavailability and ADME studies
- Pre-clinical development planning and evaluation service

7. NATIONAL INSTITUTE OF DIABETES AND DIGESTIVE AND KIDNEY DISEASES: T1D-RAID PROGRAM

Purpose: T1D-RAID is a cooperative program of the NIDDK and NCI that is designed to facilitate translation to the clinic of novel, scientifically meritorious, therapeutic interventions for the treatment of Type-1 diabetes. The goal of T1D-RAID is to support the pre-clinical tasks needed for the clinical "proof of principle" to determine if a new molecule merits expanded clinical evaluation. T1D-RAID helps bridge the resource gap between discovery and clinical testing. It is assumed that agents will be studied clinically under investigator-held INDA within the originating (or a collaborating) institution.

Websites: http://www.t1diabetes.nih.gov/T1D-RAID

http://www2.niddk.nih.gov/Research/Resources/AllServices.htm

Eligibility: Academic institutions, non-profit research institutions, biotechnology and pharmaceutical companies, U.S. and non-U.S. entities

Intellectual Property: It is anticipated that submissions will involve a potential therapeutic that either already has protected IP or will be in the public domain. Specific information can be obtained by contacting T1D-RAID Program officials.

Application Logistics: NIDDK receives requests twice per year (April 1 and November 1). Requests consist of a written description of the Request, a technology transfer form, and, if required, a letter of commitment. Once a project has been approved, NIDDK staff interact directly with the investigator. NCI contractors perform the T1D-RAID tasks under the direction of NIDDK and NCI staff. There is also a related program for agents that require additional pre-clinical testing prior to entering T1D-RAID, called Type 1 Diabetes Pre-Clinical Testing Program (T1D-PTP). Information can be found at the NIDDK Research Resources website above.

Program Capabilities: The full set of tasks for submission of an INDA. Examples of tasks that can be supported by T1D-RAID include but are not limited to

- Definition or optimization of dose and schedule for *in vivo* activity
- Development of pharmacology assays
- Pharmacology studies with a pre-determined assay
- Manufacture of bulk substance (GMP and non-GMP) and stability studies
- Development, production, and stability of dosage forms
- Range-finding initial toxicology
- IND-directed toxicology, with correlative pharmacology and histopathology
- Planning of clinical trials
- INDA filing advice

8. NATIONAL INSTITUTE ON DRUG ABUSE: MEDICATIONS DEVELOPMENT PROGRAM

Purpose: The Division of Pharmacotherapies and Medical Consequences of Drug Abuse (DPMCDA) includes a Medications Development Program (MDP). The Medications Discovery and Toxicology Branch

(MDTB), the Chemistry and Pharmaceutics Branch (CPB) within the Medication Program support a pre-clinical translational therapeutics program. Their mission is to develop medications for the treatment of drug abuse, and to facilitate the translation of basic research findings into pharmacotherapies. The program supports a full range of drug development activities from chemistry to INDA submission. The MDTB/CPB also includes the pre-clinical Addiction Treatment Discovery Program (ATDP). The goal of the ATDP is to identify agents to treat drug dependence disorders. The ATDP supports both *in vitro* and animal efficacy evaluation as well as predicative safety testing at no charge. Tasks are accomplished primarily through contract resources within DPMC, as well as grants aimed at developing and evaluating potential new pharmacotherapies to treat drug dependence.

Website: http://www.drugabuse.gov/about/organization/DPMCDA/

Eligibility: Academics, government, non-profits, small and large businesses

Intellectual Property: In general, the investigator retains ownership and development rights to their invention consistent with Bayh–Dole provisions

Background: In 1989, the U.S. Congress mandated that an MDP be established within NIDA. In 1990, NIDA created the Medications Development Division (MDD) (now the DPMCDA) to operationalize the goals of the MDP.

Program Capabilities: In addition to the *in vitro* and *in vivo* evaluations for potential efficacy performed by the ATDP, the MDP includes the following capabilities to support the full range of drug development:

- *In vitro* receptor activity profiling
- Efficacy evaluation in animal models
- *In silico* and *in vitro* predictive toxicology
- *In vitro* estimates of risk for Q-T prolongation and *in vivo* assessments of hemodynamic interactions with cocaine and methamphetamine
- Medicinal chemistry for the design and synthesis of new molecules and structural classes
- Bulk API synthesis (GMP and non-GMP) and clinical dosage forms development and manufacture
- Development of new dosage forms to improve their therapeutic effectiveness and/or to minimize the abuse potential of potential treatment medications
- Pharmacokinetics (PK) and pharmacodynamics (PD) studies
- Bioanalytical resources (including bioanalytical methods to support PK studies), biomarker studies
- Range-finding and IND-enabling toxicology
- INDA preparation and filing advice

Logistics: These resources are limited, and are used only for approved projects of high programmatic priority. They are also used to assist individuals and organizations who are developing projects based on molecular targets of interest to the program. Additional information on the Medication Development Program, its capabilities, and contact information to determine access to resources can be found at the MDTP website.

9. NATIONAL INSTITUTE OF NEUROLOGICAL DISORDERS AND STROKE

Purpose: The mission of the National Institute of Neurological Disorders and Stroke (NINDS) Office of Translational Research is to facilitate the pre-clinical discovery and development of new therapeutic interventions for neurological disorders. The Office supports pre-clinical projects from the discovery of candidate therapeutics through IND and Investigational Device Exemption (IDE) applications to the FDA. This is accomplished through both access to NINDS contract resources and direct funding mechanisms.

Website: http://www.ninds.nih.gov/funding/areas/technology_development/

Programs: The Anticonvulsant Screening Program (ASP) and the Spinal Muscular Atrophy (SMA) Project are examples of NINDS programs with goals of developing therapeutic agents. The ASP provides both *in vitro* and *in vivo* screening services to investigators, while the SMA Project is a full discovery and development program that can lead to INDA submission.

Logistics: Access to services and funding are program specific. Additional information can be found at the NINDS website above, and by contacting the appropriate program director.

Program Capabilities: Services are program specific. Additional information can be found at the NINDS website above and by contacting the appropriate Program Director.

10. ORDR TRND PROGRAM

Purpose: The Therapeutics for Rare and Neglected Diseases (TRND) program was established in 2009 to encourage and accelerate the pre-clinical development of therapeutics directed at rare and neglected diseases. These are diseases that are classified as effecting fewer than 20,000 people in the United States at a given time, and are also known as orphan diseases. The program will support drug discovery and development activities directed at submission of an INDA with the FDA. Governance and oversight of the program will be handled through the Office of Rare Diseases Research (ORDR) within the Office of the Director. Program operations will be within the intramural research program adjacent to the NIH Chemical Genomics Center (NCGC) and will be administered by the NHGRI. The program is in the formative stages and updates will be available through the ORDR website.

Websites: https://www.rarediseases.info.nih.gov (see "Research and Clinical Trials")
http://rarediseases.info.nih.gov/Resources.aspx?PageID=32
Program Capabilities:

- Optimization of lead candidates
- Animal efficacy studies
- IND-enabling toxicology studies
- Re-positioning of FDA-approved drugs for use in rare and neglected diseases

11. SBIR/STTR PROGRAMS

Purpose: The objectives of the SBIR/STTR programs are for small businesses to stimulate technological innovation, strengthen the role of small business in meeting Federal R&D needs, increase private sector commercialization of innovations developed through Federal SBIR R&D, increase small business participation in Federal R&D, and encourage participation by socially and economically disadvantaged small businesses, and women-owned business. Grants to assist biomedical research by small businesses, including drug development, have a monetary ceiling and time limit with initial feasibility Phase I awards limited to $100,000 over 6–12 months and subsequent Phase II awards limited to $750,000 over 24 months.

Website: http://grants.nih.gov/grants/funding/sbirsttr_programs.htm
Eligibility: U.S. for profit businesses of 500 or less employees
Logistics: Grant applications are accepted on April 5, August 5, and December 5, annually. Areas of interest are described in the funding opportunity announcements and solicitations available at the NIH Small Business Funding Opportunities website.

Disclaimer: The views expressed here are those of the author and do not represent the views of, nor are they an endorsement by, the U.S. Government or the National Institutes of Health.

REFERENCES

[1] See NIH Website: http://history.nih.gov/exhibits/history/index.html.

[2] V. A. Harden in *Inventing the NIH: federal biomedical research policy, 1887–1937*, Johns Hopkins University Press, Baltimore, MD, c1986.

[3] D. M. Huryn and N. D. P. Cosford, *Annu. Rep. Med. Chem.*, 42 (ed. J. E. Macor), Elsevier Press, London, 2007, p. 401.

[4] L. Nelsen, *Science*, 1998, **279** (5356), 1460.

[5] E. Liakopoulou, C. A. Blau, Q. Li, B. Josephson, J. A. Wolf, B. Founrnarakis, V. Raisys, G. Dover, T. Papayannopoulou, and G. Stamatoyannopoulos, *Blood*, 1995, **8**, 3227.

[6a] S. Perrine, M. S. Boosalis, L. Shen, G. White, A. Thomson, H. Cao, R. Fei, R. Berenson, and D. V. Faller, *Blood (ASH Annual Meeting Abstract)*, 2009, **114**, Abstract 4754.

[6b] S. Perrine, W. C. Welch, J. Keefer, R. L. Downey, M. S. Boosalis, S. Case, D. V. Faller, and R. Berenson, *Blood*, 2008, **112**, Abstract 130.

[6c] R. Bohacek, M. S. Boosalis, C. McMartin, D. V. Faller, and S. Perrine, *Chem. Biol. Drug Des.*, 2006, **67**, 318.

[7a] W. S. Messer, *Curr. Top. Med. Chem.*, 2002, **2**, 353.

[7b] W. S. Messer, W. G. Rajeswaran, Y. Cao, H. J. Zhang, A. A. El-Assadi, C. Dockery, J. Liske, J. O'Brien, F. E. Williams, X. P. Huang, M. E. Wroblewski, P. I. Nagy, and S. M. Peseckis, *Pharm. Acta. Helv.*, 2000, **74**, 135.

[8] R. H. Shoemaker, *Nat. Rev. Cancer*, 2005, **6**, 813.

[9a] H. Ueda, H. Nakajima, Y. Hori, T. Fujita, M. Nishimura, T. Goto, and M. Okuhara, *J. Antibiot. (Tokyo)*, 1994, **47**, 301.

[9b] W. Zhou and W. G. Zhu, *Curr. Cancer Drug Targets*, 2009, **9**, 91.

[10a] S. Kummar, R. Kinders, M. E. Gutierrez, L. Rubinstein, R. E. Parchment, L. R. Phillips, J. Ji, A. Monks, J. A. Low, A. Chen, A. J. Murgo, J. Collins, S. M. Steinberg, H. Eliopoulos, V. L. Giranda, G. Gordon, L. Helman, R. Wiltout, J. E. Tomaszewski, and J. H. Doroshow, *J. Clin. Oncol.*, 2009, **27**, 2705.

[10b] R. Kinders, R. E. Parchment, J. Ji, S. Kummar, A. J. Murgo, M. Gutierrez, J. Collins, L. Rubinstein, O. Pickeral, S. M. Steinberg, S. Yang, M. Hollingshead, A. Chen, L. Helman, R. Wiltrout, M. Simpson, J. E. Tomaszewski, and J. H. Doroshow, *Mol. Interv.*, 2007, **7**, 325.

PART VII:
Topics in Drug Design and Discovery

Editor: Manoj C. Desai
Medicinal Chemistry Gilead Sciences, Inc.
Foster City
California

Role of Physicochemical Properties and Ligand Lipophilicity Efficiency in Addressing Drug Safety Risks

Martin P. Edwards[*] and **David A. Price**[**]

1. INTRODUCTION

A significant challenge to drug discoverers and the pharmaceutical industry is to reduce the cost of exploring exciting new mechanisms of action and converting them into new medicines for the patient. One of the major levers the industry can use is to reduce the attrition of drug candidates through adverse safety findings at all stages in the drug discovery process and in particular in late stage development, where the cost of failure is substantial. Reducing drug candidate safety liabilities by design in order to avoid attrition is complicated given the diversity of mechanisms that can cause toxic outcomes. Adverse safety findings can have four fundamental origins and combinations thereof: (1) the primary pharmacology or mechanism of action of the compound under study; (2) the secondary pharmacology of the compound under study; (3) the presence of a well-defined structural fragment

[*] Pfizer PTx Research, Cancer Research Chemistry, San Diego, CA, USA

[**] Pfizer PTx Research, CVMED Research Chemistry, Groton, CT, USA

Annual Reports in Medicinal Chemistry, Volume 45
ISSN 0065-7743, DOI 10.1016/S0065-7743(10)45023-X

(toxicophore or structural alert); and (4) the overall physicochemical properties of the compound [1].

Extensive experimentation is often required to determine the contributing causes of toxicity within a series of compounds. The majority of safety-related attrition occurs preclinically. There is an increasing array of *in vitro* screening technologies available to identify liabilities earlier in the drug discovery process, which will lead to the selection of better drug candidates for development [2]. However, for drug discoverers, few effective strategies exist for avoiding toxicity by design prior to synthesis and screening of compounds. Many organizations are complementing *in vitro* safety assays by conducting *in vivo* repeat-dose toxicology studies as early as possible, quite often during the lead optimization phase, to help reduce attrition. Cost, speed, and ethical imperatives favor a move from a screening-only system for assessing safety to a greater use of *in silico* tools and an understanding of fundamental physicochemical liabilities to design and prepare lower risk compounds. Parallels to the evolution of drug metabolism and pharmacokinetics over the last decade are apparent. It is now commonplace to select compounds for further *in vivo* profiling on the basis of their *in vitro* profile and use physicochemical properties to help in the design of compounds with better pharmacokinetics [3]. The influence of physicochemical drug properties (in particular lipophilicity) upon the profile of compounds is well established. Lipophilicity has an impact on all parts of an oral drug's pharmacokinetic profile such as solubility, intestinal absorption, cellular permeability, binding to plasma proteins, metabolism, and eventual excretion [4].

This chapter will summarize recent developments regarding the role of physicochemical properties in safety liability and the emerging use of physicochemical property-based lead optimization using lipophilic ligand efficiency (LLE or LipE).

2. PHYSICOCHEMICAL PROPERTIES AND SAFETY LIABILITIES

Recent analyses demonstrating comparable mean values of lipophilicity, H-bond donors, and polar surface area between older (pre-1983) and newer (1983–2002) oral drugs suggest that these physicochemical attributes are fundamental characteristics of successful drugs. The noted increasing lipophilicity of compounds entering clinical development in recent years may have contributed to a commensurate higher rate of attrition [5,6].

Throughout the literature, there are many elegant studies linking physicochemical properties to *in vitro* outcomes; however, it is often difficult to make a clear link to *in vivo* safety findings. There are a number of recent publications showing a correlation between physicochemical properties, especially lipophilicity, and the likelihood of a compound to promiscuously bind to multiple known and unknown molecular targets, thus increasing the potential for pharmacologically based toxicity [6]. This concept is often referred to as the promiscuity of the compound. In contrast to having a lead with known unwanted pharmacology with a defined functional response and *in vivo* consequence, promiscuous leads with multiple off-target pharmacologies greatly complicate the task of the drug discoverer. In a recent analysis of the Cerep Bioprint™ database of drugs and reference compounds, it was clear that overall compound promiscuity was strongly influenced by lipophilicity and ionization state [6]. Regarding the ionization state, bases and quaternary bases were more promiscuous than acids, neutral compounds, or zwitterions at a given clogP and in all classes promiscuity was positively correlated with clogP. The conclusion that the risk of unwanted pharmacology increased with lipophilicity irrespective of ionization class was confirmed more recently by a publication from a second research group looking at a different data set [7]. Within this work, a trend was noted for increased promiscuity with rising pK_a in compounds with a basic center, though the trend was not statistically significant due to the small size of the dataset.

Given the expectation of a link between compound promiscuity and potential adverse safety effects, it is not surprising that evidence linking compound lipophilicity and ionization state with adverse *in vivo* toxicological outcomes was recently published [8]. A comprehensive analysis of rat and dog *in vivo* tolerability studies conducted on 280 compounds over a 3-year period was described. A statistical analysis was performed to determine if descriptors of compound physicochemical properties were correlated with a higher than average propensity for adverse outcomes. The dataset was unique in that it contained not only dose and findings in organs examined but also the exposure (either C_{max} or Area Under the Curve (AUC) in the animals for each dose described. The plasma protein binding for many of the compounds was available, so the analysis could be performed on either free or total drug concentration. Within this work, a compound was defined as "toxic" if there was a significant finding at C_{max} <10 μM (measuring total drug concentration) or <1 μM (measuring free drug concentration). The finding was categorized as significant if there was belief that it would delay development of the compound, since further work would be required to derisk the finding before progressing to humans.

Of the physicochemical descriptors, lipophilicity (as described by clogP and Topological Polar Surface Area (TPSA) gave the strongest overall correlation to incidence of adverse *in vivo* outcomes, whether analyzed in terms of free or total drug threshold concentrations. In the case of free drug threshold analysis, a Random Forest statistical method indicated that there was a higher chance of a compound with TPSA <70

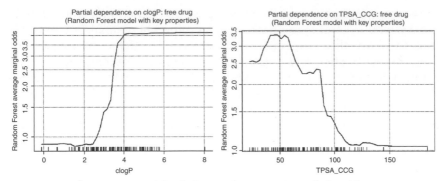

Figure 1   Random Forest models of clogP and TPSA against relative risk.

Å^2 having a significant finding (approximately 1.5:1 unfavorable odds ratio), compared to a compound with TPSA in the range 100–130 Å^2 (approximately 0.4:1 favorable odds ratio). The same analysis held true for clogP. Compounds with clogP >3 have a 4:1 unfavorable odds ratio compared to a compound with clogP <3 (Figure 1).

Having found both clogP and TPSA to be associated with toxicity, and given that these properties were not independent of each other, the researchers sought to determine whether there was any coordinated relationship in their association with toxicity. A pattern emerged and Table 1 displays the toxicity odds (ratio of toxic to nontoxic compounds) for analyses based on total-drug and free-drug exposure when compounds are separated into four categories based on high/low TPSA/clogP.

Thresholds of 3.0 and 75 Å^2, respectively, were selected for clogP and TPSA from the Random Forest analysis. From both the free drug and total drug cases, it was clear that the dominant trend involves compounds with both risk factors of high clogP and low TPSA. Compounds with low clogP and high TPSA were associated with odds of a significant finding of only 0.38 and 0.39 at a fixed concentration of 10 μM (total) or 1 μM (free), respectively. In contrast, compounds with high clogP and low TPSA were associated with odds of a significant finding of 2.4 and 2.59 at a fixed concentration of 10 μM (total) or 1 μM (free), respectively. This represents

Table 1   Observed odds for toxicity versus clogP/TPSA. Numbers of compounds in each cohort are in parenthesis

Toxicity	Total drug		Free drug	
	TPSA > 75	TPSA < 75	TPSA > 75	TPSA < 75
clogP < 3	0.39 (57)	1.08 (27)	0.38 (44)	0.5 (27)
clogP > 3	0.41 (38)	2.4 (85)	0.81 (29)	2.59 (61)

an overall odds ratio of greater than 6 when comparing compounds in these two, most and least favorable, odds buckets. A final conclusion is that it is likely that pharmacological promiscuity of the more lipophilic compounds is driving at least some of the increased likelihood of an adverse *in vivo* finding. This conclusion is in line with the data presented in the two studies mentioned above that show the most lipophilic compounds had a significantly higher likelihood of pharmacological promiscuity [6,7]. Further studies from this group using an expanded and further refined data set demonstrate that the ionization state also contributes to the likelihood of a significant finding in *in vivo* toleration studies [9]. In order to draw this conclusion, the dataset was subdivided into basic and nonbasic compounds, with the definition of a basic compound being the presence of a basic center with a pK_a >6.5 and thus a significant degree of protonation at physiological pH. For the basic compounds within this cohort, if the clogP was >3, the odds ratio for a significant finding was 36:1. For compounds that are basic and the clogP is <3, the unfavorable odds ratio is reduced to 3:1. Given that there were so few compounds in the basic and clogP>3 cohort that did not generate a significant adverse *in vivo* finding (2 out of a total of 73 compounds included in the study), 36 is not a statistically validated number, but the trend was clear. It was concluded that these odds ratios linking physicochemical properties to significant findings from *in vivo* experiments mirror the analysis of *in vitro* pharmacology datasets linking promiscuity with high lipophilicity within each ionization class [6,7].

It is hoped that the application of the conclusions from these recent publications will allow the drug discovery community to improve the chances of avoiding safety attrition by design, through an increased understanding of the links between physicochemical properties and the likelihood of an adverse event in an *in vivo* study. One significant limitation within this work was that there was no link made to the therapeutic index over the exposure required for efficacy in the species examined. For compounds with desired pharmacology of exquisite potency, for example, a finding in an *in vivo* toleration study at 1 μM free or 10 μM total drug, the thresholds used in these studies, may not be of sufficient concern to halt the clinical development of the compound.

3. LIGAND LIPOPHILICITY EFFICIENCY

Given the role of lipophilicity in influencing drug potency, pharmacokinetics, and toxicity reviewed above, it was proposed that chemists should target increased ligand lipophilicity efficiency, LLE, as an important objective for lead generation and optimization programs [6]. LLE, or Lipophilic ligand efficiency, LipE [10], was defined as in Equation 1 below and is the ratio of compound potency to lipophilicity or, put another way, the quantity of potency per unit of lipophilicity. Note that both calculated (e.g., clogP) or measured (logD) estimates of lipophilicity may be used along with the most relevant *in vitro* potency predicting *in vivo* efficacy [6,10].

$$LLE = LipE = pIC_{50}(or \ pK_i) - clogP \ (or \ logD \ or \ clogD)$$

An important point to make is that calculated and measured estimates of lipophilicity may not be in agreement, even within a given chemotype, so care should be taken to use the most appropriate lipophilicity values for the particular series of compounds under examination. For evaluation of compounds that contain both neutral and charged members, logD, calculated or measured, is the most appropriate way to assess LipE as logD is a more relevant measure of lipophilicity for charged compounds [3]. It was proposed that for drug discovery chemists targeting the average observed oral drug clogP of around 2.5 and potency in the range of 1–10 nM, LLE will need to be at least 5.5 (10 nM = pIC_{50} of 8 minus clogP of 2.5 = LLE of 5.5) and likely 7 or greater [6]. High *in vivo* potency has advantages: when the total dose in humans is low, adventitious compound-related toxicity is less of an issue. In essence, the goal for optimization is to increase potency without increasing lipophilicity at the same time [6].

The application of ligand efficiency (LE), a metric that relates the efficiency of potency per heavy atom in a ligand, is well established, in contrast to the use of LLE which is just emerging [11]. LE is a key molecular property that is important in the assessment of the attractiveness of leads and fragments. Maintaining good LE in a series is important in controlling drug candidate size [molecular weight (MW) or n heavy atoms] as potency is increased during lead optimization (e.g., a compound of around MW 450, a surrogate for n heavy atoms, with a K_i of 10 nM, has LE = 0.34).

$$LE = 1.4log \ K_i/n \ \text{heavy atoms}$$

Several publications have appeared in just the last year or so that demonstrate the application of a strategy for medicinal chemistry design that focuses on increasing LLE and targeting attractive physicochemical properties (especially lipophilicity and MW) in the lead optimization phase of a drug discovery project. This allows the project team to target a lower dose size [9] and to avoid attrition owing to adverse effects arising from high lipophilicity-driven pharmacological promiscuity or high drug burden. This section will review these publications.

The first publication describing an LLE-focused optimization was a rapid assessment of a novel series of selective CB2 agonists using parallel synthesis protocols [10]. Structural changes which increase LLE are keenly sought in this approach. During this evaluation of a series of sulfonamides, the change of a phenyl group in **1** (hCB2 EC_{50} = 7.9 nM, clogP = 3.6, LipE = 4.5) to an isoxazole in **2** (hCB2 EC_{50} = 5.1 nM, clogP = 2.4, LipE = 5.9) was a notable example.

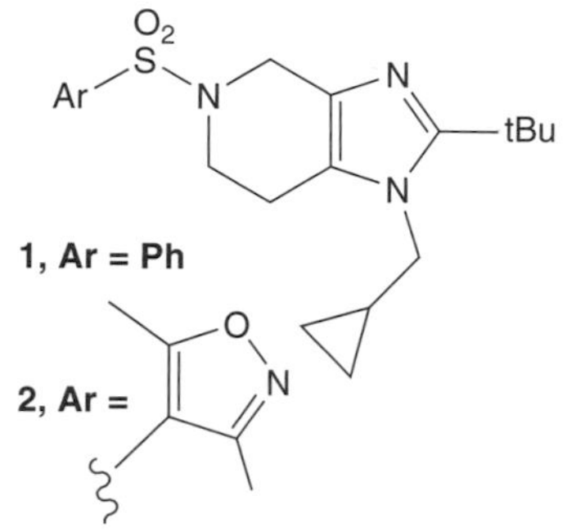

1, Ar = Ph

2, Ar =

An important point to note from this chapter is that having a design strategy targeting increased LLE in a series which shows high metabolic clearance even at lower lipophilicity (approximately clogP = 2) will not lead to compounds with good *in vivo* properties. Specifically, addressing metabolic stability issues with targeted designs, in parallel with increasing LLE, will be a more effective approach.

It is well established that high MW and high lipophilicity reduce the chances for poor oral drug-like properties. This was substantiated in a recent analysis of a diverse set of 47,018 compounds derived from drug discovery projects [12]. This analysis of the relationship of logD and MW with *in vitro* measures of compound metabolic stability and permeability concluded that the chances of achieving acceptable permeability and stability in a single molecule were highest when logD was between 1 and 2 and MW 350–450. Every drug discovery target may not be amenable to achieving high enough potency to provide the luxury of working within such a high probability physicochemical space, but it was concluded that application of a rigorous focus on increasing LLE increases the chances of discovering a compound with enough potency and a logP (or logD) low enough for good oral drug performance [12].

The author's analysis suggests that a strategy focused on maximizing both LLE and LE during lead optimization can yield highly potent compounds with logD and MW values in the range that give the highest probability of combining favorable clearance and permeability properties.

An approach that integrated structure-based drug design with physicochemical properties-based approaches to optimizing drug candidate properties in the discovery of the dual cMet/ALK inhibitor PF-02341066 (3) was recently described [13,14].

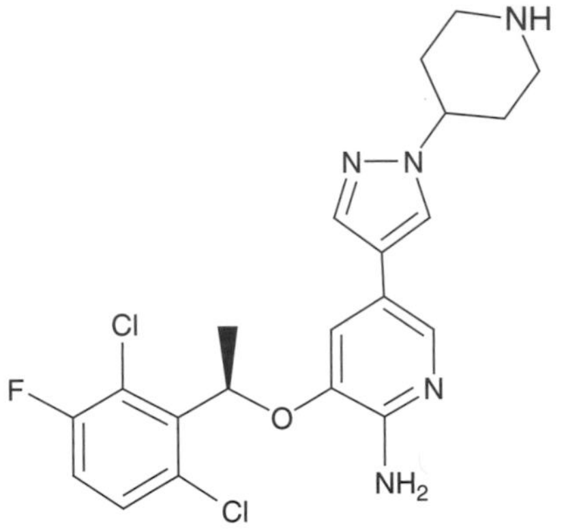

3

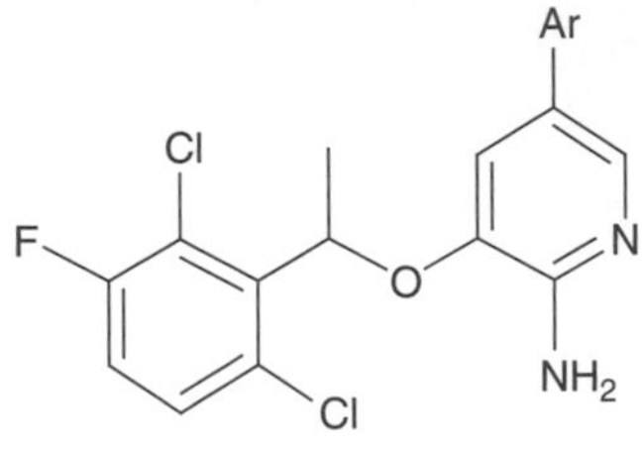

4, Ar = Ph

5, Ar = (pyrazole)

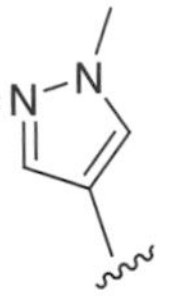

The lipophilic lead amino pyridine (4), with a clogD of 6.4, cell cMet IC$_{50}$ of 1 µM, and LLE (pIC50 – clogD) of only –0.4, was optimized to the dramatically improved PF-02341066 (3), with a clogD of 2.1, cell cMet IC$_{50}$ of 8 nM, and LLE of 6.0), which was advanced into clinical development. A critical design was replacement of the phenyl group present in 4 with an *N*-methyl pyrazole (5). Compound 5 (clogD = 4.5, cell cMet IC50 = 40 nM) had a LipE of 2.9 which was over 3 LipE units higher than the lead 4. In order to guide design of compounds in this series that would have the best chance of acceptable metabolic stability and permeability for good *in vivo* properties, the relationships between logD and both permeability and metabolic stability in *in vitro* assays were tracked. Within this series of 2-aminopyridines, compounds with logD lower than 1.5 did not demonstrate acceptable permeability in *in vitro* permeability assays. Additionally, no compounds with a logD above 2.5 were found to have acceptable metabolic stability in *in vitro* assays of metabolic stability. It was concluded that designing compounds within a logD range of 1.5–2.5 gave the highest probability of successfully combining potency, metabolic stability, and permeability in a single member of this series.

Two papers described the optimization of LLE and physicochemical properties in a series of pyrazole HIV nonnucleoside reverse transcriptase inhibitors (NNRTIs) and the selection of lersivirine (6) as a development candidate [15,16]. The early lead (7) was relatively lipophilic (clogP = 4.3), rapidly metabolized in human liver microsomes and had an LLE of only 1.9 [pIC$_{50}$ (HIV RT) – clogP] [15]. An optimization program targeting increased LLE in less lipophilic compounds of low MW (to increase metabolic stability) produced the development candidate **6**. With a clogP of 2.1 for **6**, the reduction of lipophilicity was over a 100-fold and potency was increased over 5-fold relative to **7**, giving a net dramatic increase of LLE to almost 5 for lersivirine.

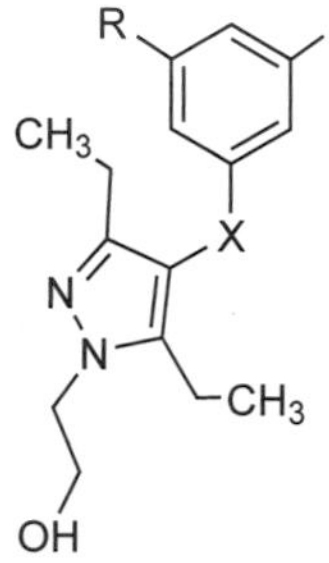

7, R = Cl, X = CH2
6, R = CN, X = O

From Figure 2 below, it can be seen that lersivirine (which was labeled 5 in this figure from the paper [16]) had the highest LLE of any compound in this series and could be considered somewhat of an outlier.

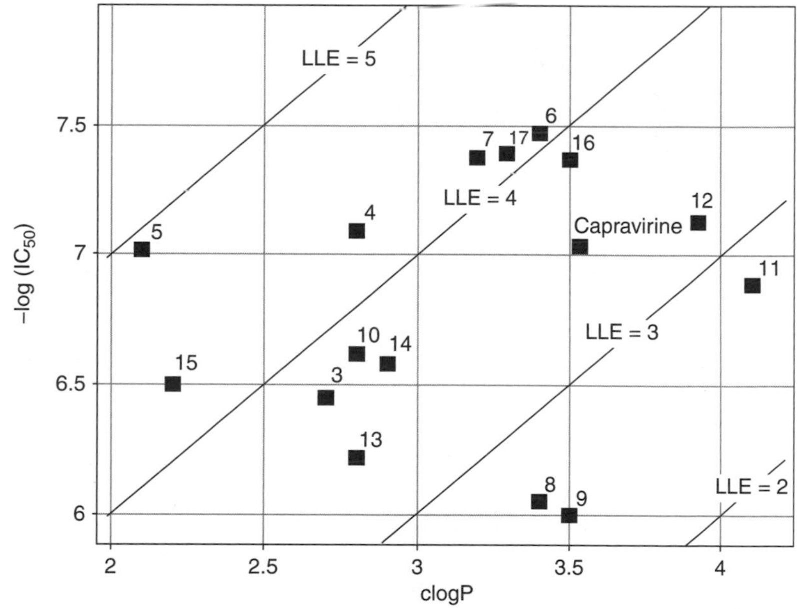

Figure 2 Plot of pIC_{50} versus clogP for the pyrazole series. The 45° lines indicate equal values of LLE.

Optimization of the large (MW = 512), lipophilic (clogP = 7.6), and very potent (K_i = 0.22 nM) Vascular Endothelial Growth Factor Receptor Tyrosine Kinase (VEGFR TK) inhibitor **8** (LLE = 2.1) to produce the development candidate axitinib (**9**) was recently reported [17].

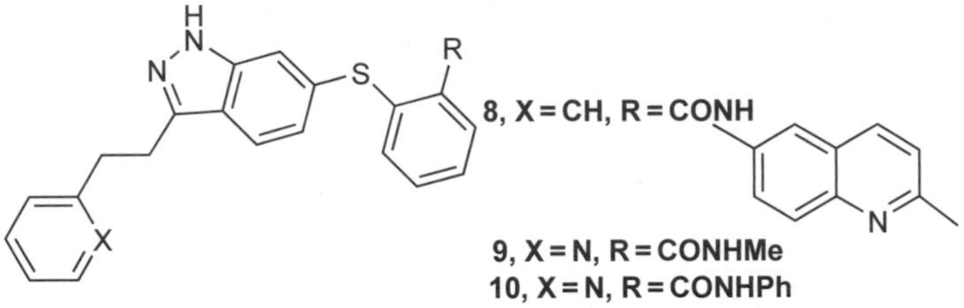

8, X = CH, R = CONH

9, X = N, R = CONHMe
10, X = N, R = CONHPh

A design strategy that targeted increased LipE through truncation, lowering lipophilicity, and heavy use of structure based drug design was successful in producing **9**, with a MW of only 386, a clogP of 3.3, and a K_i of 0.028 nM, and thus a LipE of 7.3. It was noteworthy that the greatest increase in LipE (almost a 1000-fold) was achieved when reducing MW and lipophilicity by switching methyl for phenyl in going from **10** (MW = 448, clogP = 5.1, K_i = 0.28 nM, and LipE = 4.5) to **9** [17].

Finally, a lead optimization program that utilized a strategy to maximize LipE in a series of aminopyrrolidines produced the broadly selective, potent adenosine-5′-triphosphate (ATP)-competitive Akt inhibitor **11** (Akt IC_{50} < 1 nM, clogP = 2.4, LipE = 7.2) which was nominated for clinical development [18].

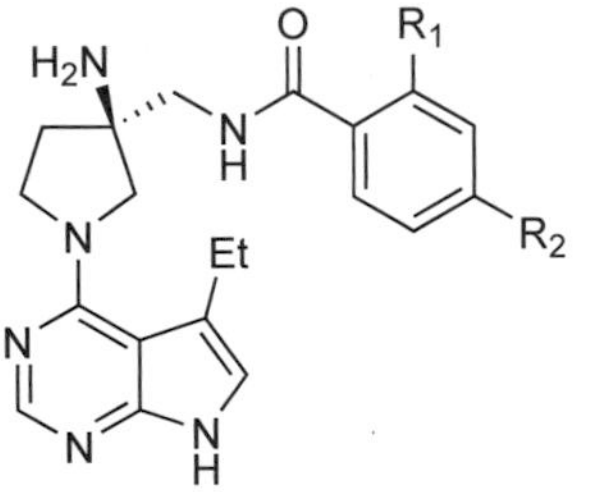

11 R^1 = F, R^2 = F, clogP 2.4, Akt IC_{50} LipE 7.2
12 R^1 = H, R^2 = F, clogP 2.6, Akt IC_{50} LipE 6.3
13 R^1 = H, R^2 = Cl, clogP 3.2, Akt IC_{50} LipE 5.6

Particularly notable structural modifications that dramatically increased LipE in this series involved optimization of the phenyl ring substituents. Changing 4-chloro (in **13**) to 4-fluoro (in **12**) increased LipE from 5.6 to 6.3 and the 2,4-difluoro analog (**11**) had the highest LipE at 7.2.

4. CONCLUSIONS

Examples of the application of strategies based on increasing LLE or LipE and optimizing physicochemical properties as key components of medicinal chemistry programs that delivered clinical development candidates have begun to be described in the last year or so. It remains to be seen whether the hoped for reduction in compound attrition that this approach seeks to achieve will be realized, as it is early days.

REFERENCES

[1] J. Blagg, *Ann. Rep. Med. Chem.*, 2006, **41**, 353.

[2] J. A. Kramer, J. E. Sagartz, and D. L. Morris, *Nat. Rev. Drug. Disc.*, 2007, **6**, 636.

[3] D. A. Smith, H. van de Waterbeemd, and D. K. Walker in *Pharmacokinetics and Metabolism in Drug Design*, 2nd Edition, Wiley-VCH, Weinheim, 2006, p. 1.

[4] H. Van de Waterbeemd, D. A. Smith, K. Beaumont, and D. K. Walker, *J. Med. Chem.*, 2001, **44**, 1313.

[5] P. D. Leeson and A. M. Davis, *J. Med. Chem.*, 2004, **47**, 6338.

[6] P. D. Leeson and B. Springthorpe, *Nat. Rev. Drug. Disc.*, 2007, **6**, 881.

[7] J. U. Peters, P. Schnider, P. Mattei, and M. Kansy, *ChemMedChem*, 2009; online publication

[8] J. D. Hughes, J. Blagg, D. A. Price, S. Bailey, G. A. DeCrescenzo, R. V. Devraj, E. Ellsworth, Y. M. Fobian, M. E. Gibbs, R. W. Gilles, N. Greene, E. Huang, T. Krieger-Burke, J. Loesel, T. Wager, L. Whiteley, and Y. Zhang, *Bioorg. Med. Chem. Lett.*, 2008, **18**, 4872.

[9] S. Bailey in *Physicochemical Drug Properties Associated with In Vivo Toxicological Outcomes*, 15th RSC-SCI Medicinal Chemistry Symposium, September 9, 2009, Cambridge, UK.

[10] T. Ryckmans, M. P. Edwards, V. A. Horne, A. M. Correia, D. R. Owen, L. R Thompson, I. Tran, M. F. Tutt, and T. Young, *Bioorg. Med. Chem. Lett.*, 2009, **19**, 4406.

[11] L. A. Hopkins, C. R. Groome, and A. Alex, *Drug Discov. Today*, 2004, **9**, 430.

[12] T. W. Johnson, K. R. Dress, and M. P. Edwards, *Bioorg. Med. Chem. Lett.*, 2009, **19**, 5560.

[13] M. P. Edwards, Discovery of cMet Kinase Inhibitors as Potential Anti-Cancer Agents, 11th Annual San Diego Med. Chem. Symposium, UCSD, California, July, 2008.

[14] J. J. Cui, I. Botrous, H. Shen, M. Tran-Dube, M. Nambu, P. Kung, L. Funk, L. Jia, J. Meng, M. Pairish, M. McTigue, N. Grodsky, K. Ryan, G. Alton, S. Yamazaki, H. Zou, J. Christensen, and B. Mroczkowski, MEDI-177, 235th ACS National Meeting, New Orleans, LA, April, 2008.

[15] C. E. Mowbray, R. Corbau, M. Hawes, L. H. Jones, J. E. Mills, M. Perros, M. D. Selby, P. A. Stupple, R. Webster, and A. Wood, *Bioorg. Med. Chem. Lett.*, 2009, **19**, 5603.

[16] C. E. Mowbray, C. Burt, R. Corbau, S. Gayton, M. Hawes, M. Perros, I. Tran, D. A. Price, F. J. Quinton, M. D. Selby, P. A. Stupple, R. Webster, and A. Wood, *Bioorg. Med. Chem. Lett.*, 2009, **19**, 5857.

[17] R. S. Kania in *Kinase Inhibitor Drugs*, (ed. R. Li and J. A. Stafford), John Wiley and Sons, 2009, p. 167.

[18] K. D. Freeman-Cook, C. Autry, G. Borzillo, D. Gordon, E. Barbacci-Tobin, V. Bernardo, D. Briere, T. Clark, M. Corbett, J. Jakubczak, S. Kakar, E. Knauth, B. Lippa, M. J. Luzzio, M. Mansour, G. Martinelli, M. Marx, K. Nelson, J. Pandit, F. Rajamohan, S. Robinson, C. Subramanyam, L. Wei, M. Wythes, and J. Morris, *J. Med. Chem.*, 2010, **53**, 4615.

CHAPTER 24

Reducing the Risk of Drug Attrition Associated with Physicochemical Properties

Paul D. Leeson[*] and **James R. Empfield**[**]

[*] AstraZeneca R&D Charnwood, Bakewell Road, Loughborough, LE11 5RH, UK

[**] AstraZeneca Pharmaceuticals, 1800 Concord Pike, Wilmington, DE 19850, USA

Annual Reports in Medicinal Chemistry, Volume 45
ISSN 0065-7743, DOI 10.1016/S0065-7743(10)45024-1

1. INTRODUCTION

1.1 Major reasons for drug failure

The productivity of pharmaceutical R&D in proportion to spend has fallen substantially to the point that its very existence is under threat [1,2]. It has been estimated that 93–96% of nominated candidate drugs fail at some stage in development [3]; all attrition, from the start of compound optimization in the discovery phase of a project to launch on the market, is probably closer to 99%. Therefore, there is an urgent imperative to directly address the reasons for attrition, which from candidate drug nomination onwards can be divided into three broad categories [3–5]:

- *Compound related*: includes drug metabolism and pharmacokinetics (DMPK), safety (off-target toxicity) and selectivity, and formulation and synthesis (cost of goods)
- *Biology related*: includes target-disease linkage, clinical efficacy, and target-based toxicity (exaggerated pharmacology)
- *Organization related*: driven by strategy, disease priorities, and costs

An analysis of 10 pharmaceutical companies [3] showed that the principal reasons for candidate failure in the year 2000 were as follows (estimated percentages from Figure 3 of Ref. [3]): toxicity/clinical safety (19% or 12%), efficacy (24%), commercial/other (19% or 6%), cost of goods (8%), DMPK (8%), and formulation (4%) (see Figure 1, adapted from

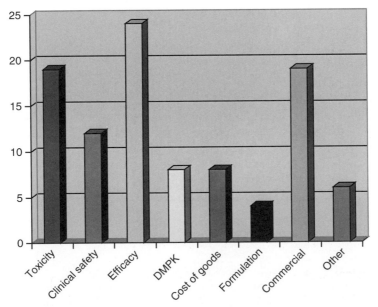

Figure 1 Percentage of clinical failures by cause in 2000 (adapted from Ref. [3]). (See Color Plate 24.1 in the Color Plate Section)

Figure 3 in Ref. [3]). An additional study is broadly consistent, showing that among 73 drugs terminated in clinical Phases I–III during the period 1992–2002, 33% failed for toxicity or clinical safety reasons, 43% for efficacy reasons, 11% for DMPK reasons, and 13% for other reasons [4,5]. Attrition is higher in the central nervous system (CNS) area compared to other therapy areas [3], partly because the need to cross the blood–brain barrier presents an additional challenge [6]. There are caveats and challenges to these studies. For example, a significant proportion of "efficacy" failures may in fact be compound related and there is no differentiation between off-target and target-based toxicities. An updated attrition study, taking into account these issues and covering all drug discovery phases, would be of great value to the drug discovery community.

Since the 1990s, there has been a significant shift in the major reason for overall compound-related attrition from DMPK to toxicity [3–5]. This is most likely the consequence of investments in screening and optimization of DMPK parameters, which are now standard practice from the lead generation phase onwards. As toxicity has become the major cause of drug attrition, attention is now increasingly focused on this issue. Significant effort has been devoted to bringing predictive toxicity screens earlier into the drug discovery process [7,8]. Toxicity attrition occurs not only within clinical trials but continues postapproval: of 548 drugs approved the US Food and Drug Administration during 1975–1999, 56 (10.2%) acquired a new black box warning or were withdrawn for unpredictable adverse drug reactions [9].

1.2 Compound-related attrition and physical properties

The quality of a candidate drug is largely a consequence of its physical properties, which significantly impact its affinity for the target, as well as pharmacokinetic properties (permeability, stability, clearance, and exposure) and nontarget-related toxicological effects, including chemistry-based toxicity (e.g., genetic toxicity and reactive metabolite formation), target promiscuity, and organ toxicity. Unlike biological efficacy, which is very hard to predict, the control of risks due to physical properties and molecular structure, and their subsequent impact on attrition, is entirely at the discretion of the medicinal chemist. This chapter focuses on these aspects.

2. USE OF PHYSICOCHEMICAL PARAMETERS

2.1 Physical properties

The importance of physical properties in drug action underlies many aspects of modern drug discovery; in particular, compound lipophilicity (as estimated by LogP, the logarithm of the 1-octanol–water partition coefficient, or LogD, the logarithm of the 1-octanol–pH7.4 buffer distribution coefficient) is recognized as a fundamentally important property influencing target affinity, solubility, *in vivo* distribution, intestinal absorption, permeability, plasma protein binding, metabolism, and toxicity (for a recent review, see [10]). The control of lipophilicity is therefore critical in drug discovery projects. The recognition that drugs occupy limited regions of physicochemical space led to the Lipinski *"rule of 5"* [11], now cited >2300 times in Scifinder®, which states that permeability is optimal when molecular weight (Mol Wt) is <500; cLogP is <5; the sum of O and N atoms is <10, and the sum of OH and NH groups is <5. More recently, polar surface area (PSA) has often been used instead of polar atom counts. It should be recognized that these simple physical properties are correlated; three-dimensional (3D) plots of related measured physical properties form a plane whose coordinates were described as "bulk" and "cohesiveness" [12], which probably better describe the underlying characteristic molecular features. Lipophilicity itself is a complex physical property, correlated with molecular size, polarity, and hydrogen-bonding characteristics [13]. From a purely physical chemistry perspective, complete quantum mechanical descriptions of molecular size and electrostatics would provide more detailed and extensive parameters than LogP/D, Mol Wt, and PSA. However, the latter are easy to calculate, use, and interpret in lead generation and optimization projects, and there is a wealth of information indicating their utility, especially in DMPK and toxicity optimization.

2.2 Oral drug profiles

The properties of oral drugs are consistent with the rule of 5, having well-defined distributions of LogP, Mol Wt, and hydrogen-bonding properties [14–16]. The concept of "drug-likeness," largely associated with the rule of 5 properties, has been expanded to "lead-likeness" [17,18] and "fragment-likeness" [19], recognizing the fact that optimization processes tend to result in increased physical size and lipophilicity. The Mol Wt of marketed drugs appears to be changing over time [16,20], with more recent oral drugs (approved since 1983) having higher Mol Wt than older oral drugs (approved prior to 1983). Among oral drugs approved since 1983, the trend toward increased Mol Wt over time persists [21]. In contrast, lipophilicity shows little or no change over time in approved oral drugs, suggesting that it is a more stable descriptor of drug-like attributes than Mol Wt. These trends are reinforced by studies showing that lipophilicity declines as compounds progress in drug development; for example, compounds reported as discontinued in Phases I–III have higher lipophilicity than compounds active in those phases [14].

2.3 Physical property "inflation"

Given the acceptance of the *rule of 5* and the substantial empirical evidence that control of fundamental physical properties, especially lipophilicity, are important for eventual success, it is perhaps a surprise to see that current chemistry from the patent literature occupies the upper ranges of the *rule of 5* drug-like spectrum [21]. Thus, the average oral drug marketed since 1983 has cLogP = 2.7 and Mol Wt of 358, whereas the average patented compound from four large pharmaceutical companies in the period 2003–2007 has cLogP of 4.1 and Mol Wt of 450. The general trend [14,21] that Log P and Mol Wt decrease in the order (research compounds > development candidates > marketed drugs) was confirmed in a larger study covering the medicinal literature [22]. These observations suggest that current drug discovery projects are carrying significant risk of attrition. Why are the physical properties, particularly Log P and Mol Wt, of current chemistry inflated compared to successful drugs? Several reasons can be put forward:

- Today's drug targets are less "druggable" than in the past [23]; the "low-hanging" fruit in the center of drug-like space has already been picked
- The search for new intellectual property often involves increasing size of existing prior art compounds, especially in "me-too," "me-better," and highly competitive projects [21]
- The use of high-throughput screening (HTS) results in hits and leads that are more lipophilic than optimized drugs [24]
- The uncontrolled use of parallel or combinatorial synthetic chemistry in hit and lead optimization (and in compound collection enhancement)
- The culture and strategy of the discovery organization [21]

Perhaps the most intriguing observation surrounds organizational strategy. In a comparison of the patent output from four large pharmaceutical companies, there were consistent differences in physical properties, which extended to target classes and even to individual targets [21]. The variation in cLogP between organizations' intellectual property is comparable in magnitude to the variation in clogP between target classes [25] and in addition there is substantial overlap of cLogP ranges between target classes [10,26]. These observations suggest that, in choosing the physical property space to work in, differences due to "behavioral" factors may be similar to the scientific challenges inherent to different targets. This is a very important conclusion relative to the ability of medicinal chemists to control compound properties and thereby influence attrition rates. Undoubtedly "medicinal prejudice" plays a key role in compound design; for example, when given the same set of compounds, chemists had inconsistent views in assessing their drug-like and lead-like attributes [27,28].

2.4 Structural properties and attrition

Recent studies comparing drugs and development compounds have extended beyond physical properties to molecular structural features. The fraction of sp3 (tetrahedral) carbon atoms (Fsp3 = sp3 atom count/ total carbon atom count) and the presence of a chiral center were shown to gradually increase from the discovery phase, through Phases I–III, to marketed drug [29]. Discovery compounds had 36% Fsp3 and drugs 47%. Of discovery compounds passing the *rule of 5*, 46% were chiral, in contrast to 61% of drugs. Higher Fsp3 reduces melting point and increases solubility. The correlation of Fsp3 with other drug properties, such as LogP, was not shown. It was proposed, quite reasonably, that increased saturation and chirality (three-dimensionality) of a molecule allowed access to greater chemical space, and that the reduction of three-dimensionality in research compounds primarily reflects the use of pre-ferred synthetic methodology, especially cross-coupling chemistry [29]. A study of the GlaxoSmithKline pipeline, the first reported Company pipeline physical property analysis we are aware of, focused on aromatic ring count [30]. In moving from precandidate selection to clinical proof of concept, the mean aromatic ring count dropped from 3.3 to 2.3, compared with 1.6 in oral drugs; this is consistent with the sp3 carbon trends. The aromatic ring count correlated positively with some devel-opability properties, namely cLogP, LogD7.4, % serum albumin binding, CyP 3A4 inhibition, and human Ether-à-go-go Related Gene (hERG) inhibition, and correlated negatively with solubility.

2.5 Biological relevance and "metabolite-likeness"

A biologically relevant score, BR, has been developed to distinguish drugs from nondrug molecules. This scoring function has been built utilizing similarity comparisons with a selected set of "biorelevant" com-pounds [31]. BR also increases in moving from preclinical compounds (BR = 0.46), through Phases I–III, to launched drugs (BR = 0.56). A related concept of "metabolite–likeness" was introduced [32] on the basis that to enter cells, many drugs use transport proteins that have naturally occurring intermediary metabolites as their substrates. It has been argued that transporter-mediated cellular uptake is much more common than generally believed and has comparable physical property determinants to passive transport [33]. There are opportunities for exploi-tation of structural features found in natural products and metabolites in the medicinal design process. These features could also be used to enrich the quality of screening collection libraries [34–36].

3. PHYSICOCHEMICAL PROPERTIES AND DMPK

3.1 Oral bioavailability

Although it has been clear for some time that physical properties drive pharmacokinetic parameters, satisfactory predictive models for bioavailability are difficult to generate because bioavailability is a multifactorial process dependent on dissolution, absorption (membrane permeability and active transport), and *in vivo* clearance by several pathways. Specific physical property dependencies vary according to the data set employed and provide only a rough guide to bioavailability prediction. Thus, a model suggesting that rotatable bond count (<10) and PSA (<140) limit from a proprietary set of GlaxoSmithKline compounds is widely cited [37] but this approach worked less effectively on other proprietary data sets [38,39]. In a set of Abbott compounds [38], the probability that bioavailability was >10% depended primarily on charge type, with acids showed declining bioavailability with increasing PSA. In neutral, basic, and zwitterionic compounds, the rule of 5 was more predictive but not impressively so, with 55% of compounds passing having bioavailability >10% and 17% of those failing having bioavailability >10%. Attempts to generate models for human bioavailability have been problematic; it appears that multiple physical properties are involved [40]. Recent studies [41] demonstrate an improved way forward by analyzing the separate contributions of intestinal absorption and first-pass elimination; thus higher Mol Wt reduces absorption, whereas elimination decreased with increasing lipophilicity, and rotatable bond increases reduced both parameters. Other review articles reinforce the overall roles of molecular size and polarity in determining bioavailability [42–44].

3.2 Permeability and stability

Three recent studies, on large proprietary databases from pharmaceutical companies, are consistent with each other and provide strong evidence for the importance of both Mol Wt and LogD/P in permeability [45], combined permeability and stability [46], and a battery of DMPK/toxicity screens [47]. In CACO-2 screens at AstraZeneca, increasing Mol Wt required coincreased LogD of 7.4, in the range LogD 0.5–3.5 and Mol Wt <300–500, to maintain a 50% chance of high permeability [45]; increasing Mol Wt to >500 substantially reduced the odds of finding acceptable permeability. Similar results were found in a Pfizer data set, using several tests for permeability and *in vitro* metabolic stability [46]. Compounds passing both permeability and clearance criteria fell into a "golden triangle" with a Mol Wt apex of ~450 and a LogD base of 0–3. Of the compounds at the center of this space (LogD = 1.5, Mol Wt = 350), 29% passed permeability and stability criteria, compared to 4% or less with Mol Wt 450 and LogD 0 or 3. When Mol Wt was >450, very few permeable and stable compounds exist; some outliers could be rationalized, for example, halogen-containing compounds with blocked sites for metabolism. The thorough studies of the GlaxoSmithKline data set, covering solubility, permeability, bioavailability, volume of distribution, clearance, hERG inhibition, Permeability-Glycoprotein (PGP) efflux, and P450 inhibition [47], led to the conclusion that, for overall lower risk, it is optimal that a compound have both Mol Wt <400 and LogP <4. This conclusion seemed to be largely independent of ion class. Further studies of proprietary databases focusing on matched-pair analyses also emphasize the importance of the change in LogD/P on a range of DMPK parameters and are useful for the medicinal chemist in substituent selection [48,49].

3.3 CNS drugs—blood–brain barrier penetration

Physicochemical parameters play a significant role in a drug candidate's ability to permeate the blood–brain barrier. In general, CNS drugs have lower PSA and Mol Wt than their non-CNS counterparts [20]. In 1999, Clark published a useful model for predicting the likelihood of drug candidates to penetrate into the brain based on the total sum of their PSA [50]. Since that time other groups have explored the relationship of other physical parameters to brain penetration of drugs. Pajouhesh and Lenz have published a review on the properties of successful CNS drug [51]. More recently, preferred ranges of physical properties for CNS drugs were proposed [52]: PSA <70, H-bond donors 0–1, cLogP 2–4, cLogD 2–4, and Mol Wt <450. Two papers from the Pfizer group provide further guidance on the design of CNS drugs based on a multiple parameter scoring function [53,54].

3.4 Nonoral routes of administration

Differences in physical property ranges were found for nonoral drugs [15]. For example, injectable drugs, which require high aqueous solubility, had lower LogP than orals (mean of 0.6 vs. 2.3) but higher Mol Wt (mean of 558 vs. 344). Treatment of respiratory disease by the inhalation route provides an opportunity to design molecules with many properties opposite to oral drugs, where ensuring lung retention and low systemic exposure is essential. Thus, low bioavailability, high clearance, and high plasma protein binding can be used to limit systemic exposure, combined with high volume of distribution or low solubility for lung retention [55].

4. PHYSICOCHEMICAL PROPERTIES AND TOXICITY/ PROMISCUITY

4.1 High-risk structures

The presence of chemically reactive structural features in potential drug candidates, especially when caused by metabolism, has been linked to idiosyncratic toxicity [56,57] although in most cases this is hard to prove unambiguously, and there is no evidence that idiosyncratic toxicity is correlated with specific physical properties *per se*. The best strategy for the medicinal chemist is *avoidance* of the liabilities associated with inherently chemically reactive or metabolically activated functional groups [58]. For reactive metabolites, protein covalent-binding screens [59] and genetic toxicity tests (Ames) of putative metabolites, for example, embedded anilines, can be employed in "risky" chemical series.

4.2 Promiscuity

The realization that drugs often interact with multiple targets has opened up a new stream of research on drug–target networks or "systems" pharmacology [60–65]. Indeed, multiple acting drugs are the cornerstone of psychiatric medicines [66,67] and the first generation of kinase inhibitors in oncology is nonspecific [68]. There is clearly a role for *designed* promiscuity in seeking efficacy but there is also a danger of binding to unwanted targets invoking toxicity. In addition, the design of compounds with multiple activities requiring more than one pharmacophore brings major challenges to compound physical property control [69].

A number of studies implicate physical properties in receptor promiscuity. A theoretical analysis suggests that the presence of charged groups or polarity, rather than hydrophobicity, increases specificity; interestingly, flexibility in polar molecules also increases specificity [70]. Increasing lipophilicity results in an increased number of hits in the Cerep Bioprint® database, whereas increasing Mol Wt reduced hit frequency in Pfizer high-throughput screens [71]. A further analysis of a larger Bioprint® data set confirmed a strong dependency on lipophilicity, with promiscuity sharply higher when cLogP > 3; there was no dependency on Mol Wt [21]. It was also demonstrated that ion class has a major role in promiscuity, with bases being notably less selective than other ion classes, although all increased promiscuity with increasing lipophilicity. Similar lipophilicity and Mol Wt trends were found in a promiscuity analysis of a set of proprietary Roche compounds [72], where it was suggested that employing amine-preferring assays (e.g., aminergic G-protein-coupled receptors (GPCRs) and transporters) increased promiscuity of bases over other ion classes. Increasing basic pK_a to >8 also increased promiscuity. In a set of Novartis compounds, the trend toward higher lipophilicity and promiscuity was also found and in this case, higher Mol Wt led to higher promiscuity [73]. In agreement with other studies, carboxylic acids had high selectivity, and bulky lipophilic amines had low selectivity. Application of Novartis' predictive model showed a clear decline in promiscuity and increase in selectivity, in moving from lead optimization, through Phases I–III, to launched drugs [73], reminiscent of similar studies with physical properties [14,29,30]. These four studies confirm the role of lipophilicity in promiscuity and show that lipophilic bases are a particular risk; the differing results with Mol Wt indicate that further research into promiscuity–molecular complexity relationships is warranted. In addition, studies to date have not looked at *margins*; while compounds may be active at other sites, margins of >100-fold in affinity above the preferred target will, in most cases, render these nonsignificant in the compound's biological action.

4.3 Phospholipidosis and hERG

Phospholipidosis is an excessive intracellular accumulation of phospholipids and drug, which is normally reversible after discontinuation of drug treatment. Currently, it is thought that phospholipidosis alone is not toxic *per se*, but because some compounds cause concurrent phospholipidosis and organ toxicity, avoidance of the issue in drug discovery projects seems prudent [74]. Fortunately, there are clear links between phospholipidosis and physical properties, especially lipophilicity, basicity, and amphiphilicity [75] which allow for good prediction of the risk.

Inhibition of the hERG ion channel is firmly associated with cardiovascular toxicity in humans, and several drugs with this liability have been withdrawn. A number of studies show that basicity, lipophilicity, and the presence of aromatic rings [76] contribute to hERG binding. The 3D models of the hERG channel [77] are potentially useful to understand more subtle structure–activity relationships. In common with receptor promiscuity, both phospholipidosis and hERG inhibition are predominantly issues with lipophilic, basic compounds, and with the predictive models available, both risks should be well controlled.

4.4 Organ toxicity

A useful review of the role of physical properties in toxicity has appeared [78]. The massive volume of literature establishing lipophilicity as an important parameter in structure–activity relationships would suggest that off-target-related toxicity should have a lipophilic component [10,79]. However, direct evidence was lacking until the study of a proprietary set of Pfizer compounds appeared [80]. This study, on the *in vivo* toxicity outcomes of 245 compounds, is valuable since the methodology used takes into account the observed exposures. A plasma exposure level of $10\,\mu M$ was chosen since this maximized the toxic/nontoxic number of observations (most or all compounds became increasingly toxic with increasing exposures). The results showed that compounds with cLogP <3 and PSA >75 were sixfold less likely to display *in vivo* toxicity versus compounds with cLogP >3 and PSA <75. A selected set of promiscuous compounds from the Bioprint® database showed the same trend with cLogP and PSA. The data are in accord with the promiscuity studies discussed earlier [21,72,73], suggesting that receptor promiscuity and organ toxicity are related to each other and to compound lipophilicity. There are some caveats; first, a proprietary data set was used [80] and we know that studies of, for example, bioavailability (see above) lead to different conclusions based on the compounds used; second, there were no statistical measures provided; third, the specific organ toxicities from the animal studies (both rats and dogs) are not separated; and fourth, interpretation of what is toxic and what is not can be challenging, even for expert toxicologists. Given the importance of improved understanding of structure–toxicity relationships in the war on attrition, we hope those scientists with access to additional compound toxicity data will pursue further studies in this area.

4.5 Ligand efficiency measures

Given the importance of lipophilicity in toxicity, promiscuity, and so many aspects of drug discovery [10], the control of this property is fundamental to eventual success. Ligand lipophilicity efficiency [21,81], LLE = pIC_{50}/pK_i – LogP/D, has been suggested as an important measure of progress for optimization projects (Figure 2). An optimized value of LLE >5 (i.e., IC_{50} < $10\,nM$ and LogP < 3) has been proposed [21] to reduce toxicity risks. The high lipophilicity of hits and leads from HTS [24] may be at least partly responsible for HTS failure. In contrast, the leads for 60 successful drugs [82] had lower mean LogP than HTS hits (Figure 2) and the optimization processes did not alter mean LogP but mean potency and LLE increased by >2 log units to a mean LLE of 5.9. The use of other ligand efficiency metrics, based on potency per heavy atom or Mol Wt [83–85] or a combination of heavy atoms and lipophilicity [24], is very useful for early stage optimization, where careful control of the effects of added size and lipophilicity to a molecule is

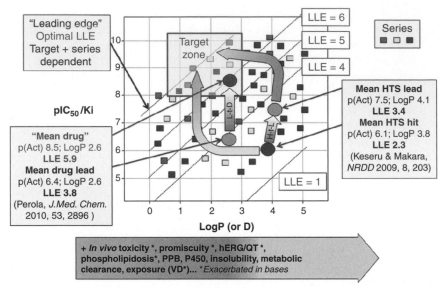

Figure 2 Use of ligand lipophilicity efficiency (LLE) in optimization. Potency versus LogP or LogD plots often show a "leading edge" where LLE is optimal. Mean high-throughput screening (HTS)-derived hits and leads [24] are more lipophilic than the mean leads successfully converted to drugs [82]. Reducing lipophilicity and increasing potency are frequently needed in lead optimization. H-t-L hit-to-lead; L-t-D lead to drug. (See Color Plate 24.2 in the Color Plate Section)

essential. Establishing lead-like [17] or fragment-like [19] starting points with high ligand efficiencies has the potential to allow optimization toward the center of drug-like space by a controlled increase in physical properties.

5. CONCLUSION

During the past few years, a significant growth in our understanding of the links of physical properties of drugs to drug attrition has occurred. A very considerable body of evidence suggests that DMPK and toxicity outcomes are related to physical properties of compounds, which are under the control of medicinal chemists and other project scientists in drug discovery projects. The component of pipeline attrition today that is compound based can therefore be substantially reduced by implementing appropriate practices and tougher drug candidate criteria. Adopting a "zero-tolerance" attitude to compound-based failure in the clinic would lead to fewer candidate nominations, but also to increased clinical success.

REFERENCES

[1] S. M. Paul, D. S. Mytelka, C. T. Dunwiddie, C. C. Persinger, B. H. Munos, S. R. Lindborg, and A. L. Schacht, *Nat. Rev. Drug Discov.*, 2010, **9**, 203−214.

[2] B. Munos, *Nat. Rev. Drug Discov.*, 2009, **8**, 959−968.

[3] I. Kola and J. Landis, *Nat. Rev. Drug Discov.*, 2004, **3**, 711−716.

[4] D. Schuster, C. Laggner, and T. Langer, *Curr. Pharm. Des.*, 2005, **11**, 3545−3559.

[5] D. Schuster, C. Laggner, and T. Langer, *Methods Princ. Med. Chem.*, 2008, **38**, 3−22.

[6] V. Nienaber, *Curr. Top. Med. Chem.*, 2009, **9**, 1688−1704.

[7] A. S. Bass, M. E. Cartwright, C. Mahon, R. Morrison, R. Snyder, P. McNamara, P. Bradley, Y. Zhou, and J. Hunter, *J. Pharmacol. Toxicol. Methods*, 2009, **60**, 69−78.

[8] J. A. Kramer, J. E. Sagartz, and D. L. Morris, *Nat. Rev. Drug Discov.*, 2007, **6**, 636−649.

[9] K. E. Lasser, P. D. Allen, S. J. Woolhandler, D. U. Himmelstein, S. M. Wolfe, and D. H. Bor, *JAMA*, 2002, **287**, 2215−2220.

[10] M. J. Waring, *Expert Opin. Drug Discov.*, 2010, **5**, 235−248.

[11] C. A. Lipinski, F. Lombardo, B. W. Dominy, and P. J. Feeney, *Adv. Drug Delivery Rev.*, 1997, **23**, 3−25.

[12] R. D. Cramer III, *J. Am. Chem. Soc.*, 1980, **102**, 1837.

[13] P. V. Oliferenko, A. A. Oliferenko, G. Poda, V. A. Palyulin, N. S. Zefirov, and A. R. Katritzky, *J. Chem. Inf. Model.*, 2009, **49**, 634−646.

[14] M. C. Wenlock, R. P. Austin, P. Barton, A. M. Davis, and P. D. Leeson, *J. Med. Chem.*, 2003, **46**, 1250−1256.

[15] M. Vieth, M. G. Siegel, R. E. Higgs, I. A. Watson, D. H. Robertson, K. A. Savin, G. L. Durst, and P. A. Hipskind, *J. Med Chem.*, 2004, **47**, 224−232.

[16] J. R. Proudfoot, *Bioorg. Med. Chem. Lett.*, 2005, **15**, 1087−1090.

[17] S. J. Teague, A. M. Davis, P. D. Leeson, and T. Oprea, *Angew. Chem. Int. Ed.*, 1999, **38**, 3743−3748.

[18] G. M. Rishton, *Curr. Op. Chem. Biol.*, 2008, **12**, 340−351.

[19] C. W. Murray, and D. C. Rees, *Nat. Chem.*, 2009, **1**, 187−192.

[20] P. D. Leeson and A. M. Davis, *J. Med. Chem.*, 2004, **47**, 6338−6348.

[21] P. D. Leeson and B. Springthorpe, *Nat. Rev. Drug Discov.*, 2007, **6**, 881−890.

[22] C. Tyrchan, N. Blomberg, O. Engkvist, T. Kogej, and S. Muresan, *Bioorg. Med. Chem. Lett.*, 2009, **19**, 6943−6947.

[23] K. Beaumont and D. A. Smith, *Curr. Opin. Drug Discov. Dev.*, 2009, **12**, 61−71.

[24] G. M. Keseru, and G.M Makara, *Nat. Rev. Drug Discov.*, 2009, **8**, 203−212.

[25] M. Vieth and J. J. Sutherland, *J. Med. Chem.*, 2006, **49**, 3451−3453.

[26] R. Morphy, *J. Med. Chem.*, 2006, **49**, 2969−2978.

[27] P. D. Leeson, A. M. Davis, and J. Steele, *Drug Discov. Today Technol.*, 2004, **1**, 189−195.

[28] M. S. Lajiness, G. M. Maggiora, and V. Shanmugasundaram, *J. Med. Chem.*, 2004, **47**, 4891−4896.

[29] F. Lovering, J. Bikker, and C. Humblet, *J. Med. Chem.*, 2009, **52**, 6752−6756.

[30] T. J. Ritchie and S. J. F. MacDonald, *Drug Discov. Today*, 2009, **14**, 1011−1020.

[31] D. Kong, W. Ren, W. Lu, and H. Zhang, *J. Chem. Inf. Model.*, 2009, **49**, 2376−2381.

[32] P. D. Dobson, Y. Patel, and D. B. Kell, *Drug Discov. Today*, 2009, **14**, 31−40.

[33] P. D. Dobson and D. B. Kell, *Nat. Rev. Drug Discov.* 2008, **7**, 205−220.

[34] J. Rosen, J. Gottfries, S. Muresan, A. Backlund, and T. I. Oprea, *J. Med. Chem.*, 2009, **52**, 1953−1962.

[35] K. Kumar and H. Waldmann, *Angew. Chem. Int. Ed.*, 2009, **48**, 3224−3242.

[36] A. Ganesan, *Curr. Opin. Chem. Biol.*, 2008, **12**, 306−317.

[37] D. F. Veber, S. R. Johnson, H. Cheng, B. R. Smith, K. W. Ward, and K. D. Kopple, *J. Med. Chem.*, 2002, **45**, 2615−2623.

[38] Y. C. Martin, *J. Med. Chem.*, 2005, **48**, 3164−3170.

[39] J. J. Lu, K. Crimin, J. T. Goodwin, P. Crivori, C. Orrenius, L. Xing, P. J. Tandler, T. J. Vidmar, B. M. Amore, A. G. E. Wilson, P. F. W. Stouten, and P. S. Burton, *J. Med. Chem.*, 2004, **47**, 6104−6107.

[40] T. Hou, J. Wang, W. Zhang, and X. Xu, *J. Chem. Inf. Model.*, 2007, **47**, 208−218.

[41] M. V. S. Varma, R. S. Obach, C. Rotter, H. R. Miller, G. Chang, S. J. Steyn, A. El-Kattan, and M. D. Troutman, *J. Med. Chem.*, 2010, **53**, 1098−1108.

[42] D. J. Livingstone and H. van de Waterbeemd, *Methods Princ. Med. Chem.*, 2009, **40**, 433−451.

[43]. J. Keldenich, *Chem. Biodiversity*, 2009, **6**, 2000−2013.

[44] M. T. Kuentz and Y. Arnold, *Pharm. Dev. Technol.*, 2009, **14**, 312−320.

[45] M. J. Waring, *Bioorg. Med. Chem. Lett.*, 2009, **19**, 2844−2851.

[46] T. W. Johnson, K. R. Dress, and M. Edwards, *Bioorg. Med. Chem. Lett.*, 2009, **19**, 5560−5564.

[47] M. P. Gleeson, *J. Med. Chem.*, 2008, **51**, 817−834.

[48] M. L. Lewis and L. Cucurull-Sanchez, *J. Comput.-Aided Mol. Des.*, 2009, **23**, 97−103.

[49] P. Gleeson, G. Bravi, S. Modi, and D. Lowe, *Bioorg. Med. Chem.*, 2009, **17**, 5906−5919.

[50] D. E. Clark, *J. Pharm. Sci,.* 1999, **88**, 815−821.

[51] H. Pajouhesh and G. R. Lenz, *J. Am. Soc. Exp. NeuroTher.*, 2005, **2**, 541−553.

[52] S. A. Hitchcock and L. D. Pennington, *J. Med. Chem.*, 2006, **49**, 1−25.

[53] T. T. Wager, X. Hou, P. R. Verhoest, and A. Villalobos, *ACS Chem Neurosci.*, 2010, **1**, 435−449.

[54] T. T. Wager, R. Y. Chandrasekaran, X. Hou, M. D. Troutman, P. R. Verhoest, A. Villalobos, and Y. Will, *ACS Chem Neurosci.*, 2010, **1**, 420−434.

[55] T. J. Ritchie, C. N. Luscombe, and S. J. F. Macdonald, *J. Chem. Inf. Model.*, 2009, **49**, 1025−1032.

[56] D. J. Antoine, D. P. Williams, and B. K. Park, *Expert Opin. Drug Metab. Toxicol.*, 2008, **4**, 1415−1427.

[57] A. S. Kalgutkar, G. Fate, M. T. Didiuk, and J. Bauman, *Expert Rev. Clin. Pharmacol.*, 2008, **1**, 515−531.

[58] A. S. Kalgutkar, I. Gardner, R. S. Obach, C. L. Shaffer, E. Callegari, K. R. Henne, A. E. Mutlib, D. K. Dalvie, J. S. Lee, Y. Nakai, J. P. O'Donnell, J. Boer, and S. P. Harriman, *Curr. Drug Metab.*, 2005, **6**, 161−225.

[59] S. Kumar and T. A. Baillie, *Drugs Pharm. Sci.*, 2009, **186**, 597−618.

[60] A. L. Hopkins, *Nat. Chem. Biol.*, 2008, **4**, 682−690.

[61] M. A. Yildirim, K. Goh, M. E. Cusick, A. Barabasi, and M. Vidal, *Nat. Biotechnol.*, 2007, **25**, 1119−1126.

[62] S. C. Janga and A. Tzakos, *Mol. BioSyst.*, 2009, **5**, 1536−1548.

[63] M. J. Keiser, V. Setola, J. J. Irwin, C. Laggner, A. I. Abbas, S. J. Hufeisen, N. H. Jensen, M. B. Kuijer, R. C. Matos, T. B. Tran, R. Whaley, R. A. Glennon, J. Hert, K. L. H. Thomas, D. D. Edwards, B. K. Shoichet, and B. L. Roth, *Nature*, 2009, **462**, 175−181.

[64] J. Mestres, E. Gregori-Puigjane, S. Valverde, and R. V. Sole, *Mol. BioSyst.*, 2009, **5**, 1051−1057.

[65] S. Renner, W. A. L. van Otterlo, M. Dominguez Seoane, S. Moecklinghoff, B. Hofmann, S. Wetzel, A. Schuffenhauer, P. Ertl, T. I. Oprea, D. Steinhilber, L. Brunsveld, D. Rauh, and H. Waldmann, *Nat. Chem. Biol.*, 2009, **5**, 585−592.

[66] B. L. Roth, D. J. Sheffler, and W. K. Kroeze, *Nat. Rev. Drug Discov.*, 2004, **3**, 353−359.

[67] M. J. Millan, *Neurotherapeutics*, 2009, **6**, 53−77.

[68] Z. A. Knight, H. Lin, and K. M. Shokat, *Nat. Rev. Cancer*, 2010, **10**, 130−137.

[69] R. Morphy and Z. Rankovic, *J. Med. Chem.*, 2006, **49**, 4961−4970.

[70] M. L. Radhakrishnan and B. Tidor, *J. Phys. Chem. B*, 2007, **111**, 13419−13435.

[71] A. L. Hopkins, J. S. Mason, and J. P. Overington, *Curr. Opin. Struct. Biol.*, 2006, **16**, 127−136.

[72] J. Peters, P. Schnider, P. Mattei, and M. Kansy, *ChemMedChem*, 2009, **4**, 680–686.

[73] K. Azzaoui, J. Hamon, B. Faller, S. Whitebread, E. Jacoby, A. Bender, J. L. Jenkins, and L. Urban, *ChemMedChem*, 2007, **2**, 874–880.

[74] L. A. Chatman, D. Morton, T. O. Johnson, and S. D. Anway, *Toxicologic Pathology*, 2009, **37**, 997–1005.

[75] A. J. Ratcliffe, *Curr. Med. Chem.*, 2009, **16**, 2816–2823.

[76] D. J. Diller, *Curr. Comput. Aided Drug Des.*, 2009, **5**, 106–121.

[77] A. Stary, S. J. Wacker, L. Boukharta, U. Zachariae, Y. Karimi-Nejad, J. Aqvist, G. Vriend, and B. L. de Groot, *ChemMedChem*, 2010, **5**, 455–467.

[78] D. A. Price, J. Blagg, L. Jones, N. Greene, and T. Wager, *Expert Opin. Drug Metab. Toxicol.*, 2009, **5**, 921–931.

[79] M. T. D. Cronin, *Curr. Comput. Aided Drug Des.*, 2006, **2**, 405–413.

[80] J. D. Hughes, J. Blagg, D. A. Price, S. Bailey, G. A. DeCrescenzo, R. V. Devraj, E. Ellsworth, Y. M. Fobian, M. E. Gibbs, R. W. Gilles, N. Greene, E. Huang, T. Krieger-Burke, J. Loesel, T. Wager, L. Whiteley, and Y. Zhang, *Bioorg. Med. Chem. Lett.*, 2008, **18**, 4872–4875.

[81] A. R. Leach, M. M. Hann, J. N. Burrows, and E. J. Griffen, *Mol. BioSyst.*, 2006, **2**, 429–446.

[82] E. Perola, *J. Med. Chem.*, 2010, **53**, 2896–2997.

[83] A. L. Hopkins, C. R. Groom, and A. Alex,*Discov. Today*, 2004, **9**, 430–431.

[84] C. Abad-Zapatero, *Expert Opinion Drug Discov.*, 2007, **2**, 469–488.

[85] C. H. Reynolds, B. A. Tounge, and S. D. Bembenek, *J. Med. Chem.*, 2008, **51**, 2432–2438.

CHAPTER 25

Compound Collection Enhancement and Paradigms for High-Throughput Screening — an Update

Stevan W. Djuric,[*] **Irini Akritopoulou-Zanze,**[*] **Philip B. Cox**[*] and **Scott Galasinski**[**]

Contents		

[*] Medicinal Chemistry Technologies, Advanced Technologies, Abbott Laboratories, Abbott Park, IL 60064, USA

[**] Lead Discovery, Advanced Technologies, Abbott Laboratories, Abbott Park, IL 60064, USA

Annual Reports in Medicinal Chemistry, Volume 45
ISSN 0065-7743, DOI 10.1016/S0065-7743(10)45025-3

1. INTRODUCTION

Several years ago in Volume 38 (2003) of *Annual Reports of Medicinal Chemistry* [1], a path forward to greater success for pharmaceutical companies was proposed based on several key principles. These included improved target identification and a high-throughput screening (HTS) collection containing an optimized, diverse set of quality compounds that would provide a greater selection of lead series for chemists to work on. Such a collection might possibly ameliorate ongoing issues of compound attrition due to off-target toxicity [2]. Given the billions of dollars spent since this time on the enhancement of corporate screening collections as a whole, a point is being reached where an assessment of the value of this exercise is possible. In this chapter, we overview the strategies that have been developed for the selection and synthesis of compounds for collection enhancement and current strategies to screen these compounds. We will also briefly assess whether these efforts are bearing fruit. Historically, corporate collections were composed of legacy compounds derived from prior program-related initiatives, many classes of which were not generally useful for HTS efforts, and compounds purchased from vendor collections that were available to everyone and potential intellectual property disasters. Notably, many compounds simply did not have drug-like properties and efforts to improve the collections with compounds having improved physicochemical properties, ligand efficiencies, and molecular complexity have been a major focus as of late [3–7]. Another problem with historic collections was that many contained large libraries of compounds around a specific scaffold (from the combinatorial chemistry days). After it was reported that scaffold substitution patterns drove the diversity of libraries, the trend moved toward the production of smaller libraries around a greater diversity of scaffolds [8,9]. Concepts of total chemical space and total biological space were developed [10–13] and considerable efforts were directed toward the synthesis of diverse libraries of compounds to target perceived pharmacologically relevant space and toward the synthesis of focused libraries targeted to particular protein families [14–18]. Relatedly, a recent publication highlighting molecular complexity (Fsp3) as a criterion for candidate selection has been propounded [19]. Additionally, several reports have attempted to deal with issues surrounding optimal size of screening collections and the number of analogs needed to screen to ensure maximal hit rates [20,21].

Another interesting facet of library development has been the extension of new chemical reactions to target novel pharmacologically relevant molecules [22–25]. In addition, an examination of potential new heterocyclic structures as starting points for novel drugs has recently been published. [26,27].

This chapter will cover recent developments in the design of diverse and protein family-targeted libraries, their physicochemical properties, and contemporary methods for screening them.

2. TARGETED/FOCUSED LIBRARIES

The terms targeted or focused libraries have been used to describe efforts to bias the outcome of HTS campaigns by predesigning/preselecting compounds based on their likelihood to interact with the targets of interest [28]. Numerous approaches are available to assess how probable it is for a real or virtual molecule to be biologically active. Some approaches focus on broad computational efforts attempting to design bioactive molecules against any target, while others concentrate on families of targets with common structural features or individual targets. Across all approaches some methodologies utilize ligand or target-based designs or combinations of both.

2.1 Kinases

The vast number of publications on kinase structures and kinase inhibitors over the last decade has enabled the development of numerous approaches to the design and selection of focused inhibitor libraries. A classical approach has been the selection of compounds from an existing pool to create a focused set for screening. The use of two-dimensional (2D) and two-dimensional (3D) filtering followed by docking to a generalized kinase model has been reported [29]. Recently, one group has thoroughly analyzed more than 250,000 published compounds and identified several structural elements most commonly found in kinase inhibitors [30]. They created a new "2–0" kinase-likeness rule that refers to a sum of heteroaromatic nitrogens greater than 2 or a sum of anilines and nitriles greater than 0. By applying this simple rule to their internal HTS campaigns, they were able to identify, on average, 89% of all their kinase hits which was comparable to other methods such as classification models and docking exercises. Another group has explored the frequencies of kinase inhibitor fragments as they relate to kinase potency and selectivity [31]. They have now reported the computational construction of new molecules by recombination of fragments deriving from known inhibitors [32]. The new molecules were scored based on a kinase pharmacophore

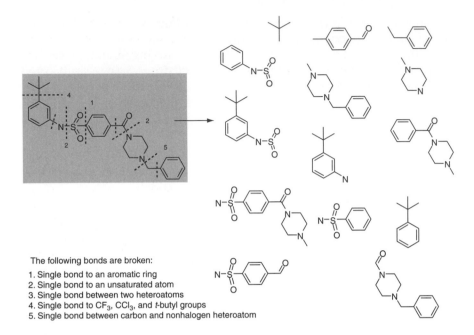

The following bonds are broken:

1. Single bond to an aromatic ring
2. Single bond to an unsaturated atom
3. Single bond between two heteroatoms
4. Single bond to CF_3, CCl_3, and t-butyl groups
5. Single bond between carbon and nonhalogen heteroatom

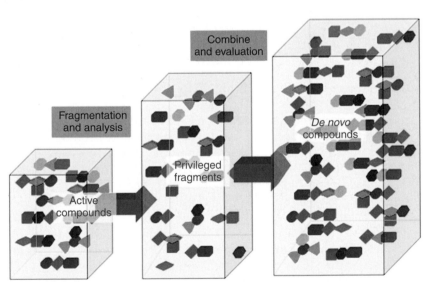

Figure 1 Schematic disassembly and reconstruction of kinase inhibitors with permission from reference 31. (See Color Plate 25.1 in the Color Plate Section)

model and the results were used for the design and synthesis of small chemical libraries which exhibited high (92%) overall hit rate toward kinase targets (Figure 1).

Although the above methodologies proved to be very successful in identifying active kinase inhibitors, they utilized "generic" kinase models and did not address selectivity issues. An interesting recent report has attempted to create quantitative structure–activity relationship (QSAR) models based on data sets of compounds tested against multiple kinases [33].

2.2 GPCRs

Recent developments in the field of G-protein-coupled receptor (GPCR) crystallography have sparked a renewed interest in identifying GPCR-targeted compounds via structural/homology modeling approaches. An investigation on which GPCRs could be modeled based on current X-ray structural data and which ones were needed for future crystallization efforts was undertaken [34].

While receptor-based approaches to GPCR ligand design are expected to bear fruit in the future, the best strategy for generating GPCR-biased libraries is based on chemical structures. The latest installment in this approach has been the identification of structural features most commonly found in GPCR ligands much like the previously cited approach for kinases. More than 40,000 GPCR ligands were compared to a set of ~16,000 general diversity compounds [35] to identify the butyl-amino moiety (often linked to an aromatic substructure) as a common feature exhibiting fourfold enrichment in GPCR ligands over random compounds.

The authors also identified the most common structural motifs unique to ligands of individual classes of GPCRs such as the adrenergic, dopamine, histamine, muscarinic, and serotonin receptors as shown in Table 1.

2.3 Ion channels

The use of similarity fingerprints to distinguish between selective and nonselective molecules has found its latest application toward ion channels. Molecular ACCess System (MACCS) and Extended Connectivity Fingerprints (ECFP4) similarity fingerprints were employed to distinguish between selective and nonselective ionotropic glutamate ligand-gated ion channel antagonists [36].

2.4 Central nervous system targets

There have been numerous efforts in library synthesis to develop compounds with central nervous system (CNS) activity [37]. Most recently, a QSAR model has been developed based on the activity of 2500 compounds against 180 assays using proprietary 3D pharmacophore descriptors [38]. The model successfully predicted 83% of a test

Table 1 Most frequent unique structural features of GPCR ligands by class

GPCR	Substructure	Frequency in class (%)	Frequency in other GPCR ligands (%)
Adrenergic	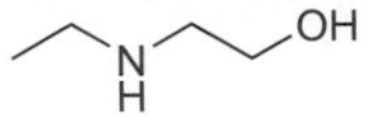	44	1
Dopamine	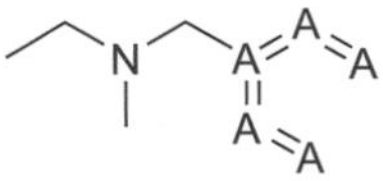 A = aromatic atom	30	8
Histamine	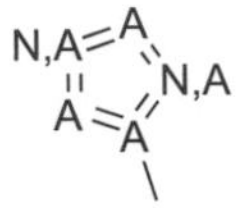 A = aromatic atom	42	6
Muscarinic		19	0
Serotonin	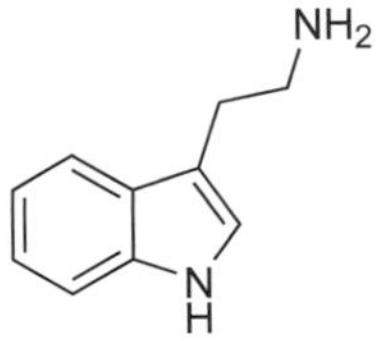	31	6

set of molecules that were active against GPCR CNS targets. Additionally, the model was successful in predicting blood–brain barrier permeability.

2.5 *De novo* design

De novo design approaches computationally generate virtual molecules based on protein active sites or simply just *de novo*-designed molecules [39].

A report of a fully integrated system of computational tools (Novo-Bench project, FlexNovo) utilized ligands from 188 reported crystal structures across eight diverse targets [40]. The suite of programs shredded ligands to basic fragments, positioned them in the active sites of the selected targets using pharmacophoric constrains, reconstructed them within the active sites, and further evaluated their physicochemical properties and ease of synthesis. The study was able to successfully reconstruct in most target cases the native ligand used for the *de novo* design. A related "all-in-one" approach involved an automated E-Novo protocol for the evaluation of lead optimization libraries [41]. Virtual molecules were built using fragmentation/enumeration tools and were subjected to core-constrained docking and physics-based rescoring. The predicted binding energies were well correlated with experimental values for seven test cases.

In addition to target-based approaches, several programs have been developed to produce *de novo*-designed molecules such as a multiobjective evolutionary graph algorithm (MEGA) framework [42] and a reaction vector approach to design synthetically feasible molecules [43]. The newly created molecules could then be interrogated in a variety of evaluation processes.

The value of *de novo* design to the identification of completely new chemotypes (scaffold hopping) was also explored [44]. Using the World Drug Index (WDI) database, 24,301 structures were selected after applying various filters. These were dissected to 10,497 unique fragments which were used in the software program Flux to generate new structures. The study explored four different targets and evaluated virtual molecules based on their similarity to one or more template structures. Three different approaches were taken by using one or five most similar or five most diverse reference ligands as templates for each target. For several targets, known reference structures, dissimilar to template or templates used, were reconstructed successfully. In addition, other new chemotypes were also obtained suggesting the utility of this approach to "scaffold hopping."

3. BIOLOGY-ORIENTED SYNTHESIS

The term biology-oriented synthesis (BIOS) [45] has been used to describe the design of compound libraries based on biologically relevant chemical space [46]. The areas in protein structures that participate in productive protein–ligand interactions have been, for the most part, already defined by natural products and drugs. Thus libraries inspired by natural products and other bioactive molecules are expected to have a higher probability of biologically activity than randomly synthesized molecules [47,48].

By using a set of natural products and metabolites to broadly define biologically active space, it was demonstrated that the majority of commercially available screening compounds were similar to natural products and metabolites and therefore biased to bind to some protein targets [49]. The study further identified almost 1300 natural product scaffolds that were missing from commercially available libraries of compounds and suggested that these scaffolds would be a fruitful starting point for new library synthesis (Figure 2).

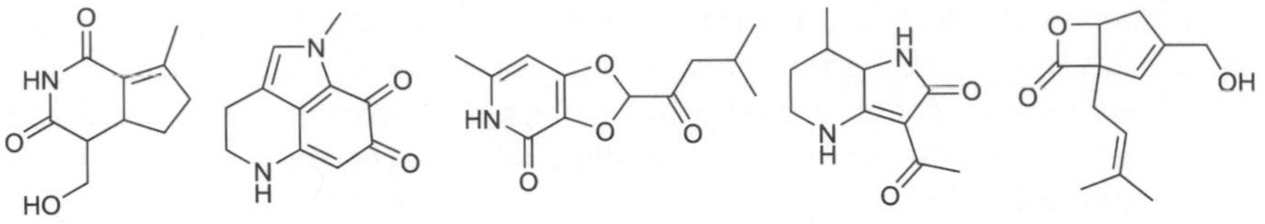

Figure 2   Examples of natural product scaffolds missing from commercial sources.

4. SCAFFOLD-ORIENTED SYNTHESIS

4.1 Privileged scaffolds

There continues to be interest in small-molecular scaffolds as good starting points for drug discovery [50]. Indoles [51], 2-oxindoles [52], 2-arylbenzothiazoles [53], rhodanines [54], quinolones [55], β-D-glucose [56], and γ-pyrones [57] have all recently been reviewed as "privileged" scaffolds.

A method to identify common scaffolds in databases by scaffold detection, alignment, and assignment has been reported [58]. This method could be used to create meaningful SAR analyses of large medicinal chemistry databases.

Although the number of potential molecules to be made is quite large, the number of small aromatic ring systems is much smaller. A recent study identified 24,847 such ring systems and found that only 1701 have been synthesized. The remaining were subjected to synthetic feasibility calculations and more than 3000 additional ring systems were identified (Figure 3) that have never been made [26].

4.2 Scaffold hopping

The concept of scaffold hopping invokes the use of computational tools that when given a reference structure can propose a different structure likely to have similar biological properties. A comprehensive scaffold database to serve for scaffold-hopping purposes has been created and is publicly available [59]. The database was based on the analysis of more than 4 million compounds to identify 241,824 unique scaffolds. In addition to the scaffold structure, the database contains information about the original molecule and its biological activity as well as its calculated physicochemical properties.

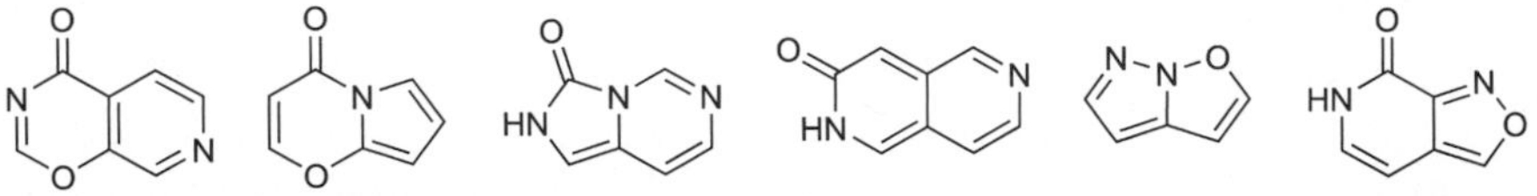

Figure 3   Examples of heteroaromatic rings generated that have never been made.

Among the various methods reported for scaffold hopping, some utilize improved virtual screening methods such as the 3D ligand similarity searches LigCSRre [60] and Ligmatch [61]. A novel scoring algorithm based on molecular interaction fingerprints has also been developed [62]. This method was compared to other scoring functions of virtual screening programs and was found to be equally successful in virtual screening and more efficient in scaffold hopping.

Another group has evaluated self-organizing maps [63] and shape/pharmacophore models [64]. They developed a new method termed SQUIRREL to compare molecules in terms of both shape and pharmacophore points. Thus from a commercial library of 199,272 compounds, 1926 were selected based on self-organizing maps trained on peroxisome proliferator-activated receptor α (PPARα) "activity islands." The compounds were further evaluated with SQUIRREL and 7 out of 21 molecules selected were found to be active in PPARα. Furthermore, a new virtual screening technique (PhAST) was developed based on representation of molecules as text strings that describe their pharmacophores [65].

In addition to similarity comparisons of entire molecules, several other methods were developed to address later stages in drug discovery where the necessary functionalities for activity are known. Such techniques, which are similar to scaffold hopping, are often termed "core hopping," "lead hopping," or "fragment hopping" and aim to replace only parts of a molecule preferably with readily accessible scaffolds/fragments [66–69].

5. DIVERSITY-ORIENTED SYNTHESIS

While focused libraries attempt to access chemical space that overlaps with biology space, diversity-oriented synthesis (DOS) attempts to explore larger and often uncovered areas of chemical space. The synthesis of libraries of azacycles, oxacycles, and carbacycles based on a build/couple/pair strategy [70] has been recently reported [71]. Each final compound was prepared in only four or five steps starting with unsaturated building blocks attached to fluorous tags, followed by reactions with propagating or capping building blocks and subsequent metathesis cascades. Only six different reaction types were employed but remarkably more than 80 new scaffolds, mostly novel, coming from 25 distinct parental scaffolds, were produced in ring sizes ranging from 5 to 15 atoms (Figure 4).

"Privileged" scaffolds are often incorporated in the design of DOSs. An unprecedented asymmetric catalytic 1,3-dipolar cycloaddition

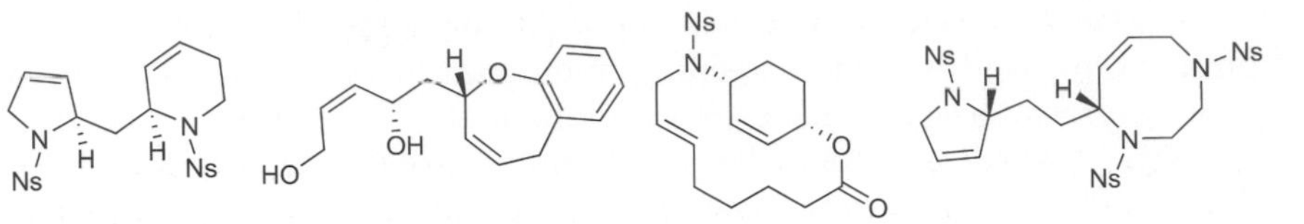

Figure 4   Examples of scaffolds produced by DOS.

reaction to suitably substituted oxindole scaffolds to produce spiro[pyrrolidin-3,3'-oxindoles] has been reported [72].

Functionalized sugars and amino acids have been used to create unique morpholine scaffolds which were further elaborated to create a second generation of morpholino derivatives following the precepts of the build/couple/pair approach [73].

Although DOS has been building an impressive amount of new chemical matter, some of the concerns regarding its overall utility are beginning to be addressed [74,75].

6. MULTICOMPONENT REACTIONS

Multicomponent reactions (MCRs) have been known to produce highly complex and diverse structures [76]. There is a considerable interest in the application of new multicomponent reactions to access biologically relevant molecules [77,78] and natural products [79]. A recent report has disclosed multicomponent Passerini and Ugi reactions to produce, rapid and efficiently, a library of redox-active selenium and tellurium compounds [80]. The compounds showed promising cytotoxicity against several cancer cell lines.

A one-pot, four-component process involving the *in situ* formation of an azadiene followed by an intermolecular or intramolecular Diels–Alder reaction for the synthesis of highly functionalized piperidone scaffolds has been reported [81]. The compounds were obtained in good yields and diastereoselectivities (Figure 5).

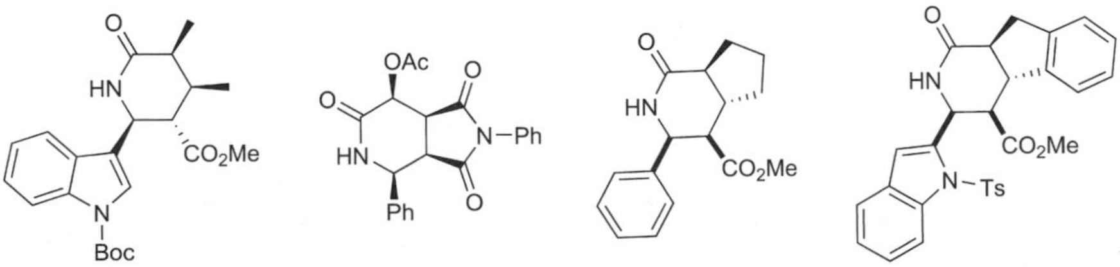

Figure 5   Examples of piperidone scaffolds.

7. FACTORS INFLUENCING COMPOUND COLLECTION DESIGN

The physicochemical guidelines or rules of thumb that are important for drug design, by inference, are also important when considering the design of compounds for corporate screening collections. In this section, we will review some recent studies focused on the importance of balancing and optimizing physicochemical properties in the context of the design of drug-like chemical matter.

7.1 Lipophilicity—"perils of logP"

Since Lipinski's seminal paper in 1997 [3], there have been many studies both vindicating and refining the "Rule of Five," as well as unearthing a multitude of liabilities associated with highly lipophilic compounds [82,83]. A more recent and very comprehensive analysis reaffirmed the negative effects of high compound lipophilicity [6]. In one of these studies, the authors analyzed promiscuity versus median clogP of 2113 compounds with promiscuity equated as the percentage of assays the compounds hit at 30% inhibition at 10 µM (Figure 6).

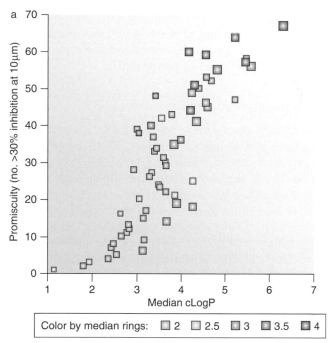

Figure 6   Plot of promiscuity versus median cLogP. (See Color Plate 25.6 in the Color Plate Section)

This analysis suggests that setting a lower threshold for cLogP is advisable particularly when designing compounds for a screening collection. In addition, this plot is consistent with the Rule of 3/75 guidelines [7] in which it was observed that compounds with a cLogP greater than 3 have an enhanced risk of toxicity (with the odds of toxicity even higher for compounds with a polar surface area less than 75). More recently, a similar analysis was performed by another industry group. Promiscuity was again shown to track with increasing lipophilicity [84].

Highly lipophilic compounds have observed liabilities on a number of fronts. For example, highly lipophilic compounds (particularly bases) have an enhanced risk of cardiovascular side effects as a result of binding to hERG (human Ether-a-go-go Related Gene) channels [85] and an increased likelihood for phospholipidosis [86,87]. Unsurprisingly, there have been many studies aimed at further understanding the effect of lipophilicity on the absorption, distribution, metabolism, and excretion (ADME) and pharmacokinetic (PK) profiles of a compound. A recent report proposed an interesting way of optimizing absorption and clearance using the "Golden Triangle" visualization tool [88]. An analysis of a set of 47,000 compounds showed that compounds with a higher probability of acceptable permeability and clearance fell within a well-defined triangular area of a plot of molecular weight (MW) versus LogD. A similar analysis [89,90] has proposed a more refined examination of the upper and lower limits of lipophilicity of drug candidates in the context of permeability. This statistical analysis showed that LogD and MW are the two most important physicochemical properties that need to be optimized within a defined range in order to achieve workable levels of permeability.

7.2 Other physicochemical roadblocks

Recently, there have been many studies associated with the negative impact of designing chemical matter outside of drug-like space. What constitutes drug-like space hinges on a number of parameters with the majority related to a broad spectrum of physicochemical properties. Indeed, many studies not satisfied with well-established physicochemical parameters have had cause to conceptualize and calculate others such as Ligand efficiency (LigE) [91], ligand Lipophilicity efficiency (LLE) [6]), aromatic proportion (AP) [92], and FSp^3 [19] in an attempt to enhance how potential drug-like chemical matter is assessed.

A recent and very comprehensive assessment of the effect of aromatic ring count on compound "developability" has been published [93]. The analysis showed a number of negative consequences of high aromatic ring count on a range of druggable factors such as solubility, cLogP, LogD, plasma protein binding, CYP inhibition and activation, and hERG channel inhibition (Table 2).

Table 2 Table of mean aromatic ring count in the Glaxo-SmithKline pipeline

	CS	FTIH	P1	P2	POC
Count	50	68	35	53	96
Mean aromatic ring count	3.3	2.9	2.5	2.7	2.3

CS, preclinical candidate selection; FTIH, first time in human; P1, phase 1; P2, phase 2; POC, proof of concept.

In all cases, increased aromatic ring count had a negative effect on these parameters. Indeed, the study concluded that compounds with more than three aromatic rings have a lower probability of "developability." The authors highlighted the fact that the average number of aromatic rings for oral drugs is less than 2. The idea of using the number of aromatic rings or the degree of aromaticity (AP) as a main parameter in developing a solubility model was recently reported by industry researchers [92]. A recursive partitioning (RP) model for aqueous solubility built by this group was developed as a quick filter for choosing compounds for their screening collection. They found that the most accurate and precise model utilized just two parameters, AP and MW out of a total of five models that included combinations of seven different physicochemical descriptors (MW, RTB, HBA, HBD, PSA, ALogP, AP). This study demonstrated the importance of considering the overall degree of aromaticity in a compound, especially in the context of designing compounds with good aqueous solubility, and should be taken into consideration when selecting or designing compounds for a screening collection.

Another concept inversely related to AP and aromatic ring count was recently proposed [19]. This study considered the fraction of Sp^3-hybridized carbons (FSp^3) as a function of total number of carbons in a molecule. The authors calculated the FSp^3 for compounds across all stages of development from initial discovery to launched drugs and found that the level of saturation (FSp^3) increased as compounds progressed. The authors also made the connection between low FSp^3 and higher melting temperatures, and therefore low solubility, thereby showing the positive effect of adding saturation to a molecule in terms of improving druggable physicochemical properties.

7.3 Assessing risk—from Rule of Five to Rule of 3/75

The large majority of the studies undertaken to understand druggability from an assessment of either *in vitro* and or *in silico* physicochemical properties are aimed at the common goal of ameliorating the risk of compound attrition. If a simple rule of thumb, such as Rule of Five, was guaranteed to deliver development candidates, with good PK profiles and a low risk of *in vivo* toxicity, then drug discovery would be relatively straightforward. While we know the truth is rather more complex, the fact remains that the Rule of Five and other studies have significantly enhanced our ability to design drug-like chemical matter. In a quest to further refine risks associated with venturing out of drug-like space, researchers have investigated correlations between physicochemical properties and *in vivo* toxicity data. The recent disclosure of the "Rule of 3/75" [7], as mentioned previously, is a classic example of this study where the odds of toxicity were correlated with high lipophilicity (cLogP > 3) and low polar surface area (<75).

As noted above, considerable research is ongoing to define the overall ideal structural and physicochemical characteristics of drug-like chemical matter. These evolving characteristics should be taken seriously in the context of compound collection design.

8. CONTEMPORARY METHODS FOR COMPOUND EVALUATION

HTS campaigns remain the starting point for most projects, even when one organization can fast follow on the results of a different organization. Assays are designed to have high sensitivity with low basal noise (high Z') [94], be highly reproducible, and come at a low cost in terms of time and money. There is usually a compromise between these factors and the size of the compound collection that can be tested, sometimes resulting in screening smaller focused collections. Because of the technical demands and the need to balance the cost of resources, many screening groups have utilized "off-the-shelf" assays for standard druggable targets, including kinases and GPCRs. These serve as excellent sources of high-quality assays that can be rapidly developed and often allow for large-scale selectivity profiling of HTS hits as well. The temptation to use these "kits" suggests that as they become more broadly adopted, a level of competitive advantage can be lost. In addition, this approach may not be viable for assay development of novel targets. Assays that meet the demands of HTS and still offer novelty in detection of mechanism of inhibitors or potential for screening new targets will be highlighted.

8.1 Improvements in biochemical activity and affinity assays

Kinases remain a highly pursued drug target for numerous therapeutic areas [95]. In addition to the primary screen, selectivity profiling of kinases has become common, providing selectivity profiles that may improve polypharmacology, for example, in oncology applications, or optimizing profiles that reduce the potential of adverse side effects [96]. Traditional radiometric flashplate assays have been compared to nonradiometric microfluidic mobility shift assays for multiple kinases [97]. Excellent agreement between both platforms was found, and in general assay development was easier for the mobility shift assay without the need for radioactivity. Between day reproducibility experiments demonstrated that the mobility shift assay was superior. This suggests that the mobility shift assay is an excellent choice for routine and repetitive kinase profiling operations. The improved precision gained by measuring the product-to-substrate ratio in this assay is not limited to kinases and should be applicable to other enzymatic target classes.

An improved time-resolved fluorescence resonance energy transfer (TR-FRET) assay has been demonstrated for kinase primary screening and broad kinome profiling [98]. The assay requires a tagged recombinant enzyme, an anti-tag antibody conjugated to europium, and a compound probe labeled with Alexa Fluor 647 moiety. Use of the tagged enzyme means that the assay signal is only produced by the kinase of interest, eliminating interference by contaminating kinases, a unique property of this assay compared to activity assays. The assay does not rely on enzyme-specific activity, thereby allowing identification of inhibitors for non-activated forms and those that would not support an activity assay. The assay is able to detect both adenosine-5'-triphosphate (ATP) competitive inhibitors and "mixed" inhibitors. The TR-FRET approach should be useful for many non-kinase applications to support both active and allosteric site inhibitor identification.

Epigenetics has identified numerous potential new therapeutic targets with uncommon enzymatic activities [99]. The types of enzymatic reactions include lysine and arginine methylation, lysine acetylation, sumolation, ubiquitination, adenosine diphosphate (ADP)-ribosylation, and phosphorylation [100]. These modifications further serve as recruitment scaffolds, thereby establishing additional protein–protein interactions for therapeutic intervention. Malignant brain tumor (MBT) domain proteins are selective for monomethyl-lysine and di-methyl-lysine and are associated with chromatin condensation and transcriptional repression. Since these proteins do not show strong sequence specificity and relatively low affinity for methylated residues, they present a unique challenge for developing an assay. An amplified luminescence proximity homogeneous assay (Alpahscreen™) has been established as high-throughput antagonist assay for an MBT protein and a cognate histone peptide [101]. This affinity format supports protein–protein interaction assays where binding partners do not necessarily have high affinity for each other.

8.2 Advances in understanding cell-based screening

GPCRs have been described as "microprocessors" of information rather than "on–off" switches [102]. The detection of these events requires functional assays and the results present both new insights and interpretation challenges. Functional assays allow for detection of agonists, antagonists, inverse agonists, partial agonists, and allosteric modulators. This latter class has become particularly popular as possible therapeutic agents because they offer the potential for increased selectivity, better control of physiology, and separation of affinity and efficacy [103]. GPCR function is generally modulated through G-protein activation and β-arrestin recruitment. Furthermore, ligands can stimulate G-protein-dependent and G-protein-independent signaling events which may not always be exclusive. These observations have given rise to the terms "cell-specific signaling," "functional selectivity," and "ligand bias" to try and capture pleiotropic GPCR signaling.

Multiple companies are now developing higher throughput platforms that measure cellular impedance in response to compound treatment, based on earlier work [104–106]. The assay transforms complex data resulting from cell morphological changes into a readout that reflects unique G-protein signatures based upon the predominant type of G protein activated. Single G-protein responses for G_{as}, $G_{ai/o}$, and G_{aq} can be discerned from cell-specific events, dual signaling, and functional selectivity imparted by ligands.

Cellular impedance systems have been recently evaluated for direct detection of different G-coupled receptors [107]. In these studies, G-protein inhibitors (e.g., Pertussis toxin) can be useful for distinguishing coupling events. Cell-specific signaling was confirmed by this methodology, demonstrating G_{as} coupling for the melanocortin-4-receptor in human embryonic kidney (HEK) cells versus G_{aq} coupling in Chinese hamster ovary (CHO) cells [108]. Dual signaling with G_{as}/G_{ai} coupling was observed for CB1 and β-adrenergic receptors. This observation suggests that dynamic signaling can be monitored simultaneously for dual coupling, and that new response profiles may indicate preferential modulation of signal transduction that can be used to bias unique pharmacological responses.

G-protein-independent signaling can be mediated by β-arrestin recruitment and receptor internalization [109]. Assays have been developed that allow for detection on standard high-throughput readers. The PathHunter assay^{TM} is a β-arrestin, cell-based, β-galactosidase protein complementation assay. Ligand binding of the receptor recruits the β-arrestin and truncated β-galactosidase fusion product to the tagged receptor and allows for full enzyme complementation that can generate a readable signal. The advantages of this assay is that it measures direct binding of β-arrestin instead of a second-messenger amplification step and is completely G-protein independent. This assay was compared to a competitive ligand-binding assay, and 3'-5' cyclic-adenosine monophosphate (cAMP) assay, for the cannabinoid receptor 2 (CB2) [110]. Assays that measure β-arrestin signaling hold special promise for detecting inverse agonists, which reduce the constitutive activity of GPCRs, and other pharmacological effects imposed by ligand bias [111,112].

9. CONCLUSIONS

The industry, as a whole, has invested well over a billion dollars on improving the quality of their corporate compound collections through the addition of compounds meeting criteria as described previously and through the removal/elimination of undesirable or impure compounds. Has this effort paid dividends? It is certainly clear that many compounds currently in lead development have arisen through HTS efforts. An interesting documented example of this is a marketed CCR5 antagonist Maraviroc (approved in 2007) whose origins resided in an HTS hit identified in 1997 [15]. On the whole, it appears that more useful lead series are being identified for druggable targets giving the chemist a choice of multiple series to work on. This can only be good. Nondruggable targets, however, remain generally recalcitrant although fragment-based screening has provided alternative options here with the Bcl-xL antagonist ABT-263, currently in clinical trials, being a notable example [113].

Having said all this, a recent report from a major pharmaceutical company with one of the largest reported screening collections revealed that just over 50% of their current lead optimization efforts invoke a "fast follower" approach [114]. Why is this? One might speculate that as most companies work on the same targets these days chemists find it easier to "patent bust" than wait for the results of an HTS campaign which may take months to develop. Possibly, when more novel targets are identified, HTS of enhanced collections will deliver upon the promise that they hold.

REFERENCES

[1] G. M. Milne, *Ann. Rep. Med. Chem.*, 2003, **38**, 383.
[2] T. Koppal, *Drug Discov. Dev.*, 2003, 22.
[3] C. A. Lipinski, F. Lombardo, B. W. Dominy, and P. J. Feeney, *Adv. Drug. Del. Res.*, 1997, **23**, 3.
[4] M. Congreve, R. Carr, C. Murray, and H. Jhoti, *Drug Discov. Today*, 2003, **8**, 876.
[5] T. I. Oprea, A. M. Davis, S. J. Teague, and P. D. Leeson, *J. Chem. Inf. Comput. Sci.*, 2001, **41** 1308.

[6] P. D. Leeson and B. Springthorpe, *Nat. Rev. Drug Discov.*, 2007, **6**, 881 and references therein.

[7] J. D. Hughes, J. Blagg, D. A. Price, S. Bailey, G. A. DeCrescenzo, R. V. Devraj, E. Ellsworth, Y. M. Fobian, M. E. Gibbs, R. W. Gilles, N. Greene, E. Huang, T. Krieger-Burke, J. Loesel, T. Wager, L. Whiteley, and Y. Zhang, *Bioorg. Med. Chem. Lett.*, 2008, **18**, 4872.

[8] W. H. B. Sauer and M. K. Schwartz, *J. Chem. Inf. Comput. Sci.*, 2003, **43**, 987.

[9] I. Akritopoulou-Zanze, J. T. Metz, and S. W. Djuric, *Drug Discov. Today*, 2007, **12**, 948.

[10] C. Lipinski and A. Hopkins, *Nature*, 2004, **432**, 855 and references therein.

[11] J. L. Medina-Franco, K. Martinez-Mayorga, M. A. Giulianotti, R. A. Houghten, and C. Pinilla, *Curr. Comput-Aided Drug Des.*, 2008, **4**, 322.

[12] G. V. Paolini, R. H. B. Shapland, W. P. van Hoorn, J. S. Mason, and A.L Hopkins, *Nat. Biotechnol.*, 2006, **7**, 805.

[13] I. Vogt and J. Bajorath, *J. Chem. Inf. Model.*, 2008, **48**, 1389.

[14] E. Jacoby, A. Schuffenhauer, M. Popov, K. Azzaoui, B. Havill, U. Schopfer, C. Engeloch, J. Stanek, P. Acklin, P. Rigollier, F. Stoll, G. Koch, P. Meier, D. Orain, R. Giger, J. Hinrichs, , K. Malagu, J. Zimmerman, and H. J. Roth, *Curr. Top. Med. Chem.*, 2005, **5**, 397.

[15] K. -H. Baringhaus and G. Hessler, *Methods Princ. Med. Chem.*, 2004, **22**, 221–242.

[16] D. M. Schnur, M. A. Hermsmeier, and A. J. Tebben, *J. Med. Chem.*, 2006, **49**, 2000.

[17] R. Heilker, M. Wolff, C. S. Tautermann, and M. Bieler, *Drug Discov. Today*, 2009, **14**, 231.

[18] J. -U. Peters, P. Schnider, P. Mattei, and M. Kansy, *ChemMedChem*, 2009, **4**, 680.

[19] F. Lovering, J. Bikker, and C. Humblet, *J. Med. Chem.*, 2009, **52**, 6752.

[20] G. Harper, S. D. Pickett, and D. V. S. Green, *Comb. Chem. High Throughput. Screen.*, 2004, **7**, 63.

[21] M. J. Lipkin, A. P. Stevens, D. J. Livingstone, and C. J. Harris, *Comb. Chem. High Throughput Screen.*, 2008, **11**, 482.

[22] I. Akritopoulou-Zanze, and S. W. Djuric, *Heterocycles*, 2007 **73**, 125.

[23] M. D. Burke and S. L. Schreiber, *Angew. Chem., Int. Ed.*, 2004, **43**, 46.

[24] D. S. Tan, *Nat. Chem. Biol.*, 2005, **1**, 74.

[25] D. R. J. Spandl, A. Bender, and D. R. Spring, *Org. Biomol. Chem.*, 2008, **6**, 1149.

[26] W. R. Pitt, D. M. Parry, B. G. Perry, and C. R. Groom, *J. Med. Chem.*, 2009, **52**, 2952.

[27] A. Wood and D. Armour, *Prog. Med. Chem.*, 2005, **43**, 239.

[28] E. Gregori-Puigjané and J. Mestres, *Curr. Opin. Chem. Biol.*, 2008, **12**, 359.

[29] H. Decornez, A. Gulyás-Forró, Á. Papp, M. Szabó, G. Sármay, I. Hajdú, S. Cseh, G. Dormán, and D. B. Kitchen, *Chem. Med. Chem.*, 2009, **4**, 1273.

[30] A. M. Aronov, B. McClain, C. S. Moody, and M. A. Murcko, *J. Med. Chem.*, 2008, **51**, 1214.

[31] J. J. Sutherland, R. E. Higgs, I. Watson, and M. Vieth, *J. Med. Chem.*, 2008, **51**, 2689.

[32] M. Vieth, J. Erickson, J. Wang, Y. Webster, M. Mader, R. Higgs, and I. Watson, *J. Med. Chem.*, 2009, **52**, 6456.

[33] R. P. Sheridan, K. Nam, V. Maiorov, D. R. McMasters, and W. D. Cornell, *J. Chem. Inf. Model.*, 2009, **49**, 1974.

[34] J. C. Mobarec, R. Sanchez, and M. Filizola, *J. Med. Chem.*, 2009, **52**, 5207.

[35] E. van der Horst, Y. Okuno, A. Bender, and A. P. Ijzerman, *J. Chem. Inf. Model.*, 2009, **49**, 348.

[36] H. E. A. Ahmed, H. Geppert, D. Stumpfe, E. Lounkine, and J. Bajorath, *Chem. Biol. Drug. Des.*, 2009, **73**, 273.

[37] P. Zajdel, M. Pawlowski, J. Martinez, and G. Subra, *Combi. Chem. High Through. Screen.*, 2009, **12**, 723.

[38] R. Gozalbes, F. Barbosa, E. Nicolaï, D. Horvath, and N. Froloff, *Chem. Med. Chem.*, 2009, **4**, 204.

[39] W. L. Jorgensen, *Acc. Chem. Res.*, 2009, **42**, 724.

[40] A. Zaliani, K. Boda, T. Seidel, A. Herwig, C. H. Schwab, J. Gasteiger, H. Claußen, C. Lemmen, J. Degen, J. Pärn, and M. Rarey, *J. Comput. Aided Mol. Des.*, 2009, **23**, 593.

[41] B. C. Pearce, D. R. Langley, J. Kang, H. Huang, and A. Kulkarni, *J. Chem. Inf. Model.*, 2009, **49**, 1797.

[42] C. A. Nicolaou, J. Apostolakis, and C. S. Pattichis, *J. Chem. Inf. Model.*, 2009, **49**, 295.
[43] H. Patel, M. J. Bodkin, B. Chen, and V. J. Gillet, *J. Chem. Inf. Model.*, 2009, **49**, 1163.
[44] B. A. Krueger, A. Dietrich, K. -H. Baringhaus, and G. Schneider, *Combi. Chem. High Throughput Screen.*, 2009, **12**, 383.
[45] A. Nören-Müller, I. Reis-Corrêa Jr., H. Prinz, C. Rosenbaum, K. Saxena, H. J. Schwalbe, D. Vestweber, G. Cagna, S. Schunk, O. Schwarz, H. Schiewe, and H. Waldmann, *Proc. Nat. Acad. Sci. USA*, 2006, **103**, 10606.
[46] D. J. Triggle, *Biochem. Pharm.*, 2009, **78**, 217.
[47] K. Hübel, T. Leßmann, and H. Waldmann, *Chem. Soc. Rev.*, 2008, **37**, 1361.
[48] M. Kaiser, S. Wetzel, K. Kumar, and H. Waldmann, *Cell. Mol. Life Sci.*, 2008, **65**, 1186.
[49] J. Hert, J. J. Irwin, C. Laggner, M. J. Keiser, and B. Shoichet, *Nature, Chem. Biol.*, 2009, **5**, 479.
[50] H. Zhao and I. Akritopoulou-Zanze, *Expert Opin. Drug Discov.*, 2010, **5**, 123.
[51] F. R. de Sá Alves, E. J. Barreiro, and C. A. M. Fraga, *Mini-Rev. Med. Chem.*, 2009, **9**, 782.
[52] S. Peddibhotla, *Curr. Bioact. Compounds*, 2009, **5**, 20.
[53] A. A. Weekes and A. D. Westwell, *Curr. Med. Chem.*, 2009, **16**, 2430.
[54] T. Tomašić and L. P. Mašič, *Curr. Med. Chem.*, 2009, **16**, 1596.
[55] C. Mugnaini, S. Pasquini, and F. Corelli, *Curr. Med. Chem.*, 2009, **16**, 1746.
[56] R. F. Hirschmann, K. C. Nicolaou, A. R. Angeles, J. S. Chen, and A. B. Smith, III, *Acc. Chem. Res.*, 2009, **42**, 1511.
[57] W. Wilk, H. Waldmann, and M. Kaiser, *Bioorg. Med. Chem.*, 2009, **17**, 2304.
[58] A. M. Clark and P. Labute, *J. Med. Chem.*, 2009, **52**, 469.
[59] B. -B. Yan, M. -Z. Xue, B. Xiong, K. Liu, D. -Y. Hu, and J. -K. Shen, *Acta Pharmacol. Sin.*, 2009, **30**, 251.
[60] F. Quintus, O. Sperandio, J. Grynberg, M. Petitjean, and P. Tuffery, *BMC Bioinformatics*, 2009, **10**, 245.
[61] S. L. Kinnings and R. M. Jackson, *J. Chem. Inf. Model.*, 2009, **49**, 2056.
[62] J. Venhorst, S. Núñez, J. W. Terpstra, and C. G. Kruse, *J. Med. Chem.*, 2008, **51**, 3222.
[63] P. Schneider, Y. Tanrikulu, and G. Schneider, *Curr. Med. Chem.*, 2009, **16**, 258.
[64] E. Proschak, H. Zettl, Y. Tanrikulu, M. Weisel, J. M. Kriegl, O. Rau, M. Schubert-Zsilavecz, and G. Schneider, *Chem. Med. Chem.*, 2009, **4**, 41.
[65] V. Hähnke, B. Hofmann, T. Grgat, E. Proschak, D. Steinhilber, and G. Schneider, *J. Comput. Chem.*, 2009, **30**, 761.
[66] F. Fontaine, S. Cross, G. Plasencia, M. Pastor, and I. Zamora, *ChemMedChem*, 2009, **4**, 427.
[67] R. Bergmann, T. Liljefors, M. D. Sørensen, and I. Zamora, *J. Chem. Inf. Model.*, 2009, **49**, 658.
[68] J. Oyarzabal, T. Howe, J. Alcazar, J. I. Andrés, R. M. Alvarez, F. Dautzenberg, L. Iturrino, S. Martínez, and I. Van der Linden, *J. Med. Chem.*, 2009, **52**, 2076.
[69] S. Senger, *J. Chem. Inf. Model.* 2009, **49**, 1514.
[70] T. E. Nielsen and S. L. Schreiber, *Angew. Chem. Int. Ed.*, 2008, **47**, 48.
[71] D. Morton, S. Leach, C. Cordier, S. Warriner, and A. Nelson, *Angew. Chem. Int. Ed.*, 2009, **48**, 104.
[72] X. -H. Chen, Q. Wei, S. -W. Luo, H. Xiao, and L. -Z. Gong, *J. Am. Chem. Soc.*, 2009, **131**, 13819.
[73] C. Lalli, A. Trabocchi, F. Sladojevitch, G. Menchi, and A. Guarna, *Chem. Eur. J.*, 2009, **15**, 7871.
[74] W. R. J. D. Galloway, A. Bender, M. Welch, and D. R. Spring, *Chem. Commun.*, 2009, **18**, 2446.
[75] A. A. Shelat and R. K. Guy, *Bioorg. Med. Chem.*, 2009, **17**, 1088.
[76] J. D. Sunderhaus and S. F. Martin, *Chem. Eur. J.*, 2009, **15**, 1300.
[77] A. Dömling, *Chem. Rev.*, 2006, **106**, 17.
[78] I. Akritopoulou-Zanze, *Curr. Opin. Chem. Biol.*, 2008, **12**, 324.
[79] B. B. Touré and D. G. Hall, *Chem. Rev.*, 2009, **109**, 4439.
[80] S. Shabaan, L. A. Ba, M. Abbas, T. Burkholz, A. Denkert, A. Gohr, L. A. Wessjohann, F. Sasse, W. Weber, and C. Jacob, *Chem. Commun.*, 2009, **31**, 4702.

[81] W. Zhu, M. Mena, E. Jnoff, N. Sun, P. Pasau, and L. Ghosez, *Angew. Chem. Int. Ed.*, 2009, **48**, 5880.

[82] D. F Veber, S. R. Johnson, H. Y. Chang, B. R. Smith, K. W. Ward, and K. D. Kopple, *J. Med. Chem.*, 2002, 45, 2615—2623.

[83] M. C. Wenlock, R. P. Austin, P. Barton, A. M. Davis, and P. D. Leeson, *J. Med. Chem.*, 2003, **46**, 1250.

[84] J. Peters, P. Schnider, P. Mattei, and M. Kansy, ChemMedChem, 2009, **4**, 680.

[85] M. J. Waring and C. Johnstone, *Bioorg. Med. Chem. Lett.*, 2007, **17**, 1759.

[86] A. J. Ratcliffe, Curr. Med. Chem., 2009, **16**, 2816.

[87] J. -P. H. T. M. Ploeman, J. Kelder, T. Hafmans, T. van de Sandt, J. A. van Burgsteden, P. J. M. Salemink, and E. van Esch, *Exp. Toxicol. Pathol.*, 2004, **55**, 347.

[88] T. W. Johnston, K. R. Dress, and M. Edwards, *Bioorg. Med. Chem. Lett.*, 2009, 19, 5560.

[89] M. J. Waring, *Bioorg. Med. Chem. Lett.*, 2009, **19**, 2844.

[90] M. J. Waring, *Expert Opin. Drug Discov.*, 2010, **5**, 235.

[91] A. L. Hopkins, C. R. Groom, and A. Alex, *DDT*, 2004, **9**, 430.

[92] C. Lamanna, M. Bellini, A. Padova, G. Westerberg, and L. Macari, *J. Med. Chem.*, 2008, **51**, 2891.

[93] T. J. Ritchie and S. J. MacDonald, *Drug Discov. Today*, 2009, **14**, 1011.

[94] J. -H. Zhang and T. D. Y. Chung, *J. Biomol. Screen.*, 1999, **4**, 67.

[95] K. Garber, *Nat. Biotechnol.*, 2006, **24**, 127.

[96] Y. Kuo, *Curr. Opin. Mol. Ther.*, 2005, **7**, 251.

[97] A. Card, C. Caldwell, H. Min, B. Lockchander, H. Xi, S. Sciabola, A. Kamath, S. Clutston, W. T. Tschantz, L. Wang, and D. Moshinsky, *J. Biomol. Screen.* 2009, **14**, 31.

[98] C. Lebakken, S. Riddle, U. Singh, W. J. Frazee, H. C. Eliason, Y. Gao, L. J. Reichling, B. D. Marks, and K. W. Vogel, *J. Biomol. Screen.*, 2009, **14**, 924.

[99] S. Berger, T. Kouzarides, R. Shiekhatter, and A. Shilatifard, *Genes Dev.*, 2009, **23**, 781.

[100] B. E. Bernstein, A. Meissner, and E. S. Lander, *Cell*, 2007, **128**, 669.

[101] E. F. Ullman, H. Kirakossian, S. Singh, Z. P. Wu, B. R. Irvin, J. S. Pease, A. C. Switchenko, J. D. Irvine, A. Dafforn, C. N. Skold, and D. B. Wagner, *Proc. Natl. Acad. Sci. USA*, 1994, **91**, 5426.

[102] T. P. Kenakin, *Nat Rev. Drug Discov.*, 2009, **8**, 617.

[103] T. P. Kenakin, *J. Biomol. Screen.*, 2010, **15**, 119.

[104] I. Giaever and C. R. Keese, *Nature*, 1993, **366**, 591.

[105] E. Verdonk, K. Johnson, R. McGuinness, G. Leung, Y. W. Chen, H. R. Tang, et al., *Assay Drug Dev. Technol.*, 2006, **4**, 609.

[106] N. Yu, J. M. Atienza, J. Bernard, S. Blanc, J. Zhu, X. Wang, X. Xu, and Y. A. Abbassi, *Anal. Chem.*, 2006, **78**, 35.

[107] M. F. Peters and C. W. Scott, *J. Biomol. Screen.*, 2009, **14**, 246.

[108] E. A. Newman, B. X. Chai, W. Zhang, J. Y. Li, J. B. Ammori, and M. W. Mulholland, *J. Surg. Res.*, 2006, **132**, 201.

[109] S. M. DeWire, S. Ahn, R. J. Lefkowitz, and S. K. Shenoy, Annu. *Rev. Physiol.*, 2007, **69**, 483.

[110] D. McGuinness, A. Malikzay, R. Visconti, K. Lin, M. Bayne, F. Monsma, and C. L. Lunn, *J. Biomol. Screen.*, 2009, **14**, 49.

[111] C. Doucette, K. Vedvik, E. Koepnick, A. Bergsma, B. Thomson, and T. C. Turek-Etienne, *J. Biomol. Screen.*, 2009, **14**, 381.

[112] S. Kredel, M. Wolff, J. Wiedenmann, B. Moepps, G. U. Nienhaus, P. Gierschik, B. Kistler, and R. Heilker, *J. Biomol. Screen.* 2009, **14**, 1076.

[113] C. Tse, A. R. Shoemaker, J. Adickes, M. G. Anderson, J. Chen, S. Jin, E. F. Johnson, K. C. Marsh, M. J. Mitten, P. Nimmer, L. Roberts, S. K. Tahir, Y. Xiao, X. Yang, H. Zhang, S. Fesik, S. H. Rosenberg, and S. W. Elmore, *Cancer Res.*, 2008, **68** 3421.

[114] F. Bertelli, Drug Discovery Leaders Summit, Montreux, June, 2009.

CHAPTER 26

The Role of Fluorine in the Discovery and Optimization of CNS Agents: Modulation of Drug-Like Properties

Kevin J. Hodgetts, Kerry J. Combs, Amy M. Elder and **Geraldine C. Harriman**

Contents

1. INTRODUCTION

Over the last five decades, researchers have utilized the unique characteristics of the fluorine atom in the creation of new drugs that have impacted the quality of life of many individuals. No other substituent has captured the imagination of medicinal chemists to the extent that

Galenea, 300 Technology Square, Cambridge, MA 02139, USA

Annual Reports in Medicinal Chemistry, Volume 45
ISSN 0065-7743, DOI 10.1016/S0065-7743(10)45026-5

fluorine has. It has a small atomic radius with the ability to be both electron withdrawing and lipophilic, while maintaining a unique ability to accept hydrogen bonds and block metabolic sites. For this reason, fluorine has been an exceptional tool for medicinal chemists to create new medicines [1–3]. An example of this and an early addition to the pool of central nervous system (CNS) drugs was haloperidol (1), which contains a 4-fluorophenyl substituent that was critical for its potency as a first-generation neuroleptic agent. Shortly after this, many new fluorine-containing CNS agents were brought to the market, including a series of atypical antipsychotic drugs that addressed both the positive and negative symptoms of schizophrenia by modulating the dopaminergic pathway [4]. This included the approval of the fluorine-containing antipsychotic, rispiridone (2), which reached the market in 1993.

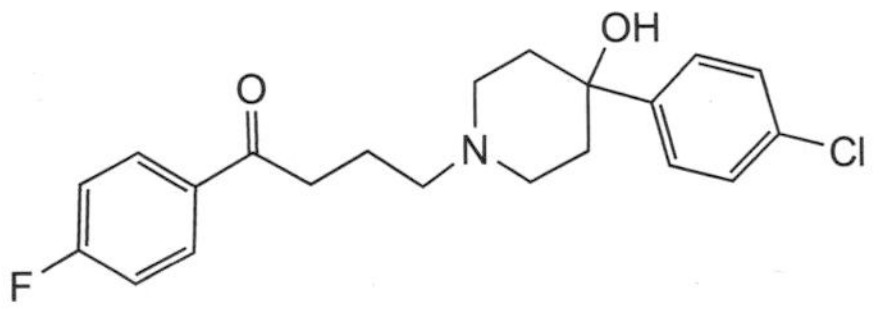

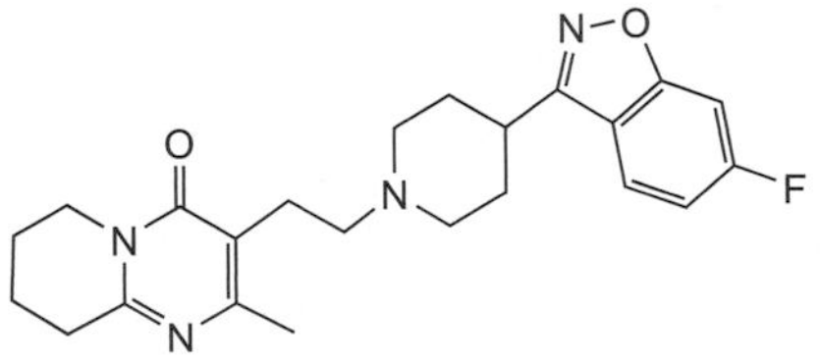

Since the introduction of fluoxetine (3) in 1987, a series of selective serotonin reuptake inhibitors (SSRIs) have been discovered that have seen broad application in many facets of mood disorders. These compounds include fluvoxamine (4) which contains a trifluoromethyl group and paroxetine (5) and citalopram (6) which contain 4-fluorophenyl groups [5,6].

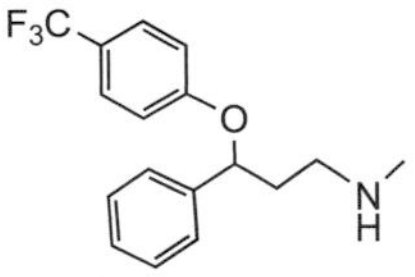

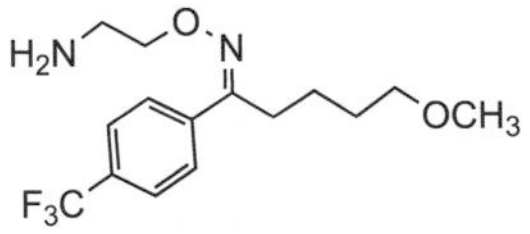

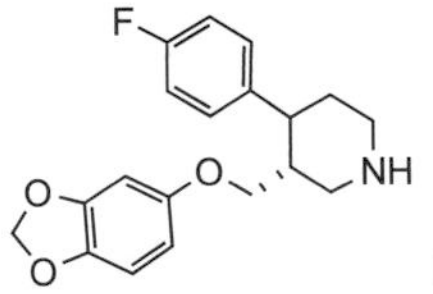

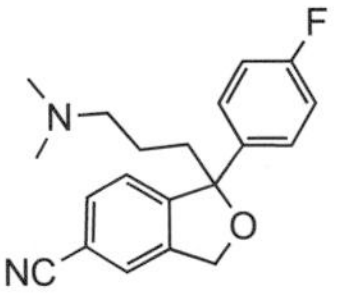

The strategic placement of an ^{18}F or ^{19}F atom has also enabled positron emission tomography (PET) [7,8] and magnetic resonance imaging (MRI) [9,10] approaches to determine receptor occupancy and perform biodistribution studies, respectively. PET has also been used to study receptor expression patterns and aid in determining efficacious brain exposures and ultimately human efficacious doses. This topic will not be covered here, as it is covered in an earlier chapter in this volume.

This chapter will highlight the strategic placement of fluorine into potential new drugs to solve medicinal chemistry challenges. These modifications would not have been possible without the synthetic advances in a variety of fluorine transformations; these will not be highlighted in this chapter, but merit attention [11,12]. Rather this chapter will focus on examples citing the utilization of fluorine and its effects on potency, selectivity (off-target effects), metabolic clearance, brain penetration, and protein binding (covalent).

2. MODULATING POTENCY AND SELECTIVITY BY INTRODUCTION OF FLUORINE

The area of CNS drug discovery offers many good examples of the utilization of the fluorine atom to modulate both potency and selectivity of bioactive molecules. Despite its small atomic radius, fluorine can have a significant effect on the conformation of rings [13] or the pK_a of neighboring groups. These effects can have profound influences on the binding affinities for the intended molecular target and subsequent off-target selectivity. For instance, in the mGluR5 field, many research groups have been in pursuit of alternative replacement series to the well-described diaryl-alkynes, such as 3-((2-methyl-4-thiazolyl)ethynyl)pyridine (MTEP) ($IC_{50} = 5$ nM) [14]. Scientists were successful in identifying a new heteroaromatic N-linked tetrazole, 7, as a potential new lead [15]. However, compound 7 lacked the desired *in vitro* potency and binding affinity. Introduction of hydrogen bond-accepting groups such as fluorine improved the potency slightly to give compound 8. However, modification of the core from a tetrazole to a pyrrole and incorporation of a fluorine provided the necessary increase in *in vitro* potency from $IC_{50} = 190$ nM (9) to $IC_{50} = 3$ nM (10). The increase in potency observed with compound 10 translated to excellent *in vivo* rat receptor occupancy.

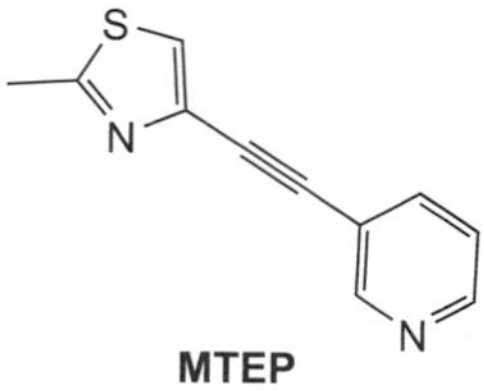

MTEP

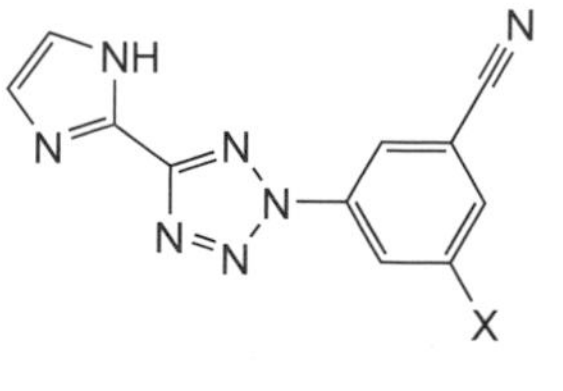

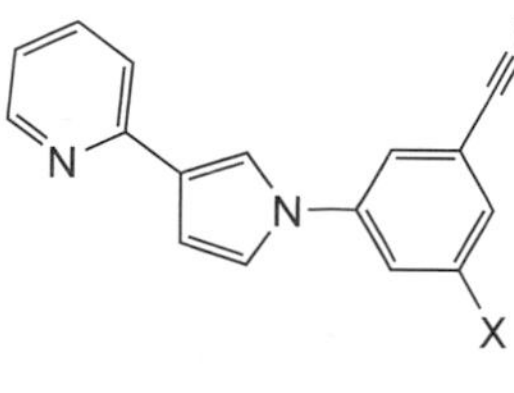

7 X = H mGlu5 Ca^{2+} flux IC_{50} = 77 nM[a]
8 X = F mGlu5 Ca^{2+} flux IC_{50} = 47 nM[a]

9 X = H mGlu5 Ca^{2+} flux IC_{50} = 190 nM[a]
10 X = F mGlu5 Ca^{2+} flux IC_{50} =   3 nM[a]

[a] Using glutamate (10 μM) as agonist

The introduction of fluorine has also been shown not only to increase the *in vitro* potency, but also to affect off-target selectivity. The identification of highly potent inhibitors of β-amyloid cleaving enzyme (BACE-1) that can demonstrate acceptable *in vivo* potency in animal models of Alzheimer's disease (AD) has been quite challenging. In the hydroxylethyl amine (HEA) series, introduction of a fluoro-substituent at the C-2 position resulted in an approximately threefold improvement in cellular potency (**11–12**) [16]. The incorporation of the C-2 fluoro, along with a six-membered sultam, and modification of the 3-methoxy benzyl (**12**) to 3-trifluoro benzyl (**13**) resulted in GSK188909 which was the first low nanomolar BACE-1 inhibitor in the cell-based assay expressing the Swedish (SWE) amyloid precursor protein (APP) substrate with good selectivity over BACE-2 and Cat-D.

11	
BACE-1 $IC_{50} = 8$ nM	
BACE-2 $IC_{50} = 4270$ nM	
Cat-D $IC_{50} = 6760$ nM	
SWE Aβ40 $IC_{50} = 928$ nM	

12
BACE-1 $IC_{50} = 4$ nM
BACE-2 $IC_{50} = 785$ nM
Cat-D $IC_{50} = 6545$ nM
SWE Aβ40 $IC_{50} = 283$ nM

13
BACE-1 $IC_{50} = 4$ nM
BACE-2 $IC_{50} = 177$ nM
Cat-D $IC_{50} = 2653$ nM
SWE Aβ40 $IC_{50} = 40$ nM

Modulation of the pK_a of basic moieties has been one strategy to improve the overall affinity for a target; this is possible by incorporation of fluorine α, β, and γ to the basic site. Recently, a number of research groups have reported antagonists of opioid receptor like-1 (ORL1) which displayed high affinity and good selectivity. For example, the novel aryl pyrazole **14** demonstrated good intrinsic potency for human ORL1 [17]. However, introduction of a fluorine beta and gamma to the two basic sites of the molecule resulted in compounds that were subnanomolar against ORL1. Modulation of the cyclopentylamine from a pK_a of 6.9 to 6.5 by introduction of a fluorine atom at the 2-position (**15**) and difluorination at the 3-position (**16**), respectively, decreased the intrinsic potency two to threefold (Table 1), but mitigated a major liability for a CNS drug, efflux by the human P-glycoprotein (P-gp). However, introduction of the 3-fluoro (compound **17**) improved the potency approximately twofold but retained the P-gp efflux liability. In contrast, modulation of the pyridine pK_a, by fluorination at the 3-position, (pK_a reduced from 4.5 to 2.1) resulted in compound **18** with improved intrinsic potency and some mitigation of the P-gp liability. Ultimately, combining these two features resulted in compound **19**, with $IC_{50} = 0.52$ nM and subsequent selectivity over the μ- and κ-opioid receptors. In addition, these

Table 1 Modulation of the pKa and P-gp ratios of ORL1 antagonists through fluorine substitution

Compound	R^1	R^2	ORL1 binding[a] IC_{50} (nM)	Antagonism[a] IC_{50} (nM)	pK_a[b] N1	pK_a[b] N2	Log $D_{7.4}$[c]	Human P-gp transport ratio[d]
14	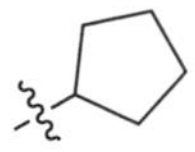	H	4.0	4.7	8.2	4.5	1.7	4.1
15	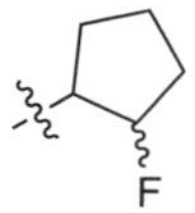	H	11	7.1	6.9	4.5	2.0	1.9
16	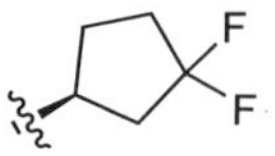	H	9.3	4.5	6.5	4.5	2.1	1.4
17[e]	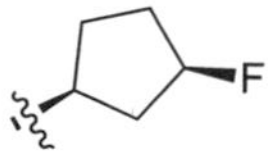	H	2.5	1.0	7.4	4.5	1.3	4.4

(Continued)

Table 1 (Continued)

Compound	R^1	R^2	ORL1 binding^a IC$_{50}$ (nM)	Antagonism^a IC$_{50}$ (nM)	pK$_a$^{b} N1	pK$_a$^{b} N2	Log D$_{7.4}$^{c}	Human P-gp transport ratio^d
18		F	1.1	1.2	8.2	2.1	2.1	2.3
19		F	0.52	0.31	7.4	2.1	1.8	1.7

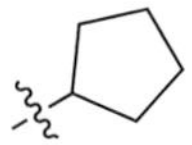

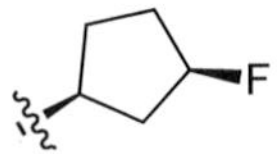

^{a} $n = 1$.
^{b} Calculated pK$_a$ values.
^{c} Measured with shake flask.
^{d} Transport ratio B–A/A–B.
^{e} More potent enantiomer.

transformations were also effective at reducing the hERG affinity and more importantly for a CNS target, eliminated the P-gp efflux liability. The below structure needs to be above table 1 because it has R substitutions.

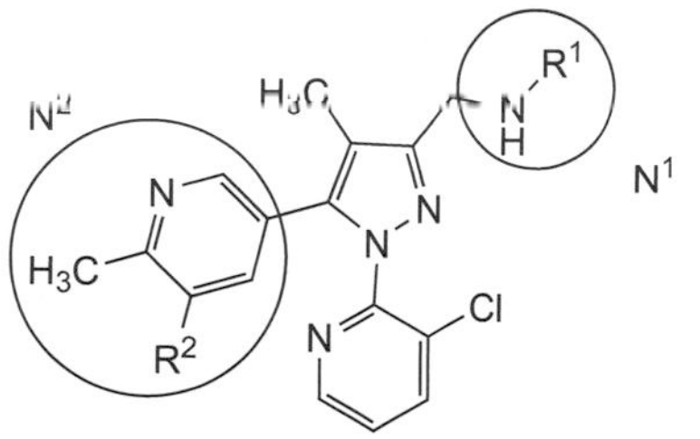

There are many examples of the introduction of fluorine enhancing on-target activity and these only represent a small number. Many more will be exemplified in the remaining sections of this chapter.

3. EFFECTS OF FLUORINE ON PHARMACOKINETIC PARAMETERS AND DRUG-LIKE PROPERTIES

The incorporation of fluorine into a molecule has been widely used to alter the pharmacokinetic properties and overall drug-like properties of compounds. This includes affecting the metabolism, oral absorption, and brain penetration of these molecules [18]. Metabolism can be affected by addition of fluorine directly at or adjacent to the site of metabolism. In addition, substitution with fluorine can increase the lipophilicity of compounds which has been shown to dramatically affect both oral absorption and brain penetration. Finally, the electron-withdrawing characteristic of fluorine has been exploited to lower the P-gp liability of compounds and modulate the pK_a of adjacent groups which resulted in increased brain exposure. In the following section, representative examples will highlight the powerful nature of fluorine to modulate overall drug-like properties.

3.1 Metabolic stability

In the $A\beta_{42}$ hypothesis of AD, $A\beta_{42}$ is produced by sequential cleavage of amyloid plaques by β- and γ-secretase-producing oligomers that are responsible for early cognitive impairment [19,20]. The γ-secretase inhibitor **20** was identified as an initial lead and subsequent optimization gave thiophene **21** with improved *in vitro* potency (Table 2) [21]. Compound **21** reduced Aβ levels in the brain of Tg2576 mice but was shown to have poor microsomal stability. The major metabolites resulted from oxidation of the side-chain methyl groups and subsequently transformed into their corresponding glucuronides. To block the site of oxidation, both methyl groups were substituted with trifluoromethyl groups. The

Table 2 Effects of fluorine on the properties of gamma-secretase inhibitors

Compound	$A\beta_{40}$ IC_{50} (nM)	Selectivity vs. notch inhibition	In vitro $T_{1/2}$ Tg2576 mice (min)	In vitro $T_{1/2}$ human (min)	% $A\beta_{40}$ reduction @ 4 h (po)	% $A\beta_{42}$ reduction @ 4 h (po)
20	5449	3.7	na	na	na	na
21	25	10	2	8	25 @ 100 mpk	25 @ 100 mpk
22	16	15	24	8	27 @ 5 mpk	22 @ 5 mpk
23	15	14	48	>90	37 @ 5 mpk	25 @ 5 mpk

na, not applicable.

resulting *bis*-trifluoromethyl analog **22** had increased microsomal stability along with improved potency and selectivity. The efficacious dose was lowered from 100 to 5 mpk. Further optimization led to a comparably potent and selective molecule **23**, which possessed enhanced microsomal stability. The γ-secretase inhibitor **23** was reported to have been selected for clinical evaluation in AD patients.

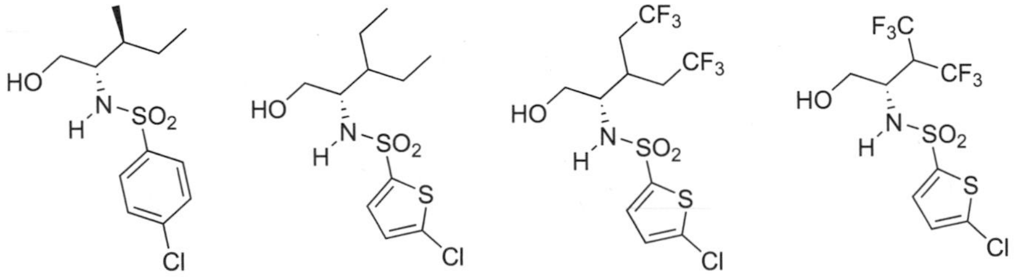

Incorporation of fluorine at a site adjacent to a "metabolic soft spot" has also been used as a strategy to increase duration of action. Linopirdine (**24**) was among the first clinical compounds that enhanced potassium-evoked release of acetylcholine in preclinical models of AD [22]. Linopirdine showed no clinical efficacy and its human pharmacokinetic profile was suggested as the reason for this lack of clinical efficacy. Specifically noted was the molecule's poor brain exposure and short half-life due to formation of the N-oxides **25** and **26** (Table 3) [23,24]. Optimization of **24** resulted in replacement of the indolone core by the anthracenone **27**, which had improved *in vitro* activity, but still exhibited a short duration of action. To improve the metabolic stability, fluorine

Table 3 Linopirdine (24) and fluorinated analogs

Compound	ACH release EC$_{50}$ (μM)	CLogP	Rat PO PK		
			$T_{1/2}$	F	B/P
24	4.5	2.97	0.5 h	30%	1/6
25	Inactive	—	—	—	—
26	Inactive	—	—	—	—
27	0.45	4.65	—	—	—
28	0.83	5.11	2.0 h	30%	1/1

was introduced at the 2-position of the pyridine rings (α- to the pyridyl nitrogen) lowering the pK_a and reducing metabolic N-oxidation. The resulting difluoro analog **28** (DMP543) had an extended duration of action and the overall increase in lipophilicity on going from **24** (clogP = 2.97) to **28** (clogP = 5.11) led to a significant improvement in brain-to-plasma ratio [25].

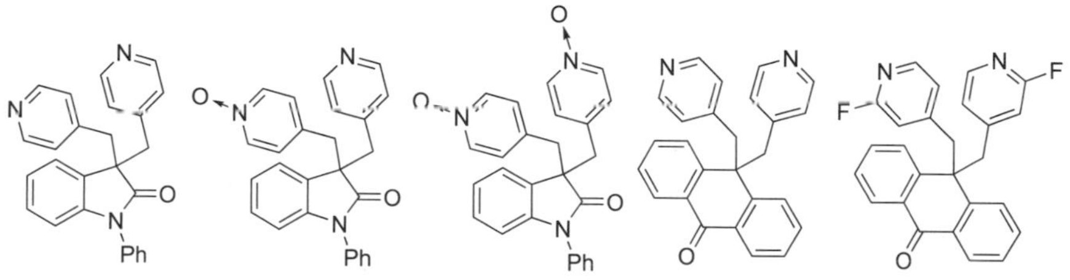

3.2 Oral exposure

Introduction of fluorine into 5-HT$_{1D}$ agonists and neurokinin 1 (NK$_1$) antagonists to improve oral exposure is highlighted in this section. Piperidine **29** and piperazine **30** were reported as 5-HT$_{1D}$ agonists with excellent selectivity over other serotonin subtypes (Table 4) [26]. Oral absorption of **29** and **30** was determined by portal vein bleed with piperazine **30** demonstrating significantly higher exposure than piperidine **29**. The lower overall exposure of **29** was reported to be due to poor absorption rather than metabolic instability. It was speculated that the more basic piperidine had a smaller proportion of uncharged species at the physiological pHs in the small intestine, thus reducing the passive diffusion across the gut wall [27,28]. Modulation of the pK_a of the piperidine was addressed by fluorine substitution at the 4-position (**31**, pK_a 8.8)

Table 4 Modulation of pKa and oral exposure of 5-HT agonists following fluorine substitution

Compound	5-HT$_{1D}$IC$_{50}$ (nM)	pK$_a$	[Drug] (ng/mL) in rat plasma	
			hpv[a]	systemic
29	0.35	9.7	25 ± 4	<2
30	0.20	8.2	178 ± 40	42 ± 24
31	1.0	8.8	570 ± 119	52 ± 21
32	0.95	8.7	781 ± 171	196 ± 60

[a] Concentration of compound in rat plasma obtained from hepatic portal vein (hpv).

and at the β-position of the side chain (**32**, pK$_a$ 8.7). The reduction in basicity resulted in higher oral exposure for both compounds **31** and **32**.

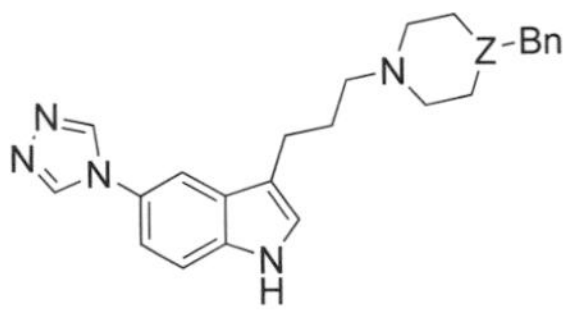

29 Z = CH
30 Z = N   **31** 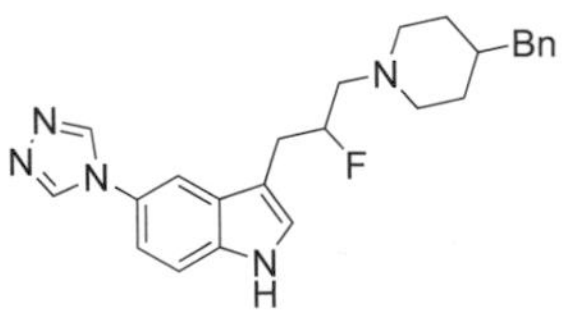   **32**

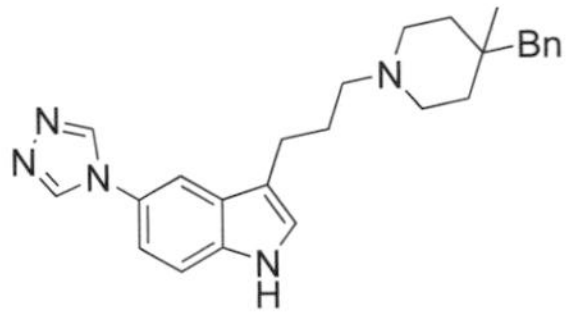

NK$_1$ antagonists have been extensively studied as treatments for diseases such as depression, anxiety, and emesis [29]. The early NK$_1$ antagonist CP-96345 (**33**, Table 5). There are no structures 33, 34, 35 and 36 in the document demonstrated excellent NK$_1$ affinity but was reported to have poor oral bioavailability in rodents. The 2-methoxybenzylamine group was identified as a potential metabolic liability within the molecule. A number of methoxybenzylamine replacements were examined leading to the identification of the 3,5-bis(trifluoromethyl)benzylether **34**. This analog was shown to have excellent affinity and improved oral bioavailability [30,31]. Compounds **33** and **34**, however, also had significant L-type calcium-channel inhibition and potential cardiovascular side effects. Replacement of the quinuclidine by a morpholine bearing an electron-withdrawing triazolinone gave **35**, which had excellent selectivity over calcium-channel activity [32,33]. Although compound **35** was efficacious in models of emesis (ID$_{50}$ = 2.3 mpk at 24 h), further efforts to improve metabolic stability and duration of action were undertaken. Addition of fluorine at the 4-position of the phenyl led to the identification of the metabolically more stable analog **36** (aprepitant), which demonstrated improvements in efficacy in both the emesis model (ID$_{50}$ = 1.8 mpk at 24 h) and the

Table 5 Fluorinated NK1 antagonists

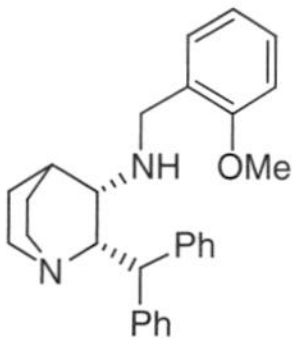

Compound	hNK$_1$ IC$_{50}$ (nM)	SYVAL^a ID$_{50}$ (mpk @ 24 h)	GR73632-induced foot tapping ID$_{50}$ (mpk @ 24 h)
33	0.6	—	—
34	2	—	—
35	0.09	2.3	1.11
36	0.09	1.8	0.33

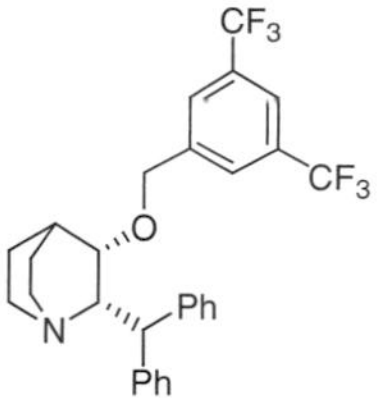

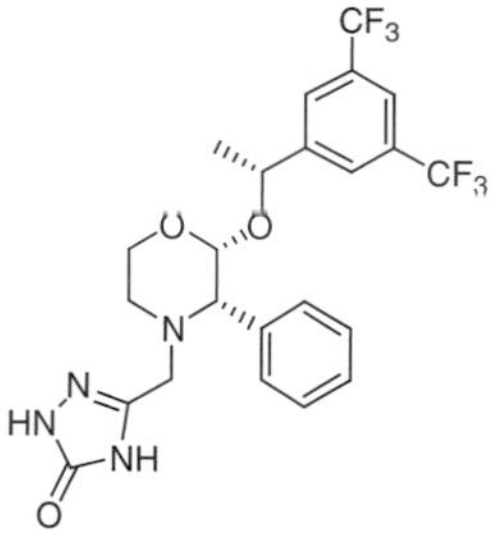

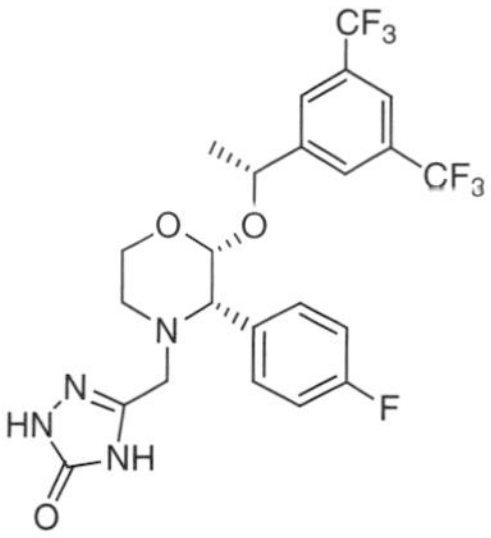

^{a} Resiniferatoxin-induced systemic vascular leakage (SYVAL).

NK1 agonist-mediated gerbil foot tapping model (ID$_{50}$ = 0.33 mpk) [34]. Aprepitant (**36**), which contains seven fluorine atoms, was subsequently approved by the Food and Drug Administration (FDA) for the prevention of acute and delayed chemotherapy-induced nausea and vomiting.

3.3 Brain exposure

The electron-withdrawing characteristics of fluorine have been successfully exploited to lower the P-gp liability within a series of bradykinin (BK) B$_1$ receptor antagonists leading to improved brain exposure. The BK B$_1$ receptor is not widely expressed peripherally in the non-diseased state, but is induced upon injury and plays a role in persistent pain and inflammation. In addition, recent studies have suggested that the receptor is also constitutively expressed in the CNS, thus a brain-penetrant BK B$_1$ receptor antagonist might have superior efficacy than a peripherally restricted agent [35]. The biaryl **37** was identified as an initial lead having moderate binding affinity for hBK B$_1$ and promising pharmacokinetic properties [36]. However, **37** proved to be a substrate for P-gp-mediated efflux and as a result would be predicted to have poor brain exposure in humans (Table 6) [37].

Table 6 Pharmacokinetic properties of fluorinated BK B1 receptor antagonists

Compound	R	hBK B1	P-gp[a]	Papp[b]	log P	Rat $T_{1/2}$ (h)	Cl (mL/ min/kg)	F (%)
37	—	63	16	18	—	9.5	9.3	25
38a	CH_2CF_3	0.81	18.4	23	2.9	4.4	6.3	79
38b	CF_2CF_3	2.95	2.2	31	—	—	—	—
38c	CF_3	1.47	4.1	23	2.7	—	—	—
39	—	0.4	1.9	34	—	0.4	40	34

[a] MDR1-directional transport ratio (B/A)/(A/B).
[b] Passive permeability (10^{-6} cm/s).

Optimization strategies of the biaryl region incorporated fluorine in both aryl rings resulting in compound **38a**, which showed a significant boost in BK B_1 potency and oral bioavailability in rat, but little change in P-gp efflux. In an attempt to reduce the P-gp liability of the series, replacement of the trifluoropropionamide group was investigated. Pentafluoropropionamide **38b** was threefold less active against hBK B_1 but had dramatically reduced efflux potential (P-gp ratio = 2.2). Replacement of the difluoromethylene for a trifluoroacetamide provided **38c** with improved binding potency, albeit with a slight gain in P-gp efflux potential (P-gp ratio = 4.1). Overall the trifluoromethyl analog **38c** exhibited a good balance between hBK B_1 potency and reduced P-gp liability. The electron-withdrawing trifluoromethyl group proved essential for lowering this P-gp liability. It was proposed that the more electron-deficient amide had less hydrogen-bonding capability, contributing to decreased recognition by P-gp [38,39]. Further optimization led to the identification of compound **39** which demonstrated minimal P-gp efflux and adequate oral efficacy in complete Freund's adjuvant (CFA) induced hyperalgesia in a humanized mouse. Compound **39** has been reported to be a clinical development candidate for the treatment of pain and inflammation.

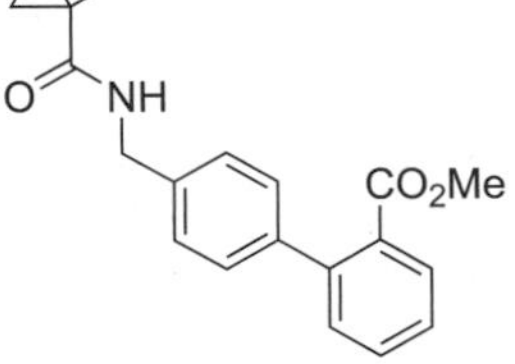

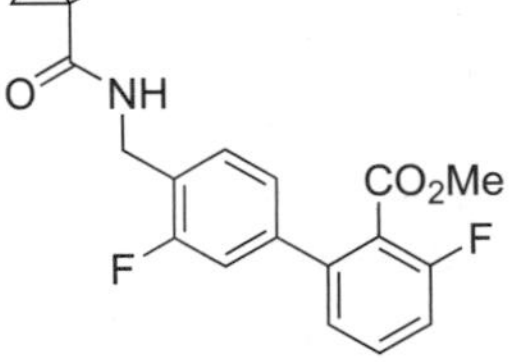

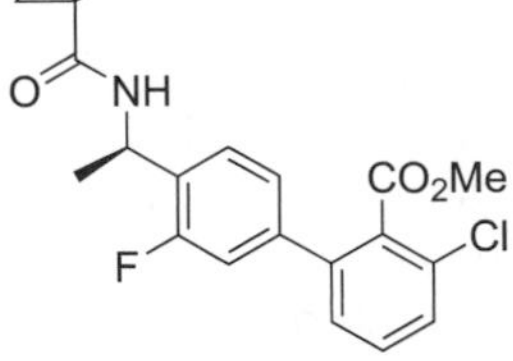

Fluorine has been used to modulate the basicity of amines which may lead to an improvement in brain exposure. Recently, the discovery of a series of α4β2 nicotinic acetylcholine receptor (nAChR) potentiators as possible treatment for Parkinson's disease and schizophrenia was were disclosed [40]. Optimization of isoxazole **40** included the bioisosteric replacement of the central amide by an imidazole ring. Introduction of a fluorine at the 6-position of the phenyl ring provided compound **41**. This compound had excellent potency but was determined to be a substrate for P-gp (efflux ratio >10). In an attempt to reduce amine basicity and decrease the efflux propensity, the 4-fluoropiperidine **42** was identified which retained potency and had significantly reduced P-gp efflux liability (efflux ratio ~1). CNS penetration of **42** was observed in rodents following intraperitoneal (IP) treatment at 5 mg/kg and showed a brain concentration of 6.5 μM.

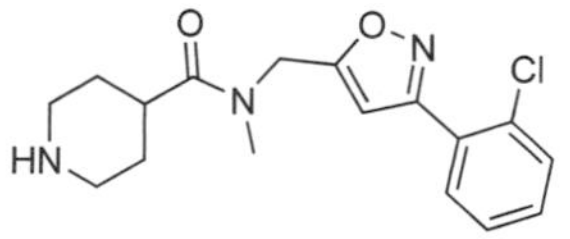

40 EC$_{50}$ = 350 nM

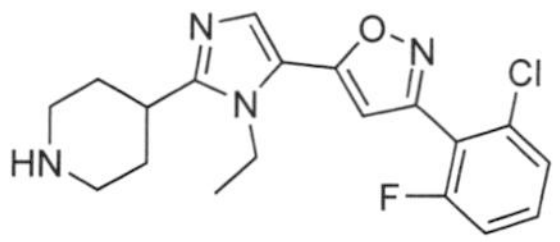

41 EC$_{50}$ = 3 nM
MDR1-LLC-PK1 efflux ratio >10

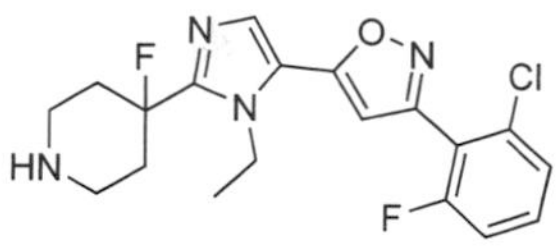

42 EC$_{50}$ = 3 nM
MDR1-LLC-PK1 efflux ratio ~1

3.4 Solubility

A simple fluorine substitution for hydrogen in an aromatic system generally increases lipophilicity and lowers aqueous solubility. The powerful electron-withdrawing effect of fluorine, however, can strongly affect the polarity of nearby functional groups, modulating properties including pK_a and hydrogen-bonding capabilities [41]. Early leads in a calcitonin gene-related peptide (CGRP) antagonist program were plagued by poor drug-like properties including high molecular weight and low solubility [42]. The 7-methylindazole **43** was a potent hCGRP antagonist but the solubility (**43**, 15 mg/mL @ pH 5) was insufficient to support intranasal dosing. Introduction of a fluorine at C-8 of the quinazolinone had a dramatic effect on increasing solubility (**44**, >500 mg/mL @ pH 6.8) while retaining potency. The authors postulated that "it is possible that the fluorine serves to polarize the NH bond of the adjacent urea, rendering it a more potent hydrogen bond donor and thereby enhancing solvation by water." Indeed, **44** exhibited a favorable intranasal pharmacokinetic profile in rabbits (F_{IN} = 59% ± 22%) and high plasma levels within 10 min, suggesting the potential for rapid migraine relief.

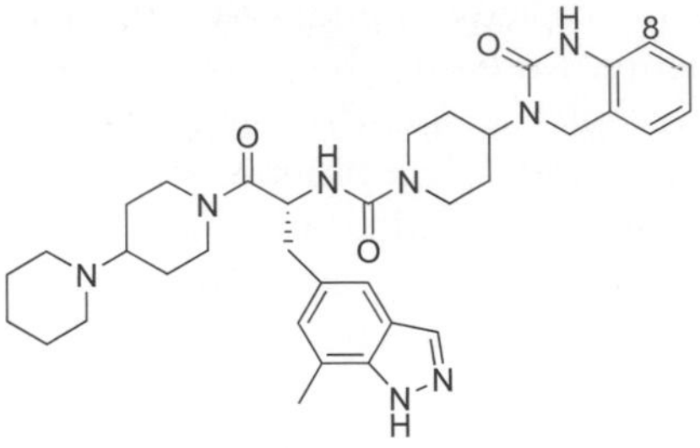

43 hCGRP = 0.01 nM
MW = 626, CLogP = 2.6
Solubility @ pH 5 = 15 mg/mL

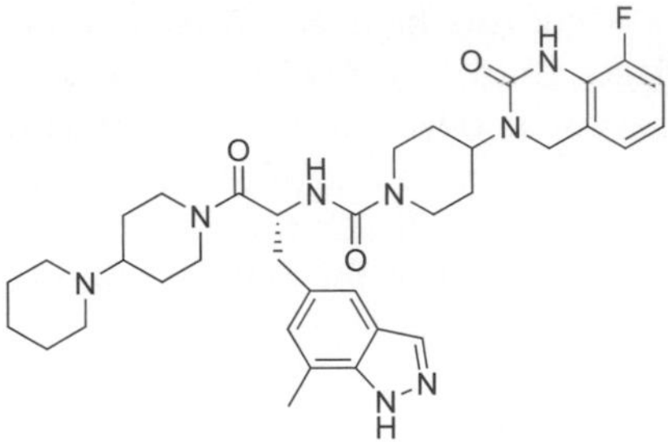

44 hCGRP = 0.013 nM
MW = 644, CLogP = 3.1
Solubility @ pH 6.8 >500 mg/mL

4. TOXICITY

4.1 hERG affinity

Evidence has linked T-type Ca^{2+} channels to a number of CNS diseases including epilepsy and movement disorders. Piperidine **45** was identified as a molecule with good potency against T-type Ca^{2+} channels and reasonable selectivity against hERG and L-type channels (Table 7) [43,44]. Strategies to improve selectivity against hERG included reduction in the basicity of the piperidine nitrogen [45] through addition of electron-with-drawing groups at the β-positions of both the piperidine and the N-alkyl chain. Although introduction of fluorine at the β-position of the alkyl chain resulted in a loss in potency for the T-type Ca^{2+} channel (**46**, EC_{50} = 169 nM), there was also a threefold reduction in hERG affinity following the modulation of the piperidine basicity. Relocation of the fluorine to the 3-position of the piperidine gave the *cis*-isomer **47** (pK_a = 7.9) and racemic *trans*-**48** (pK_a = 6.7). In the fluorescent imaging plate reader (FLIPR) assay, *cis*-**47** was twofold more potent than the starting piperidine **45**, while the *trans*-isomer **48** was less active. The more potent isomer, **47**, was also more

Table 7 Modulation of hERG and L-type calcium channel activity following fluorine substitution

Compound	T-type FLIPR IP (nM)	hERG IC_{50} (nM)	L-type IC_{50} (nM)	pK_a
45	61	1930	1190	8.7
46	169	4731	1526	nd
47	32	4114	2134	7.9
48	411	3034	1049	6.7

nd, not determined.

selective on the hERG and L-type channels and gave no adverse effects in the dog QT experiments. Overall, **47** displayed good pharmacokinetics in multiple species and efficacy in several models of movement disorder.

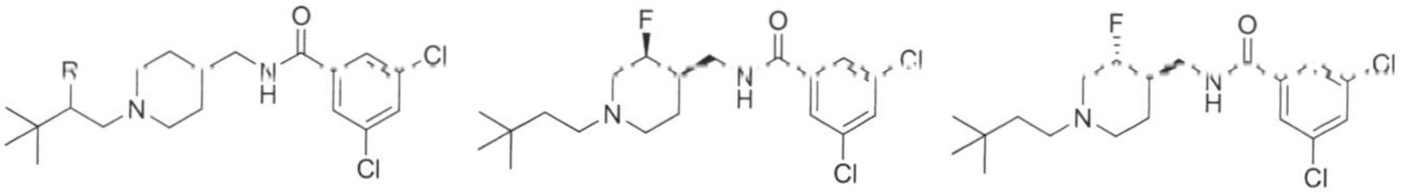

45 R = H
46 R = F   **47** (single enantiomer)   **48** (racemic)

4.2 Protein binding (covalent)

The involvement of the cannabinoid (CB) receptor in regulating feeding behavior has been demonstrated in both animal and clinical studies. Recently, researchers have described an acyclic series of CB1R inverse agonists that led to the discovery of **52** (Table 8) [46]. The initial lead **49** had poor oral bioavailability and, when incubated with human liver microsomes, displayed irreversible toxic binding [47]. The electron-rich phenoxy ring was identified as a site for oxidative metabolism which resulted in reactive metabolite formation. The phenyl ring was replaced with the more electron-deficient 2-pyridyl ring and compound **50** was found to have reduced levels of covalent binding and also higher oral bioavailability in rodents. The addition of the electron-withdrawing trifluoromethyl group at the 5-position of the pyridine gave **51** and not only improved potency but further reduced covalent binding. The residual covalent binding in **51** was thought to result from the bioactivation of the unsubstituted phenyl ring. Substitution of this ring with an electron-withdrawing cyano substituent afforded **52** with minimal covalent protein binding (27 pmol equiv/mg protein). Clinical investigation of taranabant (**52**) was discontinued following the finding that efficacy and side effects (psychiatric adverse events) were dose limiting.

Table 8   Reduction in covalent binding of CB1 inverse agonists

Compound	CB1 IC$_{50}$ (nM)	Covalent binding pmol equiv/mg protein @ 1 h incubation	% Oral bioavailability (rat)
49	2.0	3900	9
50	1.8	910	29
51	0.5	88	100
52	0.3	27	74

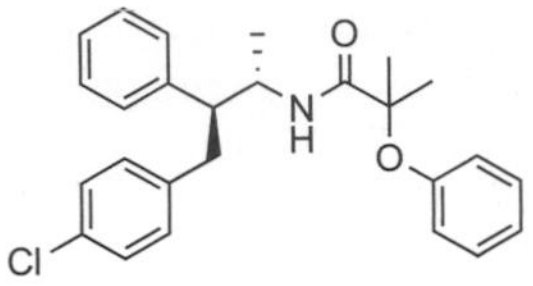

49 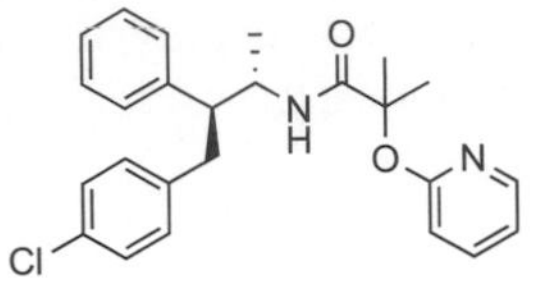   50 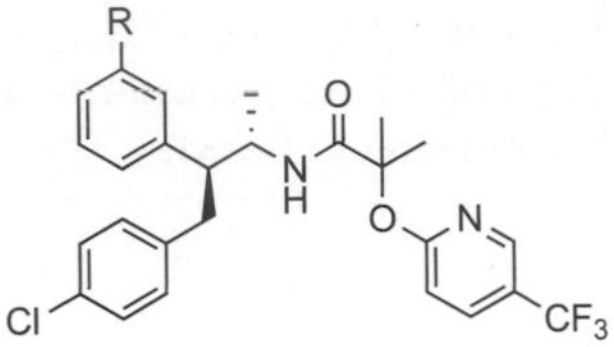   51 R=H
                       52 R=CN

5. FLUORINE AS A BIOISOSTERE

The use of fluorine to modulate properties including potency, selectivity, pharmacokinetics, and toxicity has have been highlighted. Fluorine has also been suggested as a potential bioisosteric replacement for a number of functional groups, examples of which are presented in the final section.

Gamma-aminobutryic acid (GABA) (53) is the major inhibitory neurotransmitter in the CNS. When GABA levels in the brain fall below a threshold level, convulsions begin. Since GABA does not cross the blood–brain barrier, it is not an effective oral anticonvulsant and thus lipophilic bioisosteres of the GABA carboxylic acid have been sought. As the pK_a of phenol drops from 9.8 to 7.1 following 2,6-difluoro substitution, a series of di-fluoro-substituted phenols, for example, 54 and 55 were prepared as potential inhibitors of GABA aminotransferase [48]. In this pK_a range, the 2,6-difluorophenols 54 and 55 would be completely ionized and capable of mimicking a carboxylate ion. Both 54 and 55 were found to be competitive inhibitors of GABA aminotransferase (K_i values of 6.3 and 11 mM, respectively); under the same conditions, the K_m for GABA is 2.5 mM. These results suggest that the 2,6-difluorophenol moiety can act as a bioisosteric replacement for a carboxylate anion.

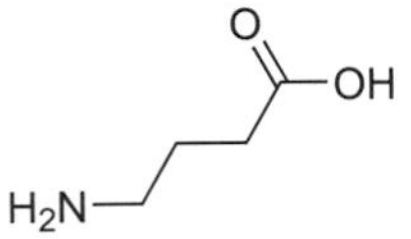

53 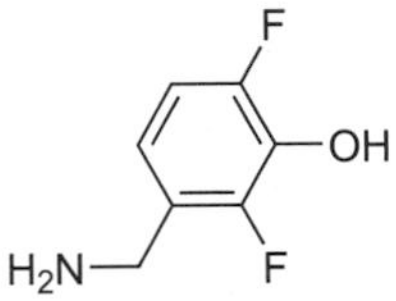   54 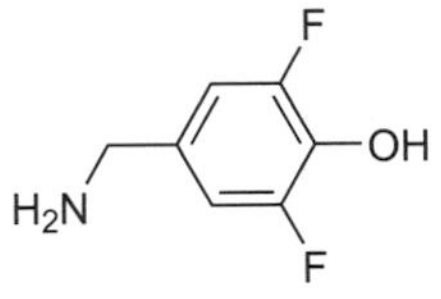   55

During the development of a series of GABA$_A$ $\alpha2/\alpha3$-selective pyridazine-based benzodiazepine site agonists, it was discovered that a ring nitrogen could be successfully replaced with a fluoro substituent; the alternative substitution was also true that a fluorine atom could be replaced by a ring nitrogen [49]. For instance, replacement of the 3-fluoropyridine of 56 by 2,6-difluorophenyl afforded agonist 57 with similar binding affinities at $\alpha1$, $\alpha3$, and $\alpha5$ (Table 9). Since 56 and 57

Table 9 Efficacy and rat pharmacokinetic data for fluorinated GABA$_A$ agonists

Compound	Ki (nM)			Efficacy				Rat PK		
	α1	α3	α5	α1	α2	α3	α5	Cl (ml/min/ kg)	$T_{1/2}$ (h)	F(%)
56	0.5	1.4	2.4	–30%	—	0.4	—	—	—	—
57	0.1	0.4	0.3	–5%	—	0.24	—	—	—	—
58	1.8	6.3	6.0	–3%	0.28	0.39	–5%	59	1.6	50
59	1.5	8.5	12.1	–7%	0.12	0.44	0.01	1.3	53	88

maintain a similar *in vitro* profile, this would suggest that a fluorine atom may be a possible replacement for a pyridine nitrogen and a possible bioisosteric replacement. Additionally, the 3,5-difluoropyridine analog **58** had high clearance and a short half-life in rat that was traced to a propensity for pyridine *N*-oxide formation. Switching of the pyridine nitrogen and an *ortho*-fluoro gave equipotent agonist **59**, which circumvented metabolism and accounted for the exceptional stability and long half-life in rat.

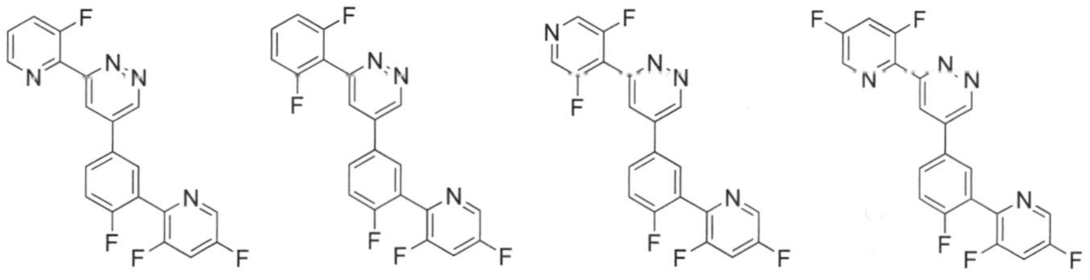

56   57   58   59

A series of imidazopyridines and imidazopyrimidines (e.g., **60–62**) were also pursued as modulators of the GABA$_A$ receptor, with the goal of minimizing the functional effects at GABA$_A$ α1 while optimizing agonist activity at GABA$_A$ α3 [50]. Introduction of a nitrogen atom into the ring gave the imidazopyrimidine **61** and a 10-fold increase in both GABA$_A$ α1 and α3 affinities (Table 10). Unfortunately, the increased affinity also resulted in undesirable enhancement of functional activity at the α1 subtype. As in the previous example, the ring nitrogen was replaced with a fluoro substituent which gave the fluoro analog **62** and a substantial increase in binding affinity for both α1 and α3. The functional *in vitro* effects of compound **62** were found to be similar to compound **60** for GABA$_A$ α1, but **62** was a far more efficacious functional agonist of GABA$_A$ α3 than both **60** and **61** with a maximal response of 104% of chlorodiazepoxide (CDZ).

Table 10 Effects of fluorine substitution on GABA$_A$ binding and efficacy

Compound	K$_i$ (nM)		Efficacy (%CDZ)	
	α1	α3	α1	α3
60	4.35	5.16	35	61
61	0.71	0.47	60	79
62	0.20	0.32	34	104

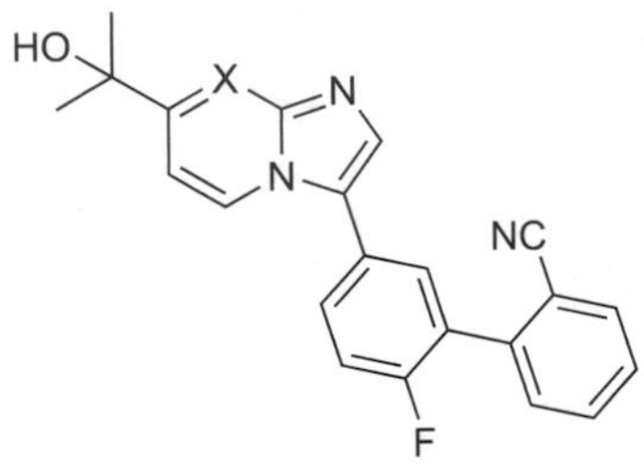

60 X = CH
61 X = N
62 X = CF

6. CONCLUSION

In summary, the addition of fluorine can have dramatic effects on the drug-like properties of compounds. The presence of this one atom can influence metabolic stability, brain exposure, off-target selectivity, protein binding, and affinity for the primary biological target. Not all effects of fluorine addition are of course positive. For example, the introduction of fluorine to increase lipophilicity and brain penetration of a molecule may in turn reduce solubility and lower oral absorption. Therefore, striking the correct balance of drug-like properties will remain a recurring challenge in lead optimization but strategies utilizing fluorine's impact on these properties will continue to influence the success of many drug discovery programs. To date, there are over 100 approved drugs that have this atom incorporated into their structure.

REFERENCES

[1] P. Shah and A. D. Westwell, *J. Enzyme Inhib. Med. Chem.*, 2007, **22**, 527.
[2] H. -J. Bohm, D. Banner, S. Bendels, M. Kansy, B. Kuhn, K. Muller, U. Obst-Sander, and M. Stahl, *ChemBioChem*, 2004, **5**, 637.
[3] K. L. Kirk, *Curr. Top. Med. Chem.*, 2006, **6**, 1447.
[4] R. Freedman, *N. Engl. J. Med.*, 2003, **349**, 1738.

[5] S. Taylor and M. B. Stein, *Med. Hypotheses*, 2005, **66**, 14.
[6] M. Vaswani, F. K. Linda, and S. Ramesh, *Prog. Neuro-Psychopharmacol. Biol. Psychiatry*, 2003, **27**, 85.
[7] F. Dolle, C. Perrio, L. Barre, M. -C. Lasne, and D. Le Bars, *Actual. Chim.*, 2006, **301–302**, 93.
[8] D. Le Bars, *J. Fluorine Chem.*, 2006, **127**, 1488.
[9] D. G. Reid and P. S. Murphy, *Drug Discov. Today*, 2008, **13**, 473.
[10] D. G. Reid and P. S. Murphy, *Drug Dev. Res.*, 2008, **69**, 279.
[11] W. K. Hagmann, *J. Med. Chem.*, 2008, **51**, 4359.
[12] S. Purser, P. R. Moore, S. Swallow, and V. Gouverneur, *Chem. Soc. Rev.*, 2008, **37**, 320.
[13] K. Isensee, M. Amon, A. Galaparti, X. Ligneau, J. C. Camelin, M. Capet, J. C. Schwartz, and H. Stark, *Bioorg. Med. Chem. Lett.*, 2009, **19**, 2172.
[14] N. D. P Cosford, L. Tehrani, J. Roppe, E. Schweiger, N. D. Smith, J. Anderson, L. Bristow, J. Brodkin, X. Jiang, I. McDonald, S. Rao, M. Washburn, and M. A. Varney, *J. Med. Chem.*, 2003, **46**, 204.
[15] L. R. Tehrani, N. D. Smith, D. Huang, S. F. Poon, J. R. Roppe, T. J. Seiders, D. F. Chapman, J. Chung, M. Cramer, and N. D. P. Cosford, *Bioorg. Med. Chem. Lett.*, 2005, **15**, 5061.
[16] P. Beswick, N. Charrier, B. Clarke, E. Demont, C. Dingwall, R. Dunsdon, A. Faller, R. Gleave, J. Hawkins, I. Hussain, C. N. Johnson, D. MacPherson, G. Maile, R. Matico, P. Milner, J. Mosley, A. Naylor, A. O'Brien, S. Redshaw, D. Riddell, P. Rowland, J. Skidmore, V. Soleil, K. J. Smith, S. Stanway, G. Stemp, A. Stuart, S. Sweitzer, P. Theobald, D. Vesey, D. S. Walter, J. Ward, and G. Wayne, *Bioorg. Med. Chem. Lett.*, 2008, **18**, 1022.
[17] K. Kobayashi, M. Uchiyama, H. Ito, H. Takahashi, T. Yoshizumi, H. Sakoh, Y. Nagatomi, M. Asai, H. Miyazoe, T. Tsujita, M. Hirayama, S. Ozaki, T. Tani, Y. Ishii, H. Ohta, and O. Okamoto, *Bioorg. Med. Chem. Lett.*, 2009, **19**, 3627.
[18] B. K. Park and N. R. Kitteringham, *Drug Metab. Rev.*, 1994, **26**, 605.
[19] M. A. Findeis, *Pharmacol. Ther.*, 2007, **116**, 266.
[20] D. M. Walsh and D. J. Selkoe, *J. Neurochem.*, 2007, **101**, 1172.
[21] S. C. Mayer, A. F. Kreft, B. Harrison, M. Abou-Gharbia, M. Antane, S. Aschmies, K. Atchison, M. Chlenov, D. C. Cole, T. Comery, G. Diamantidis, J. Ellingboe, K. Fan, R. Galante, C. Gonzales, D. M. Ho, M. E. Hoke, Y. Hu, D. Huryn, U. Jain, M. Jin, K. Kremer, D. Kubrak, M. Lin, P. Lu, R. Magolda, R. Martone, W. Moore, A. Oganesian, M. N. Pangalos, A. Porte, P. Reinhart, L. Resnick, D. R. Riddell, J. Sonnenberg-Reines, J. R. Stock, S. -C. Sun, E. Wagner, T. Wang, K. Woller, Z. Xu, M. M. Zaleska, J. Zeldis, M. Zhang, H. Zhou, and J. S. Jacobsen, *J. Med. Chem.*, 2008, **51**, 7348.
[22] L. Cook, V. J. Nickolson, G. F. Steinfels, K. W. Rohrbach, and V. J. DeNoble, *Drug Dev. Res.*, 1990, **19**, 301.
[23] H. J. Pieniaszek, Jr., W. D. Fiske, T. D. Saxton, Y. S. Kim, D. M. Garner, M. Xilinas, and R. Martz, *J. Clin. Pharmacol.*, 1995, **35**, 22.
[24] D. C. Rakestraw, D. A. Bilski, and G. N. Lam, *J. Pharm. Biomed. Anal.*, 1994, **12**, 1055.
[25] R. A. Earl, R. Zaczek, C. A. Teleha, B. N. Fisher, C. M. Maciag, M. E. Marynowski, A. R. Logue, S. W. Tam, W. J. Tinker, S. -M. Huang, and R. J. Chorvat, *J. Med. Chem.*, 1998, **41**, 4615.
[26] J. L. Castro, I. Collins, M. G. N. Russell, A. P. Watt, B. Sohal, D. Rathbone, M. S. Beer, and J. A. Stanton, *J. Med. Chem.*, 1998, **41**, 2667.
[27] D. A. Smith, B. C. Jones, and D. K. Walker, *Med. Res. Rev.*, 1996, **16**, 243.
[28] O. H. Chan and B. H. Stewart, *Drug Discov. Today*, 1996, **1**, 461.
[29] L. Quartara and M. Altamura, *Curr. Drug Targets*, 2006, **7**, 975.
[30] C. Swain and N. M. J. Rupniak, *Annu. Rep. Med. Chem.*, 1999, **34**, 51.
[31] J. M. Humphrey, *Curr. Top. Med. Chem. (Sharjah, United Arab Emirates)*, 2003, **3**, 1423.
[32] C. J. Swain, B. J. Williams, R. Baker, M. A. Cascieri, G. Chicchi, M. Forrest, R. Herbert, L. Keown, T. Ladduwahetty, S. Luell, D. E. Macintyre, J. Metzger, S. Morton, A. P. Owens, S. Sadowski, and A. P. Watt, *Bioorg. Med. Chem. Lett.*, 1997, **7**, 2959.

[33] J. J. Hale, S. G. Mills, M. MacCoss, P. E. Finke, M. A. Cascieri, S. Sadowski, E. Ber, G. G. Chicchi, M. Kurtz, J. Metzger, G. Eiermann, N. N. Tsou, F. D. Tattersall, N. M. J. Rupniak, A. R. Williams, W. Rycroft, R. Hargreaves, and D. E. MacIntyre, *J. Med. Chem.*, 1998, **41**, 4607.

[34] N. M. J. Rupniak, F. D. Tattersall, A. R. Williams, W. Rycroft, E. J. Carlson, M. A. Cascieri, S. Sadowski, E. Ber, J. J. Hale, S. G. Mills, M. MacCoss, E. Seward, I. Huscroft, S. Owen, C. J. Swain, R. G. Hill, and R. J. Hargreaves, *Eur. J. Pharmacol.*, 1997, **326**, 201.

[35] M. G. Bock, J. F. Hess, and D. J. Pettibone, *Annu. Rep. Med. Chem.*, 2003, **38**, 111.

[36] S. D. Kuduk, C. N. Di Marco, R. K. Chang, M. R. Wood, K. M. Schirripa, J. J. Kim, J. M. C. Wai, R. M. DiPardo, K. L. Murphy, R. W. Ransom, C. M. Harrell, D. R. Reiss, M. A. Holahan, J. Cook, J. F. Hess, N. Sain, M. O. Urban, C. Tang, T. Prueksaritanont, D. J. Pettibone, and M. G. Bock, *J. Med. Chem.*, 2007, **50**, 272.

[37] M. Mahar Doan Kelly, E. Humphreys Joan, O. Webster Lindsey, A. Wring Stephen, J. Shampine Larry, J. Serabjit-Singh Cosette, K. Adkison Kimberly, and W. Polli Joseph, *J Pharmacol Exp Ther*, 2002, **303**, 1029.

[38] A. Seelig, *Eur. J. Biochem.*, 1998, **251**, 252.

[39] A. Seelig and E. Landwojtowicz, *Eur. J. Pharm. Sci.*, 2000, **12**, 31.

[40] B. K. Albrecht, V. Berry, A. A. Boezio, L. Cao, K. Clarkin, W. Guo, J. -C. Harmange, M. Hierl, L. Huang, B. Janosky, J. Knop, A. Malmberg, J. S. McDermott, H. Q. Nguyen, S. K. Springer, D. Waldon, K. Woodin, and S. I. McDonough, *Bioorg. Med. Chem. Lett.*, 2008, **18**, 5209.

[41] A. Avdeef, *Curr. Top. Med. Chem. (Hilversum, The Netherlands.)*, 2001, **1**, 277.

[42] A. P. Degnan, P. V. Chaturvedula, C. M. Conway, D. A. Cook, C. D. Davis, R. Denton, X. Han, R. Macci, N. R. Mathias, P. Moench, S. S. Pin, S. X. Ren, R. Schartman, L. J. Signor, G. Thalody, K. A. Widmann, C. Xu, J. E. Macor, and G. M. Dubowchik, *J. Med. Chem.*, 2008, **51**, 4858.

[43] W. D. Shipe, J. C. Barrow, Z.-Q. Yang, C. W. Lindsley, F. V. Yang, K.-A.S. Schlegel, Y. Shu, K. E. Rittle, M. G. Bock, G. D. Hartman, C. Tang, J. E. Ballard, Y. Kuo, E. D. Adarayan, T. Prueksaritanont, M. M. Zrada, V. N. Uebele, C. E. Nuss, T. M. Connolly, S. M. Doran, S. V. Fox, R. L. Kraus, M. J. Marino, V. K. Graufelds, H. M. Vargas, P. B. Bunting, M. Hasbun-Manning, R. M. Evans, K. S. Koblan, and J. J. Renger, *J. Med. Chem.*, 2008, **51**, 3692.

[44] Z. -Q. Yang, J. C. Barrow, W. D. Shipe, K.-A.S. Schlegel, Y. Shu, F. V. Yang, C. W. Lindsley, K. E. Rittle, M. G. Bock, G. D. Hartman, V. N. Uebele, C. E. Nuss, S. V. Fox, R. L. Kraus, S. M. Doran, T. M. Connolly, C. Tang, J. E. Ballard, Y. Kuo, E. D. Adarayan, T. Prueksaritanont, M. M. Zrada, M. J. Marino, V. K. Graufelds, A. G. DiLella, I. J. Reynolds, H. M. Vargas, P. B. Bunting, R. F. Woltmann, M. M. Magee, K. S. Koblan, and J. J. Renger, *J. Med. Chem.*, 2008, **51**, 6471.

[45] J. S. Mitcheson, J. Chen, M. Lin, C. Culberson, and M. C. Sanguinetti, *Proc. Natl. Acad. Sci. USA*, 2000, **97**, 12329.

[46] L. S. Lin, T. J. Lanza, Jr., J. P. Jewell, P. Liu, S. K. Shah, H. Qi, X. Tong, J. Wang, S. S. Xu, T. M. Fong, C.-P. Shen, J. Lao, J. C. Xiao, L. P. Shearman, D. S. Stribling, K. Rosko, A. Strack, D. J. Marsh, Y. Feng, S. Kumar, K. Samuel, W. Yin, L. H. T Van der Ploeg, M. T. Goulet, and W. K. Hagmann, *J. Med. Chem.*, 2006, **49**, 7584.

[47] K. Samuel, W. Yin, R. A. Stearns, Y. S. Tang, A. G. Chaudhary, J. P. Jewell, T. Lanza, Jr., L. S. Lin, W. K. Hagmann, D. C. Evans, and S. Kumar, *J. Mass Spectrom.*, 2003, **38**, 211.

[48] J. Qiu, S. H. Stevenson, M. J. O'Beirne, and R. B. Silverman, *J. Med. Chem.*, 1999, **42**, 329.

[49] R. T. Lewis, W. P. Blackaby, T. Blackburn, A. S. R. Jennings, A. Pike, R. A. Wilson, D. J. Hallett, S. M. Cook, P. Ferris, G. R. Marshall, D. S. Reynolds, W. F. A. Sheppard, A. J. Smith, B. Sohal, J. Stanley, S. J. Tye, K. A. Wafford, and J. R. Atack, *J. Med. Chem.*, 2006, **49**, 2600.

[50] A. C. Humphries, E. Gancia, M. T. Gilligan, S. Goodacre, D. Hallett, K. J. Merchant, and S. R. Thomas, *Bioorg. Med. Chem. Lett.*, 2006, **16**, 1518.

CHAPTER 27

Patents in Drug Discovery: Case Studies, Examples, and Simple Steps Medicinal Chemists Can Take to Protect Hard-Won Intellectual Property

Charles J. Andres and **Richard L. Treanor**

Contents			

Oblon Spivak, 1940 Duke Street, Alexandria, VA 22314, USA

Annual Reports in Medicinal Chemistry, Volume 45
ISSN 0065-7743, DOI 10.1016/S0065-7743(10)45027-7

1. INTRODUCTION

Patents are of paramount importance to the pharmaceutical industry. At the discovery program level, chemotype patentability is one of the key requirements for continued work on a particular structural class. Decisions by venture capitalists to fund startup companies are based, in part, on the strength of their patent portfolios. The presence or absence of a single key patent can determine the future of even the largest pharmaceutical company. Patents thus are a critical, inseparable component of the drug discovery process.

Most medicinal chemists have a basic understanding of patents. This chapter assumes and builds upon this basic knowledge by analyzing court cases where patent owners took suspected infringers to court. It highlights a variety of defenses employed by the suspected infringers to operate outside the scope of, invalidate, or render unenforceable, the asserted patent(s).

A key theme emerging from this chapter is that actions taken during the patenting process can dramatically affect a patent's ability to survive when it is challenged in a Federal Court, the Patent Office, or before the U.S. International Trade Commission.

Another common theme is that there is really no such thing as an "ironclad" patent, and that an issued patent is subject to attack and can sometimes be defeated by a variety of mechanisms.

As noted above, in several of the case studies that follow, a patent owner has sued a suspected infringer. A patent's claims act like a property deed, setting the boundaries that separate and define what property a patent holder "owns." One important difference between real property (land) and intellectual property, however, is that a patent only defines that territory from which the patent owner can *exclude* others—*not* territory the patent owner can "occupy" with impunity. Regardless, when someone has "trespassed" (infringed) upon the patent owner's property (the claims), they can be legally forced to stop infringing and possibly be made to pay for the damage caused by infringing. These infringement actions can teach valuable lessons for going forward with patent applications.

It is important to note that this chapter, although intended to be informative, does not constitute legal advice. The authors strongly recommend that discovery chemists integrate a patent professional into every discovery team and consult with their patent professional early in the process and on a regular basis.

2. ADEQUATELY SUPPORT YOUR PATENT APPLICATION CLAIMS

2.1 In re '318 Patent Infringement Litigation (galantamine)

Case History: In an attempt to protect their galantamine franchise from generic competition, Janssen Pharmaceutica N.V., Janssen L.P., and Synaptech, Inc. (collectively Janssen) sued various generic pharmaceutical manufacturers for infringing upon Claims 1 and 4 of U.S. Patent No. 4,663,318 ('318). The claims were drawn to methods of treating Alzheimer's disease with galantamine or salts thereof. At trial, the generic manufacturers conceded that they infringed Claims 1 and 4 of the '318 patent but argued, among other things, that the '318 patent was invalid and thus had no legal effect because it did not teach how to make and use (enable) the inventions described in Claims 1 and 4, as required by law. The Federal district court looked at the issue and agreed that the '318 patent failed to teach how to make and use the claimed invention (that it lacked enablement) and further concluded that it failed to establish usefulness for the claimed invention (it lacked utility) [1]. Because enablement [2] and utility [3] are separate but interrelated requirements for patentability, the district court invalidated Claims 1 and 4. On appeal a divided Court of Appeals for the Federal Circuit, typically the highest court to which a patent case goes, agreed with the district court's decision.

Analysis: This case is an example of an issued patent being invalidated at trial and the decision being upheld upon appeal. The mechanisms used to defeat the patent were (1) lack of enablement and (2) lack of utility.

Teaching how to make and use an invention, also known as enablement, is a requirement for obtaining a patent. Establishing an invention's usefulness, also known as utility, is a separate requirement for patentability. An application can meet the utility requirement but still fail to demonstrate enablement. During the trial phase, the district court conducted a two-pronged analysis of the application and concluded that the application lacked *both* utility *and* enablement. However, on appeal, the Federal Circuit focused only on the utility requirement.

The patent application that issued into the '318 patent is slightly over a page long. Given the brevity of the application, its inventor took an unorthodox approach in teaching how to make and use the invention. The application presented no *in vivo* or *in vitro* test results showing any utility of galantamine for treating Alzheimer's disease. Rather, in attempting to meet the enablement requirement, the application relied in large part upon unintegrated summaries of six previously published scientific papers. This was enough to convince the patent examiner, who examines the application, to issue the patent.

The courts, however, are not bound by the decision of the Patent Office to issue the patent. In considering whether the application had utility, the Federal Circuit approvingly quoted the district court's determination that the application "did not provide analysis or insight connecting the [summaries of the six references] ... to galantamine's potential to treat Alzheimer's disease [4]." In conjunction with this analysis, witness testimony given at trial was also considered. For example, the sole inventor testified that "when I submitted this patent, I certainly wasn't sure, and a lot of other people weren't sure, that cholinesterase inhibitors [such as galantamine] would *ever* work [to treat Alzheimer's disease] [5]." (Emphasis added).

The Federal Circuit then briefly summarized the utility requirement. To have utility, inventions must have "substantial utility" and "specific benefit existing in currently available form [6]." The utility requirement "prevents mere ideas from being patented" and also "prevents the patenting of a *mere research proposal* ... [6]." Quoting a famous Supreme Court patent case, the court emphasized that "a patent is not a hunting license. It is not a reward for search, but compensation for its successful conclusion [7]."

The court concluded that the application was *a mere research proposal*, *lacked utility*, and therefore *was not enabled*: "at the end of the day, the [application] ... does no more than state a hypothesis and propose testing to determine the accuracy of that hypothesis [8]."

Suggested Best Practices for Chemists: Enablement requires that the patent teach how to make and use the invention. To address the "how to make" requirement of enablement:

1. *Provide* a *synthetic route* to construct each chemotype, or provide one generic scheme to the group of chemotypes, in your patent application. If the chemotype is isolated from a natural source (such as a plant part), provide a *chemotype isolation procedure*. Provide all available *characterization data* for each compound synthesized or isolated.

    To address the "how to use" requirement of enablement, and the separate utility requirement:

2. Provide either *in vitro* or *in vivo* assay results for representative compounds, *describe* how the *in vitro* or *in vivo* assay protocol is performed, and *describe* how and why the test results demonstrate that the tested compounds exhibit a useful pharmaceutical property. Ideally, *provide and link in vitro* assay results to *in vivo* assay results that in turn demonstrate that the claimed compounds can be used to treat or prevent a disease. Describe how to administer the application's compounds and intended administration recipients (e.g., humans), including dosage amounts and dosage forms (e.g., pills, tablets, capsules), possible ways of administering the dosage

forms, and what diseases are amenable to treatment or prevention via these dosage forms. Describe how each disease is connected to your assay results and would thus be a valid condition for therapeutic intervention.

3. If time or resource constraints do not allow for actual making or testing of certain compounds, *use prophetic examples*. A prophetic or "paper" example "describes a possible route to making an inventive embodiment that has not actually been carried out [9]." If prophetic examples are employed, it is very important that the verb tense of the prophetic examples *always* be in the *present tense* (e.g., acid (1) *is* reacted with amine (2) forming an amide). In the patent world, the present tense distinguishes prophetic examples from actual examples; actual examples are written in the *past tense*.

3. OWN YOUR OWN WORK

International Application No. PCT/US2009/031047 (beta-lactamase inhibitors)

Example History: International Application No. PCT/US2009/031047 published as WO 2009/091856 A2, on July 23, 2009 [10].

Analysis: Page 119 of the published application contains comments that should have been deleted but were not, before the application's filing. The comments cover a variety of topics including duplicate examples, the possibility of synergy for selected compounds, and the identification of a probable backup candidate.

Suggested Best Practices for Chemists: As an inventor, your name will be on the eventual patent, and you will have to sign a Declaration indicating, among other things, that you have reviewed the application. Careful proofreading of the final copy is critical to avoid serious pitfalls such as unintentionally giving away competitive intelligence. Towards this end:

1. When drafting or revising a patent application, keep comments, questions, and observations *separate* from the application itself.
2. Just prior to filing a patent application, scrutinize *every page* of the application to ensure completeness while avoiding extraneous and potentially damaging material.

4. SCRUTINIZE PATENT APPLICATION CLAIMS

4.1 Pfizer, Inc. v. Ranbaxy Laboratories, Ltd. (atorvastatin calcium)

Case History: In an attempt to protect Lipitor® from generic competition, Pfizer sued Ranbaxy in district court asserting, in part, that Ranbaxy infringed upon dependent Claim 6 of U.S. Patent No. 5,273,995 ('995).

Pfizer won at the district court level, but lost on appeal at the Federal Circuit, in part because Claim 6 was not written in proper form [11]. This case presents an excellent example of the "reach through" effect; how small actions early on in the patent process can have large consequences in litigation.

Analysis: The active ingredient in Lipitor[®] is atorvastatin calcium. Pfizer lost because of a deficiency in the way dependent Claim 6 related to Claims 1 and 2. The following chart is presented to help understand how this happened:

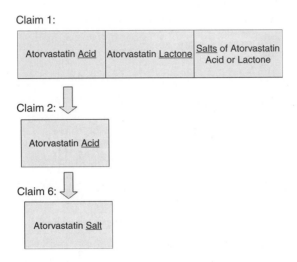

Claim 1:

| Atorvastatin <u>Acid</u> | Atorvastatin <u>Lactone</u> | <u>Salts</u> of Atorvastatin Acid or Lactone |

Claim 2:

| Atorvastatin <u>Acid</u> |

Claim 6:

| Atorvastatin <u>Salt</u> |

Claim 1 is an independent claim containing three different categories of molecules shown as boxes in the diagram. Claim 2 is a dependent claim and therefore by law must refer back to independent Claim 1, while also narrowing the scope of Claim 1 [12,13]. Claim 2 met both of these requirements by referring to Claim 1 (indicated by the arrow linking Claims 1 and 2), and by narrowing Claim 1 by excluding the subject matter of Box 2 (atorvastatin lactone) and Box 3 (salts of atorvastatin acid and atorvastatin lactone), leaving only Box 1 (atorvastatin acid). Since Claim 6 is a dependent claim depending from Claim 2, it must reference and narrow Claim 2. Claim 6 however failed to *properly* narrow Claim 2 because it reintroduced a salt *after* Claim 2 eliminated a salt from Claim 1. Put differently, the subject matter of Claims 2 and 6 do not *overlap* because Claim 2 is drawn to an acid and Claim 6 is drawn to a salt. The Federal Circuit invalidated Claim 6 because of this legal glitch. Thus, a small error in claim drafting, missed during the process of obtaining the patent, "reached through" to create the large consequence of defeating the patent in court. (Pfizer had agreed not to sue under independent Claim 1) [14].

Suggested Best Practices for Chemists: 1. Claims in a patent application tend to begin broadly and then narrow down to important individual compounds. Scrutinize application claims to make sure the most important compounds (e.g., clinical candidates and backups) are individually claimed. *One important compound = one claim.*

2. To avoid legal glitches when claiming compounds, claim important compounds individually in *independent* claims. It costs more up front, but helps avoid expensive and potentially catastrophic patent defeating legal glitches later. If you do use dependent claims, make them all dependent on a broad, independent Claim 1.

3. The "natural" language of medicinal chemists is chemical structure. When claiming important individual compounds, claim the compounds by *chemical structure* instead of, for example, International Union of Pure and applied Chemistry (IUPAC) chemical names. This minimizes the possibility that an error in a chemical structure will be missed.

4. Scrutinize all claim interrelationships to ensure, where appropriate, *overlapping subject matter*. For example, if a dependent claim is drawn to a salt, make sure the claim from which it depends includes a salt.

5. LOOK FOR SUPERIOR AND UNEXPECTED RESULTS

5.1 Sanofi-Synthelabo v. Apotex, Inc. (clopidogrel bisulfate)

Case History: Apotex received Food and Drug Administration (FDA) approval to sell, and commenced selling, generic Plavix® (i.e., the dextrorotatory enantiomer of clopidogrel bisulfate). Sanofi attempted to stop Apotex's clopidogrel bisulfate sales by suing Apotex in district court. Sanofi, in its suit, asserted that Apotex, by selling clopidogrel bisulfate, infringed upon U.S. Patent No. 4,847,265 ('265). Sanofi won at the district court level and upon appeal the decision was affirmed by the Federal Circuit [15].

Analysis: This case presents an example where superior and unexpected results associated with a molecule (one enantiomer) conveyed patentability to the enantiomer over its previously disclosed (prior art) racemate. The case also introduces another mechanism through which patents can be defeated by demonstrating that the claimed subject matter is obvious.

One of the requirements for obtaining a patent is that the claimed subject matter in the patent *not* be obvious in view of prior art (previously disclosed patents and publications). Obviousness [16] and prior art [17] are legally codified in Title 35 of the United States Code.

A basic, or *prima facie*, case of obviousness, which would otherwise defeat patentability, may in turn be defeated by one or more "secondary considerations [18]." One of the most commonly employed secondary considerations is a "superior and unexpected result [18]." At trial, Apotex argued that the dextrorotatory enantiomer of clopidogrel bisulfate, found in Plavix®, was *prima facie* obvious in view of the prior art combined with generally available knowledge. Sanofi countered in part with superior and unexpected results and won.

Apotex, in establishing its *prima facie* case of obviousness, first asserted the following facts:

1. A prior art Sanofi patent disclosed the racemate that the dextrorotatory enantiomer found in Plavix® was separated from.
2. The prior art Sanofi patent made statements concerning enantiomers and addition salts.
3. It was generally known that racemates can be separated into their component enantiomers.
4. Techniques for separating enantiomers from racemates were generally known.
5. Separated enantiomers can differ as to biological properties.
6. In the future, it was possible the FDA would require separation and testing of enantiomers.

Apotex wove these facts into a *prima facie* case of obviousness as follows: It would be obvious to separate an enantiomer from a known racemate using generally known techniques. Furthermore, there was *motivation* [19] to do so because enantiomers can have different properties from those displayed by the racemate, and because of a possible future FDA regulatory requirement for separation of enantiomers. Additionally, there was *a reasonable expectation of success* [20] in achieving the separation because techniques for separating enantiomers from racemates are known. And finally, it would be obvious to form an addition salt of the enantiomer to optimize selected physical properties. Thus, concluded Apotex, dextrorotatory clopidogrel bisulfate was obvious, it is therefore unpatentable, and this renders the '265 patent invalid.

Sanofi skillfully countered every point argued by Apotex. Sanofi first noted that their scientists had previously separated enantiomers from racemates in the same molecular class as clopidogrel, but found that the enantiomers showed *no* advantages over the racemates. For example, in one of the separations, "one of the enantiomers was more biologically active but also [was] more neurotoxic than the racemate [21]." Sanofi then presented data showing that when enantiomers of racemic clopidogrel were separated, the dextrorotatory enantiomer "provided *all* of the favorable antiplatelet activity but with *no* significant neurotoxicity, while the levorotatory enantiomer produced *no* antiplatelet activity but virtually *all* of the neurotoxicity [21]." Experts from both Sanofi and Apotex agreed that "while it was generally known that enantiomers can exhibit different biological activity, this degree and kind of stereoselectivity is *rare*, and *could not be predicted"* and that "in the *usual case,* if one enantiomer is more biologically active than the other, that activity includes the adverse as well as the beneficial properties [21]."

Thus, Sanofi had two *superior and unexpected results* that (1) the dextrorotatory enantiomer separated from racemate PCR 4099 contained substantially *all* of the antiplatelet activity and (2) was essentially *devoid* of the neurotoxicity associated with PCR 4099.

Evidence was further presented that "eventual success [in separating the enantiomers] came only after several failures using known strategies for enantiomer separation" and that it "could be ... difficult to separate the enantiomers of PC 4099 ... because it would be understood by chemists that the methyl ester substituent in PCR 4099 could make it more susceptible to re-racemizations, and thus resistant to successfully obtaining a separated product [22]." Accordingly, "neither the chemists at Sanofi nor a person of ordinary skill in the art could have reasonably expected that the separate enantiomers of PCR 4099 could be obtained ... and if obtained, by what method and configuration [22]." Thus, Apotex's contention that separating the enantiomers was obvious was also refuted.

Concerning the bisulfate salt, Sanofi presented evidence that "the prior art taught away from the use of sulfuric acid with the claimed enantiomer, for strong acids could encourage re-racemization [23]." "Teaching away" from, a separate secondary consideration trending toward patentability, is discussed in more detail in Section 7 of this chapter.

Finally, concerning the possible FDA mandate, the court concluded that "Sanofi undertook this separation [of enantiomers] in order to study adverse neurological effects of [the racemate that the dextrorotatory enantiomer found in Plavix® was separated from], and not because of a possible future regulatory requirement [24]."

In summary, Sanofi presented two superior and unexpected results, in the form of a highly significant combination of properties [associated with one enantiomer], as part of a point-by-point rebuttal of Apotex's arguments in its *prima facie* obviousness case. By doing so, Sanofi retained patent protection for Plavix®. Superior and unexpected results can thus overcome a basic case of obviousness and confer patentability, even in the close case of an enantiomer compared to a racemate.

Suggested Best Practices for Chemists: 1. Throughout the course of a discovery program, molecules are tested against multiple targets. Selectivity for a given target over other targets, degree of inhibition of key isozymes, blockage of key ion channels, toxicity, carcinogenicity, and mutagenicity are all examined. Thus, there are many opportunities to discover superior and unexpected results associated with specific discovery program molecules. Each superior and unexpected result should be assigned to its corresponding molecule and recorded in the laboratory notebook.

2. A listing of all superior and unexpected results and their associated molecules should be provided as part of each *invention disclosure* to the patent department.

3. Often, the biggest obstacle to obtaining a patent is a company's *own prior art*. Thoughtful consideration as to the content and timing of disclosures can remove or minimize self-created obstacles to obtaining patent protection. Discuss the content of proposed patent applications and publications with your patent professional.

4. For each discovery program, a company should formulate and adhere to *a patent and publication strategy*. Make sure that you have a strategy and that you understand how to apply it.

5. Molecules that are structurally similar (e.g., enantiomers) to previously known molecules (e.g., racemates) may, in a fact-dependent fashion, be patentable. Work with your patent professional early in the process to determine if such molecules are worth pursuing as program targets.

6. MAKE SURE SUPERIOR AND UNEXPECTED RESULTS ARE BOTH SUPERIOR AND UNEXPECTED

6.1 Pfizer, Inc. v. Apotex, Inc. (amlodipine besylate)

Case History: In an attempt to protect Norvasc[®] from generic competition, Pfizer sued Apotex in district court, asserting that Apotex infringed upon Claims 1–3 of U.S. Patent No. 4,879,303 ('303). Pfizer won at the district court level, but lost on appeal at the Federal Circuit [25].

Analysis: This case, in many ways, is the antithesis of *Sanofi-Synthelabo v. Apotex, Inc.*, presented in Section 5 above. Pfizer sued Apotex to prevent Apotex from selling generic Norvasc[®] (amlodipine besylate), asserting that Apotex infringed Claims 1–3 of the '303 patent. Apotex admitted that they infringed Claims 1–3, but argued, among other things, that Claims 1–3 of the '303 patent were obvious in view of prior art, that Claims 1–3 were therefore unpatentable, and that these claims were therefore not valid. Apotex successfully made a *prima facie* case of obviousness, in part by using a previous Pfizer patent as prior art. Pfizer attempted to overcome Apotex's *prima facie* case of obvious by various methods, including superior and unexpected results. The outcome was different in that, unlike Sanofi, Pfizer lost.

Pfizer's initial formulation of amlodipine relied on a maleate salt, but two problems were encountered: "(1) the chemical instability of amlodipine maleate, and (2) the stickiness of the tablet blend [26]." To solve the problem, a Pfizer scientist hypothesized that by "changing from the maleate salt to … another acid addition salt, 'many of the stability problems would disappear' [26]." Seven alternative salts were identified, three of them being "members of the same class of sulfonic acids [26]." Of these, it was discovered that the besylate salt "showed a much improved stability profile over the maleate in all cases [26]." In a subsequent communication with the FDA, Pfizer stated that "the change in salt form [to besylate] is justified since benzenesulfonate is a commercially acceptable salt, as exemplified by the tranquilizer mesoridazine (Serentil) [27]."

At trial, Pfizer and Apotex agreed that benzene sulfonate was known in the art, and that a prior art Pfizer patent "specifically states that the pharmaceutically-acceptable salts of amlodipine are those formed from acids which form non-toxic addition salts containing pharmaceutically acceptable anions [28]." Included in this list of salts is "sulfate." Building upon this, Apotex further produced a prior art publication, the Berge reference, that showed there were 53 pharmaceutically acceptable salts, including benzene sulfonate. Additional prior art disclosed that "benzene sulphonic acids considerably increase the solubility of pharmaceuticals containing one or more nitrogen atoms [a class of molecules including amlodipine]," that "besylate (benzene sulphonate) was specifically identified as the *preferred* pharmaceutically acceptable addition salt form of a pharmaceutical compound," that "an intermediate dihydropyridine compound [was found to be] useful in the form of an acid addition salt *derived from benzene sulphonate* [amlodipine is a dihydropyridine]," and that "the besylate addition salt form of a pharmaceutical composition [has] *excellent pharmacokinetic properties, near-optimal solubility, and improved stability* [29]." Thus, the court concluded that taken together, these references provide ample *motivation* to narrow the genus of 53 pharmaceutically acceptable anions to a few, including benzene sulfonate, and that there was a *reasonable expectation of success* in doing so. The court concluded that Apotex had made a *prima facie* case of obviousness.

Pfizer attempted to counter Apotex's *prima facie* case of obviousness by, among other things, arguing superior and unexpected results. Pfizer initially argued a *superior* and unexpected solubility. In considering this argument, the court first noted the trial court's finding that "the '303 patent discloses that amlodipine besylate has a solubility of 4.6 mg/mL at pH 6.6, whereas amlodipine maleate has a solubility of 4.5 mg/mL at pH 4.8" and that "[t]he district court stated that any product having a solubility greater than 1.0 mg/mL is acceptable [30]." Based on this data, the court concluded that the district court found that "the solubility of amlodipoine besylate was [not] materially superior, much less *unexpectedly* superior [compared] to the solubility of amlodipine maleate [31]."

The court next dispensed with a "nonhygroscopicity" argument.

Accordingly, the "two allegedly unexpected and superior properties remaining [were] drug stability and tablet processing [31]." Concerning these arguments, the court noted, in part, that "the prior art ... evidences that one ... would *expect* an acid addition salt made from benzene sulfonate to have *good* physicochemical properties," concluding that drug stability and tablet properties were in fact expected [32]. Finally, the court noted that "[t]he district court wrongly relied upon the fact that [for manufacturing purposes, including stickiness] the "besylate salt works" because considerable evidence shows that amlodipine maleate also *worked for its intended purpose* and even did so in human clinical trials... [32]."

Thus, in response to Pfizer's asserted superior and unexpected results, the court, at various places, reasoned a particular result (e.g., solubility, chemical stability, ease of manufacturability) was not superior, not unexpected, or both not superior and not unexpected. This analysis, combined with the court's dismissal of Pfizer's other arguments for nonobviousness, resulted in invalidation of Claims 1–3 of the '303 patent.

Suggested Best Practices for Chemists: 1. Present any suspected superior and unexpected result in an *information disclosure* to your patent department. Work with the patent department to determine if the result is likely to be considered both superior and unexpected. 2. In disclosing a superior and unexpected result, describe why the result is *both superior and unexpected*; this includes providing data and/or reasoned arguments as to what result would have been expected prior to the discovery of the superior and unexpected result. 3. *Patenting a salt* of an active compound, where the active compound is known in the prior art, may be *challenging*. Keep this in mind as part of formulating a program-specific patent strategy.

7. NOTE WHEN A REFERENCE "TEACHES AWAY"

7.1 Takeda Chem. Indus., Ltd. v. Alphapharm Pty., Ltd. (pioglitazone)

Case History: Takeda, in an attempt to protect Actos® from generic competition, sued various generic pharmaceutical manufacturers (collectively Alphapharm) for infringing the claims of U.S. Patent No. 4,687,777 ('777). At trial, Alphapharm asserted that the compounds covered by the claims of the '777 patent were obvious in view of a prior art compound. The district court found the compounds covered by the claims of the '777 patent nonobvious and held the '777 patent valid. On appeal, the Federal Circuit agreed [33].

Analysis: Pioglitazone, the active ingredient in Actos®, is a member of the thiazolidinedione (TZD) molecular class and has the structure (1):

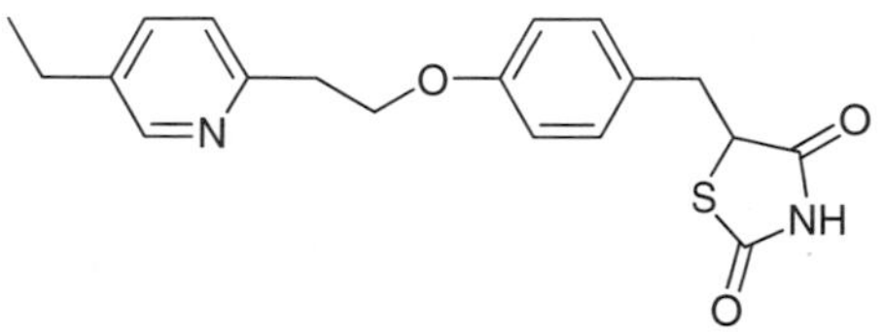

(1)

Alphapharm attempted to argue that (1) was obvious in view of known compound (2);

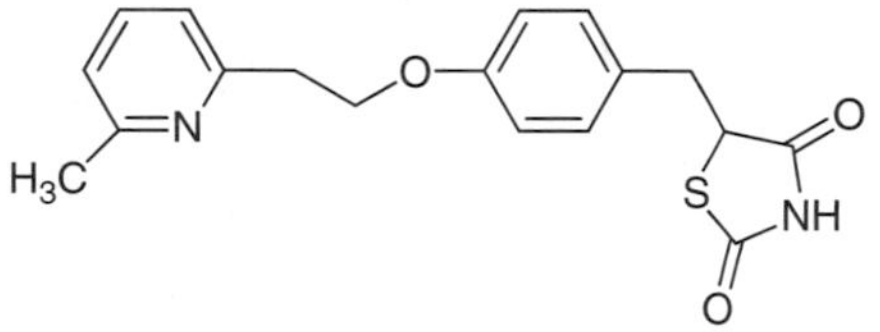

(2)

To make a case for obviousness, Alphapharm was required to show that there would be some *motivation* to select compound (2) as a "lead compound," and further, having selected compound (2), that there would be further *motivation* to both change (homologate) the methyl group on the pyridine ring to an ethyl group and "ring walk" the ethyl group to the 5 position of the pyridine ring, as shown below:

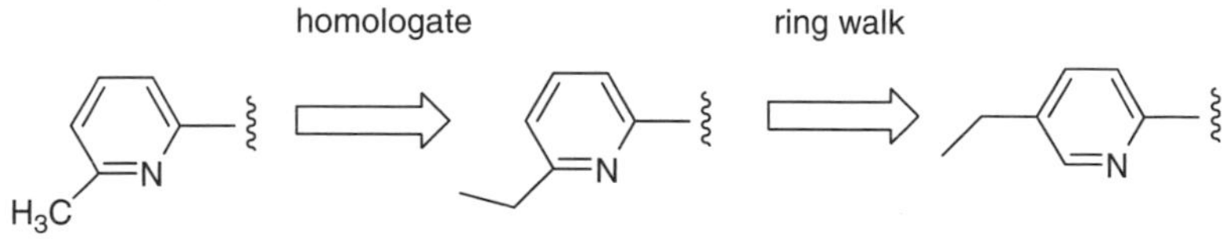

Focusing on the selection of compound (2), the district court first considered prior art U.S. Patent No. 4,287,200 ('200). The court noted that the '200 patent disclosed hundreds of *millions* of TZDs and specifically identified 54 compounds, including compound (2), but failed to disclose any experimental data or test results for the identified compounds [34]. Other data disclosed during prosecution of the application that issued into the '200 patent failed to provide motivation for selecting compound (2) as a "lead compound [34]."

The district court then turned its attention to an article entitled "Studies on Antidiabetic Agents, II. Synthesis of 5-[4-(1-Methylcyclohexylmethoxy)-benzyl]thiazoli-dine-2,4-dione (ADD-3878) and Its Derivatives" (the Sodha II reference). Sodha II, while disclosing compound (2), also identified three specific compounds that displayed the most favorable combination of toxicity, decreased side effects, and good biological activity—compound (2) was *not* one of the three favored compounds [35].

Noting that diabetes, the target disease, is a chronic condition that requires long-term treatment, the district court concluded that researchers would have been dissuaded from selecting compound (2) as a lead compound because of its suboptimal toxicity and side-effect profile [35].

One of Alphapharm's witnesses further buttressed the court's conclusion, "acknowledging that [compound (2)] had the negative side effects of increased body weight and brown fat" and that "a compound with such side effects would 'presumably not' be a suitable candidate for treatment of Type II diabetes [36]."

The district court thus concluded that selection of compound (2) as a "lead compound" was "taught away" from by the prior art, and the Federal Circuit agreed. Because, in part, selecting compound (2) as a "lead compound" was "taught away" from, Alphapharm's attempt to invalidate the '777 patent was thwarted.

Suggested Best Practices for Chemists: "Teaching away" from, like a "superior and unexpected result," is another secondary consideration that can be used to defeat a *prima facie* case of obviousness [37].

1. If a reference "teaches away" from some research activity (e.g., selecting a particular chemotype for modification, modifying a chemotype in a specific fashion, or making a particular salt form), *note* this in the laboratory notebook and in the *invention disclosure* form.

8. CONCLUSION

The fortunes of pharmaceutical companies and the careers of their employees depend on holding patents covering key molecules, their formulations, and methods of making and using them. Patents can be challenging to obtain, and once held are subject to challenge and defeat by a variety of mechanisms in the Patent Office, the Federal Courts, and before the U.S. International Trade Commission. This chapter has outlined a variety of actions that chemists can take to increase the likelihood of obtaining a strong patent.

REFERENCES

[1] *In re '318 Patent Infringement Litig.*, 583 F.3d 1317 (Fed. Cir. 2009).

[2] 35 U.S.C. § 112, para. 1.

[3] 35 U.S.C. § 101.

[4] *In re '318 Patent Infringement Litig.*, 583 F.3d 1317, 1321 (Fed. Cir. 2009).

[5] *In re '318 Patent Infringement Litig.*, 583 F.3d 1317, 1327 (Fed. Cir. 2009).

[6] *In re '318 Patent Infringement Litig.*, 583 F.3d 1317, 1324 (Fed. Cir. 2009).

[7] *In re '318 Patent Infringement Litig.*, 583 F.3d 1317, 1324 (Fed. Cir. 2009) (quoting *Brenner v. Manson*, 383 U.S. 591, 536 (1966)).

[8] *In re '318 Patent Infringement Litig.*, 583 F.3d 1317, 1327 (Fed. Cir. 2009) (citing *Rasmusson v. SmithKline Beecham Corp.*, 413 F.3d 1318, 1325 (Fed. Cir. 2005)).

[9] Manual of Patent Examining Procedure § 601.01(p).

[10] T. A. Blizzard, H. Chen, C. Gude, J. D. Hermes, J. E. Imbriglio, S. Kim, J. Y. Wu, S. Ha, C. J. Mortko, I. Mangion, N. Rivera, R. T. Ruck, and M. Shevlin, *International Patent Application Publication No. WO 2009/091856 A2*, published 2009.

[11] *Pfizer, Inc. v. Ranbaxy Labs., Ltd.*, 457 F.3d 1284 (Fed. Cir. 2006).

[12] 35 U.S.C. § 112, para. 4.

[13] 35 U.S.C. § 112, para. 4, ("...a claim in dependent form shall contain a reference to a claim previously set forth and then specify a further limitation of the subject matter claimed...").

[14] *Pfizer, Inc. v. Ranbaxy Labs., Ltd.*, 457 F.3d 1284 (Fed. Cir. 2006).

[15] *Sanofi-Synthelabo v. Apotex, Inc.*, 550 F.3d 1075 (Fed. Cir. 2008).

[16] 35 U.S.C. § 103.

[17] 35 U.S.C. § 102.

[18] Manual of Patent Examining Procedure § 2145.

[19] Manual of Patent Examining Procedure § 2143.01.

[20] Manual of Patent Examining Procedure § 2143.02.

[21] *Sanofi-Synthelabo v. Apotex, Inc.*, 550 F.3d 1075, 1081 (Fed. Cir. 2008).

[22] *Sanofi-Synthelabo v. Apotex, Inc.*, 550 F.3d 1075, 1088 (Fed. Cir. 2008).

[23] *Sanofi-Synthelabo v. Apotex, Inc.*, 550 F.3d 1075, 1089 (Fed. Cir. 2008).

[24] *Sanofi-Synthelabo v. Apotex, Inc.*, 550 F.3d 1075, 1090 (Fed. Cir. 2008).

[25] *Pfizer, Inc. v. Apotex, Inc.*, 480 F.3d 1348 (Fed. Cir. 2007).

[26] *Pfizer, Inc. v. Apotex, Inc.*, 480 F.3d 1348, 1354 (Fed. Cir. 2007).

[27] *Pfizer, Inc. v. Apotex, Inc.*, 480 F.3d 1348, 1355 (Fed. Cir. 2007).

[28] *Pfizer, Inc. v. Apotex, Inc.*, 480 F.3d 1348, 1361 (Fed. Cir. 2007).

[29] *Pfizer, Inc. v. Apotex, Inc.*, 480 F.3d 1348, 1363 (Fed. Cir. 2007).

[30] *Pfizer, Inc. v. Apotex, Inc.*, 480 F.3d 1348, 1369 (Fed. Cir. 2007).

[31] *Pfizer, Inc. v. Apotex, Inc.*, 480 F.3d 1348, 1370 (Fed. Cir. 2007).

[32] *Pfizer, Inc. v. Apotex, Inc.*, 480 F.3d 1348, 1371 (Fed. Cir. 2007).

[33] *Takeda Chem. Indus., Ltd., v. Alphapharm Pty., Ltd.*, 492 F.3d 1350 (Fed. Cir. 2007).

[34] *Takeda Chem. Indus., Ltd., v. Alphapharm Pty., Ltd.*, 492 F.3d 1350, 1357 (Fed. Cir. 2007).

[35] *Takeda Chem. Indus., Ltd., v. Alphapharm Pty., Ltd.*, 492 F.3d 1350, 1358 (Fed. Cir. 2007).

[36] *Takeda Chem. Indus., Ltd., v. Alphapharm Pty., Ltd.*, 492 F.3d 1350, 1359 (Fed. Cir. 2007).

[37] Manual of Patent Examining Procedure § 2144.05.

PART VIII:
Trends and Perspectives

CHAPTER 28

To Market, To Market—2009

Shridhar Hegde* and **Michelle Schmidt****

* Pfizer Global Research & Development, St. Louis, MO 63017, USA

** Covidien, Hazelwood, MO 63042, USA

Annual Reports in Medicinal Chemistry, Volume 45
ISSN 0065-7743, DOI 10.1016/S0065-7743(10)45028-9

The pharmaceutical market closed out the decade with a significant surge in new molecular entities (NMEs) while exploring five novel mechanisms of action in 2009. Of the 28 NMEs, the anticancer drugs reclaimed the lead with six launches, including the biologics catumaxomab, a first-in-class, trifunctional bispecific antibody, and the anti-CD20 antibody ofatumumab. Joining the anticancer biologics, three additional biologics entered the market to combat inflammatory disorders, keeping with the recent trend of garnering about 20% of the yearly introductions. While the United States consistently boasts the most active market, it monopolized 2009 with 19 out of the 28 NMEs. Germany trailed at a distant second with six, sharing some of its launches with the United Kingdom, France, and Denmark. Of the remaining NMEs, two debuted in Japan and one in Canada. Although not covered in detail in this venue, immunomodulating vaccines proved to be a prolific area fueled by the H1N1 pandemic. New combinations of existing drugs continued to be popular for providing enhanced patient benefit; however, extensive coverage of these launches is beyond the scope of this chapter. Likewise, line extensions involving new formulations and/or new indications are not elaborated extensively but are recognized for their significant contribution to the class of 2009. (The collection of new therapeutic entities first launched in 2009 originated from the following sources: (a) CIPSLINE, Prous database; (b) Iddb, Current Drugs database; (c) IMS R&D Focus; (d) Adis Business Intelligence R&D Insight; (e) Pharmaprojects; and (f) prescribing information; see also [1].)

Among the anticancer agents launched this past year, Removab® (catumaxomab) garnered significant attention as a melding of firsts. In addition to being the first approved treatment for malignant ascites, it is the only antibody on the market to target the tumor antigen Ep-CAM. The ultimate novelty, however, lies in the essence of this first trifunctional bispecific antibody that possesses an anti-CD3 arm that targets the T-cell antigen CD3 and an intact Fc region that selectively binds to FCγ receptor-positive accessory cells, in addition to its Ep-CAM recognition arm. This single antibody brings the requisite immune effector cells in close proximity to the tumor cells; catumaxomab destroys tumor cells possessing the surface antigen Ep-CAM via simultaneous activation and complex crosstalk of T cells and accessory immune cells. The other anticancer biologic debuting this year for the treatment of chronic lymphocytic leukemia (CLL) was Arzerra™ (ofatumumab), a traditional monoclonal antibody that binds to a discontinuous epitope on the CD20 molecule displayed on the surface of B cells, thereby triggering B cell ablation through the body's immune response. Although it followed the anti-CD20 mAb rituximab to the market, ofatumumab has displayed superior efficacy for CLL, attributed to the combination of a slower off-rate and targeting a different epitope on CD20. For patients with relapsed or refractory peripheral T-cell lymphoma (PTCL), Allos Therapeutics launched its first product, which targets inhibition of dihydrofolate reductase (DHFR).

PTCL is categorized as a group of aggressive blood cancers with poor prognosis, and Folotyn^{TM} (pralatrexate) is the first and only drug approved for this indication. Patients suffering from prostate cancer have a new treatment option in Firmagon® (degarelix acetate). As an antagonist of gonadotropin-releasing hormone (GnRH), degarelix acetate does not suffer from a potential flare-up of the disease, as experienced with agonists prior to down-regulation of the GnRH receptor, and it effects a rapid reduction of testosterone secretion to castration-like levels. Votrient^{TM} (pazopanib hydrochloride) is the latest angiogenesis inhibitor to hit the market. This orally active vascular endothelial growth factor receptor (VEGFR) tyrosine kinase inhibitor is indicated for the treatment of advanced renal cell carcinoma (RCC). The final NME in the anticancer category is a first-in-class small-molecule antagonist of the chemokine CXCR4 receptor. Mozobil^{TM} (plerixafor hydrochloride) received orphan drug status for use in combination with granulocyte colony-stimulating factor (G-CSF) to elicit the shuttling of stem cells out of the bone marrow and into circulating blood for collection and subsequent transplantation in patients with non-Hodgkin's lymphoma and multiple myeloma. The combination of agents significantly enhances the mobilization of stem cells over G-CSF alone, which translates to substantial patient benefit since fewer apheresis days are needed to isolate the requisite number of stem cells. The previously marketed Afinitor® (everolimus), a selective inhibitor of mTOR (mammalian target of rapamycin) that regulates tumor cell division, blood vessel growth, and cell metabolism, was launched for a new indication—the treatment of advanced RCC after failure of first-line agents sunitinib and sorafenib that both inhibit multiple kinases. In addition to serving this unmet medical need, everolimus continues to be used as an immunosuppressive drug for the prevention of rejection episodes following heart or kidney transplantation.

The CNS field was represented by the entry of three new drugs, new formulations and/or new indications for existing drugs, and a combination for a new indication. Regarding the treatment of schizophrenia, several options emerged. Saphris® (asenapine maleate) is a dual serotonin 5-HT$_2$ and dopamine D$_1$/D$_2$ receptor antagonist that was approved for acute treatment of schizophrenia in adults and for the management of manic or mixed episodes with bipolar I disorder in adults, with or without psychotic features. The dual antagonism of this atypical agent retains the antipsychotic activity with an improvement in the negative symptoms and a reduction in extrapyramidal side effects (EPS). Other previously launched atypical antipsychotics received new formulations to promote a longer duration of action. Zypadhera^{TM} (olanzapine pamoate) resurfaced as a long-acting injection formulation for the maintenance treatment of adult schizophrenic patients sufficiently stabilized during acute episodes by treatment with oral olanzapine. This depot formulation is also expected to improve patient compliance. Likewise, Invega® (paliperidone palmitate) is now available in an extended-release injectable suspension for the acute and maintenance treatment of schizophrenia and is the first atypical antipsychotic approved for once-monthly administration. While NMEs were not introduced this past year for the treatment of depression, a new combination of existing drugs was approved for treatment-resistant depression in acute situations. The product, Symbyax®, marries the atypical antipsychotic olanzapine with the selective serotonin reuptake inhibitor (SSRI) fluoxetine hydrochloride. While bupropion hydrochloride has been on the market since 1989 as an effective antidepressant therapy, this past year it was reformulated as its hydrobromide salt, hence improving patient compliance with the availability of one-tablet, daily dosing. Attention deficit hyperactivity disorder has a new approach for treatment using an existing drug. Intuniv^{TM} (guanfacine hydrochloride) is a selective α_{2A}-adrenoceptor agonist marketed as an antihypertensive agent, but it found a new application when it was realized that stimulation of postsynaptic α_{2A}-adrenoceptors in the prefrontal cortex promotes working memory, attention regulation, and behavioral inhibition; reduces susceptibility to distraction; and enhances impulse control. With its extended release formulation, the pill burden is only one per day. Epilepsy is one of the CNS conditions that welcomed an NME this past year. Zebinix® (eslicarbazepine acetate) is a third-generation antiepileptic drug (AED) that was approved as adjunctive therapy for partial-onset seizures with or without secondary generalization. As a member of the carbamazepine family (first generation), it is the prodrug of eslicarbazepine and the active metabolite of oxcarbazepine (second generation). While the exact mechanism of action has not been fully elucidated, eslicarbazepine acetate is a voltage-gated sodium channel blocker that targets the inactivated state of the ion channel and reduces repetitive firing by preventing the return of the ion channel to the active state. Another established AED Trerief® (zonisamide) secured a new indication. Its development for Parkinson's disease (PD) was based on a serendipitous finding of mitigation of PD symptoms in a patient with comorbid epilepsy. Zonisamide is approved for use with the standard of care levodopa. Neupro® (rotigotine) is a PD drug that found a new role in the treatment of restless leg syndrome. As a dopamine D_2/D_3 receptor agonist, it produced a significant improvement in symptoms in comparison to placebo that was sustained throughout the 6-month maintenance phase. Keeping with the trend of discovering new indications for CNS drugs, Savella^{TM} (milnacipran hydrochloride) is a dual norepinephrine/serotonin reuptake inhibitor that has been marketed for years as an antidepressant and has now been approved for fibromyalgia therapy. Milnacipran hydrochloride is the third drug, and second antidepressant, to be launched for this previously unmanageable disorder. Another CNS drug with a new indication is Abilify® (aripiprazole), a 5-HT$_{2A}$ receptor antagonist/5-HT$_{1a}$ receptor partial agonist/dopamine D_2 receptor partial agonist. For years, it has been used to treat various psychiatric disorders, such as schizophrenia, major depression, and bipolar disorder. Now, aripiprazole has been launched for the alleviation of irritability associated with autistic disorder in pediatric patients aged 6–17 years. The final drug in the CNS category is an NME addressing sleep disorders. Nuvigil® (armodafinil) is a longer-lasting, single-isomer formulation of the CNS stimulant modafinil, and it is indicated for the improvement of daytime wakefulness. Patients needing this assistance are struggling with excessive sleepiness as the result of obstructive sleep apnea, shift work sleep disorder (SWSD), or narcolepsy.

Concerning endocrine diseases, diabetic patients received multiple new options. Victoza® (liraglutide) is the second glucagon-like peptide 1 (GLP-1) receptor agonist to reach the market, following its predecessor Byetta®. Liraglutide possesses 97% homology to GLP-1 with only two amino acid changes and the addition of a fatty acid side chain. These modifications are implicated in the drug's resistance to dipeptidyl peptidase 4 (DPP-4) degradation by contributing to its propensity to form micelles and to bind to albumin. This incretin mimetic is the first once-daily GLP-1 analog for managing type 2 diabetes mellitus (T2DM). The other NME complements liraglutide since it blocks DPP-4, the enzyme responsible for the degradation of GLP-1 and its analogs. Onglyza™ (saxagliptin) is the third DPP-4 inhibitor to pass the market threshold. Approved as an adjunct to diet and exercise to improve glycemic control in adults, saxagliptin was the effort of a joint collaboration between Bristol-Myers-Squibb and AstraZeneca. By combining the two existing drugs repaglinide, a fast-acting secretagogue, and the insulin sensitizer metformin, Novo Nordisk created a fixed-dose combination marketed as PrandiMet™ for the treatment of T2DM. Reduction of pill burden was the driving force for this launch. For emergency contraception, EllaOne® (ulipristal acetate) offers an alternative choice to levonorgestrel, dubbed the "morning after pill." Ulipristal acetate is a selective progesterone receptor modulator (SPRM), and as such, is the first representative of this new therapeutic class. Ulipristal acetate differentiates from its predecessor by maintaining efficacy for 5 days following unprotected intercourse versus 3 days for levonorgestrel.

The metabolic sector experienced the introduction of two NMEs along with some novel combinations and a new indication for a drug fighting rare disorders. Joining a crowded market, minodronic acid hydrate is the seventh bisphosphonate to cross the finish line for osteoporosis intervention. It may have an advantage in the prevention of bone fractures. Codeveloped by Astellas and Ono, the drug is marketed as Bonoteo® and Recalbon®, respectively. Continuing with the bone remodeling theme, the Augment™ Bone Graft (GEM-OS1) is an implantable biologic composed of recombinant human platelet-derived growth factor BB (rhPDGF-BB) and β-tricalcium phosphate (β-TCP). While β-TCP donates the support system for new bone growth, rhPDGF-BB stimulates the recruitment and proliferation of new bone-forming cells and blood vessels for tissue repair. It is approved as an alternative to the use of autograft in midfoot, hindfoot, and ankle fusion indications. Touted as the first major new option for the management of hyperuricemia in gout in more than 40 years, Uloric® (febuxostat) is a xanthine oxidase inhibitor that reduces serum uric acid levels by impeding the transformation of hypoxanthine to xanthine and ultimately to uric acid; both conversions are catalyzed by the xanthine oxidoreductase enzymes xanthine oxidase and xanthine dehydrogenase. A major aim in gout management is the long-term reduction of serum uric acid concentrations below saturation levels to promote crystal dissolution and eventual disappearance. The attack of dyslipidemia has a new weapon in the fixed-dose combination of nicotinic acid and laropiprant, marketed as Tredaptive™. This blood cholesterol-lowering agent is recommended for patients with elevated levels of LDL cholesterol and triglycerides coupled with low HDL cholesterol, and it may be used in combination with statins or as monotherapy in patients for whom statins are considered inappropriate or not tolerated. While previously approved for Gaucher's disease, a rare lysosomal storage disorder, Zavesca® (miglustat) is a ceramide glucosyltransferase inhibitor that has demonstrated clinically relevant benefits in the treatment of progressive neurological manifestations in adult and pediatric patients with Niemann–Pick type C disease. Fulfilling an unmet need, miglustat is the first approved therapy for this rare neurological disorder manifested by the storage of glycosphingolipids in certain cells of the body, including the brain.

The field of anesthesia and analgesia welcomed two NMEs as well as new formulations and a combination of existing agents. As a dual μ-opioid receptor (MOR) agonist and norepinephrine reuptake inhibitor, the dual mechanism of action of Nucynta™ (tapentadol hydrochloride) provides an efficacious profile in both acute and chronic pain management with fewer side effects. It has been filed into Schedule II of the Controlled Substances Act by the U.S. Drug Enforcement Agency. While not covered in detail in this chapter, the other NME is the Schedule IV drug Lusedra™ (fospropofol disodium). This water-soluble phosphate prodrug is metabolized in the liver by alkaline phosphatase enzymes, thereby releasing the sedative-hypnotic/anesthetic drug propofol. Administered intravenously by a trained anesthesiologist, the clinical effect is more sustained with fewer side effects compared to the parent propofol. When continuous analgesia is required for an extended period, the oral extended-release Embeda™, a combination of morphine sulfate and naltrexone hydrochloride, is a new option for the management of moderate to severe pain. The novelty in this combination comes from the unique formulation that deters abuse. With a sequestered core of naltrexone, the capsules effectively release morphine for pain release while the naltrexone passes through the body with no consequence. If the capsules are chewed or crushed, however, naltrexone is liberated and reverses the effects of the opioid. On the subject of reversal, OraVerse™, a new formulation of the α-adrenoceptor antagonist and vasodilating agent phentolamine mesylate, was launched for reversing dental anesthesia. It is the first drug approved for accelerating the return to normal sensation and function following soft tissue anesthesia.

The area of anti-inflammatory agents showcased three new biologics. Stelara™ (ustekinumab) is a first-in-class fully human antibody targeting the proinflammatory cytokines IL-12 and IL-23 for the control of moderate to severe plaque psoriasis. Since overproduction of IL-12 has been implicated in several autoimmune diseases, this mAb could provide widespread benefit in patients with immunological disorders. Patients will also appreciate the attractive dosing regimen that backs down to self-injection once every 12 weeks following the starter doses at week 0 and 4. Simponi™ (golimumab) is this year's anti-TNF-α mAb that hopes to share in the success of its predecessors. It was launched for the treatment of rheumatoid arthritis (RA), psoriatic arthritis (PsA), and ankylosing spondylitis (AS). The third anti-inflammatory biologic is Ilaris® (canakinumab). As an anti-IL-1β antibody, canakinumab's mechanism of intervention in the pathogenesis of CAPS (cryopyrin-associated periodic syndrome) is the blockade of the excessive IL-1β signaling. Compared to other drugs on the market targeting IL-1β, such as anakinra and rilonocept, canakinumab possesses a less frequent dosing regimen of just once every 2 months, instigates fewer injection site reactions, and is approved in children as young as 4 years of age. While initially approved for the treatment of Crohn's disease, Cimzia® (certolizumab pegol), one of the predecessors of golimumab, is an anti-TNF-α mAb that has entered the competition for battling RA. Since Colcrys™ (colchicine) is a natural product derived from the dried seeds of *Colchicum autumnale*, it was not reviewed in detail in this article; however, it fulfills an unmet need for the management of familial Mediterranean fever (FMF). As a member of a group of disorders classified as hereditary periodic fever syndromes, FMF is characterized by recurrent bouts of fever and peritonitis with variable symptoms of pleuritis, skin lesions, arthritis, and in rare cases pericarditis. Although the precise mechanism of action has not been elucidated, it is presumed that the single-ingredient colchicine interferes with the intracellular assembly of the inflammasome complex that mediates the activation of IL-1β. Colchicine is also believed to have a role in disrupting cytoskeletal functions by inhibiting β-tubulin polymerization into microtubules. Cartilage degradation is the hallmark of osteoarthritis, and a new treatment has been approved for use in the E.U. to circumvent the limited healing capacity of cartilage. ChondroCelect® consists of chondrocytes harvested from a healthy region of the patient's cartilage, grown outside the body, and then reimplanted during an autologous surgical procedure. Not only was ChondroCelect the first cell-based product approved in Europe, but it was also the first procedure approved under the Advanced Therapy Medicinal Product regulation, providing medicinal products for human use based on gene therapy, somatic cell therapy, or tissue engineering.

In the renal-urologic arena, two new NMEs emerged coupled with a novel formulation for a previously marketed drug. Following the introduction of the dual arginine-vasopressin (AVP) V_{1a}/V_2 receptor antagonist conivaptan, Samsca^{TM} (tolvaptan) was launched as a nonpeptide selective V_2 receptor antagonist with potent aquaretic attributes for the treatment of hypervolemic and euvolemic hyponatremia, conditions associated with heart failure, cirrhosis, and the syndrome of inappropriate antidiuretic hormone (SIADH). Selective antagonism of the AVP V_2 receptor promotes water excretion without perturbing electrolyte balance, thereby making it an appealing target for preventing disease progression. Men in Sweden and Finland now have an option for coping with premature ejaculation (PE). As serotonin is believed to play a central role in the timing of ejaculation, Priligy^{TM} (dapoxetine hydrochloride) derives its efficacy by acting as a short-term SSRI. It is taken only when needed. To avoid the first-pass metabolism implicated in the side effects of dry mouth and constipation, a novel gel formulation of the muscarinic antagonist oxybutynin chloride was developed and launched. Gelnique^{TM} is the first and only topical gel formulation for the treatment of overactive bladder with symptoms of urge urinary incontinence, urgency, and frequency. While effective in both men and women, it is marketed specifically for female patients. Once-daily application to the thigh, abdomen, upper arm, or shoulder provides continuous delivery of the anticholinergic agent for 24 h.

Regarding hematology, two NMEs were introduced. Although approved in late 2008, Promacta® (eltrombopag) entered the market after its first-in-class predecessor Nplate®. Both are small-molecule agonists of the thrombopoietin (TPO) receptor for the treatment of thrombocytopenia in patients with chronic immune (idiopathic) thrombocytopenic purpura (ITP) who have an insufficient response to corticosteroids, immunoglobins, or splenectomy. Patients with acute coronary syndrome (ACS) undergoing percutaneous coronary intervention (PCI) have a new choice to prevent atherothrombotic events. Efient® (prasugrel) works by blocking the $P2Y_{12}$ receptor on the surface of blood platelets, thereby preventing platelet aggregation and clotting. By using an oral, modified-release formulation of tranexamic acid, Lysteda^{TM} emerged as a treatment for heavy menstrual bleeding; the injectable formulation (Cyklokapron) is an acceptable treatment of hemophilia. This antifibrinolytic agent is the first nonhormonal therapy for menorrhagia. Feraheme^{TM} (ferumoxytol) is a new iron replacement agent. This intravenously injected drug is indicated for the management of iron deficiency anemia in adult patients with chronic kidney disease. Compared to oral iron, ferumoxytol, comprised of superparamagnetic iron oxide particles with a proprietary semisynthetic carbohydrate coating, possesses greater bioavailability and fewer associated side effects. Replacing another product previously marketed by Biotest, the German company introduced coagulation factor IX (Haemonine®) for the acute treatment and prevention of type B hemophilia, the rarer form of the blood disorder.

The cardiovascular domain witnessed one NME. Multaq® (dronedarone) is an antiarrhythmic agent that was launched to reduce the risk of cardiovascular hospitalization in patients with paroxysmal or persistent atrial fibrillation (AF) or atrial flutter (AFL), with a recent episode of AF/AFL and associated cardiovascular risk factors, who are in sinus rhythm or who will be cardioverted. Dronedarone, a close analog of amiodarone, was structurally modified to provide improved safety and pharmacokinetic profiles. As in previous years, new combinations of existing cardiovascular drugs continued to be popular for improving patient compliance. Exforge HCT® combines the calcium channel blocker amlodipine, the angiotensin receptor blocker (ARB) valsartan, and the diuretic hydrochlorothiazide in a single tablet for once-daily administration. It is the first antihypertensive medication to incorporate three different drugs in a single pill. Valturna® is a dual-containing treatment consisting of the ARB valsartan and the direct renin inhibitor aliskiren. Its novelty lies in it being the first medication to simultaneously target two points within the renin–angiotensin–aldosterone system. Twynsta® couples telmisartan and amlodipine besylate for the treatment of hypertension. In clinical studies, Twynsta demonstrated significant blood pressure reductions of up to 40/29 mmHg, with a 24-h response rate of about 98% in hypertensive patients at risk for cardiovascular events. Two existing drugs were launched for a new indication—pulmonary arterial hypertension (PAH). Adcirca™ (tadalafil) is a phophodiesterase 5 inhibitor, also marketed as Cialis® for erectile dysfunction. While tadalafil is an oral agent for treating PAH of multiple etiologies, Tyvasco™ (trepostinil) is an inhaled option. This prostaglandin analog has previously only been available in injectable formulations.

The anti-infective category contained one NME and two combination therapies. Vibativ™ (telavancin hydrochloride) is a new approach for fighting complicated skin and skin structure infections (cSSSI) caused by susceptible Gram-positive bacteria. This once-daily injectable antibiotic has a dual mechanism of action, both inhibiting bacterial cell wall synthesis and disrupting bacterial cell membrane function. By combining acyclovir with hydrocortisone, Medivir created Lipsovir® for the purpose of reducing the likelihood of ulcerative cold sores and shortening lesion healing time in patients with recurrent herpes labialis. The combination of nifurtimox and eflornithine hydrochloride resulted in a new treatment for the life-threatening, parasitic disease of human African trypanosomiasis, also known as sleeping sickness. By simplifying the co-administration of oral nifurtimox with injectable eflornithine, the patient benefits by a reduction in the total number of intravenous eflornithine infusions from 56 to 14 and a shortening in hospitalization from 14 to 10 days.

In addition to one NME for ophthalmic disorders, new indications surfaced for existing drugs along with a novel intravitreal implant. For the treatment of bacterial conjunctivitis emanating from a wide range of ocular pathogens, Besivance™ (besifloxacin hydrochloride), the first ophthalmic fluoroquinolone with no prior systemic use, entered the market. A new formulation for bepotastine besilate (Bepreve™) generated a new indication for the treatment of ocular itching associated with allergic conjunctivitis. It is a histamine H1 receptor antagonist, and the active ingredient has been marketed since 2000 for the systemic treatment of allergic rhinitis and urticaria/pruritus. Following this trend, the prostaglandin analog bimatoprost, previously marketed for glaucoma, is now experiencing a new indication, thanks to the serendipitous discovery of eyelash growth in treating glaucoma. Latisse® is now available as a once-daily application to improve eyelash length, thickness, and darkness in patients suffering from hypotrichosis of eyelashes. As observed in arterial hypertension, combination therapy is often more effective in managing the ocular hypertension condition of glaucoma. Azarga® (brinzolamide/timolol maleate) is the latest fixed-dose combination product for elevated intraocular pressure associated with open-angle glaucoma or ocular hypertension. As a first-of-its-kind therapy, Ozurdex™ utilizes the company's proprietary Novadur™ solid polymer delivery system to administer the corticosteroid dexamethasone by intravitreal injection. It is indicated for the therapy of macular edema following branch retinal vein occlusion.

The remaining therapeutic areas are grouped together for modest market activity. Focusing first on respiratory diseases, Onbrez Breezhaler® (indacaterol maleate) is an NME for the once-daily bronchodilatory treatment of airflow obstruction in adult patients with chronic obstructive pulmonary disease (COPD). In addition to being the first new inhaled agent for COPD in 7 years, indacaterol maleate is the first and only therapy to achieve both 24-h bronchodilation and a rapid onset of action (within 5 min of inhalation). Supporting the popularity of intranasal corticosteroids as a frontline attack for allergic rhinitis by relieving the symptoms of sneezing, itching, rhinorrhea, and congestion, Japan now has its version in Erizas® (dexamethasone cipecilate). Switching to gastrointestinal diseases, Kapidex™ (dexlansoprazole) is the latest proton pump inhibitor (PPI) to join the market for the management of heartburn associated with symptomatic nonerosive gastroesophogeal reflux disease (GERD) and for the healing and maintenance of healed erosive esophagitis (EE). Dexlansoprazole differentiates from its predecessors as the first PPI with a Dual Delayed Release™ (DDR) formulation. Its unique formulation results in a concentration-time profile with two distinct peaks that translates to once-daily administration. Concluding with dermatology, Remitch® (nalfurafine hydrochloride) is a novel κ-opioid receptor agonist that was launched for the improvement of pruritis in hemodialysis patients who have not responded to conventional treatment, such as antihistamines. While its cause has not been identified, this condition is manifested by severe systemic itching without inflammation of the skin.

Bolstered by the HIN1 pandemic, new vaccines exploded onto the market. Sanofi Pasteur, CSL, and Novartis led the way in injectable H1N1 vaccines, all of which incorporated virus produced in embryonated chicken eggs. Med Immune contributed a live attenuated intranasal option. Although the FDA specified that none of the influenza A (H1N1) monovalent vaccines released in 2009 should bear a trade name in the United States, the CSL vaccine is known as Panvax® H1N1 in Australia. Regarding other countries, Celtura® (Novartis), an adjuvanted cell culture-based H1N1 vaccine, was launched in Switzerland although it was also approved in Germany. The adjuvanted egg-based vaccine Focetria® (A, H1N1 from Novartis) and the adjuvanted split influenza vaccine Pandemrix™ (H1N1 from GlaxoSmithKline) debuted in the E.U. Sanofi Pasteur provided the nonadjuvanted inactivated vaccine Panenza® to national authorities in France while the split-virion vaccine PANFLU.1™ from Sinovac Biotech was distributed in China. Hungary received the whole-virion vaccine Fluval P, containing virus propagated in Vero cells, from Omnivest. Also, using Vero Cell technology developed by Baxter, Celvapan H1N1® is the first cell culture-based and nonadjuvanted pandemic influenza vaccine to obtain marketing authorization in the E.U. It evolved from the pandemic influenza mock-up vaccine Celvapan® that was based on an inactivated H5N1 strain manufactured in Vero cells. In addition to the proliferation of H1N1 vaccines, Grippol® Neo was launched as a seasonal influenza subunit vaccine. Co-developed by Solvay and Petrovax, it utilized influenza antigens produced by Solvay in its cell-based facility and incorporated Petrovax's proprietary adjuvant polyox-idonium in a preservative-free formulation. The prevention of pneumo-coccal disease now has two new weapons. Prevenar® is a 13-valent pneumococcal conjugate vaccine developed by Wyeth (now Pfizer) for active immunization of children aged 6 weeks through 5 years for the prevention of invasive pneumonia and otitis media caused by 13 pneumo-coccal serotypes. Its first launch took place in Germany. GlaxoSmithKline launched its pneumococcal conjugate vaccine in the Czech Republic. This 10-valent vaccine protects against life-threatening diseases, such as meningitis and bacteremic pneumonia, as well as middle ear infections, and is indicated for use in children aged 6 weeks through 2 years of age. Regarding the prevention of Japanese encephalitis, a mosquito-borne arbo-virus infection with seasonal distribution predominantly in Asia, two new vaccines emerged. Jebik V® is marketed in Japan by Takeda and Mitsubishi Tanabe and is dry-cell cultured containing the Beijing strain of Japanese encephalitis. The inactivated, adsorbed Japanese encephalitis vaccine Jespect® was first launched in Australia. Developed by Intercell, it is marketed by CSL. Patients undergoing liver transplantation for hepatitis

B-induced liver failure can now turn to Niuliva®, a human antihepatitis B immunoglobulin for intravenous administration, to mitigate the high risk of HBV reinfection. Launched initially in Spain and Italy, it is approved for use during the maintenance phase in nonreplicator patients and for the immunoprophylaxis of hepatitis B.

1. Armodafinil   (Sleep disorder treatment)   [2–6]

Country of Origin:	US
Originator:	Cephalon
First Introduction:	US
Introduced by:	Cephalon
Trade Name:	Nuvigil
CAS Registry No:	112111-43-0
Molecular Weight:	273.35

Armodafinil, an α1-adrenoceptor agonist, was launched last year for the oral treatment of excessive sleepiness associated with narcolepsy, SWSD, and obstructive sleep apnea/hypopnea syndrome (OSA). It is the R-enantiomer of modafinil, which is a previously marketed wake-promoting agent. The key differentiator for armodafinil is its longer pharmacokinetic half-life as compared with the S-enantiomer (10–14 h vs. 3–4 h). At therapeutic concentrations, armodafinil does not bind to most of the potentially relevant receptors for sleep/wake regulation (e.g., serotonin, dopamine, and adenosine receptors) or transporters of neurotransmitters or enzymes involved in sleep/wake regulation (e.g., serotonin, norepinephrine, and phosphodiesterase VI transporters). Both armodafinil and modafinil block dopamine reuptake by binding to the dopamine transporter and increasing dopamine concentrations in certain regions of the brain. However, dopamine receptor antagonists (e.g., haloperidol) and dopamine synthesis inhibitors (e.g., α-methyl-p-tyrosine) do not block modafinil's action. While the α1-adrenergic receptor antagonist prazosin attenuates the wake-promoting effect of modafinil, other in vitro assays failed to show agonistic activity on α-adrenergic receptors. Thus, the precise mechanism through which armodafinil and modafinil promote wakefulness is unknown. In addition to its wake-promoting effects and ability to increase locomotor activity in animals, modafinil produces psychoactive and euphoric effects, alterations in mood, perception, thinking, and feelings typical of other CNS stimulants in humans. Modafinil has reinforcing properties, as evidenced by its self-administration in monkeys previously trained to self-administer cocaine; modafinil was also partially discriminated as stimulant-like. However, the potential for abuse and dependency appears to be lower for modafinil than amphetamine-like stimulants. Modafinil is classified as a Schedule IV drug. The recommended dosage of armodafinil in patients with excessive sleepiness associated with narcolepsy or OSA is 150 or 250 mg/day, given as a single dose in the morning. In patients with SWSD, the recommended dosage is 150 mg/day, administered ~1 h before the start of the work shift. Oral armodafinil is rapidly absorbed and exhibits linear pharmacokinetics after single and multiple 50–400 mg doses. When administered in the fasting state, peak plasma concentrations are achieved at ~2 h. While oral bioavailability is not altered significantly by food, t_{max} is delayed by 2–4 h, which can alter the time to onset and the duration of pharmacological action. The apparent terminal elimination half-life of armodafinil is ~15 h, and the oral clearance at steady state is ~33 mL/min. Armodafinil is extensively metabolized via multiple mechanisms. These include amide hydrolysis, CYP3A4- and 3A5-mediated oxidation to the corresponding sulfone, aromatic ring hydroxylation, and subsequent glucuronidation of the hydroxylated products. Less than 10% of the drug is eliminated in the urine as the unchanged parent compound. The efficacy of once-daily armodafinil in improving wakefulness in adult patients with excessive sleepiness associated with nacrolepsy, OSA, or SWSD has been demonstrated in four large ($n > 195$), double-blind, multinational trials of 12-week duration. Compared with placebo, mean sleep latency (coprimary endpoint) was significantly improved with armodafinil 150 or 250 mg once daily in patients with OSA or narcolepsy, and with armodafinil 150 mg once daily in patients with SWSD, as assessed by the Multiple Sleep Latency Test (MSLT) or the Maintenance of Wakefulness Test (MWT). Furthermore, a significantly higher proportion of armodafinil than placebo recipients achieved a response (at least a minimal improvement) on the Clinical Global Impressions of Change (CGI-C) scale at study end in these four trials. Armodafinil was generally well tolerated in all four studies. The most common adverse events associated with armodafinil included headache, nausea, dizziness, and insomnia. Armodafinil can be obtained starting from ($\pm$)-2-(diphenylmethylsulfinyl)acetic acid via optical resolution using S-(–)-α-methylbenzylamine, followed by esterification to the methyl ester and subsequent amidation with ammonium hydroxide. It can also be derived enantioselectively from 2-(diphenylmethylthio)acetamide via asymmetric oxidation with cumene hydroperoxide and titanium isopropoxide-S,S-diethyl tartrate complex.

2. Asenapine   (Antipsychotic)   [7–10]

Country of Origin:	US
Originator:	Organon
First Introduction:	US
Introduced by:	Schering-Plough (Merck)
Trade Name:	Saphris
CAS Registry No:	65576-45-6(Free base)
	85650-56-2(maleate salt)
Molecular Weight:	285.77 (Free base)
	401.84 (maleate salt)

Asenapine, a dual antagonist of dopamine D_2 and serotonin 5-HT$_2$ receptors, was launched last year for the acute treatment of schizophrenia and manic or mixed episodes associated with bipolar I disorder in adults. It is the latest entry in the class of atypical antipsychotic agents to reach the market. Schizophrenia is diagnosed by the presence of positive psychotic symptoms, such as delusional thoughts, auditory hallucinations, and irrational fears. In addition to these symptoms, many patients experience the negative symptoms of social alienation, difficulty articulating, apathy, and lack of energy. The first-generation drugs, known as typical antipsychotics, antagonize only the dopamine D_2 receptors. Chlorpromazine and haloperidol fall into this category, and while they demonstrate efficacy against psychosis, they are ineffective against the negative symptoms, and dose-limiting EPS are a concern. The second-generation antipsychotics, also known as atypical antipsychotics, antagonize both the mesolimbic pathway dopamine D_2 receptors and the serotonin 5-HT$_{2A}$ receptors in the prefrontal cortex. This dual antagonism retains the antipsychotic activity with an improvement in the negative symptoms and a reduction in EPS. Previously marketed atypical antipsychotics include olanzapine (Zyprexa®), risperidone (Risperdal®), quetiapine (Seroquel®), paliperidone (Invega®), blonanserin (Lonasen®), and ziprasidone (Geodon®). Asenapine exhibits high affinity for serotonin 5-HT$_{1A}$, 5-HT$_{1B}$, 5-HT$_{2A}$, 5-HT$_{2B}$, 5-HT$_{2C}$, 5-HT$_5$, 5-HT$_6$, and 5-HT$_7$ receptors (K_i = 2.5, 4.0, 0.06, 0.16, 0.03, 1.6, 0.25, and 0.13 nM, respectively); dopamine D_1, D_2, D_3, and D_4 receptors (K_i = 1.4, 0.42, 1.1, and 1.3 nM, respectively); α_1- and α_2-adrenergic receptors (K_i = 1.2 and 1.2 nM, respectively); and histamine H_1 receptors (K_i = 1.0 nM), and moderate affinity for H_2 receptors (K_i = 6.2 nM). In *in vitro* assays, asenapine acts as an antagonist at these receptors. Asenapine has no appreciable affinity for muscarinic cholinergic receptors (e.g., K_i = 8128 nM for M_1). Despite its potent binding affinity for multiple receptors, the efficacy of asenapine in schizophrenia is thought to be primarily mediated through a combination of antagonist activity at D_2 and 5-HT$_{2A}$ receptors. Asenapine is supplied as the maleate salt, formulated as 5- and 10-mg tablets for sublingual administration. The recommended dose is 5 mg twice daily for schizophrenia and 10 mg twice daily for bipolar disorders. Following sublingual administration, asenapine is rapidly absorbed with peak plasma concentrations occurring within 0.5–1.5 h. The absolute bioavailability of asenapine at 5 mg is 35%. The intake of water 2–5 min after asenapine administration resulted in decreased exposure. Therefore, eating and drinking should be avoided for 10 min after administration. Asenapine has high protein binding (95%), high volume of distribution (~20–25 L/kg), and high clearance (52 L/h). Asenapine is extensively metabolized, predominantly by uridine diphosphate-glucuronyltransferase (UGT) and cytochrome P450 (CYP) enzymes, with asenapine N-glucuronide and N-desmethylasenapine being two major metabolites. Both of these metabolites are largely inactive. Following an initial more rapid distribution phase, the mean terminal half-life is approximately 24 h. With multiple-doses and twice-daily dosing, steady state is attained within 3 days. Asenapine is excreted as the glucuronide derivative, with approximately 50% of the dose in urine and 40% in feces. *In vitro* studies indicate that asenapine is a substrate for UGT1A4, CYP1A2, and to a lesser extent CYP3A4 and CYP2D6. Asenapine does not cause induction of CYP1A2 or CYP3A4 in cultured human hepatocytes. The safety and efficacy of asenapine in the treatment of schizophrenia and bipolar disorder have been demonstrated in several Phase III studies. In two randomized, controlled trials in adult patients with acute schizophrenia, treatment with asenapine reduced the clinician-assessed Positive and Negative Syndrome Scale (PANSS) total score from baseline to a significantly greater extent than placebo at 6 weeks. In a third study in schizophrenic patients with predominant, persistent, negative symptoms, asenapine at 26 weeks reduced the Negative Symptom Assessment (NSA-16) total score from baseline to an extent similar to that observed with olanzapine. Similarly, in two randomized, controlled trials of asenapine monotherapy and in one randomized, controlled trial of adjunctive asenapine therapy in adult patients with bipolar I disorder, asenapine produced significantly greater reductions from baseline than placebo in clinician-assessed Young Mania Rating Scale (YMRS) total score at 3 weeks. Asenapine was generally well tolerated in clinical trials, with most treatment-related adverse events being of mild to moderate severity. Incidence rates of clinically significant weight gain, extrapyramidal symptoms, hyperprolactinemia, and alterations in glucose or lipid metabolism were generally low. The most common adverse effects associated with asenapine treatment include insomnia, somnolence, nausea, anxiety, agitation, headache, vomiting, constipation, and psychosis. In placebo-controlled trials with risperidone, aripiprazole, and olanzapine in elderly subjects with dementia, there was a higher incidence of cerebrovascular adverse reactions (cerebrovascular accidents and transient ischemic attacks) including fatalities compared to placebo-treated subjects. Asenapine is not approved for the treatment of patients with dementia-related psychosis. As with other atypical antipsychotics, the asenapine label contains a black box warning on increased risk of death associated with off-label use to treat behavioral problems in older people with dementia-related psychosis. The chemical synthesis of asenapine can be achieved by several different methods. A concise synthesis of asenapine begins with the Horner–Emmons condensation of diethyl (5-chloro-2-fluoro)benzyl phosphonate with salicylaldehyde to produce the corresponding stilbene derivative, which undergoes a 1,3-dipolar cycloaddition reaction with trimethylamine-N-oxide in the presence of lithium diisopropylamide to furnish a trans-3,4-diarylpyrrolidine intermediate. Then, a base-catalyzed intramolecular cyclization involving the nucleophilic displacement of the fluoro group by the phenolic hydroxyl group affords asenapine, which is finally treated with maleic acid in ethanol to yield the corresponding maleate salt.

3. Besifloxacin   (Antibacterial)   [11–14]

Country of Origin:	Japan
Originator:	SSP Co. Ltd. (Japan)
First Introduction:	US
Introduced by:	Bausch & Lomb
Trade Name:	Besivance
CAS Registry No:	141388-76-3 (free base)
	405165-61-9 (hydrochloride)
Molecular Weight:	393.84 (free base)
	430.3 (hydrochloride)

Besifloxacin is a new fluoroquinolone antibacterial for ophthalmic use. It is indicated for the treatment of bacterial conjunctivitis, one of the most common ocular infections encountered in the primary care setting. Although the previously marketed fluoroquinolones ciprofloxacin, levofloxacin, ofloxacin, gatifloxacin, and moxifloxacin have been widely used to treat bacterial conjunctivitis, their continued utility is hampered by the emergence of resistance among key ocular isolates such as *Staphylococcus aureus*. For instance, about 85% of isolates of multidrug-resistant (MDR) strains of methicillin-resistant *S. aureus* (MRSA) have shown resistance to ciprofloxacin, moxifloxacin, gatifloxacin, and levofloxacin. Of MRSA isolates tested in a nationwide surveillance program, 68% were resistant to ciprofloxacin, 71% to gatifloxacin, 74% to levofloxacin, and 73% to moxifloxacin. Besifloxacin is the first fluoroquinolone developed exclusively for topical ophthalmic use. Unlike the previously marketed fluoroquinolones, besifloxacin has not been used systemically and is not in development as a systemic agent. Its exclusive indication as a topical agent is expected to reduce the overall environmental exposure of bacteria to besifloxacin, which may contribute to a lower risk for the emergence of bacterial resistance. Fluoroquinolones derive their antibacterial activity via inhibition of two essential bacterial enzymes, DNA gyrase and topoisomerase IV, which regulate processes of DNA replication. Besifloxacin inhibits DNA gyrase and topoisomerase IV from *Streptococcus pneumoniae* (IC_{50} = 1 and 0.4 mg/L, respectively) and *Escherichia coli* (IC_{50} = 1 and 10 mg/L, respectively). It demonstrates potent MIC_{90} (minimum inhibitory concentration required to inhibit the growth of 90% of isolates) values against Gram-positive and Gram-negative pathogens including methicillin- and ciprofloxacin-sensitive *S. aureus* (0.06 μg/mL), methicillin- and ciprofloxacin-resistant *S. aureus* (8 μg/mL), methicillin-sensitive *Staphylococcus epidermidis* (0.5 μg/mL), methicillin-resistant *S. epidermidis* (8 μg/mL),

Haemophilus influenzae (0.06 µg/mL), beta-lactamase-positive and ampi-cillin-resistant *H. influenzae* (0.12 µg/mL), and *S. pneumoniae* (0.12 µg/mL). *In vitro* studies of spontaneous resistance development by exposing inocula of approximately 10^{10} cells to besifloxacin at a concentration $4 \times$ MIC indicate a very low frequency for the development of resistant mutants in *S. aureus* ($<3.3 \times 10^{-10}$) and *S. pneumoniae* ($<4.7 \times 10^{-10}$). Besi-floxacin is supplied as 0.6% ophthalmic suspension of the hydrochloride salt. It is formulated with DuraSite®, a mucoadhesive polymer vehicle that results in a gel-forming drop, which extends residence time of the drug at the ocular surface. The recommended dose for besifloxacin suspension is one drop in the affected eye(s) three times a day, 4–12 h apart for 7 days. The mean peak besifloxacin concentration in tear fluid after a single drop applied in healthy adult subjects is 610 µg/g. Twenty-four hours after administration, the mean concentration in tear fluid is 1.6 µg/g. Following bilateral topical ophthalmic administration three times a day (16 doses total) in adults, peak plasma besifloxacin concen-trations after the first and last dose are <1.3 ng/mL. The mean besiflox-acin peak plasma concentration is 0.37 ng/mL on day 1, and 0.43 ng/mL on day 6. The average elimination half-life of besifloxacin is about 7 h. The clinical efficacy of besifloxacin has been evaluated in a randomized, multicenter, double-blinded, vehicle-controlled study. Patients aged 1 year and older with bacterial conjunctivitis were randomized to treat-ment with besifloxacin 0.6% ophthalmic suspension or vehicle applied topically three times daily for 5 days ($n = 957$). The primary endpoints were clinical resolution and microbial eradication of baseline bacterial infection at visit 2 (day 5 ± 1). Secondary endpoints included clinical resolution and microbial eradication at visit 3 (day 8 or 9), individual clinical outcomes at follow-up visits, and safety. Clinical resolution and microbial eradication were significantly greater with besifloxacin ophthalmic suspension than with vehicle at visit 2 (45.2% vs. 33.0%, $p = 0.0084$; and 91.5% vs. 59.7%, $p < 0.0001$, respectively) and visit 3 (84.4% vs. 69.1%, $p = 0.0011$; and 88.4% vs. 71.7%, $p < 0.0001$, respectively). Results for secondary endpoints of individual clinical outcomes were consistent with primary endpoints. Fewer eyes receiving besifloxacin ophthalmic suspension experienced adverse events than those receiving vehicle (9.2% vs. 13.9%; $p = 0.0047$). In another comparative trial, besi-floxacin 0.6% ophthalmic suspension was noninferior to moxifloxacin 0.5% ophthalmic solution in bacterial conjunctivitis patients aged ≥ 1 year with regard to clinical resolution (58.3% vs. 59.4%) and microbial eradication (93.3% vs. 91.1%) rates on day 5 ± 1 of treatment. The most common adverse event reported in patients treated with besifloxacin ophthalmic suspension was conjunctival redness. Other less common adverse events were blurred vision, eye pain, eye irritation, eye pruritus, and headache.

4. Canakinumab   (Anti-inflammatory) [15–17]

Country of Origin:	US	Class:	Recombinant monoclonal antibody
Originator:	Novartis		
First Introduction:	US	Type:	Fully human IgG1, Anti-IL-1β
Introduced by:	Novartis	Molecular Weight:	~150 kDa
Trade Name:	Ilaris	Expression system:	Murine Sp2/0-Ag14 cell line
CAS Registry No:	914613-48-2	Manufacturer:	Novartis

The symptoms of inherited autoinflammatory syndromes, which include recurrent fever, joint pain, and rash, are a result of the aberrant signaling of the proinflammatory cytokine interleukin-1β (IL-1β). Patients are clinically diagnosed by an observed mutation in the NLRP3 gene (nucleotide-binding domain, leucine-rich family, pyrin domain containing 3; also known as NALP3 and CIAS1) that leads to the overproduction of the protein cryopyrin. Diseases of this type, such as Familial Cold Autoinflammatory Syndrome (FCAS), Muckle–Wells Syndrome (MWS), and neonatal onset multisystem inflammatory disease (NOMID), are also characterized as CAPS. As the central component of the cryopyrin inflammasome, excess cryopyrin leads to stimulation of the inflammasome resulting in the activation of the other major component of the inflammasome, caspase 1, which is responsible for the processing of pro-IL-1β to its functional form. As an anti-IL-1β antibody, canakinumab's mechanism of intervention in the pathogenesis of CAPS is the blockade of the excessive IL-1β signaling. Compared to other drugs on the market targeting IL-1β, such as anakinra and rilonocept, canakinumab possesses a less frequent dosing regimen of just once every 2 months, instigates fewer injection site reactions, and is approved in children as young as 4 years of age. In addition to its initial approval for CAPS, development in other indications, including COPD, type 2 diabetes, and RA, is ongoing. Canakinumab has a binding affinity of 40 pM for human IL-1β and showed no cross-reactivity with human IL-1α or mouse IL-1β. In a mouse model of IL-1β-dependent joint inflammation, canakinumab suppressed swelling with an ED_{50} value of 0.6 mg/kg. It was first discovered using Medarex's UltiMab technology where transgenic mice were genetically engineered to express the human IgGκ repertoire from introduced human antibody genes following inactivation of the mouse genes. It is ultimately produced in mouse hybridoma Sp2/0 cells by recombinant DNA technology. The lyophilized powder (150 mg of canakinumab) is reconstituted with 1 mL of preservative-free, sterile water prior to subcutaneous injection. Following a single, 150 mg/mL injection in adult CAPS patients, the peak serum concentration (C_{max} = 16 ± 3.5 µg/mL) occurred in approximately 7 days. The absolute bioavailability was estimated to be 70%, and the mean terminal half-life was 26 days. In a typical CAPS patient with a body weight of 70 kg, the clearance was about 0.174 L/day, and the volume of distribution was around 6.01 L. Both parameters varied according to body weight. Over the dose range of 0.30–10.0 mg/kg given as an intravenous infusion or from 150 to 300 mg as subcutaneous injection, both area under the curve (AUC) and C_{max} increased in proportion to dose. Following 6 months of subcutaneous administration of 150 mg canakinumab every 8 weeks, the expected accumulation ratio was 1.3-fold. The safety and efficacy of canakinumab were assessed in a three-part clinical trial in patients with the MWS phenotype of CAPS, ranging in age from 9 to 74. The first part of the trial was an 8-week, open-label, single-dose period that involved the administration of canakinumab based on weight: 150 mg for patient >40 kg and 2 mg/kg for patients weighing 15–40 kg. Patients who experienced a complete clinical response and did not relapse by week 8 were randomized into part 2, a 24-week randomized, double-blind, placebo-controlled withdrawal period. Part 3, a 16-week open-label active treatment phase, included patients that completed part 2 or suffered a disease flare. A complete response was defined based on a number of parameters, including a rating of minimal or better for physician's assessments of disease activity and skin disease, C-reactive protein (CRP) values less than 10 mg/L, and serum amyloid A (SAA) less than 10 mg/L. If CRP and SAA exceeded 30 mg/L in combination with the defined physician's assessments, a disease flare was declared. In part 1, 71% of patients achieved a complete clinical response after the first week, increasing to 97% of patients by week 8. The most compelling data was recorded in the randomized withdrawal period where 81% of patients on placebo experienced a disease flare compared to none of the patients receiving canakinumab. The most common adverse events included vertigo, injection site reaction, nasopharyngitis, upper respiratory tract infections, influenza, diarrhea, nausea, headache, weight increase, and musculoskeletal pain. Since an increased risk of serious infections and an increased risk of neutropenia have been associated with other IL-1β blockers in combination with TNF-α inhibitors, concomitant use of canakinumab with anti-TNF-α therapy is not recommended. It is also not recommended to receive live vaccinations while being treated with canakinumab due to the risk of secondary transmission of infection. Finally, as an IL-1β blocker, treatment with canakinumab is expected to normalize the formation of CYP450 enzymes suppressed by the increased level of the IL-1β cytokine in CAPS; therefore, therapeutic monitoring may be necessary with concurrent drugs that interact with the CYP450 system.

5. Catumaxomab   (Anticancer) [18–21]

Country of Origin:	Germany	Class:	Trifunctional monoclonal antibody (trAb)
Originator:	TRION Pharma		
First Introduction:	Germany	Type:	Chimeric murine IgG2a / rat IgG2b Anti-EpCAM / Anti-CD3
Introduced by:	TRION Pharma	Molecular Weight:	~180 kDa Fresenius Biotech
Trade Name:	Removab	Expression system:	Rat/mouse quadroma cell line
CAS Registry No:	509077-98-9	Manufacturer:	TRION Pharma

The human Ep-CAM protein is overexpressed on a variety of tumors, including stomach, colon, prostate, ovarian, breast, and lung cancers. This tumor-associated antigen, a type I transmembrane glycoprotein of approximately 39–42 kDa that consists of a large extracellular domain with two epidermal growth factor-like domains, a single transmembrane domain, and a short cytoplasmic tail, acts as an epithelial cell adhesion and activating molecule. Since it has been demonstrated that its inhibition is associated with a decrease in the proliferation, migration, and invasion of tumor cells, Ep-CAM has emerged as a viable target for the immunotherapy of cancer. While the recently launched catumaxomab is an anti-Ep-CAM monoclonal antibody, it has been designed to induce more efficient killing of epithelial tumor cells. More specifically, catumaxomab is a bispecific trifunctional antibody (trAb). In addition to its Ep-CAM recognition arm, it possesses an anti-CD3 arm that targets the T-cell antigen CD3; binding to the CD3 receptor activates T cells to release cytotoxic cytokines and to promote T cell-mediated lysis. In essence, catumaxomab consists of one half of an anti-Ep-CAM antibody and one half of an anti-CD3 antibody. The trifunctionality comes into play with an intact Fc region that selectively binds to FCγ receptor-positive accessory cells, such as macrophages, dendritic cells, and natural killer (NK) cells that promote phagocytosis and antibody-dependent cell cytotoxicity. As a trAb that brings the requisite immune effector cells in close proximity to the tumor cells, catumaxomab destroys tumor cells possessing the surface antigen Ep-CAM via simultaneous activation and complex crosstalk of T cells and accessory immune cells. While it has gained approval for the intraperitoneal (i.p.) treatment of malignant ascites caused by Ep-CAM-positive metastatic epithelial-derived tumors, it is also being evaluated for the treatment of ovarian and gastric cancers. Malignant ascites is defined as the accumulation of tumor cell-containing fluid in the abdominal cavity due to increased microvascular permeability of the tumor vasculature. Catumaxomab fulfills the previously unmet need for management of this condition; paracentesis (draining of peritoneal fluid via a needle) was the most common treatment to relieve the abdominal pressure, but repeated punctures were typically necessary since it did not address the cause of the symptoms. Catumaxomab recognizes a discontinuous epitope in the extracellular region of Ep-CAM with high affinity as measured by a dissociation constant of 5.6×10^{-10} M, which was comparable to the full anti-Ep-CAM mAb. Preclinical studies demonstrated the interaction of different immune effector cells in the elimination of the tumor cells by several killing mechanisms: T-cell mediated lysis, cytotoxicity by multiple released cytokines, phagocytosis, and antibody-dependent cellular cytotoxicity via activation of FcγR-positive accessory cells. Catumaxomab was generated using a hybrid hybridoma or quadroma, where two hybridoma cells producing different mAbs are fused together to form a new hybrid cell that contains sets of genes encoding the two parent antibodies. The Ep-CAM recognition arm is a mouse IgG_{2a} heavy and light chain while the CD3-binding half is a rat IgG_{2b} heavy and light chain. This combination constitutes a hybrid mouse/rat Fc domain that has been found to facilitate a single-step purification from the monospecific parental antibodies by differential binding to a Protein A column. Purified catumaxomab is formulated in a 0.1 M sodium citrate buffer solution (pH 5.6) containing 0.02% polysorbate 80, and it is supplied as a 100 µg/mL concentrate for on-site dilution with sterile isotonic saline. The pharmacokinetic parameters of catumaxomab were evaluated in a Phase II, open-label, multicenter study in 13 patients diagnosed with malignant ascites requiring paracentesis. Over a period of 11 days, four intraperitoneal infusions of 10, 20, 50, and 150 µg/mL of catumaxomab were administered. Local and systemic exposures of the mAb were determined through high-sensitivity ELISA. A bridging ELISA was employed to monitor antidrug antibody (ADA) development. During the course of the treatment, increasing concentrations of catumaxomab were detected in the ascites with a median concentration of 6121 pg/mL. Nine of the patients exhibited systemic exposure with a median C_{max} of 403 pg/mL. High interpatient variability was observed with AUC values ranging from 0 to 10,020 pg*d/mL (median = 2045); however, it was noted that there was no correlation between safety parameters and systemic concentration. The mean elimination half-life was 2.13 days with almost full immunological activity (80–100%) after 3 days of circulation in ascites. ADA responses were not significant (>100 ng/mL) until the time of the last infusion. The safety and efficacy of catumaxomab were investigated in a combined Phase I/II study in women with recurrent ascites due to Ep-CAM-positive ovarian cancer. The primary efficacy endpoints were the reduction of ascites flow rate, the need for paracentesis during the treatment period, and tumor cell elimination from the ascites. In 17 patients, the median ascites flow rate was reduced from 105 mL/h at baseline to 23 mL/h 1 day following the fourth infusion. In addition, only one patient required paracentesis during the 37-day study period. A remarkable 99.9% reduction in the number of Ep-CAM-positive malignant cells in the ascites fluid was observed on average with six patients having levels below the limit of detection. Peripheral blood analysis also demonstrated a reduced number of circulating tumor cells following i.p. administration of catumaxomab. Most of the adverse events were attributed to the proposed immunological mechanism of action, such as fever, nausea, vomiting, and abdominal pain. Occurrence of adverse events coincided with the day of or day after infusion, with resolution of symptoms within days to weeks. In other clinical studies, liver enzyme elevations were noted but were considered transient in nature. As many patients developed HAMA and HARA responses following multiple doses of catumaxomab, the clinical benefit should outweigh the risk of severe adverse events associated with its immunogenicity for prolonged treatment.

6. Dapoxetine   (Premature Ejaculation)   [22–25]

Country of Origin:	US
Originator:	Lilly
First Introduction:	Finland, Sweden
Introduced by:	Janssen-Cilag
Trade Name:	Priligy
CAS Registry No:	119356-77-3 (Free base)
	129938-20-1 (Hydrochloride)
Molecular Weight:	305.41 (Free base)
	341.88 (Hydrochloride)

PE is the most common form of male sexual dysfunction affecting 20–40% of men globally at some point in their lives. Ejaculation is a reflex comprising different sensory pathways, motor centers, and nerve pathways. This ejaculatory reflex has been shown to be controlled primarily by serotonin and dopamine. The pathophysiology associated with PE includes a reduction in the serotonergic neurotransmission, 5-HT_{2c} receptor hyposensitivity, and/or 5-HT_{1a} receptor hypersensitivity. SSRIs, commonly used in the treatment of depression, are often used to treat PE, based on the observation that delayed ejaculation is a frequent side effect of this drug class. Studies in laboratory animals have shown that the administration of SSRIs actively blocks presynaptic membranes of 5-HT transporters, resulting in higher serotonin levels in the synaptic cleft. The serotonin then binds to 5-HT_{2c} and 5-HT_{1a} receptors to delay ejaculation. The most commonly used SSRIs to treat PE include the long-acting agents fluoxetine, paroxetine, sertraline, and citalopram. Several clinical studies have shown that chronic daily dosing of long-acting SSRIs results in significant prolongation of intravaginal ejaculatory latency time (IELT) in men with PE. However, chronic dosing of SSRIs is associated with unwanted sexual side effects and withdrawal symptoms upon discontinuation. Dapoxetine hydrochloride is the newest SSRI launched last year for the oral on-demand treatment of PE in men between 18 and 64 years of age. Dapoxetine is a short-acting SSRI. It is differentiated from the existing SSRI treatments for PE by the fact that it can be administered on an as-needed basis. In healthy subjects, dapoxetine is rapidly absorbed after oral administration with a peak plasma concentration (T_{max}) occurring between 1.4 and 2 h. This is followed by a rapid decline in plasma concentration, to about 5% of peak concentration at 24 h. Both the AUC and C_{max} increase proportionally with doses up to 100 mg. The mean initial half-life of dapoxetine after a single dose is 0.5–0.8 h, and this decreases slightly to 0.4–0.6 h after multiple doses for 6 days. The terminal half-life of dapoxetine is approximately 15–19 h after a single dose and 20–24 h after multiple doses. Dapoxetine undergoes hepatic metabolism to two metabolites, desmethyldapoxetine and didesmethyldapoxetine, both of which have much lower plasma concentrations compared with dapoxetine. The efficacy of dapoxetine has been studied in five separate multicenter, randomized, double-blind, placebo-controlled trials involving more than 6000 men with PE. In four studies that evaluated IELT as an endpoint ($n = 4843$), dapoxetine 30 and 60 mg achieved statistically significant increases in IELT versus placebo relative to the baseline value of 0.9–1 min. At the week 12 endpoint, the IELT values were 2.78–3.9 min for the 30-mg group, 3.32–4.2 min for the 60-mg group, and 1.75–2.4 min for placebo. Dapoxetine also showed statistically significant improvements in perceived control over ejaculation, PE-related personal distress, and other patient-reported outcomes in all five trials. Dapoxetine treatment was generally well tolerated, with low incidences of discontinuation syndrome, sexual dysfunction, and treatment-emergent mood symptoms. The most common adverse events with dapoxetine included nausea, diarrhea, headache, dizziness, and somnolence. Dapoxetine can be synthesized in four chemical steps starting from R-1-phenyl-1,3-propanediol via selective tosylation of the primary hydroxy group with p-toluenesulfonyl chloride, triethylamine and 4-(dimethylamino)pyridine (DMAP), and subsequent condensation with 1-naphthol by means of sodium- or lithium hydroxide to yield R-3-(1-naphthyloxy)-1-phenylpropanol as the key intermediate. Conversion of the alcohol group to the corresponding mesylate with methanesulfonyl chloride, triethylamine, and DMAP, followed by treatment with dimethylamine affords dapoxetine, which is then acidified to its hydrochloride salt.

7. Degarelix Acetate   (Anticancer)   [26–28]

Country of Origin: Switzerland
Originator: Ferring Pharmaceutical
First Introduction: US
Introduced by: Ferring Pharmaceutical
Trade Name: Firmagon
CAS Registry No: 214766-78-6
Molecular Weight: 1692.31

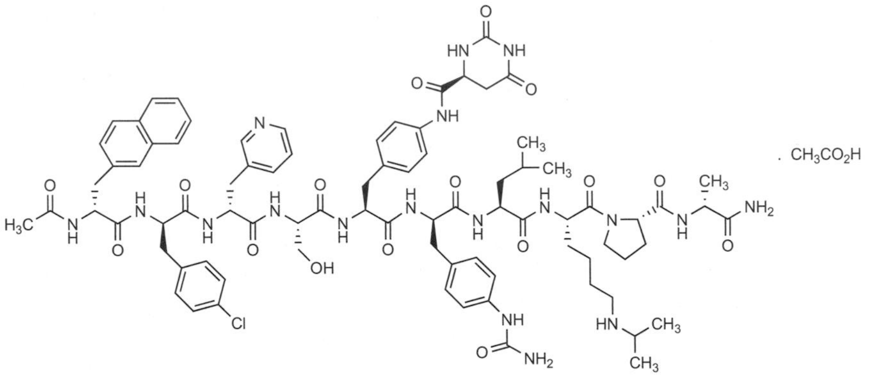

Antagonists of GnRH have proven to be an effective therapy for hormonally regulated cancers, such as prostate and some types of breast. As analogs of GnRH, they bind competitively and reversibly to GnRH receptors in the pituitary gland, thereby blocking the release of luteinizing hormone (LH) and follicle-stimulating hormone (FSH). In men, the reduction of LH triggers the ablation of testosterone secretion from the testes, and these castration-like levels have been essential in the effective management of advanced prostate cancer. In comparison to GnRH agonists, antagonists do not suffer from a potential flare of the disease as a result of an initial stimulation of the hypothalamic–pituitary–gonadal axis prior to down-regulation of the GnRH receptor. Moreover, GnRH antagonists provide beneficial effects more rapidly postdosing and result in a more efficient suppression of gonadotropin levels. With this in mind, degarelix acetate has been launched as a third-generation GnRH antagonist for the treatment of prostate cancer, and it joins other third-generation agents, ganirelix and cetronelix, on the market. Because of life-threatening allergic reactions, another GnRH antagonist abarelix was voluntarily withdrawn from the U.S. market in 2005. Relative to its predecessors, degarelix is touted as having increased aqueous solubility (>50 mg/mL), long-lasting effects, and weak histamine-releasing properties; histamine release has been an unwanted side effect with previous GnRH antagonists. Degarelix displays high potency for the GnRH receptor ($IC_{50} = 3$ nM), and in a castrated male rat assay, it exhibited a long duration of action by inhibiting LH secretion by more than 80% at 96 h following subcutaneous injection of 50 μg. Degarelix is a linear decapeptide amide containing seven unnatural amino acids. The replacement of D-Arg at position 6 by a D-ureidoalkyl residue is presumably responsible for the reduction in histamine release. One strategy for its synthesis involves Boc chemistry utilizing diisopropyl carbodiimide and 1-hydroxybenzotriazole in the coupling steps. Starting with an amide resin, the first four residues are coupled and deprotected (trifluoroacetic acid). After incorporation of N-α-Boc-D-4-(Fmoc-amino)phenylalanine, the side-chain Fmoc is cleaved by treatment with piperidine in DMF, and the liberated amine is converted to the corresponding urea by treatment with *tert*-butyl isocyanate. A similar approach is used for the incorporation of the next amino acid, N-α-Boc-L-4-(Fmoc-amino)phenylalanine, and subsequent acylation of the *p*-amino group with L-hydroorotic acid follows orthogonal deprotection. The remaining amino acids and amino acid analogs of the sequence are incorporated following the same Boc chemistry described above. Following an N-terminal acetylation with acetic anhydride, the peptide is cleaved from the resin, and benzyl and carbobenzyloxy side-chain protecting groups are removed by treatment with HF to afford crude degarelix. The drug is formulated as its acetate salt at doses of 120 or 80 mg per vial in accordance with a recommended starting regimen of two subcutaneous injections of 120 mg each with follow-up maintenance of 80-mg injections every 28 days. Following a subcutaneous injection of 240 mg at a concentration of 40 mg/mL, the mean C_{max} was 26.2 ng/mL, and the mean AUC was 1054 ng* day/mL with a T_{max} of about 2 days. Within the dose range of 120–240 mg, the pharmacokinetic parameters were linear and strongly influenced by the concentration of the injection solution. With a volume of distribution >1 L/kg, degarelix was widely distributed throughout total body water. The plasma protein binding was approximately 90%. The median terminal half-life was 53 days and was attributed to the very slow release of degarelix from the depot site. The elimination route was about 20–30% by renal excretion with the remainder excreted via the hepatobiliary system. In passing through the hepatobiliary system, degarelix was subject to peptide hydrolysis, so peptide fragments were predominantly detected in the feces. There were no quantitatively significant metabolites detected in the plasma. *In vitro* studies have shown that degarelix was not a substrate, inducer, or inhibitor of the CYP450 enzymes; therefore, drug–drug interactions are not anticipated. The safety and efficacy of degarelix were evaluated in an open-label, multicenter, randomized, parallel-group clinical study enrolling 620 patients with prostate cancer. Patients were randomized to receive degarelix at a starting dose of 240 mg followed by monthly doses of

160 mg subcutaneously, degarelix at a starting dose of 240 mg followed by monthly doses of 80 mg subcutaneously, or leuprolide at a dose of 7.5 mg intramuscularly monthly over the course of a year. The primary efficacy endpoint was the achievement and maintenance of serum testosterone suppression to castration levels. Compared to leuprolide that achieved a 96.4% castration rate from day 28 to day 364, the degarelix maintenance doses of 160 mg and 80 mg were equally successful with castration rates of 98.3 and 97.2%, respectively. In addition, patients receiving degarelix experienced a faster reduction in testosterone to castration levels. While degarelix reported a 96% castration rate by day 3, the data indicated that leuprolide was only at 18% by day 14 although it achieved a 100% rate by day 28. Prostate-specific antigen (PSA) levels were also monitored during the course of the study, and after 2 weeks, degarelix lowered PSA levels by 64%. After 1 month, PSA levels were lowered by 85%, and by the third month, levels were reduced by 95%. While this data was encouraging, it was noted that a correlation has not been established between the rapidity of PSA decline and clinical benefit. The most common adverse events included injection site reactions (pain, erythema, swelling, or induration), hot flashes, increased weight, and increases in serum levels of transaminases and gamma-glutamyltransferase. In addition to being contraindicated in patients with a previous hypersensitivity to degarelix, it should not be administered to women who are or may become pregnant as fetal harm can occur. Since long-term androgen deprivation therapy prolongs the QT interval, physicians should consider whether the benefits of degarelix outweigh the potential risks in patients with congenital long QT syndrome, electrolyte abnormalities, or congestive heart failure or in patients taking antiarrhythmic medications.

8. Dexlansoprazole　　(Gastroesophogeal reflux disease)　　[29–30]

Country of Origin:	Japan
Originator:	Takeda
First Introduction:	US
Introduced by:	Takeda
Trade Name:	Kapidex
CAS Registry No:	138530-94-6
Molecular Weight:	369.36

The mechanism of PPIs involves the irreversible binding to the hydrogen/potassium adenosine triphosphatase enzyme system, commonly referred to as the gastric proton pump, of the gastric parietal cell. As the last stage in gastric acid secretion, blockade of the gastric proton pump is an effective treatment for a variety of diseases requiring acid suppression, such as heartburn, peptic ulcers, and GERD. Dexlansoprazole is the latest PPI to hit the market, joining the ranks of omeprazole, rabeprazole, pantoprazole, esomeprazole, and lansoprazole, and is the *R*-enantiomer of the racemic lansoprazole. Compared to its predecessors, dexlansoprazole exhibits improved pharmacokinetics with slower clearance and longer terminal half-life. In addition, dexlansoprazole utilizes a novel DDR technology; drug release is optimized through the use of granules with different pH-dependent dissolution profiles, thereby providing an initial release in the proximal small intestine within 1–2 h of administration followed by a subsequent release at distal regions of the small intestine several hours later. With its longer duration of action culminating in more effective acid suppression, dexlansoprazole may have an advantage over conventional PPIs that possess single release formulations (immediate or delayed). Similar to all PPIs, dexlansoprazole is a prodrug that consists of pyridine and benzimidazole rings with a latent sulfenamide moiety. In order to form the disulfide bond with cysteine residues of the proton pump, dexlansoprazole must be activated through two protonations followed by a spontaneous rearrangement to unmask the sulfenamide. To achieve maximum efficacy, the drug must be concentrated in parietal cells. Delivery in its inactive form permits efficient entry into these cells where drug activation readily occurs. Dexlansoprazole may be obtained by optical resolution of lansoprazole (synthesis previously described) by two related methods that both involve forming a salt between the racemate and an optically active compound. The diastereomeric salt is separated by fractional crystallization and then subjected to a neutralization process to provide dexlansoprazole. Alternatively, the diastereomeric salt may be separated by chiral column chromatography. Dexlansoprazole is supplied in 30- and 60-mg capsules with the DDR formulation for oral administration and has been approved for the healing of all grades of EE (60 mg once daily for up to 8 weeks), maintenance of healed EE (30 mg once daily for up to 6 months), and treating heartburn associated with nonerosive GERD (30 mg once daily for 4 weeks). As previously mentioned, the DDR formulation results in a dexlansoprazole plasma concentration–time profile with two distinct peaks, within 1–2 h and between 4 and 5 h, following oral administration. This results in a prolonged mean residence time with increased duration of plasma concentrations, which translates to a greater percentage of time the pH is maintained above 4. The mean C_{max} was 658 ng/mL for the 30-mg dose and 1397 ng/mL for the 60-mg dose. The corresponding AUC values were 3275 ng*h/mL and 6529 ng*h/mL, respectively. After multiple doses in symptomatic GERD patients, the apparent volume of distribution was 40.3 L. Independent of concentration, the plasma protein binding ranged from 96.1 to 98.8%. Despite having a terminal half-life typical of other PPIs of approximately 1–2 h, the DDR formulation was a significant factor in prolonging the drug residence time. Dexlansoprazole was extensively metabolized in the liver by oxidation, reduction, and subsequent formation of inactive glucuronide, sulfate, and glutathione conjugate; however, dexlansoprazole was the major circulating component in plasma. Oxidative hydroxylation was mainly mediated by CYP2C19, with CYP3A4 being responsible for oxidation to the sulfone. The route of elimination was both renal and hepatic with an apparent clearance of 11.4–11.6 L/h after 5 days of 30 or 60 mg once daily. In two multicenter, double-blind, active-controlled, randomized, 8-week studies, the safety and efficacy of dexlansoprazole were compared to lansoprazole in patients with endoscopically confirmed EE, the severity of which was determined by the Los Angeles Classification Grading System (Grades A–D). The 4092 patients enrolled were randomized to receive 60 mg of dexlansoprazole, 90 mg of dexlansoprazole, or 30 mg of lansoprazole daily. The studies were designed to test noninferiority, and if noninferiority was confirmed, then superiority would be tested. While noninferiority was indeed demonstrated in both studies, the finding of superiority in one study was not replicated in the other. Regarding symptomatic nonerosive GERD, a multicenter, double-blind, placebo-controlled, randomized, 4-week study was conducted in patients that met three criteria: a history of heartburn for 6 months or longer, heartburn on at least 4 of 7 days immediately prior to randomization, and confirmed not to have EE. Patients randomized to receive 30 mg of dexlansoprazole experienced a statistically significantly greater percentage of days with heartburn-free 24-h periods (54.9%) versus patients on placebo (18.5%). The most commonly recorded adverse reactions that occurred at a higher incidence than placebo were diarrhea, abdominal pain, nausea, vomiting, flatulence, and upper respiratory tract infection. As dexlansoprazole inhibits gastric acid secretion, its use is expected to interfere with the absorption of drugs with pH-dependent oral bioavailability. Since the HIV protease inhibitor atazanavir is dependent on gastric acid for absorption, dexlansoprazole should not be co-administered with atazanavir to avoid a loss of therapeutic efficacy. While co-administration of dexlansoprazole did not affect the pharmacokinetics of warfarin or INR (international normalized ratio: the ratio of a patient's prothrombin time to a normal sample), there have been reports of increased INR and prothrombin time in patients receiving concomitant treatment with PPIs and warfarin. Since increases in INR and prothrombin time may lead to abnormal bleeding and possibly death, concomitant use of dexlansoprazole and warfarin may necessitate monitoring for increases in INR and prothrombin time. As dexlansoprazole is partly metabolized by CYP2C19 and CYP3A4, plasma levels may be affected by co-administration of drugs that inhibit or induce these enzymes. Concomitant use of dexlansoprazole with tacrolimus may lead to an increase in whole blood levels of tacrolimus, especially in patients who are intermediate or poor metabolizers of CYP2C19. As dexlansoprazole has been detected in rat milk, nursing mothers are advised to discontinue drug or nursing depending on the importance of the drug for the mother. Finally, dexlansoprazole is contraindicated in patients with known hypersensitivity to any component of the formulation.

9. Dronedarone   (Antiarrhythmic)   [31–35]

Country of Origin: US
Originator: Sanofi-Aventis
First Introduction: US
Introduced by: Sanofi-Aventis
Trade Name: Multaq
CAS Registry No: 141626-36-0
Molecular Weight: 556.76

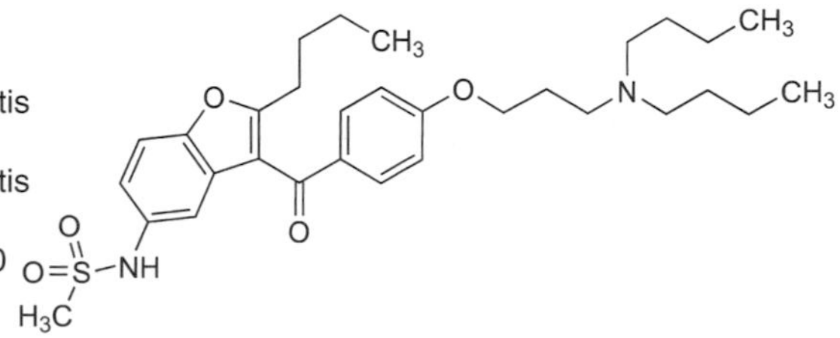

AF is the most common form of sustained cardiac arrhythmia, with an increasing prevalence in the aging population. AF accounts for 34.5% of arrhythmia-related hospital admissions in the United States. The most significant consequences of AF include congestive heart failure, a 5-fold increased risk of stroke, and increased rate of mortality. Although a 90% conversion rate from AF to normal sinus rhythm (NSR) can be achieved with electrical cardioversion, up to 70% of these patients require additional therapy with antiarrhythmic drugs in order to maintain NSR. Among the currently available antiarrhythmic agents, amiodarone (Cordarone®) is the most frequently used drug to maintain NSR. However, it is often associated with many extracardiac adverse effects, especially during chronic use. In addition, its dosing regimen is complicated due to an extremely large volume of distribution (~70 L/kg) and a long elimination half-life (30–55 days). Dronedarone, a close analog of amiodarone, is structurally modified to provide improved safety and pharmacokinetic profile. With the introduction of a sulfonamide group, dronedarone is less lipophilic, has lower tissue accumulation, and has a much shorter serum half-life (~24 h) compared with amiodarone. Additionally, dronedarone lacks the iodine moieties that are responsible for thyroid dysfunctions associated with amiodarone. Dronedarone is specifically indicated to reduce the risk of cardiovascular hospitalization in patients with paroxysmal or persistent AF or AFL, with a recent episode of AF/AFL and associated cardiovascular risk factors, who are in sinus rhythm or who will be cardioverted. Similar to amiodarone, dronedarone is a potent blocker of multiple ion currents (including the rapidly activating delayed-rectifier potassium current, the slowly activating delayed-rectifier potassium current, the inward rectifier potassium current, the acetylcholine-activated potassium current, peak sodium current, and L-type calcium current) and exhibits antiadrenergic effects. The recommended dosage of dronedarone in adults is 400 mg twice daily via oral administration. It is extensively metabolized, mainly by CYP3A4 enzyme. The initial metabolic pathway includes N-debutylation to form the active N-debutyl metabolite, oxidative deamination to form the inactive propanoic acid metabolite, and direct oxidation. The N-debutyl metabolite exhibits pharmacodynamic activity but is 3- to 10-fold less potent than dronedarone. The absolute bioavailability of dronedarone without food is low ($\sim$4%). It increases to approximately 15% when administered with a high fat meal. After oral administration under fed conditions, peak plasma concentrations of dronedarone and the main circulating active metabolite (N-debutyl derivative) are reached within 3–6 h. After repeated administration of 400 mg twice daily, steady state is reached within 4–8 days, and the mean accumulation ratio for dronedarone ranges from 2.6 to 4.5. The steady-state C_{max} and exposure of the main N-debutyl metabolite are similar to that of the parent compound. Both dronedarone and its N-debutyl metabolite have high protein binding (>98%). After intravenous administration, the volume of distribution at steady state is about 1400 L. The plasma clearance of dronedarone is 130–150 L/h. Approximately 84% of the administered dose is excreted in feces, mainly as metabolites. The efficacy and safety of dronedarone have been studied in several placebo-controlled clinical trials at an oral dose of 400 mg twice daily. In two randomized, controlled trials involving 1237 patients with AF or AFL, dronedarone was shown to be more effective than placebo in maintaining sinus rhythm and in controlling the ventricular rate during recurrences of AF or AFL. The pooled data from these two studies showed that dronedarone delayed the time to first recurrence of AF or AFL to 60–158 days compared with 5–59 days with placebo. At 12 months of follow-up, rates of pulmonary, thyroid, and hepatic adverse effects were not significantly greater with dronedarone than with placebo. In another 30-month trial involving 4628 patients with paroxysmal or persistent AF or AFL who had additional risk factors for death (such as hypertension, diabetes, or previous stroke), the primary outcome was first hospitalization for cardiovascular reasons or death from any cause. The primary outcome occurred in $\sim$32% of patients in the dronedarone group, compared with $\sim$39% of patients in the placebo group. This difference was largely attributable to the effect of dronedarone on cardiovascular hospitalization, principally hospitalization related to AF or AFL. Overall, dronedarone was well tolerated. The most common side effects were gastrointestinal in nature and included nausea, vomiting, and diarrhea.

10. Eltrombopag (Antithrombocytopenic)   [36–38]

Country of Origin: US
Originator: Ligand Pharmaceuticals
First Introduction: US
Introduced by: GlaxoSmithKline
Trade Name: Promacta
CAS Registry No: 496775-61-2
Molecular Weight: 442.47

ITP is an autoimmune disease in which antiplatelet antibodies accelerate the destruction of platelets. In addition, platelet production can be impaired because the antiplatelet antibodies can also damage megakaryocytes. Although the thrombocytopenia of ITP can be severe, signs of bleeding are usually only minor. The disease is characterized by persistently low platelet counts (<20,000 per µL), which lead to an increased risk of serious bleeding, such as intracranial hemorrhage. The goal of managing chronic ITP is to maintain platelet counts at levels that prevent bleeding, thereby reducing treatment-related toxicity. Traditional therapies of ITP include glucocorticoids and intravenous immunoglobulins, which increase platelet counts primarily by reducing the extent of platelet destruction. However, the recognition that platelet production in ITP is often suboptimal has led to the use of treatments that enhance thrombopoiesis. This approach has focused on TPO, the growth factor underlying megakaryocytopoiesis. TPO is the ligand for the TPO receptor on megakaryocytes and platelets. Eltrombopag is an oral, small-molecule, nonpeptide TPO-receptor agonist. It initiates TPO-receptor signaling by interacting with the transmembrane domain of the receptor, thereby inducing proliferation and differentiation of cells in the megakaryocytic lineage. Eltrombopag is the second TPO-receptor agonist to reach the market behind romiplostim (Nplate®), a recombinant fusion protein launched in 2008. Both drugs are specifically indicated for the treatment of thrombocytopenia in patients with chronic ITP who have had an insufficient response to corticosteroids, immunoglobulins, or splenectomy. Eltrombopag is supplied as the ethanolamine salt (also known as eltrombopag olamine). The recommended starting dose of eltrombopag is 50 mg once daily for most patients. For patients of East Asian ancestry or patients with moderate or severe hepatic insufficiency, the starting dose is 25 mg once daily. The eltrombopag dose should be adjusted to achieve platelet counts ≥ 50,000 per µL as necessary to reduce the risk of bleeding. The absorption of an orally administered dose of eltrombopag is about 50%, and the time to reach peak concentration (T_{max}) is 2–6 h. Eltrombopag is extensively metabolized, predominantly through oxidation via CYP1A2 and CYP2C8 and conjugation with glucuronic acid, glutathione, or cysteine. The predominant route of eltrombopag excretion is via feces (59%), and 31% of the dose is found in the urine. Unchanged eltrombopag in feces accounts for approximately 20% of the dose. Unchanged eltrombopag is not detectable in urine. The plasma elimination half-life of eltrombopag is approximately 21–32 h in healthy subjects and 26–35 h in ITP patients. The safety and efficacy of eltrombopag were evaluated in two double-blind, placebo-controlled clinical studies in adult patients who had failed at least one prior ITP therapy and who had baseline platelet counts <30,000 per μL. Response rate was defined as an increase in platelet count from baseline to ≥ 50,000 per μL. The first study randomized 114 patients to eltrombopag 50 mg or placebo daily. In the second study, 117 patients were randomized to either placebo or one of three doses of eltrombopag (30, 50, or 75 mg). The median duration of treatment was 6 weeks in both studies. The response rates in the 50-mg treatment groups in these two studies were 70 and 59%, compared to placebo response rates of 11 and 16% ($p < 0.001$), respectively. The 75-mg group had a response rate of 81%, and the response rate in the 30-mg group was not significantly different from placebo. In another open-label, extension study, eltrombopag was administered to 109 patients, with 74 patients receiving the drug for at least 3 months, 53 for at least 6 months, and 3 for at least 1 year. At baseline, the median platelet count was 18,000 per μL. Eltrombopag treatment resulted in median platelet counts of 74,000, 67,000, and 95,000 per μL, at 3, 6, and 9 month follow-up time points, respectively. The most common adverse reactions associated with eltrombopag were nausea, vomiting, mennorrhagia, myalgia, paresthesia and cataracts. Eltrombopag is derived in five synthetic steps starting from 2-bromo-6-nitroanisole, via Suzuki coupling reaction with 3-carboxyphenylboronic acid to a biphenyl intermediate, followed by cleavage of the methyl ether moiety with hydrobromic acid, and reduction of the nitro group to an amino group under catalytic hydrogenation conditions. Subsequently, the amino group is diazotized with sodium nitrite and hydrochloric acid, and the diazonium intermediate is condensed with 1-(3,4-dimethylphenyl)-3-methyl-2,5-dihydro-1H-pyrazol-5-one to produce eltrombopag.

11. Eslicarbazepine Acetate   (Antiepileptic)   [39–42]

Country of Origin:	Spain
Originator:	BIAL Group
First Introduction:	UK
Introduced by:	Eisai
Trade Name:	Zebinix; Exalief, Stedesa
CAS Registry No:	236395-14-5
Molecular Weight:	296.32

Although epilepsy is a neurological disorder with varying etiology and severity, the common feature is unprovoked, recurring seizures. Whether classified as generalized, involving both cerebral hemispheres, or partial, with only localized portions of brain participation at onset, effective treatment relies on accurate assessment of syndrome type to optimally decrease the frequency, duration, and severity of seizures. In particular, treatment of partial-onset seizures presents a continual challenge since nearly 60% of these patients do not achieve seizure control with current AEDs. Even when traditional AEDs, such as phenobarbital, phenytoin, carbamazepine, and valproate, control seizure activity, their effectiveness is hampered by considerable side effects and the tendency for drug–drug interactions. Designed for improved efficacy and safety, eslicarbazepine acetate is a third-generation AED that has been approved as adjunctive therapy for partial-onset seizures with or without secondary generalization. As a member of the carbamazepine family (first generation), it is the prodrug of eslicarbazepine and the active metabolite of oxcarbazepine (second generation). While the precise mechanism of action has not been elucidated, preclinical data suggests that the drug competitively interacts with site 2 of the inactivated state of a voltage-gated sodium channel, thereby preventing its return to the active state and repetitive neuronal firing. The ability of eslicarbazepine acetate to block voltage-gated sodium channels was assessed by displacement of tritiated batrachotoxinin A 20α-benzoate binding in rat cortical synaptosomes. Its IC_{50} of 138 μM was significantly improved over carbamazepine (210 μM), and it demonstrated approximately 96% inhibition of $^{22}Na^+$ uptake at a dose of 300 μM versus 64.8% with a similar dose of carbamazepine. Starting from oxcarbazepine, eslicarbazepine acetate may be prepared by related resolution methods. Reduction with sodium borohydride provides the racemic alcohol that is subsequently derivatized with either menthoxyacetyl chloride or di-O-acetyl-L-tartaric acid anhydride to afford diastereomeric esters. The desired S-diastereomer is isolated in either case by fractional crystallization, and the alcohol is liberated under alkaline conditions. The final acetylation is performed with acetyl chloride and DMAP in pyridine and dichloromethane. The active ingredient is packaged in tablets at doses of 400, 600, or 800 mg. The recommended daily dose is 800 mg once daily following a 1 week initiation phase of 400 mg once daily. Based on individual response, the dose may be increased to 1200 mg once daily; other dose adjustments may be necessary depending on renal function. As a prodrug of eslicarbazepine, the plasma levels of eslicarbazepine acetate usually remain below the limit of quantification. Since more than 90% of the eslicarbazepine acetate dose is recovered in the urine as metabolites, eslicarbazepine and its glucuronide, and more than 95% of systemic exposure is attributed to the main metabolite eslicarbazepine, oral bioavailability is classified as high. A C_{max} of 14,287 ng/mL was achieved for eslicarbazepine at a median time of 1 h following a single, 800-mg dose in healthy volunteers. The corresponding AUC was 236,834 h*ng/mL. The renal clearance was determined to be 17 mL/min while the elimination half-life was 11 h. The volume of distribution was approximately 34 L, and plasma protein binding was estimated to be less than 40%. In addition to the main metabolite eslicarbazepine and its glucuronide (inactive), minor active metabolites included oxidation to oxcarbazepine, which in turn is partly metabolized to R-licarbazepine. In a Phase II, placebo-controlled study conducted in Europe in patients with refractory partial-onset seizures, eslicarbazepine acetate was demonstrated to be efficacious and well tolerated with a 54% responder rate compared to 28% with placebo. All three of the Phase III evaluations included patients with at least four partial seizures per 4 weeks despite treatment with up to three AEDs, and all were designed as multicenter, randomized, double-blind, placebo-controlled studies. Three doses (400, 800, or 1200 mg once daily) were investigated as an adjunctive therapy and comprised an 8-week baseline period, followed by a double-blind, 2-week titration, and a double-blind, 12-week maintenance period. Of the three doses utilized, the 400-mg dose was not statistically different from placebo; however, the two higher doses recorded a significant reduction in median seizure frequency compared to placebo. Carbamazepine was the most frequently used concomitant AED, and it was used in about 60% of the patients. The most common adverse events occurring with an incidence of >2% were dizziness, somnolence, nausea, diplopia, headache, vomiting, abnormal coordination, blurred vision, vertigo, and fatigue. While most of these side effects appeared to be mild to moderate, they did display a dose dependency. Although eslicarbazepine acetate is only contraindicated in patients with a known hypersensitivity to the carbamazepine family of AEDs, caution should be exercised in patients with cardiac conduction abnormalities (potential for prolongations in PR intervals), renal impairment, and in patients of Han Chinese and Thai origin since the presence of HLA-B*1502 allele in these individuals has been shown to be strongly associated with the risk of developing Stevens–Johnson syndrome. Regarding drug interactions, eslicarbazepine acetate may decrease the effectiveness of hormonal contraceptives, so alternative or additional methods of contraception should be considered. Since eslicarbazepine acetate has been shown to be a weak inducer of CYP3A4 and UDP-glucuronyl transferases and an inhibitor of CYP2C19 in vitro, dose adjustment of concomitant drugs metabolized by these pathways may be necessary. In particular, concomitant use of the antiepileptic phenytoin may result in the need to increase the eslicarbazepine acetate dose due to induction of glucuronidation by phenytoin, and the phenytoin dose may need to be decreased due to inhibition of CYP2C19 by eslicarbazepine resulting in an increase in exposure of phenytoin. Of the other potential co-administered AEDs, concomitant carbamazepine may increase the risk of adverse reactions, particularly diplopia, abnormal coordination, and dizziness.

12. Febuxostat   (Anti-hyperuricemic)   [40–47]

Country of Origin:	Japan
Originator:	Teijin
First Introduction:	US
Introduced by:	Takeda/Teijin/Ipsen
Trade Name:	Uloric, Adenuric
CAS Registry No:	144060-53-7
Molecular Weight:	316.37

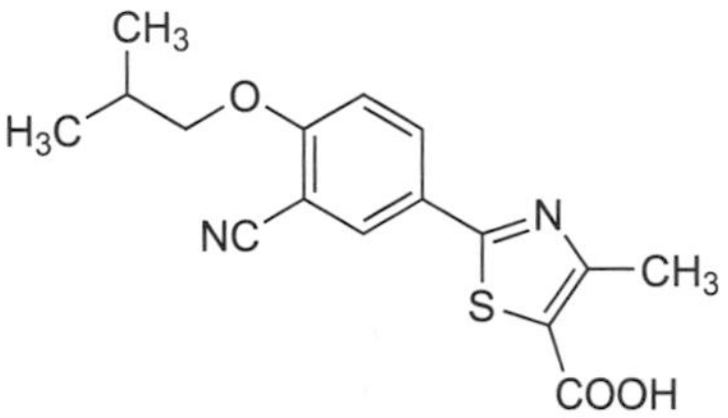

Febuxostat, a selective xanthine oxidase inhibitor, was launched last year for the chronic management of hyperuricemia in patients with gout. Hyperuricemia is defined as a serum uric acid concentration exceeding the limit of solubility. It predisposes affected persons to gout, a disease characterized by the formation of crystals of monosodium urate or uric acid from supersaturated fluids in joints and other tissues. Crystal deposition is asymptomatic, but it is revealed by bouts of joint inflammation. If left untreated, further crystals accumulate in joints and can form deposits known as tophi. A major aim in gout management is the long-term reduction of serum uric acid concentrations below saturation levels, as this results in crystal dissolution and eventual disappearance. Xanthine oxidase inhibitors reduce serum uric acid levels by impeding the transformation of hypoxanthine to xanthine and of xanthine to uric acid; both conversions are catalyzed by the xanthine oxidoreductase enzymes xanthine oxidase and xanthine dehydrogenase. The active sites of both enzymes are structurally equivalent; xanthine dehydrogenase readily converts to xanthine oxidase in mammals. Allopurinol, a xanthine oxidase/xanthine dehydrogenase inhibitor, has been the mainstay of uric-acid-lowering therapy for more than three decades. However, a significant proportion of patients receiving allopurinol do not achieve the desired reduction in serum uric acid levels, and the side effects of the drug, although uncommon, can be severe and are more frequent in patients with renal impairment. Another key drawback of allopurinol is the fact that it inhibits other enzymes besides xanthine oxidase in the purine and pyrimidine metabolism pathways. Febuxostat is a nonpurine derivative with higher potency and selectivity than allopurinol for inhibiting xanthine oxidase. It completely inhibits human xanthine oxidase activity in the lung cancer cell line A549, whereas the activities of other enzymes involved in purine or pyrimidine metabolism (e.g., purine nucleoside phosphorylase, adenosine deaminase, and pyrimidine nucleoside phosphorylase) are affected by <4%. The IC_{50} for xanthine oxidase derived from bovine milk, mouse liver, or rat liver are 1.4, 1.8, and 2.2 nM, respectively, for febuxostat as compared with 1700, 380, and 100 nM, respectively, for allopurinol. The recommended starting dosage of febuxostat is 40 mg once daily taken orally. For patients who do not achieve a serum uric acid concentration of less than 6 mg/dL after 2 weeks at the 40 mg dose, febuxostat 80 mg once daily is recommended. Following oral administration of febuxostat, the time to peak concentration is 0.5–1.5 h. The plasma concentrations increase proportionally over a range of doses from 10 to 120 mg. At doses from 120 to 240 mg, peak plasma concentrations remain dose proportional, whereas the increase in the AUC is more than dose proportional. Febuxostat is highly bound to albumin (approximately 99%). The volume of distribution at steady state is approximately 0.7 L/kg. Febuxostat is primarily metabolized in the liver to acyl-glucuronide derivatives via uridine diphosphate glucuronosyltransferase (UGT) enzymes and, to a lesser extent, to oxidative metabolites via cytochrome P450 enzymes. The renal clearance of unchanged drug is less than 6%. The mean elimination half-life of febuxostat is about 5–8 h. The safety and efficacy of oral once-daily febuxostat has been studied in two randomized controlled Phase III trials involving 1832 patients with hyperuricemia (serum uric acid level ≥ 8.0 mg/dL) and gout. The primary efficacy endpoint in each study was the proportion of patients whose serum uric acid levels were <6.0 mg/dL at the last three monthly measurements. In both trials, significantly more febuxostat-treated gout patients met the primary endpoint (48 and 53% with 80 mg; 65 and 62% with 120 mg), compared with those receiving allopurinol 300 mg (22 and 21%; $p < 0.001$ in both studies). Febuxostat was more effective than allopurinol in the subset with impaired renal function, with no requirement of dose adjustment if the renal impairment is mild to moderate. Long-term extension studies confirmed the efficacy and tolerability of febuxostat. In patients who achieved the serum uric acid target of 6 mg/dL, the incidence of gout flares fell steadily and tophi resolved in many patients. The incidence of adverse events such as dizziness, diarrhea, headache, and nausea with febuxostat was similar to allopurinol. Febuxostat is contraindicated in patients being treated with the xanthine oxidase substrates such as azathioprine, mercaptopurine, and theophylline. Febuxostat can be synthesized in a multistep sequence from 2,4-dicyanophenol, starting with the alkylation of the phenolic hydroxyl group with isobutyl bromide and potassium carbonate, followed by treatment with thioacetamide in hot dimethyl formamide to yield 3-cyano-4-isobutoxythiobenzamide. Cyclization of the thioamide group with 2-chloroacetoacetic acid ethyl ester in refluxing ethanol affords 2-(3-cyano-4-isoutoxyphenyl)-4-methylthiazole-5-carboxylic acid ethyl ester, which is hydrolyzed with sodium hydroxide to produce febuxostat.

13. Golimumab   (Anti-inflammatory) [48–52]

Country of Origin:	US	Class:	Recombinant monoclonal antibody
Originator:	Centocor Ortho Biotech		
First Introduction:	US, Canada	Type:	Fully human IgG1, Anti-TNF-
Introduced by:	Centocor Ortho	Molecular Weight:	~150 kDa Biotech / Schering-Plough
Trade Name:	Simponi	Expression system:	Murine Sp2/0-cell line
CAS Registry No:	476181-74-5	Manufacturer:	Centocor Ortho Biotech

The proinflammatory cytokine tumor necrosis factor-alpha (TNF-α) has been implicated as the primary mediator of articular inflammation in diseases such as RA, PsA, and AS. Targeting TNF-α has been a successful strategy in the intervention of a range of immunoinflammatory disorders. Biologics have risen to the forefront of TNF-α blockade with infliximab, a chimeric monoclonal antibody (mAb), and etanercept, a fusion protein comprised of the ligand-binding segment of the soluble TNF receptor, reaching the market in the late 1990s for the initial indication of RA. Since then, their use has expanded to other diseases of inflammatory etiology, and two additional TNF-α inhibitors have joined the competition. As TNF-α-specific mAbs, adalimumab (fully human) and certolizumab pegol (PEGylated humanized) cover the above indications in addition to Crohn's disease. With the success of these groundbreaking TNF-α inhibitors, the quest has continued for the discovery of agents with improved dosing regimens or activity in refractory patient populations. Golimumab, a human anti-TNF-α mAb that binds to both soluble and transmembrane forms of TNF-α, is the first once-monthly subcutaneous agent to enter the market and is currently approved for the treatment of RA in combination with methotrexate (MTX), PsA alone or in combination with MTX, and AS. Like canakinumab, it was first discovered using Medarex's UltiMab technology. Although the amino acid sequences of the constant regions (heavy and light chain) are identical to infliximab, the heavy- and light-chain variable regions of golimumab are human. By surface plasmon resonance, the affinity of golimumab for human TNF-α is 17 pM, and in solution, it is 1.4 pM with an apparent dissociation constant of 0.14 nM. The recommended dosage is 50 mg administered subcutaneously once a month. Following a 50-mg subcutaneous injection in healthy patients, a C_{max} of approximately 2.5 µg/mL was achieved with a T_{max} ranging from 2 to 6 days, and the absolute bioavailability was estimated at about 53%. The pharmacokinetics of golimumab in patients with RA were dose-proportional in the range of 0.1–10.0 mg/kg following a single intravenous dose. Over the same dose range, the mean systemic clearance was estimated to be 4.9–6.7 mL/day/kg, and the mean volume of distribution varied from 58 to 126 mL/kg, suggesting limited extravascular distribution. The median terminal half-life was approximately 2 weeks in all patient populations. The safety and efficacy of golimumab in treating RA were evaluated in three multicenter, randomized, double-blind, controlled trials (RA-1, RA-2, and RA-3) comprising a total of 1542 patients ≥ 18 years of age. For admission, patients had to be diagnosed according to the American College of Rheumatology (ACR) criteria with moderate to severe RA with an endurance of at least 3 months prior to first administration of study agent. In addition, patients had to possess at least four swollen and four tender joints. In RA-1, patients who were previously treated with one or more doses of a biologic TNF-α blocker without adverse reaction were randomized to receive either 50 or 100 mg of golimumab or placebo. While patients were permitted to continue stable doses of MTX, sulfasalazine (SSZ), and/or hydroxychloroquine (HCQ) during the trial, the use of other DMARDs (disease-modifying antirheumatic drugs) including cytotoxic agents or other biologics was prohibited. Study RA-2 consisted of patients with active RA despite prior stable dosing with MTX. These patients were also biologic TNF-α-blocker-naive. Patients in this group were randomized to MTX alone, 50 or 100 mg of golimumab plus background MTX, or 100 mg of golimumab monotherapy. In study RA-3, patients were both MTX- and biologic TNF-α-blocker-naive, and dosing randomization was identical to RA-2. The primary efficacy endpoint for RA-1 and RA-2 was the percentage of patients achieving an ACR-20 response at week 14 while RA-3 measured the percentage of patients achieving an ACR-50 response at week 24. In these three trials, a greater percentage of patients treated with a combination of golimumab and MTX met the defined primary efficacy endpoint versus those on MTX alone or placebo. For example, 35% of patients receiving 50 mg of golimumab in RA-1 achieved an ACR-20 response at week 14 compared to 18% in the placebo group. In support of the recommended dose, there was no clear evidence of improved ACR response in the higher 100-mg golimumab group compared to the lower 50-mg dose group. The most common adverse reactions, occurring with an incidence >5%, were upper respiratory tract infection and nasopharyngitis. Regarding drug interactions, live vaccines should not be administered while being treated with golimumab. As an increased risk of serious infection has been associated with concomitant use of abatacept and anakinra, combination of these drugs is not recommended. Also, golimumab treatment should not be initiated in patients with an active infection. Furthermore, patients should be warned about the higher incidence of malignancies observed with anti-TNF α therapy and increased risks of worsening or new onset of heart failure, of exacerbation or new onset of demyelinating disease, and of hepatitis B reactivation.

14. Indacaterol   (Chronic obstructive pulmonary disease)   [53–57]

Country of Origin:	United Kingdom
Originator:	Novartis
First Introduction:	Germany
Introduced by:	Novartis / Skye Pharma
Trade Name:	Onbrez Breezhaler
CAS Registry No:	312753-06-3
Molecular Weight:	392.50

Inhaled β2 adrenoceptor agonists are effective in the management of asthma and COPD, primarily through their bronchodilating properties. These drugs induce bronchodilation by causing direct relaxation of airway smooth muscle through activation of adenylate cyclase, which in turn increases intracellular cAMP levels. Salbutamol is an inhaled β2 agonist that provides rapid bronchodilation and has been widely used over the past 30 years. However, its major drawback is its short duration of action (4–6 h), requiring the drug to be administered several times a day. Two longer-acting inhaled β2 agonists, formoterol and salmeterol, are now available and are used in the management of asthma and COPD. These two drugs have a bronchodilating effect lasting for 12 h after a single inhalation and are, therefore, given twice daily. Indacaterol is the newest β2 agonist to reach the market. It is an ultra-long-acting agent with a duration of action suitable for once-a-day dosing. Indacaterol is supplied as an aerosol formulation of its maleate salt and is administered via a dry powder inhaler device. It is specifically approved for once-daily maintenance treatment of airflow obstruction in adult patients with COPD. In preclinical models, indacaterol is close to a full agonist at the human β2 adrenoceptor ($E_{max} = 73 \pm 1\%$ of isoprenaline's maximal effect, $pEC_{50} = 8.06 \pm 0.02$) while salmeterol displays only partial efficacy ($38 \pm 1\%$). The functional selectivity profile of indacaterol over β1 human adrenoceptors is similar to that of formoterol, whereas its β3 adrenoceptor selectivity profile is similar to that of formoterol and salbutamol. In isolated superfused guinea pig trachea, indacaterol has a fast onset of action (30 ± 4 min) similar to formoterol and salbutamol, and a long duration of action (529 ± 99 min) comparable to salmeterol. In the conscious guinea pig, when given intratracheally as a dry powder, indacaterol inhibits serotonin-induced bronchoconstriction for at least 24 h while salmeterol, formoterol, and salbutamol have durations of action of 12, 4, and 2 h, respectively. When given via nebulization to anesthetized rhesus monkeys, all compounds dose-dependently inhibit methacholine-induced bronchoconstriction, although indacaterol produces the most prolonged bronchoprotective effect and induces the lowest increase in heart rate for a similar degree of antibronchoconstrictor activity. In a Phase I study of patients with mild-to-moderate COPD, administration of 800 µg/day indacaterol for 14 days showed that it had a $t_{1/2}$ of 48 h, with a median T_{max} of between 30 and 60 min. $AUC_{0-24 h}$ analysis demonstrated increased systemic exposure to indacaterol by approximately 2-fold from day 1 to day 14. In a Phase III study ($n = 1732$), indacaterol 300 and 600 µg once daily produced clinically significant bronchodilation (>120 mL in trough FEV1 (forced expiratory volume in one second)) at week 12 postdose, compared to placebo (indacaterol 300 and 600 µg, 1.48 L; placebo, 1.31 L; $p < 0.001$). This improvement was seen as early as 5 min postdose and at every subsequent time point measured in each study. Indacaterol 300 and 600 µg also produced significantly greater differences in trough FEV1 versus placebo, compared to that seen with formoterol 12 µg twice daily, at week 12 (170 and 170 mL vs. 70 mL, $p < 0.001$) and at week 52 (160 and 150 mL vs. 50 mL, $p < 0.001$). In another trial ($n = 416$), after 12 weeks of treatment, indacaterol at a dose of 150 µg once daily resulted in 9% fewer days of poor control in COPD patients compared to placebo ($p < 0.001$). The most commonly reported adverse events associated with indacaterol treatment were nasopharyngitis, upper respiratory tract infection, and headache and cough following inhalation. Adverse events were mild or moderate in most cases, and became less frequent with continued treatment. The chemical synthesis of indacaterol begins with α-chlorination of 5-acetyl-8-benzyloxy-2-quinolone with benzyltrimethylammonium dichloro-iodate. The resultant chloroketone is reduced with borane in tetrahydrofuran in the presence of the chiral boron catalyst R-tetrahydro-1-methyl-3,3-diphenyl-1H,3H-pyrrolo[1,2-c][1,3,2]oxazaborole to produce the corresponding chlorohydrin intermediate in high enantiomeric excess. The chlorohydrin intermediate is cyclized to the corresponding epoxide by treatment with potassium carbonate, the epoxide is condensed with 5,6-diethylindan-2-amine, and the benzyl protecting group is removed by hydrogenolysis to produce indacaterol. The 5,6-diethylindan-2-amine intermediate is derived from 1,2-diethylbenzene via Friedel–Crafts acylation with 3-chloropropionyl chloride, cyclization of the resultant 3-chloro-1-(3,4-diethylphenyl)-1-propanone by means of concentrated sulfuric acid to 5,6-diethylindan-1-one, oximation with butyl nitrite, and reduction of the oxime to an amine via treatment with hydrogen over palladium-carbon.

15. Liraglutide   (Antidiabetic)   [58–64]

Country of Origin:	US
Originator:	Massachusetts General Hospital
First Introduction:	UK, Germany, Denmark
Introduced by:	Novo Nordisk
Trade Name:	Victoza
CAS Registry No:	204656-20-2
Molecular Weight:	3751.20

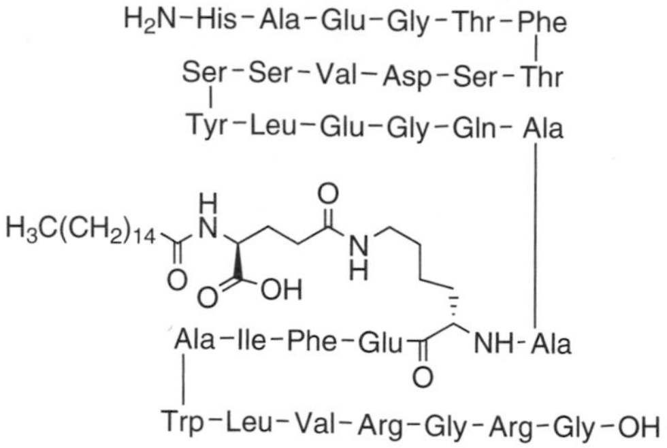

H_2N–His–Ala–Glu–Gly–Thr–Phe

Ser–Ser–Val–Asp–Ser–Thr

Tyr–Leu–Glu–Gly–Gln– Ala

$H_3C(CH_2)_{14}$

Ala–Ile–Phe–Glu   NH–Ala

Trp–Leu–Val–Arg–Gly–Arg–Gly–OH

Approaches to treating T2DM, a disease characterized by the dual defect of islet cell dysfunction and insulin resistance, include agents that increase the secretion of insulin by the pancreas (secretagogues), agents that increase the sensitivity of target organs to insulin (sensitizers), and agents that decrease the glucose absorption rate from the gastrointestinal tract. Focusing on the secretagogues, the incretins, GLP-1, and gastric inhibitory peptide (also known as glucose-dependent insulinotropic peptide) are endogenous peptides that fulfill this role by provoking postprandial, glucose-induced pancreatic insulin secretion. In addition to stimulating insulin secretion, GLP-1 also decreases glucagon secretion, improves beta-cell function, and slows gastric emptying. Unfortunately, in patients with T2DM, GLP-1 production is reduced, and exogenous administration of GLP1-1 is ineffective due to rapid inactivation by the enzyme DPP-4. The first marketed GLP-1 mimic exenatide has a 53% homology with GLP-1, and it exerts its action through binding to the membrane GLP receptor, thereby promoting the increased release of insulin from pancreatic beta cells. Liraglutide, the most recent GLP-1 receptor agonist to reach the market, possesses a 97% homology to GLP-1 with only two amino acid changes and the addition of a fatty acid side chain. Specifically, the lysine in position 34 has been replaced with an arginine, and the lysine in position 26 has been modified with a C16 acyl chain via a glutamoyl spacer. Liraglutide derives its resistance to DPP-4 degradation from its propensity to form micelles and to bind to albumin. Unlike its predecessor exenatide, which requires two daily subcutaneous injections before the first and last meals of the day, liraglutide is approved as a once-daily treatment regimen and may be used in combination with metformin or a sulfonylurea in patients with insufficient glycemic control with either monotherapy or combined dual therapy. It is also approved in combination with the dual therapy of metformin and a thiazolidinedione in patients with insufficient glycemic control. Liraglutide displayed a binding potency of 61 pM ($EC_{50} = 55$ pM for GLP-1) for the cloned human GLP-1 receptor. While the length of the fatty acid side chain did not significantly impact its potency for the human GLP-1 receptor, there was a strong correlation to half-life improvement with increasing fatty acid chain length. The parent chain of liraglutide may be prepared by standard solid-phase methodology employing Fmoc chemistry. After cleavage from the resin and side chain deprotection with a trifluoroacetic acid scavenging cocktail, the peptide is purified by reverse phase chromatography before treatment with N-α-palmitoyl-L-glutamic acid α-*tert*-butyl ester γ-succinimidyl ester in an equal mixture of water and N-methylpyrrolidone in the presence of N,N-diisopropylethylamine. Final purification affords liraglutide. Another method describes its production by recombinant DNA technology in *Saccharomyces cerevisiae*. Liraglutide is formulated to contain 6 mg of peptide per 1 mL of solution; one prefilled pen contains 18 mg of liraglutide in 3 mL. Following a single subcutaneous injection, absorption of liraglutide was slow, achieving maximum concentration 8–12 h after dosing with a bioavailability of approximately 55%. At a dose of 0.6 mg, the maximum plasma concentration was 9.4 nmol/L, and exposure was found to increase linearly with dose. As designed, plasma protein binding was high (98%), and the mean volume of distribution was 0.07 L/kg. Within the first 24 h following administration, intact liraglutide was the predominant component in plasma. It was metabolized similarly to other peptides without a preference for a major route of elimination. The mean clearance of liraglutide was 1.2 L/h with an elimination half-life of about 13 h. In clinical studies involving patients with type 2 diabetes, liraglutide was found to stimulate insulin secretion in a glucose-dependent manner while suppressing glucagon secretion. It also slowed gastric emptying and led to a loss in body weight in obese patients. Furthermore, liraglutide administered once daily significantly lowered fasting plasma glucose (FPG) and hemoglobin A1c (HbA1c) in a dose-dependent manner with a low risk of hypoglycemia. The most frequent adverse events associated with liraglutide therapy were gastrointestinal in nature, in accordance with GLP-1 receptor activation, and included nausea, diarrhea, vomiting, constipation, abdominal pain, and dyspepsia. To circumvent these gastrointestinal effects, a gradual increase in dose is suggested starting with 0.6 mg of liraglutide daily during the first week and increasing to 1.2 mg the second week. After another week, the dose may be increased to 1.8 mg, but this is the maximum recommended dose. Since liraglutide does not inhibit or induce the cytochrome P450 enzymes, drug–drug interactions are unlikely. In combination with a sufonylurea, patients may be at a higher risk for hypoglycemia, but lowering the dose of the sulfonylurea diminishes the potential for this side effect. Finally, concomitant use of other GLP-1 analogs has been associated with a risk of pancreatitis. If pancreatitis is suspected, both medications should be suspended.

16. Minodronic Acid   (Osteoporosis)   [65–67]

Country of Origin:	Japan
Originator:	Astellas Pharma
First Introduction:	Japan
Introduced by:	Ono / Astellas Pharma
Trade Name:	Recalbon, Bonoteo
CAS Registry No:	180064-38-4
	155648-60-5 (hydrate)
Molecular Weight:	322.15
	340.16 (hydrate)

Minodronic acid hydrate, a third-generation bisphosphonate, was approved and launched last year in Japan for the oral treatment of osteoporosis. Osteoporosis is a skeletal disorder characterized by low bone mass and structural deterioration of bone tissue resulting in bone fragility and increased susceptibility to fractures, particularly of the hip, spine, and wrist. It is the most common type of metabolic bone disease and a major health concern affecting 200 million women worldwide. The majority of the agents currently available for the treatment of osteoporosis decrease bone resorption (e.g., estrogens, selective estrogen modulators, calcitonin, and bisphosphonates), whereas other agents such as fluoride and parathyroid hormone increase bone formation. Bisphosphonates are synthetic analogs of the endogenous mineral deposition inhibitor pyrophosphate and are the most widely used agents for the treatment of osteoporosis. They prevent bone resorption and increase bone mineral density (BMD), although their exact mechanism of action can vary depending on the structure. Nitrogen-containing bisphosphonates, such as pamidronate, ibandronate, alendronate, and zoledronate induce osteoclast apoptosis by inhibiting farnesyl pyrophosphate (FPP) synthase, an enzyme in the mevalonate pathway, thereby preventing prenylation of small GTPase signaling proteins that are required for normal cellular function. Bisphosphonates that lack a nitrogen in the chemical structure (e.g., clodronate, etidronate) do not inhibit protein prenylation and have a different mode of action that seems to involve primarily the formation of cytotoxic metabolites in osteoclasts. Minodronic acid is the newest nitrogen-containing bisphosphonate to reach the market. *In vitro*, it inhibits recombinant human FPP synthase activity with similar potency as zoledronic acid ($IC_{50} = 3\,nM$ for both agents). The recommended dosage of oral minodronic acid in adults is 1 mg daily. The efficacy of oral minodronic acid has been assessed in a placebo-controlled Phase III trial that included 704 postmenopausal Japanese women with established osteoporosis and 1–5 vertebral fractures. After treatment with 1 mg of minodronate or placebo daily for 26 months, the cumulative incidence of new vertebral fractures was reduced by 58.9% with minodronate compared to placebo. In addition, the incidence of new vertebral fractures from months 6 to 26 was reduced by 74.1% with minodronate. Minodronate treatment also durably suppressed bone turnover markers by approximately 50% beginning at 6 months. Safety was similar between treatment and placebo groups. Another Phase III trial compared the efficacies of once-daily minodronate 1 mg and alendronate 5 mg when given over 12 months in a similar patient population ($n = 270$). At 12 months, lumbar spine BMD increased by 5.86 and 6.29% in these groups, respectively, from baseline. Total hip BMD increased by 3.47 and 3.27% in these groups, respectively. Safety was also comparable between treatments. The most common adverse events associated with minodronic acid included stomach and abdominal discomfort, abdominal pain, decrease in blood calcium level, and gastritis. The chemical synthesis of minodronic acid starts with a condensation reaction of 2-aminopyridine with ethyl 4-bromo-3-oxobutyrate by means of sodium bicarbonate to produce 2-(imidazo[1,2-*a*]pyridin-3-yl)acetic acid ethyl ester, which is hydrolyzed to the corresponding carboxylic acid with potassium hydroxide. Treatment of the carboxylic acid intermediate with phosphorus acid and phosphorus trichloride followed by hydrolysis with hydrochloric acid gives minodronic acid.

17. Nalfurafine Hydrochloride   (Pruritus)   [68–70]

Country of Origin:	Japan
Originator:	Toray industries
First Introduction:	Japan
Introduced by:	Toray / Japan Tobacco
Trade Name:	Remitch
CAS Registry No:	152657-84-6 (free base)
	152658-17-8 (hydrochloride)
Molecular Weight:	476.56 (free base)
	513.03 (hydrochloride)

Pruritus (chronic itching) is a common symptom seen in 25–90% of uremic patients, especially those with chronic renal failure requiring hemodialysis. Although the precise mechanism of induction of pruritus is still unclear, accumulating evidence points to the activation of MORs and the ensuing imbalance in the endogenous opioid system as significant contributors. On the other hand, activation of κ opioid receptors is known to control or inhibit the signals activated through MORs. Hence, κ-opioid receptor agonists have potential therapeutic value in treating pruritus. However, κ-opioid agonism is also associated with CNS side effects such as psychotomimetic alterations and dysphoria. Although several κ-opioid agonists have been explored previously, insufficient therapeutic margin between their beneficial effects and the CNS side effects have hampered clinical development. Nalfurafine hydrochloride is a new member of this class that exhibits an improved safety profile as compared with its predecessors in preclinical studies. It is a potent agonist for the κ-opioid receptor ($K_i = 0.24$ nM, $EC_{50} = 0.008$ nM, $I_{max} = 91\%$), with substantially lower binding and agonism of the μ- or δ-opioid receptors ($K_i = 2.24$ and 484 nM, $EC_{50} = 1.66$ and 21.3 nM, $I_{max} = 53$ and 78%, respectively). In vivo, nalfurafine hydrochloride demonstrates potent antipruritic activity against histamine-sensitive as well as histamine-resistant itch in mouse pruritogen-induced scratching models. For example, it dose-dependently inhibits scratching behavior induced by histamine or substance P with ED_{50} values of 7.3 and 19.6 µg/kg, respectively. In addition, it does not produce suppression of locomotor activity or alterations of general behavior in rodent models at doses up to 100 µg/kg. Nalfurafine hydrochloride was launched last year in Japan as an oral treatment for refractory pruritis in hemodialysis patients. The recommended dosage of nalfurafine hydrochloride in adults is 2.5 µg once daily. The dosage may be adjusted according to symptoms up to a maximum of 5 µg once daily. The efficacy and safety of oral nalfurafine hydrochloride have been demonstrated in a double-blind, placebo-controlled Phase III trial in 337 hemodialysis patients with pruritus that were resistant to currently available treatments, such as antihistamines. Patients were randomized to receive either 5 µg ($n = 114$) or 2.5 µg ($n = 112$) of nalfurafine hydrochloride or placebo ($n = 111$) for 14 days. Changes in itch severity on a 100 mm visual analog scale (VAS) were compared between the nalfurafine and the placebo groups as the primary endpoint of the study. The mean decrease in VAS was significantly larger in the 5-µg nalfurafine hydrochloride group than in the placebo group (22 mm vs. 13 mm, $p = 0.0002$). The decrease in the VAS in the 2.5-µg group was also significantly larger than that in the placebo group (23 mm vs. 13 mm, $p = 0.0001$). The most common adverse event associated with nalfurafine hydrochloride was insomnia or sleep disturbance, seen in ~10% of the treated patients. Nalfurafine is structurally related to naltrexone (Revia®), an opioid receptor antagonist marketed for treating alcohol dependence. Nalfurafine is synthesized in two steps starting from naltrexone, via reductive amination with methylamine under catalytic hydrogenation conditions, and subsequent acylation with 3(E)-(3-furyl) acryloyl chloride.

18. Ofatumumab   (Anticancer)   [71–74]

Country of Origin:	Denmark	Class:	Recombinant monoclonal antibody
Originator:	Genmab		
First Introduction:	US	Type:	Fully human IgG1κ, Anti-CD20
Introduced by:	Genmab / GSK	Molecular Weight:	~149 kDa
Trade Name:	Arzerra	Expression system:	Murine NS0 cell line
CAS Registry No:	679818-59-8	Manufacturer:	Lonza

Considered as the most common form of leukemia in adults, CLL is characterized by the accumulation of abnormal B lymphocytes and manifests itself as lymph node enlargement and bone marrow failure in symptomatic patients. The disease afflicts primarily older individuals with a median age of diagnosis of 70 years for men and 74 years for women. As this type of leukemia varies significantly in its course of progression with survival ranging from months to decades, treatment conforms to an individualized approach with some doctors taking a "watch and see" attitude to assess disease progression. For patients receiving treatment, there has been an evolution from first-generation DNA-alkylating agents, such as chlorambucil and cyclophosphamide to nucleoside analogs (fludarabine) to a combination of these approaches. In 2008, a second-generation DNA-alkylating drug, bendamustine, hit the market. More recently, biologics that target B lymphocytes have received considerable interest. Alemtuzumab, a humanized mAb that was approved as a single-agent treatment for CLL in 2007, targets the surface antigen CD52, which is ubiquitously expressed on lymphocytes and monocytes. While alemtuzumab has demonstrated efficacy in relapsing patients and can eliminate residual disease in bone marrow, this agent has been plagued by hematological and immunological toxicity issues due to the widespread expression of CD52. Since rituximab, a chimeric mAb, is specific for the B-cell surface antigen CD20, toxicity issues have been minimized in its application to treat CLL; however, its combination with fludarabine seems essential for maximal effectiveness. Ofatumumab is the second marketed mAb to target CD20. Compared to its predecessor, it is a fully human mAb that was generated via transgenic mouse and hybridoma technology. In addition to the Fab portion that recognizes CD20, the Fc domain mediates immune effector functions to lyse cells via complement-dependent cytotoxicity (CDC) and antibody-dependent, cell-mediated cytotoxicity (ADCC). Ofatumumab targets a different, discontinuous epitope on CD20 than rituximab; it binds specifically to both the small and large extracellular loops of the CD20 molecule expressed on normal B lymphocytes and on B-cell CL, and its binding does not induce receptor internalization nor shedding from the cell surface. In addition to targeting a different epitope on CD20, ofatumumab also exhibits a slower off-rate. These characteristics have been proposed as reasons for its ability to lyse rituximab-resistant cells that express low levels of CD20 and for its greater *in vitro* activity against CLL cells versus rituximab. Ofatumumab is supplied as a 20 mg/mL (100 mg/5 mL) solution in single-use vials for intravenous infusion. The recommended dosage for the treatment of CLL refractory to fludarabine and alemtuzumab consists of the following 12-dose regimen: a 300-mg initial dose, followed 1 week later by 2000-mg doses weekly for seven doses, followed by 2000-mg doses every 4 weeks for four doses. Premedication with oral acetaminophen, oral or intravenous antihistamine, and intravenous corticosteroid is recommended to mitigate infusion reactions. The pharmacokinetic parameters were evaluated in patients with refractory CLL following an initial dose of 500 mg and three subsequent doses of 2000 mg on a weekly basis. After the first and fourth doses, the AUC increased from 7654 to 392,418 μg/mL/h while C_{max} increased from 131 to 1087 mg/L. The half-life increased from 30 to 3054 h, and the volume of distribution increased from 3054 to 3642 mL, whereas the clearance values decreased from 65 to 10 mL/h. Ofatumumab was eliminated through both a target-independent route and a B cell-mediated route. The safety and efficacy of ofatumumab were evaluated in 154 patients with relapsed or refractory CLL in a single-arm, multicenter study. Drug refractoriness was defined as failure to achieve at least a partial response to fludarabine or alemtuzumab or disease progression within 6 months of the last dose of fludarabine or alemtuzumab. The main efficacy outcome was durable objective tumor response rate, and the results of the study indicated an overall response rate of 42% with a median duration of response of 6.5 months. No complete responses were recorded. The most common adverse events were neutropenia, pneumonia, pyrexia, cough, diarrhea, anemia, fatigue, dyspnea, rash, nausea, bronchitis, and upper respiratory tract infections. As mentioned, a premedication routine is requested to avoid infusion reactions. Patients should also be monitored at regular intervals for drug-induced cytopenias and progressive multifocal leukoencephalopathy, and use should be discontinued if suspected. Since hepatitis B reactivation is an additional concern, ofatumumab administration should be terminated if the viral infection is confirmed. Regarding drug–drug interactions, formal studies have not been conducted.

19. Pazopanib   (Anticancer)   [75–79]

Country of Origin:	US
Originator:	GlaxoSmithKline
First Introduction:	US
Introduced by:	GlaxoSmithKline
Trade Name:	Votrient
CAS Registry No:	444731-52-6 (free base)
	635702-64-6 (hydrochloride)
Molecular Weight:	437.52 (free base)
	473.98 (hydrochloride)

The growth of solid tumors is dependent on angiogenesis, the process wherein new capillaries are formed from existing blood vessels. VEGF is one of the most important inducers of angiogenesis and expressed at high levels by most tumors. Hence, the inhibition of VEGF or its receptor signaling system is an attractive target for cancer therapeutics. The most studied and developed inhibitors are monoclonal antibodies that neutralize VEGF (e.g., bevacizumab), anti-VEGF ribozymes (e.g., angiozyme), and small-molecule VEGFR kinase inhibitors (e.g., sunitinib, sorafenib). Pazopanib is the latest VEGFR kinase inhibitor to reach the market. It is indicated for the oral treatment of advanced RCC. The biological functions of the VEGF family are mediated by activation of three structurally homologous tyrosine kinase receptors, VEGFR-1, VEGFR-2, and VEGFR-3. *In vitro*, pazopanib inhibits VEGFR-1, VEGFR-2, and VEGFR-3 with IC_{50} values of 10, 30, and 47 nM, respectively. In addition, it inhibits several of the closely related tyrosine receptor kinases, including platelet-derived growth-factor receptor β (PDGFR-β), c-kit, and fibroblast growth factor receptor-1 (FGFR1) with IC_{50} values of 84, 74, and 140 nM, respectively. In human umbilical vein endothelial cells (HUVEC), pazopanib inhibits VEGF-induced proliferation more potently than basic fibroblast growth factor (bFGF)-stimulated proliferation ($IC_{50} = 21$ nM vs. 721 nM) and concentration-dependently inhibits VEGF-induced VEGFR-2 phosphorylation ($IC_{50} = 7$ nM). It also potently inhibits angiogenesis in Matrigel plug and corneal micropocket assays. In preclinical *in vivo* studies, orally administered pazopanib (10, 30, or 100 mg/kg, qd or bid) demonstrated dose-dependent antitumor activities in mice bearing human xenografts of HT29 colon, Caki-2 RCC, A375P melanoma, BT474 breast, PC3 prostate, and H322 lung cancers. The recommended dose of pazopanib is 800 mg orally once daily without food. The median time to achieve peak plasma concentrations is 2–4 h after the oral dose. Systemic exposure to pazopanib is increased when administered with food. Administration of pazopanib with a high-fat or low-fat meal results in an approximately 2-fold increase in AUC and C_{max}. Therefore, pazopanib should be administered at least 1 h before or 2 h after a meal. Pazopanib is highly bound to human plasma protein *in vivo* (>99%). *In vitro* studies suggest that pazopanib is a substrate for P-glycoprotein (Pgp) and breast cancer-resistant protein (BCRP). It is primarily metabolized by CYP3A4 with a minor contribution from CYP1A2 and CYP2C8. Pazopanib has a mean half-life of 31 h after administration of the recommended dose of 800 mg. Elimination is primarily via feces with renal elimination accounting for <4% of the administered dose. The safety and efficacy of pazopanib have been evaluated in a randomized, double-blind, placebo-controlled study in 435 subjects with locally advanced and/or metastatic RCC who had received either no prior therapy or one prior cytokine-based systemic therapy. The subjects were randomized to receive pazopanib 800 mg once daily or placebo once daily. The primary endpoint of efficacy was progression-free survival (PFS). The secondary endpoints included overall survival (OS), overall response rate (RR), and duration of response. The median PFS was 9.2 months in the pazopanib arm versus 4.2 months in the placebo group ($p <$ 0.001). In the pazopanib arm, there was a 30% response rate versus 3% in the placebo arm. The median duration of response was 58.7 weeks in the pazopanib arm. The most common adverse events associated with pazopanib were diarrhea, hypertension, hair depigmentation, nausea, anorexia, and vomiting. Pazopanib is synthesized in five chemical steps starting from 3-methyl-6-nitroindazole, which is converted to the corresponding 2,3-dimethylindazole analog via *N*-methylation with trimethyloxonium tetrafluoroborate. Subsequent reduction of the nitro group to the amino group using tin chloride followed by condensation with 2,4-dichloropyrimidine yields a chloropyrimidinylaminoindazole intermediate. The final two steps leading up to pazopanib consist of an *N*-methylation reaction using iodomethane and cesium carbonate followed by condensation with 5-amino-2-methylbenzenesulfonamide.

20. Plerixafor Hydrochloride   (Stem cell mobilizer)   [80–84]

Country of Origin:	Canada
Originator:	AnorMED
First Introduction:	US
Introduced by:	Genzyme
Trade Name:	Mozobil
CAS Registry No:	110078-46-1 (free base)
	155148-31
	(octahydrochloride)
Molecular Weight:	502.78 (free base)
	794.47 (octahydrochloride)

.8HCl

Autologous hematopoietic stem cell (HSC) transplantation, a standard treatment for hematological malignancies, such as non-Hodgkin's lymphoma or multiple myeloma, involves collection of HSCs from the patient, chemo- or radiotherapy of the patient to eliminate malignant cells, and retransplantation of the stored HSCs. This process requires infusion of a minimum number of HSCs for successful engraftment and proliferation in the bone marrow. HSCs are normally bound to CD34+ cells and reside in the bone marrow matrix. Because stem cell concentrations are usually low in the peripheral blood, mobilization is required for optimal stem cell collection during apheresis prior to autologous transplantation. Currently, mobilization of HSCs for removal and storage is initiated by G-CSF, a cytokine that promotes the proliferation and differentiation of HSCs and induces the release of enzymes that act to prevent them from anchoring to bone marrow stroma. Although G-CSF therapy has been widely used, it has some limitations. For example, up to 30% of patients may not mobilize HSCs in sufficient numbers. In addition, several days of G-CSF treatment are required for adequate mobilization. Plerixafor is a potent antagonist of the CXCR4 chemokine receptor and a first-in-class drug for stem cell mobilization. CXCR4 is specific for stromal-derived-factor-1 (SDF-1), a molecule endowed with potent chemotactic activity for lymphocytes. Because the interaction between SDF-1 and CXCR4 plays an important role in holding HSCs in the bone marrow, drugs that block the activity of CXCR4 receptor are capable of mobilizing HSCs into the bloodstream. *In vitro*, plerixafor potently blocks SDF-1 binding of CXCR4 and inhibits SDF-1-induced calcium flux ($IC_{50} = 0.01-0.13$ µg/mL) and chemotaxis ($IC_{50} = 0.13$ µg/mL) in several different cell types. Plerixafor, in combination with G-CSF, is specifically indicated to mobilize HSCs to the peripheral blood for collection and subsequent autologous transplantation in patients with non-Hodgkin's lymphoma and multiple myeloma. Plerixafor is supplied as a solution of the hydrochloride salt for subcutaneous injection. Treatment with plerixafor should begin after the patient has received G-CSF once daily for 4 days. It should be administered approximately 11 h prior to initiation of apheresis for up to 4 consecutive days. The recommended initial dose of plerixafor is 0.24 mg/kg body weight. Plerixafor exhibits linear pharmacokinetics over the dose range from 0.04 to 0.32 mg/kg. Median systemic absorption after subcutaneous injection is 87%, and the time to reach peak plasma concentration is approximately 30–60 min. Distribution of plerixafor is mostly confined to extravascular space, with a volume of distribution of 0.2–0.3 L/kg and a distribution half-life of 0.3–0.4 h. Plasma protein binding of plerixafor is about 58%. Plerixafor is not metabolized to any significant extent. Its primary route of excretion is by renal clearance. Approximately 70% of the dose is excreted in the urine as the parent drug during the first 24 h after administration. The terminal half-life is 5.3 h in patients with healthy renal function. Plerixafor elimination is reduced in patients with renal impairment. As a consequence, AUC values can increase by up to 39% in patients with severe renal impairment versus those in healthy volunteers, and elimination half-lives are approximately tripled. Hence, in patients with moderate and severe renal impairment, the dose should be reduced by one-third to 0.16 mg/kg of body weight. The therapeutic efficacy of plerixafor in conjunction with G-CSF has been evaluated in two separate placebo-controlled clinical trials in patients with either non-Hodgkin's lymphoma ($n = 298$) or multiple myeloma ($n = 302$). In each study, the patients were randomized to receive either plerixafor (0.24 mg/kg by subcutaneous injection) or placebo on each evening prior to apheresis. Patients received daily morning doses of G-CSF (10 µg/kg) for 4 days prior to the first dose of plerixafor or placebo, and on each morning prior to apheresis. In the first study in patients with non-Hodgkin's lymphoma, $>5 \times 10^6$ CD34+ cells per kg from the peripheral blood were collected in ≤4 apheresis sessions from 59% of patients treated with plerixafor and G-CSF, compared with 20% of patients who were treated with placebo and G-CSF. The median number of days required to reach $\geq 5 \times 10^6$ CD34+ cells per kg was 3 days for the plerixafor group and was not evaluable in the placebo group. In the second study in patients with multiple myeloma, $\geq 6 \times 10^6$ CD34+ cells per kg from the peripheral blood were collected in ≤2 apheresis sessions from 72% of patients treated with plerixafor and G-CSF, compared with 34% of patients who were treated with placebo and G-CSF. The median number of days needed to reach $\geq 6 \times 10^6$ CD34+ cells per kg was 1 day for the plerixafor group and 4 days for the placebo group. The most common adverse reactions (occurring in more than 10% of patients) observed during plerixafor use include diarrhea, nausea, fatigue, injection-site reactions, headache, arthralgia, dizziness, and vomiting. Plerixafor can be synthesized by several related procedures starting from tetraazacyclotetradecane, the macrocyclic tetraamine cyclam. The general synthetic strategy entails the masking of three of the four ring nitrogens, followed by alkylation with *para*-xylylene dibromide, and subsequent removal of the masking groups. In one approach, tetraazacyclotetradecane is protected as the phosphorotriamide derivative by reaction with tris (dimethylamino)phosphine followed by oxidation with carbon tetrachloride and sodium hydroxide. After condensation with xylylene dibromide, the dimeric bis-phosphoramide intermediate is hydrolyzed to plerixafor by treatment with dilute hydrochloric acid solution.

21. Pralatrexate   (Anticancer)   [85–88]

Country of Origin:	US
Originator:	SRI International/ Southern Research Institute/ Sloan-Kettering
First Introduction:	US
Introduced by:	Allos
Trade Name:	Folotyn
CAS Registry No:	146464-95-1
Molecular weight:	477.47

Pralatrexate, an injectable DHFR inhibitor, was launched last year for the treatment of patients with relapsed or refractory PTCL. PTCL is an aggressive form of non-Hodgkin's lymphoma (NHL) characterized by the proliferation of abnormal T-lymphocytes that circulate in the peripheral bloodstream. The inhibition of the folate enzymes DHFR and thymidylate synthase is a well-validated method of cancer treatment. Several launched antifolate anticancer agents, including MTX, pemetrexed, and raltitrexed, act via inhibition of these enzymes. DHFR reduces dihydrofolic acid to tetrahydrofolic acid—a compound that is required by dividing cells for the synthesis of thymine. Reduced thymine levels following DHFR inhibition prevent cell division in rapidly dividing cancer cells. Pralatrexate is marketed as a 1:1 mixture of epimers at the C10 position bearing the propargyl group. *In vitro*, pralatrexate is slightly less potent than MTX in inhibiting DHFR derived from murine leukemia L1210 cells ($K_i = 18.2\,pM$ vs. $5.75\,pM$) and human leukemia CCRF-CEM cells ($K_i = 13.4\,pM$ vs. $5.4\,pM$). However, it is transported into both types of cells with ~10-fold higher efficiency than MTX, thereby providing a more potent inhibition of cell growth as compared with MTX. *In vivo*, intraperitonally administered pralatrexate at 60 mg/kg twice weekly for three or four doses caused complete lymphoma regressions in 89, 56, and 30% of HT, RL, and SKI-DLBCL-1 xenografted mice, respectively, whereas a similar dosing of MTX at 40 mg/kg twice weekly did not produce complete regression. The posttreatment tumor diameter was also smaller in pralatrexate-treated animals. The recommended dose of intravenous pralatrexate is 30 mg/m^2 administered once weekly for 6 weeks in 7-week cycles until progressive disease or unacceptable toxicity. In PTCL patients, the total systemic clearance of pralatrexate diastereomers is 417 mL/min for the *S,S*-diastereomer and 191 mL/min for the *S,R*-diastereomer. The terminal elimination half-life is about 12–18 h. The total systemic exposure (AUC) and maximum plasma concentration (C_{max}) of pralatrexate increase dose-proportionately within a dose range of 30–325 mg/m^2, with no significant change in pharmacokinetics over multiple treatment cycles and no accumulation of drug. Pralatrexate is approximately 67% bound to plasma proteins. The steady-state volume of distribution is 105 L for the *S,S*-diastereomer and 37 L for the *S,R*-diastereomer. Pralatrexate is not significantly metabolized by hepatic CYP450 enzymes or glucuronidases. Its primary mechanism of clearance is renal. The mean fraction of unchanged drug excreted in urine following an intravenous dose of 30 mg/m^2 is 31% for the *S*-diastereomer and 38% for the *R*-diastereomer. The efficacy and safety of pralatrexate have been evaluated in an open-label, single-arm, multicenter trial in patients with relapsed or refractory PTCL ($n = 111$). The subjects received pralatrexate at 30 mg/m^2 once weekly by i.v. push over 3–5 min for 6 weeks in 7-week cycles until disease progression or unacceptable toxicity. Of the 111 subjects treated, 109 were evaluable for efficacy. The primary efficacy endpoint was overall response rate (complete response, complete response unconfirmed, and partial response) as assessed by International Workshop Criteria (IWC). The key secondary efficacy endpoint was duration of response. Response assessments were scheduled at the end of cycle 1 and then every other cycle (every 14 weeks). The overall response rate was 27%, and the median duration of response was 287 days (9.4 months). Of the responders, 66% responded within cycle 1. The median time to first response was 45 days. The most common adverse reactions associated with pralatrexate are mucositis, thrombocytopenia, nausea, and fatigue. Folic acid and vitamin B12 supplements are administered as adjunct therapies to potentially reduce pralatrexate-related hematological toxicity and mucositis. The chemical synthesis of pralatrexate starts with the alkylation of the anion of dimethyl homoterephthalate with propargyl bromide, promoted by potassium hydride in dimethylformamide, to afford the corresponding α-propargyl diester. Further alkylation of the potassium salt of α-propargyl diester with 2,4-diamino-6-(bromomethyl)pteridine followed by saponification with sodium hydroxide yields a diacid intermediate (2,4-diamino-4-deoxy-10-propargyl-10-deazapteroic acid). Mono-decarboxylation of the diacid intermediate by heating in dimethylsulfoxide at 120°C, followed by coupling with diethyl L-glutamate, and subsequent ester hydrolysis with sodium hydroxide yields pralatrexate.

22. Prasugrel　　　(AntiplateletTherapy)　　　[89–96]

Country of Origin:	Japan
Originator:	Daiichi Sankyo
First Introduction:	UK
Introduced by:	Daiichi Sankyo & Eli Lilly
Trade Name:	Effient
CAS Registry No:	150322-43-3
Molecular Weight:	373.4

Prasugrel is a third-generation thienopyridine that has been developed and launched for the prevention of atherothrombotic events in patients with ACS or following PCI. While the second-generation agent clopidogrel was an improvement over the first-generation ticlopidine, which suffered from gastrointestinal adverse effects and the risk of neutropenia with prolonged use, its delayed onset of action and considerable interpatient variability prompted the search for the next-generation thienopyridine. The mechanism of action of these platelet inhibitors involves initial biological activation to a sulfhydryl metabolite that irreversibly binds to the $P2Y_{12}$ receptor on platelets via disulfide formation, thereby preventing platelet activation and aggregation by the endogenous agonist adenosine diphosphate (ADP). The advantage of prasugrel over its predecessors is its more efficient and consistent absorption and rapid conversion to its active metabolite. Co-administration of thienopyridines with acetylsalicylic acid (aspirin), an inhibitor of the synthesis of the platelet aggregation mediator thromboxane A_2, is an effective antiplatelet strategy and joins antagonists of glycoprotein IIb/IIIa, which target the final step in platelet aggregation, in the medical arsenal combating atherothrombotic events. Regarding prasugrel's effects on ADP-induced platelet aggregation in rats, its ED_{50} of 1.2 mg/kg was 10 times lower than clopidogrel (16 mg/kg) and 300 times lower than ticlopidine (>300 mg/kg). In addition, inhibition of platelet aggregation occurred with an onset of less than 0.5 h and a duration persisting longer than 3 days, thereby spanning the life of the circulating platelets. Prasugrel also reduced thrombus formation dose-dependently in rat arteriovenous shunt models with an ED_{50} of 0.68 mg/kg, exhibiting greater potency than clopidogrel ($ED_{50} = 6.2$ mg/kg) and ticlopidine (ED_{50} > 300 mg/kg). One route to prepare prasugrel begins with the reaction of 2-fluorobenzyl bromide with cyclopropyl cyanide in diethyl ether in the presence of magnesium to afford cyclopropyl 2-fluorobenzyl ketone. Alpha bromination is then accomplished with bromine in carbon tetrachloride, and the resulting bromide is displaced with 5,6,7,7a-tetrahydro-thieno[3,2-c]pyridin-2(4H)-one hydrochloride to generate 5-[2-cyclopropyl-1-(2-fluorophenyl)-2-oxoethyl]-5,6,7,7a-tetrahydro-thieno[3,2-c]pyridin-2 (4H)-one. The final step involves treatment with acetic anhydride and sodium hydride in DMF with pure prasugrel isolated following column chromatography. It is formulated into 5-mg, yellow, elongated hexagonal tablets or 10-mg, beige, elongated hexagonal tablets. The recommended dosage is initiation with a single 60-mg oral loading dose followed by 10-mg once daily, with or without food. Following oral administration, $\geq 79\%$ of the dose is absorbed, and prasugrel is rapidly converted to its active metabolite, so no parent compound is detected in plasma. Unlike clopidogrel, which requires a two-step metabolic activation mediated primarily by CYP3A4, prasugrel is initially hydrolyzed by carboxylesterases in the intestine to a thiolactone followed by a rapid conversion to the active metabolite via CYP enzymes, predominately CYP3A4 and CYP2B6. Over the dose range of 5–60 mg, AUC increased in a slightly more than proportional fashion. With a singe daily dose of 15 mg, the AUC of the active metabolite was unaffected by a high-fat, high-calorie meal; however, T_{max} was increased from 0.5 to 1.5 h, and C_{max} was decreased by 49%. The estimated volume of distribution ranged from 44 to 68 L while estimates of clearance varied from 112 to 166 L/h in healthy subjects. The active metabolite was approximately 98% bound to human serum albumin. Two inactive metabolites generated from the active metabolite by S-methylation or conjugation to cysteine were also highly bound to plasma proteins. Of the administered prasugrel dose, about 68% was eliminated as inactive metabolites via the renal route and 27% by the fecal route. The average elimination half-life was 7.4 h. In one Phase II study labeled TRITON-TIMI 38 (Trial to Assess Improvement in Therapeutic Outcomes by Optimizing Platelet Inhibition with Prasugrel) involving 13,608 patients with ACS who were to be managed by PCI, the efficacy and safety of prasugrel were compared with that of clopidogrel in a randomized, double-blind, parallel group manner. Both drugs were co-administered with aspirin and other therapies, such as heparin and GP IIb/IIIa inhibitors, at the discretion of the physician. The primary efficacy endpoint was the composite of cardiovascular death, nonfatal myocardial infarction, or nonfatal stroke while the key safety endpoint was major bleeding. In the prasugrel population, the primary efficacy endpoint occurred in 9.9% of patients compared to 12.1% receiving clopidogrel. A significant decrease in nonfatal myocardial infarction was the contributing factor to the more favorable outcome of prasugrel treatment. With regard to the safety endpoint, an increase in bleeding was observed in 2.4% of patients receiving prasugrel compared to 1.8% with clopidogrel. In addition to the hemorrhagic side effect, other serious adverse events included AF, bradycardia, leucopenia, severe thrombocytopenia, angiodema, anemia, and abnormal hepatic function with hypertension, headache, back pain, dyspnea, nausea, dizziness, and diarrhea as less severe complaints. Prasugrel is contraindicated in patients with active pathological bleeding, such as peptic ulcers or intracranial hemorrhage, and in patients with a history of prior transient ischemic attack or stroke. In addition, in patients ≥ 75 years old, <60 kg, or likely to undergo urgent coronary artery bypass graft surgery, the risk may not outweigh the benefit. When possible, prasugrel treatment should be discontinued at least 7 days prior to any surgery. While warfarin and non-steroidal anti-inflammatory drugs (NSAIDS) may increase the risk of bleeding with co-administration of prasugrel, no drug interactions are anticipated with concomitant use of drugs that are inducers or inhibitors of the cytochrome P450 enzymes. Prasugrel may also be administered with aspirin (75–325 mg per day), heparin, GP IIb/IIIa inhibitors, statins, digoxin, and drugs that elevate gastric pH, including PPIs and H_2 blockers.

23. Saxagliptin   (Antidiabetic)   [97–100]

Country of Origin:	US
Originator:	Bristol-Myers-Squibb
First Introduction:	US
Introduced by:	BMS / AstraZeneca
Trade Name:	Onglyza
CAS Registry No:	361442-04-8
	945667-22-1 (monohydrate)
Molecular Weight:	315.41

While liraglutide targets the incretin system as a GLP-1 analog for therapeutic intervention in T2DM, saxagliptin enhances GLP-1 plasma concentrations by inhibiting the DPP-4 enzyme responsible for degradation of GLP-1 by cleavage of the two N-terminal amino acids adjacent to alanine (or proline), thereby rendering the hormone inactive. It joins the previously marketed, reversible DDP-4 inhibitors, sitagliptin and vildagliptin. With a K_i of 1.3 ± 0.3 nM, saxagliptin is 10-fold more potent than either sitagliptin (K_i = 18 ± 2 nM) or vildagliptin (K_i = 13 ± 3 nM) and appears to possess greater selectivity versus other DPP family members. Similar to vildagliptin, saxagliptin is a cyanopyrrolidine substrate-based inhibitor. The proline mimetic occupies the small S1-pocket while the nitrile's trajectory aligns with the scissile bond of the substrate. The slow dissociation kinetics of saxagliptin is attributed to a reversible, covalent reaction between the nitrile and the catalytically active serine hydroxyl (Ser630). The steric bulk provided by the adamantyl moiety and the constrained cyclopropyl ring stabilizes the *trans*-rotamer of the amide, thereby preventing the problematic intramolecular cyclization, via the amino group attacking the nitrile, which is favored in the *cis*-conformation. In its protonated state, the amino functionality is involved in salt bridges to negatively charged glutamic acid residues of the protein. One efficient route to saxagliptin starts with *S*-adamantylglycine. Following Boc-protection of the amine, the adamantyl ring is hydroxylated using potassium permanganate in hot aqueous potassium hydroxide. The pyrrolidine ring is installed by coupling with (1*S*,3*S*,5*S*)-2-azabicyclo[3.1.0]hexane-3-carboxamide employing 1-ethyl-3-(3-dimethylamino-propyl)carbodiimide hydrochloride and *N*-hydroxybenzotriazole. After protection of the hydroxyl group with triethylsilyl triflate, the primary amide is dehydrated to the corresponding nitrile by treatment with phosphorus oxychloride and imidazole in pyridine. Final deprotection with trifluoroacetic acid provides saxagliptin. Since the recommended daily dose is 2.5 or 5 mg taken once daily, saxagliptin is formulated into 2.5- and 5-mg tablets. The pharmacokinetic properties were evaluated in healthy subjects and in patients with T2DM, and the parameters were similar. Saxagliptin has an active metabolite, 5-hydroxy saxagliptin, which is approximately half as potent as its parent. Following a single 5-mg dose, the mean AUC for saxagliptin and its active metabolite were 78 and 214 ng*h/mL, respectively. The corresponding C_{max} values were 24 and 47 ng/mL, respectively. Both parameters were found to increase proportionally to dose in the 2.5–400-mg dose range. The median T_{max} was 2 h for saxagliptin and 4 h for its active metabolite. The oral bioavailability was estimated to be around 70%. Elimination occurred by both renal and hepatic routes with about 22% of the dose excreted from the latter pathway. The average renal clearance was 230 mL/min, which is greater than the average glomerular filtration rate (~120 mL/min), suggesting active renal excretion. The mean terminal half-life was 2.5 h for saxagliptin and 3.1 h for its active metabolite. Since metabolism of saxagliptin is primarily mediated by CYP3A4/5, there is a potential for drug interactions with concomitant administration of strong inducers or inhibitors of these cytochrome P450 enzymes. With strong inhibitors (ketoconazole), the recommendation is to limit the dose of saxagliptin to 2.5 mg once daily. Dose adjustment has been deemed unnecessary with concomitant use of inducers, such as rifampin. The safety and efficacy of saxagliptin as both monotherapy and combination therapy have been evaluated in multiple clinical trials involving T2DM patients. In one 24-week, double-blind, placebo-controlled study with saxagliptin as monotherapy, 401 patients were randomized to 2.5, 5, or 10 mg of saxagliptin or placebo after a 2-week, single-blind diet, exercise, and placebo lead-in period. If patients failed to meet specified glycemic controls, they were treated with metformin rescue therapy; however, the last recorded measurement was obtained prior to rescue. Doses of 2.5 and 5 mg once daily were sufficient to elicit significant improvements in hemoglobin A1C (HbA1c), FPG, and postprandial glucose (PPG) compared to placebo. In the placebo group, 26% of individuals had to discontinue for lack of glycemic control relative to 16% in the 2.5-mg and 20% in the 5.0-mg saxagliptin treatment populations. Regarding saxagliptin's add-on combination therapy with metformin, a total of 743 patients with T2DM experiencing inadequate glycemic control with metformin alone were enrolled in a 24-week, randomized, double-blind, placebo-controlled trial. Following a similar lead-in period as described in the monotherapy study, patients were randomized to 2.5, 5, or 10 mg of saxagliptin or placebo in addition to their prestudy dose of metformin (1500–2550 mg daily). Patients who failed to meet specific glycemic goals were rescued with pioglitazone since titration of saxagliptin or metformin doses was not permitted. Placebo-subtracted changes in HbA1c were −0.73, −0.83, and −0.71%, respectively, at the end of 24 weeks. FPG and PPG were also lower. Similar results were observed in studies examining saxagliptin as add-on combination therapy with thiazolidinedione or glyburide. Overall, saxagliptin was well tolerated, not associated with an increased incidence of hypoglycemia compared with placebo, and considered to be weight-neutral. The most common adverse events (>5% of patients) included upper respiratory tract infection, urinary tract infection, and headache

24. Tapentadol Hydrochloride   (Analgesic)   [101–103]

Country of Origin:	Germany
Originator:	Grunenthal GmbH
First Introduction:	US
Introduced by:	Ortho-McNeil-Janssen
Trade Name:	Nucynta
CAS Registry No:	175591-09-0
Molecular Weight:	257.80

Agonism of the MOR is a common strategy for moderate to severe pain intervention. Opioid drugs, such as morphine, that modulate this receptor have demonstrated efficacy in acute situations; however, chronic conditions, particularly those of neuropathic or inflammatory etiology, suffer from inadequate pain management with this treatment. With a narrow therapeutic window, traditional MOR agonists flirt with side effects at optimal analgesia, and prolonged use increases the potential for physical dependency. Since extensive efforts to design activators of MOR have failed to dissociate the undesirable adverse effects from the analgesic properties, the focus has been on enhancing the analgesic efficacy through a dual mechanism of action. Tapentadol hydrochloride brings this concept to fruition; MOR agonism is coupled with noradrenaline reuptake inhibition in a combinatory contribution to analgesia. Compared to morphine, it is about 50-fold less potent for MOR ($K_i = 100\,nM$ for tapentadol versus $2\,nM$ for morphine). The K_i for inhibition of noradrenaline reuptake was $500\,nM$ while reuptake of serotonin was only weakly inhibited ($K_i = 2.5\,\mu M$). Despite its lower affinity for MOR, the dual mechanism has provided an efficacious profile in both acute and chronic conditions with fewer side effects. The synthesis of tapentadol hydrochloride begins with diethyl ketone, which is subjected to a standard Mannich condition to provide a β-dimethylamino intermediate. A Grignard reaction with 3-bromoanisole was followed by the separation of diastereomers and racemic resolution. Treatment with thionyl chloride converts the hydroxy group to its corresponding chloride, which is removed by treatment with zinc borohydride (generated from zinc chloride and sodium borohydride), with overall retention of stereochemistry. The methyl ether is cleaved in refluxing HBr to afford the tapentadol salt that is ultimately converted to the hydrochloride for formulation into 50-, 75-, and 100-mg immediate-release (IR) oral tablets. The recommended dosage is individualized based on the severity of the pain, previous drug experience, and the ability to monitor the patient. As per the formulated amounts, the initial dose ranges from 50 to 100 mg. While the dosing frequency is typically every 4–6 h, the second dose may be administered as soon as 1 h after the initial dose if adequate analgesia is not obtained, and the tolerability is acceptable. Following a single oral dose of 100 mg, a C_{max} of 78 ng/mL was achieved after 1.2 h. The oral bioavailability of 32% was comparable to morphine. The terminal half-life was 4.9 h, and the clearance was 5007 mL/min. Over the 50–150-mg dose range, dose-proportional increases in C_{max} and AUC values were observed. Tapentadol had low plasma protein binding (approximately 20%) and was widely distributed throughout the body. Metabolism was extensive with only 3% of intact drug being excreted in urine; however, none of the metabolites are active. Phase 2 glucuronidation was the predominant route of metabolism with Phase 1 oxidative pathways being minor contributors. Of the 70% conjugate drug excreted in urine, 55% has been identified as the O-glucuronide and 15% as sulfate. CYP2C9 and CYP2C19 are accountable for 13% formation of N-desmethyl tapentadol and CYP2D6 for 2% hydroxy tapentadol, both of which are further metabolized by conjugative mechanisms. Since the cytochrome P450 enzymes play a minor role in the metabolism of tapentadol, clinically relevant interactions with concomitant drugs that are metabolized by or that inhibit or induce these enzymes are not anticipated. The efficacy and safety of tapentadol were evaluated in both postoperative pain and end-stage joint disease settings. In a randomized, double-blind Phase III bunionectomy study, three different doses of tapentadol IR (50, 75, and 100 mg) were compared with oxycodone IR (15 mg) and placebo, each given every 4–5 h for 72 h. Efficacy was determined by evaluating the sum of pain intensity difference over the first 48 h (SPID48) versus placebo. Based on these SPID48 values, tapentadol provided a statistically significant reduction in pain compared to placebo. While tapentadol IR 100 mg produced similar analgesia to oxycodone HCl IR 15 mg, a lower incidence of nausea and vomiting was associated with tapentadol, suggesting an improved safety profile. The common adverse effects of tapentadol were nausea, vomiting, somnolence, dizziness, and itching. As with all opioid medications, constipation was also an issue. Tapentadol is contraindicated in patients taking monoamine oxidase inhibitors because of the potential for adverse cardiovascular events due to additive effects on norepinephrine levels. As with other MOR agonists, it is also contraindicated in patients with paralytic ileus. In patients with a history of epilepsy or seizure, tapentadol may induce seizures. Since tapentadol causes somnolence, its combination with other sleep aids could dangerously affect breathing. Similarly, patients with existing breathing or lung problems are cautioned about using tapentadol. Furthermore, patients with past or present substance abuse or drug addiction should consult the doctor prior to use since physical dependency and addiction is a risk with tapentadol. Alcohol should be avoided due to the potential additive effect on CNS depression.

25. Telavancin   (Antibiotic)   [104–111]

Country of Origin:	US
Originator:	Theravance
First Introduction:	US
Introduced by:	Theravance / Astellas Pharma
Trade Name:	Arbelic, Vibativ
CAS Registry No:	372151-71-8 (free base)
CAS Registry No:	380636-75-9 (hydrochloride)
Molecular Weight:	1729.11

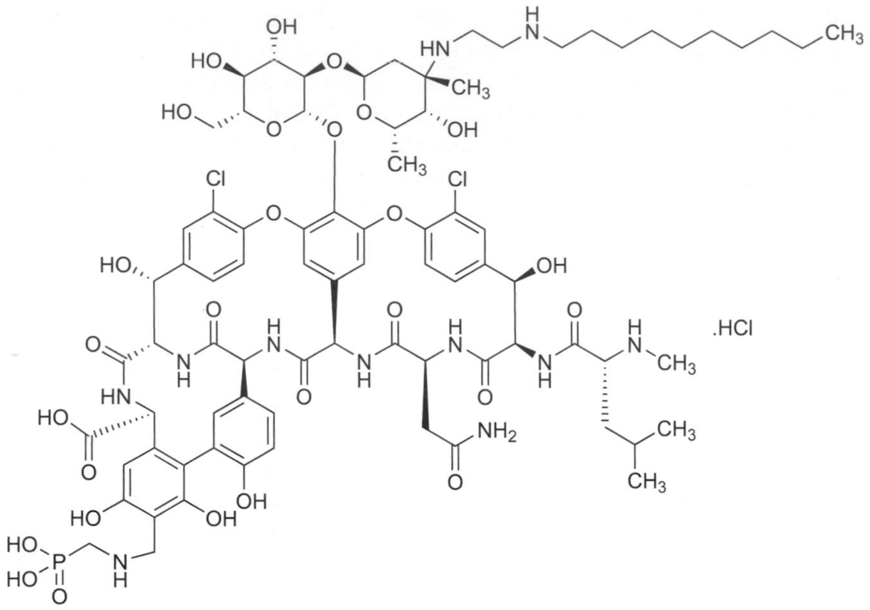

Due to the increased emergence of resistant Gram-positive pathogens, there is a persistent need for novel antibacterial agents with enhanced broad-spectrum activity. While never relegated to first-line treatment, vancomycin, a glycopeptide antibiotic derived from fermentations of *Amycolatopsis orientalis*, has experienced a resurgence in use. Because of low oral absorption, the primary route of administration is intravenous, and monitoring of vancomycin plasma levels is deemed necessary to circumvent adverse events. In addition, its slow bacterial kill rate raises concerns about treatment outcomes, and increasing rates of resistance to vancomycin among *S. aureus* and *Enterococcus* species further supports the urgent desire for improved analogs. Fulfilling this need, telavancin, a semisynthetic derivative of vancomycin with a hydrophobic side chain on the vancosamine sugar, has been developed and launched for the treatment of adults with cSSSI caused by susceptible Gram-positive bacteria. Unlike vancomycin, this lipoglycopeptide displays a dual mode of action with a significantly enhanced bactericidal killing rate. By binding to peptidoglycan precursors terminating in D-alanyl-D-alanine, similar to vancomycin, telavancin interferes with the polymerization and cross-linking of peptidoglycan, effectively inhibiting bacterial cell wall synthesis. Telavancin, however, is also capable of disrupting membrane barrier function by binding to the bacterial membrane through targeted interaction with the cell wall precursor lipid II, and its rapid bactericidal killing profile is attributed to this trait. Its dual mechanism of action provides telavancin with broad-spectrum activity against Gram-positive organisms, including vancomycin-resistant species. Against methicillin-susceptible *S. aureus* (MSSA) and MRSA, telavancin exhibits low MICs of 0.03 and 1 μg/mL, respectively. These values are approximately 2- to 4-fold more potent than vancomycin, 4- to 8-fold more potent than linezolid, and 2-fold more potent than daptomycin. Telavancin also performs better than these same antibiotics against penicillin-resistant *Streptococcus pneumoniae* with MIC values in the range of 0.016–0.063 μg/mL. Regarding anaerobic Gram-positive bacteria, telavancin displays broad-spectrum activity that is 2- to 4-fold more potent than vancomycin against most strains except for *Clostridium clostridioforme* (MIC_{90} = 8 μg/mL for telavancin and 1 μg/mL for vancomycin). In addition to the lipid tail whose length ensures optimal activity against MSRA and VanA enterococci, telavancin also differs from vancomycin by the introduction of a negatively charged phosphonomethyl aminomethyl group at the 4′-position of the resorcinol-like residue. This modification improves the ADME properties. The synthesis of telavancin from vancomycin starts with installation of the lipid tail by reductive alkylation of vancomycin with N-fluorenylmethoxycarbonyl-N-decylaminoacetaldehyde and sodium cyanoborohydride. The Fmoc protecting group is subsequently removed by treatment with piperidine, and a Mannich reaction is then utilized to incorporate the phosphonomethyl aminomethyl moiety to afford telavancin. Telavancin is formulated as a lyophilized powder (250 or 750 mg per vial) with cyclodextrin to enhance its solubility, and the mode of administration is intravenous infusion over 60 min at a dose of 10 mg/kg. The dosing regimen is every 24 h for a period of 7–14 days. In healthy volunteers, the pharmacokinetics were linear at doses ≤12.5 mg/kg after a single intravenous infusion. At the 10 mg/kg dose, a C_{max} of 93.6 ± 14.2 μg/mL, with a corresponding AUC of 747 ± 129 μg*h/mL, was achieved. The clearance was 13.9 ± 2.9 mL/h/kg while the elimination half-life was 8.0 ± 1.5 h. The primary route of excretion was renal. Although telavancin was not appreciably metabolized *in vitro*, three hydroxylated metabolites were detected *in vivo* with radiolabeled telavancin, but the predominant metabolite accounted for <10% of the radioactivity in the urine and <2% in plasma; however, the metabolic pathway has not been identified. The volume of distribution at steady state was 145 ± 23 mL/kg, and the plasma protein binding was approximately 90%. The safety and efficacy of telavancin (10 mg/kg administered intravenously every 24 h) were compared to vancomycin (1 g administered intravenously every 12 h) in a randomized, double-blind trial involving adult patients with cSSSI. The primary cause of infection in these cSSSI patients was suspected or confirmed MRSA. Patients who received any amount of study medication according to their randomized treatment group and were evaluated for efficacy were classified as the all-treated efficacy (ATE) population. Within this group was a subset of patients who had sufficient adherence to the protocol to be considered clinically evaluable (CE). The primary efficacy endpoint was the clinical cure rate at a follow-up visit in the ATE and CE populations. With telavancin treatment, the clinical cure rate was 84.3% in the CE cluster and 72.5% in the ATE population; the corresponding rates with vancomycin therapy were 82.8 and 71.6%, respectively. The most common adverse events were taste disturbance, nausea, vomiting, and foamy urine. While there were no contraindications, precautionary warnings included potential nephrotoxicity, infusion-related reactions, and *Clostridium difficile*-associated diarrhea common with many antibacterial agents. Since telavancin caused developmental defects in three animal species at clinically relevant doses, women of child-bearing age should have a serum pregnancy test prior to administration and should use effective contraception during treatment. Caution should also be used in patients with a prior history of prolongation of the QTc interval or concomitantly taking drugs known to prolong the QTc interval. Regarding drug interactions, telavancin interferes with coagulation test parameters, by binding to the artificial phospholipid surfaces commonly added to these tests, and some urine protein tests.

26. Tolvaptan (Hyponatremia)   [112–116]

Country of Origin:	US
Originator:	Otsuka Pharmaceutical
First Introduction:	US
Introduced by:	Otsuka America Pharmaceutical, Inc.
Trade Name:	Samsca
CAS Registry No:	150683-30-0
Molecular Weight:	448.95

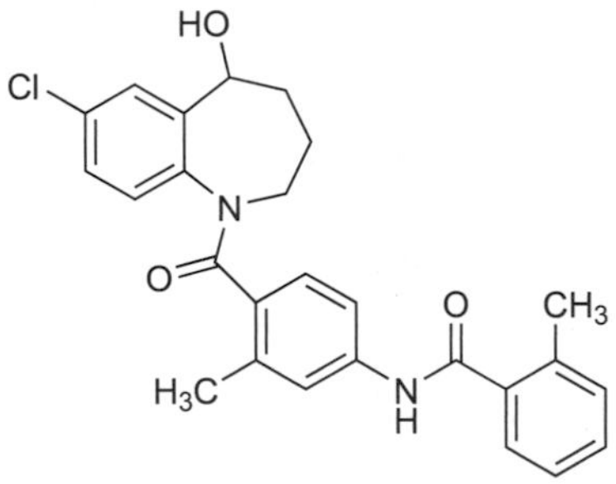

The potent antidiuretic hormone AVP orchestrates the regulation of free water absorption, body fluid osmolality, cell contraction, blood volume, and blood pressure through stimulation of three G-protein-coupled receptor subtypes: V_1-vascular types a and b, V_2-renal, and V_3-pituitary. Increased AVP secretion is the trademark of several pathophysiological disorders, including heart failure, impaired renal function, liver cirrhosis, and SIADH. As a consequence, these patients experience excess water retention or inadequate free-water excretion, which results in the dilution of sodium concentrations, frequently manifesting as clinical hyponatremia (serum sodium concentration <135 mmol/L). This electrolyte imbalance increases mortality rates by 60-fold. Selective antagonism of the AVP V_2 receptor promotes water excretion without perturbing electrolyte balance making it an appealing target for preventing disease progression. Following the introduction of the dual AVP V_{1a}/V_2 receptor antagonist conivaptan, tolvaptan has recently been launched as a nonpeptide, selective V_2 receptor antagonist with potent aquaretic attributes for the treatment of hypervolemic and euvolemic hyponatremia (serum sodium concentration of <125 mmol/L or less distinct hyponatremia that is symptomatic and has resisted correction with fluid restriction). As a more potent and selective V_2 receptor antagonist, tolvaptan is a follow-up to mozavaptan, which possesses weak V_1 receptor antagonism and was approved for the treatment of SIADH in Japan. With a K_i of 0.43 nM for human recombinant V_2 (hV2) and 12.3 nM for hV_1, tolvaptan exhibits a 29-fold selectivity for hV_2 compared to its predecessors, which were equipotent at hV_1 and hV_2. Tolvaptan may be prepared in 11 steps starting from 5-chloro-2-nitrobenzoic acid. Following esterification, reduction of the nitro moiety with tin(II) chloride, subsequent protection (tosylation) and alkylation of the resulting aniline, and a Dieckmann cyclization with potassium *tert*-butoxide generate the benzazepinone core that is ultimately decarboxylated by heating with hydrochloric acid in acetic acid. After deprotection of the amine group, condensation with 2-methyl-4-nitrobenzoyl chloride affords another nitro handle that is reduced with tin(II) chloride. This aniline is coupled with 2-methylbenzoyl chloride to give the penultimate intermediate. In the final step, reduction of the ketone functionality with sodium borohydride provides racemic tolvaptan that is formulated into 15- and 30-mg tablets for oral administration. The pharmacokinetics of tolvaptan have been evaluated in healthy individuals. While the AUC increases proportionally with dose, C_{max} increases are not proportional at doses above 60 mg. Peak concentrations are observed between 2 and 4 h postdose, and the steady-state ratio of S-(−) and R-(+) is 3:1, respectively. Intersubject variations in peak and average exposure to tolvaptan are an issue with a coefficient of variation ranging from 30 to 60%. The plasma protein binding is high at 99%, and the apparent volume of distribution is 3 L/kg. The clearance is 4 mL/min/kg with a terminal phase half-life of 12 h. While the absolute bioavailability of tolvaptan is unknown, at least 40% of the dose is absorbed as either the parent or its metabolites. Tolvaptan is predominantly metabolized by CYP3A4 and is eliminated primarily in the feces. Since CYP3A4 is the primary metabolic pathway, drug interactions are a concern. Concomitant administration with the strong CYP3A4 inhibitor ketoconazole results in a 5-fold increase in exposure for tolvaptan; therefore, strong CYP3A4 inhibitors should not be co-administered. Tolvaptan is also a P-glycoprotein (P-gp) substrate and inhibitor; as a result, a reduction in dose may be required in patients concomitantly treated with other P-gp inhibitors, such as cyclosporine. In the presence of CYP3A4 and P-gp inducers, the dose of tolvaptan may need to be increased. For example, the CYP3A4 and P-gp inducer rifampin reduces the exposure of tolvaptan by 85%. In two randomized, double-blind, placebo-controlled trials including patients with euvolemic or hypervolemic hyponatremia with various underlying causes (heart failure, liver cirrhosis, and SIADH), the efficacy and safety of tolvaptan were investigated. A total of 424 patients were randomized to receive either placebo or tolvaptan over the course of 30 days with monitoring for another 7 days following withdrawal. In the tolvaptan group, the initial daily oral dose was 15 mg with the dose being increased to 30 mg and then 60 mg daily if warranted by serum sodium concentrations. The primary endpoint was the average daily AUC for change in serum sodium concentration from baseline to day 4 and baseline to day 30. During both periods, tolvaptan elicited a statistically significant increase in serum sodium concentration compared to placebo, regardless of etiology. By the end of the 7-day withdrawal period, sodium levels in the tolvaptan patients declined to values similar to those of placebo-treated patients. In an open-label follow-up trial, patients were given the option of initiating or reinitiating tolvaptan treatment. Serum sodium concentrations increased to approximately the same levels as experienced for those previously treated with tolvaptan, and these levels were sustained for at least a year. The most common adverse events were thirst, dry mouth, asthenia, constipation, pollakiuria or polyuria, and hyperglycemia. The recommended starting dose is 15 mg daily with a daily 15-mg adjustment to a maximum of 60 mg daily to raise serum sodium concentration. Initiation should be in a hospital setting where serum sodium and volume status may be monitored since too rapid correction of hyponatremia (>12 mEq/L/24 h) can cause osmotic demyelination resulting in dysarthria, mutism, dysphagia, lethargy, spastic quadriparesis, seizures, coma, and death. In addition to avoiding concomitant use of strong CYP3A4 inhibitors, tolvaptan is contraindicated in settings of urgent need to raise serum sodium acutely, in patients with an inability to sense or appropriately respond to thirst, in hypovolemic hyponatremia conditions, and in anuric patients.

27. Ulipristal acetate   (Contraceptive)   [117–121]

Country of Origin:	US
Originator:	Research Triangle Institute
First Introduction:	US
Introduced by:	HRA Pharma
Trade Name:	ellaOne
CAS Registry No:	126784-99-4
Molecular Weight:	475.62

Ulipristal acetate is a synthetic steroid derived from 19-nor-progester-one that exerts potent progesterone receptor (PR) antagonist activity at the transcriptional level. It is an orally administered drug indicated for emergency contraception (i.e., postcoital contraception) up to 120 h following unprotected intercourse. Ulipristal acetate is the second oral progestogen marketed for this indication behind levonorgestrel. As PR antagonists, both drugs are believed to act via delay of ovulation and inhibition of follicular development. Levonorgestrel 1.5 mg is approved for contraception up to 72 h following unprotected intercourse and is widely accessible via prescription, and directly from clinics and community pharmacies. Prior to ulipristal acetate, the only approved postcoital contraceptive option between 72 and 120 h was the insertion of an intra-uterine device (IUD). The recommended dose of ulipristal acetate is one 30-mg tablet taken as soon as possible and no more than 120 h following intercourse. The vast majority of PR antagonists belonging to the mifepristone (RU486) family also bind to the glucocorticoid receptor (GR) with high affinity and thereby exert antiglucocorticoid activity. When administered over a short time period, the antiglucocorticoid properties are unlikely to cause long-term consequences. However, long-term administration for contraception may be expected to cause undesirable side effects. In receptor-binding studies using extracts from human cell lines, ulipristal shows similar affinity as mifepristone for the human PR (K_d = 0.31 nM vs. 0.5 nM, respectively) and a slightly lower affinity for human GR (K_d = 1.68 nM vs. 0.68 nM, respectively, at [^{3}H]dexamethasone binding sites). Ulipristal exhibits potent *in vitro* antiprogestin activity in T47D cells transfected with the PR-responsive reporter genes MMTV-LUC or PRE2-tk-LUC. Ulipristal is less effective than mifepristone in inhibiting the GR-mediated transcription of MMTV-LUC and PRE2-tk-LUC reporter genes in HeLa and HepG2 cells, respectively. Following oral administration of a single 30-mg dose, ulipristal acetate is rapidly absorbed, with peak plasma concentration occurring ~1 h after ingestion. Administration of ulipristal acetate together with a high-fat breakfast resulted in ~45% lower mean C_{max}, a delayed T_{max} (from a median of 0.75–3 h), and 25% higher mean AUC compared with administration in the fasted state. Serum levels of ulipristal acetate are dose-proportional from 1 to 50 mg but dose-independent at higher doses. Ulipristal acetate is highly bound (>98%) to plasma proteins. It is extensively metabolized by CYP3A4, primarily via N-mono-demethylation, N-bis-demethylation, and hydroxylation. The mono-demethylated metabolite is pharmacologically active. The terminal half-life of ulipristal acetate in plasma following a single 30-mg dose is estimated to be 32.4 ± 6.3 h, with a mean oral clearance (CL/F) of 76.8 ± 64.0 L/h. The efficacy and safety of ulipristal acetate at 30-mg oral dose have been evaluated in two Phase III studies. In a noncomparative open study, 1334 women aged between 18 and 35 years were treated with ulipristal acetate 48–120 h following intercourse. Twenty-nine pregnancies were observed against an expected number of 75, yielding an efficacy rate of 61%. In another active-comparator study, women were randomized to contraception with ulipristal acetate 30 mg ($n = 939$) or levonorgestrel 1.5 mg ($n = 954$) at between 0 and 120 h postcoitus. There were 15 pregnancies in the ulipristal acetate group and 25 in the levonorgestrel group, yielding pregnancy rates of 1.6 and 2.6%, respectively. The difference between groups was not significant ($p > 0.05$). The most common adverse effects with ulipristal acetate were headache, dysmenorrhea, nausea, abdominal pain, dizziness, fatigue, and upper abdominal pain.

28. Ustekinumab   (Anti-psoriatic)   [122–126]

Country of Origin:	US	Class:	Humanized IgG1κ monoclonal antibody
Originator:	Medarex	Type:	anti-IL12 and anti-IL-23
First Introduction:	Canada	Molecular Weight:	149 kDa
Introduced by:	Janssen-Ortho	Expression system:	Unspecified recombinant cell line
Trade Name:	Stelara	Manufacturer:	Cilag AG
CAS Registry No:	815610-63-0		

Psoriasis is a chronic inflammatory skin disorder that affects approximately 2–3% of the world's population. The most common form of the disease is plaque psoriasis, which is a result of dysregulated cell growth. Conventional treatment options for psoriasis include topical corticosteroids, phototherapy, and systemic drugs (oral retinoids, cyclosporine, and MTX), but all of these therapies have limitations. Topical therapies are of limited use in patients with moderate to severe disease because of the extent of body surface involved; phototherapy may increase patients' risk of developing skin cancer; and systemic therapies with retinoids, cyclosporine, and MTX are limited by safety issues. In recent years, novel biologics targeting the underlying immunopathogenesis of the disease have been introduced, such as etanercept, infliximab, adalimumab, alefacept, and efalizumab, which inhibit either the inflammatory cytokine tumor necrosis factor-α (TNF-α) or the activation of T-cells. Ustekinumab is the latest biologic to be approved for the treatment of plaque psoriasis. It is a human IgG1k monoclonal antibody that binds with high affinity and specificity to the p40 protein subunit used by both the interleukin-12 (IL-12) and interleukin-23 (IL-23) cytokines. IL-12 and IL-23 are naturally occurring cytokines that are involved in inflammatory and immune responses, such as NK cell activation and CD4+ T-cell differentiation and activation. In *in vitro* models, ustekinumab is shown to disrupt IL-12- and IL-23-mediated signaling and cytokine cascades by disrupting the interaction of these cytokines with a shared cell-surface receptor chain, IL-12ß1. Ustekinumab is indicated for the treatment of adult patients (18 years or older) with moderate to severe plaque psoriasis who are candidates for phototherapy or systemic therapy. Using DNA recombinant technology, ustekinumab is produced in a well-characterized recombinant cell line and is purified using standard bio-processing technology. It is comprised of 1326 amino acids and has an estimated molecular mass of 149 kDa. Ustekinumab is administered by subcutaneous injection. For patients weighing ≤100 kg, the recommended dose is 45 mg initially and 4 weeks later, followed by 45 mg every 12 weeks. For patients weighing >100 kg, the recommended dose is 90 mg initially and 4 weeks later, followed by 90 mg every 12 weeks. In psoriasis subjects, the median time to reach the maximum serum concentration (T_{max}) is 13.5 and 7 days, respectively, after a single subcutaneous administration of 45 and 90 mg of ustekinumab. In healthy subjects, the median T_{max} value (8.5 days) following a single subcutaneous administration of 90 mg of ustekinumab is comparable to that observed in psoriasis subjects. Following multiple subcutaneous doses of ustekinumab, the steady-state serum concentrations are achieved by week 28. There is no apparent accumulation in serum ustekinumab concentration over time when given subcutaneously every 12 weeks. The metabolic pathway of ustekinumab has not been characterized. As a human IgG1κ monoclonal antibody, ustekinumab is expected to be degraded into small peptides and amino acids via catabolic pathways in the same manner as endogenous IgG. The mean half-life of ustekinumab ranges from 15 to 46 days across all psoriasis studies following intravenous and subcutaneous administration. After a single subcutaneous dose of ustekinumab, the median apparent volume of distribution ranges from 0.076 to 0.161 L/kg, and the clearance ranges from 2.7 to 5.3 mL/kg/day. The safety and efficacy of ustekinumab have been evaluated in two placebo-controlled studies involving 1996 patients with moderate to severe plaque psoriasis who were candidates for phototherapy or systemic therapy. The primary endpoint in both studies was the proportion of patients who achieved Psoriasis Area and Severity Index (PASI) 75 response (i.e., PASI improvement of at least 75% relative to baseline) from baseline at week 12. In both studies, subjects were randomized to the following groups: placebo at weeks 0 and 4 followed by crossover to ustekinumab (either 45 or 90 mg) at weeks 12 and 16; or ustekinumab 45-mg or 90-mg doses, regardless of weight, at weeks 0, 4, and 16. In the first study ($n = 766$), the PASI 75 response was reached by 3, 67, and 66% of the placebo and ustekinumab 45- and 90-mg doses, respectively. In the second study ($n = 1230$), the PASI 75 response was reached by 4, 67, and 76% of the placebo and ustekinumab 45- and 90-mg doses, respectively. In subjects who weighed <100 kg, response rates were similar with both the 45- and 90-mg doses; however, in subjects who weighed >100 kg, higher response rates were seen with 90-mg dosing compared with 45-mg dosing. The long-term efficacy of ustekinumab was also evaluated in an extended arm of the first study. Patients who were originally randomized to ustekinumab and who achieved a PASI 75 response at both weeks 28 and 40 were re-randomized at week 40 to maintenance ustekinumab or withdrawal from treatment until loss of response. All patients were followed for up to 76 weeks. Maintenance of PASI 75 was significantly superior with continuous ustekinumab treatment compared with treatment withdrawal; at week 76, 84% of patients re-randomized to maintenance treatment achieved a PASI 75 response compared with 19% of patients re-randomized to placebo. Adverse events associated with the use of ustekinumab included nasopharyngitis, upper respiratory tract infection, headache, fatigue, and diarrhea.

REFERENCES

[1] A. I. Graul, L. Sorbera, P. Pina, M. Tell, E. Cruces, E. Rosa, M. Stringer, R. Castañer, and L. Revel, *Drug News Perspect.*, 2009, **23**, 1.
[2] H. Y. Meltzer, A. Dritselis, U. Yasothan, and P. Kirkpatrick, *Clin. Med. Ther.*, 2009, **1**, 415.
[3] K. P. Garnack-Jones, S. Dhillon, and L. J. Scott, *CNS Drugs*, 2009, **23**, 793.
[4] B. Boyd and J. Castañer, *Drugs Future*, 2006, **31**, 17.
[5] S. Nishino and M. Okuro, *Drugs Today*, 2008, **44**, 395.
[6] D. A. Lankford, *Expert Opin. Invest. Drugs*, 2008, **17**, 565.
[7] H. Y. Meltzer, A. Dritselis, U. Yasothan, and P. Kirkpatrick, *Nat. Rev. Drug Discov.*, 2009, **8**, 843.
[8] J. Weber and P. L. McCormack, *CNS Drugs*, 2009, **23**, 781.
[9] M. Shahid, G. B. Walker, S. H. Zorn, J. A. Ward, and E. H. F Wong, *J. Psychopharmacol.*, 2009, **23**, 65.
[10] M. van der Linden, T. Roeters, R. Harting, E. Stokkingreef, A. S. Gelpke, and G. Kemperman, *Org. Process Res. Dev.*, 2008, **12**, 196.
[11] N. J. Carter and L. J. Scott, *Drugs*, 2010, **70**, 83.
[12] J. S. Bertino and J. Zhang, *Expert Opin. Pharmacother.*, 2009, **10**, 2545.
[13] W. Haas, C. M. Pillar, G. E. Zurenko, J. C. Lee, L. S. Brunner, and T. W. Morris, *Antimicrob. Agents Chemother.*, 2009, **53**, 3552.
[14] A. E. Harms, *WO Patent Application 08045673*, 2008.
[15] S. Savic and M. F. McDermott, *Nat. Rev. Rheum.*, 2009, **5**, 529.
[16] G. M. Walsh, *Drugs Today*, 2009, **45**, 731.
[17] L. D. Church and M. F. McDermott, *Current Opin. Mol. Ther.*, 2009, **11**, 81.
[18] P. Revill, N. Mealy, and J. Bozzo, *Drugs Future*, 2008, **33**, 385.
[19] J. Shen and Z. Zhu, *Current Opin. Mol. Ther.*, 2008, **10**, 273.
[20] M. Sebastian, A. Kuemmel, M. Schmidt, and A. Schmittel, *Drugs Today*, 2009, **45**, 589.

[21] H. Lindhofer, R. Mocikat, B. Steipe, and S. Thierfelder, *J. Immunol.*, 1995, **155**, 219.

[22] L. A. Sorbera, J. Castañer, and R. M. Castañer, *Drugs Future*, 2004, **29**, 1201.

[23] B. Feret, *Formulary*, 2005, **40**, 227.

[24] W. J. G Hellstrom, *Neuropsychiatr. Dis. Treat.*, 2009, **5**, 37.

[25] F. Giuliano and W. J. G. Hellstrom, *BJU Int.*, 2008, **102**, 668.

[26] L. A. Sorbera, J. Bolós, N. Serradell, and M. Bayés, *Drugs Future*, 2006, **31**, 755.

[27] L. Klotz, *Drugs Today*, 2009, **45**, 725.

[28] G. Deraux, A. Soupart, and G. Vassail, *Lancet*, 2008, **371**, 1624.

[29] K. M. Olsen and M. L. Hitzeman, *Clin. Med. Ther.*, 2009, **1**, 1641.

[30] N. Aslam and R. Wright, *Expert Opin. Pharmacother.*, 2009, **10**, 2329.

[31] K. M. Zareba, *Drugs Today*, 2006, **42**, 75.

[32] S. M. Hoy and S. J. Keam, *Drugs*, 2009, **69**, 1647.

[33] R. L. Page, B. Hamad, and P. Kirkpatrick, *Nat. Rev. Drug Discov.*, 2009, **8**, 769.

[34] K. M. Dale and C. M. White, *Ann. Pharmacother.*, 2007, **41**, 599.

[35] S. Krishnamoorthy and G. Y. H Lip, *Expert Rev. Cardiovasc. Ther.*, 2009, **7**, 473.

[36] P. Revill, N. Serradell, and J. Bolós, *Drugs Future*, 2006, **31**, 767.

[37] J. B. Bussel, G. Cheng, M. N. Saleh, B. Psaila, L. Kovaleva, B. Meedeb, J. Kloczko, H. Hassani, B. Mayer, N. L. Stone, M. Arning, D. Provan, and J. M. Jenkins, *New Engl. J. Med.*, 2007, **357**, 2237.

[38] J. M. Jenkins, D. Williams, Y. Deng, J. Uhl, V. Kitchen, D. Collins, and C. L. Erickson-Miller, *Blood*, 2007, **109**, 4739.

[39] P. L. McCormack and D. M. Robinson, *CNS Drugs*, 2009, **23**, 71.

[40] C. Dulsat, N. Mealy, R. Castañer, and J. Bolós, *Drugs Future*, 2009, **34**, 189.

[41] S. S. Chung, *Eur. Neurol. J.*, 2009, **1**, 1.

[42] R. Talati, C. M. White, and C. I. Coleman, *Formulary*, 2009, **44**, 357.

[43] E. Pascual, F. Sivera, U. Yasothan, and P. Kirkpatrick, *Nat. Rev. Drug Discov.*, 2009, **8**, 101.

[44] P. I. Hair, P. L. McCormack, and G. M. Keating, *Drugs*, 2008, **68**, 1865.

[45] R. T. Keenan and M. H. Pillinger, *Drugs Today*, 2009, **45**, 247.

[46] N. L. Edwards, *Rheumatology*, 2009, **48** (suppl 2), ii15.

[47] Y. Takano, K. Hase-Aoki, H. Horiuchi, L. Zhao, Y. Kasahara, S. Kondo, and M. A. Becker, *Life Sci.*, 2005, **76**, 1835.

[48] C. Dulsat and N. Mealy, *Drugs Future*, 2009, **34**, 352.

[49] G. Hutas, *Current Opin. Mol. Ther.*, 2008, **10**, 393.

[50] O. J. Phung and C. I. Coleman, *Formulary*, 2009, **44**, 264.

[51] D. A. Pappas, J. M. Bathon, D. Hanicq, U. Yasothan, and P. Kirkpatrick, *Nat. Rev. Drug Discov.*, 2009, **8**, 695.

[52] L. K. McCluggage and J. M. Scholtz, *Ann. Pharmacother.*, 2010, **44**, 135.

[53] S. L. Davies and J. Castañer, *Drugs Future*, 2005, **30**, 1219.

[54] K. M. Beeh and J. Beier, *Adv. Ther.*, 2009, **26**, 691.

[55] B. Cuenoud, I. Bruce, R. A. Fairhust, and D. Beattie, *WO Patent Application 00075114*, 2000.

[56] O. Bauwens, V. Ninane, B. Van de Maele, R. Firth, F. Dong, R. Owen, and M. Higgins, *Curr. Med. Res. Opin.*, 2009, **25**, 463.

[57] C. Battram, S. J. Charlton, B. Cuenoud, M. R. Dowling, R. A. Fairhurst, D. Farr, J. R. Fozard, J. R. Leighton-Davies, C. A. Lewis, L. McEvoy, R. J. Turner, and A. Trifilieff, *J. Pharmacol. Exp. Ther.*, 2006, **317**, 762.

[58] K. F. Croom and P. L. McCormack, *Drugs*, 2009, **69**, 1985.

[59] B. Gallwitz, *Drugs Future*, 2008, **33**, 13.

[60] T. Vilsbøll, *Expert Opin. Invest. Drugs*, 2007, **16**, 231.

[61] S. D. Sakauye and S. A. Shah, *Formulary*, 2009, **44**, 136.

[62] L. B. Knudson, P. F. Nielsen, P. O. Huusfeldt, N. L. Johansen, K. Madsen, F. Z. Pedersen, H. Thøgersen, M. Wilken, and H. Agersø, *J. Med. Chem.*, 2000, **43**, 1664.

[63] K. Madsen, L. B. Knudson, H. Agersoe, P. F. Nielsen, H. Thøgersen, M. Wilken, and N. L. Johansen, *J. Med. Chem.*, 2007, **50**, 6126.

[64] J. Meece, *Pharmacotherapy*, 2009, **29**, 33S.

[65] L. A. Sorbera, J. Castañer, and P. A. Leeson, *Drugs Future*, 2002, **27**, 935.

[66] M. Hiroshi, T. Makoto, K. Ryoji, M. Taisei, O. Yasuo, Y. Hiroyuki, K. Katsuya, I. Masako, and N. Toshitaka, *Bone*, 2008, **43**, 840.

[67] T. Makoto, M. Hiroshi, K. Ryoji, O. Yasuo, K. Naoki, Y. Hiroyuki, and K. Katsuya, *Bone*, 2008, **43**, 894.

[68] L. A. Sorbera, J. Castañer, and P. A. Leeson, *Drugs Future*, 2003, **28**, 237.

[69] K. Nakao and H. Mochizuki, *Drugs Today*, 2009, **45**, 323.

[70] H. Kumagai, T. Ebata, K. Takamori, T. Muramatsu, H. Nakamoto, and H. Suzuki, *Nephrol. Dial. Transplant*, 2010, **25**, 1251.

[71] J. Castillo, C. Milani, and D. Mendez-Allwood, *Expert Opin. Invest. Drugs*, 2009, **18**, 491.

[72] T. Robak, *Curr. Opin. Mol. Ther.*, 2008, **10**, 294.

[73] B. Zhang, *mAbs*, 2009, **1**, 326.

[74] M. J. Keating, A. Dritselis, U. Yasothan, and P. Kirkpatrick, *Nat. Rev. Drug Discov.*, 2010, **9**, 101.

[75] R. M. Bukowski, U. Yasothan, and P. Kirkpatrick, *Nat. Rev. Drug Discov.*, 2010, **9**, 17.

[76] P. A. Harris, A. Boloor, M. Cheung, R. Kumar, R. M. Crosby, R. G. Davis-Ward, A. H. Epperly, K. W. Hinkle, R. N. Hunter, III, J. H. Johnson, V. B. Knick, C. P. Laudeman, D. K. Luttrell, R. A. Mook, R. T. Nolte, S. K. Rudolph, J. R. Szewczyk, A. T. Truesdale, J. M. Veal, L. Wang, and J. A. Stafford, *J. Med. Chem.*, 2008, **51**, 4632.

[77] S. Limvorasak and E. M. Posadas, *Expert Opin. Pharmacother.*, 2009, **10**, 3091.

[78] K. Podar, G. Tonon, M. Sattler, Y. Tai, S. LeGouill, H. Yasui, K. Ishitsuka, S. Kumar, R. Kumar, L. N. Pandite, T. Hideshima, D. Chauhan, and K. C. Anderson, *Proc. Natl. Acad. Sci. USA*, 2006, **103**, 19478.

[79] L. A. Sorbera, J. Bolós, and N. Serradell, *Drugs Future*, 2006, **31**, 585.

[80] J. F. DiPersio, G. L. Uy, U. Yasothan, and P. Kirkpatrick, *Nat. Rev. Drug Discov.*, 2009, **8**, 105.

[81] A. F. Cashen, *Drugs Today*, 2009, **45**, 497.

[82] A. J. Wagstaff, *Drugs*, 2009, **69**, 319.

[83] G. L. Uy, M. P. Rettig, and A. F. Cashen, *Expert Opin. Biol. Ther.*, 2008, **8**, 1797.

[84] S. L. Davies, N. Serradell, J. Bolós, and M. Bayés, *Drugs Future*, 2007, **32**, 123.

[85] O. A. O'Connor, *Curr. Opin. Oncol.*, 2006, **18**, 591.

[86] J. R. Molina, *IDrugs*, 2008, **11**, 508.

[87] E. Izbicka, A. Diaz, R. Streeper, M. Wick, D. Campos, R. Steffen, and M. Saunders, *Cancer Chemother. Pharmacol.*, 2009, **64**, 993.

[88] J. I. DeGraw, M. W. T. Colwell, J. R. Piper, and F. M. Sirotnak, *J. Med. Chem.*, 1993, **36**, 2228.

[89] U. S. Tantry, K. P. Bliden, and P. A. Gurbel, *Expert Opin. Invest. Drugs*, 2006, **15**, 1627.

[90] S. T. Duggan and G. M. Keating, *Drugs*, 2009, **69**, 1707.

[91] J. M. Schillig and J. S. Kalus, *Formulary*, 2008, **43**, 402.

[92] P. P. Dobesh, *Pharmacotherapy*, 2009, **29**, 1089.

[93] P. A. Gurbel and U. S. Tantry, *Current Opin. Invest. Drugs*, 2008, **9**, 324.

[94] D. J. Angiolillo, S. Suryadevara, P. Capranzano, and T. A. Bass, *Expert Opin. Pharmacother.*, 2008, **9**, 2893.

[95] W. L. Baker and C. M. White, *Am. J. Cardiovasc. Drugs*, 2009, **9**, 213.

[96] P. R. Padi, S. R. S. Peri, M. R. Ganta, S. Polavarapu, P. Cherukupally, B. Ireni, S. Padamata, K. Jonnada, K. Vinigari, and K. Nerella, *WO Patent 062044*, 2009.

[97] A. A. Tahrani, M. K. Piya, and A. H. Barnett, *Adv. Ther.*, 2009, **26**, 249.

[98] P. Cole, N. Serradell, J. Bolós, and R. Castañer, *Drugs Future*, 2008, **33**, 577.

[99] J. H. Shubrook, R. A. Colucci, and F. L. Schwartz, *Expert Opin. Pharmacother.*, 2009, **10**, 2927.

[100] S. A. Miller, E. L. St. Onge, and J. R. Taylor, *Formulary*, 2008, **43**, 122.

[101] T. M. Tzschentke, J. De Vry, R. Terlinden, H. -H. Hennies, C. Lange, W. Strassburger, M. Haurand, J. Kolb, J. Schneider, H. Buschmann, M. Finkam, U. Jahnel, and E. Friderichs, *Drugs Future*, 2006, **31**, 1053.

[102] T. M. Tzschentke, U. Jahnel, B. Kögel, T. Christoph, W. Englberger, J. De Vry, K. Schiene, A. Okamoto, D. Upmalis, H. Weber, C. Lange, J. -U. Stegmann, and R. Kleinert, *Drugs Today*, 2009, **45**, 483.

[103] T. M. Tzschentke, T. Christoph, B. Kögel, K. Schiene , H.-H. Hennies, W. Englberger, M. Haurand, U. Jahnel , T. I. F. H. Cremers, E. Friderichs, and J. De Vry, *J. Pharm. Exp. Ther.*, 2007, **323**, 265.

[104] L. A. Sorbera and J. Castañer, *Drugs Future*, 2004, **29**, 1211.

[105] E. C. Nannini and M. E. Stryjewski, *Expert Opin. Pharmacother.*, 2008, **9**, 2197.

[106] Z. A. Kanafani, *Expert. Rev. Anti-Infect. Ther.*, 2006, **4**, 743.

[107] S. Laohavaleeson, J. L. Kuti, and D. P. Nicolou, *Expert Opin. Invest. Drugs*, 2007, **16**, 347.

[108] G. R. Corey, M. E. Stryjewski, W. Weyenberg, U. Yasothan, and P. Kirkpatrick, *Nat. Rev. Drug Discov.*, 2009, **8**, 929.

[109] S. A. Knechtel, C. Jacobs, and M. E. Klepser, *Formulary*, 2007, **42**, 545.

[110] L. Charmeski, P. N. Patel, and D. Sym, *Ann. Pharmacother.*, 2009, **43**, 928.

[111] S. N. Leonard and M. J. Rybak, *Pharmacotherapy*, 2008, **28**, 458.

[112] T. Miyazaki, H. Fujiki, Y. Yamamura, S. Nakamura, and T. Mori, *Cardiovasc. Drug Rev.*, 2007, **25**, 1.

[113] L. A. Sorbera, J. Castañer, and M. Bayés, *Drugs Future*, 2002, **27**, 350.

[114] S. E. Shoaf, Z. Wang, P. Bricmont, and S. Mallikaarjun, *J. Clin. Pharmacol.*, 2007, **47**, 1498.

[115] U. C. Rangasetty, M. Gheorghiade, B. F. Uretsky, C. Orlandi, and A. Barbagelata, *Expert Opin. Invest. Drug*, 2006, **15**, 533.

[116] D. L. Jennings and J. S. Kalus, *Formulary*, 2008, **43**, 236.

[117] P. A. Orihuela, *Curr. Opin. Invest. Drugs*, 2007, **8**, 859.

[118] G. Benagiano, C. Bastianelli, and M. Farris, *Expert Opin. Pharmacother.*, 2008, **9**, 2473.

[119] B. J. Attardi, J. Burgenson, S. A, Hild, and J. R. Reel, *J. Steroid Biochem. Mol. Biol.*, 2004, **88**, 277.

[120] D. L. Blithe, L. K. Nieman, R. P. Blye, P. Stratton, and M. Passaro, *Steroids*, 2003, **68**, 1013.

[121] L. Dancsi, G. Visky, Z. Tuba, J. Csoergei, C. Molnar, and E. Magyari, *WO Patent 07144674*, 2007.

[122] A. L. Chien, J. T. Elder, and C. N. Ellis, *Drugs*, 2009, **69**, 1141.

[123] K. Reich, U. Yasothan, and P. Kirkpatrick, *Nat. Rev. Drug Discov.*, 2009, **8**, 355.

[124] J. V. Scanlon, B. P. Exter, M. Steinberg, and C. I. Jarvis, *Ann. Pharmacother.*, 2009, **43**, 1456.

[125] B. Cuenoud, I. Bruce, R. A. Fairhust, and D. Beattie, *WO Patent 00075114*, 2000.

[126] O. Bauwens, V. Ninane, B. Van de Maele, R. Firth, F. Dong, R. Owen, and M. Higgins, *Curr. Med. Res. Opin.*, 2009, **25**, 463.

CUMULATIVE NCE INTRODUCTION INDEX, 1983–2009

GENERAL NAME	INDICATION	YEAR INTRO.	ARMC VOL., PAGE
abacavir sulfate	antiviral	1999	35, 333
abarelix	anticancer	2004	40, 446
abatacept	rheumatoid arthritis	2006	42, 509
acarbose	antidiabetic	1990	26, 297
aceclofenac	anti-inflammatory	1992	28, 325
acemannan	wound healing agent	2001	37, 259
acetohydroxamic acid	hypoammonuric	1983	19, 313
acetorphan	antidiarrheal	1993	29, 332
acipimox	hypolipidemic	1985	21, 323
acitretin	antipsoriatic	1989	25, 309
acrivastine	antihistamine	1988	24, 295
actarit	antirheumatic	1994	30, 296
adalimumab	rheumatoid arthritis	2003	39, 267
adamantanium bromide	antiseptic	1984	20, 315
adefovir dipivoxil	antiviral	2002	38, 348
adrafinil	psychostimulant	1986	22, 315
AF-2259	anti-inflammatory	1987	23, 325
afloqualone	muscle relaxant	1983	19, 313
agalsidase alfa	fabry's disease	2001	37, 259
alacepril	antihypertensive	1988	24, 296
alclometasone dipropionate	topical anti-inflammatory	1985	21, 323
alefacept	plaque psoriasis	2003	39, 267
alemtuzumab	anticancer	2001	37, 260
alendronate sodium	osteoporosis	1993	29, 332
alfentanil HCl	analgesic	1983	19, 314
alfuzosin HCl	antihypertensive	1988	24, 296
alglucerase	enzyme	1991	27, 321
alglucosidase alfa	Pompe disease	2006	42, 511
aliskiren	antihypertensive	2007	43, 461
alitretinoin	anticancer	1999	35, 333
alminoprofen	analgesic	1983	19, 314
almotriptan	antimigraine	2000	36, 295
anakinra	antiarthritic	2001	37, 261
anidulafungin	antifungal	2006	42, 512
alosetron hydrochloride	irritable bowel syndrome	2000	36, 295
alpha-1 antitrypsin	protease inhibitor	1988	24, 297
alpidem	anxiolytic	1991	27, 322
alpiropride	antimigraine	1988	24, 296
alteplase	thrombolytic	1987	23, 326
alvimopan	post-operative ileus	2008	44, 584
ambrisentan	pulmonary arterial hypertension	2007	43, 463
amfenac sodium	anti-inflammatory	1986	22, 315
amifostine	cytoprotective	1995	31, 338
aminoprofen	topical anti-inflammatory	1990	26, 298
amisulpride	antipsychotic	1986	22, 316
amlexanox	antiasthmatic	1987	23, 327

GENERAL NAME	INDICATION	YEAR INTRO.	ARMC VOL., PAGE
balsalazide disodium	ulcerative colitis	1997	33, 329
bambuterol	bronchodilator	1990	26, 299
barnidipine HCl	antihypertensive	1992	28, 326
beclobrate	hypolipidemic	1986	22, 317
befunolol HCl	antiglaucoma	1983	19, 315
belotecan	anticancer	2004	40, 449
benazepril HCl	antihypertensive	1990	26, 299
benexate HCl	antiulcer	1987	23, 328
benidipine HCl	antihypertensive	1991	27, 322
beraprost sodium	platelet aggreg. inhibitor	1992	28, 326
besifloxacin	antibacterial	2009	45, 482
betamethasone butyrate prospinate	topical anti-inflammatory	1994	30, 297
betaxolol HCl	antihypertensive	1983	19, 315
betotastine besilate	antiallergic	2000	36, 297
bevacizumab	anticancer	2004	40, 450
bevantolol HCl	antihypertensive	1987	23, 328
bexarotene	anticancer	2000	36, 298
biapenem	antibacterial	2002	38, 351
bicalutamide	antineoplastic	1995	31, 338
bifemelane HCl	nootropic	1987	23, 329
bimatoprost	antiglaucoma	2001	37, 261
binfonazole	hypnotic	1983	19, 315
binifibrate	hypolipidemic	1986	22, 317
biolimus drug-eluting stent	anti-restenotic	2008	44, 586
bisantrene HCl	antineoplastic	1990	26, 300
bisoprolol fumarate	antihypertensive	1986	22, 317
bivalirudin	antithrombotic	2000	36, 298
blonanserin	antipsychotic	2008	44, 587
bopindolol	antihypertensive	1985	21, 324
bortezomib	anticancer	2003	39, 271
bosentan	antihypertensive	2001	37, 262
brimonidine	antiglaucoma	1996	32, 306
brinzolamide	antiglaucoma	1998	34, 318
brodimoprin	antibiotic	1993	29, 333
bromfenac sodium	NSAID	1997	33, 329
brotizolam	hypnotic	1983	19, 315
brovincamine fumarate	cerebral vasodilator	1986	22, 317
bucillamine	immunomodulator	1987	23, 329
bucladesine sodium	cardiostimulant	1984	20, 316
budipine	anti-Parkinsonian	1997	33, 330
budralazine	antihypertensive	1983	19, 315
bulaquine	antimalarial	2000	36, 299
bunazosin HCl	antihypertensive	1985	21, 324
bupropion HCl	antidepressant	1989	25, 310
buserelin acetate	hormone	1984	20, 316
buspirone HCl	anxiolytic	1985	21, 324
butenafine HCl	topical antifungal	1992	28, 327
butibufen	anti-inflammatory	1992	28, 327
butoconazole	topical antifungal	1986	22, 318

GENERAL NAME	INDICATION	YEAR INTRO.	ARMC VOL., PAGE
butoctamide	hypnotic	1984	20, 316
butyl flufenamate	topical anti-inflammatory	1983	19, 316
cabergoline	antiprolactin	1993	29, 334
cadexomer iodine	wound healing agent	1983	19, 316
cadralazine	hypertensive	1988	24, 298
calcipotriol	antipsoriatic	1991	27, 323
camostat mesylate	antineoplastic	1985	21, 325
canakinumab	anti-inflammatory	2009	45, 484
candesartan cilexetil	antihypertension	1997	33, 330
capecitabine	antineoplastic	1998	34, 319
captopril	antihypertensive	1982	13, 086
carboplatin	antibiotic	1986	22, 318
carperitide	congestive heart failure	1995	31, 339
carumonam	antibiotic	1988	24, 298
carvedilol	antihypertensive	1991	27, 323
caspofungin acetate	antifungal	2001	37, 263
catumaxomab	anticancer	2009	45, 486
cefbuperazone sodium	antibiotic	1985	21, 325
cefcapene pivoxil	antibiotic	1997	33, 330
cefdinir	antibiotic	1991	27, 323
cefditoren pivoxil	oral cephalosporin	1994	30, 297
cefepime	antibiotic	1993	29, 334
cefetamet pivoxil HCl	antibiotic	1992	28, 327
cefixime	antibiotic	1987	23, 329
cefmenoxime HCl	antibiotic	1983	19, 316
cefminox sodium	antibiotic	1987	23, 330
cefodizime sodium	antibiotic	1990	26, 300
cefonicid sodium	antibiotic	1984	20, 316
ceforanide	antibiotic	1984	20, 317
cefoselis	antibiotic	1998	34, 319
cefotetan disodium	antibiotic	1984	20, 317
cefotiam hexetil HCl	antibiotic	1991	27, 324
cefozopran HCl	injectable cephalosporin	1995	31, 339
cefpimizole	antibiotic	1987	23, 330
cefpiramide sodium	antibiotic	1985	21, 325
cefpirome sulfate	antibiotic	1992	28, 328
cefpodoxime proxetil	antibiotic	1989	25, 310
cefprozil	antibiotic	1992	28, 328
ceftazidime	antibiotic	1983	19, 316
cefteram pivoxil	antibiotic	1987	23, 330
ceftibuten	antibiotic	1992	28, 329
ceftobiprole medocaril	antibiotic	2008	44, 589
cefuroxime axetil	antibiotic	1987	23, 331
cefuzonam sodium	antibiotic	1987	23, 331
celecoxib	antiarthritic	1999	35, 335
celiprolol HCl	antihypertensive	1983	19, 317
centchroman	antiestrogen	1991	27, 324
centoxin	immunomodulator	1991	27, 325
cerivastatin	dyslipidemia	1997	33, 331

GENERAL NAME	INDICATION	YEAR INTRO.	ARMC VOL., PAGE
certolizumab pegol	Crohn's disease	2008	44, 592
cetirizine HCl	antihistamine	1987	23, 331
cetrorelix	female infertility	1999	35, 336
cetuximab	anticancer	2003	39, 272
cevimeline hydrochloride	anti-xerostomia	2000	36, 299
chenodiol	anticholelithogenic	1983	19, 317
CHF-1301	antiparkinsonian	1999	35, 336
choline alfoscerate	nootropic	1990	26, 300
choline fenofibrate	dyslipidemia	2008	44, 594
cibenzoline	antiarrhythmic	1985	21, 325
ciclesonide	asthma, COPD	2005	41, 443
cicletanine	antihypertensive	1988	24, 299
cidofovir	antiviral	1996	32, 306
cilazapril	antihypertensive	1990	26, 301
cilostazol	antithrombotic	1988	24, 299
cimetropium bromide	antispasmodic	1985	21, 326
cinacalcet	hyperparathyroidism	2004	40, 451
cinildipine	antihypertensive	1995	31, 339
cinitapride	gastroprokinetic	1990	26, 301
cinolazepam	hypnotic	1993	29, 334
ciprofibrate	hypolipidemic	1985	21, 326
ciprofloxacin	antibacterial	1986	22, 318
cisapride	gastroprokinetic	1988	24, 299
cisatracurium besilate	muscle relaxant	1995	31, 340
citalopram	antidepressant	1989	25, 311
cladribine	antineoplastic	1993	29, 335
clarithromycin	antibiotic	1990	26, 302
clevidipine	antihypertensive	2008	44, 596
clevudine	hepatitis B	2007	43, 466
clobenoside	vasoprotective	1988	24, 300
cloconazole HCl	topical antifungal	1986	22, 318
clodronate disodium	calcium regulator	1986	22, 319
clofarabine	anticancer	2005	41, 444
clopidogrel hydrogensulfate	antithrombotic	1998	34, 320
cloricromen	antithrombotic	1991	27, 325
clospipramine HCl	neuroleptic	1991	27, 325
colesevelam hydrochloride	hypolipidemic	2000	36, 300
colestimide	hypolipidaemic	1999	35, 337
colforsin daropate HCl	cardiotonic	1999	35, 337
conivaptan	hyponatremia	2006	42, 514
crotelidae polyvalent immune fab	antidote	2001	37, 263
cyclosporine	immunosuppressant	1983	19, 317
cytarabine ocfosfate	antineoplastic	1993	29, 335
dabigatran etexilate	anti-coagulant	2008	44, 598
dalfopristin	antibiotic	1999	35, 338
dapiprazole HCl	antiglaucoma	1987	23, 332
dapoxetine	premature ejaculation	2009	45, 488

GENERAL NAME	INDICATION	YEAR INTRO.	ARMC VOL., PAGE
drospirenone	contraceptive	2000	36, 302
drotrecogin alfa	antisepsis	2001	37, 265
droxicam	anti-inflammatory	1990	26, 302
droxidopa	antiparkinsonian	1989	25, 312
duloxetine	antidepressant	2004	40, 452
dutasteride	5a reductase inhibitor	2002	38, 353
duteplase	anticougulant	1995	31, 342
eberconazole	antifungal	2005	41, 449
ebastine	antihistamine	1990	26 302
ebrotidine	antiulcer	1997	33, 333
ecabet sodium	antiulcerative	1993	29, 336
eculizumab	hemoglobinuria	2007	43, 468
edaravone	neuroprotective	2001	37, 265
efalizumab	psoriasis	2003	39, 274
efavirenz	antiviral	1998	34, 321
efonidipine	antihypertensive	1994	30, 299
egualen sodium	antiulcer	2000	36, 303
eletriptan	antimigraine	2001	37, 266
eltrombopag	antithrombocytopenic	2009	45, 497
emedastine difumarate	antiallergic/antiasthmatic	1993	29, 336
emorfazone	analgesic	1984	20, 317
emtricitabine	antiviral	2003	39, 274
enalapril maleate	antihypertensive	1984	20, 317
enalaprilat	antihypertensive	1987	23, 332
encainide HCl	antiarrhythmic	1987	23, 333
enfuvirtide	antiviral	2003	39, 275
enocitabine	antineoplastic	1983	19, 318
enoxacin	antibacterial	1986	22, 320
enoxaparin	antithrombotic	1987	23, 333
enoximone	cardiostimulant	1988	24, 301
enprostil	antiulcer	1985	21, 327
entacapone	antiparkinsonian	1998	34, 322
entecavir	antiviral	2005	41, 450
epalrestat	antidiabetic	1992	28, 330
eperisone HCl	muscle relaxant	1983	19, 318
epidermal growth factor	wound healing agent	1987	23, 333
epinastine	antiallergic	1994	30, 299
epirubicin HCl	antineoplastic	1984	20, 318
eplerenone	antihypertensive	2003	39, 276
epoprostenol sodium	platelet aggreg. inhib.	1983	19, 318
eprosartan	antihypertensive	1997	33, 333
eptazocine HBr	analgesic	1987	23, 334
eptilfibatide	antithrombotic	1999	35, 340
erdosteine	expectorant	1995	31, 342
erlotinib	anticancer	2004	40, 454
ertapenem sodium	antibacterial	2002	38, 353
erythromycin acistrate	antibiotic	1988	24, 301
erythropoietin	hematopoetic	1988	24, 301
escitalopram oxolate	antidepressant	2002	38, 354

GENERAL NAME	INDICATION	YEAR INTRO.	ARMC VOL., PAGE
flutrimazole	topicalantifungal	1995	31, 343
flutropium bromide	antitussive	1988	24, 303
fluvastatin	hypolipaemic	1994	30, 300
fluvoxamine maleate	antidepressant	1983	19, 319
follitropin alfa	fertility enhancer	1996	32, 307
follitropin beta	fertility enhancer	1996	32, 308
fomepizole	antidote	1998	34, 323
fomivirsen sodium	antiviral	1998	34, 323
fondaparinux sodium	antithrombotic	2002	38, 356
formestane	antineoplastic	1993	29, 337
formoterol fumarate	bronchodilator	1986	22, 321
fosamprenavir	antiviral	2003	39, 277
fosaprepitant dimeglumine	anti-emetic	2008	44, 606
foscarnet sodium	antiviral	1989	25, 313
fosfosal	analgesic	1984	20, 319
fosfluconazole	antifungal	2004	40, 457
fosinopril sodium	antihypertensive	1991	27, 328
fosphenytoin sodium	antiepileptic	1996	32, 308
fotemustine	antineoplastic	1989	25, 313
fropenam	antibiotic	1997	33, 334
frovatriptan	antimigraine	2002	38, 357
fudosteine	expectorant	2001	37, 267
fulveristrant	anticancer	2002	38, 357
gabapentin	antiepileptic	1993	29, 338
gadoversetamide	MRI contrast agent	2000	36, 304
gallium nitrate	calcium regulator	1991	27, 328
gallopamil HCl	antianginal	1983	19, 319
galsulfase	mucopolysaccharidosis VI	2005	41, 453
ganciclovir	antiviral	1988	24, 303
ganirelix acetate	female infertility	2000	36, 305
garenoxacin	anti-infective	2007	43, 471
gatilfloxacin	antibiotic	1999	35, 340
gefitinib	antineoplastic	2002	38, 358
gemcitabine HCl	antineoplastic	1995	31, 344
gemeprost	abortifacient	1983	19, 319
gemifloxacin	antibacterial	2004	40, 458
gemtuzumab ozogamicin	anticancer	2000	36, 306
gestodene	progestogen	1987	23, 335
gestrinone	antiprogestogen	1986	22, 321
glatiramer acetate	Multiple Sclerosis	1997	33, 334
glimepiride	antidiabetic	1995	31, 344
glucagon, rDNA	hypoglycemia	1993	29, 338
GMDP	immunostimulant	1996	32, 308
golimumab	anti-inflammatory	2009	45, 503
goserelin	hormone	1987	23, 336
granisetron HCl	antiemetic	1991	27, 329
guanadrel sulfate	antihypertensive	1983	19, 319
gusperimus	immunosuppressant	1994	30, 300
halobetasol propionate	topical anti-inflammatory	1991	27, 329

GENERAL NAME	INDICATION	YEAR INTRO.	ARMC VOL., PAGE
lopinavir	antiviral	2000	36, 310
loprazolam mesylate	hypnotic	1983	19, 321
loprinone HCl	cardiostimulant	1996	32, 312
loracarbef	antibiotic	1992	28, 333
loratadine	antihistamine	1988	24, 306
lornoxicam	NSAID	1997	33, 337
losartan	antihypertensive	1994	30, 302
loteprednol etabonate	antiallergic ophthalmic	1998	34, 324
lovastatin	hypocholesterolemic	1987	23, 337
loxoprofen sodium	anti-inflammatory	1986	22, 322
lulbiprostone	chronic idiopathic constipation	2006	42, 525
luliconazole	antifungal	2005	41, 454
lumiracoxib	anti-inflammatory	2005	41, 455
Lyme disease	vaccine	1999	35, 342
mabuterol HCl	bronchodilator	1986	22, 323
malotilate	hepatoprotective	1985	21, 329
manidipine HCl	antihypertensive	1990	26, 304
maraviroc	anti-infective — HIV	2007	43, 478
masoprocol	topical antineoplastic	1992	28, 333
maxacalcitol	vitamin D	2000	36, 310
mebefradil HCl	antihypertensive	1997	33, 338
medifoxamine fumarate	antidepressant	1986	22, 323
mefloquine HCl	antimalarial	1985	21, 329
meglutol	hypolipidemic	1983	19, 321
melinamide	hypocholesterolemic	1984	20, 320
meloxicam	antiarthritic	1996	32, 312
mepixanox	analeptic	1984	20, 320
meptazinol HCl	analgesic	1983	19, 321
meropenem	carbapenem antibiotic	1994	30, 303
metaclazepam	anxiolytic	1987	23, 338
metapramine	antidepressant	1984	20, 320
methylnaltrexone bromide	opioid-induced constipation	2008	44, 612
mexazolam	anxiolytic	1984	20, 321
micafungin	antifungal	2002	38, 360
mifepristone	abortifacient	1988	24, 306
miglitol	antidiabetic	1998	34, 325
miglustat	gaucher's disease	2003	39, 279
milnacipran	antidepressant	1997	33, 338
milrinone	cardiostimulant	1989	25, 316
miltefosine	topical antineoplastic	1993	29, 340
minodronic acid	osteoporosis	2009	45, 509
miokamycin	antibiotic	1985	21, 329
mirtazapine	antidepressant	1994	30, 303
misoprostol	antiulcer	1985	21, 329
mitiglinide	antidiabetic	2004	40, 460
mitoxantrone HCl	antineoplastic	1984	20, 321
mivacurium chloride	muscle relaxant	1992	28, 334
mivotilate	hepatoprotectant	1999	35, 343
mizolastine	antihistamine	1998	34, 325
mizoribine	immunosuppressant	1984	20, 321

GENERAL NAME	INDICATION	YEAR INTRO.	ARMC VOL., PAGE
moclobemide	antidepressant	1990	26, 305
modafinil	idiopathic hypersomnia	1994	30, 303
moexipril HCl	antihypertensive	1995	31, 346
mofezolac	analgesic	1994	30, 304
mometasone furoate	topical anti-inflammatory	1987	23, 338
montelukast sodium	antiasthma	1998	34, 326
moricizine HCl	antiarrhythmic	1990	26, 305
mosapride citrate	gastroprokinetic	1998	34, 326
moxifloxacin HCL	antibiotic	1999	35, 343
moxonidine	antihypertensive	1991	27, 330
mozavaptan	hyponatremia	2006	42, 527
mupirocin	topical antibiotic	1985	21, 330
muromonab-CD3	immunosuppressant	1986	22, 323
muzolimine	diuretic	1983	19, 321
mycophenolate mofetil	immunosuppressant	1995	31, 346
mycophenolate sodium	immunosuppressant	2003	39, 279
nabumetone	anti-inflammatory	1985	21, 330
nadifloxacin	topical antibiotic	1993	29, 340
nafamostat mesylate	protease inhibitor	1986	22, 323
nafarelin acetate	hormone	1990	26, 306
naftifine HCl	antifungal	1984	20, 321
naftopidil	dysuria	1999	35, 344
nalfurafine hydrochloride	pruritus	2009	45, 510
nalmefene HCl	dependence treatment	1995	31, 347
naltrexone HCl	narcotic antagonist	1984	20, 322
naratriptan HCl	antimigraine	1997	33, 339
nartograstim	leukopenia	1994	30, 304
natalizumab	multiple sclerosis	2004	40, 462
nateglinide	antidiabetic	1999	35, 344
nazasetron	antiemetic	1994	30, 305
nebivolol	antihypertensive	1997	33, 339
nedaplatin	antineoplastic	1995	31, 347
nedocromil sodium	antiallergic	1986	22, 324
nefazodone	antidepressant	1994	30, 305
nelarabine	anticancer	2006	42, 528
nelfinavir mesylate	antiviral	1997	33, 340
neltenexine	cystic fibrosis	1993	29, 341
nemonapride	neuroleptic	1991	27, 331
nepafenac	anti-inflammatory	2005	41, 456
neridronic acide	calcium regulator	2002	38, 361
nesiritide	congestive heart failure	2001	37, 269
neticonazole HCl	topical antifungal	1993	29, 341
nevirapine	antiviral	1996	32, 313
nicorandil	coronary vasodilator	1984	20, 322
nifekalant HCl	antiarrythmic	1999	35, 344
nilotinib	anticancer — CML	2007	43, 480
nilutamide	antineoplastic	1987	23, 338
nilvadipine	antihypertensive	1989	25, 316
nimesulide	anti-inflammatory	1985	21, 330
nimodipine	cerebral vasodilator	1985	21, 330

GENERAL NAME	INDICATION	YEAR INTRO.	ARMC VOL., PAGE
nimotuzumab	anticancer	2006	42, 529
nipradilol	antihypertensive	1988	24, 307
nisoldipine	antihypertensive	1990	26, 306
nitisinone	antityrosinaemia	2002	38, 361
nitrefazole	alcohol deterrent	1983	19, 322
nitrendipine	hypertensive	1985	21, 331
nizatidine	antiulcer	1987	23, 339
nizofenzone	fumaratenootropic	1988	24, 307
nomegestrol acetate	progestogen	1986	22, 324
norelgestromin	contraceptive	2002	38, 362
norfloxacin	antibacterial	1983	19, 322
norgestimate	progestogen	1986	22, 324
OCT-43	anticancer	1999	35, 345
octreotide	antisecretory	1988	24, 307
ofatumumab	anticancer	2009	45, 512
ofloxacin	antibacterial	1985	21, 331
olanzapine	neuroleptic	1996	32, 313
olimesartan Medoxomil	antihypertensive	2002	38, 363
olopatadine HCl	antiallergic	1997	33, 340
omalizumab allergic	asthma	2003	39, 280
omeprazole	antiulcer	1988	24, 308
ondansetron HCl	antiemetic	1990	26, 306
OP-1	osteoinductor	2001	37, 269
orlistat	antiobesity	1998	34, 327
ornoprostil	antiulcer	1987	23, 339
osalazine sodium	intestinal antinflamm.	1986	22, 324
oseltamivir phosphate	antiviral	1999	35, 346
oxaliplatin	anticancer	1996	32, 313
oxaprozin	anti-inflammatory	1983	19, 322
oxcarbazepine	anticonvulsant	1990	26, 307
oxiconazole nitrate	antifungal	1983	19, 322
oxiracetam	nootropic	1987	23, 339
oxitropium bromide	bronchodilator	1983	19, 323
ozagrel sodium	antithrombotic	1988	24, 308
paclitaxal	antineoplastic	1993	29, 342
palifermin	mucositis	2005	41, 461
paliperidone	antipsychotic	2007	43, 482
palonosetron	antiemetic	2003	39, 281
panipenem/betamipron carbapenem	antibiotic	1994	30, 305
panitumumab	anticancer	2006	42, 531
pantoprazole sodium	antiulcer	1995	30, 306
parecoxib sodium	analgesic	2002	38, 364
paricalcitol	vitamin D	1998	34, 327
parnaparin sodium	anticoagulant	1993	29, 342
paroxetine	antidepressant	1991	27, 331
pazopanib	anticancer	2009	45, 514
pazufloxacin	antibacterial	2002	38, 364
pefloxacin mesylate	antibacterial	1985	21, 331

GENERAL NAME	INDICATION	YEAR INTRO.	ARMC VOL., PAGE
pegademase bovine	immunostimulant	1990	26, 307
pegaptanib	Age-related macular degeneration	2005	41, 458
pegaspargase	antineoplastic	1994	30, 306
pegvisomant	acromegaly	2003	39, 281
pemetrexed	anticancer	2004	40, 463
pemirolast potassium	antiasthmatic	1991	27, 331
penciclovir	antiviral	1996	32, 314
pentostatin	antineoplastic	1992	28, 334
pergolide mesylate	antiparkinsonian	1988	24, 308
perindopril	antihypertensive	1988	24, 309
perospirone HCL	neuroleptic	2001	37, 270
picotamide	antithrombotic	1987	23, 340
pidotimod	immunostimulant	1993	29, 343
piketoprofen	topical anti-inflammatory	1984	20, 322
pilsicainide HCl	antiarrhythmic	1991	27, 332
pimaprofen	topical anti-inflammatory	1984	20, 322
pimecrolimus	immunosuppressant	2002	38, 365
pimobendan	heart failure	1994	30, 307
pinacidil	antihypertensive	1987	23, 340
pioglitazone HCL	antidiabetic	1999	35, 346
pirarubicin	antineoplastic	1988	24, 309
pirfenidone	idiopathic pulmonary fibrosis	2008	44, 614
pirmenol	antiarrhythmic	1994	30, 307
piroxicam cinnamate	anti-inflammatory	1988	24, 309
pitavastatin	hypocholesterolemic	2003	39, 282
pivagabine	antidepressant	1997	33, 341
plaunotol	antiulcer	1987	23, 340
plerixafor hydrochloride	stem cell mobilizer	2009	45, 515
polaprezinc	antiulcer	1994	30, 307
porfimer sodium	antineoplastic adjuvant	1993	29, 343
posaconazole	antifungal	2006	42, 532
pralatrexate	anticancer	2009	45, 517
pramipexole HCl	antiParkinsonian	1997	33, 341
pramiracetam H2SO4	cognition enhancer	1993	29, 343
pramlintide	anti-diabetic	2005	41, 460
pranlukast	antiasthmatic	1995	31, 347
prasugrel	antiplatelet therapy	2009	45, 519
pravastatin	antilipidemic	1989	25, 316
prednicarbate	topical anti-inflammatory	1986	22, 325
pregabalin	antiepileptic	2004	40, 464
prezatide copper acetate	vulnery	1996	32, 314
progabide	anticonvulsant	1985	21, 331
promegestrone	progestogen	1983	19, 323
propacetamol HCl	analgesic	1986	22, 325
propagermanium	antiviral	1994	30, 308
propentofylline propionate	cerebral vasodilator	1988	24, 310
propiverine HCl	urologic	1992	28, 335
propofol	anesthetic	1986	22, 325
prulifloxacin	antibacterial	2002	38, 366

GENERAL NAME	INDICATION	YEAR INTRO.	ARMC VOL., PAGE
pumactant	lung surfactant	1994	30, 308
quazepam	hypnotic	1985	21, 332
quetiapine fumarate	neuroleptic	1997	33, 341
quinagolide	hyperprolactinemia	1994	30, 309
quinapril	antihypertensive	1989	25, 317
quinfamide	amebicide	1984	20, 322
quinupristin	antibiotic	1999	35, 338
rabeprazole sodium	gastric antisecretory	1998	34, 328
raloxifene HCl	osteoporosis	1998	34, 328
raltegravir	anti-infective — HIV	2007	43, 484
raltitrexed	anticancer	1996	32, 315
ramatroban	antiallergic	2000	36, 311
ramelteon	insomnia	2005	41, 462
ramipril	antihypertensive	1989	25, 317
ramosetron	antiemetic	1996	32, 315
ranibizumab	age-related macular degeneration	2006	42, 534
ranimustine	antineoplastic	1987	23, 341
ranitidine bismuth citrate	antiulcer	1995	31, 348
ranolazine	angina	2006	42, 535
rapacuronium bromide	muscle relaxant	1999	35, 347
rasagiline	parkinson's disease	2005	41, 464
rebamipide	antiulcer	1990	26, 308
reboxetine	antidepressant	1997	33, 342
remifentanil HCl	analgesic	1996	32, 316
remoxipride HCl	antipsychotic	1990	26, 308
repaglinide	antidiabetic	1998	34, 329
repirinast	antiallergic	1987	23, 341
retapamulin	anti-infective	2007	43, 486
reteplase	fibrinolytic	1996	32, 316
reviparin sodium	anticoagulant	1993	29, 344
rifabutin	antibacterial	1992	28, 335
rifapentine	antibacterial	1988	24, 310
rifaximin	antibiotic	1985	21, 332
rifaximin	antibiotic	1987	23, 341
rilmazafone	hypnotic	1989	25, 317
rilmenidine	antihypertensive	1988	24, 310
rilonacept	genetic autoinflammatory syndromes	2008	44, 615
riluzole	neuroprotective	1996	32, 316
rimantadine HCl	antiviral	1987	23, 342
rimexolone	anti-inflammatory	1995	31, 348
rimonabant	anti-obesity	2006	42, 537
risedronate sodium	osteoporosis	1998	34, 330
risperidone	neuroleptic	1993	29, 344
ritonavir	antiviral	1996	32, 317
rivaroxaban	anticoagulant, venous thromboembolism	2008	44, 617
rivastigmin	anti-Alzheimer	1997	33, 342
rizatriptan benzoate	antimigraine	1998	34, 330

GENERAL NAME	INDICATION	YEAR INTRO.	ARMC VOL., PAGE
tenofovir disoproxil fumarate	antiviral	2001	37, 271
tenoxicam	anti-inflammatory	1987	23, 344
teprenone	antiulcer	1984	20, 323
terazosin HCl	antihypertensive	1984	20, 323
terbinafine HCl	antifungal	1991	27, 334
terconazole	antifungal	1983	19, 324
tertatolol HCl	antihypertensive	1987	23, 344
thrombin alfa	hemostat	2008	44, 627
thrombomodulin (recombinant)	anticoagulant	2008	44, 628
thymopentin	immunomodulator	1985	21, 333
tiagabine	antiepileptic	1996	32, 319
tiamenidine HCl	antihypertensive	1988	24, 311
tianeptine sodium	antidepressant	1983	19, 324
tibolone	anabolic	1988	24, 312
tigecycline	antibiotic	2005	41, 468
tilisolol HCl	antihypertensive	1992	28, 337
tiludronate disodium	Paget's disease	1995	31, 350
timiperone	neuroleptic	1984	20, 323
tinazoline	nasal decongestant	1988	24, 312
tioconazole	antifungal	1983	19, 324
tiopronin	urolithiasis	1989	25, 318
tiotropium bromide	bronchodilator	2002	38, 368
tipranavir	HIV	2005	41, 470
tiquizium bromide	antispasmodic	1984	20, 324
tiracizine HCl	antiarrhythmic	1990	26, 310
tirilazad mesylate	subarachnoid hemorrhage	1995	31, 351
tirofiban HCl	antithrombotic	1998	34, 332
tiropramide HCl	antispasmodic	1983	19, 324
tizanidine	muscle relaxant	1984	20, 324
tolcapone	antiParkinsonian	1997	33, 343
toloxatone	antidepressant	1984	20, 324
tolrestat	antidiabetic	1989	25, 319
tolvaptan	hyponatremia	2009	45, 528
topiramate	antiepileptic	1995	31, 351
topotecan HCl	anticancer	1996	32, 320
torasemide	diuretic	1993	29, 348
toremifene	antineoplastic	1989	25, 319
tositumomab	anticancer	2003	39, 285
tosufloxacin tosylate	antibacterial	1990	26, 310
trabectedin	anticancer	2007	43, 492
trandolapril	antihypertensive	1993	29, 348
travoprost	antiglaucoma	2001	37, 272
treprostinil sodium	antihypertensive	2002	38, 368
tretinoin tocoferil	antiulcer	1993	29, 348
trientine HCl	chelator	1986	22, 327
trimazosin HCl	antihypertensive	1985	21, 333
trimegestone	progestogen	2001	37, 273

GENERAL NAME	INDICATION	YEAR INTRO.	ARMC VOL., PAGE
trimetrexate glucuronate	Pneumocystis carinii pneumonia	1994	30, 312
troglitazone	antidiabetic	1997	33, 344
tropisetron	antiemetic	1992	28, 337
trovafloxacin mesylate	antibiotic	1998	34, 332
troxipide	antiulcer	1986	22, 327
ubenimex	immunostimulant	1987	23, 345
udenafil	erectile dysfunction	2005	41, 472
ulipristal acetate	contraceptive	2009	45, 530
unoprostone isopropyl ester	antiglaucoma	1994	30, 312
valaciclovir HCl	antiviral	1995	31, 352
vadecoxib	antiarthritic	2002	38, 369
ustekinumab	anti-psoriatic	2009	45, 532
vaglancirclovir HCL	antiviral	2001	37, 273
valrubicin	anticancer	1999	35, 350
valsartan	antihypertensive	1996	32, 320
vardenafil	male sexual dysfunction	2003	39, 286
varenicline	nicotine-dependence	2006	42, 547
venlafaxine	antidepressant	1994	30, 312
verteporfin	photosensitizer	2000	36, 312
vesnarinone	cardiostimulant	1990	26, 310
vigabatrin	anticonvulsant	1989	25, 319
vildagliptin	antidiabetic	2007	43, 494
vinorelbine	antineoplastic	1989	25, 320
voglibose	antidiabetic	1994	30, 313
voriconazole	antifungal	2002	38, 370
vorinostat	anticancer	2006	42, 549
xamoterol fumarate	cardiotonic	1988	24, 312
ximelagatran	anticoagulant	2004	40, 470
zafirlukast	antiasthma	1996	32, 321
zalcitabine	antiviral	1992	28, 338
zaleplon	hypnotic	1999	35, 351
zaltoprofen	anti-inflammatory	1993	29, 349
zanamivir	antiviral	1999	35, 352
ziconotide	severe chronic pain	2005	41, 473
zidovudine	antiviral	1987	23, 345
zileuton	antiasthma	1997	33, 344
zinostatin stimalamer	antineoplastic	1994	30, 313
ziprasidone hydrochloride	neuroleptic	2000	36, 312
zofenopril calcium	antihypertensive	2000	36, 313
zoledronate disodium	hypercalcemia	2000	36, 314
zolpidem hemitartrate	hypnotic	1988	24, 313
zomitriptan	antimigraine	1997	33, 345
zonisamide	anticonvulsant	1989	25, 320
zopiclone	hypnotic	1986	22, 327
zuclopenthixol acetate	antipsychotic	1987	23, 345

CUMULATIVE NCE INTRODUCTION INDEX, 1983–2009 (BY INDICATION)

GENERIC NAME	INDICATION	YEAR INTRO.	ARMC VOL, (PAGE)
gemeprost	ABORTIFACIENT	1983	19 (319)
mifepristone		1988	24 (306)
lanreotide acetate	ACROMEGALY	1995	31 (345)
pegvisomant		2003	39 (281)
lisdexamfetamine	ADHD	2007	43 (477)
pegaptanib	AGE-RELATED MACULAR DEGENERATION	2005	41 (458)
ranibizumab		2006	42 (534)
nitrefazole	ALCOHOL DETERRENT	1983	19 (322)
omalizumab	ALLERGIC ASTHMA	2003	39 (280)
tacrine HCl	ALZHEIMER'S DISEASE	1993	29 (346)
quinfamide	AMEBICIDE	1984	20 (322)
tibolone	ANABOLIC	1988	24 (312)
mepixanox	ANALEPTIC	1984	20 (320)
alfentanil HCl	ANALGESIC	1983	19 (314)
alminoprofen		1983	19 (314)
dezocine		1991	27 (326)
emorfazone		1984	20 (317)
eptazocine HBr		1987	23 (334)
etoricoxib		2002	38 (355)
flupirtine maleate		1985	21 (328)
fosfosal		1984	20 (319)
ketorolac tromethamine		1990	26 (304)
meptazinol HCl		1983	19 (321)
mofezolac		1994	30 (304)
parecoxib sodium		2002	38 (364)
propacetamol HCl		1986	22 (325)
remifentanil HCl		1996	32 (316)
sufentanil		1983	19 (323)
suprofen		1983	19 (324)
tapentadol hydrochloride	ANALGESIC	2009	45 (523)
desflurane	ANESTHETIC	1992	28 (329)
propofol		1986	22 (325)
ropivacaine		1996	32 (318)
sevoflurane		1990	26 (309)
levobupivacaine hydrochloride	ANESTHETIC, LOCAL	2000	36 (308)
ivabradine	ANGINA	2006	42 (522)
ranolazine		2006	42 (535)
azelaic acid	ANTIACNE	1989	25 (310)
betotastine besilate	ANTIALLERGIC	2000	36 (297)
emedastine difumarate		1993	29 (336)
epinastine		1994	30 (299)
fexofenadine		1996	32 (307)
fluticasone furoate		2003	39 (274)

GENERIC NAME	INDICATION	YEAR INTRO.	ARMC VOL, (PAGE)
nedocromil sodium		1986	22 (324)
olopatadine hydrochloride		1997	33 (340)
ramatroban		2000	36 (311)
repirinast		1987	23 (341)
suplatast tosilate		1995	31 (350)
tazanolast		1990	26 (309)
lodoxamide tromethamine		1992	28 (333)
rupatadine fumarate		2003	39 (284)
loteprednol etabonate	OPHTHALMIC	1998	34 (324)
donepezil hydrochloride	ANTI-ALZHEIMERS	1997	33 (332)
rivastigmin		1997	33 (342)
gallopamil HCl	ANTIANGINAL	1983	19 (319)
cibenzoline	ANTIARRHYTHMIC	1985	21 (325)
dofetilide		2000	36 (301)
Dronedarone		2009	45(495)
encainide HCl		1987	23 (333)
esmolol HCl		1987	23 (334)
ibutilide fumarate		1996	32 (309)
landiolol		2002	38 (360)
moricizine hydrochloride		1990	26 (305)
nifekalant HCl		1999	35 (344)
pilsicainide hydrochloride		1991	27 (332)
pirmenol		1994	30 (307)
tiracizine hydrochloride		1990	26 (310)
anakinra	ANTIARTHRITIC	2001	37 (261)
celecoxib		1999	35 (335)
etoricoxib		2002	38 (355)
meloxicam		1996	32 (312)
leflunomide		1998	34 (324)
rofecoxib		1999	35 (347)
valdecoxib		2002	38 (369)
amlexanox	ANTIASTHMATIC	1987	23 (327)
emedastine difumarate		1993	29 (336)
ibudilast		1989	25 (313)
levalbuterol HCl		1999	35 (341)
montelukast sodium		1998	34 (326)
pemirolast potassium		1991	27 (331)
seratrodast		1995	31 (349)
zafirlukast		1996	32 (321)
zileuton		1997	33 (344)
balofloxacin	ANTIBACTERIAL	2002	38 (351)
besifloxacin		2009	45 (482)
biapenem		2002	38 (351)
ciprofloxacin		1986	22 (318)
enoxacin		1986	22 (320)

GENERIC NAME	INDICATION	YEAR INTRO.	ARMC VOL, (PAGE)
ertapenem sodium		2002	38 (353)
fleroxacin		1992	28 (331)
gemifloxacin		2004	40 (458)
norfloxacin		1983	19 (322)
ofloxacin		1985	21 (331)
pazufloxacin		2002	38 (364)
pefloxacin mesylate		1985	21 (331)
pranlukast		1995	31 (347)
prulifloxacin		2002	38 (366)
rifabutin		1992	28 (335)
rifapentine		1988	24 (310)
rufloxacin hydrochloride		1992	28 (335)
sitafloxacin hydrate		2008	44 (621)
teicoplanin		1988	24 (311)
temafloxacin hydrochloride		1991	27 (334)
tosufloxacin tosylate		1990	26 (310)
arbekacin	ANTIBIOTIC	1990	26 (298)
aspoxicillin		1987	23 (328)
astromycin sulfate		1985	21 (324)
azithromycin		1988	24 (298)
aztreonam		1984	20 (315)
brodimoprin		1993	29 (333)
carboplatin		1986	22 (318)
carumonam		1988	24 (298)
cefbuperazone sodium		1985	21 (325)
cefcapene pivoxil		1997	33 (330)
cefdinir		1991	27 (323)
cefepime		1993	29 (334)
cefetamet pivoxil hydrochloride		1992	28 (327)
cefixime		1987	23 (329)
cefmenoxime HCl		1983	19 (316)
cefminox sodium		1987	23 (330)
cefodizime sodium		1990	26 (300)
cefonicid sodium		1984	20 (316)
ceforanide		1984	20 (317)
cefoselis		1998	34 (319)
cefotetan disodium		1984	20 (317)
cefotiam hexetil hydrochloride		1991	27 (324)
cefpimizole		1987	23 (330)
cefpiramide sodium		1985	21 (325)
cefpirome sulfate		1992	28 (328)
cefpodoxime proxetil		1989	25 (310)
cefprozil		1992	28 (328)
ceftazidime		1983	19 (316)
cefteram pivoxil		1987	23 (330)
ceftibuten		1992	28 (329)

GENERIC NAME	INDICATION	YEAR INTRO.	ARMC VOL, (PAGE)
ceftobiprole medocaril		2008	44 (589)
cefuroxime axetil		1987	23 (331)
cefuzonam sodium		1987	23 (331)
clarithromycin		1990	26 (302)
dalfopristin		1999	35 (338)
dirithromycin		1993	29 (336)
doripenem		2005	41 (448)
erythromycin acistrate		1988	24 (301)
flomoxef sodium		1988	24 (302)
flurithromycin ethylsuccinate		1997	33 (333)
fropenam		1997	33 (334)
gatifloxacin		1999	35 (340)
imipenem/cilastatin		1985	21 (328)
isepamicin		1988	24 (305)
lenampicillin HCl		1987	23 (336)
levofloxacin		1993	29 (340)
linezolid		2000	36 (309)
lomefloxacin		1989	25 (315)
loracarbef		1992	28 (333)
miokamycin		1985	21 (329)
moxifloxacin HCl		1999	35 (343)
quinupristin		1999	35 (338)
rifaximin		1985	21 (332)
rifaximin		1987	23 (341)
rokitamycin		1986	22 (325)
RV-11		1989	25 (318)
sparfloxacin		1993	29 (345)
sultamycillin tosylate		1987	23 (343)
telavancin		2009	45 (525)
telithromycin		2001	37 (271)
temocillin disodium		1984	20 (323)
tigecycline		2005	41 (468)
trovafloxacin mesylate		1998	34 (332)
meropenem	ANTIBIOTIC, CARBAPENEM	1994	30 (303)
panipenem/betamipron		1994	30 (305)
mupirocin	ANTIBIOTIC, TOPICAL	1985	21 (330)
nadifloxacin		1993	29 (340)
abarelix	ANTICANCER	2004	40 (446)
alemtuzumab		2001	37 (260)
alitretinoin		1999	35 (333)
arglabin		1999	35 (335)
azacitidine		2004	40 (447)
belotecan		2004	40 (449)
bevacizumab		2004	40 (450)
bexarotene		2000	36 (298)
bortezomib		2003	39 (271)
catumaxomab		2009	45 (486)
cetuximab		2003	39 (272)

GENERIC NAME	INDICATION	YEAR INTRO.	ARMC VOL, (PAGE)
clofarabine		2005	41 (444)
dasatinib		2006	42 (517)
degarelix Acetate		2009	45 (490)
denileukin diftitox		1999	35 (338)
erlotinib		2004	40 (454)
exemestane		2000	36 (304)
fulvestrant		2002	38 (357)
gemtuzumab ozogamicin		2000	36 (306)
ibritumomab tiuxetan		2002	38 (359)
ixabepilone		2007	43 (473)
lapatinib		2007	43 (475)
letrazole		1996	32 (311)
nelarabine		2006	42 (528)
nimotuzumab		2006	42 (529)
OCT-43		1999	35 (345)
ofatumumab		2009	45 (512)
oxaliplatin		1996	32 (313)
panitumumab		2006	42 (531)
pazopanib		2009	45 (514)
pemetrexed		2004	40 (463)
pralatrexate		2009	45 (517)
raltitrexed		1996	32 (315)
SKI-2053R		1999	35 (348)
sorafenib		2005	41 (466)
sunitinib		2006	42 (544)
talaporfin sodium		2004	40 (469)
tamibarotene		2005	41 (467)
tasonermin		1999	35 (349)
temozolomide		1999	35 (350)
temsirolimus		2007	43 (490)
topotecan HCl		1996	32 (320)
tositumomab		2003	39 (285)
trabectedin		2007	43(492)
valrubicin		1999	35 (350)
vorinostat		2006	42 (549)
angiotensin II	ANTICANCER ADJUVANT	1994	30 (296)
nilotinib	ANTICANCER — CML	2007	43 (480)
chenodiol	ANTICHOLELITHOGENIC	1983	19 (317)
dabigatran etexilate	ANTICOAGULANT	2008	44 (598)
duteplase		1995	31 (342)
lepirudin		1997	33 (336)
parnaparin sodium		1993	29 (342)
reviparin sodium		1993	29 (344)
rivaroxaban		2008	44 (617)
thrombomodulin (recombinant)		2008	44 (628)
ximelagatran		2004	40 (470)
lacosamide	ANTICONVULSANT	2008	44 (610)

GENERIC NAME	INDICATION	YEAR INTRO.	ARMC VOL, (PAGE)
lamotrigine		1990	26 (304)
oxcarbazepine		1990	26 (307)
progabide		1985	21 (331)
rufinamide		2007	43 (488)
vigabatrin		1989	25 (319)
zonisamide		1989	25 (320)
bupropion HCl	ANTIDEPRESSANT	1989	25 (310)
citalopram		1989	25 (311)
desvenlafaxine		2008	44 (600)
duloxetine		2004	40 (452)
escitalopram oxalate		2002	38 (354)
fluoxetine HCl		1986	22 (320)
fluvoxamine maleate		1983	19 (319)
indalpine		1983	19 (320)
medifoxamine fumarate		1986	22 (323)
metapramine		1984	20 (320)
milnacipran		1997	33 (338)
mirtazapine		1994	30 (303)
moclobemide		1990	26 (305)
nefazodone		1994	30 (305)
paroxetine		1991	27 (331)
pivagabine		1997	33 (341)
reboxetine		1997	33 (342)
setiptiline		1989	25 (318)
sertraline hydrochloride		1990	26 (309)
tianeptine sodium		1983	19 (324)
toloxatone		1984	20 (324)
venlafaxine		1994	30 (312)
acarbose	ANTIDIABETIC	1990	26 (297)
epalrestat		1992	28 (330)
exenatide		2005	41 (452)
glimepiride		1995	31 (344)
insulin lispro		1996	32 (310)
liraglutide		2009	45 (507)
miglitol		1998	34 (325)
mitiglinide		2004	40 (460)
nateglinide		1999	35 (344)
pioglitazone HCl		1999	35 (346)
pramlintide		2005	41 (460)
repaglinide		1998	34 (329)
rosiglitazone maleate		1999	35 (347)
saxagliptin		2009	45 (521)
sitagliptin		2006	42 (541)
tolrestat		1989	25 (319)
troglitazone		1997	33 (344)
vildagliptin		2007	43 (494)
voglibose		1994	30 (313)
acetorphan	ANTIDIARRHEAL	1993	29 (332)
anti-digoxin polyclonal	ANTIDOTE	2002	38 (350)

GENERIC NAME	INDICATION	YEAR INTRO.	ARMC VOL, (PAGE)
antibody			
crotelidae polyvalent immune fab		2001	37 (263)
fomepizole		1998	34 (323)
aprepitant	ANTIEMETIC	2003	39 (268)
dolasetron mesylate		1998	34 (321)
fosaprepitant dimeglumine		2008	44 (606)
granisetron hydrochloride		1991	27 (329)
indisetron		2004	40 (459)
ondansetron hydrochloride		1990	26 (306)
nazasetron		1994	30 (305)
palonosetron		2003	39 (281)
ramosetron		1996	32 (315)
tropisetron		1992	28 (337)
eslicarbazepine Acetate	ANTIEPILEPTIC	2009	45 (498)
felbamate		1993	29 (337)
fosphenytoin sodium		1996	32 (308)
gabapentin		1993	29 (338)
levetiracetam		2000	36 (307)
pregabalin		2004	40 (464)
tiagabine		1996	32 (320)
topiramate		1995	31 (351)
centchroman	ANTIESTROGEN	1991	27 (324)
anidulafungin	ANTIFUNGAL	2006	42 (512)
caspofungin acetate		2001	37 (263)
eberconazole		2005	41 (449)
fenticonazole nitrate		1987	23 (334)
fluconazole		1988	24 (303)
fosfluconazole		2004	40 (457)
itraconazole		1988	24 (305)
lanoconazole		1994	30 (302)
luliconazole		2005	41 (454)
micafungin		2002	38 (360)
naftifine HCl		1984	20 (321)
oxiconazole nitrate		1983	19 (322)
posaconazole		2006	42 (532)
terbinafine hydrochloride		1991	27 (334)
terconazole		1983	19 (324)
tioconazole		1983	19 (324)
voriconazole		2002	38 (370)
amorolfine hydrochloride	ANTIFUNGAL, TOPICAL	1991	27 (322)
butenafine hydrochloride		1992	28 (327)
butoconazole		1986	22 (318)

GENERIC NAME	INDICATION	YEAR INTRO.	ARMC VOL, (PAGE)
cloconazole HCl		1986	22 (318)
liranaftate		2000	36 (309)
flutrimazole		1995	31 (343)
neticonazole HCl		1993	29 (341)
sertaconazole nitrate		1992	28 (336)
sulconizole nitrate		1985	21 (332)
apraclonidine HCl	ANTIGLAUCOMA	1988	24 (297)
befunolol HCl		1983	19 (315)
bimatroprost		2001	37 (261)
brimonidine		1996	32 (306)
brinzolamide		1998	34 (318)
dapiprazole HCl		1987	23 (332)
dorzolamide HCl		1995	31 (341)
latanoprost		1996	32 (311)
levobunolol HCl		1985	21 (328)
tafluprost		2008	44 (625)
travoprost		2001	37 (272)
unoprostone isopropyl ester		1994	30 (312)
acrivastine	ANTIHISTAMINE	1988	24 (295)
astemizole		1983	19 (314)
azelastine HCl		1986	22 (316)
cetirizine HCl		1987	23 (331)
desloratadine		2001	37 (264)
ebastine		1990	26 (302)
levocabastine hydrochloride		1991	27 (330)
levocetirizine		2001	37 (268)
loratadine		1988	24 (306)
mizolastine		1998	34 (325)
setastine HCl		1987	23 (342)
alacepril	ANTIHYPERTENSIVE	1988	24 (296)
alfuzosin HCl		1988	24 (296)
aliskiren		2007	43 (461)
amlodipine besylate		1990	26 (298)
amosulalol		1988	24 (297)
aranidipine		1996	32 (306)
arotinolol HCl		1986	22 (316)
azelnidipine		2003	39 (270)
barnidipine hydrochloride		1992	28 (326)
benazepril hydrochloride		1990	26 (299)
benidipine hydrochloride		1991	27 (322)
betaxolol HCl		1983	19 (315)
bevantolol HCl		1987	23 (328)
bisoprolol fumarate		1986	22 (317)
bopindolol		1985	21 (324)
bosentan		2001	37 (262)

GENERIC NAME	INDICATION	YEAR INTRO.	ARMC VOL, (PAGE)
budralazine		1983	19 (315)
bunazosin HCl		1985	21 (324)
candesartan cilexetil		1997	33 (330)
captopril		1982	13 (086)
carvedilol		1991	27 (323)
celiprolol HCl		1983	19 (317)
cicletanine		1988	24 (299)
cilazapril		1990	26 (301)
cinildipine		1995	31 (339)
clevidipine		2008	44 (596)
delapril		1989	25 (311)
dilevalol		1989	25 (311)
doxazosin mesylate		1988	24 (300)
efonidipine		1994	30 (299)
enalapril maleate		1984	20 (317)
enalaprilat		1987	23 (332)
eplerenone		2003	39 (276)
eprosartan		1997	33 (333)
felodipine		1988	24 (302)
fenoldopam mesylate		1998	34 (322)
fosinopril sodium		1991	27 (328)
guanadrel sulfate		1983	19 (319)
imidapril HCl		1993	29 (339)
irbesartan		1997	33 (336)
isradipine		1989	25 (315)
ketanserin		1985	21 (328)
lacidipine		1991	27 (330)
lercanidipine		1997	33 (337)
lisinopril		1987	23 (337)
losartan		1994	30 (302)
manidipine hydrochloride		1990	26 (304)
mebefradil hydrochloride		1997	33 (338)
moexipril HCl		1995	31 (346)
moxonidine		1991	27 (330)
nebivolol		1997	33 (339)
nilvadipine		1989	25 (316)
nipradilol		1988	24 (307)
nisoldipine		1990	26 (306)
olmesartan medoxomil		2002	38 (363)
perindopril		1988	24 (309)
pinacidil		1987	23 (340)
quinapril		1989	25 (317)
ramipril		1989	25 (317)
rilmenidine		1988	24 (310)
spirapril HCl		1995	31 (349)
telmisartan		1999	35 (349)
temocapril		1994	30 (311)
terazosin HCl		1984	20 (323)

GENERIC NAME	INDICATION	YEAR INTRO.	ARMC VOL, (PAGE)
tertatolol HCl		1987	23 (344)
tiamenidine HCl		1988	24 (311)
tilisolol hydrochloride		1992	28 (337)
trandolapril		1993	29 (348)
treprostinil sodium		2002	38 (368)
trimazosin HCl		1985	21 (333)
valsartan		1996	32 (320)
zofenopril calcium		2000	36 (313)
febuxostat	ANTI-HYPERURICEMIC	2009	45 (501)
daptomycin	ANTI INFECTIVE	2003	39 (272)
garenoxacin		2007	43 (471)
retapamulin		2007	43 (486)
maraviroc	ANTI-INFECTIVE — HIV	2007	43 (478)
raltegravir		2007	43 (484)
aceclofenac	ANTI-INFLAMMATORY	1992	28 (325)
AF-2259		1987	23 (325)
amfenac sodium		1986	22 (315)
ampiroxicam		1994	30 (296)
amtolmetin guacil		1993	29 (332)
butibufen		1992	28 (327)
canakinumab		2009	45 (484)
deflazacort		1986	22 (319)
dexibuprofen		1994	30 (298)
droxicam		1990	26 (302)
etodolac		1985	21 (327)
flunoxaprofen		1987	23 (335)
fluticasone propionate		1990	26 (303)
golimumab		2009	45 (503)
interferon, gamma		1989	25 (314)
isofezolac		1984	20 (319)
isoxicam		1983	19 (320)
lobenzarit sodium		1986	22 (322)
loxoprofen sodium		1986	22 (322)
lumiracoxib		2005	41 (455)
nabumetone		1985	21 (330)
nepafenac		2005	41 (456)
nimesulide		1985	21 (330)
oxaprozin		1983	19 (322)
piroxicam cinnamate		1988	24 (309)
rimexolone		1995	31 (348)
sivelestat		2002	38 (366)
tenoxicam		1987	23 (344)
zaltoprofen		1993	29 (349)
fisalamine	ANTI-INFLAMMATORY, INTESTINAL	1984	20 (318)
osalazine sodium		1986	22 (324)
alclometasone dipropionate	ANTI-INFLAMMATORY, TOPICAL	1985	21 (323)
aminoprofen		1990	26 (298)
betamethasone butyrate propionate		1994	30 (297)

GENERIC NAME	INDICATION	YEAR INTRO.	ARMC VOL, (PAGE)
butyl flufenamate		1983	19 (316)
deprodone propionate		1992	28 (329)
felbinac		1986	22 (320)
halobetasol propionate		1991	27 (329)
halometasone		1983	19 (320)
hydrocortisone aceponate		1988	24 (304)
hydrocortisone butyrate propionate		1983	19 (320)
mometasone furoate		1987	23 (338)
piketoprofen		1984	20 (322)
pimaprofen		1984	20 (322)
prednicarbate		1986	22 (325)
pravastatin	ANTILIPIDEMIC	1989	25 (316)
arteether	ANTIMALARIAL	2000	36 (296)
artemisinin		1987	23 (327)
bulaquine		2000	36 (299)
halofantrine		1988	24 (304)
mefloquine HCl		1985	21 (329)
almotriptan	ANTIMIGRAINE	2000	36 (295)
alpiropride		1988	24 (296)
eletriptan		2001	37 (266)
frovatriptan		2002	38 (357)
lomerizine HCl		1999	35 (342)
naratriptan hydrochloride		1997	33 (339)
rizatriptan benzoate		1998	34 (330)
sumatriptan succinate		1991	27 (333)
zolmitriptan		1997	33 (345)
dronabinol	ANTINAUSEANT	1986	22 (319)
amrubicin HCl	ANTINEOPLASTIC	2002	38 (349)
amsacrine		1987	23 (327)
anastrozole		1995	31 (338)
bicalutamide		1995	31 (338)
bisantrene hydrochloride		1990	26 (300)
camostat mesylate		1985	21 (325)
capecitabine		1998	34 (319)
cladribine		1993	29 (335)
cytarabine ocfosfate		1993	29 (335)
docetaxel		1995	31 (341)
doxifluridine		1987	23 (332)
enocitabine		1983	19 (318)
epirubicin HCl		1984	20 (318)
fadrozole HCl		1995	31 (342)
fludarabine phosphate		1991	27 (327)
flutamide		1983	19 (318)
formestane		1993	29 (337)
fotemustine		1989	25 (313)

GENERIC NAME	INDICATION	YEAR INTRO.	ARMC VOL, (PAGE)
geftimib		2002	38 (358)
gemcitabine HCl		1995	31 (344)
idarubicin hydrochloride		1990	26 (303)
imatinib mesylate		2001	37 (267)
interferon gamma-1a		1992	28 (332)
interleukin-2		1989	25 (314)
irinotecan		1994	30 (301)
lonidamine		1987	23 (337)
mitoxantrone HCl		1984	20 (321)
nedaplatin		1995	31 (347)
nilutamide		1987	23 (338)
paclitaxal		1993	29 (342)
pegaspargase		1994	30 (306)
pentostatin		1992	28 (334)
pirarubicin		1988	24 (309)
ranimustine		1987	23 (341)
sobuzoxane		1994	30 (310)
temoporphin		2002	38 (367)
toremifene		1989	25 (319)
vinorelbine		1989	25 (320)
zinostatin stimalamer		1994	30 (313)
porfimer sodium	ANTINEOPLASTIC ADJUVANT	1993	29 (343)
masoprocol	ANTINEOPLASTIC, TOPICAL	1992	28 (333)
miltefosine		1993	29 (340)
dexfenfluramine	ANTIOBESITY	1997	33 (332)
rimonabant		2006	42 (537)
orlistat		1998	34 (327)
sibutramine		1998	34 (331)
atovaquone	ANTIPARASITIC	1992	28 (326)
ivermectin		1987	23 (336)
budipine	ANTI-PARKINSONIAN	1997	33 (330)
CHF-1301		1999	35 (336)
droxidopa		1989	25 (312)
entacapone		1998	34 (322)
pergolide mesylate		1988	24 (308)
pramipexole hydrochloride		1997	33 (341)
ropinirole HCl		1996	32 (317)
talipexole		1996	32 (318)
tolcapone		1997	33 (343)
lidamidine HCl	ANTIPERISTALTIC	1984	20 (320)
prasugrel	ANTIPLATELET THERAPY	2009	45 (519)
gestrinone	ANTIPROGESTOGEN	1986	22 (321)
cabergoline	ANTIPROLACTIN	1993	29 (334)
tamsulosin HCl	ANTIPROSTATIC HYPERTROPHY	1993	29 (347)
acitretin	ANTIPSORIATIC	1989	25 (309)
calcipotriol		1991	27 (323)
efalizumab		2003	39 (274)
tazarotene		1997	33 (343)

GENERIC NAME	INDICATION	YEAR INTRO.	ARMC VOL, (PAGE)
tacalcitol		1993	29 (346)
ustekinumab		2009	45 (532)
amisulpride	ANTIPSYCHOTIC	1986	22 (316)
asenapine		2009	45 (479)
blonanserin		2008	44 (587)
paliperidone		2007	43 (482)
remoxipride hydrochloride		1990	26 (308)
zuclopenthixol acetate		1987	23 (345)
biolimus drug-eluting stent	ANTI-RESTENOTIC	2008	44 (586)
actarit	ANTIRHEUMATIC	1994	30 (296)
diacerein		1985	21 (326)
octreotide	ANTISECRETORY	1988	24 (307)
adamantanium bromide	ANTISEPTIC	1984	20 (315)
drotecogin alfa	ANTISEPSIS	2001	37 (265)
cimetropium bromide	ANTISPASMODIC	1985	21 (326)
tiquizium bromide		1984	20 (324)
tiropramide HCl		1983	19 (324)
eltrombopag	ANTITHROMBOCYTOPENIC	2009	45 (497)
romiplostim		2008	44 (619)
argatroban	ANTITHROMBOTIC	1990	26 (299)
bivalirudin		2000	36 (298)
defibrotide		1986	22 (319)
cilostazol		1988	24 (299)
clopidogrel hydrogensulfate		1998	34 (320)
cloricromen		1991	27 (325)
enoxaparin		1987	23 (333)
eptifibatide		1999	35 (340)
ethyl icosapentate		1990	26 (303)
fondaparinux sodium		2002	38 (356)
indobufen		1984	20 (319)
limaprost		1988	24 (306)
ozagrel sodium		1988	24 (308)
picotamide		1987	23 (340)
tirofiban hydrochloride		1998	34 (332)
flutropium bromide	ANTITUSSIVE	1988	24 (303)
levodropropizine		1988	24 (305)
nitisinone	ANTITYROSINAEMIA	2002	38 (361)
benexate HCl	ANTIULCER	1987	23 (328)
dosmalfate		2000	36 (302)
ebrotidine		1997	33 (333)
ecabet sodium		1993	29 (336)
egualen sodium		2000	36 (303)
enprostil		1985	21 (327)
famotidine		1985	21 (327)
irsogladine		1989	25 (315)
lansoprazole		1992	28 (332)
misoprostol		1985	21 (329)

GENERIC NAME	INDICATION	YEAR INTRO.	ARMC VOL, (PAGE)
nizatidine		1987	23 (339)
omeprazole		1988	24 (308)
ornoprostil		1987	23 (339)
pantoprazole sodium		1994	30 (306)
plaunotol		1987	23 (340)
polaprezinc		1994	30 (307)
ranitidine bismuth citrate		1995	31 (348)
rebamipide		1990	26 (308)
rosaprostol		1985	21 (332)
roxatidine acetate HCl		1986	22 (326)
roxithromycin		1987	23 (342)
sofalcone		1984	20 (323)
spizofurone		1987	23 (343)
teprenone		1984	20 (323)
tretinoin tocoferil		1993	29 (348)
troxipide		1986	22 (327)
abacavir sulfate	ANTIVIRAL	1999	35 (333)
adefovir dipivoxil		2002	38 (348)
amprenavir		1999	35 (334)
atazanavir		2003	39 (269)
cidofovir		1996	32 (306)
delavirdine mesylate		1997	33 (331)
didanosine		1991	27 (326)
efavirenz		1998	34 (321)
emtricitabine		2003	39 (274)
enfuvirtide		2003	39 (275)
entecavir		2005	41 (450)
etravirine		2008	44 (602)
famciclovir		1994	30 (300)
fomivirsen sodium		1998	34 (323)
fosamprenavir		2003	39 (277)
foscarnet sodium		1989	25 (313)
ganciclovir		1988	24 (303)
imiquimod		1997	33 (335)
indinavir sulfate		1996	32 (310)
interferon alfacon-1		1997	33 (336)
lamivudine		1995	31 (345)
lopinavir		2000	36 (310)
nelfinavir mesylate		1997	33 (340)
nevirapine		1996	32 (313)
oseltamivir phosphate		1999	35 (346)
penciclovir		1996	32 (314)
propagermanium		1994	30 (308)
rimantadine HCl		1987	23 (342)
ritonavir		1996	32 (317)
saquinavir mesylate		1995	31 (349)
sorivudine		1993	29 (345)
stavudine		1994	30 (311)
tenofovir disoproxil fumarate		2001	37 (271)

GENERIC NAME	INDICATION	YEAR INTRO.	ARMC VOL, (PAGE)
valaciclovir HCl		1995	31 (352)
zalcitabine		1992	28 (338)
zanamivir		1999	35 (352)
zidovudine		1987	23 (345)
influenza virus live	ANTIVIRAL VACCINE	2003	39 (277)
cevimeline hydrochloride	ANTI-XEROSTOMIA	2000	36 (299)
alpidem	ANXIOLYTIC	1991	27 (322)
buspirone HCl		1985	21 (324)
etizolam		1984	20 (318)
flutazolam		1984	20 (318)
flutoprazepam		1986	22 (320)
metaclazepam		1987	23 (338)
mexazolam		1984	20 (321)
tandospirone		1996	32 (319)
ciclesonide	ASTHMA, COPD	2005	41 (443)
atomoxetine	ATTENTION DEFICIT HYPERACTIVITY DISORDER	2003	39 (270)
flumazenil	BENZODIAZEPINE ANTAG.	1987	23 (335)
bambuterol	BRONCHODILATOR	1990	26 (299)
doxofylline		1985	21 (327)
formoterol fumarate		1986	22 (321)
mabuterol HCl		1986	22 (323)
oxitropium bromide		1983	19 (323)
salmeterol hydroxynaphthoate		1990	26 (308)
tiotropium bromide		2002	38 (368)
APD	CALCIUM REGULATOR	1987	23 (326)
clodronate disodium		1986	22 (319)
disodium pamidronate		1989	25 (312)
gallium nitrate		1991	27 (328)
ipriflavone		1989	25 (314)
neridronic acid		2002	38 (361)
dexrazoxane	CARDIOPROTECTIVE	1992	28 (330)
bucladesine sodium	CARDIOSTIMULANT	1984	20 (316)
denopamine		1988	24 (300)
docarpamine		1994	30 (298)
dopexamine		1989	25 (312)
enoximone		1988	24 (301)
flosequinan		1992	28 (331)
ibopamine HCl		1984	20 (319)
loprinone hydrochloride		1996	32 (312)
milrinone		1989	25 (316)
vesnarinone		1990	26 (310)
amrinone	CARDIOTONIC	1983	19 (314)
colforsin daropate HCL		1999	35 (337)
xamoterol fumarate		1988	24 (312)
cefozopran HCL	CEPHALOSPORIN, INJECTABLE	1995	31 (339)
cefditoren pivoxil	CEPHALOSPORIN, ORAL	1994	30 (297)
brovincamine fumarate	CEREBRAL VASODILATOR	1986	22 (317)

GENERIC NAME	INDICATION	YEAR INTRO.	ARMC VOL, (PAGE)
nimodipine		1985	21 (330)
propentofylline		1988	24 (310)
succimer	CHELATOR	1991	27 (333)
trientine HCl		1986	22 (327)
fenbuprol	CHOLERETIC	1983	19 (318)
lulbiprostone	CHRONIC IDIOPATHIC CONSTIPATION	2006	42 (525)
deferasirox	CHRONIC IRON OVERLOAD	2005	41 (446)
arformoterol	CHRONIC OBSTRUCTIVE PULMONARY DISEASE	2007	43 (465)
indacaterol		2009	45 (505)
auranofin	CHRYSOTHERAPEUTIC	1983	19 (314)
taltirelin	CNS STIMULANT	2000	36 (311)
aniracetam	COGNITION ENHANCER	1993	29 (333)
pramiracetam H2SO4		1993	29 (343)
carperitide	CONGESTIVE HEART	1995	31 (339)
nesiritide	FAILURE	2001	37 (269)
drospirenone	CONTRACEPTIVE	2000	36 (302)
norelgestromin		2002	38 (362)
ulipristal acetate		2009	45 (530)
nicorandil	CORONARY VASODILATOR	1984	20 (322)
certolizumab pegol	CROHN'S DISEASE	2008	44 (592)
dornase alfa	CYSTIC FIBROSIS	1994	30 (298)
neltenexine		1993	29 (341)
amifostine	CYTOPROTECTIVE	1995	31 (338)
nalmefene HCL	DEPENDENCE TREATMENT	1995	31 (347)
ioflupane	DIAGNOSIS CNS	2000	36 (306)
azosemide	DIURETIC	1986	22 (316)
muzolimine		1983	19 (321)
torasemide		1993	29 (348)
atorvastatin calcium	DYSLIPIDEMIA	1997	33 (328)
cerivastatin		1997	33 (331)
choline fenofibrate		2008	44 (594)
naftopidil	DYSURIA	1999	35 (343)
silodosin		2006	42 (540)
alglucerase	ENZYME	1991	27 (321)
udenafil	ERECTILE DYSFUNCTION	2005	41 (472)
erdosteine	EXPECTORANT	1995	31 (342)
fudosteine		2001	37 (267)
agalsidase alfa	FABRY'S DISEASE	2001	37 (259)
cetrorelix	FEMALE INFERTILITY	1999	35 (336)
ganirelix acetate		2000	36 (305)
follitropin alfa	FERTILITY ENHANCER	1996	32 (307)
follitropin beta		1996	32 (308)
reteplase	FIBRINOLYTIC	1996	32 (316)
esomeprazole magnesium	GASTRIC ANTISECRETORY	2000	36 (303)
lafutidine		2000	36 (307)
rabeprazole sodium		1998	34 (328)

GENERIC NAME	INDICATION	YEAR INTRO.	ARMC VOL, (PAGE)
dexlansoprazole	GASTROESOPHOGEAL REFLUX DISEASE	2009	45 (492)
cinitapride	GASTROPROKINETIC	1990	26 (301)
cisapride		1988	24 (299)
itopride HCL		1995	31 (344)
mosapride citrate		1998	34 (326)
imiglucerase	GAUCHER'S DISEASE	1994	30 (301)
miglustat		2003	39 (279)
rilonacept	GENETIC AUTOINFLAMMATORY SYNDROMES	2008	44 (615)
somatotropin	GROWTH HORMONE	1994	30 (310)
somatomedin-1	GROWTH HORMONE INSENSITIVITY	1994	30 (310)
factor VIIa	HAEMOPHILIA	1996	32 (307)
levosimendan	HEART FAILURE	2000	36 (308)
pimobendan		1994	30 (307)
anagrelide hydrochloride	HEMATOLOGIC	1997	33 (328)
erythropoietin	HEMATOPOETIC	1988	24 (301)
eculizumab	HEMOGLOBINURIA	2007	43 (468)
thrombin alfa	HEMOSTAT	2008	44 (627)
factor VIII	HEMOSTATIC	1992	28 (330)
telbivudine	HEPATITIS B	2006	42 (546)
clevudine		2007	43 (466)
malotilate	HEPATOPROTECTIVE	1985	21 (329)
mivotilate		1999	35 (343)
icatibant	HEREDITARY ANGIODEMA	2008	44 (608)
darunavir	HIV	2006	42 (515)
tipranavir		2005	41 (470)
buserelin acetate	HORMONE	1984	20 (316)
goserelin		1987	23 (336)
leuprolide acetate		1984	20 (319)
nafarelin acetate		1990	26 (306)
somatropin		1987	23 (343)
zoledronate disodium	HYPERCALCEMIA	2000	36 (314)
cinacalcet	HYPERPARATHYROIDISM	2004	40 (451)
sapropterin hydrochloride	HYPERPHENYL-ALANINEMIA	1992	28 (336)
quinagolide	HYPERPROLACTINEMIA	1994	30 (309)
cadralazine	HYPERTENSIVE	1988	24 (298)
nitrendipine		1985	21 (331)
binfonazole	HYPNOTIC	1983	19 (315)
brotizolam		1983	19 (315)
butoctamide		1984	20 (316)
cinolazepam		1993	29 (334)
doxefazepam		1985	21 (326)
eszopiclone		2005	41 (451)
loprazolam mesylate		1983	19 (321)
quazepam		1985	21 (332)
rilmazafone		1989	25 (317)

GENERIC NAME	INDICATION	YEAR INTRO.	ARMC VOL, (PAGE)
zaleplon		1999	35 (351)
zolpidem hemitartrate		1988	24 (313)
zopiclone		1986	22 (327)
acetohydroxamic acid	HYPOAMMONURIC	1983	19 (313)
sodium cellulose PO4	HYPOCALCIURIC	1983	19 (323)
divistyramine	HYPOCHOLESTEROLEMIC	1984	20 (317)
lovastatin		1987	23 (337)
melinamide		1984	20 (320)
pitavastatin		2003	39 (282)
rosuvastatin		2003	39 (283)
simvastatin		1988	24 (311)
glucagon, rDNA	HYPOGLYCEMIA	1993	29 (338)
acipimox	HYPOLIPIDEMIC	1985	21 (323)
beclobrate		1986	22 (317)
binifibrate		1986	22 (317)
ciprofibrate		1985	21 (326)
colesevelam hydrochloride		2000	36 (300)
colestimide		1999	35 (337)
ezetimibe		2002	38 (355)
fluvastatin		1994	30 (300)
meglutol		1983	19 (321)
ronafibrate		1986	22 (326)
conivaptan	HYPONATREMIA	2006	42 (514)
tolvaptan		2009	45 (528)
mozavaptan		2006	42 (527)
modafinil	IDIOPATHIC HYPERSOMNIA	1994	30 (303)
pirfenidone	IDIOPATHIC PULMONARY FIBROSIS	2008	44 (614)
bucillamine	IMMUNOMODULATOR	1987	23 (329)
centoxin		1991	27 (325)
thymopentin		1985	21 (333)
filgrastim	IMMUNOSTIMULANT	1991	27 (327)
GMDP		1996	32 (308)
interferon gamma-1b		1991	27 (329)
lentinan		1986	22 (322)
pegademase bovine		1990	26 (307)
pidotimod		1993	29 (343)
romurtide		1991	27 (332)
sargramostim		1991	27 (332)
schizophyllan		1985	22 (326)
ubenimex		1987	23 (345)
cyclosporine	IMMUNOSUPPRESSANT	1983	19 (317)
everolimus		2004	40 (455)
gusperimus		1994	30 (300)
mizoribine		1984	20 (321)
muromonab-CD3		1986	22 (323)
mycophenolate sodium		2003	39 (279)
mycophenolate mofetil		1995	31 (346)
pimecrolimus		2002	38 (365)

GENERIC NAME	INDICATION	YEAR INTRO.	ARMC VOL, (PAGE)
tacrolimus		1993	29 (347)
ramelteon	INSOMNIA	2005	41 (462)
defeiprone	IRON CHELATOR	1995	31 (340)
alosetron hydrochloride	IRRITABLE BOWEL, SYNDROME	2000	36 (295)
tegaserod maleate		2001	37 (270)
sulbactam sodium	β-LACTAMASE INHIBITOR	1986	22 (326)
tazobactam sodium		1992	28 (336)
nartograstim	LEUKOPENIA	1994	30 (304)
pumactant	LUNG SURFACTANT	1994	30 (308)
sildenafil citrate	MALE SEXUAL DYSFUNCTION	1998	34 (331)
gadoversetamide	MRI CONTRAST AGENT	2000	36 (304)
telmesteine	MUCOLYTIC	1992	28 (337)
laronidase	MUCOPOLYSACCARIDOSIS	2003	39 (278)
galsulfase	MUCOPOLYSACCHARIDOSIS VI	2005	41 (453)
idursulfase	MUCOPOLYSACCHARIDOSIS II (HUNTER SYNDROME)	2006	42 (520)
palifermin	MUCOSITIS	2005	41 (461)
interferon X-1a	MULTIPLE SCLEROSIS	1996	32 (311)
interferon X-1b		1993	29 (339)
glatiramer acetate		1997	33 (334)
natalizumab		2004	40 (462)
afloqualone	MUSCLE RELAXANT	1983	19 (313)
cisatracurium besilate		1995	31 (340)
doxacurium chloride		1991	27 (326)
eperisone HCl		1983	19 (318)
mivacurium chloride		1992	28 (334)
rapacuronium bromide		1999	35 (347)
tizanidine		1984	20 (324)
decitabine	MYELODYSPLASTIC SYNDROMES	2006	42 (519)
lenalidomide	MYELODYSPLASTIC SYNDROMES, MULTIPLE MYELOMA	2006	42 (523)
naltrexone HCl	NARCOTIC ANTAGONIST	1984	20 (322)
tinazoline	NASAL DECONGESTANT	1988	24 (312)
aripiprazole	NEUROLEPTIC	2002	38 (350)
clospipramine hydrochloride		1991	27 (325)
nemonapride		1991	27 (331)
olanzapine		1996	32 (313)
perospirone hydrochloride		2001	37 (270)
quetiapine fumarate		1997	33 (341)
risperidone		1993	29 (344)
sertindole		1996	32 (318)
timiperone		1984	20 (323)
ziprasidone hydrochloride		2000	36 (312)
rocuronium bromide	NEUROMUSCULAR BLOCKER	1994	30 (309)
edaravone	NEUROPROTECTIVE	1995	37 (265)

GENERIC NAME	INDICATION	YEAR INTRO.	ARMC VOL, (PAGE)
fasudil HCL		1995	31 (343)
riluzole		1996	32 (317)
varenicline	NICOTINE-DEPENDENCE	2006	42 (547)
bifemelane HCl	NOOTROPIC	1987	23 (329)
choline alfoscerate		1990	26 (300)
exifone		1988	24 (302)
idebenone		1986	22 (321)
indeloxazine HCl		1988	24 (304)
levacecarnine HCl		1986	22 (322)
nizofenzone fumarate		1988	24 (307)
oxiracetam		1987	23 (339)
bromfenac sodium	NSAID	1997	33 (329)
lornoxicam		1997	33 (337)
methylnaltrexone bromide	OPIOID-INDUCED CONSTIPATION	2008	44 (612)
OP-1	OSTEOINDUCTOR	2001	37 (269)
alendronate sodium	OSTEOPOROSIS	1993	29 (332)
ibandronic acid		1996	32 (309)
incadronic acid		1997	33 (335)
minodronic acid		2009	45 (509)
raloxifene hydrochloride		1998	34 (328)
risedronate sodium		1998	34 (330)
strontium ranelate		2004	40 (467)
fesoterodine	OVERACTIVE BLADDER	2008	44 (604)
imidafenacin		2007	43 (472)
tiludronate disodium	PAGET'S DISEASE	1995	31 (350)
rasagiline	PARKINSON'S DISEASE	2005	41 (464)
rotigotine		2006	42 (538)
tadalafil	PDE5 INHIBITOR	2003	39 (284)
vardenafil		2003	39 (286)
temoporphin	PHOTOSENSITIZER	2002	38 (367)
verteporfin		2000	36 (312)
alefacept	PLAQUE PSORIASIS	2003	39 (267)
beraprost sodium	PLATELET AGGREG.	1992	28 (326)
epoprostenol sodium	INsHIBITOR	1983	19 (318)
iloprost		1992	28 (332)
sarpogrelate HCl	PLATELET ANTIAGGREGANT	1993	29 (344)
trimetrexate glucuronate	PNEUMOCYSTIS CARINII PNEUMONIA	1994	30 (312)
solifenacin	POLLAKIURIA	2004	40 (466)
alglucosidase alfa	POMPE DISEASE	2006	42 (511)
alvimopan	POST-OPERATIVE ILEUS	2008	44 (584)
histrelin	PRECOCIOUS PUBERTY	1993	29 (338)
dapoxetine	PREMATURE EJACULATION	2009	45 (488)
atosiban	PRETERM LABOR	2000	36 (297)
gestodene	PROGESTOGEN	1987	23 (335)
nomegestrol acetate		1986	22 (324)
norgestimate		1986	22 (324)

GENERIC NAME	INDICATION	YEAR INTRO.	ARMC VOL, (PAGE)
promegestrone		1983	19 (323)
trimegestone		2001	37 (273)
alpha-1 antitrypsin	PROTEASE INHIBITOR	1988	24 (297)
nafamostat mesylate		1986	22 (323)
Nalfuraline hydrochloride	PRURITUS	2009	45 (510)
adrafinil	PSYCHOSTIMULANT	1986	22 (315)
dexmethylphenidate HCl		2002	38 (352)
dutasteride		2002	38 (353)
ambrisentan	PULMONARY ARTERIAL HYPERTENSION	2007	43 (463)
sitaxsentan	PULMONARY HYPERTENSION	2006	42 (543)
finasteride	5a-REDUCTASE INHIBITOR	1992	28 (331)
surfactant TA	RESPIRATORY SURFACTANT	1987	23 (344)
sugammadex	REVERSAL OF NEUROMUSCULAR BLOCKADE	2008	44 (623)
abatacept	RHEUMATOID ARTHRITIS	2006	42 (509)
Adalimumab		2003	39 (267)
dexmedetomidine hydrochloride	SEDATIVE	2000	36 (301)
ziconotide	SEVERE CHRONIC PAIN	2005	41 (473)
kinetin	SKIN PHOTODAMAGE/ DERMATOLOGIC	1999	35 (341)
Armodafinil	SLEEP DISORDER TREATMENT	2009	45 (478)
Plerixafor hydrochloride	STEM CELL MOBILIZER	2009	45 (515)
tirilazad mesylate	SUBARACHNOID HEMORRHAGE	1995	31 (351)
APSAC	THROMBOLYTIC	1987	23 (326)
alteplase		1987	23 (326)
balsalazide disodium	ULCERATIVE COLITIS	1997	33 (329)
darifenacin	URINARY INCONTINENCE	2005	41 (445)
tiopronin	UROLITHIASIS	1989	25 (318)
propiverine hydrochloride	UROLOGIC	1992	28 (335)
Lyme disease	VACCINE	1999	35 (342)
clobenoside	VASOPROTECTIVE	1988	24 (300)
rivaroxaban	VENOUS THROMBOEMBOLISM	2008	44 (617)
falecalcitriol	VITAMIN D	2001	37 (266)
maxacalcitol		2000	36 (310)
paricalcitol		1998	34 (327)
doxercalciferol	VITAMIN D PROHORMONE	1999	35 (339)
prezatide copper acetate	VULNERARY	1996	32 (314)
acemannan	WOUND HEALING AGENT	2001	37 (257)
cadexomer iodine		1983	19 (316)
epidermal growth factor		1987	23 (333)

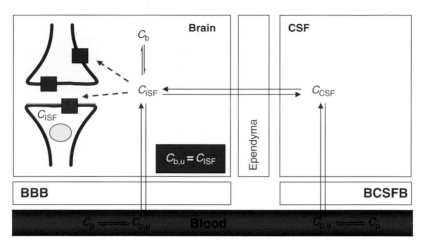

Plate 4.1 A schematic conceptualization of the three-compartment model of CNS penetration demonstrating the importance of intercompartmental unbound compound concentration relationships to target pharmacology interactions [21,22,25–28]. An exaggerated synapse is shown in the brain compartment to emphasize the locale of transmembrane proteins (squares) versus intracellular (oval) targets, and the matrix compound concentrations dictating their respective ligand–target interactions.

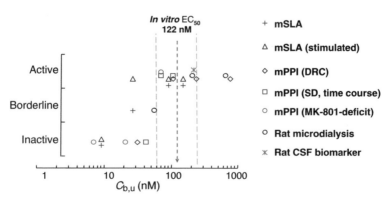

Plate 4.3 Rodent pharmacodynamic effects versus $C_{b,u}$ for **6**. Dashed lines represent a twofold separation from the *in vitro* functional assay EC_{50} (122 nM, dashed arrow). mSLA, mouse spontaneous locomotor activity; mPPI, mouse prepulse inhibition; DRC, dose–response curve; SD, single dose.

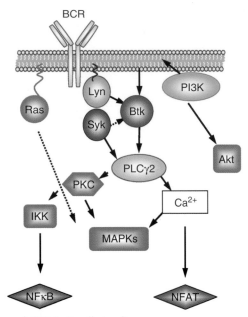

Plate 11.1   Btk, Syk, and PI3K in B-cell signaling.

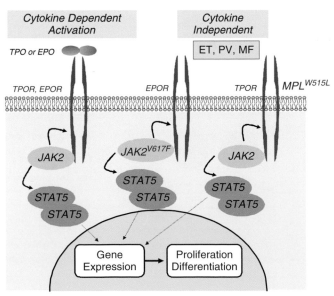

Plate 13.1   JAK2-STAT Pathway Activation in MPN.

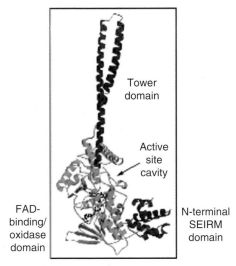

Plate 15.4 The X-ray crystal structure of LSD1.

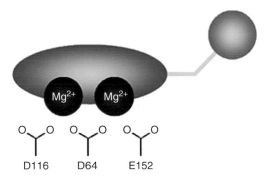

Plate 16.2 Two-metal-binding pharmacophore.

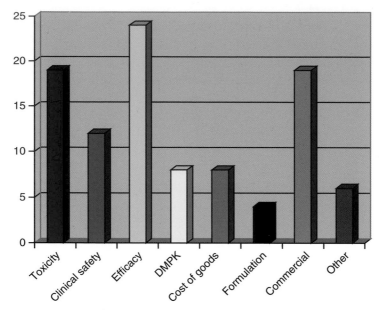

Plate 24.1 Percentage of clinical failures by cause in 2000 (adapted from Ref. [3]).

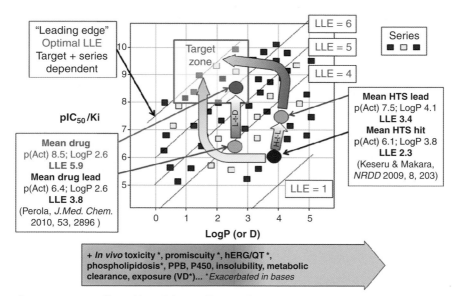

Plate 24.2 Use of ligand lipophilicity efficiency (LLE) in optimization. Potency versus LogP or LogD plots often show a "leading edge" where LLE is optimal. Mean high-throughput screening (HTS)-derived hits and leads [24] are more lipophilic than the mean leads successfully converted to drugs [82]. Reducing lipophilicity and increasing potency (red and green arrow strategies) are frequently needed in lead optimization. H-t-L hit-to-lead; L-t-D lead to drug.

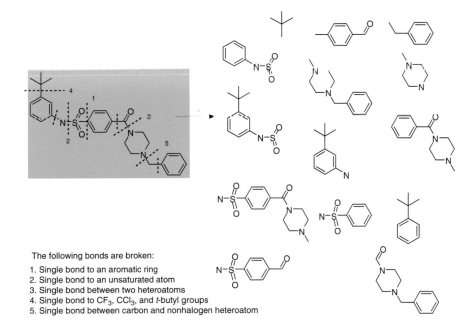

The following bonds are broken:

1. Single bond to an aromatic ring
2. Single bond to an unsaturated atom
3. Single bond between two heteroatoms
4. Single bond to CF$_3$, CCl$_3$, and *t*-butyl groups
5. Single bond between carbon and nonhalogen heteroatom

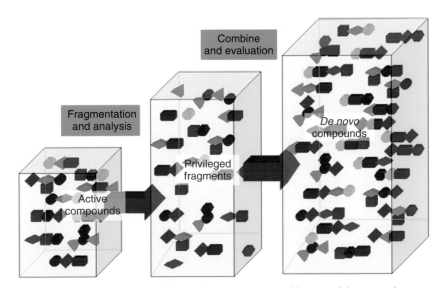

Plate 25.1 Schematic disassembly and reconstruction of kinase inhibitors with permission from reference 31.

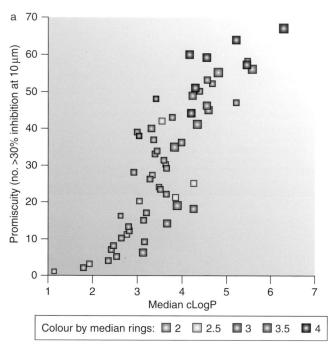

Plate 25.6 Plot of promiscuity versus median cLogP.